Nonequilibrium Phonon Dynamics

NATO ASI Series

Advanced Science Institutes Series

A series presenting the results of activities sponsored by the NATO Science Committee, which aims at the dissemination of advanced scientific and technological knowledge, with a view to strengthening links between scientific communities.

The series is published by an international board of publishers in conjunction with the NATO Scientific Affairs Division

A Life Sciences **B Physics**	Plenum Publishing Corporation New York and London
C Mathematical and Physical Sciences	D. Reidel Publishing Company Dordrecht, Boston, and Lancaster
D Behavioral and Social Sciences **E Engineering and Materials Sciences**	Martinus Nijhoff Publishers The Hague, Boston, and Lancaster
F Computer and Systems Sciences **G Ecological Sciences**	Springer-Verlag Berlin, Heidelberg, New York, and Tokyo

Recent Volumes in this Series

Volume 118—Regular and Chaotic Motions in Dynamic Systems
edited by G. Velo and A. S. Wightman

Volume 119—Analytical Laser Spectroscopy
edited by S. Martellucci and A. N. Chester

Volume 120—Chaotic Behavior in Quantum Systems: Theory and Applications
edited by Giulio Casati

Volume 121—Electronic Structure, Dynamics, and Quantum Structural Properties of Condensed Matter
edited by Jozef T. Devreese and Piet Van Camp

Volume 122—Quarks, Leptons, and Beyond
edited by H. Fritzsch, R. D. Peccei, H. Saller, and F. Wagner

Volume 123—Density Functional Methods in Physics
edited by Reiner M. Dreizler and Jõao da Providencia

Volume 124—Nonequilibrium Phonon Dynamics
edited by Walter E. Bron

Series B: Physics

Nonequilibrium Phonon Dynamics

Edited by

Walter E. Bron
Indiana University
Bloomington, Indiana

Plenum Press
New York and London
Published in cooperation with NATO Scientific Affairs Division

Proceedings of a NATO Advanced Study Institute on
Nonequilibrium Phonon Dynamics,
held August 27-September 7, 1984,
in Les Arcs, France

Library of Congress Cataloging in Publication Data

NATO Advanced Study Institute on Nonequilibrium Phonon Dynamics (1984: Les Arcs, France)
Nonequilibrium phonon dynamics,

(NATO ASI series. Series B, Physics; v. 124)
"Proceedings of a NATO Advanced Study Institute on Nonequilibrium Phonon Dynamics, held August 27-September 7, 1984, in Les Arcs, France"—T.p. verso.
"Published in cooperation with NATO Scientific Affairs Division."
Bibliography: p.
Includes index.
1. Phonons—Congresses. 2. Lattice dynamics—congresses. I. Bron, Walter E. II. North Atlantic Treaty Organization. Scientific Affairs Division. III. Title. IV. Title: Phonon dynamics. V. Series.
QC176.8.P5N38 1984 530.4'1 85-6449

ISBN-13: 978-1-4612-9513-6 e-ISBN-13: 978-1-4613-2501-7
DOI: 10.1007/978-1-4613-2501-7

Softcover reprint of the hardcover 1st edition 1985
A Division of Plenum Publishing Corporation
233 Spring Street, New York, N.Y. 10013

PREFACE

Phonons are always present in the solid state even at an absolute temperature of 0 K where zero point vibrations still abound. Moreover, phonons interact with all other excitations of the solid state and, thereby, influence most of its properties.

Historically experimental information on phonon transport came from measurements of thermal conductivity. Over the past two decades much more, and much more detailed, information on phonon transport and on many of the inherent phonon interaction processes have come to light from experiments which use nonequilibrium phonons to study their dynamics. The resultant research field has most recently blossomed with the development of ever more sophisticated experimental and theoretical methods which can be applied to it. In fact, the field is moving so rapidly that new members of the research community have difficulties in keeping up to date.

This NATO Advanced Study Institute (ASI) was organized with the objective of overcoming the information barrier between those expert in the field and those who are new to it. Thus it was decided to (i) organize a set of tutorially based lectures covering most of the important facets in the field, and (ii) to produce an Institute proceedings which would serve both as the first general textbook, as well as a valuable reference book, for this field of knowledge.

The set of lectures and this text are organized into two main subjects. These are Nonequilibrium Phonon Dynamics in the Bulk, and Phonons and Phonon Dynamics at (and near) Solid Surfaces. This also represents, for the most part, the order in which the material was presented during the ASI. Each lecturer presented essentially all of his material during a single 3 1/4 hour period. Although such a lecture period is quite an imposition on the lecturer, it results in a more coherent form of presentation than when the material is presented over fragmented periods of time. After each three lecture periods, a full period of 3 1/4 hours was dedicated to a general and completely open discussion period. The discussion periods were under the leadership of a chairman and the three previous lecturers. These periods were designed to offer to the participants ample time to ask detailed questions on the previous lectures (in a format useful to all the participants) and to make comments and contributions of their own. All of the discussion periods took up all their allotted time and would have taken more if time had been available.

It became obvious as the ASI progressed that although much new knowledge in the field of nonequilibrium phonon dynamics has come to light in the past decade, much remains to be done before the field can be considered to be fully mature. Perhaps the most obvious missing links are theoretical formulations which are readily reducible to comparison with experimental results. We expect progress in this area in the future.

Some comments about the manuscripts. Modern word processors have become an obvious tool for speeding up the typing of camera-ready manuscripts. However, they lead to a wide spread of typing fonts and in the general appearance of manuscripts, as compared to the past when the IBM selectric typewriter, or its equivalent, were the norm. I have, however, decided to accept this spread in manuscript appearance so as not to undergo both the expense and the consequent delay in retyping a number of the manuscripts. Content is more important than form.

I also regretfully take note that the manuscripts by Professor Bilz on the Dynamical Aspects of Nonlinear Oscillators and that by Professor Maneval on the Thermal Response of Thin Metal Films were not available at the time of publication.

The ASI was held in the Hotel du Golf, Les Arcs (Savoy), France from August 27 through September 7, 1984. Situated some 1800 meters above sea level, and overlooking Mont Blanc and other exceptionally beautiful scenery, the Hotel and its surroundings became a welcomed relief from the ASI's full and, at times, hectic program.

It goes almost without saying that an Institute of this scope and size can not occur without the financial and organizational help from a variety of sources. Most important, of course, was the funding and organizational advice from the Scientific Affairs Office of NATO. We are all grateful for their help. I would also wish to thank the Max Planck Institut fuer Festkoerperforschung, Stuttgart which helped to produce and mail the ASI's poster, and the Physics Department at Indiana University who provided the personnel for the second and subsequent mailings. The U.S. National Science Foundation provided travel funding for three young U.S. participants. The same is so for the science foundations and universities of the countries of many of the other lecturers and participants. We thank them all for their help and trust that their aid to future ASIs will continue to assist scientific communication.

I have also to thank many individuals whose aid helped to make the ASI the success it turned out to be. Among these persons, I would particularly like to acknowledge the help and advice of the members of the Organizing Committee; namely, Professors L. Challis and K. Renk, and of Professor J.-P. Maneval who undertook to be the local coordinator and who made the original contact with the hotel and, most importantly, battled with the vagaries of the international banking system to balance the hotel phase of the budget.

Many thanks go to Kathryn Crouch, my excellent secretary at Indiana University. By far the greatest graditude goes to my wife, Ann, who took on the secretarial and treasurers job during our prolonged stay in Stuttgart during the summer of 1984, plus during and after the ASI. Her help and organizational skills have been acknowledged by lecturers and participants alike. Finally, my thanks go to Mrs. J. Challis and Mrs. A.-M. Maneval who helped my wife to staff the secretariat at Les Arcs, and to Mr. T. Gazaud and Mr. P. Fabing of the staff of the Hotel du Golf for making our stay there so pleasant.

Walter E. Bron
Institute Director and
Professor of Physics
Department of Physics
Indiana University
Bloomington, IN 47405 USA

CONTENTS

BULK NONEQUILIBRIUM PHONONS

SURFACE NONEQUILIBRIUM PHONONS

STUDIES OF NONEQUILIBRIUM DYNAMICS IN THE TIME DOMAIN

W.E. Bron

Department of Physics
Indiana University
Bloomington, Indiana 47405, U.S.A.

I. INTRODUCTION

With but one exception, the material to be covered in this lecture concerns the excitation, and subsequent dynamics in the time domain, of nonequilibrium phonons in solids. This being the case, one needs to be aware of the natural time-scale associated with phonon dynamics; namely, the "lifetime" over which the excitation exists. It is now known that the lifetime of optical phonons in a simple solid is of the order of picoseconds, whereas those of acoustic phonons varies strongly with phonon frequency (ν) and phonon branch. Typically, the lifetimes (τ) of 1 THz longitudinal acoustic (LA) phonons is of the order of microseconds and varies with the inverse fifth power of the frequency (see e.g., Baumgartner et al., 1981), such that $\tau(\nu) \approx 10^{54}\ \nu^{-5}$. On the other hand, in the absence of any impurity scattering it has been postulated that transverse acoustic (TA) phonons in the dispersive regime may have lifetimes of the order of seconds (Orbach and Vredevoe, 1964). I consider here only those cases in which even TA phonons possess lifetimes $\leq 10^{-6}$ s; hence to a regime $10^{-12} < \tau < 10^{-6}$ which may be called the short to ultrashort time domain. I will further restrict myself to the transport of so-called high frequency phonons, i.e., $\nu \gtrsim 200$ GHz. Low-frequency ($\nu < 0.2$ THz) phonon propagation in most crystal systems occurs in an environment which differs considerably from that inherent in higher frequency phonon transport. Among the differentiating properties are dispersion, defect and boundary scattering, and phonon-phonon interactions, all of which become more pronounced at higher frequencies.

In an earlier review (Bron, 1980a) I discussed the spectrographic techniques necessary to investigate high-frequency nonequilibrium phonon dynamics. Since that time many new techniques have been discovered to take the search for knowledge in this field several steps forward. It has been tempting in this lecture to simply bring my earlier review up to date. However, in doing so I would in part be discussing work which will be presented in much greater detail in this volume in the lecture notes of Professors Challis, Goodstein, Kinder, Renk, Wolfe, and Ulbrich. Accordingly, I am restricting myself to the techniques and results obtained by my own research group and our collaborators and, moreover, to our efforts to elucidate new transport mechanisms and to determine their relevant parameters.

II. THEORETICAL BACKGROUND

The theoretical basis for the various phonon generation and detection schemes will be discussed in the relevant sections below. It is assumed that the reader has sufficient knowledge of the general theoretical description of phonons so that only that part of the description, which is pertinent to the discussion of later sections, needs to be briefly reviewed here. (For a general introductory treatment of the subject see Ashcroft and Mermin (1976). For a more detailed discussion see, for example, Born and Huang (1956), Ziman (1960), or Tucker and Rampton (1972).)

If the ions at one end of a crystal lattice are somehow suddenly uniformly displaced from their equilibrium positions, as a consequence of interionic forces, information on the sudden displacement propagates through the lattice in the form of a displacement wave. Such a displacement wave can always be described in terms of linear combinations of plane wave Fourier components of the form

$$\vec{\eta}_\alpha(\vec{k}) = A_\alpha(\vec{k})\exp[i(\vec{k}.\vec{r}_\alpha-\omega t)] \qquad (2.1)$$

in which $\vec{\eta}_\alpha$ is the displacement of the αth ion (located at $\vec{r}_\alpha$ in the lattice) due to the Fourier component with frequency ω and wavevector $\vec{k}$ evaluated after the elapsed time t. Since the lattice masses are not continuous but discrete, the vector $\vec{k}$ must be one of the discrete set of allowed vectors of the reciprocal lattice. The amplitudes $|A_\alpha|$ are at this point arbitrary except that the sum over all Fourier components must yield the initial displacement wave.

It is often more convenient to describe the displacement wave instead in terms of an orthonormal set of displacement vectors of the form

$$\eta_{\alpha}^{i}(\vec{k},\lambda) = N^{-3/2}\ \varepsilon_{\mu}^{i}(\vec{k},\lambda)\exp[i(\vec{k}.\vec{r}_{\alpha}-\omega t)] \tag{2.2}$$

in which i refers to the ith Cartesian component, ε is called the polarization vector, it is assumed that there are N^3 unit cells in the lattice, and we have introduced the branch index λ. (The total number of branches equals three times s, the number of ions in the unit cell, and μ runs over 1, 2,..., s.) The relations (2.2) can also be obtained by solving the 3s-dimensional dynamical equations of motion of a harmonic lattice and are thereby identified as the normal modes of vibrations of the lattice. As is the case in electromagnetic displacements, it is often convenient to replace the normal modes by an equivalent corpuscular description of the lattice displacements. As a result it is usual to associate a phonon with each of the normal modes.

The solutions of the dynamical equation also yield the relationship between ω and $\vec{k}$ along the 3s branches, the so-called dispersion relations. The density of allowed phonon states, $g(\omega)$, with the same energy $\hbar\omega$ associated with any one branch s turns out to be

$$g^{s}(\omega) = \frac{1}{2(\pi)^3}\int \frac{dS}{\nabla_k \omega^{s}} \tag{2.3}$$

in which $\nabla_k \omega^s$ is the gradient of ω with respect to $\vec{k}$ on the branch

s, and the integration is over all points in the dispersion relation for which the phonon freqency equals ω. For an idealized isotropic solid with linear dependence of ω on $\vec{k}$ (a dispersionless solid) and with a Debye cut-off frequency ω_D, the total density of one phonon state is:

$$g(\omega) = \frac{3}{2\pi^2}\frac{\omega^2}{(\nabla_k \omega)^3} \qquad \text{for } \omega<\omega_D$$

and (2.4)

$$g(\omega) = 0 \qquad \text{for } \omega>\omega_D \ .$$

The phonon states are eigenstates of the pure harmonic lattice, i.e., the lifetime of a phonon once created is infinitely long (except for reflections at boundaries). All real crystals, however, have an anharmonic component in the vibrational potential in whose presence phonon-phonon interactions can take place limiting, thereby, phonon lifetimes. Several attempts have been made to calculate these lifetimes (see e.g., Orbach and Vredevoe 1964, Kwok and Miller 1966, Klemens 1967, and Tua and Mahan 1982). It will be of interest for later discussions to note here a few simplified analytical expressions for the lifetime of phonons in the presence of simple phonon-phonon interactions. A case in point is a phonon of given frequency ω and belonging to the longitudinal acoustic, LA, dispersion branch decaying into two phonons of lower frequency, one belonging to the longitudinal acoustic, and the other to the transverse acoustic branch, TA. The analytic expression for the lifetime against decay, $\tau(\omega)$, most useful for comparison with experimental data is that formulated by Orbach and Vredevoe from a perturbative treatment of the anharmonic lattice. In this formulation the anharmonic interaction is described in terms of linear combinations of second- and higher-order elastic constants, C_i, and in terms of the density of states available to the decay phonons which is taken to be that of an isotropic dispersionless Debye solid. It is found that

$$\tau^{-1}(\omega) = \frac{\hbar[(C_1+C_4) + \frac{3}{2}(C_2+C_5)]^2}{16\pi\, \rho^3 v_\ell^{\,6} v_t^{\,3}} (1-\beta^2)(\beta^2-\beta+\frac{3}{8})\omega^5 \ . \qquad (2.5)$$

In equation (2.5), ρ is the mass density of the crystal in which the phonons propagate, v_ℓ and v_t are, respectively, the propagation velocity of phonons belonging to the LA and TA branches, and $\beta = v_t/v_\ell$. It is important to note at this point that the LA phonon lifetime varies with frequency as ν^{-5}, i.e., high-frequency LA phonons are predicted to decay much more rapidly than low-frequency phonons. The frequency dependence indicated in equation (2.5) has recently been partially verified for CaF_2 for phonons with $\nu \lesssim 3.5$ THz by Baumgartner et al. (1981). (See also the lecture notes of Professor Renk below.)

A related expression has been formulated by Klemens (1967) for the spontaneous decay of a LA phonon into two TA phonons. In this case it is found, again for $T_A \approx 0$, that

$$\tau^{-1}(\omega) = \frac{3\pi}{4\sqrt{2}} \gamma^2 \omega \left(\frac{\hbar\omega}{M v_\ell^2}\right)\left(\frac{\omega}{\omega_o}\right)^3 \beta \qquad (2.6)$$

in which M is the atomic mass, and γ is the Grueneisen constant. It is important to note at this point that the LA phonon lifetime against spontaneous decay varies as ν^{-5}, i.e., high frequency LA phonons are predicted to decay much more rapidly than low frequency phonons. In simple systems near $T_A \approx 0$, the spontaneous three-phonon decay of LA phonons, described above, should dominate the observed temporal three phonon interactions (Orbach and Vredevoe, 1964).

The above expressions describe the decay of nonequilibrium LA phonons in an ambient environment in which the temperature $T_A \approx 0$ K. In a technique, to be described in section 6, it is possible to lift this restraint on T_A which can then take on any value. Moreover, in this case it is the spontaneous decay of optical phonons into pairs of lower energy acoustic phonons which is probed. An approximate analytic expression for the decay time for this case has been obtained by Klemens (1966), and is

$$\tau^{-1} \approx \omega_o \frac{J}{24\pi} \gamma^2 \frac{\hbar\omega_o}{Mv_o^2} \frac{a^3\omega_o^3}{v_g^3} \times [(n_o+1)(n'n'')-n_o(n'+1)(n''+1)] \qquad (2.7)$$

in which the subscript o stands for the optical phonon which decays into two acoustic phonons denoted by superscript prime and double prime, v_g is the group velocity of the acoustic phonons, a^3 is the volume per atom, the n is the phonon occupation number, and J is a number between 1 and 6. One word of caution. Expression (2.7) was derived for a monoatomic lattice, such as crystalline silicon. The expression cannot be expected to be directly applicable for diatomic lattices as, for example, GaAs, etc.

Phonons also scatter elastically from crystal imperfections such as isotopic and foreign-ion impurities. Theoretical expressions for the scattering rates at point imperfections are in general quite complicated (see, for example, Klein 1968). A qualitative insight into the scattering probability can, however, be gained through the assumption that the scattering is of the Rayleigh type (Klemens 1958), that the crystal is an isotropic dispersionless Debye solid, and that the disturbance is limited to the local change in mass, ΔM. Under these assumptions the elastic scattering rate, τ^{*-1}, becomes

$$\tau^{*-1} = (N/\eta)(\Delta M/M)^2\ \omega^4(4\pi v^3)^{-1} \qquad (2.8)$$

in which N is the defect concentration, η is the number of atoms per unit volume, and v is the average group velocity of the acoustic phonons. Note should be taken that high-frequency phonons are predicted to be also elastically scattered more strongly than low-frequency phonons.

At this point we make note of a number of additional general features which govern the experimental methods and observations. With but the exception noted in section 7, the nominal ambient temperature of the crystal system is near 0 K. At this temperature only zero point vibrations remain. In reality the crystals are held at ambient temperatures, T_A, between 2 and 10 K and short pulses of additional nonequilibrium phonons are introduced at a repetition rate such that the entire crystal returns to the ambient temperature before the next pulse of nonequilibrium phonons is injected. Since we are here interested in injecting primarily high frequency, ω_I, phonons, the inequality $\hbar\omega_I > k_B T_A$ holds. This describes the so-called "collisionless" regime (Niklasson 1972) for which $\omega_I > 2\pi/\tau_T$ where τ_T represents the time for the system to return to local thermal equilibrium. (For the theory of the transport in the hydrodynamic regime, for which $\hbar\omega_I < k_B T_A$, see the lecture by Professor Klein.) Moreover, it follows that the mean free path against thermalization, ℓ^*, in reasonably pure materials, will exceed the (injected) phonon wavelength, i.e., $\ell^* > \lambda_I$. Accordingly, the displacement amplitude undergoes several complete oscillations before the phonon is scattered in any way. It is this fact which allows us to carry over the phonon picture into the scattering regime.

It is worth noting further that, according to equation (2.8), elastic impurity scattering of acoustic phonons can be approximated by $\tau^* \approx A^*\omega^{-4}$, whereas equations (2.5) and (2.6) predict that spontaneous anharmonic scattering of LA phonons can be approximated by $\tau \approx A\omega^{-5}$. Average values of A^*, obtainable from the literature for crystals containing $\approx$ 100 ppm imperfections, are of the order of 10^{39} s^{-3}. As has already been noted above, A is found in insulators to be typically of the order of 10^{54} s^{-4}. On the assumption that the phonon propagation velocity is approximately constant and equal to 5×10^3 ms^{-1}, one concludes that the mean free path against impurity and anharmonic scattering is much less than the dimensions of typically sized samples only if the phonon frequency $\nu \gtrsim 1$ THz.

Optical phonons, in contrast, normally possess low group velocities, so that impurity dominated elastic scattering times are long compared to those of most acoustic phonons. As noted earlier, lifetimes against anharmonic decay typically lie in the picosecond regime, and thereby dominate the observable transport properties of these phonons.

III. NONEQUILIBRIUM TRANSPORT OF A HEAT PULSE

A particularly important step forward in the spectroscopy of phonon transport was taken by von Gutfeld and Nethercot (1964) (see also a review article by von Gutfeld (1968)). They introduced the "heat pulse" technique which differs from thermal conductivity measurements in that a nonequilibrium short-duration pulse of phonons is transported across the solid rather than a steady-state thermalized phonon distribution. The technique's primary advantage is that the "time of flight" for phonons to

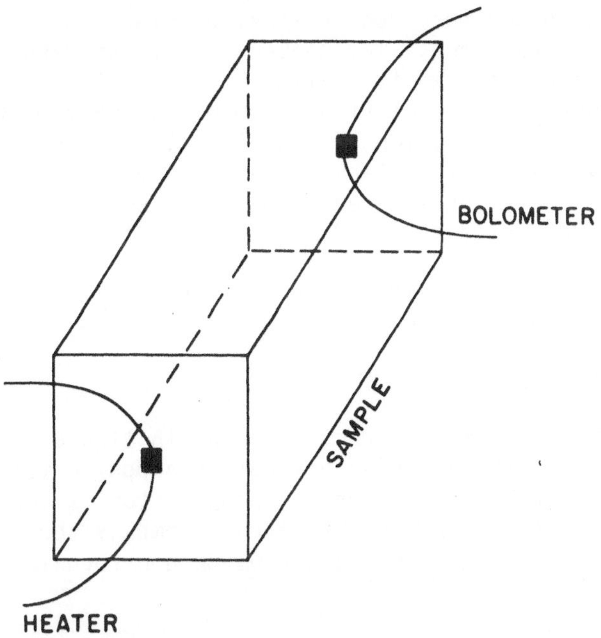

Fig. 1. Sample arrangement in the heat pulse method. The sample is maintained near the superconducting transition temperature of the bolometer.

traverse a solid sample can be directly measured and which, since the length of a given sample is constant, translates into a measurement of the effective phonon group velocity, v_g. Moreover, since v_g is normally different for longitudinally and transversely polarized phonons, the technique permits a separation of the phonon distribution into polarization (dispersion) branches.

A typical experimental arrangement is illustrated in figure 1. This figure shows a thin film metallic heater strip which has been evaporated onto one face of a single crystal. The film acts as a Joule-heated source of a phonon pulse when it is subjected to a short-duration voltage pulse. On the opposing face another small thin metallic film is evaporated which is superconducting at the temperature of the experiment. The latter film acts as a fast response phonon bolometric detector (von Gutfeld and Nethercot 1964). Somewhat more sophisticated detectors in the form of superconducting tunneling junctions may be substituted for the bolometer.

Two types of "time of flight" spectra were distinguished in the early experiments. Since the inherent transport phenomena also play a role in the more recent work, a brief description of these spectra is given here. The heater film is thought (Perrin and Budd 1972) to thermalize in a time short ($\lesssim$ 0.1 ns) compared to the experimental time resolution ($\approx$ 10 ns) thereby producing an approximate Planck distribution of phonons which are transported across the heater-sample interface. In nearly perfect crystals these phonons propagate unhindered across the sample and reach the detector in a time period governed by their group velocity, v_g, and the length, L, of the sample such that the arrival time, t_A, of the phonons at the detector is

$$t_A = L/v_g \tag{2.9}$$

A typical "time of flight" spectrum is shown in figure 2 which displays data taken with a "pure" crystal of sapphire (O'Connor 1977). This type of unhindered phonon propagation is termed "ballistic" flow. (The minor peaks which accompany the major ones arise from side wall reflections and from phonon focusing effects (Maris 1970).)

If instead of a nearly pure crystal, phonons are sent by the same technique, through a crystal of sapphire containing impurity ions, the time of flight spectrum changes to that shown in figure 3. Here the sharp peaks associated with ballistic flow are diminished relative to a strong, very broad, signal which persists

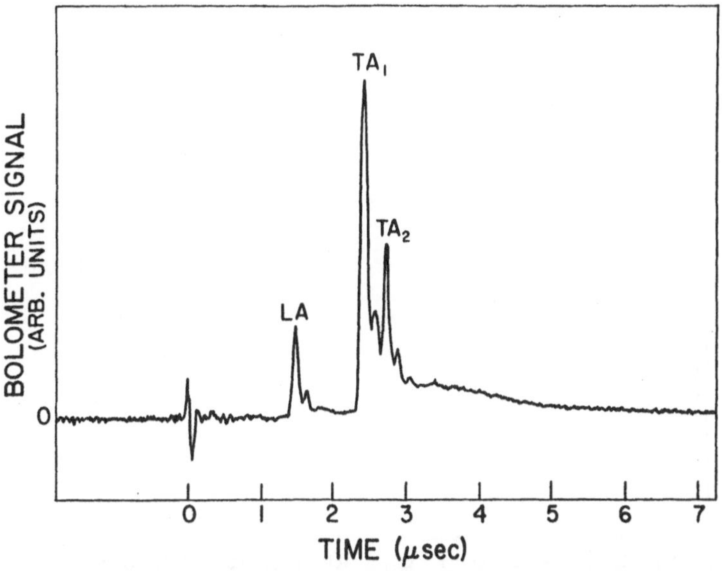

Fig. 2. Time of flight spectrum of a heat pulse injected into a "pure" sample of Al_2O_3.

to long times of flight. This latter behavior, which is termed "diffusive flow," arises (as we shall see in sections 4 and 5) from strong elastic scattering of the phonons at the lattice disturbance produced by the presence of the impurities. Other disturbances such as vacancies, isotopic impurities, dislocations, etc., produce similar effects.

If the impurity concentration is high enough then the phonons undergo random walk diffusion and the mean square displacement, $\overline{X}^2$, from the injection surface approaches

$$\overline{X}^2 \propto Dt \tag{2.10}$$

where D is the diffusion constant which, based on assumptions similar to those of the kinetic theory of gases, is proportional to

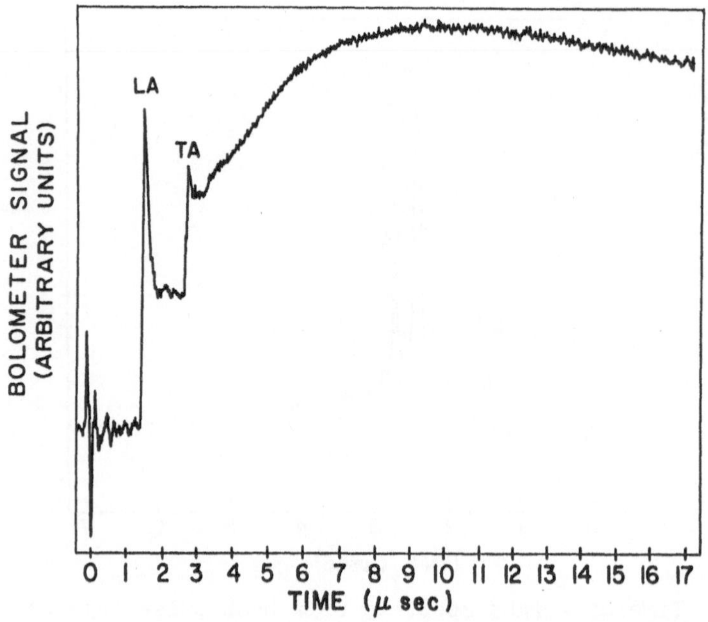

Fig. 3. The time of flight spectrum of a heat pulse generated through a Al_2O_3 sample containing vanadium probe ions. Figures 2 and 3 are taken from O'Connor (1977).

$$D \propto v^2\tau^* \propto v\ell^* \tag{2.11}$$

where ℓ^* is the mean free path, and v the average velocity between scattering events and τ^* is the lifetime against elastic scattering with a frequency dependence approximated in equation (2.8). Since high-frequency phonons scatter more strongly than low frequency ones, we may suspect that in figure 3 the major component of the diffusive signal stems from high-frequency phonons and that the ballistic signal involves primarily low frequency phonons.

The actual frequency dependence of the scattering rate cannot, however, be obtained experimentally by these means, nor can any spatial resolution, since no frequency resolving element is present and since the source and detector are fixed to the surfaces and their relative distances cannot be altered during the experiment. Limited spectral information is obtained by replacing the source and the detector by superconducting junctions again fixed to two opposing surfaces (Eisenmenger and Dayem 1967, Eisenmenger 1976).

In the early investigations the anharmonic scattering was mostly neglected as a contributor to the phonon dynamics, i.e., it was tacitly assumed that $\tau^* \ll \tau$. I shall demonstrate in section 4 through 6 that this assumption is not always valid.

The introduction of optical methods for either the phonon source and/or the phonon detector allows one to vary the source to detector distance, thereby gaining spatial resolution of the phonon dynamics. Moreover, as will be shown in the next section, it becomes possible in most cases to gain spectral information through optical detection the of electron-phonon interaction.

In the earlier experiments in this field the entire sample plus the source and the detector were immersed in the coolant. More recently, and for the cases described herein, the sample is cooled through conduction by attaching it to a cold finger on one (or more) surfaces, but with the source and detector areas remaining under vacuum.

IV. VIBRONIC SIDEBAND PHONON SPECTROSCOPY (VSPS)

A major innovation to the study of nonequilibrium phonon dynamics is the application of the so-called vibronic sideband phonon spectrometer to the problem. The basic physics underlying the spectrometer can best be understood in terms of the following arguments.

The electronic energy levels of a single atom, molecule, or ion in free space are modified upon solution into a solid (for simplicity here taken to be an insulator). We limit consideration to those cases in which the solute ion differs from those of the solid in that it possesses electronic states at energies less than the excited electronic states of the pure solid. Since upon solution the local symmetry is lowered, one effect is to remove some degeneracies in the electronic states of the solute which exist in the higher symmetry of free space, and another is to shift the energies of the states. These effects, and the corresponding spectral lines, are illustrated in figure 4a and 4b. In addition, the local solute's electronic states are modulated, through electron-phonon coupling, by the time-varying potential field which arises from the vibrational motion of the ions in the solid, i.e., from the presence of phonons. As a consequence of the modulation, the electronic transitions between the localized

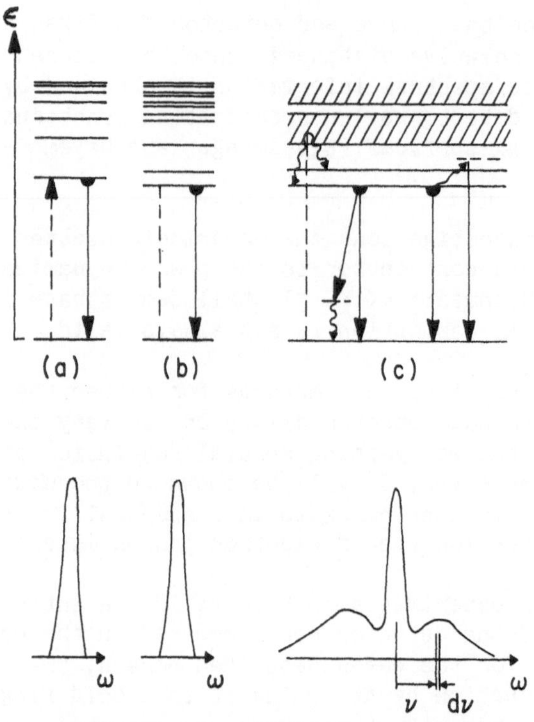

Fig. 4. a) Energy level scheme (upper diagram) and luminescence spectral lines (lower diagram) for a hypothetical probe ion in free space. The dashed lines indicate absorptive-, the solid lines indicate luminescent-, and the wavy lines nonradiative-, transitions. b) The effect of placing the probe ion into a static crystal field. c) Inclusion of a vibrational crystal field. The Stokes and antiStokes vibronic sidebands are shown, respectively, to lower and higher luminescent frequencies relative to that of the "zero phonon" spectral line.

states are accompanied by phonon sidebands, the so-called "vibronic sidebands," which appear in the observed absorption and emission spectra in addition to the normal spectral line (the so-called "zero-phonon" line). If the phonon coupling to a particular electronic state is very strong, multiphonon vibronic sidebands dominate the transition and broad absorption (or emission) bands appear in the spectrum (Bron and Wagner 1965).

These bands are convenient for the absorption of excitation energy as indicated in figure 4c. In the weak coupling case, contributions to the one-phonon sideband amplitude from all phonons of a given frequency interval, between ν and $\nu+d\nu$, are observed displaced by ν from the center frequency of the zero-phonon spectral line (see figure 4c). If a suitable zero-phonon line can be found somewhere in the visible spectrum, the vibronic interaction shifts, to this experimentally convenient range, the information on phonons which would otherwise be in the far-infrared region of the spectrum.

In case the electrons of the solute are strongly localized, the resultant vibronic sidebands contain information which readily yields the spectral distribution of the phonons which can interact with a given electronic transition. The vibronic spectra associated with the solute are, accordingly, effective probes of the phonon states in the crystal.

Vibronic sidebands can be used to gain detailed descriptions of the frequency distribution of the allowed phonon states, and of the electron-phonon interaction in a variety of solids. Moreover, it has been demonstrated that they may be used to test theoretical models of the electron-phonon interaction, the lattice dynamics of the host crystal, and the modification of the lattice dynamics which results from the presence of the probe ion (Bron 1972, Bron 1975, Timusk and Buchanan 1967, Wagner 1968).

The application of vibronic sideband spectroscopy of special interest here involves the spectral, spatial, and temporal resolution of nonequilibrium phonon distributions. It is this form of phonon spectroscopy which we discuss in some detail below.

Analytical expressions for the amplitude of the vibronic sidebands can be written (Wagner 1968) in the following form for the (one-phonon) "Stokes sideband," which appears to the low-energy side of the zero-phonon luminescence line,

$$I_S(\omega) = \frac{e^2\hbar}{4\omega}(n(\omega)+1)\sum_\Gamma F(\Gamma)\rho(\omega,\Gamma)F'(\Gamma) \tag{3.1}$$

whereas the amplitude of the (one-phonon) "antiStokes sideband," which appears to the high-energy side of the zero-phonon luminescence line, can be written as

$$I_{AS}(\omega) = \frac{e^2\hbar}{4\omega}n(\omega)\sum_\Gamma F(\Gamma)\rho(\omega,\Gamma)F'(\Gamma) \quad . \tag{3.2}$$

In equations (3.1) and (3.2), $n(\omega)$ is the phonon distribution function, $\rho(\omega,\Gamma)$ is the phonon density of states projected onto the irreducible group representations, Γ, which are consistent with the electronic transition, and $F(\Gamma)$ is an electron-phonon interaction operator.

The Stokes sideband represents the simultaneous emission of phonons during the electronic transition and the antiStokes sideband represents the simultaneous absorption of phonons. In the above expressions $n(\omega)$ follows a Planck distribution if a temperature, T, can be assigned to the phonon distribution. It is, accordingly, clear that in equation (3.1) $n(\omega)$ vanishes as $T\to 0$ and no phonons are injected into the solid by other means. In this case only the sum on Γ (times $e^2\hbar/4\omega$) survives. The surviving part is the spectral response function, $R(\omega)$, of the electronic system to the electron-phonon interaction. It is then clear that when phonons are present

$$I_{AS}(\omega)/R(\omega) = n(\omega) \qquad (3.3)$$

such that the antiStokes sideband, when corrected by the response function, directly yields the spectral distribution function for the phonons in the solid. This is the basis of the vibronic sideband phonon spectrometer (VSPS).

The capability of the vibronic sideband phonon spectrometer has been extended to simultaneous spectral, spatial, and temporal resolution of nonequilibrium phonon distribution (Bron and Grill 1977a,b). In these experiments the luminescence from Eu^{2+} probe ions is used to determine the evolution of an injected phonon distribution propagating in a single crystal of SrF_2. The luminescence is excited in a focal column (beam waist) of a N_2 laser beam. The focal column can be placed anywhere inside the crystal gaining, thereby, spatial resolution. Temporal resolution is obtained by gating the luminescence signal from a photomultiplier tube and delaying the onset of the luminescence together with the gate by varying amounts relative to the time of phonon injection.

As a first application of their vibronic sideband phonon spectrometer (VSPS), Bron and Grill carried out a comparison between theoretical predictions and the experimentally determined spectral content of the phonon distribution which enters the crystal after it has been generated by Joule heating of a thin-film resistive element evaporated onto one of the crystal faces. A sketch of the sample and the optical arrangement is shown in figures 5 and 6.

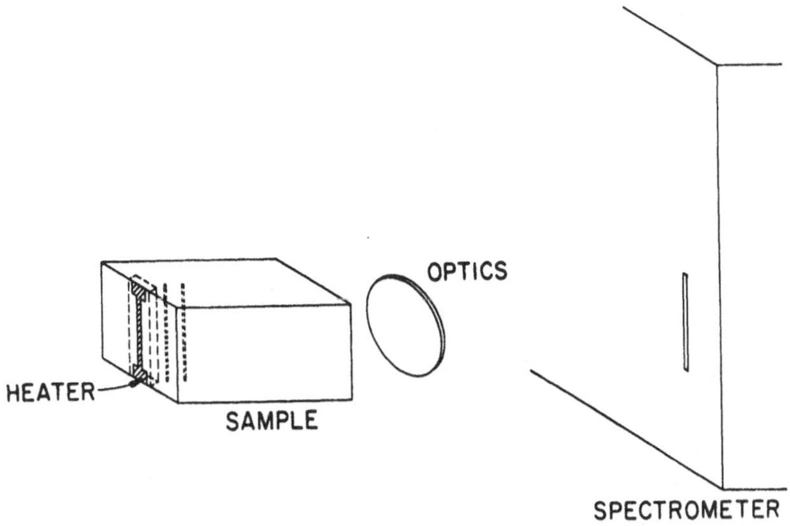

Fig. 5. Sample arrangement for the VSPS. The shaded area on one face of the sample indicates an evaporated heater film. The broken lines indicate regions of the sample which were illuminated by focused laser beams causing, thereby, luminescence which was optically collected and focused on the slit of a high resolution grating spectrometer.

The theoretical models of the phonon distribution emitted from "heater" films include the so-called acoustic mismatch (AM) model (Little 1959, Weis 1969) and a somewhat more advanced treatment by Perrin and Budd (1972). The acoustic mismatch model leads to the prediction that the spectral energy flux across the film-crystal interface follows from the film acting as a thermal radiator, i.e., that the spectral function is that of a Planck distribution. In both models it is tacitly assumed that the transport through the interface is frequency independent and determinable from acoustic mismatch theory in terms of simple continuum mechanics.

The experimental results indeed show that, for input power

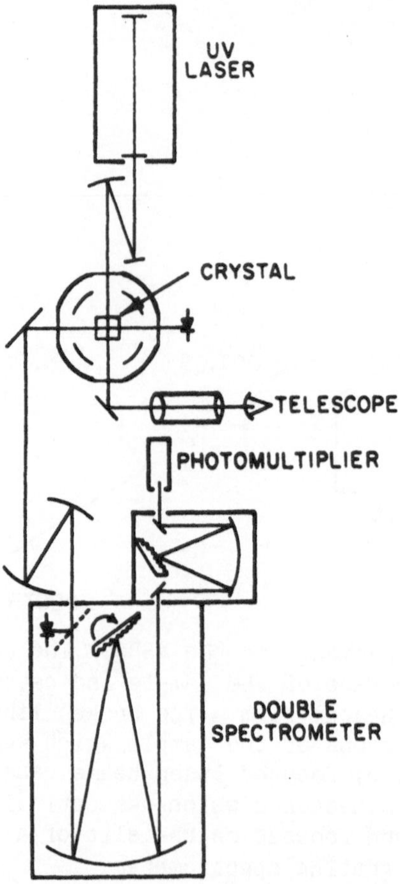

Fig. 6. Experimental apparatus for the VSPS. The sample is attached to the cold finger of an optical Dewar.

densities of the order of tens of W/mm^{-2}, the observed spectral energy flux is in rough agreement with theoretical predictions. However, a predicted increase in the proportion of high-frequency phonons as the input power, and hence the effective film temperature, is increased is not observed. In fact, the spectral distribution of the flux is found to be roughly independent of input power. Essentially only the amplitude of the distribution changes. The effect was described by Bron and Grill in terms of a phenomenological frequency-dependent reflective filter which serves to retain phonons with frequencies in excess of 1 THz in the heater film for periods longer than those considered in the theoretical models. An empirically obtained filtering function is shown to require a decrease in the transmission efficiency of four orders of magnitude between 1 and 3 THz. Although these early experiments shed no further light on the origin of the filtering action, the results are similar to those previously reported. Indeed, under conditions in which the phonon distribution contains primarily phonons of frequency less than 1 THz (Herth and Weis 1970) the acoustic mismatch model has been found to adequately describe experimental results. In contrast, in several other experiments (Herth and Weis 1970, Cheeke et al. 1972, Kappus and Weis 1973, Martinon and Weis 1979) in which heater films are pulsed to higher powers, and hence contain phonons of higher frequencies, deviations from theory have also been observed.

V. PHONON TRANSPORT IN DISPERSIVE MEDIA

In the discussion of heat pulse transport presented so far, we have not fully taken into account two major contributors to the phonon transport; namely dispersive and anharmonic effects. We consider dispersive effects next. A dispersive medium in the present context is a medium in which some parameters of the transport depend on the phonon frequency. It had already been postulated by Bron and Grill (1977b) that the phonons entering a substrate of SrF_2 containing Eu^{2+} probe ions are diffusively scattered with a probability which increases with increasing phonon frequency, ν. As a result, high frequency phonons ($\nu > 0.6$ THz) spend long periods of time ($> 10^{-6}$ sec) very close to the film-substrate interface. These observations led to the development of a model treatment (Schaich 1978) of the distribution of phonons emitted by a metallic film heater into a diffusive medium. The treatment indicates, for experimental observations over time periods of the order of microseconds (Bron and Grill 1977a,b), that the phonon distributions in both the film and the substrate do not reach a steady state. As a result of the frequency dependent diffusive scattering, the higher frequency phonons are readily scattered back into the heater film, where they rapidly thermalize (Perrin and Budd 1972), and eventually emerge from the film as lower frequency phonons which alone propagate in the substrate with relative ease (Bron and Keilmann 1975). It is further demonstrated, with parameters which correspond to the experimental conditions of Bron and Grill (1977a,b), that the spectral distributions which are calculated to arise in the substrate are similar to those observed in Bron and Grill (1977a). A necessary consequence of the back scattering of the high frequency phonons is that the instantaneous "temperature," and therefore the instantaneous electrical resistance, of the metal film increases with time while the film is electrically driven, and decays thereafter at rates consistent with the frequency dependent phonon transport in the substrate.

In contrast, under the AM model a steady state between the electrical power into the film and the phonon energy flux out of the film (into the substrate) is reached within a period, τ', comparable to the electron-phonon interaction time and the transit time of phonons across the film (Perrin and Budd 1972). For the samples of $SrF_2:Eu^{2+}$ used $\tau' < 0.5 \times 10^{-9}$ sec. Under these assumptions, no time dependent changes in the resistivity of the heater film during or after the Joule heating pulse should be observed in experiments with temporal resolution of the order of nanoseconds.

In order to check these predictions an experimental measurement was undertaken by Bron, et al. (1979). In the experiment of the instantaneous resistance of ≈ 3000 Å x 5 mm x 0.22 mm nickel films was measured during and after Joule heating pulses of approximately 85 nsec (halfwidth) duration and up to 5 KW electrical power absorbed by the film. Results were reported for three crystalline substrates: i) SrF_2: 0.1 mole % Eu, ii) "pure" SrF_2 containing <0.01 mole % foreign impurities, and iii) high purity Z-cut $A\ell_2O_3$. The instantaneous resistance of the film during the heat pulse was determined by measuring the voltage reflection coefficient from a 50 Ω transmission line terminated by the heater film. After the heating pulse is terminated, the same measurement was accomplished by sending down the transmission line, at variable times after the Joulse pulse, a 40 nsec duration probe pulse whose peak power was less than one-tenth that of the Joule heating pulse (see insert in figure 8 below). Temporal resolution is achieved with a gated boxcar integrater and is approximately 15 nsec throughout.

The time dependence of the film temperature <u>during</u> the Joule pulse as a function of absorbed electrical power, P_A, is shown in figure 7. The temperature is defined as that which corresponds to the same resistance which an equivalent film exhibits at thermal equilibrium during an independent measurement of its thermal coefficient of resistivity. Below 30 K (shaded area) the film temperature could not be accurately determined because in this region the resistivity is dominated by scattering at imperfections and is practically independent of temperature. The film temperature <u>after</u> the Joule pulse is turned off is plotted for two cases in figure 8. The results shown in figure 7 indicate that nickel films on both "pure" and Eu doped SrF_2 substrates exhibit time dependent increases in temperature during the Joule pulse which depend on P_A. No time dependent effects could be detected for $P_A \lesssim 10^3$ W. Higher absorbed powers are required to reach the

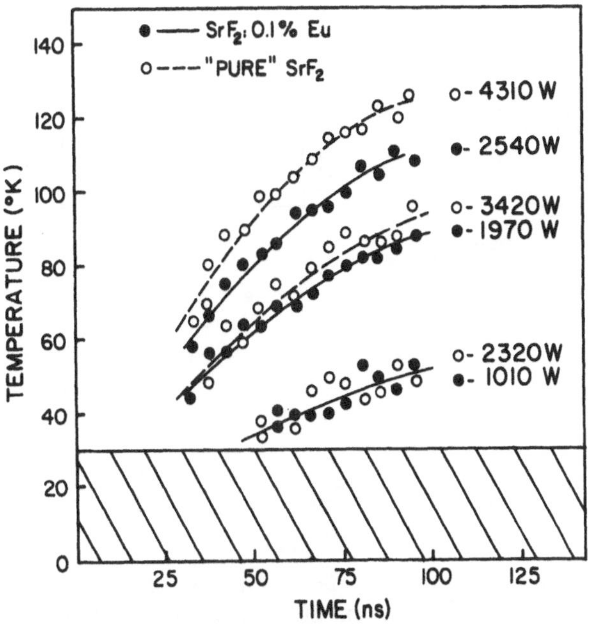

Fig. 7. Time dependent temperature of nickel thin film heaters during the time that a Joule voltage pulse is on. The solid and dashed lines are smooth line fits to the data points. See text for explanation of shaded area.

same terminal temperature in the "pure" SrF_2 than in the doped SrF_2. After the Joule pulse the temperature of the film slowly decreases (nonexponentially) and eventually returns to 4.2°K. Neither of these time dependent effects was observed in nickel films on the $A\ell_2O_3$ sample for P_A up to 4980 W, although a slight heating effect during the Joule pulse was observed for $P_A > 5$ KW.

Since both the pure and Eu doped SrF_2 exhibit time dependent resistances it is clear that the Eu probe ions are not the sole source of the observed effects. That these effects are not found with $A\ell_2O_3$ substrates, up to even higher absorbed electrical power than for SrF_2 substrates, argues against the possibility that the observed effects arise from phonon-phonon anharmonic scattering in

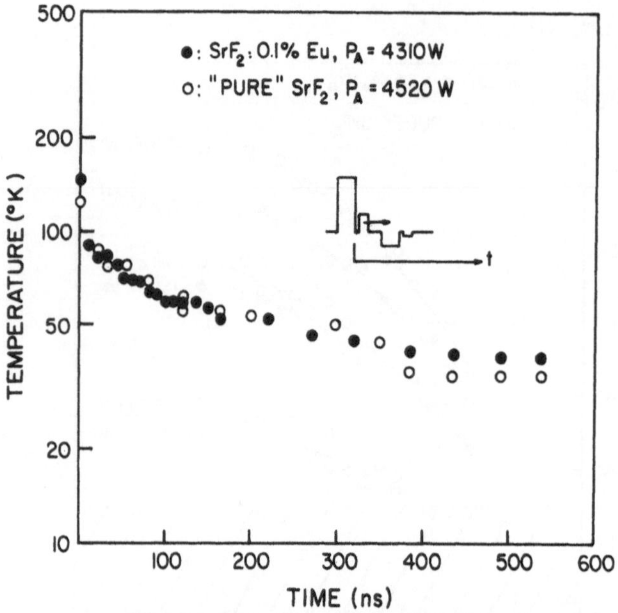

Fig. 8. Time dependent temperature of nickel thin film heaters on SrF_2 crystals after the Joule pulse is turned off. The insert illustrates the main and probe pulse and their reflections.

these crystals. Moreover, since the scale of the time dependent effects is of the order of tens of nanoseconds, it is unlikely that the source is in the metal film or the film-substrate interface for which we would expect characteristic times of the order $\tau' < 0.5 \times 10^{-9}$ sec.

The main difference between "pure" $A\ell_2O_3$ and "pure" SrF_2 appears to be their relative natural isotopic impurity content which is a total of ≈ 0.24% of the oxygen in sapphire and 17.26% of the Sr in SrF_2. The effects of these isotopes has to be compared to that of the Eu probe ions. A rough estimate can be made of the relative scattering strengths on the basis of equation (2.8). Using handbook values for the various isotopic impurities, one obtains relative scattering rates for $A\ell_2O_3/SrF_2/SrF_2$:Eu to be roughly 1/23/173, which is in qualitative agreement with the experiment.

These results were compared to the predictions of the model proposed by Schaich (1978). For simplicity it was assumed that the transport of the real phonon distribution can be adequately represented by a two-frequency Einstein spectrum. The high frequency, diffusive, mode is placed at 3 THz, with an elastic scattering rate of 2.5×10^9/sec; while the low frequency, ballistic, mode is at .6 THz and has a relative weight of 1/125 in order to correspond to a Debye density of states. With these parameters good overall agreement is obtained with the observed heating in the doped SrF_2 for the case of $P_A = 2540$ W. Some disagreement arises, however, between observed and calculated temperatures at lower P_A with, for example, the calculated temperatures for $P_A = 1010$ W being some 10 K higher than the observed temperatures, although the rates of change are quite similar. The ratio of the scattering probabilities of Eu versus isotopic impurities in SrF_2 is determined by the model to be 5, which compares well with the value of 7.5 cited above. The decay of the film temperature after the voltage pulse is also in qualitative agreement with the model treatment. The slow decay arises from diffusive flow of high frequency phonons back into the heater at the same time that their spatial distribution diffuses further and further into the substrate.

As will, however, be demonstrated in the next section on "quasidiffusion," that for experimental observations in the microsecond domain, anharmonic process must also be taken into account; particularly in transport in dispersive media.

In order to probe the effect of anharmonic processes on the heater "temperature" Schaich (1984) has revised his earlier formulation to include a model treatment of anharmonicity, albeit limited to three phonon interactions. The full application of this formalism will be discussed in the next section. Here we merely note that a comparison of these theoretical results (see figure 1 in Schaich (1984)) and those shown in figures 7 and 8 is insufficient to decide whether or not anharmonic interactions play an important role. For this purpose more sophisticated experimental techniques are required.

VI. QUASIDIFFUSIVE PHONON DYNAMICS

The term "quasidiffusive" transport was coined by Kazakovtsev

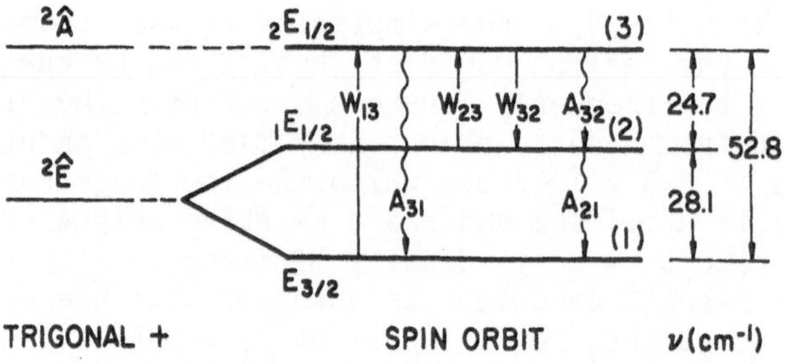

Fig. 9. Energy level diagram and group representational assignments for the lowest three levels of the d^1 configuration of V^{4+} in Al_2O_3 and transition rates for electromagnetic absorption and emission (solid lines with arrows) and phonon emission (wavy lines).

and Levinson (1978, 1979, and 1981) to describe a new transport mode which nonequilibrium phonons undergo in a dispersive medium in the presence of anharmonic interactions. The special requirement of "quasidiffusive" vs. "quasiballistic" transport is that in the former the rate of elastic scattering τ^{*-1} is faster than that for anharmonic scattering, τ^{-1} and vica versa. In this section, I show two separate experimental observations which demonstrate the existence of quasidiffusion.

From a tutorial perspective it is advantages to interrupt at this point the discussion of the transport of a broadband (heat) pulse and consider instead first the transport of an originally monochromatic phonon distribution.

In an experiment by Bron et al. (1982) monochromatic phonons were injected into a sample of sapphire containing V^{4+} ions. Quadravalent vanadium when dissolved in $A\ell_2O_3$ has a lowest lying set of electronic energy levels as illustrated in figure 9. The ground state is numbered 1 in the figure. The first excited state

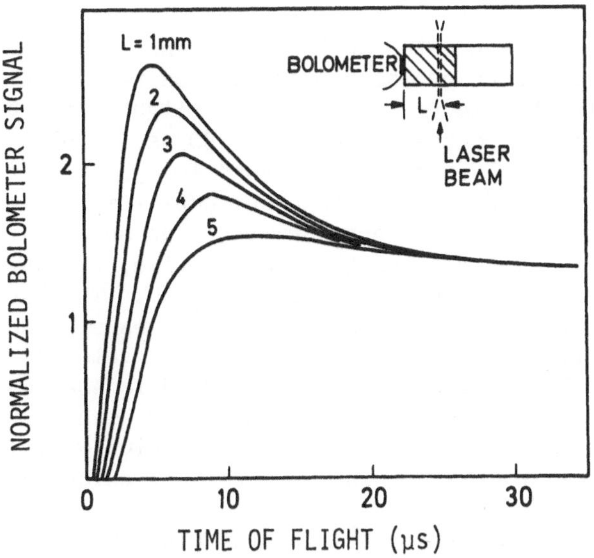

Fig. 10. Time of flight spectrum for various propagation distances, L. The hatched area in the inset represents the vanadium doped part of the Al_2O_3 crystal.

(numbered 2) appears at 28.1 cm^{-1} (0.84 THz) above the ground state and the next excited state (numbered 3) appears at 52.8 cm^{-1} (1.58 THz) above the ground state. Electric dipole absorptive transitions from the ground state (1) to either the (2) or the (3) excited states have been observed by Wong et al. (1968).

We limit here attention to the 1 → 2 absorptive transition followed by spontaneous phonon emission during the relaxation from the excited to the ground state (2 → 1). (See Bron and Grill (1978) for a discussion of stimulated phonon emission from the 3 → 2 transition.) The sample was exposed to a focused beam from a CO_2-laser-pumped CH_3NH_3 FIR laser operating at 28.8 cm^{-1}. The 1 → 2 transition, peaked at 28.1 cm^{-1}, is sufficiently broad so that resonant absorption of the incident light of roughly 30% occurs. Since at this frequency the density of one-phonon states greatly exceeds that in the photon field, the most likely de-excitation of the electronic system is the emission of phonons with $\omega_o = 28.1\ cm^{-1}$. A superconducting tin bolometer, 1 x 1 mm² in cross section, was evaporated on the doped end of the crystal and the center of the focused laser beam was positioned at distances L from the bolometer. The time-of-flight spectrum (arbitrarily normalized to the value at t = 80 μs) as obtained at various fixed values of L is shown in figure 10. If a diffusion profile is forced to fit the data, then the empirically obtained diffusion constant is found to increase with L as shown in figure 11, which also contains a plot of the arrival time, t_A, as determined from the leading edge of each spectrum. The ratio L/t_A is found to be 2.3×10^5 cm/s which is some 5 times smaller than the LA and some 3 times smaller than the TA sound velocity. No signal is detected if the laser light is focused in the undoped region of the crystal, and the signal is observed with $28.8\ cm^{-1}$ light, provided the electric component is perpendicularly polarized to the c-axis in agreement with the selection rules governing the transitions of the V^{4+} ion (Bron and Grill 1978). It is clear from these results that the observed detector signal is the result of phonons which were at least originally monochromatic and which have traversed a volume of a crystal which contains resonant and nonresonant elastic scattering sites. The situation is in fact analogous to the resonant trapping phenomenon as discussed in the lecture by Professor Renk in connection with $29\ cm^{-1}$ phonons in ruby. This being the case, one would expect strong diffusive transport. However, the experimental observations shown in figure 11 are inconsistent with this hypothesis, since for simple diffusion of monochromatic phonons the diffusion constant, D, should be a constant of the motion and not depend on L. Moreover, for simple diffusive motion the arrival time, t_A, should scale as L^2 (see equation (2.10)) and not linearly with L as it appears in figure 11. In fact, the only observation consistent with diffusive motion is that the time of flight spectrum has a broad peak to which the diffusion equation can be readily fit.

It has, however, been pointed out (Kazakovtzev and Levinson 1978, 1979) that, if during the time of observation the original

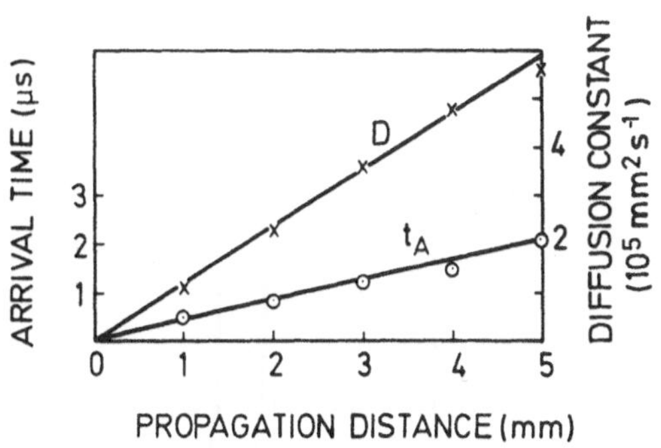

Fig. 11. Arrival time, t_A, and the empirically obtained diffusion constant, D, as a function of propagation distance, L.

phonon packet, in addition to elastic scattering, experiences anharmonic decay, then a new transport mode termed "quasidiffusion" can arise in the phonon dynamics.

In order to discuss quasidiffusive transport we limit ourselves to the regime for which the elastic scattering time τ^* is shorter than the anharmonic scattering time τ. For simplicity, we further assume that all anharmonic processes obey equation (2.5) or (2.6) and that the two phonons generated each have half the energy of the original phonon, i.e., $\hbar\omega_0/2$. It follows that

$$\tau(\omega_0/2) \sim \tau(\omega_0)/32 \qquad (6.1)$$

Moreover, we assume that equation (2.8) describes the controlling elastic scattering processes, (for all but the resonantly scattered species) so that

$$\tau^*(\omega_0/2) \sim \tau^*(\omega_0)/16 \qquad (6.2)$$

The temporal, spatial, and spectral evolution of the originally monochromatic phonon distribution is illustrated in figure 12. In the figure we sketch the spatial extent of the phonon distribution at a set of specific times in the temporal evolution. (The distribution in the interim between such special times is implied but not specifically shown.)

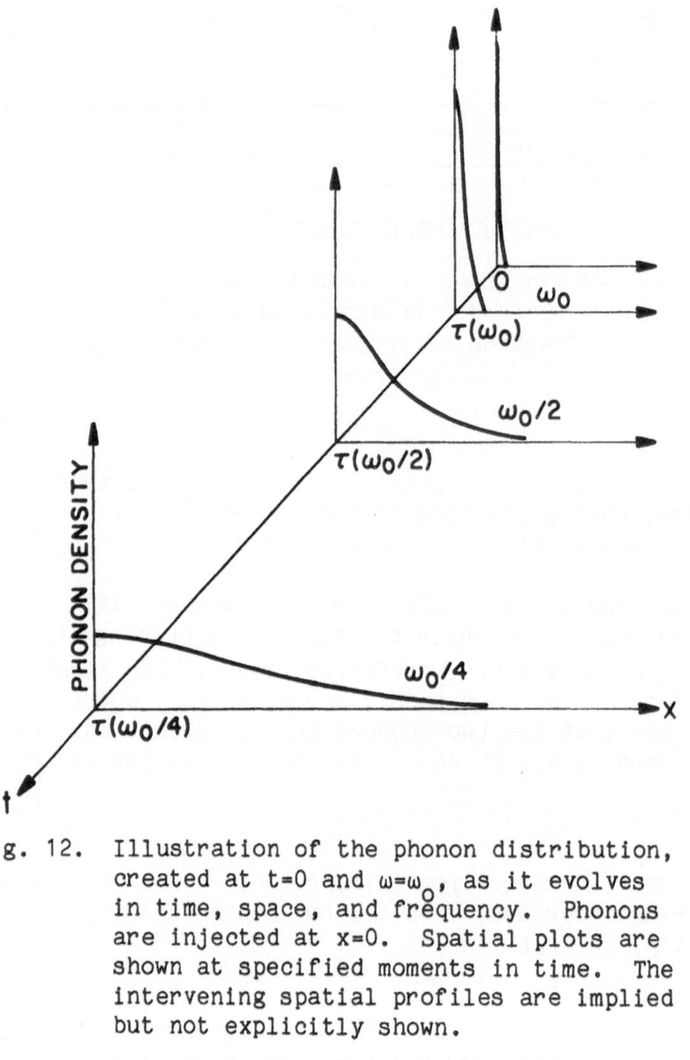

Fig. 12. Illustration of the phonon distribution, created at t=0 and $\omega=\omega_o$, as it evolves in time, space, and frequency. Phonons are injected at x=0. Spatial plots are shown at specified moments in time. The intervening spatial profiles are implied but not explicitly shown.

At the zero of time, t=0, a monochromatic nonequilibrium distribution of phonons is injected into the crystal at some distance from the position of the detector which we define as x=0. Since $\tau^* < \tau$ these phonons execute strong diffusive propagation which limits their location to a volume not much bigger than the excitation volume. However, in a time $t = \tau(\omega_o)$, 1/e of the original phonons have decayed to $\omega_o/2$ phonons. By virtue of the strong dependence of ω on τ^* and τ these "children" (at $\omega_o/2$) of the "parent" distribution (at ω_o) occupy a much larger volume of the crystal during their increased lifetime than that of the parents. This process may well repeat itself producing successive generations of children at $\omega_o/4$, $\omega_o/8$,... with each successive "generation" of phonons living much longer and occupying vastly greater space during its lifetime than its predecessors. In fact, to each generation it looks as though all previous generations occupy delta functions in space and time compared to its own domain.

Accordingly, a measurement of the temporal evolution of the phonon population at a distance L_A from the site of injection reflects almost exclusively the transport of the last generation. That is, for all intents and purposes, the contributions of all previous generations can be neglected. Thus the distance traveled to the detector $L_A \approx C(D_A t_A)^{1/2}$ and $t_A \approx \tau(\omega_A)$, where D_A is the effective diffusion constant of the generation which arrives at the detector, and C is a factor of order unity. But from equation (2.11), $D_A \approx (v^2)\tau^*(\omega_A)$, such that $(L_A)^2 \approx (v^2)\tau^*(\omega_A)t_A$. Accordingly, $L_A \approx v[\tau^*(\omega_A)\tau(\omega_A)]^{1/2}$. Moreover, from equations (2.5) and (2.6), $\omega_A \approx \omega_o(t_A/\tau_o)^{-1/5}$. By direct substitution, one finds that

$$t_A = C' t_B (t_B \tau_o{}^4/\tau_o{}^{*5})^{1/9} \quad , \tag{6.3}$$

where $t_B = L_A/v$ (the ballistic arrival time (see equation (2.9)) and C' is another factor, of order unity, which contains all the geometric and other proportionality factors which have so far been neglected, and we have shortened the symbol $\tau(\omega_o)$ to τ_o, etc. It is assumed that $\tau_o{}^* < \tau_o < t_B$. It follows from equation (6.3) that $t_A \propto L^{10/9}$ and $t_A > t_B$.

To summarize, the distinguishing features which characterize quasidiffusive propagation are that t_A is nearly linearly related to L_A and not to $L_A{}^2$ as in pure diffusion. On the other hand, $t_A > t_B = L/v$. Moreover, the time-of-flight spectrum has a long tail similar to pure diffusion, but the empirically observed diffusion constant increases as L increases.

These are precisely the characteristics observed

experimentally. This experiment was the first to observe quasidiffusive transport, and serves to illustrate the power of the modern optical techniques in comparison to the heat pulse and thermal conductivity measurements.

We now turn to quasidiffusive transport of a broadband phonon distribution (Wilson et al. 1984). In fact, we return to the case, discussed earlier, of phonons generated by the "heat pulse" method and propagating in a sample of $SrF_2:Eu^{2+}$. Moreover, we again apply the vibronic sideband spectrometer to obtain a more quantitative evaluation of the phonon transport than was available to Bron and Grill (1977a,b).

Before proceeding we briefly review the characteristics of quasidiffusive transport already recognized above. Among these characteristics are; (i) as a result of the strong elastic scattering, the high frequency component of the phonon distribution undergoes random deflections and, accordingly, propagates diffusively, hence (ii) the high frequency component propagates only slowly away from the phonon source, and (iii) since anharmonic processes are present, the spectral composition of the phonon distribution changes with time, and consequently since Γ^* is a strong function of ν, the effective diffusion rate changes with time and displacement from the phonon source.

The experimental arrangement employed is similar to the one used by Grill and Bron (1977a,b) and aleady discussed in section 4 above. Major refinements were achieved, however, in the signal to noise recovery by changing from the original analog signal recovery to a digital signal recovery system, and by carefully normalizing the intensity of the zero-phonon line and the anti-Stokes sideband structure $I_{AS}(\omega)$, as well as that associated with the detectivity function $R(\omega)$. This procedure (see equation (2) of Bron and Grill 1977a) makes it possible to report the absolute (rather than relative) values of the phonon distribution function $n(\omega,R,t)$, for phonons of a given frequency ω, at various positions R relative to $z=0$ and at any time t after $t=0$. Alternatively, we may transform to the mode temperature $T(\omega,R,t)$ since

$$T(\omega,R,t) = \frac{\hbar\omega}{k_B} [\ell n(n(\omega,R,t)^{-1} + 1)]^{-1} \quad . \tag{6.4}$$

As in the previous work, the actual signal recovered is the vibronic sideband intensity when the broadband phonon distribution

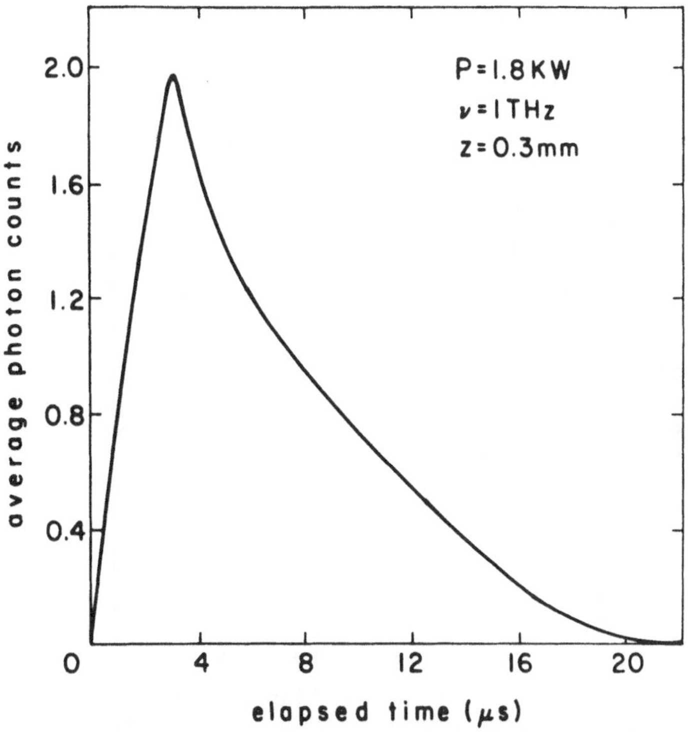

Fig. 13. Vibronic sideband intensity as a function of elapsed time, t.

is present less that when only ambient phonons are present. Thus to within linear effects, only the signal due to the nonequilibrium injected phonon distribution is observed here.

Figure 13 is an example of the observed vibronic sideband signal intensity plotted as a function of the elapsed time after the heater pulse has been injected at t=0, z=0. This "time of flight" spectrum was measured for ν = 1 THz, at a detector column located at z=0.3 mm, after a voltage pulse of 40 ns duration, such that a peak power, P, of 1.8 KW, was passed through the heater film. The lack of sharp structure plus the broad temporal distribution of this signal have already been assigned to "diffusive" phonon propagation (see e.g., von Gutfeld 1968) brought on by elastic scattering at lattice imperfections. We now

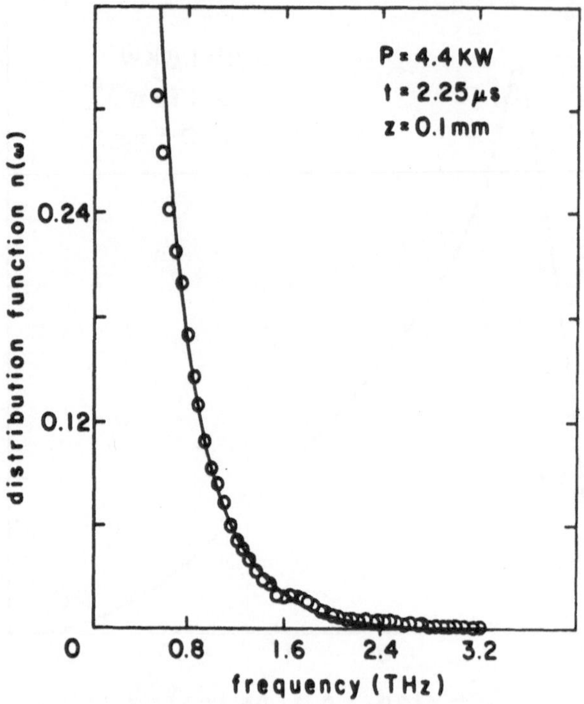

Fig. 14. The variation of the observed phonon distribution function, n(ω,R,t), with phonon frequency. Circles are data points. The solid line is a best fit Planck distribution to the data.

show, however, that other experimental evidence requires that anharmonic scattering processes are also present. Figure 14 is an example of the observed phonon distribution function, n(ω,R,t), here presented as a function of ν, as observed immediately behind the heater after 80 ns have elapsed after a 4.3 KW (40 ns duration) pulse was passed through the heater film. The solid line is a best fit to a Planck distribution (T=22.2 K). Note n(ω) for small phonon frequencies reaches values greater than 0.3. Also note that the overall agreement with a black body distribution is good, but that definite deviations from that distribution exist. These deviations can be more readily observed in figure 15a which is a plot of T(ω,R,t) as a function of ν for t=80 ns and various values of z. In this representation a true Planck distribution would be a straight horizontal line.

Deviations from the Planck distribution are easily recognized particularly when z=0.1 mm. These deviations are always observed under our experimental conditions and are fully repeatable in every detail. After increasing periods of time (figures 15a to 15d) the deviations are seen to disappear. Another observation is that at any fixed time, for $z > 0.01$ the low frequency component of the distribution tends to be at a slightly higher mode temperature than the high frequency component. Moreover, the high frequency component is almost at a constant mode temperature: i.e., phonons of $\nu \gtrsim 1.6$ THz tend to form a local quasiequilibrium among themselves. Finally, at early times the highest mode temperatures occur near z=0, but as time progresses this difference disappears such that at t = 4.5 μs a more complete spatial quasiequilibrim is forming albeit at a mode temperature of nearly 3 times the ambient temperature of 5 K. A true thermal equilibrium at the ambient temperature of 5 K is reached at a much slower rate and is completed only in times of the order of milliseconds (Bron, et al. 1979). This result reflects the slow heat conduction from the sample to the cold finger to which it is attached. It is for this and other reasons that the heater pulse repetition rate is kept to less than 50 Hz.

To summarize the observed experimental features of the transport of the broadband phonon distribution we note; (i) the "time of flight" spectrum follows that usually assigned to purely diffusive phonon transport, (ii) at the time of injection immediately behind the injection surface, the phonon distribution is almost a black body source but with marked deviations, (iii) the deviations from the Planck distribution disappear as the elapsed time after injection increases, (iv) in the interior of the sample and for $t \lesssim 2.25$ μs the mode temperature of the low frequency component of the phonon distribution is slightly higher than the high frequency component, with the latter component tending to form a local quasiequilibrium, and (v) as time progresses a more complete spatial quasiequilibrium begins to form.

The origin of the deviations from a Planck distribution remain unknown. In the earlier work it was described in terms of an <u>ad hoc</u> interface filtering function. We suspect these features arise from residual pump oil deposits which lie at the interface between the evaporated heater film and the SrF_2 crystal. The disappearance of this structure with time, plus the formation of a quasiequilibrium imply that thermalization, through anharmonic processes, takes place within the experimental time span. This conclusion is in contrast with the earlier interpretation by Bron and Grill (1977a,b) that only elastic scattering processes, at the Eu^{2+} and isotopic impurities, play a role and give rise to the

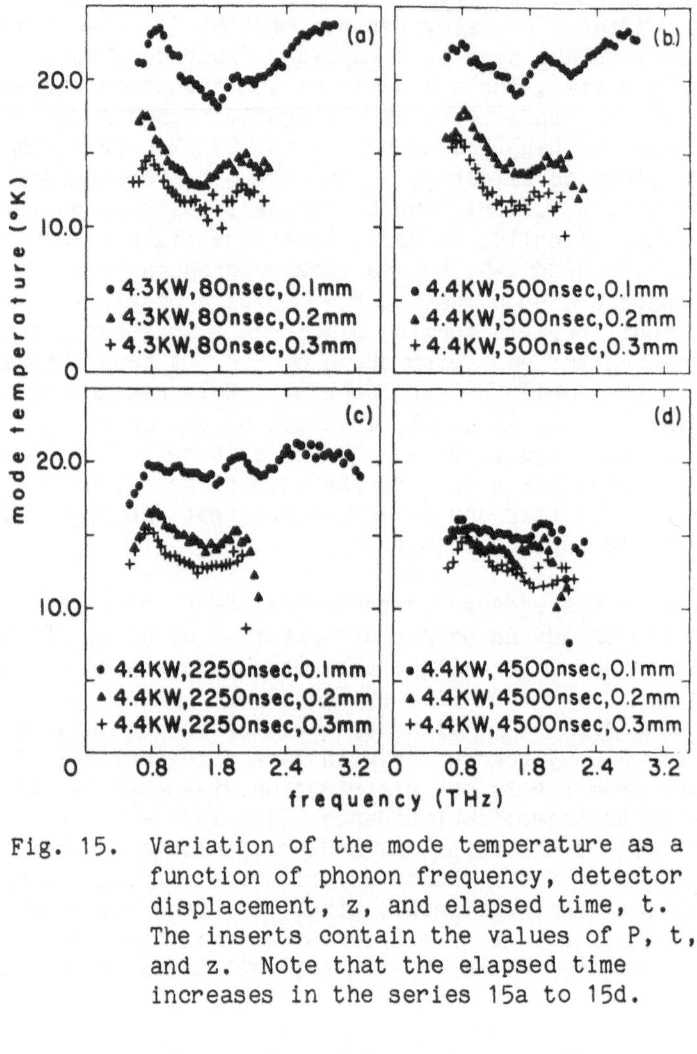

Fig. 15. Variation of the mode temperature as a function of phonon frequency, detector displacement, z, and elapsed time, t. The inserts contain the values of P, t, and z. Note that the elapsed time increases in the series 15a to 15d.

"time of flight" spectrum noted above. The earlier interpretation was, however based in part, on inconclusive knowledge (Bron 1980b) of acoustic phonon lifetimes and their dependence on phonon frequency. However, the work of Baumgartner, et al. (1981), clearly shows that, in simple crystal systems, the effective anharmonic scattering rate, Γ, varies as ν^5 and is for 1 THz phonons of the order of 10^6 s^{-1}, i.e., well within the time span of the present experiment. Our previous measurements (Bron and Grill 1977) imply that the scattering rate, Γ*, resulting from the presence of 0.1 mole % of Eu^{2+} (plus isotopic impurities) dissolved in SrF_2, is of the order of $10^8\ s^{-1}$ for 1 THz phonons. The condition for quasidiffusive transport that $\tau^* < \tau$ is, accordingly, met for the present crystal system.

In order to make headway in interpreting these experimental results it is necessary to return to a discussion of the underlining theoretical formalism. It has, already been pointed out that the theory of transport of nonequilibrium phonons in dispersive media under the presence of anharmonic processes has been discussed by Kazakovtsev and Levinson (1978, 1979, and 1981) and by Schaich (1984). (See also Wilson and Schaich 1984.)

The basic starting point in the theoretical formulations is the diffusion equation modified by anharmonic collision terms. At low ambient and mode temperatures such that $\hbar\omega < kT$, we expect three-phonon interaction processes to dominate the anharmonic interactions. Hence, for this regime, an expression for quasidiffusive phonon transport is

$$\frac{\partial n(\omega,z,t)}{\partial t} - D(\omega)\frac{\partial^2 n}{\partial z^2} = -\frac{1}{2}\int d\omega'\rho'\int \partial\omega''\rho'' A(\omega;\omega',\omega'')$$

$$\times\ \delta(\hbar\omega - \hbar\omega' - \hbar\omega'')[n(1+n')(1+n'')$$

$$-(1+n)n'n''] - \int d\omega'\rho'\int d\omega'\rho'' A(\omega';\omega',\omega)$$

$$\times\ \delta(\hbar\omega'' - \hbar\omega' - \hbar\omega)[nn'(1+n'')$$

$$-(1+n)(1+n')n''] \quad . \tag{6.5}$$

In expression (6.5), the coefficients A describe the strengths of the three-phonon creation and annihilation processes. Recombination processes involving n', n" may not be neglected as is clear from the high values of n indicated in figure 14 for low frequency phonons.

Two attempts have been made to solve equation (6.5) in analytical form. The first by Levinson (1980) considers an instantaneous temperature increment located somewhere inside an infinite medium. The second approach by Wilson and Schaich (1984), which is closer to our experimental conditions, considers an arbitrary time-dependent temperature perturbation at the boundary of a semi-infinite medium. However, in order to obtain an analytical solution in either approach, it is necessary to further restrict the temperature increment of the source to a small fraction of the ambient temperature. This restriction is not obeyed in the present experiment since it is clear from the results displayed in figure 15, that the mode temperature excursion is of the order of 20 K compared to an ambient temperature of about 5 K.

A similar, albeit numerical, approach to the solution of equation (6.5) has been taken by Schaich (1984). Here the integral equation (6.5) is supplanted by a series of equations which describe the response of a small set of Einstein modes,

$\ell = 0,1,2,3,4$, with frequencies $\nu_\ell = 2^\ell \nu_0$ with $\nu_0 = 1/4$ THz.

These modes are weighted by a factor ρ_ℓ which is a measure of the integrated number of states for the complete set of lattice modes assuming that $\rho(\nu) \propto \nu^2$ hence $\rho_\ell \propto \nu_\ell{}^3$. A further

restriction to processes such that $\nu_\ell \rightleftarrows 2\nu_{\ell-1}$, leads to

$$\frac{\partial n_\ell}{\partial t} + D_\ell \frac{\partial^2 n_\ell}{\partial x^2} = -R_\ell^-[n_\ell(1+n_{\ell-1})^2 - (1+n_\ell)n_{\ell-1}^2] \tag{6.6}$$

$$+ bR_\ell^+[n_{\ell+1}(1+n_\ell)^2 - (1+n_{\ell+1})n_\ell^2] \quad .$$

where R^+ and R^- are now the relevant rate constants. The factor b=16 ensures that anharmonic scattering conserves total energy. A numerical solution of expression (6.6), with boundary conditions which reflect our experiment, has been shown by Schaich to predict a temporal variation of the heater film which is similar to that previously observed experimentally (Bron et al. 1979). A number of other properties of the transport which are predicted in this way are discussed by Schaich. Those which bear most directly on the results of the present experiment are; (i) components of the phonon distribution, which can interact anharmonically with each other during the experimental time span, reach a common mode temperature, (ii) only the lowest frequency component of the phonon distribution (which elastically scatters the least) can penetrate rapidly into the sample, but since low frequency phonons interact anharmonically only infrequently during the time span of the experiment their mode temperatures remain high compared to that of the quasiequilibrium, (iii) the high frequency component, because of strong elastic scattering, penetrates into the sample

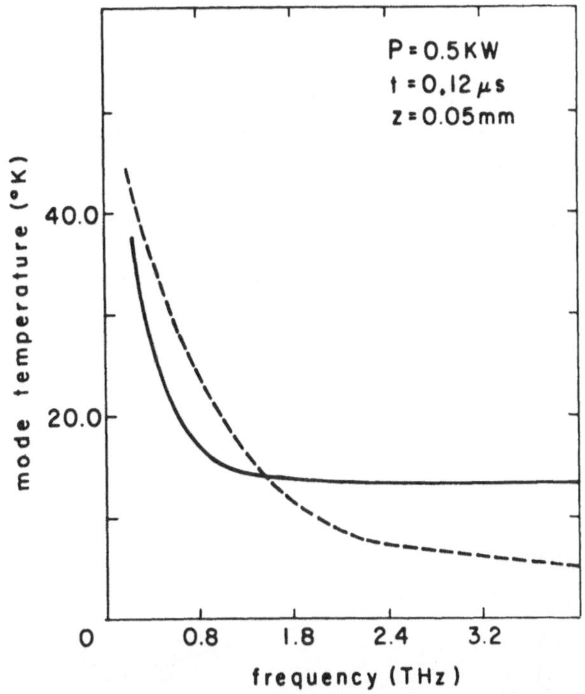

Fig. 16. Comparison between the theoretically predicted mode temperature as a function of phonon frequency with only elastic scattering (dashed line) and with quasidiffusive transport (solid line).

rather slowly at a rate determined by the diffusion constant, D_ℓ, of essentially the lowest frequency component of the quasiequilibrium. The existence of (i) and (ii) is illustrated in figure 16 which displays the predicted mode temperature as a function of phonon frequency as calculated at z = 0.05 mm, t = 0.12 μs after a pulse of 0.5 KW has passed through the heater film. In the absence of anharmonic processes (dashed line) no quasiequilibrium forms (for all t and z) whereas in the presence of anharmonic processes (solid line) a quasiequilibrium forms among phonons with $\nu \gtrsim 1$ THz. In contrast, the low frequency component remains "hot" and does not form a quasiequilibrium. This situation corresponds qualitatively to the experimented results indicated in figures 15a to 15c for $z \geq 0.2$ mm.

A further comparison between data and the predicted properties (i) to (iii) above is illustrated in figure 17. The symbols represent the observed values of the normalized phonon distribution function $n^o = n(\omega,z,t=4.5\ \mu s)/n(\omega,z=0,t=4.5\ \mu s)$ as a function of z for $\nu = 1/2$, 1 and 2 THz with P = 1 KW. For comparison figure 17 contains the theoretically predicted values obtained via equation (6.6) for quasidiffusive transport (solid lines) and those obtained for transport under elastic scattering only (dashed lines). As a result of limitations on computational time and on the temporal resolution of the experiment, the theoretically predicted values of n^o and a part of the experimental data needed to be scaled to correspond to the same

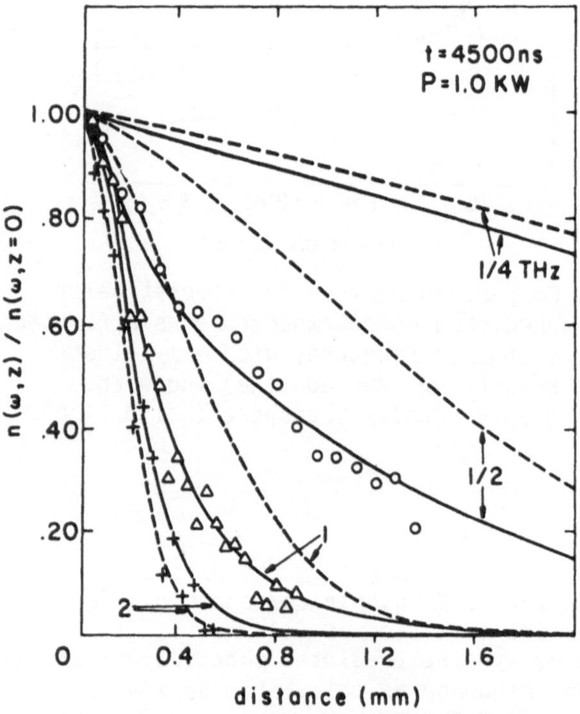

Fig. 17. Normalized distribution function as a function of the detector displacement distance, z, and phonon frequency $\nu = 1/4$, 1/2, 1, and 2 THz. The symbols represent experimental data, the dashed lines are theoretically predicted values for elastic scattering only. The solid lines are theoretically predicted values for quasidiffusive transport.

elapsed time of 4.5 μs. Based on the results noted above, it is assumed that phonons with frequencies $\nu \leq 1$ THz (because of their long lifetimes) travel essentially diffusively according to $z \propto (D(\nu)t)^{1/2}$. Moreover, to a sufficient approximation, the high frequency component, $\nu \geq 1$ THz, also propagates diffusively but at a rate determined by the lowest frequency in the quasiequilibrium. Hence once n^0 (as a function of z) is calculated under quasidiffusion or is obtained experimentally, the displacement coordinate can be scaled as $z_2/z_1 = (t_2/t_1)^{1/2}$. Accordingly, the theoretical values, which were initially calculated for t = 80 ns, were scaled by a common factor of 7.5 so that they correspond to a time of 4500 ns. Similarly, the experimental results for 2 THz phonons, observed at t = 1000 ns were scaled by a common factor of 2.1 to make them correspond to t = 4500 ns. One final additional scaling needed to be done to arrive at the results shown in figure 17. That is to multiply the distance scale of the theoretically predicted displacement z uniformly for all phonon frequencies by a factor of 3. (If purely diffusive transport is appropriate, this would correspond to $D(\nu = 1 \text{ THz}) = 2.4 \times 10^4 \text{ mm}^2 \text{ s}^{-1}$ rather than $D(\nu = 1 \text{ THz}) = 2.62 \times 10^5 \text{ mm}^2 \text{ s}^{-1}$ as chosen by Schaich.) Although this scaling procedure is clearly only an approximate one, it does yield quite good agreement between theory and data, particularly for the low frequency components for which it should be most applicable.

The agreement, illustrated by figure 17, between experiment and theory when both elastic and anharmonic processes are taken into account, and the lack of agreement when anharmonic processes are neglected, together with the qualitative agreement between the observed and predicted properties of the transport, yield strong evidence that the quasidiffusive process is present in the transport of the broadband phonon distribution reported here.

Since even most "pure" crystals contain elastic scattering centers in the form of isotopic impurities we suspect that some level of quasidiffusion occurs in most systems. If this is the case, then much of the earlier work on "heat-pulse" propagation may need to be reexamined. A case in point is our earlier work (Bron et al. 1979 and Bron and Grill 1977a,b) on the $SrF_2:Eu^{2+}$ system. Here we were led by the observed temporal evolution of the high-frequency component of a heat pulse distribution to conclude that the lifetimes of high-frequency acoustic phonons are anomalously long (Bron 1980b). However, it is now clear that the temporal evolution of the high-frequency phonon component does not reflect pure anharmonic decay, but rather the rate at which energy is extracted from the quasiequilibrium. The extraction is controlled by the low frequency phonon component, which alone can readily leave the area adjacent to the phonon source. However, since the density of states of these phonons is low, the amount of energy they can carry away is small, and hence the apparent slow decay of the high-frequency component.

VII. DEPHASING TIME OF LONGITUDINAL OPTICAL PHONONS

It has been demonstrated above that the appropriate time scale in the transport of acoustic phonons is hundreds of nanoseconds to microseconds. In contrast, optical phonons normally possess low group velocities, so that impurity dominated elastic scattering times are long compared to those of most acoustic phonons. On the other hand, lifetimes against anharmonic decay typically lie in the picosecond regime, and thereby dominate the observable transport properties of these phonons.

One of the most exciting prospects in the study of phonon dynamics in the picosecond time domain is the use of nonlinear optical phenomena to generate monochromatic, coherent phonon distributions and to detect their subsequent evolution.

Of the various methods which fall under the category of nonlinear optical techniques (Demtroder 1981, Levenson and Song 1980, Laubereau and Kaiser 1978, von der Linde 1981), the one which so far appears to have the most potential is the generation of short duration (subnanosecond) coherent phonon packets through coherent excitation using mode locked laser systems, and the detection of the packet by pulsed coherent antiStokes Raman scattering (CARS).

A qualitative understanding of this rather complicated experimental technique may be gained from the following semiclassical treatment. According to Placzek (1934), in a Raman active medium the important terms in the optical polarizability, α, can be written as

$$\alpha = \alpha_0 + (\partial\alpha/+\partial Q_v)Q_v \tag{7.1}$$

in which Q_v is the coordinate corresponding to some normal mode of vibration (phonon) of the solid, and α_0 refers to a static polarizability if one is present.

An electromagnetic field, E, can interact with this polarization. The pertinent term in the interaction Hamiltonian, H_I, corresponding to the lowest order Raman process, is

$$H_I = -\frac{1}{2}(\partial\alpha/\partial Q_v)Q_vE^2 \quad . \tag{7.2}$$

It follows that as a result a force

$$F = -\frac{\partial H_I}{\partial Q_v} = \frac{1}{2}(\partial\alpha/\partial Q_v)E^2 \tag{7.3}$$

acts on the vibrational system. Moreover, under the action of both the vibrational and the em field, a polarization, P, is induced with

$$P = -\frac{\partial H_I}{\partial E} = N(\partial\alpha/\partial Q_v)Q_v E \tag{7.4}$$

in which N is the number density of the vibrational modes. For simplicity we associate with each vibrational mode a damped oscillator of mass m, and write an equation of motion,

$$\ddot{Q}_v + 2\Gamma\dot{Q}_v + \omega_v^2 Q_v = \frac{1}{2m}(\partial\alpha/\partial Q_v)E^2 \tag{7.5}$$

in which Γ describes the damping of Q. Under coherent excitation, Q becomes the coherent amplitude and 1/Γ is related to the dephasing time T_2.

In the actual experiment an electromagnetic field is chosen which contains two frequency components, ω_ℓ and ω_s, produced by two well defined synchronously pumped coherent laser beams propagating with wavevector $\vec{k}_\ell$ and $\vec{k}_s$. The em field amplitude of these beams is

$$E(\vec{x},t) = \frac{1}{2}E_\ell \exp[i(\omega_\ell t - \vec{k}_\ell\cdot\vec{x})] + \frac{1}{2}E_s \exp[i(\omega_s t - \vec{k}_s\cdot\vec{x})] + cc \quad , \tag{7.6}$$

in which $\vec{x}$ is the position vector and t is the temporal coordinate, respectively. It is to be understood that the laser beam frequencies ω_ℓ and ω_s will be in the visible region, whereas ω_v is in the far infrared. Hence ω_ℓ, $\omega_s >> \omega_v$.

In order to solve (7.5) for Q_v, we use the trial function

$$Q_v = \frac{1}{2}q_v \exp(i\omega t) + cc \tag{7.7}$$

and pick the terms which are consistent with the inequalities in ω_ℓ, ω_s and ω_v, i.e.,

$$(\omega_v^2 - \omega^2 - i2\Gamma\omega)q_v \exp(i\omega t) = \frac{1}{2m}(\partial\alpha/\partial Q_v)E_\ell E_s \tag{7.8}$$

$$\exp[i(\omega_\ell-\omega_s)t-(\vec{k}_\ell-\vec{k}_s)\cdot x]$$

Note should be taken that the temporal terms in equation (7.8) require that $\omega = \omega_\ell-\omega_s$, i.e., that the vibrational system is being driven at the difference frequency of the two lasers.

Solving equation (7.8) for Q_v yields

$$Q_v = \frac{1}{2m}\frac{\partial\alpha/\partial Q_v}{[\omega_v^2-(\omega_\ell-\omega_s)^2+i2\Gamma(\omega_\ell-\omega_s)]} \qquad (7.9)$$

$$\exp[-i(\omega_\ell-\omega_s)t-(\vec{k}_\ell-\vec{k}_s)\cdot\vec{x}]$$

It is clear from equation (7.9) that a strong resonant excitation can occur if $\omega_\ell-\omega_s$ equals the frequency of a Raman mode, ω_v, providing that the phase matching condition $\vec{k}_v = \vec{k}_\ell-\vec{k}_s$ can be met.

The discussion so far has neglected to explicitly include the fact that the laser output is not CW but rather consists of trains of picosecond duration pulses. If the durations of the exciting laser pulses Δt_ℓ and Δt_s are short compared to the relaxation time of the phonon modes and to the time between successive laser pulses, then the excitation undergoes the equivalence of "free fall decay." This decay process can be monitored by a Raman interaction between the vibrational mode and a time delayed third (pulsed) laser beam. This interaction can also be driven coherently as described below.

The presence of the coherent phonon packet causes an oscillatory polarizability at frequency ω_v, as indicated in equation (7.1), which can in turn interact with the third laser at frequency ω_3 to yield a nonlinear polarization, P^{NL}, (see equation (7.4))

$$P^{NL} = \chi^{(3)}E_\ell E_s E_3 \qquad (7.10)$$

where $\chi^{(3)}$ is the so-called total third-order nonlinear susceptibility which contains (as we shall see) contributions from the response of the electronic and the lattice dynamical system. The polarization acts as a source term in Maxwell's equation (see e.g., Armstrong et al. 1962 and Maker and Terhune 1965) to produce a strong coherent output beam with $\omega_4 = \omega_3+(\omega_\ell-\omega_s)$ provided that

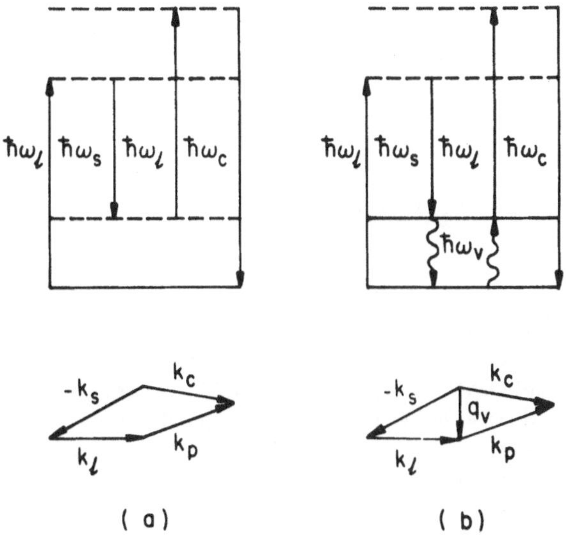

Fig. 18. Energy level system and electromagnetic and lattice transitions for the nonresonant (a) and resonant (b) CARS technique.

$\vec{k}_4 = \vec{k}_\ell - \vec{k}_s + \vec{k}_3$ can be satisfied. It is usual to set $\omega_3 = \omega_\ell$ by taking a small amplitude component of the incident "ℓ" laser beam. Therefore, in practice, $\omega_4 = 2\omega_\ell - \omega_s$. The corresponding energy and

wavevector diagrams are shown in figure 18. It is clear from the figure that the interaction leads to coherent antiStokes Raman scattering, CARS. Accordingly, we identify ω_{AS} with ω_4 and $\vec{k}_{AS}$ with $\vec{k}_4$.

It follows from the above discussion that the intensity of the CARS signal as a function of the delay time between the pump and the probe laser pulses can be written as

$$I(\Delta t) = AS(\Delta k) \int_{-\infty}^{+\infty} dt \, |\vec{E}_p(t-\Delta t)[NR_a Q(t) + 3\chi_E^{(3)} \vec{E}_\ell(t) \vec{E}_s^*(t)]|^2 \quad (7.11)$$

where

$$S(\Delta k) = \sin^2(\Delta k L/2)/(\Delta k L/2)^2 \quad (7.12)$$

and

$$\Delta\vec{k} = \vec{k}_c - [\vec{k}_\ell + \vec{k}_p - \vec{k}_s] \tag{7.13}$$

and

$$\ddot{Q} + \Gamma\dot{Q} + \omega_v^2 Q = R_a m^{-1} \vec{E}_\ell(t)\vec{E}_s^*(t) \quad . \tag{7.14}$$

In these expressions R_a is the Raman tensor, and the coherent amplitude, Q, of the excited phonon packet is given by equation (7.14). The wavevector mismatch is given by Δk; it is set experimentally to approximately zero (hence, $S(\Delta k) \approx 1$). Finally, Δt is the temporal delay between the pulses of the probe laser, p, and those of the lasers ℓ and s; m is the reduced lattice mass, N is the number of primitive cells per cm³ and $A = 2\pi\omega_c^2 L^2/c\varepsilon_c$. In the expression for A, c is the speed of light, L is the effective length within the medium over which spatial overlap of the three-wave mixing exists, and ε_c is the dielectric constant of the medium at ω_c.

According to equation (7.11), $I(\Delta t)$ has two components. If $\omega_\ell - \omega_s$ is detuned from the LO phonon frequency, ω_v, Q becomes very small and only that part of equation (7.11) controlled by $\chi_E^{(3)}$ remains. Thus a purely "electronic" part of $I(\Delta t)$ (i.e., independent of any coherent excitation) appears under these circumstances. If, on the other hand, $\omega_v = \omega_\ell - \omega_s$, the temporal evolution of the coherent excitation also contributes to $I(\Delta t)$.

Application of the nonlinear response of solids to the determination of the lifetimes of optical phonons from time domain studies has received recent attention (Gale and Laubereau 1983, Kuhl and Bron 1984, Schosser and Dlott 1984). I present here as model cases results for high purity GaP and ZnSe. These measurements are facilitated by recent rapid advances in commercially available mode-locked lasers and synchronously driven dye lasers. Such a system generates picosecond duration pulses at high repetition rates. As a result it becomes possible to observe decay processes which are typically in the 1 to 10 psec range.

The experimental apparatus is depicted in figure 19. A mode-locked, argon-ion laser is used to pump synchronously two matched dye lasers. The dye laser output contains an autocorrelation pulse which typically has durations of 2 to 3 ps FWHM, a repetition rate of 76 MHz and an average power in each output beam of 30-40 mW. The dye laser outputs are tuned to the frequencies ω_ℓ and ω_s such that $\omega_\ell - \omega_s$ equals the frequency of the

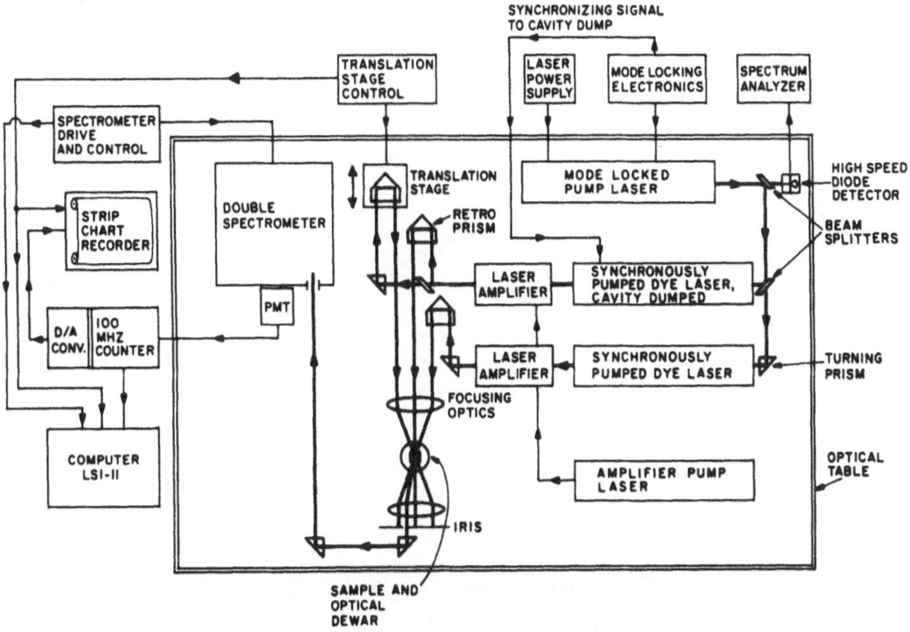

Fig. 19. CARS apparatus.

LO phonons, $\omega_v = \omega_{LO}$. When these two beams are spatially and temporally overlapped in the GaP crystals, such that the wavevector conservation is obeyed, a coherent packet of LO phonons propagating along $\vec{k}_v$ is generated.

The subsequent temporal variation of the coherent LO phonon packet is measured through the CARS technique. For this purpose a small part of the ω_ℓ beam is extracted and delayed by an optical delay line, so that pulses from it appear at the focal point at any delayed time, Δt, of between 0 and 0.3 ns, with a resolution of fractions of picoseconds. As described above, the delayed beam scatters from the component of the packet which remains coherent after Δt and produces a coherent beam propagating along $\vec{k}_{AS}$.

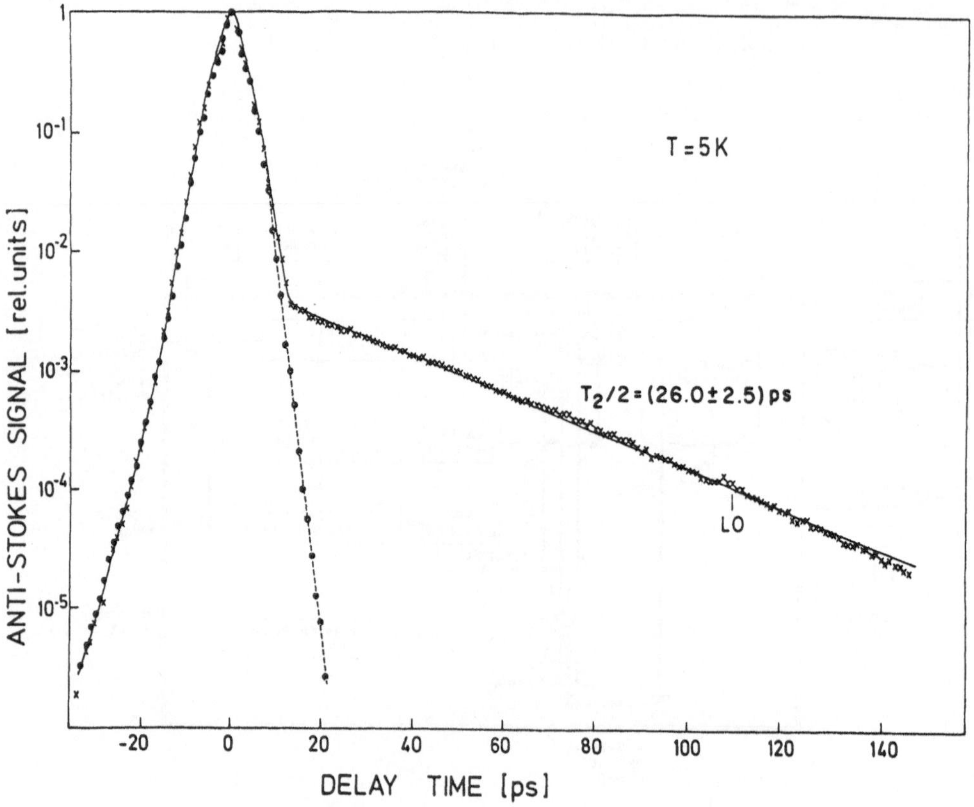

Fig. 20. Time dependence of the CARS signal for GaP at 5 K.

Figure 20 is an illustration of the typical temporal variation of the CARS signal intensity, $I(\Delta t)$, from LO phonons with $\omega_v = 403\ cm^{-1}$ and an ambient temperature of 5 K for a GaP crystal with a carrier concentration of less than $10^{16}\ cm^{-3}$. The component of the signal from ≈ -40 to ≈ +20 ps is primarily the result of nonlinear excitation of the electronic system. The full shape of this signal (dashed curve) can be obtained by slightly detuning $\omega_\ell-\omega_s$ away from ω. This part of the signal is controlled by $\chi_E^{(3)}$ and will be discussed in the next section.

The signal beyond 20 ps is a measure of the temporal evolution of the coherent intensity $\langle Q\rangle^2 \propto \exp[-2t/T_2]$, in which

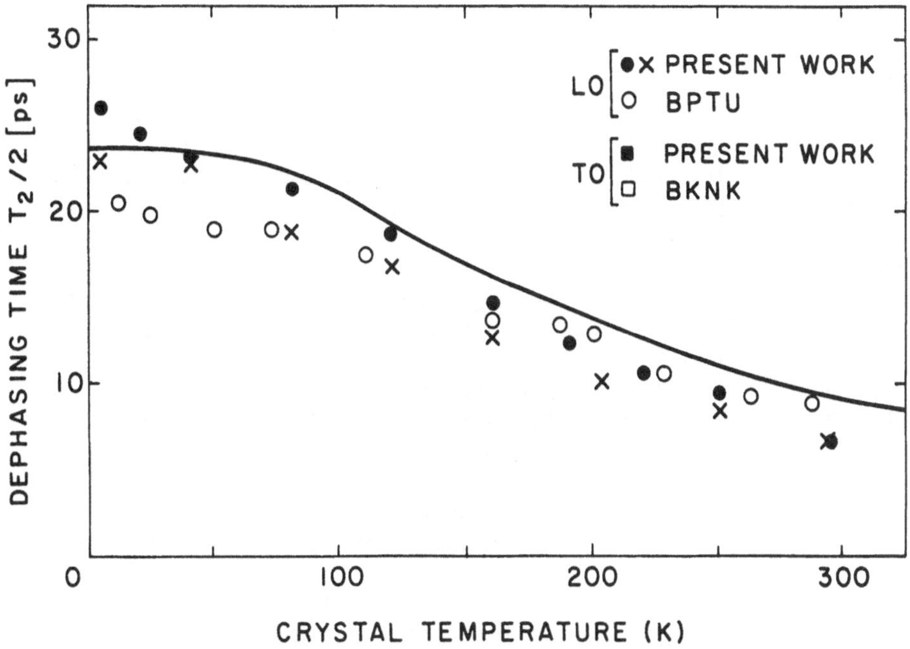

Fig. 21. Temperature dependence of $T_2/2$ for GaP. Symbols represent the data, and the solid line represents theory.

<Q> is the coherent amplitude of the excited LO mode (Laubereau and Kaiser 1978). The intensity is seen from figure 19 to decay exponentially over several orders of magnitude with a time constant $\tau = T_2/2$. Here T_2 is the traditional dephasing time.

The temperature dependence of τ (and the uncertainty in its measurement), as observed over a range from 5 to 300 K, is shown in figure 21 (full circles).

It is generally accepted that the linewidth $\delta\nu$ (in cm^{-1}) of the same Raman mode, ω_v, as observed from spontaneous Raman

scattering is related to T_2 by $\delta\nu = (\pi c T_2)^{-1}$ provided that the Raman line is homogeneously broadened (see, e.g., Laubereau and Kaiser 1978). Linewidth measurements of the LO phonon (at $\vec{k} \approx 0$) have previously been reported by Bairamov et al. (1974, 1979) who used a high resolution Fabry-Perot interferometer. The corresponding dephasing times are designated as BPTU in figure 21 (open circles). We have repeated these measurements using our sample and a standard Raman double monochromator providing

0.15 cm^{-1} resolution. Phonon dephasing times are obtained by deconvoluting the spectral function of the instrument from the measured Raman spectrum assuming a Gaussian profile for the spectrometer function and a Lorentzian for the Raman line. The results are displayed as crosses in figure 21.

The time, T_2, includes contributions from all dephasing phenomena such as elastic scattering from surfaces, from impurities, from imperfections, and from electronic carriers, etc., and also includes phonon-phonon scattering which leads to phonon population decay. The population decay time is conventionally referred to as T_1. Since T_2 contains T_1 it follows that $T_2 \leq T_1$.

In the absence of any detailed knowledge of the strength of the various scattering processes it is not possible to calculate the magnitude of T_2, although it is possible to predict theoretically its temperature dependence. Contributions to the temperature dependence of T_2 can arise from scattering from thermally activated carriers and from the temperature dependence of the various depopulation processes. We concentrate here on the latter processes. At low ambient temperature, such that $\hbar\omega_v >> k_BT$, the spontaneous three-phonon decay of $\vec{q} \approx 0$ optical phonons into two acoustic phonons (plus the corresponding recombination process) should dominate all other phonon-phonon interactions (Orbach and Vredevoe 1964).

The possible three-phonon decay channels are illustrated in the inset to figure 20. Channel (b) is not active since it would produce a phonon of higher energy than the LO phonon which is not possible in GaP. Accordingly, the temperature dependent part of the expression for the inverse lifetimes of the LO phonons is (Klemens 1966)

$$\tau^{-1} = \tau_0^{-1}[1+n'+n''] \qquad (7.15)$$

in which τ_0 is the lifetime at T=0°K and the n's are the phonon occupation numbers of the acoustic phonons. Equation (7.15) also reflects the fact that the topography of the dispersion relations of GaP (see for e.g., Bilz and Kress 1979) suggest that the decay proceeds primarily as $LO(\omega,o) \rightarrow LA(\omega/2,\vec{k}) + LA(\omega/2,-\vec{k})$. Expression (7.15), with its magnitude fitted to the low temperature data, appears as the solid line in figure 21. Comparison with the data shows indeed that the low temperature dependence of T_2 is dominated by three-phonon interactions, i.e.,

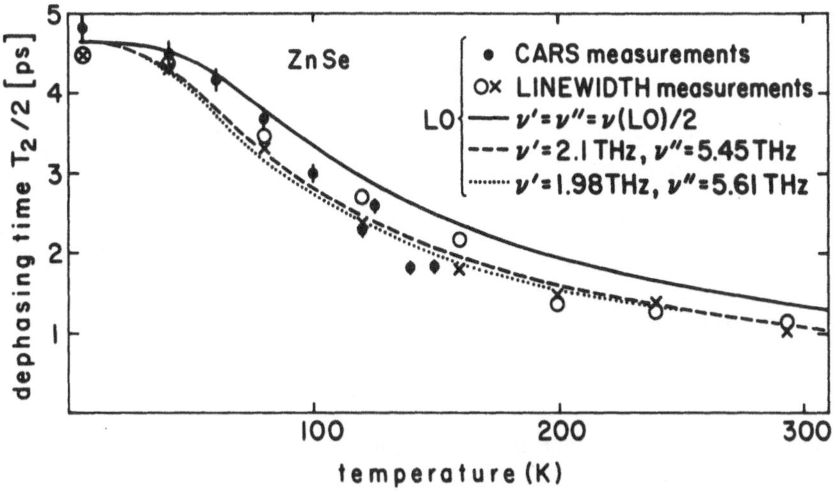

Fig. 22. Temperature dependence of $T_2/2$ for ZnSe. Symbols represent the data; solid lines represent the theoretically obtained values.

by T_1. Beyond this temperature additional processes, including higher order phonon-phonon interactions and scattering from thermally activated carriers, must also be considered.

Quite similar results and agreement have been obtained for the temperature dependence of the LO phonon in ZnSe even though the topography of the dispersion relations for this crystal allows decay to various sets of acoustic phonons. The fit of equation (7.15) for this case is given in figure 22 for the decay routes shown in the insert to the figure.

We conclude, therefore, that it is possible to measure the temperature dependence of the dephasing time of optical phonons and to account for the observation in terms of the simple Bose-Einstein dependence on phonon occupation numbers. Although this procedure does indeed yield quite good agreement between experiment and theory, no theoretial apparatus exists for the prediction of the magnitude of T_2 at T=0 K.

The only readily usable formulation for T_1 is that by Klemens (1966) which does not directly apply to diatomic lattices, and if forced to apply to such lattices yields values of T_1 one to two orders of magnitude smaller than those observed.

VIII. THIRD ORDER NONLINEAR ELECTRONIC SUSCEPTIBILITY

The third order nonlinear susceptibility, $\chi^{(3)}$, has already been defined in terms of equation (7.10). I will not attempt to arrive at a theoretical description of $\chi^{(3)}$ to the same level as was done above for T_2. This is not due to any special difficulty in the derivation but rather due to the fact that the nomenclature becomes rather cumbersome and, moreover, a number of excellent reviews of the theory appear elsewhere (see, for example, Flytzanis 1978, or Flytzanis and Bloembergen 1976).

Accordingly, I will present only the result which is applicable to the experimental conditions to be presented below. The pertinent expression for the components of the $\chi^{(3)}$ for the case of three interacting electromagnetic fields $(\vec{E}_\ell(\omega_\ell),\ \vec{E}_s(\omega_s),\ \vec{E}_p(\omega_p=\omega_\ell))$ which produce a CARS signal at $\omega_{AS} = 2\omega_\ell-\omega_s$ is

$$\chi^{(3)}_{xyxy}(-\omega_c,\omega_\ell,\omega_\ell,-\omega_s) = \chi^{(3)}_{xyxyE}$$

$$- \frac{4\pi(\Delta\hat{k}_z)^2}{3\varepsilon(\Delta\omega)}[\chi^{(2)}_{xyzE}]^2 + \frac{4\pi(1-(k_z/k'))^2}{3[(ck/\omega)^2-\varepsilon(2\omega_\ell)]}[\chi^{(2)}_{xyzE}]^2$$

$$+ (\Delta\hat{k}\ )^2 \frac{N\alpha^2_{LO}}{3M(\omega^2_{LO}-\Delta\omega^2+i\Delta\omega\Gamma)} \qquad (8.1)$$

$$+ \frac{4\pi(1-\Delta\hat{k}_z^{\ 2})}{3[(c\Delta k/\Delta\omega)^2-\varepsilon(\Delta\omega)]}\{[\chi^{(2)}_{xyzE}]^2$$

$$+ \frac{N\chi^{(2)}_{xyzE}\ \ell^*_{TO}\alpha^{(1)}_{TO}}{M(\omega^2_{TO}-\Delta\omega^2+i\Delta\omega\Gamma)}$$

$$+ \frac{N\alpha^2_{TO}}{4M(\omega^2_{TO}-\Delta\omega^2+i\Delta\omega\Gamma)}[(c\Delta k/\Delta\omega)^2-\varepsilon(\Delta\omega)]\} \ .$$

In this expression $\chi^{(i)}_{ijk\ell E}$ refers to the purely electronic contribution, $\vec{\Delta k} = \vec{k}_\ell - \vec{k}_s$, $\hat{\Delta k_z}$ is a unit vector in the direction of the phonon wavevector, $\varepsilon(\nu)$ is the dielectric constant at the frequency ν, $[\chi^{(2)}_{xyzE}]^2$ are contributions to the total $\chi^{(3)}_E$ from so-called two-step nonlinear processes, k' is the wavevector for the frequency $2\omega_\ell$, $\Delta\omega = \omega_\ell - \omega_s$, α_{TO}, α_{LO} are, respectively, the Raman tensor at the TO and LO frequencies, M is the reduced lattice mass, and $\vec{k}'$ is the wavevector of the field at $2\omega_\ell$, and N is the number of unit cells per unit volume.

When $\omega_\ell - \omega_s = \omega_{LO}$, and in the absence of any contributions from two-step two wave nonlinear processes, equation (8.1) reduces to

$$\chi^{(3)}_{xyxy}(-\omega_c,\omega_\ell,\omega_\ell,-\omega_s) = \chi^{(3)}_{xyxyE} - (\hat{\Delta k_z})^2 \frac{iN\alpha^2_{LO}}{3M\Delta\omega\Gamma} \qquad (8.2)$$

$\chi^{(3)}_{xyxyE}$ is the purely electronic contribution to the third-order nonlinear response, whereas the second term describes the damped response of the LO phonons. The relation of these terms to the two components of $I(\Delta t)$ as given in equation (7.11) is straightforward. Consequently, if by experimental means, the two contributions to $I(\Delta t)$ can be separated from each other, the contributions from the lattice and that of the electronic response can be separately determined. This is the procedure we have carried out for the example case of GaP.

It follows from equations (7.11) and (7.14) that, once Γ is determined, by the method described in section 6, a further analysis of the full $I(\Delta t)$ becomes possible in terms of the integration over the temporal profiles of the laser fields $F_i(t)$, plus an independent evaluation of the Raman tensor elements R_a. The result of this procedure leads to a determination of the components of $\chi^{(3)}_{ijk\ell E}$. In the analysis presented here (Rhee, Bron and Kuhl 1984), the actual form of the temporal profiles of the

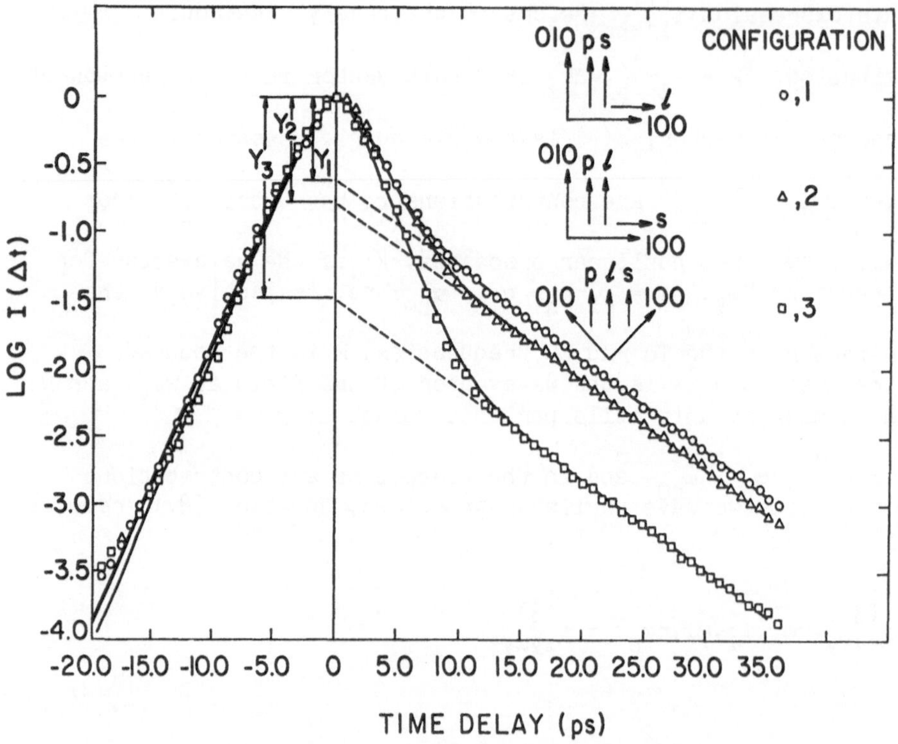

Fig. 23. Experimental data (symbols) and fitted I(Δt) (solid lines) for each of the polarization configurations indicated in the inset to the figure.

laser pulses are obtained from I(Δt) and from the auto- and cross-correlation signals which in the experiment are obtained by mixing the ℓ and s lasers in a KDP crystal. The observed autocorrelation signal can be fitted well by the slightly asymmetric exponential function $f(t) = \exp(\delta\gamma t)$ for $t \leq 0$ and $f(t) = \exp(-\gamma t)$ for $t \geq 0$. It is further assumed that the observed broadening of the cross-, compared to the auto-correlation signal can be accounted for by a Gaussian temporal fluctuation (jitter) between the ℓ and s lasers. Since the final I(Δt) involves the product of the three fields E_ℓ, E_s, E_p, it is possible to shift

for convenience the Gaussian broadening totally on to E_s and write its temporal profile as $f'(t) = [\exp(\delta\gamma t)-(1/z)\exp(z\delta\gamma t)]/(1-1/z)$ for $t \leq 0$ and $f'(t) = [\exp(-\gamma t)-(1/z)\exp(-z\gamma t)]/(1-1/z)$ for $t \geq 0$. Here we have taken advantage of the expansion for small values of α of the Gaussian profile $\exp(-\alpha t^2)$.

Every effort is made before each experimental run to maximize the temporal overlap (at $\Delta t=0$) between the ℓ and s laser. All the same, a temporal mismatch, $\Delta\tau$, of the order of $\lesssim$ 0.5 ps is often found between the two lasers. (The mismatch can be extracted from the slight asymmetry which it causes in the CARS signal in the vicinity of the peak at $\Delta t=0$.) Finally, the relative contributions of the two components of $I(\Delta t)$ can be experimentally specified by a quantity, Y, defined as the difference between the peak of the normalized CARS intensity at zero time delay, $I(\Delta t=0)$ and the intensity of the phonon dephasing component extrapolated back to $\Delta t=0$. This procedure is illustrated in figure 23 for a subset of possible laser field polarizations.

At this point the entire function, $I(\Delta t)$, can be specified (for each polarization condition) in terms of Y, Γ, γ, δ, z, and $\Delta\tau$. The first two quantities are obtained directly as described above. The remaining quantities are determined from a best fit to each $I(\Delta t)$ as shown by the solid lines in figure 23.

It is important to recognize that R_a and $\chi_E^{(3)}$ as they appear in equations (7.11) and (7.14) are each effective quantities whose magnitudes depend on the actual polarization (relative to the crystal axes) of the fields $\vec{E}_\ell$, $\vec{E}_s$, and $\vec{E}_p$. At resonance the phonon part of $I(\Delta t)$ is purely imaginary (see equation (8.2)). Moreover, $\chi_E^{(3)}$ is in general complex; particularly, in the present case (Levenson and Bloembergen 1974) for which $2\hbar\omega_\ell$ exceeds the fundamental gap energy, i.e., in the case that two photon absorptive processes may occur.

The complex effective $\chi_E^{(3)}$ can be obtained for various polarizations of the laser beams relative to the crystallographic axes of the crystal. Three of the polarization conditions used are illustrated in the insert to figure 23. A grand average over seven such conditions, together with the published value of R_a for GaP (Calleja et al. 1982), leads to the following values of the magnitude of the real, χ', and of the imaginary, χ'', parts of the active components of $\chi_E^{(3)}$ (in units of 10^{-10} e.s.u.):

$$\chi'_{1221} = 2.1\pm0.15,\ \chi''_{1221} = -0.48\pm0.24;\ \chi'_{1122} = 1.8\pm0.44\ ,$$

$$\chi''_{1122} = -0.7\pm0.48; \; \chi'_{1111} = 2.1\pm0.7, \; \chi''_{1111} = -0.73\pm0.81 \quad .$$

The imaginary part of $\chi_E^{(3)}$ has its main effect on $I(\Delta t)$ through crossterms (see e.g., (7.11)) with the phonon part. For values of $\chi'' < 10^{-10}$ e.s.u. these crossterms effect only slightly the curvature of $I(\Delta t)$ in the region (Zinth et al. 1978) $0 < \Delta t < 10$ ps; hence, the relatively large uncertainty in χ''. It is observed that the sign of χ'' is opposite to that of χ' and to that of the (positive definite) phonon part.

In order to avoid complications arising from high lying conduction bands, most previous experimental determinations of $\chi_E^{(3)}$, and most theoretical evaluations (see e.g., Flytzanis 1978) have been limited to the "zero frequency limit," i.e., to the case that $2\hbar\omega_\ell$ is much less than the gap energy, E_g, of the medium so that two-photon absorptive processes can be neglected. This constraint means that $\chi_E^{(3)}$ is real, and also leads to the so-called Kleinman symmetry (Kleinman 1962) for which $\chi'_{1122} = \chi'_{1221}$. This condition is not met in the present experiment on GaP for which $E_g = 2.26$ eV whereas $2\hbar\omega_\ell = 4.30$ eV. It is, accordingly, not unexpected that we find $\chi_E^{(3)}$ to be complex and that $\chi'_{1122} \neq \chi'_{1221}$; the observed ratio $\chi'_{1221}/\chi'_{1122} = 1.2$. Moreover, for an isotropic medium we expect the anisotropy factor $\sigma = [2\chi'_{1122} + \chi'_{1221} - \chi'_{1111}]/\chi'_{1111}$ to equal zero whereas 1.7 is observed here. This implies coupling to considerably more anisotropic electronic states than has previously been reported for other tetrahedrally bonded crystals such as diamond (σ = 0.17), Ge (σ = 0.83), Si (σ = 0.44), and GaAs (σ = 0.56) (see Levenson and Bloembergen 1974, Wynne 1969, and Yablonovitch et al. 1972).

Unfortunately, the need to take higher excited electronic bands into account makes it unlikely that the mostly phenomenologically theoretical approaches used so far are applicable to the results reported here (Levenson and Bloembergen 1974). It is particularly unlikely that these models would be able to account for the observed anisotropy in $\chi_E^{(3)}$. On the other hand, it is possible that the more microscopic, ab initio, calculations of semiconductor systems currently being developed (see, for example, Martin 1981) will in time become theoretical bases for a comparison to the experimentally observed components of $\chi_E^{(3)}$.

Clearly it is possible to determine experimentally the dephasing times of optical phonons and their temperature dependence, and that it is equally possible to determine experimentally all components of the third order nonlinear electronic susceptibility. The measurements reported here were taken in the temporal domain. Similar measurements of T_2 and $\chi_E^{(3)}$ taken in the spectral domain have previously been reported for other crystal systems.

Although it has been possible to account for the temperature dependence of T_2 for LO phonons in GaP and ZnSe, full quantitative comparisons between experiment and theory of the magnitude of T_2 and $\chi_E^{(3)}$ has not been possible. In the case of T_2, the effects of elastic scattering events can be limited by suitable experimental conditions, such that the measured T_2 reflects mainly the effects of anharmonic phonon-phonon scattering time T_1. At very low temperatures three-phonon interactions should dominate the possible anharmonic processes. Indeed, experimental and theoretical evidence on the temperature dependence of T_1 substantiate this claim. However, the only readily usable theoretical formulations for T_1 are based on crude simplifications of the anharmonic crystal; namely, a crystal for which the effect of the anharmonic component of the vibrational potential is obtainable from the Grueneisen parameter or from higher order elastic constants, and for which the density of phonon states stems from an isotropic, dispersionless Debye solid. Clearly, such formulations can be relied upon to yield only qualitative comparisons to experimental results. More detailed, and more self-consistent, formulations exist but their application to real crystal systems require detailed knowledge of the anharmonic potential and a self-consistent lattice density of states. Again these requirements often lead to simplifying approximations. In any event, attempts along these lines have been limited to only a few crystal systems. Clearly, as the wealth and detail of the experimental results accumulate, an effort toward a fully self-consistent, possibly ab initio, treatment will become necessary.

A quite similar case can be made for the comparison of the observed components of $\chi_E^{(3)}$ and theory. The theoretical basis for

$\chi_E^{(3)}$ (see e.g., Flytzanis and Bloembergen 1976) is straightforward though rather cumbersome, and requires for its application to real cases knowledge of various orders of the crystal polarizabilities. Traditionally, these parameters have been obtained in terms of the response of simplifying models of the interionic bonds. This approach has been qualitatively successful for simple cases providing the nonlinear excitation energy $2\hbar\omega_\ell$ is much less than the energy of the fundamental gap, E_g, of the crystal (Levenson and Bloembergen 1974). In the experimental case presented here $2\hbar\omega_\ell > E_g$, so that the response of electronic states both at and above E_g need to be taken into account. Moreover, our observation of the considerable anisotropy in the observed $\chi_E^{(3)}$, suggests that the theoretical treatment will need to take the microscopic details of the participating wavefunctions into account. Again, it is hoped that the ab initio, large scale, calculations currently being attempted will in time yield a sufficiently detailed theoretical basis with which a comparison to the experimental results can be made.

IX. CONCLUSIONS

I have reviewed a small number of experiments in order to illustrate recent progress in the field of nonequilibrium phonon dynamics.

The field has in the past few years experienced a rebirth spurred on by ever more sophisticated experimental techniques which have made it possible to delve deeper into the complex interactions which control nonequilibrium phonon dynamics in real systems. In this lecture I have used as examples the interplay of impurity based scattering and anharmonic interactions to demonstrate quasidiffusive transport dynamics. I have also emphasized phonon dephasing effects. The depth of the knowledge so gained cannot be obtained by the more traditional experimental methods of phonon transport; namely, thermal conductivity or even with time resolved heat pulse measurements.

I trust that I have demonstrated both the vitality and the complexity of this field of research. Many of the lectures which are to follow in this volume will underscore this observation.

I wish to make a final point. The material which I presented involves experimental results which have outpaced their theoretical basis, not in the sense that a theoretical basis does not exist, but rather that its application to the experimental situation is either too cumbersome to carry out or some components of the analysis, such as crystal potential or electronic polarizabilities, are not sufficiently known. Thus, I further trust that this lecture will stimulate new attacks on the theoretical basis of nonequilibrium phonon dynamics.

Acknowledgments

I would like to acknowledge the contributions of the many collaborators in the research program which I have described. I would like to specifically acknowledge the work of J. Kuhl, F.M. Lurie, B.K. Rhee, and T. Wilson. The work presented here has been carried out in part through the support of ARO Contracts DAAG 29-80-C-0085 and DAAG 29-83-K-0091.

REFERENCES

Armstrong, J.A., Bloembergen, N., Ducuing, J., and Pershan, P.S., 1962, Phys. Rev. 127:1918.
Ashcroft, N.W. and Mermin, N.D., 1976, "Solid State Physics," Holt, Rinehardt, and Winston, New York, Chap. 25.
Bairamov, B.Kh., Kitaev, Yu.E., Negoduiko, V.K., and Kashkotzev, Z.M., 1974, Fiz. Toerd. Tela. 16:2036, Engl. translation: Sov. Phys., Sol. State 16:1323.
Bairamov, B.Kh., Parshin, V.V., Toporov, and Ubaidullav, Sh.B., 1979, Pis'ma Zh. Tekh. Fiz. 5:1116, Engl. translation: Sov. Tech. Phys. Lett. 5:466.
Baumgartner, P., Englehardt, M., and Renk, K.F., 1981, Phys. Rev. Lett. 47:1403.
Bilz, H. and Kress, W., 1979, "Phonon Dispersion Relations in Insulators," Springer, Berlin.
Born, M. and Huang, H., 1956, "Dynamical Theory of Crystal Lattices," Clarendon, Oxford.
Bron, W.E., 1972, in: Physics of Impurity Centers in Crystals, G.S. Savt, ed., Estonian Academy of Sciences, Tallin, pg. 343.
Bron, W.E., 1975, Phys. Rev. B 11:3951.
Bron, W.E., 1980a, Rep. Prog. Phys. 43:301.
Bron, W.E., 1980b, Phys. Rev. B 21:2627.
Bron, W.E. and Grill, W., 1977a, Phys. Rev. B 16:5303.
Bron, W.E. and Grill, W., 1977b, Phys. Rev. B 16:5315.
Bron, W.E. and Keilmann, F., 1975, Phys. Rev. 21:2627.
Bron, W.E., Levinson, Y.B., and O'Connor, J.M., 1982, Phys. Rev. Lett. 49:209.
Bron, W.E., Patel, J.L., and Schaich, W.L., 1979, Phys. Rev. B 20:5394.
Bron, W.E. and Wagner, M., 1965, Phys. Rev. 139:A233.
Calleja, J.M., Vogt, H., and Cardona, M., 1983, Phil. Mag. A45:239.
Cheeke, J.D.N., Hebral, B., and Martinon, C., 1972, J. Phys. (Paris) 33:C4-57.

Demtroder, W., 1981, "Laser Spectroscopy," Springer, Berlin.
Eisenmenger, W. and Dayem, H.H., 1967, Phys. Rev. Lett. 18:125.
Eisenmenger, W., 1976, in: Physical Acoustics, vol. 12, Mason, W.P. and Thurston, R.N., eds., Academic, New York, pg. 79.
Flytzanis, C., 1978, in: "Quantum Electronics: A Treatise," vol. 1, part A, Rabin, H. and Tang, C.L., eds., Academic, New York, pg. 9.
Flytzanis, C. and Bloembergen, N., 1976, Proc. Quant. Elect. 4:271.
Gale, G.M. and Laubereau, A., 1983, Opt. Comm. 44:273.
von Gutfeld, R.G. and Nethercot, 1964, Phys. Rev. Lett. 12:641.
von Gutfeld, R.J., 1976, in: "Physical Acoustics," vol. 5, Mason, W.P., ed., (Academic, New York) pg. 9.
Herth, P. and Weis, O., 1970, Z. Angew. Phys. 29:101.
Kappus, W. and Weis, O., 1973, J. Appl. Phys. 44:1947.
Kazakovtzev, D.V. and Levinson, Y.B., 1978, Pis'ma Zh. Eksp. Teor. Fiz. 27:194 [Engl. Translation JETP Lett. 27:181].
Kazakovtzev, D.V. and Levinson, Y.B., 1979, Phys. Status Solidi (b) 96:117.
Kazakovtzev, D.V. and Levinson, Y.B., 1981, J. Low Temp. Phys. 45:49.
Klein, M.V., 1968, in: "Physics of Color Centers," Fowler, W.B., ed., Academic, New York, pg. 429.
Kleinman, D.A., 1969, Phys. Rev. 126:1977.
Klemens, P.G., 1958, in: "Solid State Physics," vol. 1, Seitz. F. and Turnbull, D., eds., Academic, New York.
Klemens, P.G., 1966, Phys. Rev. 148:845.
Klemens, P.G., 1967, J. Appl. Phys. 38:4573.
Kuhl, J. and Bron, W.E., 1984, Sol. State Comm. 48:935.
Kwok, P.C. and Miller, P.B., 1966, Phys. Rev. 146:592.
Laubereau, A. and Kaiser, W., 1978, Rev. Mod. Phys. 50:607.
Levenson, M.D. and Bloembergen, 1974, Phys. Rev. B 10:4447.
Levenson, M.D. and Song, J.J., 1980, in: "Coherent Nonlinear Optics," Feld, M.S. and Letokhov, V.S., eds. Springer, Berlin.
Levinson, Y.B., 1980, Zh. Eksp. Teor. Fiz. 79:1394, [Sov. Phys. JETP 52:704]. Also see Sol. State Comm. 36:73.
von der Linde, D., 1981, in: "Defects in Insulating Crystals," Tuckevich, V.M. and Shvartz, eds., Springer, Berlin, pp. 542.
Little, W., 1959, Can. J. Phys. 37:334.
Maker, P.D. and Terhune, R.W., 1965, Phys. Rev. 137:A801.
Maris, H.J., 1970, J. Acoust. Soc. Am. 50:812.
Martinon, C. and Weis, O., 1979, Z. Phys. B32:259.
Niklasson G., 1972, Phys. kondens. Materie 14:138.
O'Connor, J., 1972, Unpublished Ph.D. thesis submitted to Indiana University.
Orbach, R. and Vredevoe, L.A., 1964, Physics 1:91.
Perrin, N. and Budd, H., 1972, Phys. Rev. Lett. 28:1701.
Placzek, G., 1934, in: "Handbuck der Radiologie," Marx, E., ed., Akademische Verlagsgesellschaft, Leipzig.

Rhee, B.K., Bron, W.E., and Kuhl, J., to be published in the November 15, 1984 issue of Phys. Rev. B.

Schaich, W.L., 1978, J. Phys. C 11:4341.

Schaich, W.L., 1984, Sol. State Comm. 49:55.

Schosser, C.L. and Dlott, D.D., 1984, submitted to the Physical Review.

Timush, T. and Buchanan, M., 1967, Phys. Rev. 164:345.

Tua, P. and Mahan, G., 1982, Phys. Rev. B 26:2208.

Tucker, J.W. and Rampton, V.W., 1972, "Microwave Ultrasonics in Solid State Physics," North Holland, Amsterdam.

Wagner, M., 1968, Z. Phys. 214:78.

Weis, O., 1969, Z. Angew. Phys. 26:325.

Wilson, T.E., Lurie, F.M., and Bron, W.E., to be published in the November 15, 1984 issue of Phys. Rev. B.

Wilson, T.E. and Schaich, W.L., 1984, Sol. State Comm. 50:3.

Wong, J.Y., Berggren, M.J., and Schawlow, A.L., 1968, J. Chem. Phys. 49:835.

Wynn, J.J., 1969, Phys. Rev. 178:1295.

Yablonovitch, E., Flytzanis, C., and Bloembergen, N., 1972, Phys. Rev. Lett. 29:865.

Ziman, J., 1960, "Electrons and Phonons," Oxford Univ. Press, Oxford.

Zinth, W., Laubereau, A., and Kaiser, W., 1978, Opt. Comm. 26:457.

STUDIES OF NONEQUILIBRIUM PHONONS BY OPTICAL TECHNIQUES

Karl F. Renk

Institut für Angewandte Physik
Universität Regensburg
8400 Regensburg, W.Germany

1 INTRODUCTION

1.1 General Survey

Optical techniques of phonon generation and detection and the application for the study of phonon scattering processes are described in this article. Special emphasis is put on the discussion of high-frequency acoustic phonons in the terahertz frequency range. As important phonon scattering processes, spontaneous decay of acoustic phonons by splitting processes, caused by the crystal anharmonicity, and elastic resonance scattering and inelastic scattering at impurity ions are treated.

High-frequency phonons can be generated by nonradiative transitions of optically excited electronic states. This technique has been used in many studies of high-frequency phonons. Recently, phonon generation by infrared multiphonon absorption was introduced as another method. The optical detection techniques discussed in this article are based on the observation of phonon-induced fluorescence from impurity probe ions.

Phonon scattering processes are partly known from heat conductivity measurements. The use of the optical techniques of phonon generation and detection permits studies of high-frequency phonons of well defined frequencies and at low crystal temperature, $k_B T \ll h\nu$ (k_B = Boltzmann's constant, h = Planck's constant, T = crystal temperature, ν = phonon frequency). Thus, it is possible to eliminate scattering of nonequilibrium phonons with thermal phonons. The optical techniques were used to investigate anharmonic phonon decay by phonon splitting processes, elastic defect-mass scattering at impurities, elastic resonance scattering at impurities that have electronic states resonant with phonons, and inelastic scattering at impurities that have low lying electronic energy levels. Because of strong frequency dependences of the anharmonic decay rate (typically ν^5), of the defect-mass scattering cross section (ν^4) and of the inelastic scattering cross section (ν^4), these scattering processes are most important for the propagation of phonons at terahertz frequencies. Since the one-phonon spin-lattice relaxation rate is also increasing strongly with frequency (as ν^3 or ν^5), homogeneous broadening of electronic resonance lines is characteristic for $\nu \gtrsim 1$ THz and therefore strong resonance scattering of high-frequency phonons at electronic two-level states is possible.

1.2 Historical

Nonequilibrium phonons at frequencies above 10^{11} Hz were first studied by use of the heat pulse technique (von Gutfeld and Nethercot 1964). Eisenmenger and Dayem (1967) introduced superconducting tunnel junctions which are well suited for phonon spectroscopy in the range of 10^{11} Hz to 10^{12} Hz.

A first optical method for the study of phonons above 10^{11} Hz was developed by Anderson and Sabisky (1968). In $SrF_2:Tm^{2+}$ the Kramers ground state of Tm^{2+} was magnetically split and phonons were detected by determing phonon-induced population changes in the magnetic sublevel system. The changes were measured by observation of transmission changes for circularly polarized visible radiation. The technique was applied for tunable phonon detection up to a frequency of $5 \cdot 10^{11}$ Hz (Anderson and Sabisky 1971).

Detection of high-frequency phonons by phonon-induced fluorescence was first reported by Renk and Deisenhofer (1971). Phonons at an energy of 29 cm^{-1} (frequency 0.87 THz) were detected in ruby by R_2 fluorescence arising from optically excited Cr^{3+} ions; phonons were generated by the heat pulse technique. With this method, transient spatial distributions of 29 cm^{-1} phonons within ruby crystals were determined and spatial trapping due to resonance scattering at excited Cr^{3+} ions was demonstrated (Renk and Deisenhofer 1971, Renk 1971, 1972). Due to resonance scattering at excited Cr^{3+} ions, that is caused by electronic transitions between the $\overline{E}$ and $2\overline{A}$ states, 29 cm^{-1} phonons can be spatially trapped in an optically excited volume for times much larger than the ballistic time of flight through the volume. Renk and Peckenzell (1972) showed that 29 cm^{-1} phonons can be totally spatially trapped due to the resonance scattering and that the trapped phonons decayed by inelastic scattering (first attributed to anharmonic phonon decay).

Using the same technique, Kaplyanskii et al. (1975a) performed time-of-flight experiments for large distances between heater film and detector volume and studied the anisotropy of the $\bar{E} \rightarrow 2\bar{A}$ phonon absorption by excited Cr^{3+} ions, Kaplyanskii et al. (1975b) investigated resonant phonon trapping under the influence of an external magnetic field and Akimov et al. (1977) and Kaplyanskii et al. (1981a) observed phonon focusing in Al_2O_3.

Dijkhuis et al. (1976) and Meltzer and Rives (1977) generated 29 cm^{-1} phonons by optical excitation of ruby crystals with continuous or pulsed laser radiation, respectively; by optical excitation, nonradiative transitions lead to population of $\bar{E}$ and $2\bar{A}$ levels and, by $2\bar{A} \rightarrow \bar{E}$ relaxation, 29 cm^{-1} phonons were produced; phonons were detected by R_2 fluorescence. The technique was applied to study resonant phonon trapping under different experimental conditions. At large concentrations of excited Cr^{3+} ions, total spatial trapping was found (Dijkhuis et al. 1976, Meltzer and Rives 1977), in agreement with the heat pulse experiments. Dijkhuis and de Wijn (1979a,b) studied diffusive phonon escape from small crystal volumes. In connection with different trapping experiments, various inelastic scattering processes for resonantly trapped phonons were discussed. Dijkhuis and de Wijn (1979a,b) suggested that inelastic scattering at excited Cr^{3+} ions that are under the influence of random internal magnetic fields can be responsible for the decay of strongly trapped phonons. Meltzer and Rives (1982) proposed inelastic scattering at exchange-coupled Cr^{3+} ions as another inelastic scattering process for trapped phonons. Extended studies using optical techniques did not yet lead to a clear answer to the question by which processes strongly trapped phonons escape from a resonance volume. Depending on experimental conditions, different mechanisms may be responsible for the phonon decay.

In experiments with optical excitation, monochromatic phonons were generated by $2\bar{A} \rightarrow \bar{E}$ relaxation, but in addition, broadband phonons were produced by nonradiative transitions between the optical-pump bands and the $\bar{E}$ and $2\bar{A}$ levels. It seems possible that these broadband phonons influenced the trapping very strongly. Basun et al. (1982a) suggested that broadband phonons at high frequencies were responsible for additional generation of 29 cm^{-1} phonons because of $2\bar{A} \rightarrow \bar{E}$ Raman scattering processes at excited Cr^{3+} ions. Renk (1984) suggested that broadband phonons around 29 cm^{-1}, generated by the nonradiative transitions, can dominate the trapping at large concentrations of excited Cr^{3+} ions, rather than monochromatic 29 cm^{-1} phonons generated by $2\bar{A} \rightarrow \bar{E}$ relaxation.

Lengfellner and Renk (1977) reported generation of phonons in optically excited ruby by $\bar{E} \rightarrow 2\bar{A}$ far infrared excitation and $2\bar{A} \rightarrow \bar{E}$ relaxation; for detection, R_2 fluorescence was observed. Using this technique, Retzer et al. (1984) found evidence that phonons, resonantly trapped in a ruby crystal (containing 0.05 mol% Cr^{3+}), remained almost monochromatic during the trapping time even at very large concentrations of excited Cr^{3+} ions where strong multiple resonance scattering occurred. Engelhardt and Renk (1984) showed that Cr^{2+} ions, produced by X-ray irradiation of a ruby crystal, are strong inelastic scattering centers for 29 cm^{-1} phonons. It was suggested (Renk 1984) that Cr^{2+} ions which are present as impurities in unirradiated ruby crystals are also responsible for the decay of phonons in unirradiated ruby crystals. Accordingly, a 29 cm^{-1} phonon resonantly trapped in optically excited ruby decays by a single inelastic scattering process.

In a magnetic field the $\bar{E}$ and $2\bar{A}$ levels split into $\bar{E}_{\pm}$ and $2\bar{A}_{\pm}$ sublevels. Therefore, the resonance scattering at excited Cr^{3+} ions in a magnetic field is not purely elastic, but Raman scattering processes are possible; in a Raman scattering process a phonon looses or gains the Zeeman splitting energy that corresponds to the separation between $\bar{E}_-$ and $\bar{E}_+$. This Raman scattering process was studied experimentally and theoretically (Kaplyanskii et al. 1975b, Dijkhuis et al. 1976, Dijkhuis and de Wijn 1979a, Basun et al. 1979, 1981, Kaplyanskii et al. 1981b, Basun et al. 1982b); the theoretical treatments are based on photon trapping theories (Holstein 1947) and phonon trapping theories (Levinson 1978, Malyshev and Shektman 1978).

It should be mentioned that nonthermal 29 cm^{-1} phonons were first observed by Geschwind et al. (1965) and Adde et al. (1969) using a microwave technique. It was found that 29 cm^{-1} phonons produced by optical pumping with a mercury lamp influenced Orbach relaxation processes between the magnetic sublevels of excited Cr^{3+} ions.

Resonance scattering of phonons near 1 THz was also studied for Al_2O_3 containing V^{4+} ions. Bron et al. (1982) generated phonons by far infrared excitation and relaxation of V^{4+} ions and detected them bolometrically. Engelhardt et al. (1983) reported a different technique: In an X-ray irradiated Al_2O_3 crystal containing V^{4+} and Cr^{3+} ions, phonons were generated monochromatically at one frequency by far infrared excitation and relaxation of V^{4+} ions and detected at another frequency by observation of R_2 fluorescence. It was found that phonon propagation in this system is governed by both multiple elastic resonance scattering at V^{4+} ions and multiple inelastic scattering at Cr^{2+} ions. Engelhardt et al. (1984) studied the dynamics of the inelastic scattering using pulsed far infrared excitation at one frequency and optical detection at another frequency and presented a theory for propagation of phonons that undergo both multiple elastic resonance scattering and multiple inelastic scattering at impurities. The experiment and the theory are discussed in section 3.

Further experiments with monochromatic phonon generation and bolometric detection should also be mentioned. Bron and Grill (1978) reported evidence for stimulated phonon generation by far infrared excitation of V^{4+} ions in Al_2O_3. Hu (1980) reported evidence for stimulated generation of 29 cm^{-1} phonons in ruby by $2\bar{A} \rightarrow \bar{E}$ relaxation; the $2\bar{A}$ level was directly optically excited with radiation of a pulsed dye laser.

Tunable phonon detection was achieved by vibronic sideband spectroscopy and by use of tunable two-level states. The vibronic sideband spectroscopy was applied for detection of phonons in diamond (Colles and Giordmaine 1971), CdS (Shah et al. 1974a,b), SrF_2 (Bron and Keilmann 1975, Bron and Grill 1977a,b), CaF_2 (Kaplyanskii 1977, Akimov et al. 1979), and LaF_3 (Meltzer et al. 1983). Phonon-induced fluorescence from stress-split energy levels was applied to detect phonon in CaF_2 (Eisfeld and Renk 1979,1980, Akimov et al. 1980, Abramov et al. 1980) and SrF_2 (Eisfeld and Renk 1979). Phonon-induced fluorescence from magnetically split excited state levels was used to study phonons in LaF_3 (Will et al. 1983, Meltzer et al. 1984). Using these techniques, transient frequency distributions of terahertz phonons generated by the heat pulse technique were studied (Shah et al. 1974a,b, Bron and Grill 1977a,b, Eisfeld and Renk 1979).

It was demonstrated by Baumgartner et al. (1981a,b,c) that the anharmonic lifetime of acoustic phonons in CaF_2 which have wave vectors in the linear part of the dispersion curves decreases with the fifth power of frequency; this is characteristic for spontaneous decay by anharmonic phonon splitting processes. Theoretical studies of this phonon decay have been performed by Slonimskii (1937), Orbach and Vredevoe (1964), Klemens (1967), Tua and Mahan (1981, 1982), Okubo and Tamara (1983), and Tamara and Okubo (1984).

Baumgartner et al. (1983) investigated nonequilibrium phonons generated by nonradiative transitions of optically excited impurity ions and showed that the frequency distributions are non-Planckian. Happek et al. (1984a,b) generated phonons by infrared multiphonon excitation and studied phonon distributions for late times after excitation. The decay of a population of high-frequency phonons in a crystal is governed by anharmonic decay of phonons at very high frequencies and by diffusive escape from the crystal of phonons at lower frequencies (section 2.4).

The method of phonon generation by nonradiative transitions, which is now widely used for broadband generation of high-frequency phonons, was applied by Benderskii et al. (1974) and Broude et al. (1977) to generate phonons in anthracene crystals. Phonons were also detected by phonon-induced fluorescence. In connection with these experiments, propagation of phonon mixtures was studied theoretically (Kazakovtsev and Levinson 1978,1979,

Guseinov and Levinson 1983a,b) taking into account anharmonic phonon decay and elastic scattering at impurities.

Several other optical techniques, not treated in this article, are in development in various laboratories. Such techniques are phonon generation by far infrared piezoelectric surface excitation (Grill and Weis 1975, Bron et al. 1983), phonon detection by use of shallow traps in ZnO (Baumgartner et al. 1978), phonon detection by far infrared phonon-difference absorption (Lengfellner and Renk 1981), by two-phonon Raman scattering (Tsen et al. 1982) and by X-ray Brillouin scattering (McWhan et al. 1982) and level crossing spectroscopy (Challis et al. 1982).

Several previous surveys contain information on optical techniques of generation and detection of high-frequency phonons (Renk 1972, Kaplyanskii et al. 1979, Renk 1979, Bron 1980b, Kaplyanskii et al. 1981c, Renk and Lengfellner 1981, Renk 1984).

2 ANHARMONIC DECAY OF HIGH-FREQUENCY ACOUSTIC PHONONS

2.1 Tunable Detection of Terahertz Phonons

High-frequendy phonons can be detected by the vibronic sideband spectroscopy. As an example, detection of phonons in CaF_2 containing Eu^{2+} ions is discussed. A crystal, surrounded by liquid helium of 2 K, is excited with UV radiation. The fluorescence spectrum shows a sharp zero phonon line at a wavelength of 413 nm due to transitions from the lowest excited state Γ_8^+ $(4f^6 5d)$ to the $4f^7$ ground state of Eu^{2+}, with a phonon sideband (upper curve in Fig. 1) that extends over a large frequency range. In the vicinity of the zero phonon line there is a strong one-phonon contribution (Ignatev and Ovsyankin 1980) indicated by the lower curve in Fig. 1. When phonons are excited in the crystal, anti-Stokes fluorescence occurs. The occupation number $p(\nu)$ of these phonons is obtained from the relation

$$p(\nu) = I_{AS}(\nu)/I_S(\nu) \tag{2.1}$$

where I_{AS}/I_S is the ratio of the anti-Stokes and Stokes fluorescence intensities, measured at frequency distances ν relative to the frequency of the zero phonon line. Eq.(2.1) is applicable for small phonon occupation numbers, $p \ll 1$. From the phonon occupation number the spectral energy density $u(\nu)$ is obtained using the relation

$$u(\nu) = p(\nu)\rho(\nu)h\nu \tag{2.2}$$

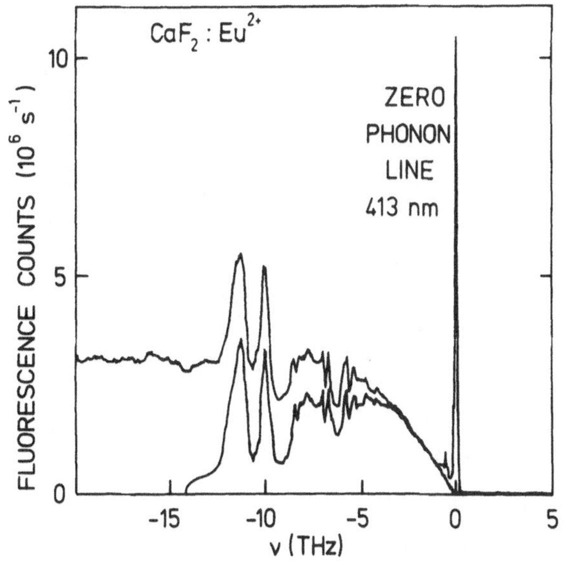

Fig. 1. Fluorescence spectrum of $CaF_2 : Eu^{2+}$ at low crystal temperature; upper curve, multiphonon spectrum and lower curve, one-phonon Stokes sideband.

where $\rho(\nu) = 12\pi\nu^2/v^3$ is the phonon density of state, v an average velocity of sound ($v = 4\cdot10^5$ cm/s for CaF_2) and h Planck's constant.

As an alternative technique, a stress technique was used for phonon detection. The Γ_8^+ ($4f^65d$) level can be split by uniaxial stress (Fig. 2a). The splitting energy corresponds to terahertz frequencies; at a stress of S = 8 kbar ($8\cdot10^8$ N/m^2), a splitting frequency of 3.2 THz is obtained. This system is suitable for tunable detection of terahertz phonons. The principle is shown in Fig. 2b. The lower of the stress split energy levels (W_1) is optically excited. When phonons are injected, emission from the upper level (W_2) occurs. The occupation number p(ν) of nonequilibrium phonons is obtained from the ratio

$$p(\nu) = I_2(\nu)/I_1(\nu) \qquad (2.3)$$

where ν is the frequency separation between the two levels, determined by the stress, and $I_2(\nu)/I_1(\nu)$ is the ratio of the fluorescence intensities for emission from the upper and lower level, respectively. The stress spectrometer is more sensitive (by an order of magnitude at 3 THz) then the vibronic sideband spectrometer. The high sensitivity was important for the study of anharmonic phonon decay at high frequencies (section 2.2).

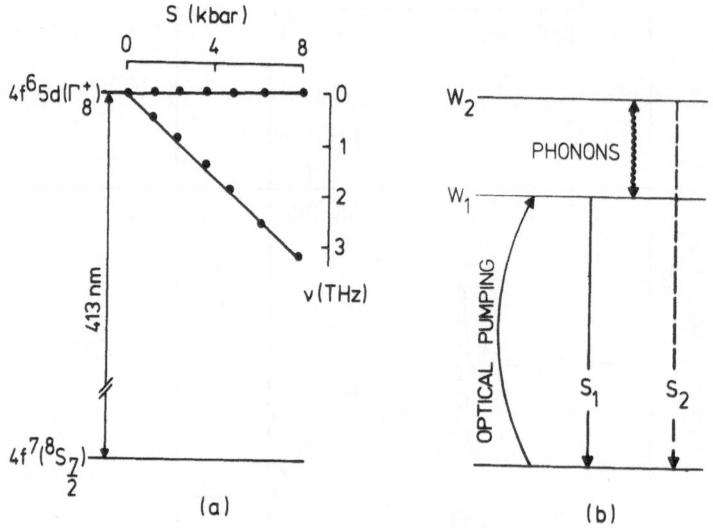

Fig. 2. (a) Stress-splitting of the Γ_8^+ ($4f^65d$) level of CaF_2:Eu^{2+} and (b) principle of phonon detection.

First experiments were performed with heat pulses as phonon sources. The principle is shown in Fig. 3. A CaF_2:Eu^{2+} crystal, immersed in liquid helium, is irradiated in a small part with optical or uv laser radiation. Fluorescence radiation from the crystal is focused to the entrance slit of a double monochromator (D.M.) and detected with a photomultiplier (P.M.). The photomultiplier pulses are analyzed using multichannel averaging techniques. The electronic device is triggered by a reference pulse (R) that is delivered from a pulse generator (G) synchronously with electric pulses P_H. These generate in the heater (H) phonons which propagate into the crystal.

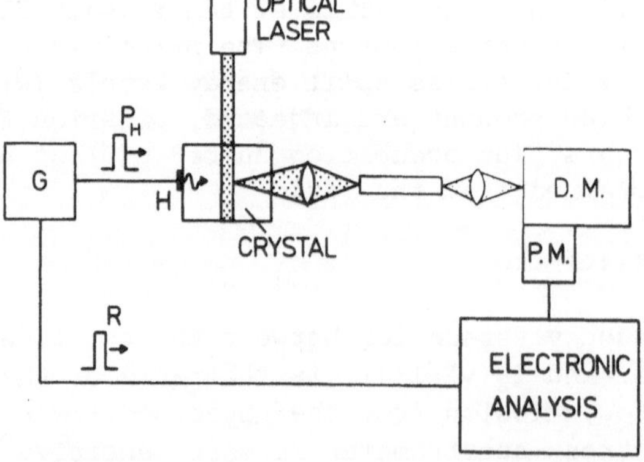

Fig. 3. Experimental setup for phonon detection by phonon-induced fluorescence.

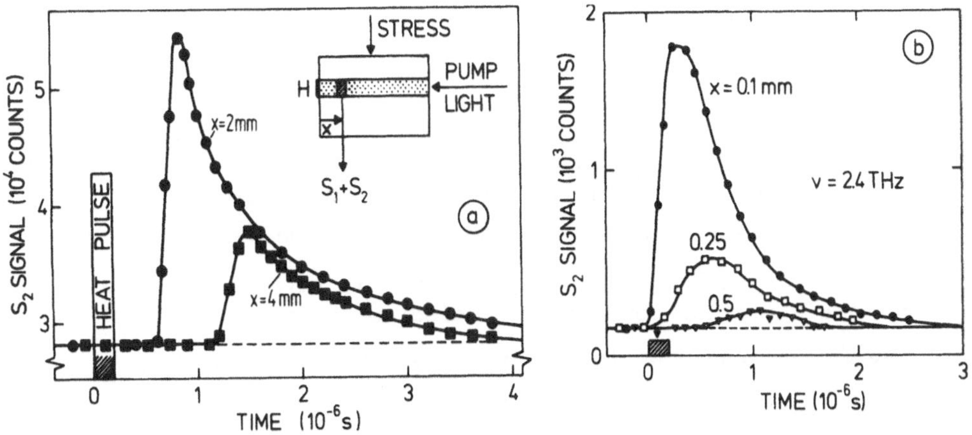

Fig. 4. Time of flight experiment with 600 GHz phonons (a) and with 2.4 THz phonons (b) in CaF_2 containing 0.01 mol% Eu^{2+}.

Results (Eisfeld and Renk 1980) of a time of flight experiment with heat pulse phonons propagating in CaF_2 along the [100] direction are shown in Fig. 4a. The heat pulses (repetition rate 10^3 Hz) had a pulse power of 20 W/mm^2 and a duration of $2 \cdot 10^{-7}$ s; averaging was performed over 10^6 pulses. The arrival times correspond to propagation of transverse acoustic phonons that have a velocity of sound of $3.3 \cdot 10^5$ cm/s. The phonons, at a frequency of 600 GHz, propagated nearly ballistically. From the pulse broadening a mean free path against defect-mass scattering at Eu^{2+} ions of about 1 cm is estimated. From the mean free path λ a cross section for defect-mass scattering is obtained according to the relation $\sigma_{def} = N_i^{-1} \lambda^{-1}$ where N_i is the impurity concentration. With $N_i \simeq 2 \cdot 10^{18}$ cm^{-3}, σ_{def} $(6 \cdot 10^{11}$ Hz$) \simeq 10^{-18}$ cm^2 is estimated. Experimental results (Eisfeld 1979) for propagation of phonons at 2.4 THz are shown in Fig. 4b; in this experiment the heat pulse power was 100 W/mm^2. For a distance $x \simeq 0.1$ mm between heater and detector volume a relative fast temporal increase of the signal is observed. For larger distances however, the increase occurs at late times after phonon generation and the signals are much smaller than for small distances. The result indicate that high-frequency phonons propagate diffusively because of strong defect-mass scattering.

According to Klemens (1955), an average cross section for elastic scattering at an isotopic impurity in a monoatomic lattice is given by

$$\sigma_{def} = (4\pi)^{-1} \, N_L^{-1} \Omega (\Delta m/m)^2 \, q^4 \tag{2.4}$$

where N_L is the number of lattice atoms per unit volume, Ω the unit cell volume, m the mass of the lattice atoms, $m+\Delta m$ the mass of the impurity atom, $q = 2\pi\nu/v$ the phonon wave vector and v the velocity of sound. For an estimate, the expression (2.4) can be applied to describe elastic scattering of Eu^{2+} defects in CaF_2. With $v = 4\cdot10^5$ cm s^{-1}, $N_L = 2.6\cdot10^{22}$ cm^{-3}, $\Omega = 1.6\cdot10^{-22}$ cm^3, $\Delta m/m \simeq 1$ (m_{CaF_2} = 78, m_{Eu} = 152) a cross section for defect-mass scattering $\sigma_{def} = B'\nu^4$ where $B' \simeq 3\cdot10^{-65}$ cm^2 s^4 is obtained. It follows that the corresponding diffusion constant for defect-mass scattering $D(\nu) = \frac{1}{3} v N_i^{-1} \sigma_{def}^{-1}$ where N_i is the impurity concentration can be written as

$$D(\nu) = B\,\nu^{-4} \tag{2.5}$$

where $B = 2\cdot10^{51}$ cm^2 s^{-5} for $N_i = 2\cdot10^{18}$ cm^{-3} (0.01 mol%). The result of the time-of-flight-experiment at 600 GHz (Fig. 4a) is within an order of magnitude consistent with the theoretical estimate. The result of the propagation experiment with 2.4 THz phonons (Fig. 4b) is in accordance with a strongly frequency dependent diffusion constant (2.5). A quantitative analysis is however difficult since the time-of-flight spectrum is determined by both diffusive phonon propagation and anharmonic decay.

Instead of time-of-flight-experiments, spatial distributions of phonons, at fixed times t after phonon injection, can be determined by the use of the experimental setup shown in the inset of Fig. 4a; fluorescence is observed at different distances x relative to the heater (H). Experimental results (Eisfeld and Renk 1980) obtained for phonons in SrF_2 injected by the heat pulse technique are shown in Fig. 5a. After t = 1 μs, phonons at 1.4 THz have propagated over a larger distance than phonons at 2.4 THz. The diffusion equation for the phonon occupation number p (x,y,z,ν,t) has the form

$$\frac{\partial p}{\partial t} = D(\nu)\,\nabla^2 p - \tau_a^{-1}(\nu)\,p \tag{2.6}$$

where $\tau_a^{-1}(\nu)$ is an anharmonic decay rate of phonons at frequency ν. In the expression (2.6) it is neglected that high-frequency phonons can contribute by their decay to the phonon occupation number at frequency ν. For low-power heat pulses ($\lesssim$ 50 W/mm^2) used in the experiment the approximation was justified. The solution of (2.6) is

$$p(x,0,0,\nu,t) = \frac{K}{D(\nu)t^d}\,\exp\left[-\frac{t}{\tau(\nu)}\right]\exp\left[-\frac{x^2}{4D(\nu)t}\right] \tag{2.7}$$

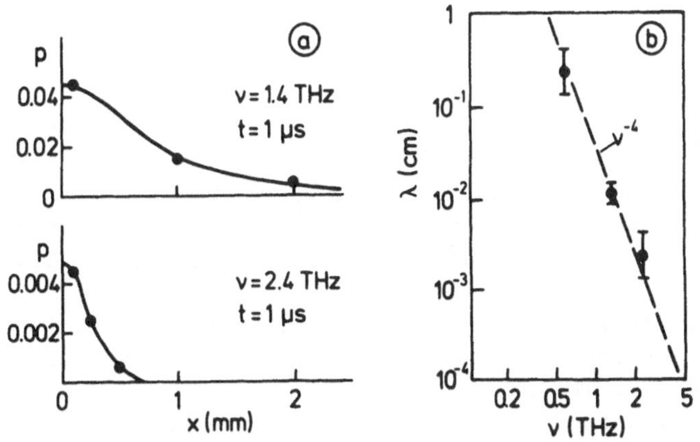

Fig. 5. Defect-mass scattering of phonons in SrF_2 containing 0.05 mol% Eu^{2+}; (a), spatial phonon distributions and, (b), mean free path of phonons at different frequencies.

where K is a constant that depends on the actual heat pulse power, d is the dimension of the diffusion ($d \simeq 2$ for small x and $\simeq 3$ for x large compared to the dimensions of the $1 \times 1\ mm^2$ heater film) and where x is the direction perpendicular to the heater film (inset of Fig. 4a). For fixed times, the last term in (2.7) describes the spatial distribution. The analysis of experimental distributions (Fig. 5a) allows therefore to determine $D(\nu)$ and according to $D(\nu) = \frac{1}{3} v\, \lambda(\nu)$ also the mean free path $\lambda(\nu)$. Results obtained from phonon distribution measurements (Renk 1980) are shown in Fig. 5b. For comparison a theoretical curve obtained from (2.4) is also shown. In the analysis, isotopic scattering at Sr isotopes was also included (Renk 1980). Though there are only few data available, the experiment gives evidence for a strong frequency dependence of the defect-mass scattering as it is predicted by theory.

2.2 Spontaneous Decay of Acoustic Phonons

In this section an experiment (Baumgartner et al. 1981a,b,c) is discussed that gives evidence for spontaneous decay of high-frequency acoustic phonons. The principle of the experiment is shown in Fig. 6a. Pulsed radiation of a uv nitrogen laser is absorbed by transitions from the $4f^7$ ground state to levels of the $4f^6 5d$ manifold (which gives rise to a broad uv absorption band). By fast nonradiative transitions the W_1 and W_2 levels are populated and, by $W_2 \rightarrow W_1$ one-phonon relaxation, monochromatic phonons are generated at the splitting frequency ν. These phonons are monitored by S_2 fluorescence (section 2.1). It should be noted that in addition to the monochromatic phonons also a background of broadband phonons is created due to the nonradiative transitions. It was found (Baumgartner et al. 1983) that the number of these phonons within the detection bandwidth has a value of about $\frac{1}{3}$ of the number of directly generated phonons. The method of phonon generation is the same as used for generation of 29 cm^{-1} phonons in ruby at fixed level splitting (Dijkhuis et al. 1976, Meltzer and Rives 1977).

The experimental arrangement is shown in the inset of Fig. 6b. A crystal under stress was homogeneously optically excited in a volume of 3×4×4 mm^3. The nitrogen laser pulse had a pulse duration of 2 ns and a pulse energy of 10^{-6} J. A phonon-induced signal obtained for a frequency of 1 THz is shown in Fig. 6b (upper curve); the curve is an average curve through the experimental points. The signal decay is only slightly shorter than the lifetime (670 ns) of the excited Eu^{2+} ions. The result shows that phonons at 1 THz have a lifetime longer than 600 ns. At higher frequency, a much faster decay is found (lower curve). The decay curve indicates that phonons at 2.4 THz have a lifetime of about 200 ns. The decay is attributed to anharmonic decay of the high-frequency phonons. It is interesting that an increase of the signal at low frequency (upper curve) is found; this is most likely due to the decay of

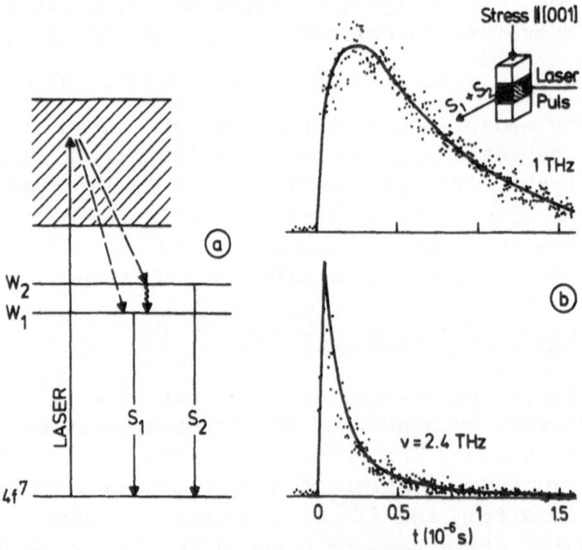

Fig. 6. Generation of phonons in $CaF_2:Eu^{2+}$ under stress by optical excitation and spin-lattice relaxation; (a) principle; (b) phonon-induced S_2 fluorescence signals; inset, arrangement.

high-frequency phonons created by the nonradiative transitions. The phonon occupation numbers in the signal maxima, obtained according to (2.3) from the ratio of S_2 and S_1 fluorescence, were 10^{-2} for the 1 THz signal and 10^{-3} for the 2.4 THz signal.

Experimental lifetimes obtained for different detector frequencies are shown in Fig. 7. The lifetime is almost constant up to a frequency of 1.5 THz and then decreases very strongly. The lifetime at the higher frequencies is consistent with a ν^{-5} dependence (solid line). This dependence gives strong evidende for anharmonic phonon splitting processes. From the experiment an anharmonic decay rate

$$\tau_a^{-1} = A\,\nu^5 \tag{2.8}$$

with $A = 5.5 \cdot 10^{-56}\ s^4$ for CaF_2 can be concluded. At the smaller frequencies (Fig. 7), the phonon lifetime is larger than the escape times τ_F^T and τ_F^L for ballistically propagating transverse and longitudinal phonons, respectively. This indicates that the phonons leave the crystal volume in which the phonons are generated by diffusion that is caused by defect-mass scattering at the Eu^{2+} impurity ions.

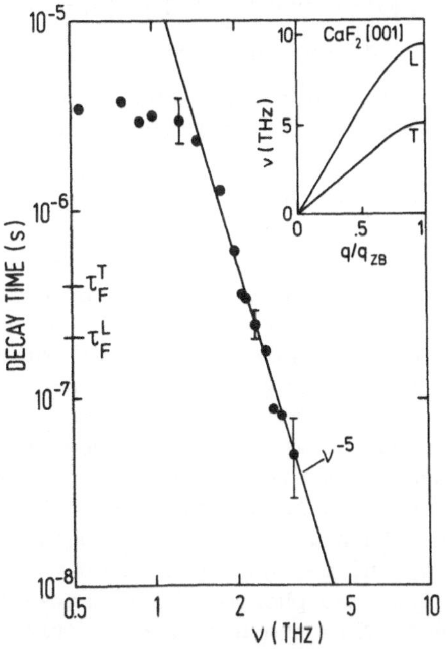

Fig. 7. Experimental lifetimes (points) and theoretical curve (solid line) describing anharmonic decay for phonons in CaF_2; inset, dispersion curves.

Akimov et al. (1981) determined also lifetimes of phonons in CaF_2. The results, obtained in a smaller frequency range, are consistent with the results of Fig. 7 if it is taken into account that the experiment was performed with a smaller crystal volume and that phonons decayed faster due to faster spatial escape from the detector volume. Evidence for the anharmonic decay of high-frequency acoustic phonons was also found for SrF_2 (Baumgartner et al. 1981a) and for LaF_3 (Will et al. 1983).

For an analysis it has to be taken into account that a phonon mixture consisting of phonons of different polarization and propagation directions is observed in the experiment. Because of elastic scattering at impurities, phonons at one frequency but belonging to different modes can be in an equilibrium. Assuming full equilibration between the phonon modes at fixed frequency, the experimental lifetime τ_a is related to the lifetime τ_L of the longitudinal phonons by the expression

$$\tau_a(\nu) = \left[1 + \frac{\rho_T(\nu)}{\rho_L(\nu)}\right]\tau_L(\nu) \tag{2.9}$$

where $\rho_T(\nu)$ is the density of states of transverse phonons that are assumed as non-decaying and $\rho_L(\nu)$ the density of states of longitudinal phonons. With $\rho_T(\nu) = 8\pi\nu^2/v_T^3$ and $\rho_L(\nu) = 4\pi\nu^2/v_L^3$ where $v_T \simeq 4\cdot10^5$ cm/s and $v_L \simeq 7\cdot10^5$ cm/s are average velocities of sound in CaF_2 for transverse and longitudinal phonons, respectively, a ratio $\rho_T(\nu)/\rho_L(\nu) \simeq 13$ is obtained. It follows therefore from the solid curve in Fig. 7 that the anharmonic decay rate for longitudinal phonons is given by $\tau_L^{-1} = A'\nu^5$ where $A' \simeq 8\cdot10^{-55}$ s^4.

The phonons for which lifetimes were determined have wave vectors in the linear part of the dispersion curves of CaF_2 (Elcombe and Pryor 1970). Dispersion curves for phonons propagating along the [001] direction are shown in the inset of Fig.7; transverse phonons at the zone boundary (ZB) have frequencies at 5 THz and longitudinal phonons at 10 THz. For a further discussion of the anharmonic phonon decay at low crystal temperature, the crystal is described in the model of an isotropic solid, with a degenerate transverse phonon branch and a longitudinal branch. In this model, transverse acoustic phonons have a negligible decay rate (Slonimskii 1937); because of energy and momentum selection rules, a transverse phonon can decay only into two phonons which propagate collinear with the decaying phonon. This process is characterized by a negligible joint density of states. If dispersion of the transverse branches is additionally taken into account, transverse acoustic phonons in low lying branches cannot decay at low crystal temperature because of energy and momentum selection rules as has been shown theoretically by Orbach and Vredevoe (1964), Maris (1965), and Lax et al. (1981).

Longitudinal acoustic phonons propagating in a crystal at low temperature ($h\nu \ll kT$) can spontaneously decay by splitting processes due to the crystal anharmonicity. A phonon decays either into two transverse phonons (Fig. 8a) or into a transverse and a longitudinal phonon (Fig. 8b). In each decay process, energy and momentum conservation laws have to be fulfilled

$$h\nu = h\nu_1 + h\nu_2 \quad \text{and} \quad \vec{q} = \vec{q}_1 + \vec{q}_2 \tag{2.10}$$

where ν is the frequency and $\vec{q}$ the wave vector of a decaying longitudinal phonon and ν_1, ν_2, $\vec{q}_1$, $\vec{q}_2$ the frequencies and wave vectors of the phonons created by the decay. The potential energy density, including anharmonic terms, can be written in the form (Slonimskii 1937, Landau and Lifshitz 1970, Orbach and Vredevoe 1964, Tucker and Rampton 1972)

$$V = C_1\, \varepsilon_{ii}^2 + C_2\, \varepsilon_{ik}^2 + P'\varepsilon_{ii}^3 + Q'\varepsilon_{ii}\, \varepsilon_{ik}^2 + R'\varepsilon_{ik}^3 \tag{2.11}$$

where ε_{ik} is the deformation tensor ($i,k = 1,2,3$); double-indices indicate that a summation has to be performed [e.g. $\varepsilon_{ii}^2 = (\varepsilon_{11} + \varepsilon_{22} + \varepsilon_{33})^2$]. The constants are related to second order elastic constants c_{ik} and to the third order elastic constants c_{ikl} by the relations $C_1 = \frac{1}{2} c_{12}$, $C_2 = c_{44}$ and $P' = \frac{1}{6} c_{111} - c_{155} + \frac{2}{3} c_{456}$, $Q' = c_{155} - 2\, c_{456}$, $R' = \frac{4}{3} c_{456}$.

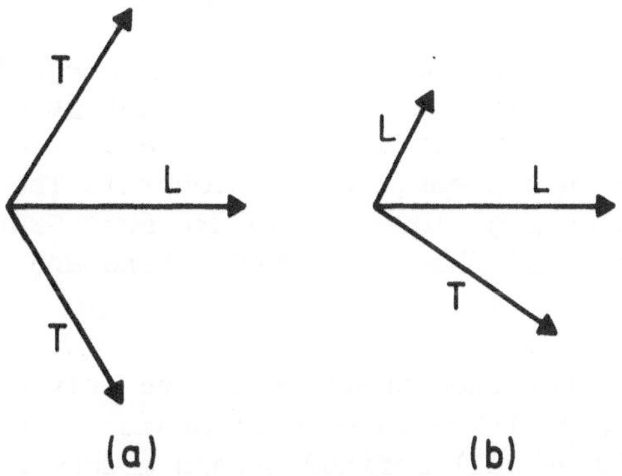

Fig. 8. Phonon splitting processes; a longitudinal (L) phonon splits in two transverse (T) phonons (a) or in a longitudinal and a transverse phonon (b).

The expression (2.11) is of third order in the elongations, and therefore vibrations at different frequencies are coupled with each other. Slonimskii (1937) has performed a perturbation calculation for the quantized system and calculated the decay of a longitudinal phonon by summing over all decay processes allowed according to (2.10), and taking into account the influence of the phonon polarizations. The result can be summarized by a simple relation for the anharmonic lifetime τ_L for longitudinal phonons,

$$\tau_L^{-1} = \frac{h\varphi^2}{128\,\rho_m}\, q^5 \tag{2.12}$$

where φ is a dimensionless anharmonicity parameter, ρ_m the mass density of the crystal and q the wave vector of the decaying longitudinal phonons. The parameter φ depends in a complicated way on the linear and nonlinear elastic constants. It can be shown by taking into account experimental values for the second and third order elastic constants that the anharmonicity parameter φ is of the order of 1; φ has the same structure as a Grüneisen parameter, but is related to a different combination of elastic constants. The factor ρ_m in the denumerator of (2.12) indicates that the phonon splitting is caused by the zero point motion of the atoms in the crystal lattice. The factor q^5 is composed of a factor q^2 that is measure of the joint phonon density of states in which a longitudinal phonon can decay and of a factor q^3 that occurs in the decay probability because three phonons participate in a decay process.

From the experimental decay time for CaF_2 ($\rho_m = 3.4$ g/cm^3, $q = 2\pi\nu/v_L$, $v_L \simeq 7\cdot 10^5$ cm/s) an anharmonicity parameter $\varphi \simeq 1$ is found. This value is by a factor of 2.4 smaller than the theoretical value that was derived from a point charge model calculation (Tua and Mahan 1982) and that was also estimated by use of the isotropic model (Slonimskii 1937) taking into account experimental (room temperature) third-order elastic constants. The theoretical anharmonic lifetimes are therefore about an order of magnitude smaller than the experimental lifetimes. A better agreement is obtained with values from other theoretical estimates (Orbach and Vredevoe 1967, Klemens 1964), however, this seems to be accidental. The parameter φ is comparable with an average low-temperature mode Grüneisen parameter ($\simeq 1.2$) for longitudinal phonons in CaF_2 (Wong and Schuele 1966).

It should be noted that the experiments give only a first survey on the phonon decay. It would be most desiderable to develop techniques for the study of well defined phonon modes. A further unsolved problem concerns the decay of transverse acoustic phonons which do not belong to the lowest branches.

2.3 Phonon Generation by Nonradiative Transitions

The principle of phonon generation by nonradiative transitions is shown in Fig. 9a. Pulsed radiation of a uv nitrogen laser is absorbed by transitions from the $4f^7$ ground state to $4f^65d$ levels 5000 cm^{-1} (170 THz) above the lowest excited state level $4f^65d$ (Γ_8^+). Subsequent nonradiative transitions within the 5d multifold lead to generation of phonons at very high frequencies, mainly optical phonons. These decay fast and therefore acoustic phonons at terahertz frequencies are obtained. In an experiment by Baumgartner et al. (1983) phonons were detected by vibronic sideband spectroscopy. After excitation, similar decay curves as shown in Fig. 6b were obtained. From the signals at fixed times after phonon generation, but measured at different detection frequencies, phonon occupation numbers were determined using (2.1). A phonon distribution, observed immediately (10 ns) after pulsed laser excitation (with laser pulses of 1 ns duration and a pulse energy of 10^{-6} J) is shown in Fig. 9b. For comparison with the experimental curve (points and solid line), the dashed line of Fig.9b shows a Planckian distribution calculated from the pulse energy that is transformed into vibrational energy for the case of fast thermalization. The comparison shows that the experimental spectrum is strongly nonthermal; the phonon concentration at 3 THz exceeds the thermal value by two orders of magnitude and at 0.5 THz the experimental concentration is smaller than for a thermalized spectrum.

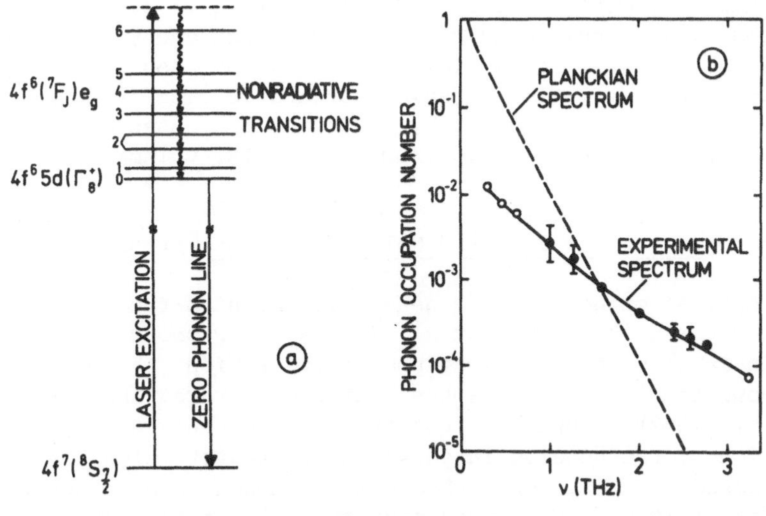

Fig. 9. Generation of phonons by nonradiative transitions in $CaF_2:Eu^{2+}$; (a) principle and (b) phonon occupation number immediately after generation; for comparison a Planckian spectrum is shown.

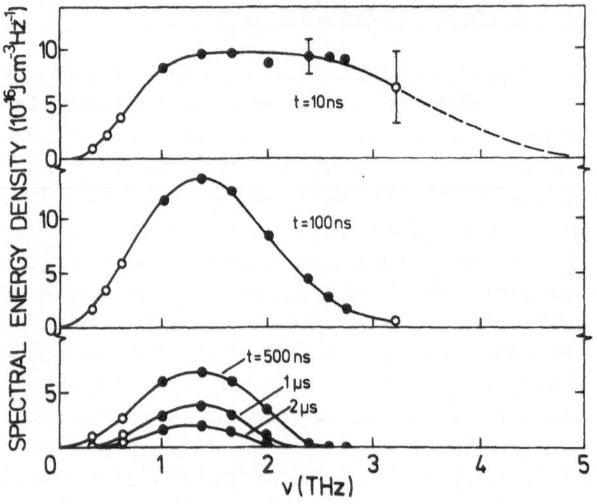

Fig. 10. Spectral phonon distributions in $CaF_2:Eu^{2+}$ for different times t after phonon generation.

From the phonon occupation number p(ν) of Fig.9b the spectral energy density of nonequilibrium phonons is obtained according to (2.2). Spectral energy densities obtained for different times t after phonon generation are shown in Fig. 10. Immediately after phonon generation (upper curve) the phonon distribution is very broad. At a time (t = 100 ns) at which phonons did not yet leave the crystal volume, the concentration of high-frequency phonons has decreased while the concentration of phonons at lower frequencies has increased. This result demonstrates the phonon decay by splitting processes. At later times (lower curves), the phonon distributions become narrower, with a maximum around 1.3 THz. Phonons at higher frequencies decayed mainly by splitting processes and phonons at lower frequencies escaped from the crystal volume by diffusion.

2.4 Phonon Generation by Infrared Multiphonon Absorption

A new optical method for the generation of high-frequency phonons was reported by Happek et al. (1984a,b). Phonons were generated in CaF_2 by infrared multiphonon absorption. The principle is shown in Fig. 11a. CaF_2 has a strong infrared absorption (Kaiser et al. 1962) around the reststrahlen frequency at 10 THz. The absorption at frequencies larger than the reststrahlen frequency is caused by multiphonon absorption. Phonons were generated by absorption of 10.6 μm CO_2 laser radiation at a frequency $\nu_L \approx$ 28 THz (arrow in Fig. 11a). The absorption coefficient has a value of 1 cm^{-1} and therefore a large crystal volume can be excited homogeneously. By the absorption, mainly optical phonons at the zone boundary are generated by three-phonon absorption processes. These

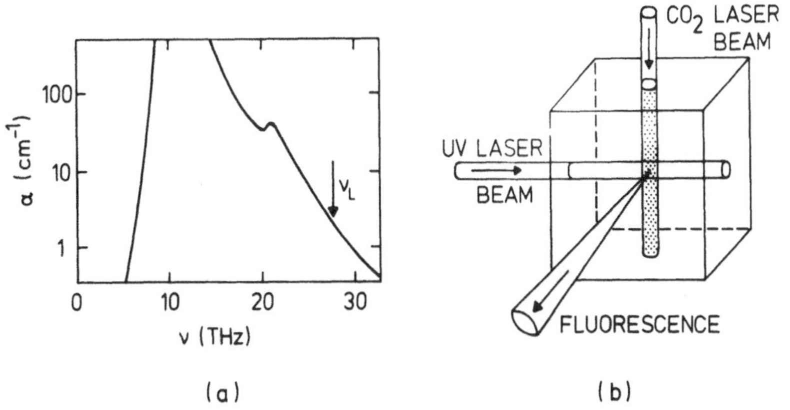

Fig. 11. Phonon generation by multiphonon infrared absorption; (a) infrared absorption spectrum of CaF_2 and (b) arrangement.

phonons decay fast into phonons of lower frequencies. Phonons were detected, using Eu^{2+} ions, by vibronic sideband spectroscopy in the range around 1 THz.

The arrangement is shown in Fig. 11b. Radiation of a Q-switched CO_2 laser (pulse duration $2\cdot10^{-7}$ s, pulse energy 10^{-4} J, repetition rate 100 cps) was focused to a CaF_2 crystal. Accordingly, a cylindrical phonon source of 1 mm diameter within the large crystal (size $8\times8\times10$ mm^3) was obtained. The crystal was also weakly doped with Eu^{2+} ions (0.01 mol%) which were used for phonon detection. For detection, continuous uv laser radiation was guided through the crystal and anti-Stokes fluorescence radiation was observed from the center of the crystal.

Experimental signal curves (Happek et al. 1984b) are shown in Fig. 12. After CO_2 laser excitation (upper curve), phonon signals are found; averaging was performed over $3\cdot10^5$ pulses. The signal behavior is similar as for phonons generated by nonradiative transitions, i.e. phonons at high frequencies decay fast (lower curve) and phonons at lower frequencies have a much larger lifetime. The upper curve shows furthermore, that the signal at 1.1 THz reaches its maximum only after 3 μs which gives evidence for long-lived high-frequency phonons.

From signal curves at different detection frequencies, frequency distributions were determined using (2.1) and (2.2). There are two interesting time regimes. Shortly after phonon generation phonon redistribution is observed (Fig. 13a) and at late times (Fig. 13b) the distribution keeps its shape. Immediately after phonon generation the distribution (lower curve in Fig. 13a) has

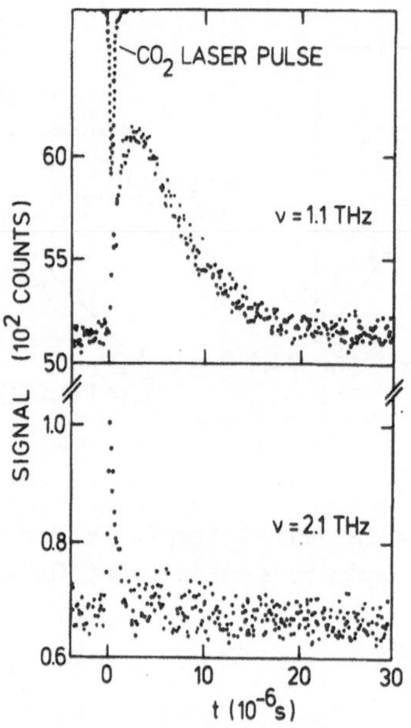

Fig. 12. Phonon-induced fluorescence for phonons generated by multiphonon absorption of CO_2 laser radiation.

two maxima, near 1 THz and 2 THz. With increasing time the maximum at the larger frequency decreases while the maximum at the smaller frequency increases up to a time of about 3 μs. The result shows again that phonons at very high frequencies decay by splitting processes. From the increase a lifetime of the order of 10^{-6} s can be concluded for very high frequency phonons. A comparison with the

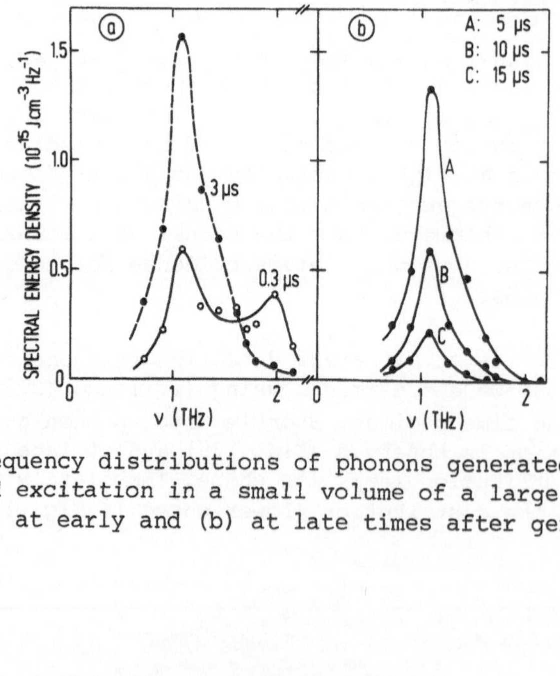

Fig. 13. Frequency distributions of phonons generated by infrared excitation in a small volume of a large CaF_2 crystal; (a) at early and (b) at late times after generation.

dispersion curves and with the density of states (Bilz and Kress 1979) shows that long-lived high-frequency phonons are expected around a frequency of 4.4 THz where the density of states of low lying transverse acoustic branches has a maximum. Decay of these phonons via splitting processes of longitudinal phonons can mainly lead to phonons around 2 THz, corresponding to the maximum of the lower curve in Fig. 13a, and decay of these phonons can be responsible for the maximum near 1 THz.

The second interesting regime concerns the behavior of the frequency distribution at large times after the phonon generation (Fig. 13b). The time-dependence of the distribution is consistent with anharmonic decay of high-frequency phonons and diffusive escape from the detector volume of phonons at lower frequencies. The maximum of the distribution indicates a maximum lifetime for phonons at a frequency $\nu_{max} \simeq 1.1$ THz. The experiment shows that the distribution remains nonthermal even at very large times after phonon generation.

The dynamics of the phonon system can be described by the equation

$$\frac{\partial f(\nu)}{\partial t} = D(\nu)\nabla^2 f(\nu) - \tau_a^{-1}(\nu)f(\nu) + 4\,\tau_a^{-1}(2\nu)f(2\nu) \qquad (2.13)$$

where $f(r,\nu,t)$ is the phonon distribution function, i.e. the number of phonons per unit volume and per unit frequency interval, r the distance from the axis of the cylindrical source, $D(\nu)$ the diffusion constant for spatial diffusion and $\tau_a(\nu)$ the anharmonic

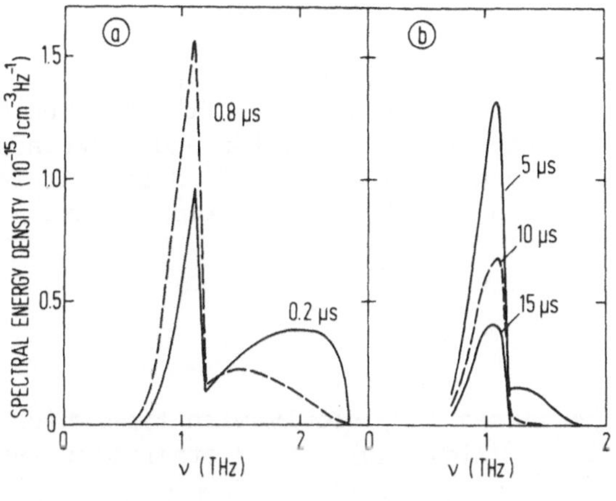

Fig. 14. Calculated frequency distributions; (a) for early times and (b) for late times after phonon generation, according to (2.13).

lifetime. For simplicity it is assumed that a phonon at frequency 2ν decays into two phonons of frequency ν. A numerical solution (Happek et al. 1984b) of (2.13) is shown in Fig. 14. As an initial condition, it was assumed that phonons were spatially distributed for $t = 0$ in a cylindrical volume (diameter 1 mm) around an axis (with coordinate $r = 0$) and that the spectral distribution was proportional to a Debye density of state, with a maximum frequency of 4.4 THz. For the anharmonic lifetime $\tau_a^{-1}(\nu) = A\nu^5$ with $A = 5.5 \cdot 10^{-56}\,s^4$ (section 2.2) and for the diffusion constant $D(\nu) = B\nu^{-4}$ with $B = 2 \cdot 10^{52} cm\ s^{-5}$ (section 2.1) were used. The calculation was performed for a cylindrical source in an infinite crystal. The calculated distributions (Fig. 14) show qualitatively the main features of the experimental distributions, namely strong redistribution for small times and a nearly time-independent shape of the distribution for late times. There is a difference between theory and experiment concerning the redistribution at small times (Fig. 14a). While the experiment showed a signal increase up to 3 µs, the calculation showed an increase up to 0.8 µs only and then the whole spectrum decreased. This discrepancy is probably due to long-lived zone-boundary phonons which are not included in the theoretical description. In order to account for these long-lived phonons, the curves of Fig. 14b are stretched in the vertical direction by a factor of 2 compared to the curves of Fig. 14a. The behavior at late times can alternatively be described by an approximate solution of (2.13), given by

$$f(0,\nu,t) = \frac{f(0,\nu,t')}{D(\nu)t} \exp\left[-\frac{t-t'}{\tau_a(\nu)}\right] \tag{2.14}$$

where t is the time after phonon generation (at $t = 0$), t' a time at which the final distribution was established and $f(0,\nu,t')$ the phonon distribution on the axis of the cylindrical source at time t'. In the solution (2.14) phonons at half frequencies are neglected; it is assumed that the corresponding phonons at lower frequencies spread fast, and leave the crystal fast. The distribution decreases as t^{-1} because of diffusion and shows additionally an exponential decrease because of anharmonic decay. It follows from (2.14) that the maximum of the distribution occurs at a frequency

$$\nu_m = \left(\frac{4}{5At}\right)^{1/5} \tag{2.15}$$

which shows that ν_m decreases only slowly with time which is in agreement with experiment (Fig. 13b). The distribution function at ν_m has the value

$$f(0,\nu_m,t) = 0.42\ B^{-1}\ A^{-\frac{4}{5}}\ f(0,\nu,t')\,t^{-\frac{9}{5}}. \tag{2.16}$$

The maximum of the distribution decreases for large times ($t > t'$) as $t^{-9/5}$. In the time-range of the experiment (Fig. 13b), the signal maximum decreases indeed according to (2.16). It was found (Happek et al. 1984b) by experiments with different diameters of the laser beam and also by solving (2.13) that the intermediate distribution $f(0,\nu,t')$ depends on the diameter of the cylindrical source.

The behavior of high-frequency phonons as found for CaF_2 crystals is most likely characteristic for many crystals and also for amorphous materials. An unsolved problem is the behavior of terahertz phonons at crystal surfaces.

The technique of infrared excitation of phonons by multiphonon absorption presented in this section should be universally applicable because all nonmetallic solids show multiphonon absorption.

3 ELASTIC-RESONANCE AND INELASTIC SCATTERING OF PHONONS

3.1 Observation of Spectral Diffusion of Resonantly Trapped Phonons

In this part of the lecture two new processes of phonon scattering will be treated, namely resonance scattering at impurities due to electronic transitions between electronic energy levels that are resonant with phonons and inelastic scattering of phonons (phonon Raman scattering) at impurities that have low lying energy levels.

Both resonance scattering and inelastic scattering will be discussed in connection with a recent experiment (Engelhardt et al. 1984). The experiment was performed with an Al_2O_3 crystal that contained the three ion types V^{4+}, Cr^{2+} and Cr^{3+}. The V^{4+} ions were used for phonon generation and acted also as resonance scattering centers, Cr^{2+} ions were responsible for inelastic scattering and Cr^{3+} ions were used for monochromatic phonon detection. V^{4+} has a resonance line that is due to electronic transitions from the $E_{3/2}$ ground state to the lowest excited state $E_{1/2}$. It is known from far infrared absorption measurements (Wong et al. 1968) that the resonance frequency for the transition is $\nu_o \simeq 843$ GHz (energy 28.1 cm^{-1}). The halfwidth $\Gamma \simeq 20$ GHz (Engelhardt et al. 1983) is most likely partly due to inhomogeneous broadening and partly due to homogeneous broadening, caused by one-phonon relaxation. Neglecting for simplicity inhomogeneous broadening, the line profile can be described by a Lorentzian shape

$$g(\nu) = \frac{1}{1 + 4\ (\nu-\nu_o)^2/\Gamma^2} \tag{3.1}$$

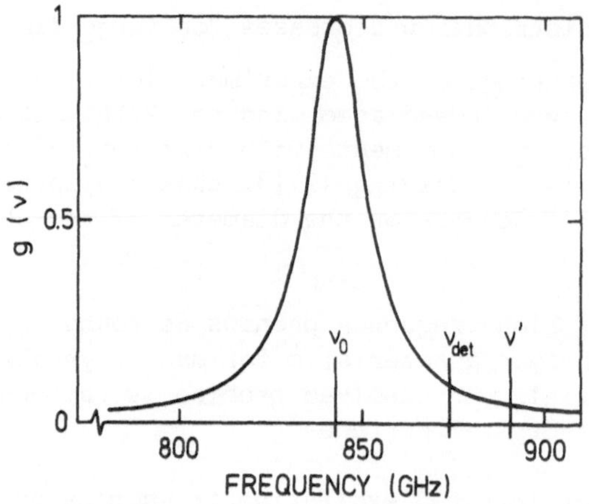

Fig. 15. Resonance line of V^{4+} in Al_2O_3.

where ν_o is the resonance frequency and Γ the halfwidth of the line. The normalization of the lineshape function (3.1) is chosen so that $g(\nu_o) = 1$ and therefore

$$\int_0^\infty g(\nu)d\nu = \frac{2}{\pi\Gamma} \quad . \tag{3.2}$$

The halfwidth Γ is related with the lifetime T_1^s of the $E_{1/2}$ level against spontaneous one-phonon emission by the relation

$$\Gamma\, T_1^s = \frac{1}{2\pi} \quad . \tag{3.3}$$

The lineshape function (3.1) is shown in Fig. 15. In the experiment (Engelhardt et al. 1984) phonons were generated at frequency ν' (= 891 GHz) and the "same" phonons were detected at the frequency ν_{det} (= 874 GHz); both frequencies were different from the resonance frequency ν_o (Fig. 15). The phonons were generated by pulsed far infrared laser excitation and detected by observation of phonon-induced R_2 fluorescence.

An Al_2O_3 crystal doped with both Cr_2O_3 (0.05 mol%) and V_2O_3 (0.3 mol%) had been X-ray irradiated. Because of charge transfer processes a part of the V^{3+} ions were converted into V^{4+} ions; an estimated concentration of V^{4+} ions was $N_{V^{4+}} \simeq 10^{18}\ cm^{-3}$. By the X irradiation, also a part of the Cr^{3+} ions were converted into Cr^{2+} ions; an estimated concentration was $N_{Cr^{2+}} \simeq 10^{18}\ cm^{-3}$ corresponding to a relative concentration of 5 % of the chromium ions (Brown and Brown 1981). The crystal contained therefore V^{4+},

Cr^{3+} and Cr^{2+} ions which were important for phonon generation and resonant trapping (V^{4+} ions), for monochromatic phonon detection (Cr^{3+} ions) and for inelastic scattering (Cr^{2+} ions).

The Al_2O_3 crystal (size of 3×3×5 mm^3) was immersed in liquid helium of 2 K. The crystal was homogeneously illuminated with far infrared radiation pulses (duration 1.8 μs, pulse energy 10^{-7} Joule) from a Q-switched HCN laser (emission frequency $\nu' = 891$ GHz, bandwidth about 5 MHz). A few per cent of the radiation was absorbed in the crystal by excitation of V^{4+} ions. By the far infrared excitation and the subsequent one-phonon relaxation, phonons at the laser frequency ν' were excited (Fig. 15); the estimated phonon energy density was about 10^{-8} J/cm^3. The crystal was continuously irradiated with visible radiation of an argon ion laser and contained excited Cr^{3+} ions. By observation of phonon-induced fluorescence from Cr^{3+} ions, phonons at the detection frequency ν_{det} (874 GHz) were monitored (Fig. 15).

An R_2 signal curve is shown in Fig. 16. Compared with the infrared radiation pulse (dashed), the signal is delayed and shows a slower decrease. There are two important results. (1) A signal is observed though there was an energy mismatch (of about 17 GHz) between generation and detection frequencies. This gives evidence for inelastic phonon scattering with small frequency shifts $\nu_R \ll \nu_o$. The delay time t_m (≈1 μs) is a measure for the average time a phonon needs to be inelastically scattered from a frequency interval at ν' to a frequency interval at ν_{det}. (2) The signal decay time (≃ 4 μs) is much larger than the ballistic escape time (≈0.3 μs) from the

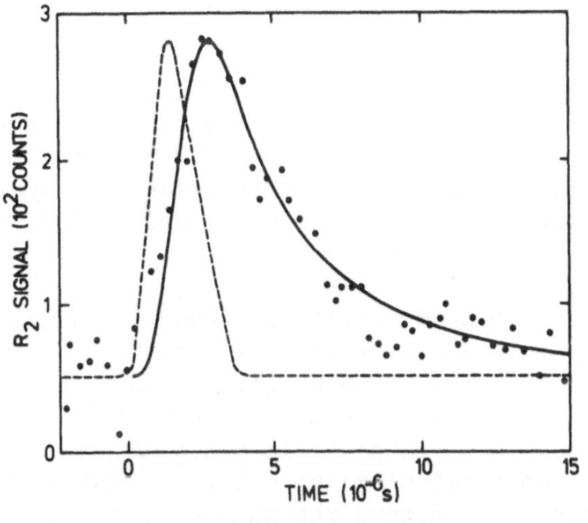

Fig. 16. Phonon-induced R_2 signal from an $Al_2O_3{:}V^{4+}$ crystal (points) and calculated curve (solid line); dashed, far infrared pulse.

crystal. This shows that phonon trapping occurred and, furthermore, that the anharmonic lifetime of the phonons near 1 THz is larger or comparable to the decay time of 4 μs.

The last result is consistent with an estimate of the anharmonic decay time for phonons in Al_2O_3. Using (2.12) and assuming that the anharmonic phonon-phonon interaction is similar as in CaF_2, it is expected that longitudinal phonons at 0.9 THz in Al_2O_3 have a spontaneous lifetime of about $2\cdot10^{-5}$ s and that a phonon mixture of longitudinal and transverse phonons has a lifetime against spontaneous decay of about $3\cdot10^{-4}$ s (for Al_2O_3, $\rho_m = 3g/cm^3$, $v_L \simeq 11\cdot10^5$ cm/s and $v_T \simeq 6\cdot10^5$ cm/s). According to this estimate, the observed phonon decay is most likely not caused by anharmonic decay due to the anharmonicity of the Al_2O_3 crystal. It is suggested (Engelhardt et al. 1984) that a combined effect of elastic and inelastic phonon scattering at impurities is responsible for the observed signal decay.

In the next sections the principle of phonon detection and generation are described in more detail and a theoretical analysis of the experiment is presented. The analysis contains a discussion of elastic resonance scattering and of inelastic scattering of high-frequency phonons.

3.2 Monochromatic Detection of Phonons in Ruby

In this section, monochromatic detection of 29 cm^{-1} phonons in ruby by phonon-induced fluorescence is discussed. The principle (Renk and Deisenhofer 1971) has been applied in various studies (section 1.2).

In Fig. 17a, the energy level scheme of Cr^{3+} in Al_2O_3 is shown. There are broad absorption bands $^4T_1$ and $^4T_2$. The lowest excited state is the ^2E state. Due to the trigonal electric field at Cr^{3+} ion sites, and because of spin-orbit interaction, the quartett ^2E level is split into two Kramers levels $\bar{E}(^2E)$ and $2\bar{A}(^2E)$, with a splitting energy of 29 cm^{-1} (Fig. 17b).

The principle of phonon detection is shown in Fig. 17b. A ruby crystal is optically pumped at low crystal temperature and contains excited Cr^{3+} ions in the $\bar{E}$ state which give rise to R_1 fluorescence radiation (at a wavelength of 693.4 nm). When 29 cm^{-1} phonons are injected, $\bar{E}\rightarrow 2\bar{A}$ transitions are possible and a $2\bar{A}$ population and therefore R_2 fluorescence emission (at 692.2 nm) are obtained. The R_2 fluorescence intensity monitors the number of 29 cm^{-1} phonons. The ratio of the R_2 and R_1 fluorescence intensities is a direct measure of the relative electronic population N/N^* and therefore of the phonon occupation number p_{det} at the detection frequency

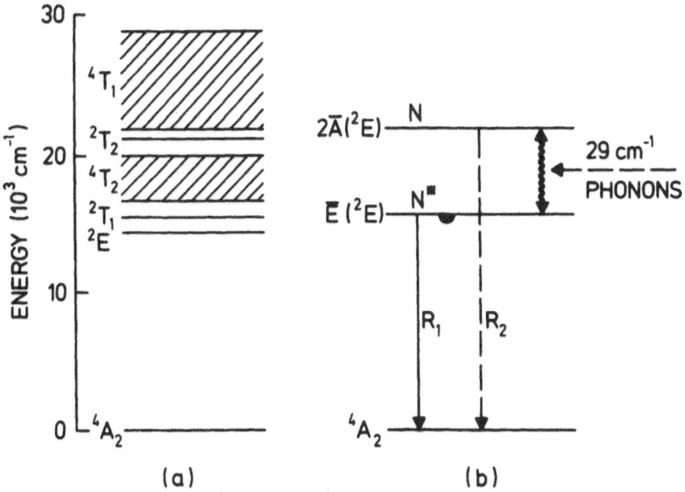

Fig. 17. (a) Energy levels of Cr^{3+} in Al_2O_3 and (b) principle of phonon detection.

ν_{det}. For $p_{det} \ll 1$, the occupation number is therefore obtained from the relation

$$p_{det} = \frac{N}{N^*} \simeq \frac{R_2\text{-intensity}}{R_1\text{-intensity}} \tag{3.4}$$

where N^* is the population of the $\bar{E}$ level and N the population of the $2\bar{A}$ level. If a phonon pulse is excited, then an equilibrium is established between the 29 cm^{-1} phonons and the relative electronic population within a time T_1^S ($\simeq$0.5 nsec) which is the one-phonon spin-lattice relaxation time (Retzer et al. 1983). The detector has therefore a fast response time. The detector is a narrow-band detector, only phonons in a frequency band of about 0.4 GHz are detected; the bandwidth corresponds to the halfwidth of the $\bar{E} \rightarrow 2\bar{A}$ resonance line which is known from far infrared absorption measurements (Retzer et al. 1983). In the experiment shown in Fig. 16, the maximum phonon occupation number had a value $p_{det} \simeq 3 \cdot 10^{-5}$.

3.3 Monochromatic Phonon Generation by Far Infrared Excitation

The phonon generation by far infrared excitation and relaxation of V^{4+} ions can be described as resonance photon-phonon conversion. Under the influence of a monochromatic far infrared radiation field at frequency ν', a V^{4+} ion performs a forced oscillation at the frequency ν' and therefore phonons of the same frequency ν' are created by the one-phonon relaxation process. Supposing that the V^{4+} ions are isolated ions without interaction, energy conservation requires that the phonons have the same quantum energy $h\nu'$ as the incident photons. Therefore, it is possible to generate monochromatic phonons at a frequency ν' which is different from the resonance frequency ν_o (Fig. 15).

Almost each absorbed photon is converted into a phonon. This follows from a discussion of the Einstein relations. For simplification, all phonon modes (transverse and longitudinal) in the Al_2O_3 crystal are assumed to have the same velocity v ($\approx 7 \cdot 10^5$ cm/s) and all interactions are assumed to be isotropic. Then, the phonon density of states $\rho(\nu)$ is given by the Debye density of states

$$\rho(\nu) = \frac{12\pi\nu^2}{v^3} . \tag{3.5}$$

Accordingly, the photon density of states $\rho^*(\nu)$ is given by

$$\rho^*(\nu) = \frac{8\pi\nu^2}{c^3} \tag{3.6}$$

where c is an average velocity of the far infrared radiation in Al_2O_3. The difference in the prefactors of (3.5) and (3.6) follows from the fact that phonons have three and electromagnetic radiation two independent polarization directions. The resonance interaction of phonons with the $E_{1/2}$ and $E_{3/2}$ states of V^{4+} can be described by the Einstein relation

$$A_{21} = \rho(\nu_o)\, h\nu_o\, B_{12} \tag{3.7}$$

where A_{21} is the Einstein coefficient for spontaneous phonon emission, $\rho(\nu_o)$ the phonon density of states at the resonance frequency ν_o and B_{12} the Einstein coefficient for phonon absorption. The $E_{3/2}$ and $E_{1/2}$ states have equal degeneracy (of two) and therefore

$$B_{12} = B_{21} \tag{3.8}$$

where B_{21} is the Einstein coefficient for induced phonon emission. The appropriate Einstein relations for photons are

$$A^*_{21} = \rho^*(\nu_o)\, h\nu_o\, B^*_{12} \tag{3.9a}$$

and

$$B^*_{12} = B^*_{21} \tag{3.9b}$$

where A^*_{21} is the Einstein coefficient for spontaneous photon emission by $E_{1/2} \to E_{3/2}$ transitions of V^{4+}, $\rho^*(\nu_o)$ the photon density of states at ν_o, B^*_{12} the Einstein coefficient for photon absorption and B^*_{21} the Einstein coefficient for induced photon emission. Assuming that the Einstein coefficient for absorption of a phonon is not much different from that for absorption of a photon, $B_{12} \simeq B^*_{12}$, it follows by dividing (3.9a) through (3.7) that

$$\frac{A^*_{21}}{A_{21}} \simeq \frac{\rho^*(\nu_o)}{\rho(\nu_o)} = \frac{v^3}{c^3} \simeq 10^{-12} . \tag{3.10}$$

Emission of photons can therefore be neglected.

3.4 Resonance Scattering and Trapping of Phonons

By far infrared excitation and relaxation of V^{4+} ions in Al_2O_3, monochromatic phonons are generated (section 3.3). These phonons can be resonantly scattered at the V^{4+} ions. In a scattering process a phonon is absorbed by a $E_{3/2} \to E_{1/2}$ transition of a V^{4+} ion and reemitted by the converse process. Since the V^{4+} ions are isolated "two-level atoms", the resonance scattering is purely elastic (Heitler 1954). The resonance scattering cross section is given by

$$\sigma(\nu) = \sigma_o \, g(\nu) \tag{3.11}$$

where σ_o is the cross section in the line center at frequency ν_o and $g(\nu)$ the Lorentzian line shape function (3.1). The maximum cross section can be derived by use of the Füchtbauer-Ladenburg relation (Lengyel 1971)

$$\int \sigma(\nu) \, d\nu = v^{-1} \, h\nu_o \, B_{12} . \tag{3.12}$$

Taking into account that $A_{21} = 1/T_1^s$ and using (3.2), (3.3), (3.5), (3.7), and (3.8), it follows from (3.12) that the cross section in the line-center is given by

$$\sigma_o = \frac{4\pi}{3} q_o^{-2} = \frac{1}{3\pi} \Lambda_o^2 \tag{3.13}$$

where $q_o = 2\pi\nu_o/v$ is the wave vector and Λ_o the wavelength of the phonons at frequency ν_o. The expression (3.13) shows that the cross section in the line center depends only on the phonon wavelength and not on the strength of the electron-phonon interaction; the strength of the electron-phonon interaction determines only the halfwidth Γ of the Lorentzian line. The maximum cross section (3.13) is characteristic for resonance scattering of radiation at two-level atoms which have an upper level that has a finite lifetime which is only due to emission of the resonance radiation. It should be noted that the expression (3.13) can also be derived by use of scattering theory (Goldberger and Watson 1964).

In comparison, if relaxation of a two-level atom occurs not by phonons but mainly by photons, then the maximum cross section for resonance scattering of electromagnetic radiation is given by the expression

$$\sigma_o^* = 2\pi\, q_o^{*-2} = \frac{1}{2\pi}\, \Lambda_o^{*2} \tag{3.14}$$

where q_o^* and Λ_o^* are wave vector and wavelength of the radiation at the resonance frequency ν_o. The result (3.14) can be obtained from an expression that corresponds to (3.12) taking into account the Einstein relations for photons (3.9a) and (3.9b). The difference in the prefactors of (3.13) and (3.14) follows from the different density of states for phonons and photons, according to (3.5) and (3.6). However, because of (3.10), elastic resonance scattering of electromagnetic radiation plays no role in the Al_2O_3:V^{4+} system.

The phonon scattering is purely elastic. The lifetime $\tau_r(\nu)$ of a phonon against resonance scattering is given by the expression

$$\tau_r^{-1}(\nu) = v\, N_{V^{4+}}\, \sigma_o\, g(\nu) \tag{3.15}$$

where $N_{V^{4+}}$ is the concentration of V^{4+} scattering centers.

Phonons at the frequency $\nu_o = 0.84$ THz in Al_2O_3 have an average wavelength $\Lambda_o \simeq 0.8 \cdot 10^{-6}$ cm and a maximum scattering cross section $\sigma_o \simeq 0.7 \cdot 10^{-13}$ cm^2. For $N_{V^{4+}} = 10^{18}$ cm^{-3} the mean free path of phonons in the line center is $\Lambda_o = (N\sigma_o)^{-1} \simeq 10^{-5}$ cm and the lifetime against resonance scattering is $\tau_r(\nu_o) \simeq 2 \cdot 10^{-11}$ s. The resonance scattering as it occurs in the time scale is illustrated in Fig. 18 for phonons in the center of the resonance line. For the lifetime of the $E_{1/2}$ level a value $T_1^s \simeq (2\pi\Gamma)^{-1} \simeq 10^{-11}$ s is estimated. For phonons in the wing of the line the lifetime $\tau_r(\nu)$ is, according to (3.15), larger than $\tau_r(\nu_o)$. Elastic resonance scattering at V^4 ions leads to spatial diffusion of the phonons, described by the diffusion constant $D(\nu) = \frac{1}{3} v^2\, \tau_r(\nu)$, or

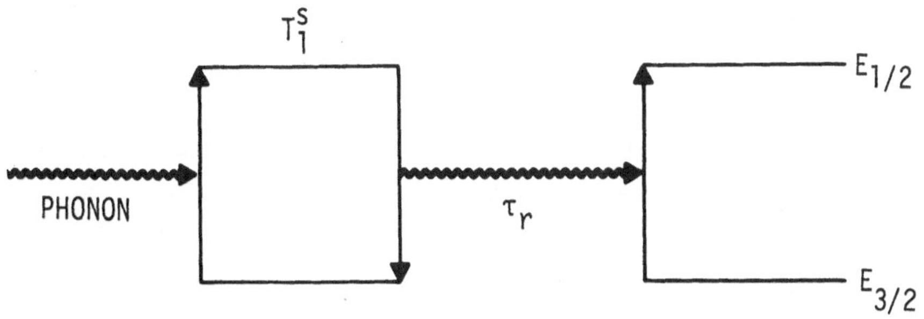

Fig. 18. Resonance scattering of phonons by V^{4+} ions in Al_2O_3.

$$D(\nu) = D_o \, [1 + 4 \, (\nu - \nu_o)^2/\Gamma^2] \tag{3.16}$$

where $D_o = \frac{1}{3} v \, (N_{V^{4+}} \, \sigma_o)^{-1}$ is the diffusion constant in the line center, with an estimated value $D_o \simeq 3 \; cm^2/s$ at $N_{V^{4+}} \simeq 10^{18} \; cm^{-3}$. According to (3.16), phonons with frequencies at line-center undergo strong resonance scattering and are therefore strongly spatially trapped, i.e. a phonon escapes from a crystal by diffusion in a time that is much larger than the ballistic time of flight through the crystal. Phonons with frequencies in the wing of the resonance line, (3.1), have according to (3.16) larger diffusion constants and are less strongly spatially trapped than phonons in the line center.

3.5 Spectral-Spatial Phonon Diffusion

In Al_2O_3 containing V^{4+} ions and Cr^{2+} ions, phonons undergo both elastic resonance scattering at V^{4+} ions and inelastic scattering (Raman scattering) at Cr^{2+} ions. Because of multiple elastic and multiple inelastic scattering, phonons diffuse in real space as well as in frequency space and perform therefore spatial diffusion as well as spectral diffusion. This spectral-spatial diffusion can be described by the diffusion equation

$$\frac{\partial f}{\partial t} = D_\nu \frac{\partial^2 f}{\partial \nu^2} + D_o \, [1 + 4 \, (\nu - \nu_o)^2/\Gamma^2] \, \nabla^2 f \tag{3.17}$$

where $f(\vec{r},\nu,t)$ is the phonon distribution function, i.e. the number of phonons per unit volume and unit frequency interval, and where $D_\nu = \nu_R^2/\tau_{in}$ is the diffusion constant in frequency space assumed to be independent of frequency. The spectral-diffusion constant is given by

$$D_\nu = \nu_R^2/\tau_{in} \tag{3.18}$$

where ν_R is an average Raman shift in an inelastic scattering process and τ_{in} the inelastic lifetime of a phonon against a Raman scattering process. To find a solution of (3.17), a procedure developed by Engelhardt et al. (1984) is presented here. A mode analysis, setting

$$f = \sum_n f_n(\nu,t)\ \chi_n(\vec{r}) \tag{3.19}$$

leads to separation of the variables. The spatial distribution is then described by

$$\nabla^2\chi_n + k_n^2\ \chi_n = 0 \tag{3.20}$$

where k_n are appropriate wave vectors of the diffusive modes (n= 1, 2,...). The frequency and time dependencies are described by

$$\frac{\partial f_n}{\partial t} = D_\nu \frac{\partial^2 f_n}{\partial \nu^2} - \tau_{o,n}^{-1}\ [1 + 4\ (\nu-\nu_o)^2\ \Gamma^{-2}]f_n \tag{3.21}$$

where $\tau_{o,n}^{-1} = D_o\ k_n^2$ are diffusive escape rates for phonons at line center. With $f_n = g_n \exp(-t/\tau_{o,n})$ the following equation is obtained,

$$\frac{\partial g_n}{\partial \tau_n} = \frac{\partial^2 g_n}{\partial \xi_n^2} - \xi_n^2\ g_n \tag{3.22}$$

where $\tau_n = t/T_n$ and $\xi_n = (\nu-\nu_o)/\gamma_n$ are dimensionless time and frequency variables and

$$T_n = \left(\frac{1}{4}\ \Gamma^2\ \tau_{o,n}/D_\nu\right)^{1/2} \tag{3.23}$$

a characteristic time and γ_n a characteristic frequency given by

$$\gamma_n^2 = \left(\frac{1}{4}\ \Gamma^2\ \tau_{o,n}\ D_\nu\right)^{1/2}. \tag{3.24}$$

The equation (3.22) is solved by the Green's function

$$G_n\,(\tau_n,\xi_n,\xi_n') = \left[2\pi\,\sinh\,(2\tau_n)\right]^{\frac{1}{2}} \exp\left\{-\left[\frac{1}{4}\,(\xi_n-\xi_n')^2\,\coth\,\tau_n + \frac{1}{4}\,(x_n+x_n')^2\,\mathrm{tangh}\,\tau_n\right]\right\}. \tag{3.25}$$

$f_n = G_n\,\exp(-\tau_n/T_n\,\tau_{o,n})$ is the probability to find a phonon at frequency ξ_n that was originally generated at frequency ξ_n'. The solution at small time, $t < T_n$ $(\tau_n<1)$, is approximately

$$f_n = \left(\frac{T_n}{4\pi t}\right)^{\frac{1}{2}} \exp\left[-\frac{(\nu-\nu')^2}{4D_\nu t}\right] \exp\left[-\tau_{d,n}^{-1}(\nu)\,t\right] \tag{3.26}$$

where $\tau_{d,n}$ is an effective diffusion escape time given by

$$\tau_{d,n}(\nu) = \tau_{o,n}\,[1+(\nu+\nu'-2\nu_o)^2/\Gamma^2]^{-1}\,. \tag{3.27}$$

The expression (3.26) shows that spectral diffusion from a frequency ν' to a frequency ν takes a time

$$t_m \simeq (\nu-\nu')^2/D_\nu. \tag{3.28}$$

The phonon distribution at large times, $t > T_n$ $(\tau>1)$, is given by

$$f_n = (2\pi)^{-\frac{1}{2}} \exp\,[-\tfrac{1}{2}\,(\nu'-\nu_o)^2\gamma_n^{-2}]\exp\,[-(T_n^{-1}+\tau_{o,n}^{-1})]\ \exp\left[-\tfrac{1}{2}\,(\nu-\nu_o)^2\gamma_n^{-2}\right]. \tag{3.29}$$

The distribution of the mode n decays with a rate $T_n^{-1}+\tau_{o,n}^{-1} \simeq T_n^{-1}$ which is independent of the frequency at which phonons were originally generated. The corresponding stationary distribution has Gaußian shape $\exp\,[-\tfrac{1}{2}\,(\nu-\nu_o)^2/\gamma_n^2]$.

Using (3.25) an analysis of the experimental result (Fig. 16) was performed (Engelhardt et al. 1984) by taking into account the finite duration of the phonon excitation pulse; for simplification the analysis was restricted to the $n = 1$ mode. Best agreement between the experimental points and the theory (solid curve in Fig. 16) was obtained for the values $T_1 = 6 \cdot 10^{-6}$ s and $\gamma_1 = 45$ GHz; after the time T_1, phonons are distributed over the width γ_1. It follows from (3.23) and (3.24) that the spectral-diffusion constant is given by $D_\nu = \gamma_1^2/T_1 \simeq 3 \cdot 10^{26}$ s^{-3}. The value of D_ν is, according to (3.28), consistent with an experimental delay time between excitation pulse and signal pulse of $t_m \simeq 10^{-6}$ s. The inelastic scattering is most likely caused by Cr^{2+} ions which are produced by the X-ray irradiation of the crystal. Anticipating a result of the next section, the inelastic scattering at Cr^{2+} ions leads to an inelastic scattering rate $\tau_{in}^{-1} \simeq 10^7$ s^{-1}. It follows therefore from the spectral-diffusion constant that an average Raman shift is given by $\nu_R \simeq (D_\nu \tau_{in})^{1/2} \simeq 6$ GHz. This value is consistent with low lying electronic levels of Cr^{2+} which are known from hypersonic absorption measurements (Anderson et al. 1972) and from theoretical studies (Bates and Wardslaw 1980). The coupling to phonons is very strong; the Cr^{2+} ion is a Jahn Teller ion with a strong electron-phonon interaction (Al'tshuler and Kozyrev 1974, Bates and Wardslaw 1980).

From the analysis of the experiment it follows furthermore, by multiplying (3.23) with (3.24), that $\frac{1}{4}\Gamma^2\tau_{0,1} \simeq \gamma_1^2 T_1 \simeq 1.2 \cdot 10^{16}$s^{-1}. By assuming that the phonons decay at the crystal surfaces, delivering their energy to the helium bath, (3.20) has the solution (for the main mode)

$$\chi_1 = A \cos k_x x \cos k_y y \cos k_z z \qquad (3.26)$$

where A is determined by the pulse energy of the generated phonons, $k_x = \pi/d_x$, $k_y = \pi/d_y$, $k_z = \pi/d_z$ are the wave vectors of the $n = 1$ diffusive mode, d_x, d_y, d_z are the crystal dimensions and x,y,z the cartesian coordinates relative to the center of the crystal. With $d_x = d_y = 0.3$ cm and $d_z = 0.5$ cm, a value $k_1 = (k_x^2 + k_y^2 + k_z^2)^{1/2} \simeq 16$ cm^{-1} and a diffusive escape rate $\tau_{0,1}^{-1} \simeq k_1^2 D_o \simeq 10^2$ s^{-1} are obtained. It follows for the halfwidth $\Gamma \simeq 2\gamma_1 (T_1 \tau_{0,1}^{-1})^{1/2} \simeq 8$ GHz. The experiment gives therefore evidence that the V^{4+} resonance line has a homogeneous width that is about half the experimental halfwidth.

3.6 Inelastic Scattering Centers for High-Frequency Phonons

In the preceding sections it was suggested that Cr^{2+} ions in Al_2O_3 are possible inelastic scattering centers for high-frequency phonons. Direct evidence for inelastic phonon scattering at Cr^{2+} ions was reported by Engelhardt and Renk (1984). The experiment was performed with phonons in optically excited ruby containing excited Cr^{3+} ions. Due to resonance scattering at excited Cr^{3+} ions, phonons were totally spatially trapped. The trapped phonons decayed most likely because of inelastic scattering at centers other than Cr^{3+} ions as has recently been concluded from trapping experiments performed with monochromatic phonons generated by $\bar{E} \rightarrow 2\bar{A}$ far infrared excitation and $2\bar{A} \rightarrow \bar{E}$ relaxation (Retzer et al. 1984, Renk 1984). In the experiment by Engelhardt and Renk (1984), inelastic scattering centers have been produced by X-ray irradiating a ruby crystal. It was found that the inelastic phonon lifetime was reduced by the X irradiation by a factor of 5 and that the inelastic lifetime in the irradiated ruby crystal had a value of $\tau_{in} \simeq 10^{-7}$ s.

Excited Cr^{3+} in Al_2O_3 has an extremely narrow resonance line. In high-quality ruby crystals the resonance line has Lorentzian shape, (3.1), with $\nu_0 = 874$ GHz (29 cm^{-1}) and a halfwidth $\Gamma \simeq 0.4$ GHz (Retzer et al. 1983). Because of the narrow resonance line, an inelastic scattering process with a Raman shift of several GHz transfers a phonon far away from resonance and the inelastically scattered phonon is no longer resonantly scattered but leaves the crystal by ballistic flight. While inelastically scattered phonons in Al_2O_3 containing V^{4+} ions remain in the range of the resonance and undergo therefore multiple inelastic scattering, resonant phonons in Al_2O_3 containing excited Cr^{3+} ions (and no V^{4+} ions) are inelastically scattered only once.

The principle of the experiment (Engelhardt and Renk 1984) is shown in Fig. 19. By applying a magnetic field (parallel to the c axis of ruby) the $\bar{E}$ level is split into the $\bar{E}_-$ and $\bar{E}_+$ levels and the $2\bar{A}$ level is split into the $2\bar{A}_-$ and $2\bar{A}_+$ levels. The crystal is

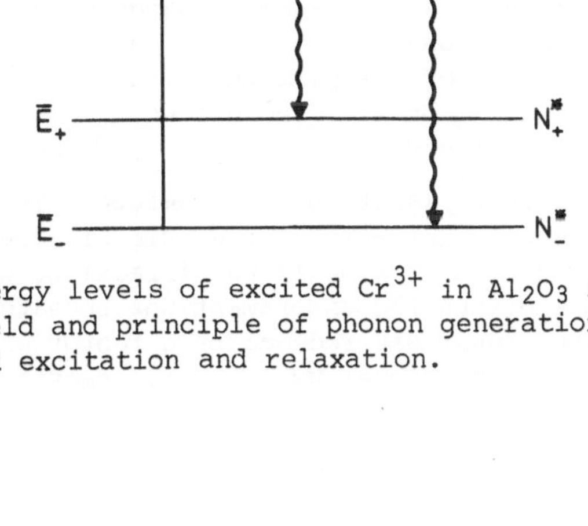

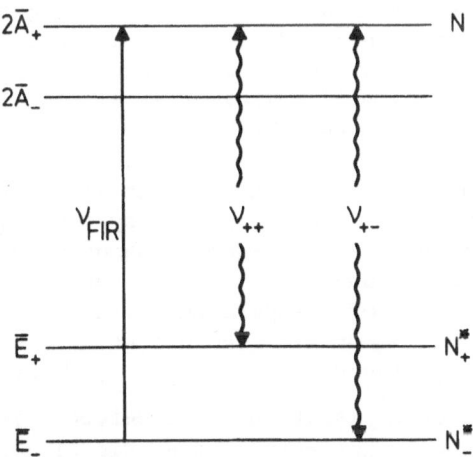

Fig. 19. Energy levels of excited Cr^{3+} in Al_2O_3 in a magnetic field and principle of phonon generation by far infrared excitation and relaxation.

optically pumped and contains excited Cr^{3+} ions, with nearly equal populations in the $\bar{E}_{\pm}$ levels, $N_-^* = N_+^* \simeq \frac{1}{2} N^*$ where N^* is the total concentration of excited Cr^{3+} ions. The $\bar{E}_- \rightarrow 2\bar{A}_+$ resonance line is tuned to resonance with far infrared radiation at fixed frequency ν_{FIR}. Far infrared absorption leads to generation of two phonon packets, at the frequencies $\nu_{+-} = \nu_{FIR}$ and $\nu_{++}(B) = \nu_{FIR} - g_{\bar{E}}\, \mu_B B$ where $g_{\bar{E}}$ is the Landé factor for the splitting of the $\bar{E}$ level (μ_B Bohr's magneton) and B the magnetic field strength. At large concentrations of excited Cr^{3+} ions both phonon packets are totally spatially trapped and decay by inelastic scattering processes. It can be shown (Engelhardt and Renk 1984, Renk 1985) that for continuous far infrared excitation the stationary $2\bar{A}_+$ population N is given by the expression

$$N/N^* = \frac{1}{4} R\, T_{++}\, v\, \sigma_o\, \tau_{in}\, N^* \tag{3.27}$$

where R is the phonon generation rate by the far infrared excitation, T_{++} the lifetime of the $2\bar{A}_+$ level against one-phonon relaxation by $2\bar{A}_+ \rightarrow \bar{E}_+$ non-spinflip transitions, v an average velocity of sound, $\sigma_o = \frac{1}{3\pi} \Lambda^2$ the cross section for phonon absorption by $\bar{E}_+ \rightarrow 2\bar{A}_+$ transitions and τ_{in} the inelastic lifetime assumed to be equal for both phonon packets. Though spinflip ($2\bar{A}_+ \rightarrow \bar{E}_-$) transitions occur with a weaker rate, the phonons at frequency ν_{+-} are (at large N^*) in an equilibrium with the non-spinflip phonons at ν_{++} due to multiple absorption and reemission processes. Using (3.27), the inelastic phonon lifetime τ_{in} can be determined; the relative population N/N^* is obtained from the fluorescence emission according to (3.4), R can be estimated from the far infrared radiation intensity reaching the crystal and from the far infrared absorption cross section ($10^{-18} cm^2$ at $N^* \simeq 10^{18}$ cm^{-3}), $T_{++} \simeq 5 \cdot 10^{-10}$ s (Retzer et al. 1983), $\sigma_0 \simeq 10^{-13}$ cm^2 and N^* can be estimated from the optical pump intensity.

In the experiment by Engelhardt and Renk (1984) two identical samples (3×3×4 mm^3) were cut from a large ruby crystal that contained 0.05 mol% Cr^{3+}. By use of a sample change mount, phonons in both crystals were studied under identical conditions. The crystals were located in the center of a superconducting magnet and were surrounded by liquid helium of 2 K. Phonons were generated by use of radiation of a cw HCN laser (ν_{FIR} = 891 GHz, power 10^{-1} W) and detected by R_2 fluorescence (section 3.2). Before X irradiation, both crystals showed the same phonon-induced R_2 signal curve (upper curve in Fig. 20). One of the crystals was irradiated with X-rays for several minutes and then the experiment was repeated. It was found that the phonon-signal was reduced by a factor of 5 (lower curve in Fig. 20). From the signal an inelastic scattering time

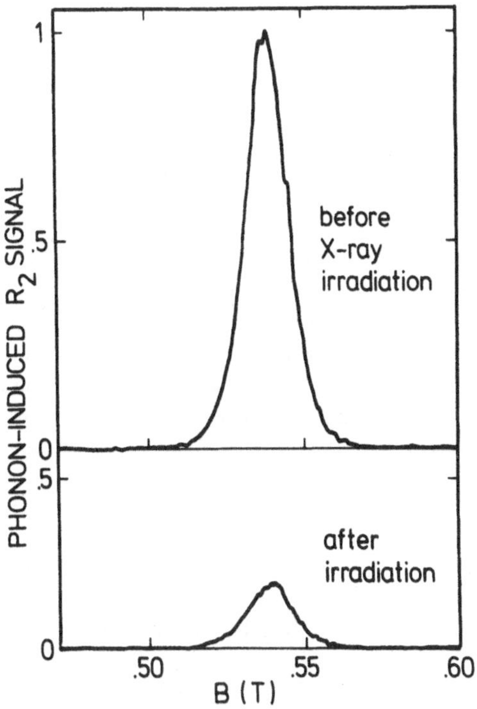

Fig. 20. Phonon-induced R_2 signal before and after X-ray irradiation of a ruby crystal.

$\tau_{in} \approx 10^{-7}$ s is estimated for the X irradiated crystal. The reduction was attributed to a shortening of the phonon lifetime because of production of inelastic scattering centers, most likely Cr^{2+} ions. By the X irradiation about 5% of the Cr^{3+} ions were converted into Cr^{2+} ions; this conversion is known from other studies (Brown and Brown 1981). From the corresponding concentration of Cr^{2+} ions, $N_{Cr^{2+}} \approx 10^{18}$ cm^{-3}, and using the relation $\tau_{in}^{-1} = N_{Cr^{2+}} \, v \, \sigma_{in}$, a cross section $\sigma_{in} \approx 5 \cdot 10^{-18}$ cm^2 can be estimated for the inelastic scattering.

It is most likely that unirradiated ruby crystals contain also Cr^{2+} ions and that these ions are responsible for the decay of trapped phonons. A concentration of $2 \cdot 10^{17}$ cm^{-3} Cr^{2+} ions, i.e. 1% of the chromium ions, would be sufficient to explain the experimental lifetime of $\tau_{in} \approx 5 \cdot 10^{-7}$ s. Different values for the inelastic lifetime, found in various experiments may be explained by different Cr^{2+} impurity concentrations in the crystals. A detailed analysis of different experiments with 29 cm^{-1} phonons in ruby will be published elsewhere (Renk 1985).

Acknowledgements

I would like to thank my coworkers for their cooperation in the field of phonon physics and T. Holstein for very stimulating discussions.

References

Abramov, A.P., Akramova, I.N., Gerlovin, I.Ya., and Razumova, I.K., 1980, Sov. Phys. Solid State, 22:556.

Adde, R., Geschwind, S., and Walker, L.R., 1969, The Observation of a Phonon Bottleneck in the Orbach Relaxation of the $\overline{E}(^2E)$ State in Ruby, in: Proc. Colloque Ampère XV, ed. P. Averbuch, North Holland, Amsterdam, p. 460.

Akimov, A.V., Basun, S.A., Kaplyanskii, A.A., Rachin, V.A., and Titov, R.A., 1977, JETP Lett., 25:461.

Akimov, A.V., Basun, S.A., Kaplyanskii, A.A., and Titov, R.A., 1979, Sov. Phys. Solid State, 21:136.

Akimov, A.V., Kaplyanskii, A.A., and Syrkin, A.L., 1980, JETP Lett., 32:124.

Akimov, A.V., Kaplyanskii, A.A., and Syrkin, A.L., 1981, JETP Lett.., 33:393.

Al'tshuler, S.A., and Kozyrew, B.M., 1974, Electron Paramagnetic Resonance in Compounds of Transition Elements, Wiley, New York, p. 221.

Anderson, C.H. and Sabisky, E.S., 1968, Appl.Phys.Lett., 13:214.

Anderson, C.H. and Sabisky, E.S., 1971, Spin-Phonon Spectrometer, in: Physical Acoustics, vol. 8, W.P. Mason and R.N. Thurston, ed., Academic, New York, p. 1.

Anderson, R.S., Bates, C.A., and Jaussaud, P.C., 1972, J. Physique C5:3397.

Basun, S.A., Kaplyanskii, A.A., and Shekhtman, V.L., 1979, JETP Lett., 30:255.

Basun, S.A., Kaplyanskii, A.A., and Shekhtman, V.L., 1981, Sov. Phys. Solid State, 23:2149.

Basun, S.A., Kaplyanskii, A.A., and Shekhtman, V.L., 1982a, Sov. Phys. Solid State, 24:1093.

Basun, S.A., Kaplyanskii, A.A., and Shekhtman, V.L., 1982b, Sov. Phys. JETP, 55:1119.

Baumgartner, R., Eisfeld, W., Pauli, G., Renk, K.F., and Riehl, N., 1978, Solid State Commun., 27:1105.

Baumgartner, R., Engelhardt, M., and Renk, K.F., 1981a, Phys.Rev. Lett., 47:1403.

Baumgartner, R., Engelhardt, M., Renk, K.F., and Orbach, R., 1981b, Physica, 107B:109.

Baumgartner, R., Engelhardt, M., and Renk, K.F., 1981c, J. Physique C6, 42:119.

Baumgartner, R., Engelhardt, M., and Renk, K.F., 1983, Physics Lett. 94A:55.

Benderskii, V.A., Brikenstein, V.Kh., Broude, V.L., and Lavrushko, A.G., 1974, Solid State Commun., 15:1235.
Bron, W.E., 1980a, Phys. Rev. B, 21:2627.
Bron, W.E., 1980b, Rep. Progr. Phys., 43:301.
Bron, W.E. and Grill, W., 1977a, Phys. Rev. B, 16:5303.
Bron, W.E. and Grill, W., 1977b, Phys. Rev. B, 16:5315.
Bron, W.E. and Grill, W., 1978, Phys. Rev. Lett., 40:1459.
Bron, W.E. and Keilmann, F., 1975, Phys. Rev. B, 12:2496.
Bron, W.E., Levinson, Y.B., and O'Connor, J.M., 1982, Phys. Rev. Lett., 49:209.
Bron, W.E., Rossinelli, M., Bai, Y.H., and Keilmann, F., 1983, Phys. Rev. B, 27:1370.
Broude, V.L., Vidmont, N.A., Korshunov, V.V., Levinson, Y.B., Maksimov, A.A., and Tartakovskii, I.I., 1977, JETP Lett., 25:261.
Brown, I.J. and Brown, M.A., 1981, Phys. Rev. Lett., 46:835.
Challis, L.J., Ghazi, A.A., and Wybourne, M.N., 1982, Phys. Rev. Lett., 48:759.
Colles, M.J. and Giordmaine, J.A., 1971, Phys. Rev. Lett., 27:670.
Dijkhuis, J.I. and de Wijn, H.W., 1979a, Phys. Rev. B, 20:1844.
Dijkhuis, J.I. and de Wijn, H.W., 1979b, Solid State Commun., 31:39.
Dijkhuis, J.I., van der Pal, A., and de Wijn, H.W., 1976, Phys. Rev. Lett., 37:1554.
Eisenmenger, W. and Dayem, A.H., 1967, Phys. Rev. Lett., 18:125.
Eisfeld, W., 1979, thesis, Universität Regensburg, unpublished.
Eisfeld, W. and Renk, K.F., 1979, Appl. Phys. Lett., 34:481.
Eisfeld, W. and Renk, K.F., 1980, Tunable Optical Detector and Generator for Terahertz Phonons in CaF_2 and SrF_2, in: Proc. Conf. Phonon Scattering in Condensed Matter, H.J. Maris, ed., Plenum Publishing, New York, p. 329.
Elcombe, M.M. and Pryor, A.W., 1970, J. Phys. C: Solid State Phys., 3:492.
Engelhardt, M. and Renk, K.F., 1984, Phonon Decay in X-Ray Irradiated Ruby Crystals, in: Phonon Scattering in Condensed Matter, W. Eisenmenger, K. Laßmann, S. Döttinger, eds., Springer-Verlag, Berlin-Heidelberg-New York, p. 124.
Engelhardt, M., Happek, U., and Renk, K.F., 1983, Phys. Rev. Lett., 50:116.
Engelhardt, M., Fritsch, U., Happek, U., Holstein, T., and Renk, K.F., 1984, to be published.
Geschwind, S., Devlin, G.E., Cohen, R.L., and Chinn, S.R., 1965, Phys. Rev. A, 137:1087.
Goldberger, M.L. and Watson, K.M., 1964, Collision Theory, Wiley.
Grill, W. and Weis, O., 1975, Phys. Rev. Lett., 35:588.
Guseinov, N.M. and Levinson, Y.B., 1983a, Solid State Commun., 45:371.
Guseinov, N.M. and Levinson, Y.B., 1983b, Sov. Phys. JETP, 58:452.
Happek, U., Baumgartner, R., and Renk, K.F., 1984a, Infrared Excitation of High-Frequency Phonons by Multiphonon Absorption, in: Phonon Scattering in Condensed Matter, W. Eisenmenger,

K. Laßmann, S. Döttinger, eds., Springer-Verlag, Berlin-Heidelberg-New York, p. 37.
Happek, U., Netter, H., and Renk, K.F., 1984b, to be published.
Heitler, W., 1954, The Quantum Theory of Radiation, Oxfored.
Holstein, T., 1947, Phys. Rev., 72:1212.
Hu, P., 1980, Phys. Rev. Lett., 44:417.
Ignatev, I.A. and Ovsyankin, V.V., 1980, Opt. Spectrosk. (USSR), 49:538.
Kaiser, W., Spitzer, W.G., Kaiser, R.H., and Howarth, L.E., 1962, Phys. Rev., 127:1950.
Kaplyanskii, A.A., 1977, Coll. Int. CNRS No. 255:137.
Kaplyanskii, A.A., Basun, S.A., Rachin, V.A., and Titov, R.A., 1975a, JETP Lett., 21:200.
Kaplyanskii, A.A., Basun, S.A., Rachin, V.A., and Titov, R.A., 1975b, Sov. Techn. Phys. Lett., 1:281.
Kaplyanskii, A.A., Basun, S.A., and Shekhtman, V.L., 1979, Resonant Scattering and Trapping of 29 cm^{-1} Acoustic Phonons in Ruby Crystals, in: Light Scattering in Solids, J.L. Birman, H.Z. Cummins, K.K. Rebane, eds., Plenum Publishing, New York, p. 95.
Kaplyanskii, A.A., Rachin, V.A., Akimov, A.V., and Basun, S.A., 1981a, Sov. Phys. Solid State, 23:274.
Kaplyanskii, A.A., Basun, S.A., and Shekhtman, V.L., 1981b, J.Physique, C6:461.
Kaplyanskii, A.A., Basun, S.A., and Shekhtman, V.L., 1981c, J. Physique, C6:439.
Kazakovtsev, D.V. and Levinson, Y.B., 1978, JETP Lett., 27, 181.
Kazakovtsev, D.V. and Levinson, Y.B., 1979, phys. stat. sol. (b), 96:117.
Klemens, P.G., 1955, Proc. Phys. Soc. A,68:1113.
Klemens, P.G., 1967, J. Appl. Phys.,38:4573.
Landau, L.D. and Lifshitz, E.M., 1970, Theory of Elasticity, Pergamon Press.
Lax, M., Hu, P., and Narayanamurti, V., 1981, Phys. Rev. B, 23:3095.
Lengfellner, H. and Renk, K.F., 1977, IEEE J. Quantum Electronics, QE-13:421.
Lengfellner, H. and Renk, K.F., 1981, Phys. Rev. Lett., 46:1210.
Lengyel, B.A., 1971, Lasers, Wiley-Interscience, New York.
Levinson, I.V., 1978, Sov. Phys. JETP, 48:117.
Malyshev, V.A. and Shekhtman, V.L., 1978, Sov. Phys. Solid State, 20:1684.
Maris, H.J., 1965, Phys. Lett. A, 17:228.
Mc Whan, D.B., Hu, P., Chin, M.A., and Narayanamurti, V., 1982, Phys. Rev. B, 26:4774.
Meltzer, R.S. and Rives, J.E., 1977, Phys. Rev. Lett., 38:421.
Meltzer, R.S. and Rives, J.E., 1982, Phys. Rev. B, 25:3026.
Mletzer, R.S., Rives, J.E., and Dixon, G.S., 1983, Phys. Rev. B., 28:4786.

Meltzer, R.S., Rives, J.E., Sox, D.J., and Dixon, G.S., 1984, High-Frequency Phonon Dynamics in LaF_3 Using Monoenergetic Optical Detection Methods, in: Phonon Scattering in Condensed Matter, W. Eisenmenger, K. Laßmann, S. Döttinger, eds., Springer-Verlag, Berlin-Heidelberg-New York, p. 115.

Okubo, K. and S. Tamura, 1983, Phys. Rev. B, 28:4847.

Orbach, R. and Vredevoe, L.A., 1964, Physics, 1:91.

Renk, K.F., 1971, Optical Detection of Resonant Phonon Trapping, in: Proc. II. Int. Conf. on Light Scattering in Solids, Paris, 1971, M. Balkanski, ed., Flammarion, Paris, p. 12.

Renk, K.F., 1972, Phononpulse, in: Festkörperprobleme XII, O. Madelung, ed., Pergamon, Vieweg, p. 107.

Renk, K.F., 1979, A Survey on the Optical Detection of Terahertz Phonons, in: Proc. 1979 Ultrasonic Symposium of the IEEE, New York, 1979, J. de Klerk, E.R. McAvoy, eds., IEEE, New York, p. 427.

Renk, K.F., 1984, Far Infrared Volume Generation and Detection of Phonons, in: Phonon Scattering in Condensed Matter, W. Eisenmenger, K. Laßmann, S. Döttinger, eds., Springer-Verlag, Berlin-Heidelberg-New York, p. 10.

Renk, K.F., 1985, Optical Generation and Detection of 29 cm^{-1} Phonons in Ruby, in: Non-equilibrium Phonons in Nonmetallic Crystals, W. Eisenmenger and A.A. Kaplyanskii, eds. (North-Holland, Amsterdam).

Renk, K.F. and Deisenhofer, J., 1971, Phys. Rev. Lett., 26:764.

Renk, K.F. and Lengfellner, H., 1981, Generation, Detection and Resonant Trapping of Very High Frequency Phonons in Ruby, in: Proc. Int. Conf. on Lasers '80, New Orleans, 1980, C.B. Collins, ed., STS Press, McLean, Va., p. 398.

Renk, K.F. and Peckenzell, J., 1972, J. Physique, C4:103.

Retzer, N., Lengfellner, H., and Renk, K.F., 1983, Phys. Lett. A, 96:487.

Retzer, N., Lengfellner, H., and Renk, K.F., 1984, to be published.

Shah, J., Leheny, R.F., and Dayem, A.H., 1974a, Phys. Rev. Lett., 33:818.

Shah, J., Leheny, R.F., and Brinkman, W.F., 1974b, Phys. Rev. B, 10:659.

Slonimskii, G.L., 1937, JETP, 7:1457.

Tamura, S. and Okubo, K., 1984, Anharmonic Decay of High-Energy LA Phonons, in: Phonon Scattering in Condensed Matters, W. Eisenmenger, K. Laßmann, S. Döttinger, eds., Springer-Verlag, Berlin-Heidelberg-New York, p. 109.

Tsen, K.T., Abramsohn, D.A., and Bray, R., 1982, Phys. Rev. B., 26:4770.

Tua, P.F. and Mahan, G.D., 1981, J. Physique C6, 42:122.

Tua, P.F. and Mahan, G.D., 1982, Phys. Rev. B, 26:2208.

Tucker, J.W. and Rampton, V.W., 1972, Microwave Ultrasonics in Solid State Physics, North-Holland Publ., Amsterdam.

von Gutfeld, R.J. and Nethercot, A.H., 1964, Phys. Rev. Lett., 12:641.
Will, J.M., Eisfeld, W., and Renk, K.F., 1983, Appl. Phys. A, 31:191.
Wong, C. and Schuele, D.E., 1966, J. Phys. Chem. Solids, 28:1225.
Wong, J.Y., Berggren, M.J., and Schawlow, A.L., 1968, J. Chem. Phys., 49:835.

GENERATION, PROPAGATION AND DETECTION OF TERAHERTZ PHONONS IN GALLIUM ARSENIDE

R.G.Ulbrich

Institut für Physik der Universität Dortmund
46 Dortmund 50, Fed.Rep.Germany

I. INTRODUCTION

Tera-Hertz phonons in crystalline solids have attracted increasing attention in recent years. A variety of quite different methods has been developed to study the generation and decay mechanisms, and the scattering of dispersive high-frequency phonons in insulators and semiconductors[1]. The field is far from being mature; it is driven by the principal interest in excitations far from equilibrium. In addition there is the practical interest in studies of "hot" phonon phenomena, because of their close connection with the problem of non-equilibrium heat transport in electronic and electro-optic semiconductor devices[2,3,4]. The potential for probing bulk semiconductor materials and structures with a Tera-Hertz acoustic phonon microscope possibly down to 10 Å resolution is another fascinating challenge.

Recent progress in the field of high-frequency acoustic phonons came essentially from experimental studies of structurally well-defined, high-purity crystalline solids – mostly the technologically important semiconductors – with the application of optical methods and the use of visible and IR laser light sources. The observation of very long energy and momentum relaxation times of Tera-Hertz TA phonons in GaAs and InP[5], optical excitation and decay studies in CaF_2[6], direct IR excitation of zone-edge phonons in TlCl[7], and detailed phonon focussing studies in Ge[8] are among the new experimental results in the last four years.

The following article deals with the generation, the transport properties, and the detection of non-equilibrium acoustic phonons in the Tera-Hertz frequency range in semiconductors at low temperatures. For the generation process we restrict ourselves on non-radiative

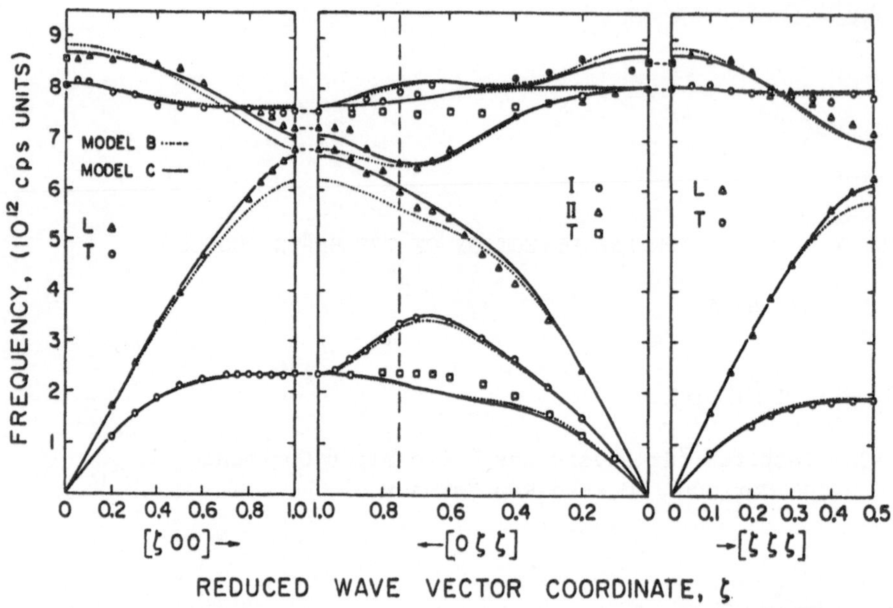

Fig.1 Phonon dispersion $\omega(\vec{q})$ for GaAs (after Waugh and Dolling, Ref.9). Of specific interest is the lowest acoustical branch: dispersive TA phonons span the frequency range 0.7 ... 2.2 THz.

recombination of optically excited electron-hole (e-h) pairs. (Direct optical excitation of phonons with IR light is covered in the article by K.F.Renk). The propagation of these phonons, their lifetimes with respect to energy and momentum, their transport properties in specific time-of-flight configurations, and finally phonon detection schemes are discussed. Special emphasis is laid on the dispersive TA phonon regime, with frequencies ranging from 0.7 ... 2.2 THz, in the technologically relevant III-V semiconductor GaAs. Figure 1 shows the phonon dispersion $\omega(\vec{q})$ of GaAs for reference[9]. Noteworthy is the flatness and low cut-off frequency of the lowest TA branch, a common feature of all zinc-blende type crystals[9,10]. As a consequence the TA phonon group velocity $v_{gr} = |\nabla\ \omega(\vec{q})|$ shown in Fig.2 decreases rapidly with increasing q and ω, and eventually reaches zero already around 0.8 q_{max}, or $\hbar\omega \approx 2.2$ THz. The term "dispersive" will be used here for the range of frequencies where v_{gr} changes appreciably with q.

II. PHONON GENERATION

It is well-known that the spontaneous recombination of e-h pairs in common bulk semiconductors is dominated by so-called "non-radiative" processes which do not involve light emission[11]. Because of the small phase space volume available in the photon field, the e-h pairs loose

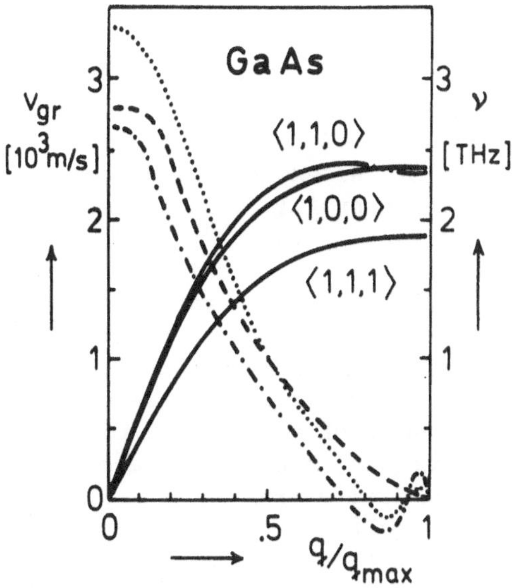

Fig.2 The lowest TA phonon branch (——) in the three principal directions together with the corresponding group velocities $v_{gr} = |\nabla\, \omega(\vec{q})|$. .... (1,0,0), -·-·- (1,1,0) , and ---- (1,1,1). Data taken from Ref.9.

their energy preferentially through coupling to the lattice, i.e. by emission of incoherent phonons. The various electron-phonon coupling mechanisms differ appreciably from each other, and depend strongly on the initial kinetic energy of the carriers. In addition, the omnipresence of specific shallow and deep defect levels in real crystals — even the so-called "highest-purity" crystals contain life-time controlling amounts of recombination centers — plays a crucial role in the localization and interband recombination process of the optically excited e-h pairs[12]. Nonradiative e-h-pair recombination has been known to be an efficient source of high-frequency phonons for some time. Among the first quantitative experiments were electron-hole

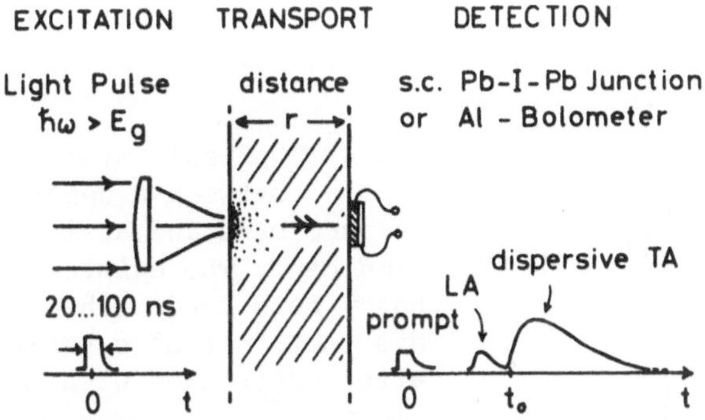

Fig.3 Phonon generation and propagation in a time-of-flight arrangement. Tera-Hertz phonons are generated in the process of non-radiative recombination of optically excited electron-hole pairs (dotted region).

plasma studies in Ge[13,14]. The effect was later used in experimental studies of the decay of optical into acoustical phonons in InSb[15]. Only recently full advantage was taken of the flexibility which is inherent in the optical excitation method[5]. Variation of the distance between source and detector, controlled scans of the light spot over a crystal surface[16], and the adjustments of the excitation photon energy to obtain a desired absorption depth profile can be easily performed with photoexcitation. A typical set-up for time-of-flight experiments involving high frequency acoustical phonons is shown in Fig.3. A bunch of incoherent Tera-Hertz phonons is generated in the process of intraband energy relaxation and the recombination of e-h pairs after photoexcitation in a well-confined volume within the crystal (dotted region). This method has obviously the virtue of both spatial confinement (down to a few μm) and high temporal resolution (better than 10^{-9}s) which exceeds, in practice, the available detector response speed. The subsequent phonon propagation and damping is studied with superconducting 2Δ-threshold Pb junction detectors or Al bolometers. Crystals of different thickness may be used as well as varying angles of propagation direction. The detector response usually limits the time resolution in the 10 ns range. Typical parameters for pulse widths and a schematic line shape for TA dispersive phonon transport are also given in Fig.3. In the following paragraph we will discuss details of the optical absorption process, the e-h pair energy relaxation, and the phonon decay in the excitation volume.

A. Optical Excitation

The absorption of light with photon energies close to or above the band gap energy is schematically shown in Fig.4. In the single-particle description, optical interband transitions connect states in the doubly degenerate valence band of symmetry Γ_8, represented by the

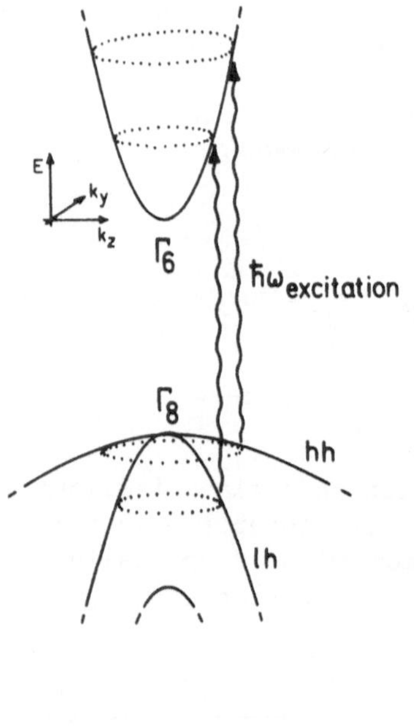

Fig.4 Optical interband transition between a degenerate Γ_8 valence band and a single Γ_6 conduction band. The existence of light and heavy hole bands (lh,hh) causes two distinct initial carrier distributions after excitation with monochromatic light.

heavy and light hole bands (hh,lh), with the single Γ_6 conduction band[17]. One obtains two distinct initial energy distributions for electrons and holes, with mean kinetic energies ΔE_e, ΔE_h determined by the excess energy $\hbar\omega_{excitation} - E_g$ of the exciting photons and by the mass ratios m_e/m_{hh} and m_e/m_{lh}. The anisotropic nature of the band structure ("warping"), which is especially pronounced in the valence band, broadens the energetic width of both electron and hole distributions, and the angular variation of the corresponding momentum matrix elements, which connect the valence band and conduction band wave functions, determines the initial carrier momentum distributions of e, lh and hh[18]. All parameters in this description of the fundamental absorption process of light in the common group IV and III-V semiconductors are known with high accuracy[11,19].

Much less is known about the processes which follow the e-h pair generation. They provide the principal mechanisms for the relaxation of initial pair momentum and energy, and represent in a phenomenological sense the "damping" of e-h pair motion. The kinetics of these carrier interactions in energy and wavevector space is summarized in Fig.5. Most efficient in the randomization of energy and quasi-momentum of the initially almost mono-energetic, highly anisotropic distributions are collisions caused by the long range Coulomb interaction

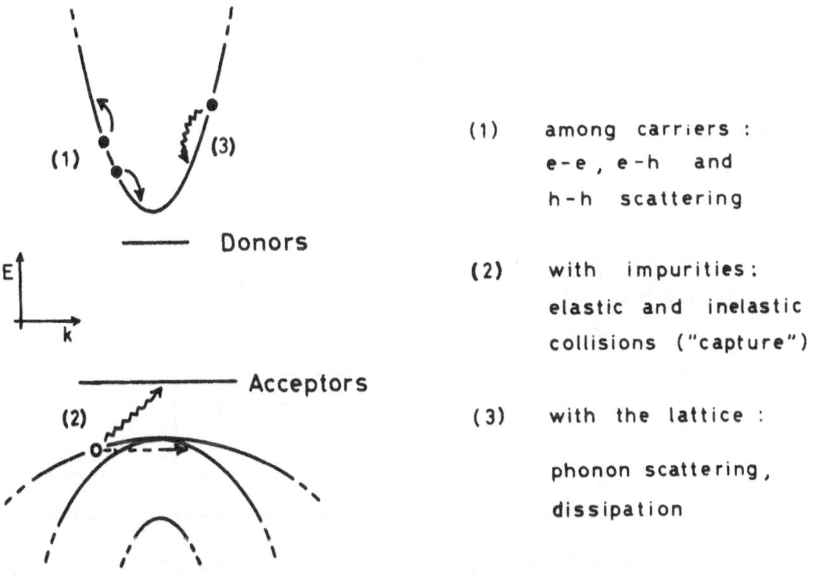

Fig.5 Fundamental interactions of electrons and holes in semiconductors. These processes contribute to the relaxation of carrier momentum and energy.

among the photoexcited carriers themselves: e-e, e-h and h-h scattering provides extremely fast channels towards quasi-equilibrium within the carrier system. The scattering times are of the order of 10^{-13}s for typical carrier densities of $n_e = n_h = 10^{14}$ cm^{-3}. The collisions establish a carrier temperature T_c within a few scattering times, and for a large range of e-h pair concentrations (10^{11} ... 10^{16} cm^{-3}) this occurs before the interaction with the lattice (see below) sets in. The total kinetic energy and the total quasimomentum of the carrier system remain, of course, constant for this type of scattering.[20]

The presence of impurities, neutral or charged, provides another mechanism for momentum relaxation of mobile carriers. When the ionized impurity concentration N_{A-}, N_{D+} exceeds the optically excited e-h pair density, it takes over the momentum randomization of the mobile carriers. Carrier-impurity collisions may be either elastic or inelastic. The latter processes involve Auger-type (impact ionization) or vibronic excitations of the impurity atoms and contribute to the net transfer of kinetic energy from the e-h-pairs to the crystal lattice. In general the bulk electron-phonon interaction dominates the energy relaxation of mobile e-h pairs. The exchange of energy between the two subsystems of elementary excitations, electrons and phonons, is shown schematically in Fig.6

B. Intraband Energy Relaxation

A charge carrier (e or h) with kinetic energy $\Delta E > 3/2 \cdot kT_L$, where T_L is the crystal lattice temperature, will loose energy by emission of phonons until equilibrium $\frac{2}{3k}\langle\Delta E\rangle = T_e = T_h = T_L$ is

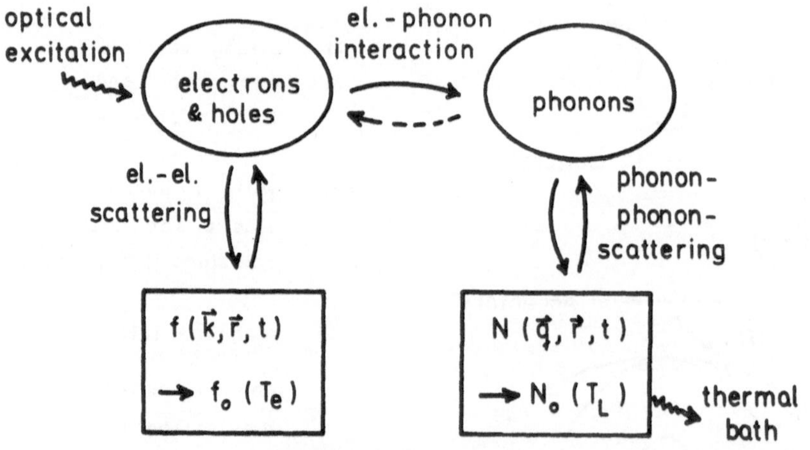

Fig. 6 The exchange of energy and momentum between the subsystems of elementary excitations. Thermal equilibrium, represented by the temperature of the external bath, is reached through a chain of interactions.

reached or the lifetime τ of the carrier is terminated (see Fig.7). For a discussion of the relevant electron-phonon coupling constants and transition rates W , it is convenient to set first $T_L = 0$ ("cold lattice"approximation) and $\tau = \infty$,i.e. one carrier in an otherwise unperturbed, ideal crystal[20]. The generalization to finite lattice temperatures, or even non-equilibrium phonon distributions, and finite carrier lifetimes is straightforward: the quantum mechanical phonon emission rates W are modified by a Bose factor N_q+1 accounting for the occupancy of phonon modes and the branching ratio.

Two basic types of electron-phonon interaction exist in polar semiconductors like GaAs: the short-range deformation potential and the long-range dielectric polarization mechanism. The former term describes the change of the electronic eigen-energies due to the local deformation of the crystal elementary cell in the presence of a lattice vibration, either acoustical or optical. The latter mechanism couples the charge of the electron (or hole) to the dielectric polarization which accompanies the displacement field of polar acoustic or optical lattice vibrations. Both couplings have been studied extensively, experimentally and theoretically, in the context of electric transport, especially in high electric fields[20,21]. The wavevectors $\vec{q}$ of the phonons involved depend on the initial electron and hole excess energies ΔE roughly like $q_{typ} \sim \hbar^{-1} \cdot \sqrt{2\, m^* \Delta E}$. For electrons with $\Delta E = 10$ meV in GaAs, the deformation potential coupling generates LA phonons up to $\hbar\omega_{max} = 0.2$ meV energy, according to the isotropic matrix element at a rate of $W = 3 \cdot 10^9\ s^{-1}$, with $W \propto \Delta E$, and $\hbar\omega_{max} \propto \sqrt{\Delta E}$. Piezoelectric coupling, on the other hand, is highly anisotropic, and crates LA and TA phonons at a similar average rate, but with another energy dependence, $W \propto E^{-0.5}$. Above $\Delta E = \hbar\omega_{opt}$,

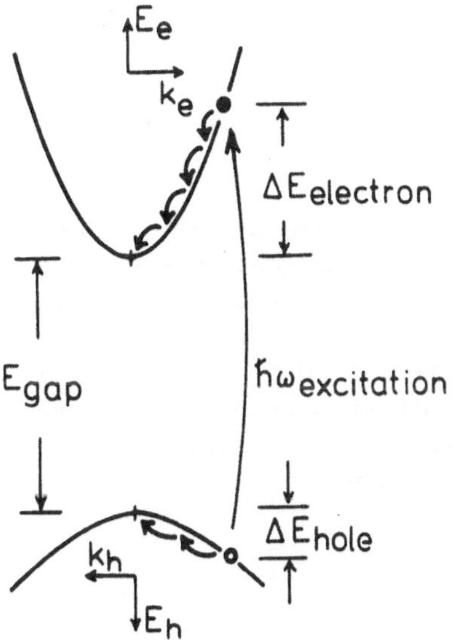

Fig.7 Intraband energy relaxation of photoexcited electron-hole pairs is an efficient source of both optical and acoustical phonons with wavevectors and frequencies depending on the initial electron and hole excess energies ΔE.

the threshold energy for optical phonon emission, the so-called polar optical ("Fröhlich") and the optical deformation potential coupling provide very efficient energy relaxation for the carriers via LO and TO phonon emission[20], with transition rates of the order of $2.10^{12}s^{-1}$.

When the carrier concentration exceeds $n_c \sim 10^{12}$ cm^{-3}, Coulomb collisions between the carriers occur far more frequent than e-phonon collisions. As a consequence the carrier dsitributions are quasi-thermal, with carrier temperature T_c and a high energy tail reaching up to the optical phonon threshold, so that above the (small) critical carrier temperature $T_c \approx 30$ K the intraband energy relaxation is dominated by optical phonon emission[22]. A good survey on the relevant intraband energy relaxation mechanisms of electrons and holes has been given by Yoffa[23].

C. Interband Energy Relaxation

Mobile electrons and holes in the bands loose kinetic energy by phonon em ission until they reach the band minima. Further energy relaxation across the band gap, i.e. interband e-h pair recombination, proceeds via different paths: radiative recombination would be the dominant channel in ideal, intrinsic crystals. For a direct gap band structure like that of GaAs, the spontaneous transition rate for this process is $\sim 10^9 s^{-1}$, and depends somewhat on the sample geometry ("reabsorption" effects), on carrier temperature and carrier density. In real crystals with defects like substitutional shallow or deep impurities, vacancies, line defects, etc., there is the other, usually much more efficient channel via phonon-assisted capture and subsequent recombination involving these localized states, see Fig.8. Typical time constants for trapping and recombination depend strongly on the

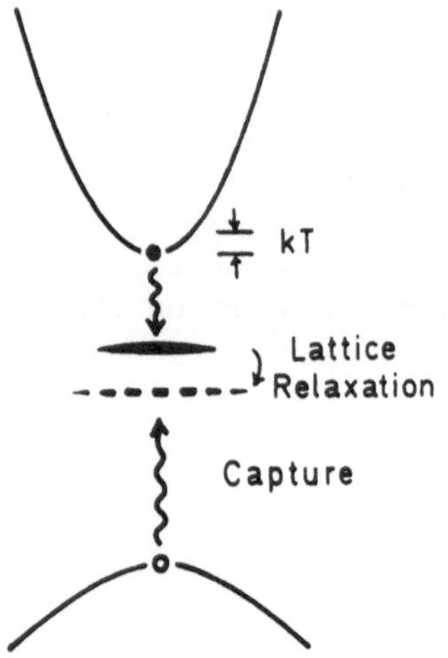

Fig.8 Interband energy relaxation proceeds typically via multi-phonon emission mediated by shallow and/or deep impurity levels. The energy transfer is dominated by optical phonon emission when a deep level with strongly localized lattice distortion is involved.

specific channel, the charge state of the defect, and the temperature of the lattice[24]. Only very few actual systems have been investigated in such detail that reliable time constants and explicit dynamical studies of carrier and lattice dynamics are available[24-28].

Total energy calculations have shown that the static lattice deformation around apoint defect (substitutional or interstitial impurity; vacancy) is of short range and involves only nearest and second-nearest neighbors[25]. A change in the charge state of the defect, caused e.g. by capture of an electron or a hole, may release part of the energy stored in the deformation and excite lattice vibrations. The decomposition of such local excitations into free phonon modes gives, in diatomic lattices, preferential population of LO phonon modes (see Fig.9). For dispersionless optical phonons their distribution in q-space will follow roughly the density-of-states factor $D(q) \sim q^2$.

Actual defects in semiconductors, like the so-called EL2 center in GaAs[26], have a more complicated structure, and hence may couple to more phonon branches. A direct coupling of the EL2 center to zone-edge TA phonons has been found empirically in optical excitation spectra[29]. From these observations it may be concluded that recombination via the EL2 center generates as well zone-edge TA phonons.

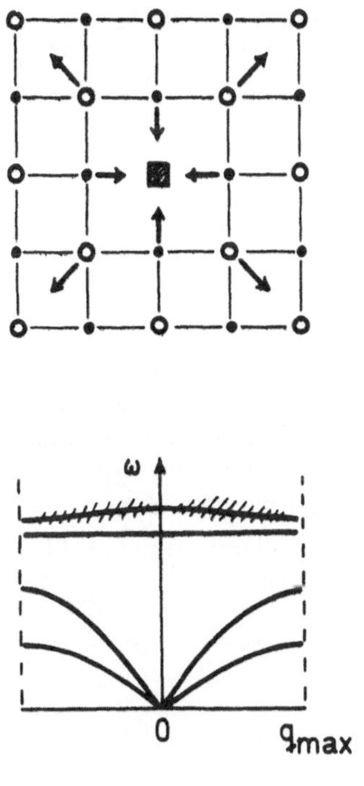

Fig.9 Static lattice deformation around a substitutional impurity in a diatomic lattice is of short range, involving nearest and second-nearest neighbor atoms (above). After electron capture part of the potential energy is released and decomposed into the "free" LO phonon modes indicated on the dispersion curve (below).

D. Fast Phonon Decay

The energy relaxation of e-h pairs in GaAs is dominated by the emission of optical phonons: the intraband processes create LO and TO phonons within a relatively narrow band of allowed wavevectors, $q \approx 0$, whereas the interband non-radiative recombination creates optical phonons over a broad range of $0 < q \leq q_{max}$. All optical phonon branches have short lifetimes of the order of 10^{-11}s. Raman linewidth measurements[30], time-resolved Raman scattering measurements[31], and theoretical estimates based on known Grüneisen parameters give consistently this short value. The most probable three-phonon decay process which conserves energy and momentum of an initial LO phonon with finite wave vector is shown in Fig.10. The peculiar phonon dispersion of GaAs with relatively low-lying transverse acoustic branches and much stiffer LA modes allows only one decay channel in the first step: LO $\rightarrow$ LA + LA [32]. The density of final states in the LA branch is high at the zone edge and therefore favors the asymmetric pattern indicated in Fig.10. According to theory, the next step in the phonon decay cascade, LA $\rightarrow$ LA + TA (dashed lines in Fig.10), is also relatively fast[33]. No direct experimental data in GaAs are available at present, the only indirect information is the prompt and practically complete conversion of initial LO phonon total energy into high-lying, dispersive TA phonons on the lowest TA branch in less than 100 ns[34].

Anharmonic decay of LA phonons with $\hbar\omega \leq 11$ meV may involve either LA or TA mode final states, both being compatible with kinematic requirements and the selection rules. In this case the TA branch final states are favored, because of their much higher density-of-states. Experimental studies of anharmonic processes involving non-

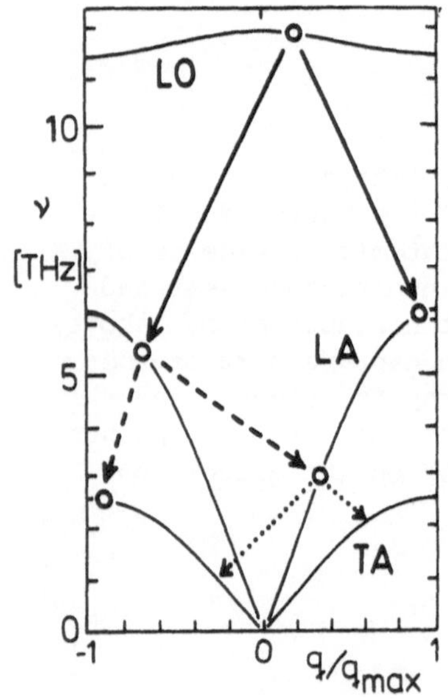

Fig.10 Anharmonic decay of a non-equilibrium LO phonon in GaAs. The most probable three-phonon process which conserves energy and momentum is shown in the first (—), second (---), and third (...) step of the cascade.

equilibrium LA phonons in the dispersive regime are very rare at present[6]. The empirical time constants show the expected strong temperature dependence, theoretical calculations based on static Grüneisen parameters are in reasonable agreement with the experiment[6,35].

In the second and all higher cascade steps (see Fig.10) the lowest TA phonon branch is populated. It is important to note that there is no further anharmonic decay for this branch because of kinematic restrictions[36,37]. The usual convex curvature of $\omega(\vec{q})$ for the lowest branch does not allow simultaneous conservation of momentum and energy, even for higher-order anharmonic decay, and including Umklapp-processes[38]. This "bottle-neck" behavior of the lowest phonon branch has been demonstrated in recent experiments in solid He^4 (Ref. 39). The only mechanism which provides further energy relaxation of TA phonons is eleastic scattering of the type STA $\rightarrow$ LA or STA $\rightarrow$ FTA, so-called mode conversion.

In the common group IV and III-V semiconductors the dominant "intrinsic" source for elastic scattering and mode conversion is the isotopic disorder of the crystal lattice[40]. In GaAs the two stable nuclei Ga^{69} and Ga^{71} have a natural abundance of 36% and 64%, and thus represent a dense ensemble of weak scatterers with $(\delta M/M)^2 \sim 8 \cdot 10^{-4}$ mass defect. The theory for elastic phonon scattering on dilute isotopic impurities in otherwise perfect crystals has been developed by Klemens[41] and recently adapted to the dispersive regime and to diatomic lattices[42,43]. A general feature of isotope scattering is the dependence of the scattering rate $W \sim \omega^4$. For frequencies above 3 THz one expects from theory W 10^9 s^{-1} and therefore extremely rapid momentum randomization of the LA phonons after the first cascade step (see Fig.10). For dispersive TA phonons of 1 THz theory predicts for GaAs a momentum relaxation time of 80 nsec[43], much too short to explain the empirical time-of-flight data in the 0.8 ... 1.5 THz range (see below). We shall discuss this discrepancy, which may be due to a correlation effect not properly accounted for in theory, in connection with the data presented here.

E. Spatial and Temporal Evolution

The optical absorption process, the subsequent expansion of the e-h pairs into the bulk crystal, and their loss of kinetic energy by phonon emission are occurring on vastly different length and time scales (see Fig.11): the optical absorption constant in the continuum directly above the gap energy of GaAs is $\alpha \cong 10^4$ cm^{-1}, so that the penetration depth of the light is 1um. The subsequent ambipolar diffusion of hot electron-hole pairs injected with excess kinetic energies of $\Delta E \gg kT_L$ may carry the excitation energy as far as 100 μm within 10 ns into the crystal. The transport of e-h pairs into the bulk crystal is usually limited by the final trapping, i.e. localization on defects and recombinations centers. This leads to a strong dependence on crystal purity, carrier densities, and tempe-

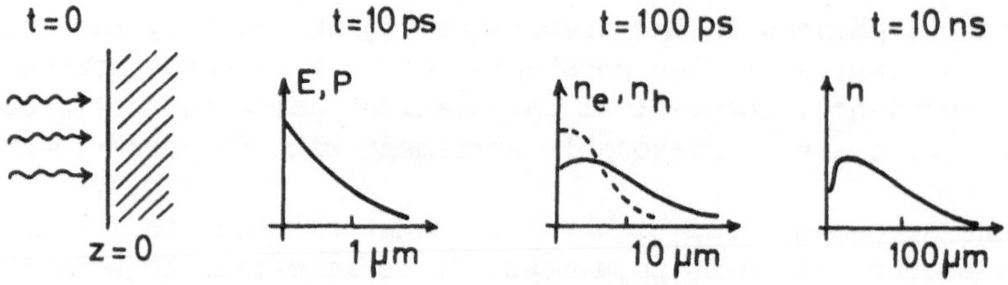

Fig.11 The temporal and spatial evolution of the optical absorption and subsequent phonon emission process is determined by the expansion of the mobile e-h pairs, their loss of kinetic energy and final localization into traps.

rature. Empirical data at $T_L = 2$ K in high-purity LPE GaAs with $N_D < N_A = 2\cdot10^{14}$ cm^{-3}, $\hbar\omega_{excitation} = 1.59$ eV, excitation density 10^{-7} j/cm^2, repetition rate 10 kHz, yielded the following results: within 10 ps after excitation with a psec light pulse the e-h pairs are localized in a depth not much larger than 1 µm (Fig.11, left). After 100 ps most of the intraband energy relaxation by LO phonon emission has occurred, the excited volume reaches now to 10 µm depth (Fig.11, middle). The final trapping resp. localization of the carriers is completed after 10 ns within a depth of $\lesssim 100$ µm (Fig.11, right). Total lifetimes of carriers in shallow traps of the given concentration do not exceed 100 ns. In the bulk-grown GaAs crystals of somewhat lesser quality, which were used for the time-of-flight experiments (see below) we found much smaller penetration depths of ~ 5 µm, and corresponding shorter lifetimes of ~ 20 ns.

III. PHONON TRANSPORT IN THE DISPERSIVE REGIME

A pulse of incoherent acoustic phonons with a given frequency distribution will propagate according to (i) the wave equation which describes the lattice vibrations, and (ii) the scattering processes acting on the phonons. If there is no scattering, i.e. in the limit of purely ballistic transport, the pulse propagation and broadening in space r and time t is determined solely by the different group velocities $v_{gr} = \partial\omega/\partial q$ contained in its spectrum: the shape of the propagating and expanding pulse merely reflects, apart from a factor r/t, the preserved frequency distribution[34]. The propagation of heat pulses with only low-frequency phonons usually gives sharp pulses with no broadening or tails, and pulse shapes independent of r and t over a wide range. It is characteristic of ballistic transport in the non-dispersive regime, where v_{gr} is constant[44].

A phonon pulse with a frequency distribution extending into the dispersive range behaves differently: the initially sharp pulse

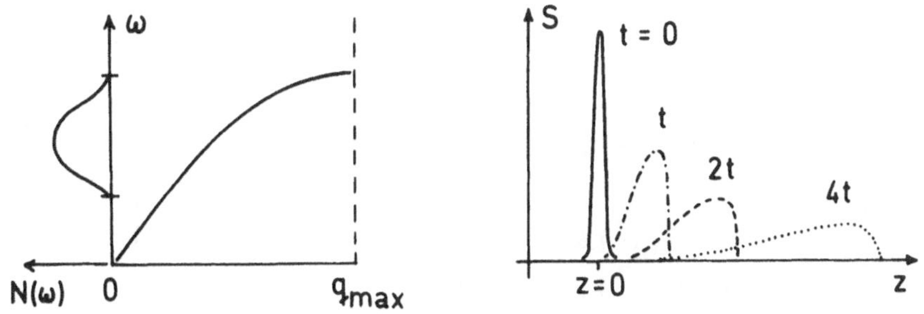

Fig.12 Dispersive transport: a pulse of incoherent phonons with a frequency distribution N(ω) well up into the dispersive regime (left) propagates and broadens in space and time according to the dispersion of the group velocity v_{gr}.

developes a halfwidth proprtional to the time elapsed, resp. the distance travelled from the origin, and its peak propagates with a velocity smaller than the low-frequency group velocity v_{max}. Under such conditions, the peak energy density in a given crystal direction will, in addition to the usual $1/r^2$ behavior (Lambert's law) fall off with an extra factor 1/r. The overall pulse shape scales linearly in r and t, i.e. remains affine when r/t is used as a variable. This special form of pulse propagation is shown schematically in Fig.12. If the pulse had started at t=0 with a given shape of finite width, its shape would evolve in a more complicated way, and would be no more strictly affine in r/t. For sufficiently large distances, however, the behavior is qualitatively similar to Fig.12.

Characteristic shapes of dispersive heat pulses were calculated with the actual dispersion of TA phonons in quartz[45]. The verification of this effect turned out to be difficult: generation of sufficiently short pulses of high-frequency phonons with heater films is a subtle problem in itself[46,47], and diffusive tails due to elastic (or inelastic) scattering in the bulk often tend to mask the effect. The first experiments with clear-cut evidence for dispersive broadening of heat pulses were performed in InSb[48]. Photo-excited TA phonon pulses in InSb showed similar behavior[15].

The next step was the discovery of very long-lived TA phonons above 0.7 THz in high-purity and structurally perfect crystals of InP and GaAs[5]. Time-of-flight experiments with photo-excited acoustic phonons showed for the first time linear scaling of pulsewidth with propagation distance up to several mm, and the above-mentioned $1/r^3$ signal peak height dependence[34]. Based on (i) the observation of affine pulse shapes, (ii) the $1/r^2$ - dependence of time-integrated phonon signals, and (iii) the threshold characteristics of the lead junction detectors it was concluded that the signals observed in GaAs

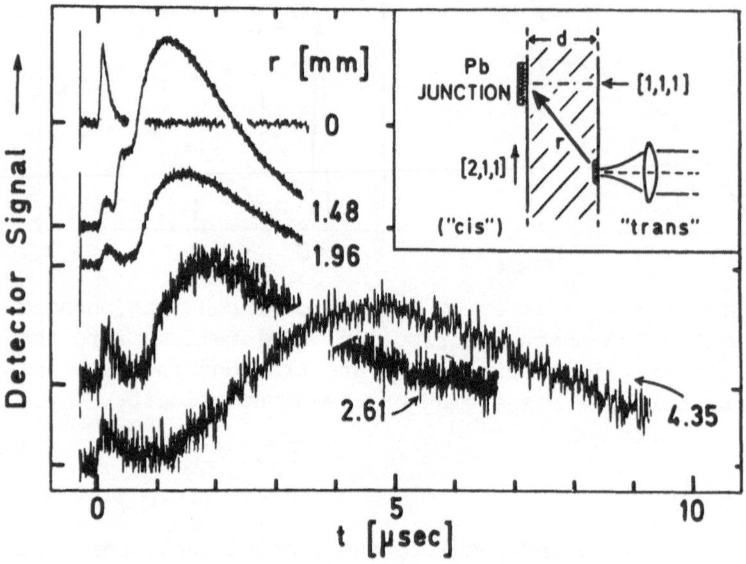

Fig.13 Phonon time-of-flight spectra for different source to detector distances r. The signals were measured with a superconducting lead junction detector in the geometrical configuration shown in the insert. (After Ref.5).

were the signature of dispersive ballistic transport of TA phonons with frequencies ranging from 0.7 ... 1.5 Thz, corresponding to group velocities from 0.95 down to 0.5 v_{max} in (1,1,1) direction[34]. The interpretation implies a mean free path of ~ 2 mm for 1.5 THz TA phonons (see Fig.14), much larger than theory predicts for isotope scattering[41].

In order to explain the apparent contradiction, Levinson et al.[49] proposed a model of cascaded phonon relaxation and diffusive propaga-

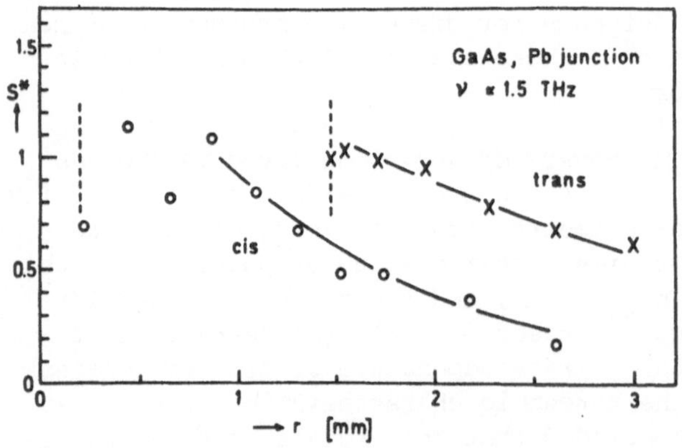

Fig.14 Time-integrated detector signal, corrected for the $1/r^2$ dependence, as a function of r. (After Ref.5).

tion, the so-called "quasi-diffusion" model, which reproduces the empirical linear scaling behavior, but incorporates very short isotope scattering times. Their basic idea is to assume (i) a chain of anharmonic decay events in the phonon population during the propagation, and (ii) a random distribution of phonon quasi-momenta $\vec{q}$, imposed by the condition $\tau_{elastic} \ll \tau_{anharmonic}$. A careful analysis of the transport equation leads indeed to an almost linear scaling between mean travelled distance and pulse width, accompanied by a formidable reduction of mean propagation velocity. One key feature of the model is down-conversion of the mean phonon energy in each cascade step by a factor 1/2. With a given starting energy of 2 THz, phonon energy would drop below 0.7 THz, the experimental detector threshold, already after the second relaxation step. This was not observed in experiment, so that the quasi-diffusion model is not applicable in our case[50].

Further experiments with a blocking slit, shown in Fig.15, designed for the distinction between ballistic and diffusive propagation were performed in (1,1,1) GaAs[51]. The comparison of time-of-flight signals for direct, intermediate, and blocked phonon propagation from source to detector (positions 1,2,3 in Fig.15 a) corroborated the earlier results: the major portion of the Tera-Hertz phonon flux reaches the detector unscattered. In the same experiment (Fig.15b right traces) heater generated low-frequency phonon signals were used to check the response of the arrangement. It became also clear that

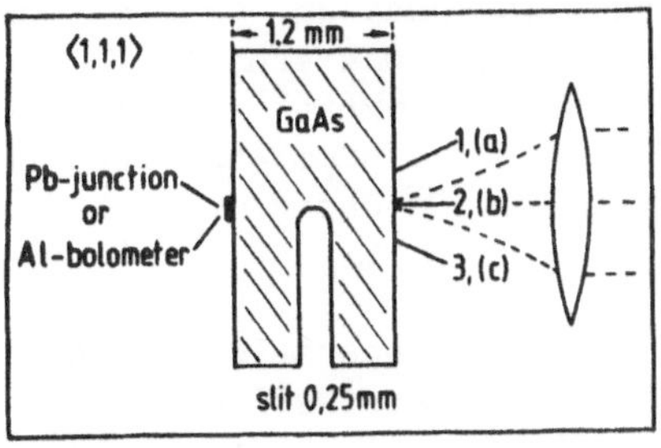

Fig.15a Experimental configuration with blocking slot for measurements of direct "ballistic" versus indirect "diffusive" propagation of photo-excited Tera-Hertz and heater-generated low-frequency TA phonons in GaAs.

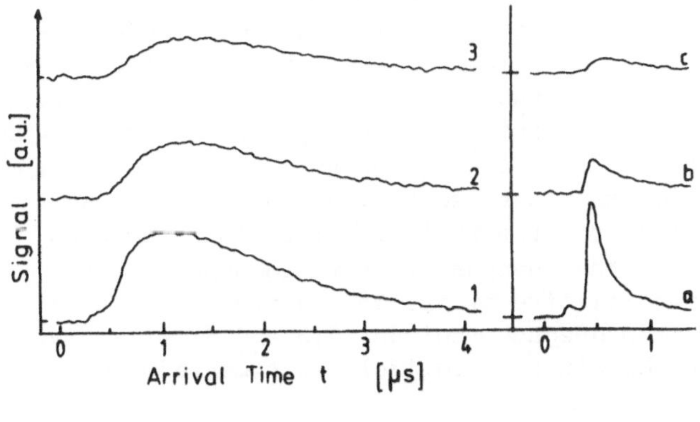

Fig.15b Signals for direct (1), intermediate (2) and blocked (3) transport. of THz phonons. Traces a,b,c show corresponding heater-generated low-frequency phonon signals. (Ref.51).

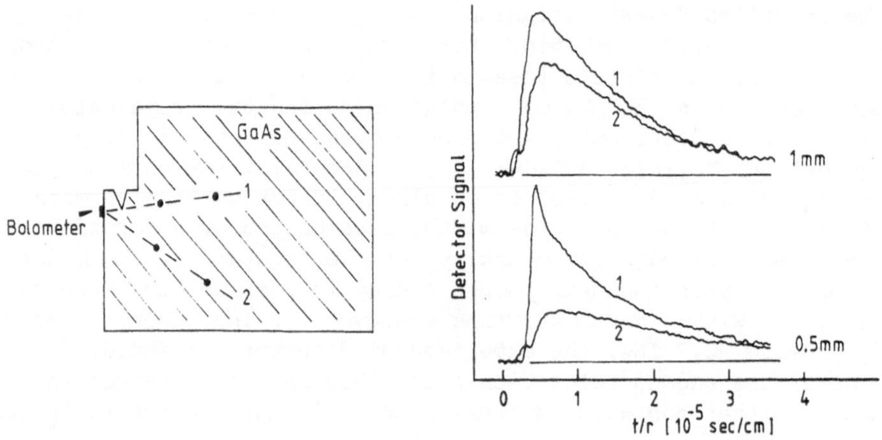

Fig.16 Down-conversion of high-frequency phonons in the damaged region close to the wedge-cut produces "extra" signal (1) in the leading portion of the usual, unperturbed time-of-flight signal (2). (After Ref.52).

the trailing portion of the signal, i.e. signal arriving later than 2 t_o, contains an increasing contribution from scattered phonons.

A variant of the blocking geometry is shown in Fig.16. A fine v-shaped groove was cut into a crystal surface, and propagation and down-conversion close to this damaged region was studied in edge-on configuration with photo-excitation[52]. The "extra" signal at early times ($t < 0.2$ µs) and short distances ($r < 0.5$ mm) has been tentatively ascribed to damaged-induced anharmonic decay of high-frequency LA phonons[52].

Very detailed information on Tera-Hertz acoustic phonon transport could finally be extracted from highly angular-resolved time-of-flight data ("phonon focussing") in GaAs[53]. The principal effect has been known for some time[54], it was quantified in detailed application to Ge[55], and recently refined into a diagnostic tool[56]. Fig.17 shows schematically the principles of phonon focussing and the generalization necessary for application in the dispersive regime. We start with a monochromatic distribution of TA phonons localized in r-space, and distributed homogeneously on the constant frequency surface $\omega(\vec{q}) = \omega_o$ of Fig.17a. The mapping of the resulting group velocities $v_{gr} = \nabla\omega(\vec{q})$ leads in most solids to singular lines, so-called caustics, and as a result the phonon energy flux from the initial point source becomes highly anisotropic[56]. In fact the phonons are not "focussed", but rather channeled in preferential directions. In the given direction $\vec{u}$, monochromatic phonons from q-space regions 2,3 and 4 will travel with three different group velocities (Fig.17c). If time-resolved

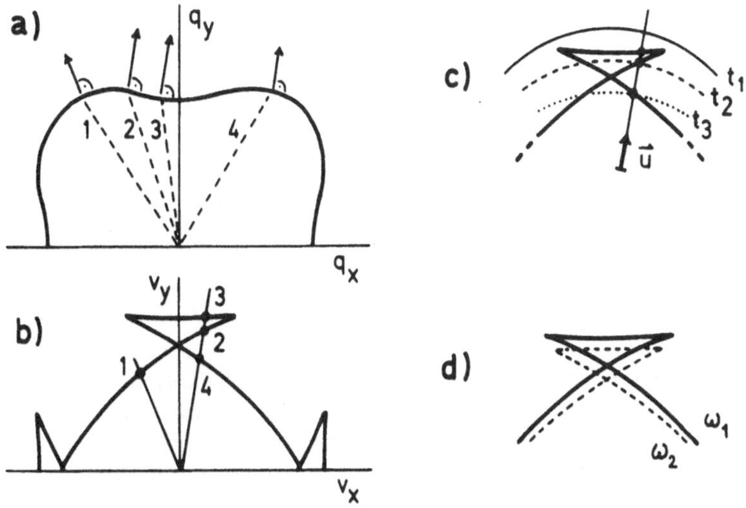

Fig.17 a) Surface of constant phonon energy $\omega(\vec{q}) = \omega_0$ and group velocities (indicated by arrows) for four different $\vec{q}$-directions. b) Mapping of the group velocities leads to singular lines (caustics). c) In the given direction $\vec{u}$ monochromatic phonons from regions 2,3,4 in q-space will travel with three different group velocities. Fixed arrival times t' are represented by spherical shells with radii inversely proportional to t'. d) In the dispersive regime, each phonon frequency generates another mapping and other singular direction lines.

measurements are performed with a fixed gate delay t' such that $|\vec{v}_{gr}| = |\vec{r}| / t' = \text{const.}$, this condition is described by spherical shells with radii inversely proportional to t'. The intersections of these shells with the v_{gr} - mapping of Fig.17c represent the ideal time-of-flight signal pattern in space and time for a monochromatic point source. When $\nabla\omega(\vec{q})$ is independent of ω, i.e. in the non-dispersive regime, the resulting signal patterns become singular for certain directions of v_{gr} and certain values of t', of course. This directional pattern consists of closed critical lines and eventually singularities[56], which are directly connected with the turning points of Fig.17b. In the dispersive regime, however, each phonon frequency generates another mapping of v_{gr} (see Fig.17d), and this will, in general, smear out the sharp critical lines and singular cusps . One cannot expect singular patterns for polychromatic, dispersive phonons.

Fig.18 shows the experimental arrangement for the measure ments of dispersive phonon focussing patterns[53]. The boxcar time gate position t' is adjusted continuously by the computer to fulfil the condition $v_{gr} = r/t' = \text{const.}$, in accordance to Fig.17c. The aspect angle and Lambert's law were taken into account in the quantity $S^*(t')$

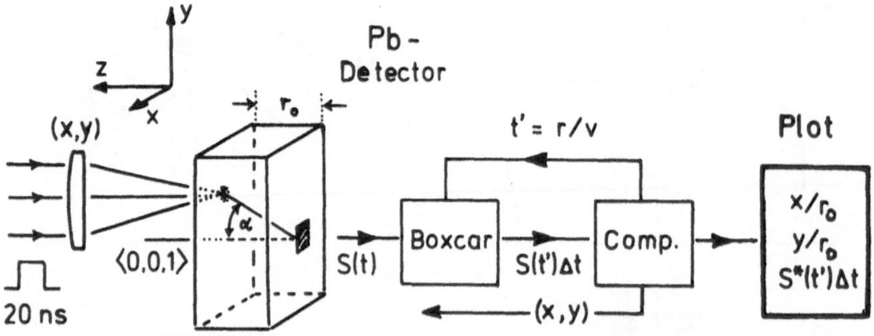

Fig.18 Experimental arrangement for measurement of dispersive phonon focussing patterns. The boxcar time gate position t is adjusted continuously by the computer to fulfil the condition $v = r/t$ = constant.

which is plotted finally, $S^*(t')\cdot\Delta t = S(t')\cdot\Delta t \cdot r^2/\cos\alpha$. The dispersed time-of-flight signal in the (1,0,0) direction is shown in Fig.19a, together with five especially chosen time gate positions t' (arrows), which correspond to the different propagation velocities $v = r/t'$ indicated in the plots 19b-f. The time gate was kept very small (< 20 ns), to avoid broadening of the critically time-dependent dispersive focussing patterns. The sharp, crown-like structure in Fig.19b is due to STA phonons, the v-shaped ridges pointing towards

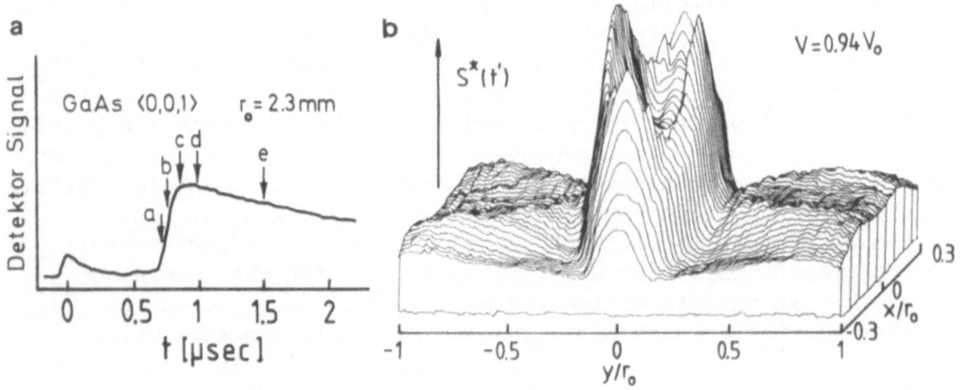

Fig.19 a) Time-of-flight signal and time gate positions t' (arrows) corresponding to five different propagation velocities $v = r/t'$. b) Phonon focussing pattern centered around the (1,0,0) direction, which is at the origin $x = 0$, $y = 0$. See text. (After Ref.53).

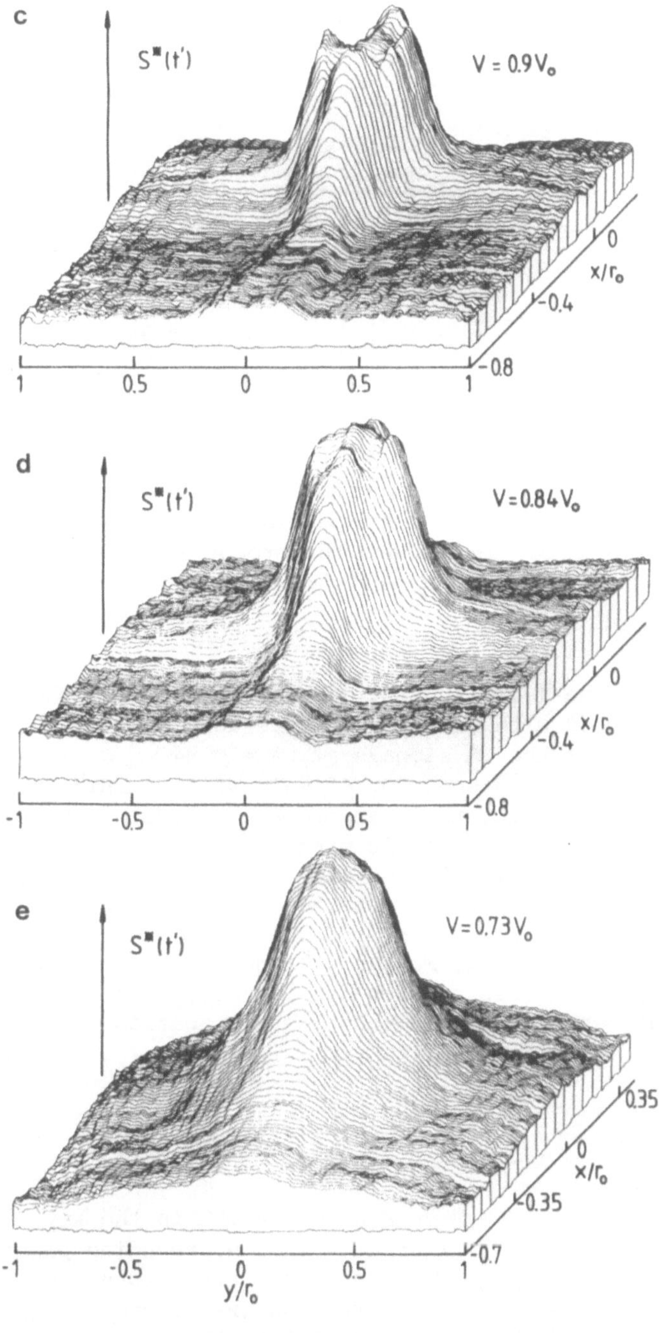

Fig.19 c), d), e) Dispersive phonon focussing patterns centered around the (1,0,0) direction, with propagation velocity v as indicated. See text.(After Ref.53).

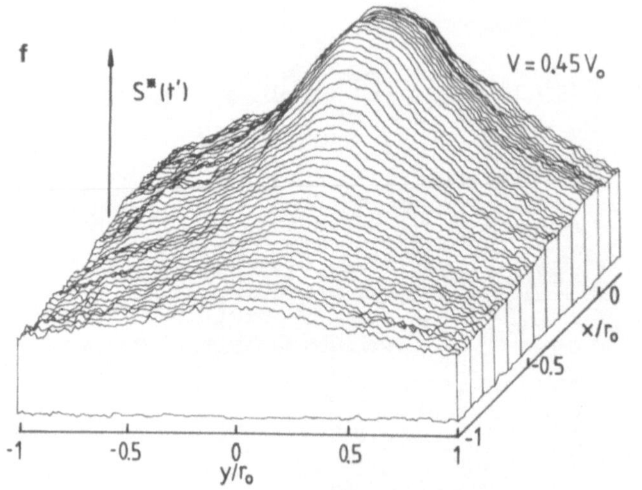

Fig.19f Dispersive phonon focussing pattern for very late arrival time t', corresponding to $v = 0.45\ v_{max}$. Note the increased background and the broadening of the central peak structure. (After Ref.53).

the corners, i.e. the (1,1,1) direction, are due to FTA phonons. The signals were measured with a lead 2Δ-threshold junction detector, the observed onset of the TA signals at $v = 0.94\ v_{max}$ is in good agreement with the known dispersion of GaAs at 0.7 THz[9,10]. With increasing t', corresponding to lower propagation velocity v, the patterns change in a characteristic manner: the center region around the (1,0,0) direction fills up (see Figs.19c,d,e), the background does not increase and, most important, the overall directionality of the pattern, i.e. the angular width of the central peak structure, does not change up to $t' = 1.5\ t_o$. For very large t', corresponding to $v = 0.45\ v_{max}$, the pattern does broaden considerably (see Fig.19f), and gains background, as one would expect for propagation with one or

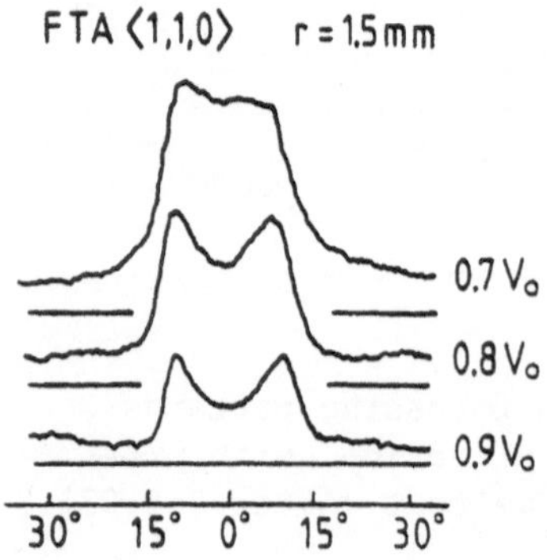

Fig.20 A cut in the x,y plane of the preceding plots (Figs. 19a-f) centered around the (1,0,1) direction ($x=r_o$, $y=0$, $z=r_o$) and parallel to the y-axis. The charactersitic FTA double ridge singular structure changes gradually into a mesa-like shape, but does not broaden its halfwidth. (After Ref.53).

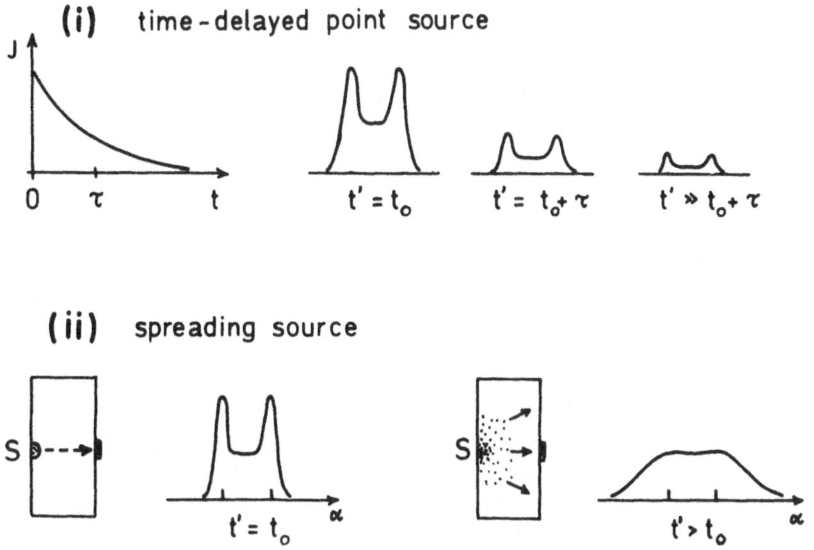

Fig.21 The spatial and temporal confinement of the phonon source influences the angular focussing pattern detected at some distance r : (i) a stationary point source with a spatial extension smaller than the detector size, but with a storage effect in time ("delayed source") will generate sharp, affine focussing patterns at all times $t'>t_o$. (ii) a spatially expanding source ("spreading source") will cause a corresponding broadening of the focussing pattern for increasing t'.

more scattering events. A cut in the x,y plane of Fig.19 parallel to the y-axis and containing the point $x=r_o$, $y=0$, $z=r_o$ is shown in Fig.20. The characteristic FTA ridge structure changes gradually into a mesa-like shape, but does not broaden its angular halfwidth[52]. This behavior is consistent with the evolution of the central structure in Fig.19b-e.

It is clear from the foregoing discussion, that the spatial and temporal confinement of the phonon source may as well influence the angular focussing pattern detected at some distance r (see Fig.21). We expect that (i) a stationary point source with an extension smaller than the detector size, but with a delayed emission J(t), would generate sharp, affine focussing patterns for all times; (ii) a spatially extending source region with a lateral spread would cause a corresponding broadening of the focussing pattern for larger times t'. Both broadening mechanisms are obviously not prominent in the observed patterns of Fig.19 and Fig.20. The dispersive phonon focussing patterns are a clear indication for ballistic transport of photoexcited TA phonons in the dispersive regime in GaAs, with frequencies from 0.7 to 1.5 THz.

IV. Detection

Various schemes for the detection of phonons have been developed. Most phonon detectors do not resolve single quanta of energy, but some come close to it: energy fluxes as small as 10^{-13} W are detectable in 1 Hz bandwidth[57]. Some of the techniques are also applicable in the Tera-Hertz frequency range.[58]

At first sight, inelastic light scattering, i.e. first-order Raman or Brillouin scattering seems to be an ideal probe for high-frequency phonons in the bulk crystal. Kinematic restrictions, the small wavevector of visible light ($k_{light} \sim 10^{-3}\, q_{max}$), and the smallness of the scattering cross section limit its practical use for phonon spectroscopy, however. When the incident photon energy is close to a (direct) band gap energy, and falls in the range of the associated excitonic resonances, there is a tremendous enhancement of the phonon scattering efficiency, but the resonance effects reduce the penetration depth of the probing light as well[59]. Second-order light sacttering spectra contain rich information on two-phonon density-of-states, and the sum and difference frequencies of phonon branches at critical points in the Brillouin zone[60]. Recent attempts to detect non-equilibrium populations of zone-edge phonons by an analysis of overtone Raman spectra are promising, but require careful elimination of spurious signal due to other inelastic scattering mechanisms in the same energy range, like e.g. acceptor impurity excitations[61].

A variant of inelastic light scattering is phonon sideband spectroscopy of the sharp emission lines which originate from specific impurity atoms doped into the host crystal. This method has the virtue of inherent energy resolution combined with good sensitivity and spatial resolution, and can be performed in the visible part of the spectrum[58]. The presence of impurity atoms "in situ", with potentially strong electron-phonon coupling, will in general affect the propagation characteristics of the phonons which are investigated. This undesired effect can be eliminated by a separation of undoped propagation and doped detection regions in the crystal[58].

The second class of energy resolving phonon detectors is based on Cooper-pair breaking in thin superconducting metal films evaporated on the crystal surface[57]. High resolution phonon spectroscopy with superconducting tunnel junctions as generators and detectors, and bias modulation techniques has been performed up to $\sim$0.9 THz[57]. The extension of this method to higher phonon frequencies is dificult, because of problems with the preparation of superconducting thin films with larger 2Δ. Phonon wavelengths of the order of 10 Å clearly require smooth interfaces on an atomic scale. For phonon frequencies above 1 THz, and wavevectors comparable with q_{max}, the question of phonon transmission through the boundary, i.e. the acoustic mismatch problem between semiconductor crystal and metal film, and the question of mode-conversion and anharmonic decay at the interface are open at present.

The dependence of the quantum efficiency η of superconducting tunnel diodes on phonon energy around the 2Δ-threshold has recently been reexamined in a series of experiments[62]. The analysis of well-characterized heater phonon spectra with suitably biased lead junctions on very smooth GaAs surfaces has confirmed the sharp step in η at 2Δ, and the plateau above[57]. All lead junctions with sufficiently good characteristics (current ratios of 300:1 and more, at 1.4 K) did show this behavior. Phonons with frequencies below 2Δ did not contribute to the signal[62]. The high frequency response ($\hbar\omega \gg 2\Delta$) of the same lead junctions was also tested with photo-excited LA and TA phonons[52]. There is evidence that the high-frequency cut-off in lead, $\hbar\omega_{max}$ = 2.2 THz, restricts the efficient detection of phonons with higher energies because of total reflection on the crystal-lead film interface. The exponentially damped phonon field inside the lead probably contributes not much to the signal in the tunnel current. As described above, the LO phonon relaxation cascade in GaAs generates LA phonons with 4...6 THz in the first step, and 2...3 THz in the second. It is not clear at present what fraction of these LA phonons with $\hbar\omega > \hbar\omega^{Pb}_{max}$ are down-converted at the interface, so that their decay fragments are indirectly detected by the junction. The analysis of time-of-flight signals for very small r (0.2 mm), and the dependence of integrated signal strength on r suggest that a rather substantial fraction of high-frequency LA phonons, $\hbar\omega > 2.2$ THz, is not detected by the Pb junctions[52].

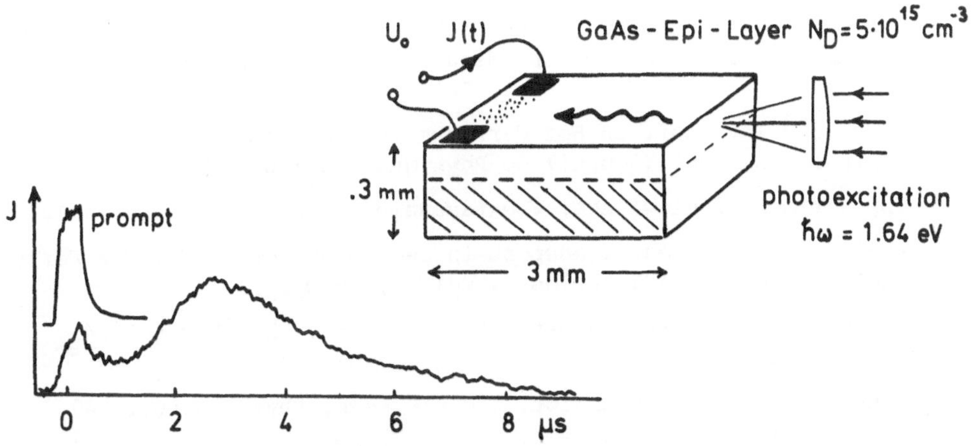

Fig.22 Novel detection scheme for Tera-Hertz phonons in GaAs based on phonon-induced ionization of shallow, effective-mass-like donors with 1s → 2s threshold energy of 4.4 meV (1.05 THz). Sample and contacts (above) and typical ionization current (below) are shown.

We finally describe a novel scheme for Tera-Hertz phonon detection based on the ionization of impurity bound electrons. The effect is the inverse process of phonon-assisted capture of mobile carriers into localized impurity states[63]. In GaAs, the usual shallow donors have a ground state bindung energy of 5.8 meV [64], a hydrogen-like spectrum of excited states, and a 1s-2s separation of 4.4 meV. The absorption of a 4.4 meV phonon by the 1s donor electron leaves the donor in the 2s excited state, which has a high probability for subsequent thermal ionization even at low temperatures ($E_{2s\rightarrow\infty}$ = 1.4 meV). The resulting free conduction band electrons can be measured as a current with the help of an applied electric field. It must not exceed the critical value for avalanche breakdown ($\sim$1 V/cm) . With increasing electric field the sensitivity improves. Fig.21 shows the principal set-up and a phonon signal measured at 1.4 K. Extreme care has to be taken to shield the exciting light pulse from the detector region, where it might create e-h pairs which lead to a strong "prompt" photo-conductive response at t=0 (see Fig.21). Comparison of the donor-ionization signals with lead junction signals measured in the identical geometrical configuration at the same distances r give strong evidence that indeed high-frequency phonons above 1.05 THz are detected by the shallow donors. Heater-generated low-frequency phonons gave no significant signal. The sensitivity of the detector for TA and for LA phonons can be estimated from the known piezoelectric and deformation potential constants in GaAs, and empirical data of the inverse capture process[63]. It compares favorably with the sensitivity of superconducting tunnel junctions.

REFERENCES

1. Proc. 4th Int.Conf. on Phonon Scattering in Condensed Matter, Stuttgart, 1983 (eds. W.Eisenmenger, K.Laßmann,S.Döttinger, Springer Series in Solid State Sciences Vol.51, 1984) Part III.
2. Proc. 3rd Int.Conf. on Hot Carriers in Semiconductors, Montpellier, 1981 (Journal de Physique, Colloque C-7,1981).
3. For a timely review, see: V.Narayanamurti, Science 213, 717(1981).
4. Physics of Nonlinear Transport in Semiconductors (eds. D.K.Ferry and J.R.Barker, NASI Series B, Vol.52, 1980).
5. R.G.Ulbrich, V.Narayanamurti, and M.A.Chin, Phys.Rev.Lett.45, 1432 (1980).
6. R.Baumgartner, M.Engelhardt, and K.F.Renk, Phys.Rev.Lett. 47, 1403 (1981).
7. H.Lengfellner and K.F.Renk, Phys.Rev.Lett. 46, 1210 (1981).
8. W.Dietsche, G.A.Northrop, and J.P.Wolfe, Phys.Rev.Lett.47, 660 (1981); G.A.Northrop, Phys.Rev. B26, 903 (1982).

9. J.T.Waugh and G.J.Dolling, Phys.Rev. 132, 2410 (1963).

10. K.C.Rustagi and W.Weber, Solid State Commun. 18, 673 (1976).

11. J.I.Pankove, Optical Processes in Semiconductors (Dover Publications, New York, 1975) Ch.7.

12. G.D.Watkins, in: Proc.16th Int.Conf. Physics of Semiconductors, Montpellier 1982 (Physica 117B & 118B, p.9,1983).

13. V.S.Bagaev, L.V.Keldysh, N.N.Sibeldin, and V.A.Tsvetkov, Zh.Eksp.Teor.Fiz.70, 702 (1976); transl. Sov.Phys.-JETP 43,362 (1976).

14. J.C.Hensel, T.G.Phillips, and G.A.Thomas, in: Solid State Physics Vol.32 (eds.H.Ehrenreich, F.Seitz, D.Turnbull, Academic Press, New York, 1977) p.152.

15. D.Huet and J.P.Maneval, in: Proc. 3rd Int.Conf. on Phonon Scattering in Condensed Matter, Providence,1979 (ed. H.J.Maris, Plenum,1980) p.145.

16. P.Hu, V.Narayanamurti, and M.A.Chin, Phys.Rev.Lett.46, 192(1981).

17. J.Friedel, in: Optical Properties of Solids (ed.F.Abelès, North Holland, Amsterdam, 1972) p.1.

18. R.G.Ulbrich, Solid-St.Electron. 21, 51 (1978).

19. F.Bassani and G.Pastori Parravicini, Electronic States and Optical Transitions in Solids (Pergamon, Oxford, 1975) Ch.4.

20. E.M.Conwell, High Field Transport in Semiconductors (Academic Press, New York, 1967) Ch.9.

21. W.Zawadzki, in: Handbook of Semiconductors, Vol.1 (ed.T.Moss, North Holland, Amsterdam, 1982) p.713.

22. R.G.Ulbrich, Phys.Rev. B8, 5719 (1973).

23. E.J.Yoffa, Phys.Rev. B21, 2415 (1980).

24. P.J.Dean, in: Topics in Applied Physics, Vol.17 (ed.J.I.Pankove, Springer, Berlin, 1977)p.63.

25. S.Pantelides, Rev.Mod.Phys. 50, 797, (1978).

26. D.Lang, in: Proc.15th Int.Conf. Physics of Semiconductors, Kyoto,1980 (J.Phys.Soc.Japan 49,1980) p.215.

27. C.H.Henry, in:Relaxation of Elementary Excitations (ed.R.Kubo, E.Hanamura, Springer, Berlin,1980) p.19.

28. U.Kaufmann, J.Schneider, in: Advances in Solid State Physics, Vol.20 (ed.J.Treusch, Vieweg, Wiesbaden,1980) p.87.

29. M.Kaminska, M.Skowronski, J.Lagowski, J.M.Parsey, and H.C.Gatos, Appl.Phys.Lett43, 302 (1983).

30. A.Mooradian, in: Laser Handbook Vol.2 (ed.F.Arrecchi, E.O.Schulz-Dubois, North Holland,1972) p.1409.

31. D.von der Linde, J.Kuhl, and H.Klingenberg, Phys.Rev.Lett.44, 1505 (1980); see also Ref.23, p.653.

32. R.Orbach, Phys.Rev.Lett. 16, 15 (1966).

33. S.Tamura and K.Okubo, in: Ref1, p.109.

34. R.G.Ulbrich, V.Narayanamurti, and M.A.Chin, J.Physique (Paris) 42, c-6, 226 (1981).

35. P.T.Tua and G.D.Mahan, Phys.Rev. B26,2208 (1982).

36. H.J.Maris, Phys.Rev.Lett. 17, 228 (1965).

37. R.Orbach and L.A.Vredevoe, Physics 1, 91 (1964).

38. M.Lax, P.Hu, and V.Narayanamurti, Phys.Rev. B23, 3095 (1981).

39. T.Haavasoja, V.Narayanamurti, and M.A.Chin, Phys.Rev.B27, 2767 (1983).

40. T.H.Geballe, G.W.Hull, Phys.Rev.110, 773 (1958).

41. P.G.Klemens, in: Solid State Physics Vol.7 (ed.F.Seitz, D.Turnbull, Academic Press, New York, 1958) Ch.1.

42. S.Tamura, Phys.Rev B27, 858 (1983).

43. S.Tamura, Phys.Rev B28, (1984).

44. R.J.von Gutfeld, in: Physical Acoustics (ed. W.P.Mason, Academic Press, New York, 1965) Vol.V, p.233.

45. J.M.Andrews and M.W.P.Strandberg, Proc.IEEE 54, 523 (1966).

46. F.Rösch and O.Weis, Z.Physik B27, 33 (1977).

47. T.E.Wilson, W.E.Bron, F.M.Lurie, and W.L.Schaich, Ref.1, p.34.

48. D.Huet, J.P.Maneval, and A.Zylberstejn, Phys.Rev.Lett.29,1092 (1972).

49. D.V.Kazakovtsev, Y.B.Levinson, JETP Lett. 27, 181(1978); N.M.Guseinov and Y.B.Levinson, Solid State Commun.45, 371 (1983).

50. M.Lax, V.Narayanamurti, and R.G.Ulbrich, Ref.1, p.103.

51. B.Stock, M.Fieseler, and R.G.Ulbrich, ibid. p.97.

52. B.Stock, PhD thesis, University Dortmund, 1984 (unpublished).

53. B.Stock, M.Fieseler, and R.G.Ulbrich, Proc. 17th Int.Conf. Physics of Semiconductors, 1984, San Francisco (in print).

54. B.Taylor, H.J.Maris, and C.Elbaum, Phys.Rev. B3, 1462 (1971).

55. J.C.Hensel and R.C.Dynes, in: Proc.14th Int.Conf.Physics of Semiconductors, 1978, Edinburgh (ed.B.L.H.Wilson, London)p.371.

56. G.A.Northrop and J.P.Wolfe, Phys.Rev. B22, 6196 (1980).

57. W.Eisenmenger, in: Physical Acoustics, Vol.XII (ed.W.P.Mason, R.N.Thurston, Academic Press, New York, 1976) Ch.IV.

58. W.Bron, Rep.Prog.Phys. 43, 301 (1980).

59. C.Weisbuch and R.G.Ulbrich, in: Topics in Applied Physics Vol.51 (eds.M.Cardona, G.Güntherodt, Springer, 1982) Ch.7.

60. R.Trommer, M.Cardona, Phys.Rev.B17, 1856 (1978).

61. K.T.Tsen, D.A.Abramson, and R.Bray, Phys.Rev. B26, 4770 (1982); see also Proc. IUPAP Symposium on High Excitation and Short Pulse Phenomena, Trieste 1984 (J.Luminescence, in press).

62 M.Fieseler, Diploma thesis, University Dortmund, 1984 (unpubl.).

63. R.G.Ulbrich, Proc.12th Int.Conf.Physics of Semiconductors, Stuttgart, 1974 (ed.M.H.Pilkuhn, Teubner, Stuttgart, 1974) p.376.

64. see, e.g.: A.M.White, in Ref.55, p.123.

MONOCHROMATIC PHONON GENERATION

BY SUPERCONDUCTING TUNNEL JUNCTIONS

Helmut Kinder

Physik Department
der Technischen Universität München
D-8046 Garching, Federal Republic of Germany

INTRODUCTION

Superconducting tunneling junctions are metal-oxide-metal structures where the insulating oxide is thin enough so that electrons can tunnel quantum mechanically from one side to the other. These are extremely versatile devices, and were used, among other things, for fast switching computer elements, low noise microwave mixers, ultra sensitive magnetic flux detectors (SQUID), and highly repoducible voltage standards. The junctions are also well suited for the generation and the detection of phonons. This will be the subject of the present lecture.

Presently we have two different ways of tunable monochromatic phonon generation by superconducting tunnel junctions: (i)Tunneling of single particles leads to a nonequilibrium distribution of quasiparticles in the superconductors which relaxes back to equilibrium by phonon emission. A monochromatic part of the emitted phonon spectrum is then isolated by a modulation technique. This way of phonon generation was applied to many different problems in condensed matter physics. Examples of this will be given here. (ii)Tunneling of Cooper pairs generates electromagnetic waves with Josephson frequency 2eV/h inside the junction. These electromagnetic waves are then dissipated by dielectric loss in the oxide layer of the junction. The dielectric loss mechanism converts directly photons into phonons of the same frequency. Thus, their frequency resolution should be ultimately as high as that of the Josephson effect itself. This way of phonon generation is quite recent. The hitherto existing experiments will be reviewed at the end of this lecture.

Tunneling junctions serve also as phonon detectors in three different ways: (i) Phonons incident in homogenous junctions can break Cooper pairs if their energy exceeds the threshold set by the energy gap 2Δ of the superconductors. The resulting single electrons give rise to an extra tunneling current at bias voltages V less than $2\Delta/e$. This is the most common use. (ii) Spatial resolution can be achieved by laser scanning a homogenous junction when biased just at $V = 2\Delta/e$. (iii) Frequency resolution is obtained by using hetero-junctions made of two superconductors with different energy gaps. These junctions exhibit a variable threshold, $(\Delta_1+\Delta_2-eV)/h$ which can be used for spectrum analysis. These three ways of phonon detection will be also presented.

PHONON GENERATION BY SINGLE PARTICLE TUNNELING

Junction Fabrication

As it is well known, superconducting tunneling junctions consist of two overlapping metal films which are separated by a thin oxide layer. These films are usually evaporated in a vacuum of 10^{-5} to 10^{-6} mbar. The oxide layer is grown on the bottom film before the top film is evaporated. On aluminium, this is done by introducing oxygen at 5 mbar for 5 minutes, on tin by applying an oxygen glow discharge at 1 mbar for 10 min, and on lead at ambient air in an oven at 60°C for 20 minutes. These conditions vary, not only from one evaporator to the other, but also from time to time in the same evaporator.

Typical film thicknesses are 150 nm each, a typical junction area, as defined by the overlap of both films, is 1x1 mm^2. For junctions of this size, the shapes of the films can be defined by metallic masks, and photolithography is not required. Indium pads are pressed directly onto the films for electrical contact.

Tunneling

If the oxide layer is assumed to provide a potential barrier of rectangular shape with thickness d and height B above the Fermi energy E_F (see Fig.1), the Schrödinger equation for electrons at E_F is, inside the barrier

$$(-\hbar^2/2m + B + E_F)\,\psi(x) = E_F\,\psi(x)$$

which is solved by evanescent waves $\psi(x) \propto \exp(-\sqrt{(2mB)}x/\hbar)$. An electron which has a plane wave state on the left side of the barrier (see Fig. 1) then has the finite probability $|\psi(d)|^2$ of tunneling to the right side if energy can be conserved. On tunneling, a hole is created on the left and an electron on the right. Because of the energy conservation, the excitation energies of the hole and the electron must add up to the voltage applied to the

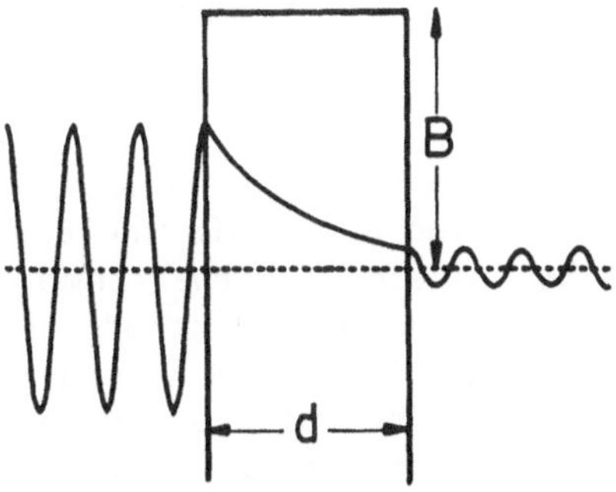

Fig. 1. Wave function of an electron tunneling through a barrier of rectangular shape.

junction, i.e. $E_h + E_e = eV$. The larger the voltage, the larger is the range of energies and accordingly, the number of states that can tunnel. Hence, the tunneling current is directly proportional to the voltage, if the density of states is constant as in normal conductors. Ohm's law is then obeyed. Yet in the case of superconductors, the density of states is not constant. This will be discussed next.

Density of States of Superconductors

In the ground state of a superconductor, the electrons form a condensate of correlated "Cooper pairs" in order to make use of a weak attractive force between each other which is mediated by the electron-phonon interaction. Due to the correlation, the condensate occupies the k-states close to the Fermi energy only partially, even at T=0. But if a state with given k and spin direction happens to be

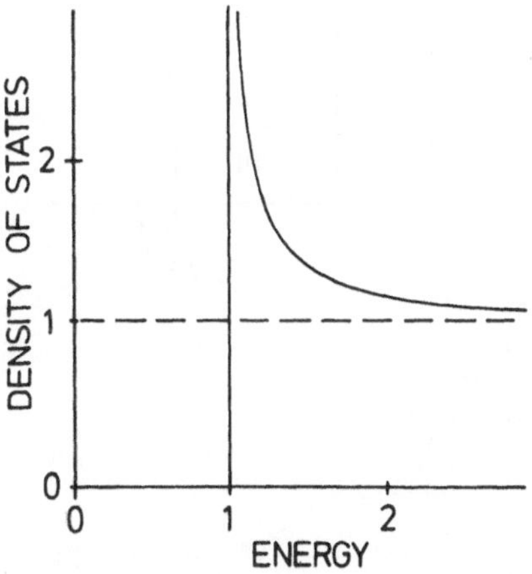

Fig 2. BCS-density of quasiparticle states in superconductors.

occupied, then also the conjugate state with -k and opposite spin is occupied with certainty. And so, if one of two conjugate states is empty, the other one is also empty, of course.

An excited state of a superconductor (=quasiparticle) consists of an electron in one k-state and a hole in the conjugate state. Thus, quasiparticles can be created by either removing one electron from an occupied pair of k-states, or by introducing an electron into an empty pair of k-states. Since each quasiparticle blocks a pair of k-states which would be otherwise accessible to Cooper pairs, it destroys a small part of the correlation and hence requires some extra excitation energy in addition to the kinetic energy of the electron or hole. Thus, the density of states of quasiparticles in superconductors is, in comparison with that of the normal state, always shifted up to higher energies, and a gap opens usually. This is shown in Fig. 2 for the well known BCS case, where there is a singularity at the gap edge.

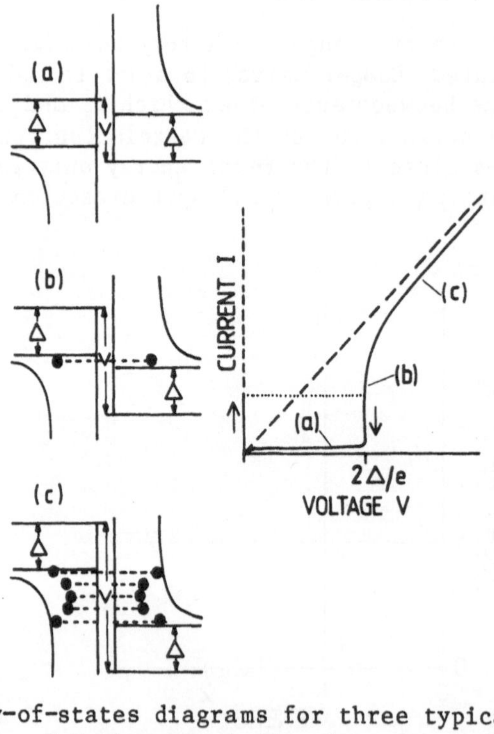

Fig. 3. Density-of-states diagrams for three typical voltages, and resulting I-V characteristic of a superconducting tunneling junction.

Single particle tunneling between superconductors

If a single electron tunnels out of an occupied pair of k-states of one superconductor into an empty pair of k-states of another superconductor, then two quasiparticles are created. One is the partner left behind, and the other one is the electron itself. The energies of both particles must again add up to the voltage, $E_1 + E_2 = eV$, and the number of states that can actually tunnel depends on the densities of states on both sides. This can be visualized by the diagrams of Fig. 3.

The energy scales are displayed vertically here, and the E_1 scale is upside down so as to automatically conserve energy, $E_2 = eV - E_1$, for "horizontal tunneling", i.e. creation of both quasiparticles at the same height. For diagram (a), the voltage is less than 2Δ, and in this regime tunneling is not possible at T=0. In diagram (b), the density of states is large on both sides, resulting in a steep rise of the tunneling current at $eV = 2\Delta$. Diagram (c) shows a case of even higher voltage, where the number of tunneling states again increases almost linearly with the voltage, as in normal conductors. Consequently, the superconducting junction exhibits the well known I-V-characteristic shown on the right of Fig. 3. The "thermal current" below 2Δ will be discussed in the section on homogenous junction detectors. The dc Josephson current due to pair tunneling at V=0 is also included in Fig. 3, but it will be discussed in the section on Josephson phonons.

Emitted Phonon Spectra

As can be readily seen from Fig. 2 (c), the tunneling current injects a nonequilibrium distribution of quasiparticles into an energy band of width $eV-2\Delta$ on either side. On both, top and bottom of this band, the injection strength has singularities due to those of the densities of states. This nonequilibrium distribution relaxes back to equilibrium by electron-phonon interaction, i.e. under phonon emission (Eisenmenger and Dayem 1967).

In a first step, see Fig. 4(a), the quasiparticles slow down by transitions towards the gap edge. This results in a broad band "bremsstrahlung" or "relaxation" spectrum, see Fig. 4(b). The spectrum exhibits a discontinuous cutoff at the maximum frequency which corresponds to transitions from the top of the injected distribution down to the gap edge in a single step. This process is highly probable because of the convolution of the singularity in the injection strength with the one in the density of final states. The discontinuity is absent in normal conductors.

In a second step, Fig. 4(c), the quasiparticles recombine into the pair condensate. This step is usually slower than the first one, because it depends on the availability of partners. The resulting spectrum of "recombination-" or "2Δ-" phonons is sketched in

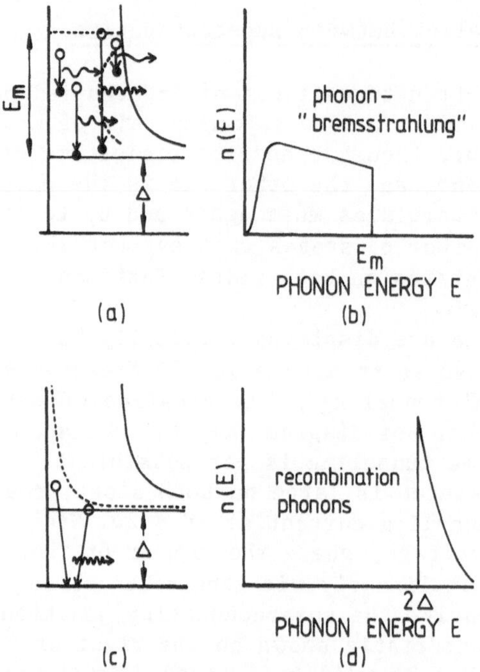

Fig. 4. Relaxation (a) and recombination (c) of injected quasiparticles yields "bremsstrahlung-" (b) and "2Δ-" (d) phonons.

Fig. 4(d). Experimentally measured spectra will be presented in the section on applications.

Modulation Technique

Because the discontinuous cutoff of the "bremsstrahlung" part of the emitted phonon spectrum depends on the voltage, it lends itself to a modulation technique which allows to isolate a narrow band of quasi monochromatic phonons (Kinder 1972). In Fig. 5(a), the tunneling characteristic is again displayed with an operation point V_o indicated. Fig. 5(b) shows the corresponding (total) emitted phonon spectrum. If now the operation point is shifted to $V_o+\delta V$, the cutoff frequency shifts accordingly, as indicated in Fig. 5(b). In addition, the number of recombination phonons increases because of the increased current. The difference of the two spectra is given by the hatched area. This difference spectrum can be detected alone, when δV is a time varying voltage, i.e. an ac modulation or small

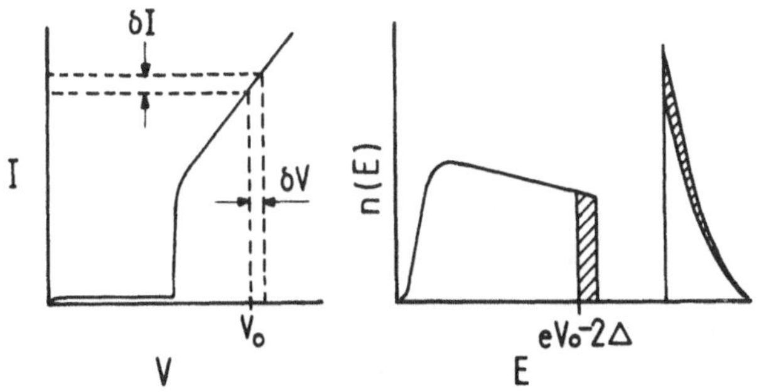

Fig. 5. The change of the emitted "bremsstrahlung" spectrum on a small voltage change is quasi monochromatic.

pulses, and the phonon detector is ac coupled to the preamplifier, as shown in Fig. 6.

If pulses are used for δV, then the time of flight can be used to separate phonons having different polarizations or travelling on different paths, as shown in Fig. 7. A boxcar averager is then commonly used to measure the height of one of the pulses as a function of phonon frequency which is continuously varied by sweeping the bias voltage V_0.

In all superconductors but aluminum, there is an upper frequency limit set to the "bremsstrahlung" phonons by the energy gaps of the superconductors used. Above this limit, the phonons are reabsorbed inside the junction, and their energy is released only as 2Δ-phonons and low frequency phonons. These frequency limits are 290GHz for Sn, 650GHz for Pb, and 870GHz for $Pb_{.7}Bi_{.3}$. Only in the case of aluminum, the electron-phonon interaction is weak enough so as to let the phonons above 2Δ (~100GHz) escape (Forkel et al. 1973). Frequencies of up to 1.5THz have been so far reached by using extremely thin Al junctions (Eisenmenger 1978).

The frequency resolution attainable with this technique depends on the modulation amplitude, and eventually on the sharpness of the energy gap which is usually somewhat smeared by anisotropy (Sn, Pb)

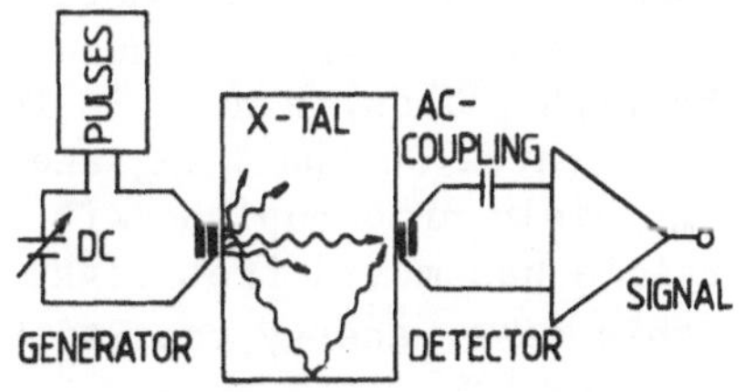

Fig. 6. Schematic circuit diagram of the modulation technique.

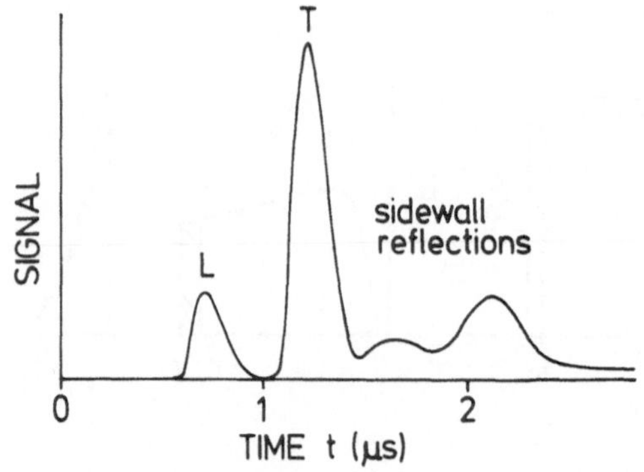

Fig. 7. Phonon pulse pattern reveals polarizations and transits.

or varying oxygen content (Al). A resolution of 5GHz was reached with Al, of 10GHz with Sn, and of 2GHz with PbBi (Kinder and Dietsche 1976).

Before we can go on to describe the many experiments done with this technique, we need to discuss the operation of tunnel junctions as detectors.

HOMOGENOUS JUNCTION DETECTOR

The most commonly used detector junctions consist of two equal superconductors, and they operate at voltages below 2Δ (Eisenmenger and Dayem 1967). In this regime, single electrons cannot tunnel out of the pair condensate as discussed above. However, quasiparticles that do already exist, e.g. at finite temperature, can lead to electron tunneling also in this regime. This should be considered in some detail to understand the detector operation.

Electrons can go from left to right by two different processes, because of the two ways of quasiparticle creation (and annihilation) as discussed in the previous section. One way is tunneling out of a quasiparticle state on the left. Then, an empty pair of k-states is left behind and another one on the right side is occupied by the electron, forming a new quasiparticle there. Energy conservation thus requires $E_2 = E_1 + eV$. The other way is tunneling into a quasiparticle state on the right. Then, a hole is left behind in a formerly occupied pair of k-states on the left, and another hole is filled on the right side to yield an occupied pair of k-states there. Thus, the quasiparticle has moved from right to left, opposite to the electron. In this case, energy conservation requires $E_1 = E_2 + eV$.

Both processes can be combined into one density-of-states dia-

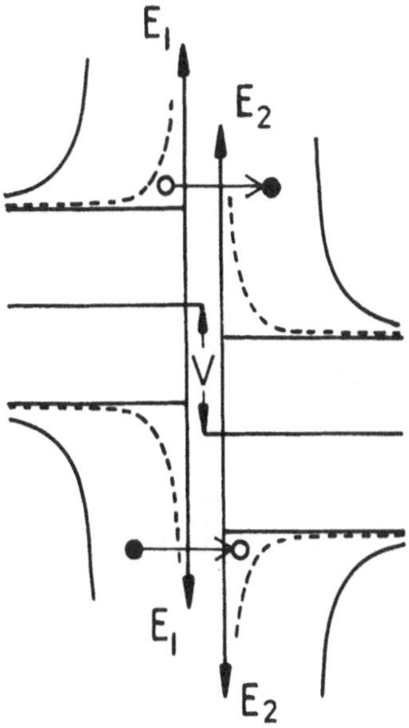

Fig. 8. Electron tunneling from left to right by thermally excited quasiparticles changing over from left to right (upper half) and from right to left (lower half).

gram for "horizontal tunneling", if two energy scales, one up and one down, are used on either side, see Fig. 8. This is similar to the so called "semiconductor pictiure", but is strictly consistent with the quasiparticle description of superconductors. Note that the two-fold energy scale represents the same states twice, and merely stands for two possible ways of tunneling. From the same diagram, we see that quasiparticles with energies above Δ+eV also lead to electron tunneling opposite to the applied voltage. The tunneling current in the regime below eV = 2Δ thus measures the number of quasiparticles between E = Δ and E = Δ + eV.

If now phonons of energy $\hbar\omega > 2\Delta$ are incident into the junction, they can be absorbed by pair breaking, i.e. scattering of electrons out of pair states into two quasiparticles at a time. These extra quasiparticles give rise to an extra tunneling current as discussed above. The extra current leads to the signal in the way indicated in Fig. 9. Since phonons with $\hbar\omega < 2\Delta$ are not absorbed at sufficiently low temperatures, the homogenous junction detector has a fixed frequency threshold of $\hbar\omega = 2\Delta$ which is often very useful. Also, the sensitivity of junctions is roughly 10 times higher than

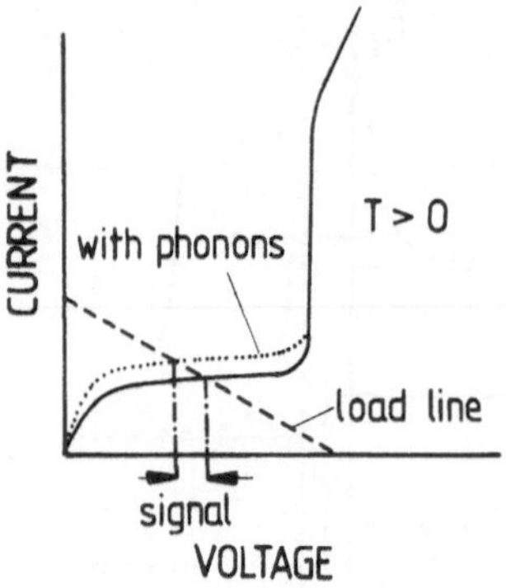

Fig. 9. Detector characteristic. The extra current due to phonons shifts the operation point and so produces the signal.

that of superconducting bolometers.

A "standard" phonon spectroscopy experiment consists of a "bremsstrahlung" generator junction on one face of the sample and a detector junction on the far face, as indicated above in Fig.6. The useful frequency range is then spanned by the gaps $2\Delta_D$ and $2\Delta_G$ of detector and generator, respectively, because of the threshold of the former and the reabsorption limit of the latter.

APPLICATIONS

The "bremsstrahlung" method has been applied to a variety of different problems. The purpose of this lecture is not a complete review, however. Rather, one or two typical examples will be given for each subject. We begin with studies of localized excitations in solids.

$Al_2O_3:V^{3+}$

Sapphire (Al_2O_3) containing V^{3+} impurities was the first system studied (Kinder 1972, 1973). The V^{3+} ion has two electrons in its 3d-shell. According to Hund's first rule, the spins are parallel, i.e. S=1. If the ion is placed into a solid, the electric field of the nearest neighbor atoms (crystal field) causes a Stark effect which is much stronger than the spin-orbit coupling. This mixes the orbital angular momentum states such that they no longer precess, but rather lock into the crystal field, with the expectation value <L>=0, to first order. In other words, the orbital angular momentum is quenched by the crystal field.

The ground state is therefore characterized by the spin only, and forms an S=1 triplet. To second order the spin still magnetizes the orbit to some extent by the L·S-coupling. This results, e.g. in effective g-factors deviating from 2. The magnetizability of the orbit depends on the orientation, however. Therefore the $S_Z = 0$ state happens to be at somewhat lower energy than the $S_Z = \pm 1$ doublet). The latter exhibits a further tiny hyperfine splitting which will be of importance in the context of Josephson phonons. Here the V^{3+} ion will be simply viewed as an effective two-level-system.

The system is coupled to phonons by the so called van Vleck mechanism. The strain field of the phonon modulates the nearest neighbor distances and hence, the cystal field which then distorts the orbit which eventually L·S-couples to the spin. The phonons

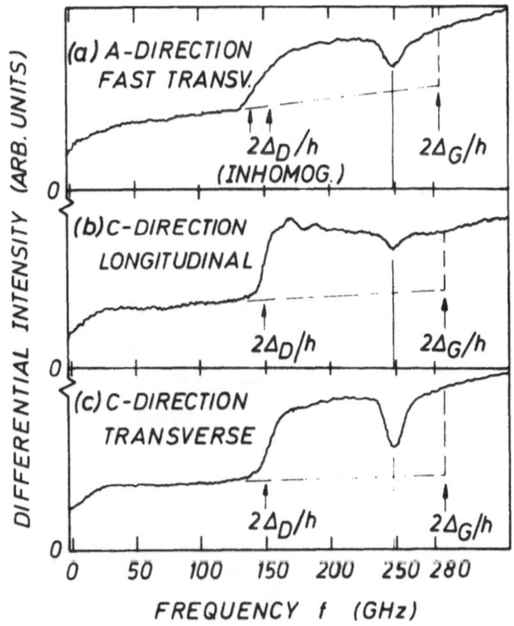

Fig. 10. Resonance scattering of monochromatic phonons at 248 GHz by V^{3+} ions in sapphire.

cause transitions if their energy is coincident with the level splitting, and hence are resonantly scattered. By this, the ballistic phonons are attenuated and the detector signal decreases at this frequency.

The experimental results are shown in Fig.10 (above) for some phonon directions and/or polarizations. The detected intensity is plotted as a function of the phonon frequency (eV-2Δ)/h. Besides the resonance scattering line at the splitting frequency (248GHz), there is also a step at the detector threshold frequency (150GHz). Below this, only background phonons of mainly 2Δ-frequency are detected. Thus, the background signal is known and can be fairly well subtracted at bremsstrahlung frequencies above the detector threshold. The next example is a much more complicated system.

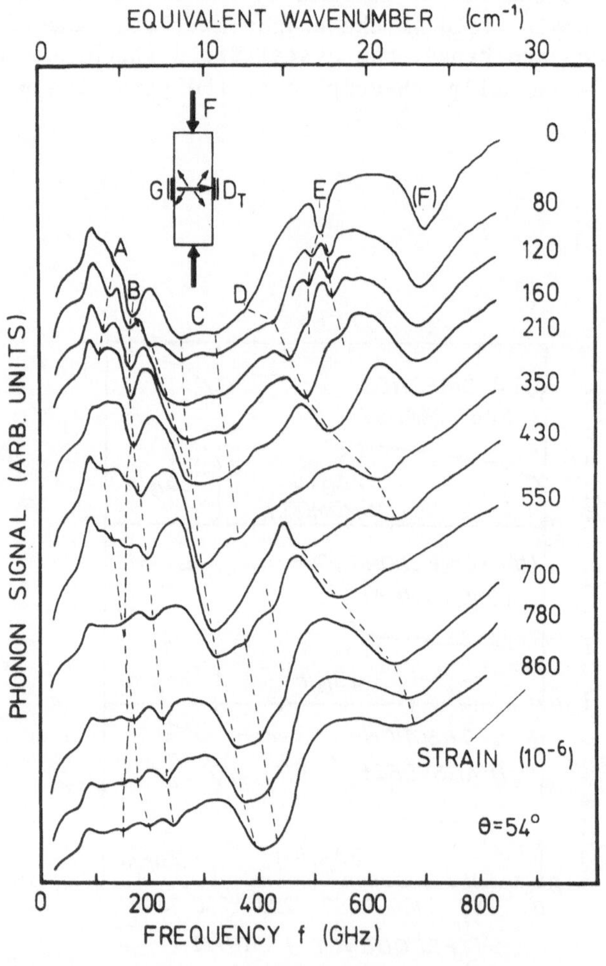

Fig. 11. Phonon spectroscopy of Mn^{3+} in sapphire with uniaxial stress.

$Al_2O_3:Mn^{3+}$

The Mn^{3+} ion has four electrons in its d-shell, with S=2. Since ⟨L⟩ is quenched again, this results, to first order, in a quintet ground state. This quintet is then further split by coupling to a dynamical distortion of the vicinity of the ion with lower than trigonal symmetry (dynamic Jahn-Teller effect). Because of the 3-fold axial symmetry of the ion site, this results in 5x3 = 15 vibronic (vibrational+electronic) states. Fig. 11 gives the results of phonon spectroscopy under varying uniaxial stress (Zoller et al. 1980). The many lines do confirm the Jahn-Teller effect, but are inconsistent with the theory of Bates and Wardlaw (1980).

Tunneling States in Alkali Halides

A simple model of tunneling states is a particle in a double well potential, see Fig. 12(a). Localized states in either one of the wells are time dependent if the tunneling probability is finite. Rather, the true eigenstates are the symmetric and antisymmetric linear combinations of the localized states, see Fig. 12(b). Since the antisymmetric mode has a node, it has slightly more kinetic energy from which results the well known tunnel splitting. The system couples to phonons if the strain modulates either the relative depth or the relative distance of the two wells (Sussmann 1964). Phonons with the splitting energy are then resonantly scattered.

As an example of a real system, we now consider OH^- molecules on Cl^- sites in NaCl. As a function of the orientation, the OH^- dipole has six equivalent potential wells along the cubic axes. Here, the tunneling wave functions are more complicated linear combinations of the (orientationally) localized states. The resulting level scheme for the cubic case and for tetragonal distortion under uniaxial stress is shown in the inset of Fig. 13, and the phonon allowed transitions are indicated.

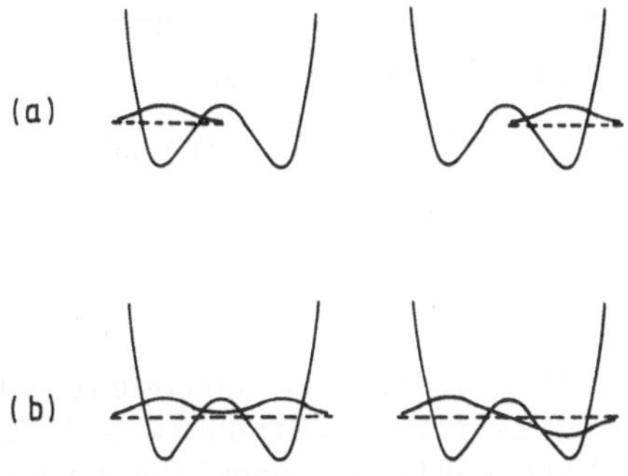

Fig. 12. Double well potential with localized (a) and tunneling (b) wave functions.

The result of phonon spectroscopy is also shown in Fig. 13, for various stresses (Windheim and Kinder 1975). The line position as function of stress fits the theory very well in this case. In this system, the interaction with phonons is so strong that a very low concentration of OH^- had to be used, less than 0.1 ppm. For many

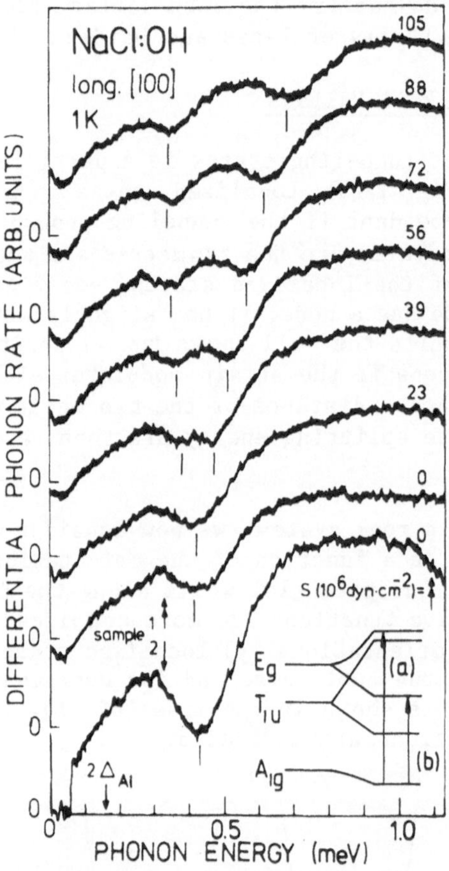

Fig. 13. Spectroscopy of OH^- in NaCl with uniaxial stress.

other orientational tunneling states the agreement between theory and experiment was also satisfactory (Windheim and Kinder 1976), while "off center ions" with position tunneling turned out to agree less (Bridges and Zoller 1979). The most complicated system studied was $KI:OH^-$ which is supposed to exhibit both, orientational and positional tunneling (Zoller and Bridges 1981).

Bending Mode of Si:O

Oxygen in silicon goes on interstitial sites where it has a bond bending mode with a frequency of 870GHz. This was studied using the method of very thin Al junctions, as mentioned above (Forkel et al. 1973; Eisenmenger 1976).

As the bending mode frequency is more than 10 times higher than the energy gap, the "bremsstrahlung" spectrum resembles very nearly that of a normal conductor where there is a kink instead of a discontinuity at the cutoff frequency. Therefore, second derivative modulation was used to isolate the latter. Fig.14 shows the result (Forkel et al. 1973, Eisenmenger 1981). Above 870 GHz up to about 1.5 THz, a rich variety of further lines is seen, probably due to oxygen pairs of various relative distances.

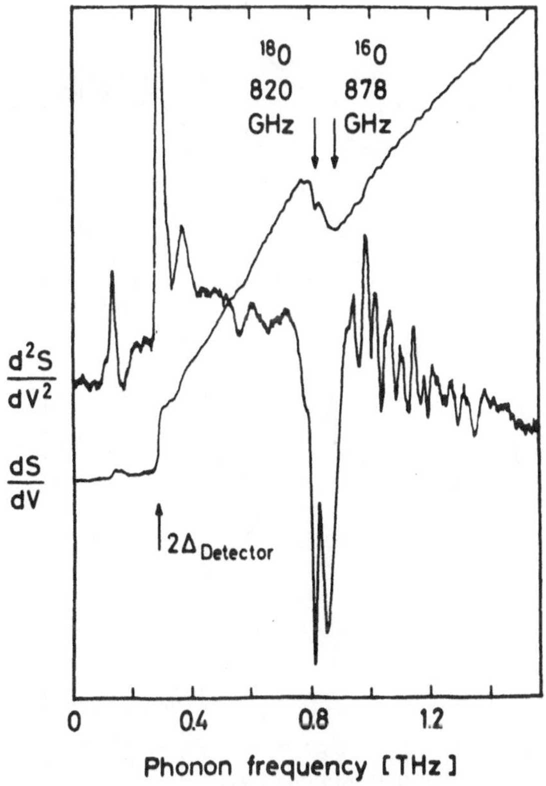

Fig. 14. Bending modes of oxygen isotopes in silicon. Structure up to 1.5THz due to oxygen pairs (Eisenmenger 1981).

Shallow Donors and Acceptors in Semiconductors.

Shallow donors and acceptors introduce n- or p-conductivity into semiconductors at room temperature, and hence are all-important for electronics. At low temperatures, the electrons or holes are captured by the charged donors or acceptors and form hydrogen-like bound states. The binding energy is small, however (~10meV), because of the high dielectric constant and low effective mass in semiconductors like Si, Ge or GaAs. Many of such semiconductor:dopant systems have been studied by phonon spectroscopy. As an example, we consider here Si:B.

Si:B. The valence band edge of silicon contains two kinds of holes which differ by their effective masses, the "light" and "heavy" holes. If bound to the B^- ions, both form bound states with s-character and spin up or down. Thus, the total ground state is

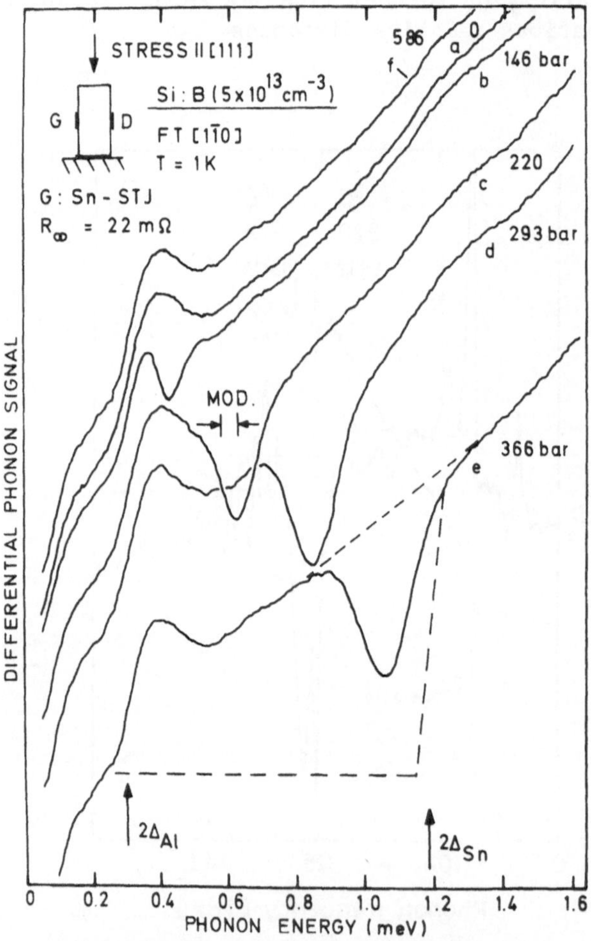

Fig. 15. Boron acceptors in silicon exhibit a stress tunable absorption line (Schwarte and Berberich 1985).

fourfold degenerate. Lowering the symmetry by uniaxial stress results in a splitting of the quartet into two doublets of time reversal symmetric states (Kramers doublets). This splitting is known to be perfectly linear as a function of stress. Phonons interact with these states due to the "deformation potential" which causes a very strong resonant scattering at the actual splitting frequency. Fig.15 shows the results of phonon spectroscopy for varying stress (Schwarte and Berberich 1985). The line strength varies approximately linearly with the resonance frequency, and the line width is mainly due to life-time broadening. Because this system tunes so nicely with stress, it has been used as a phonon spectrometer.

Application of Si:B as spectrometer. Consider any type of phonon generator that emits an unknown spectrum into the crystal. Without stress, the phonons travel ballisticly to the detector where they produce a signal. At any finite stress, the corresponding resonant phonons are partially blocked and hence, the signal is

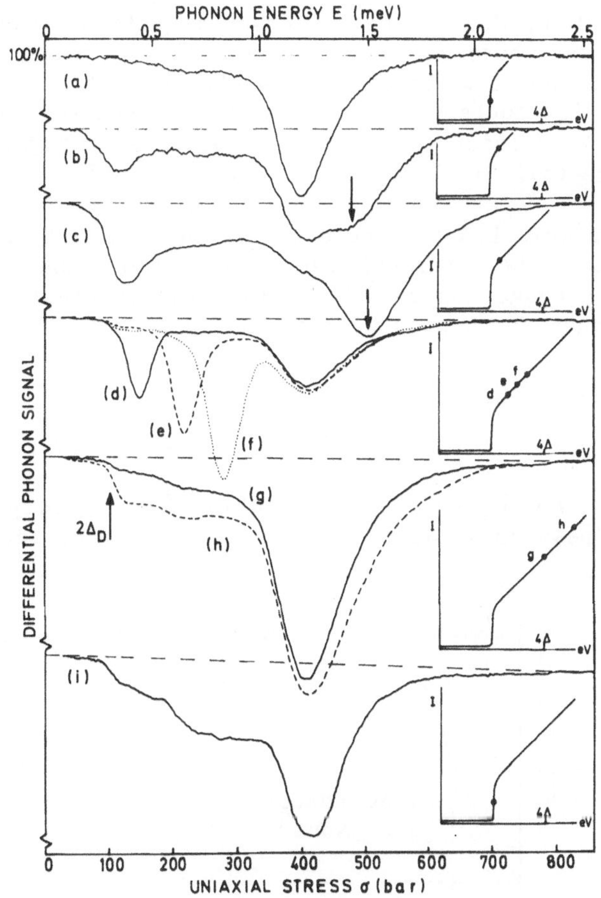

Fig. 16. Spectra (upside down) of Sn generator junction with modulation technique at indicated bias points.

reduced by an amount which depends on the contribution of these phonons to the total spectrum. In other words, the difference from 100% signal as a function of stress reproduces the emitted phonon spectrum proper.

This spectrometer was applied to investigate experimentally the real spectra emitted by superconducting junctions in the single particle mode, see Fig.16 (Berberich and Kinder 1981). Traces (a) to (h) were obtained from the same Sn junction by using the modulation technique at various bias voltages as indicated by the insets. At $V_0 = 2\Delta$, trace (a), a pure recombination phonon spectrum is obtained which exhibits a Boltzmann tail corresponding to an effective temperature of T^*=1.1K, slightly above the bath temperature of T=1.0K. For somewhat higher voltage, Traces (b) and (c) the 2Δ-peak shifts to higher frequencies, obviously due to direct recombination from the injection energies. For still higher voltages, traces (d) to (f), the relaxation of quasiparticles takes over which is seen from the "bremsstrahlung" peaks at eV-2Δ, and also from the recombination phonon peak which has shifted back to 2Δ. The Boltzmann tails now correspond to T^*=1.5K, due to the increased injection. Traces (g) and (h) at V_0=4Δ and above demonstrate the effect of reabsorption on the bremsstrahlung phonons. Trace (i) was obtained from a junction with higher impurity content in the metal films, due to the preparation conditions. This leads to a strongly increased background, as compared to trace (a).

Altough more work has been done so far on localized excitations, one can study the interaction of monochromatic phonons with extended states (propagating modes) as well. The simplest of these interactions is that of phonons with themselves.

Phonon-Phonon Interaction

Phonons can interact with each other in lowest order by three-phonon processes (3PP) which must conserve energy and quasi-momentum, $\omega_1 + \omega_2 = \omega_3$, and $\vec{q}_1 + \vec{q}_2 = \vec{q}_3 + \vec{G}$, where $\vec{G}$ is a reciprocal lattice vector. These constraints are more or less restrictive

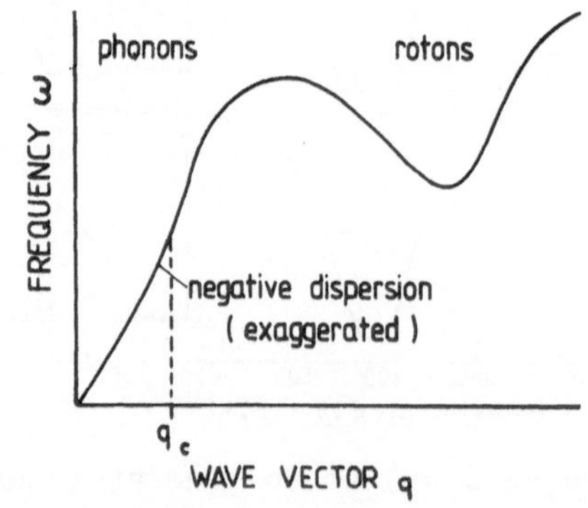

Fig. 17. Dispersion curve of superfluid helium (schematic).

depending on the phonon dispersion curves. For example, the lowest phonon branch cannot fulfill the constraints at all, and hence cannot decay by 3PP, if the dispersion of this branch has downward curvature. In this case, the smaller frequencies, ω_1 and ω_2, have larger phase velocities v_1 and v_2, than the combination frequency ω_3 and hence,

$$|\vec{q}_1 + \vec{q}_2| \leq |\vec{q}_1| + |\vec{q}_2| = \omega_1/v_1 + \omega_2/v_2 < \omega_1/v_3 + \omega_2/v_3 = \omega_3/v_3 = |\vec{q}_3|$$

i.e. momentum cannot be conserved. In contrast, for upward curvature there is always a range of allowed angles between $\vec{q}_1$ and $\vec{q}_2$, and hence the phonons decay much more rapidly. This is the case for liquid helium.

The superfluid helium dispersion curve is shown schematically in Fig. 17. Besides the well known roton minimum at large wave

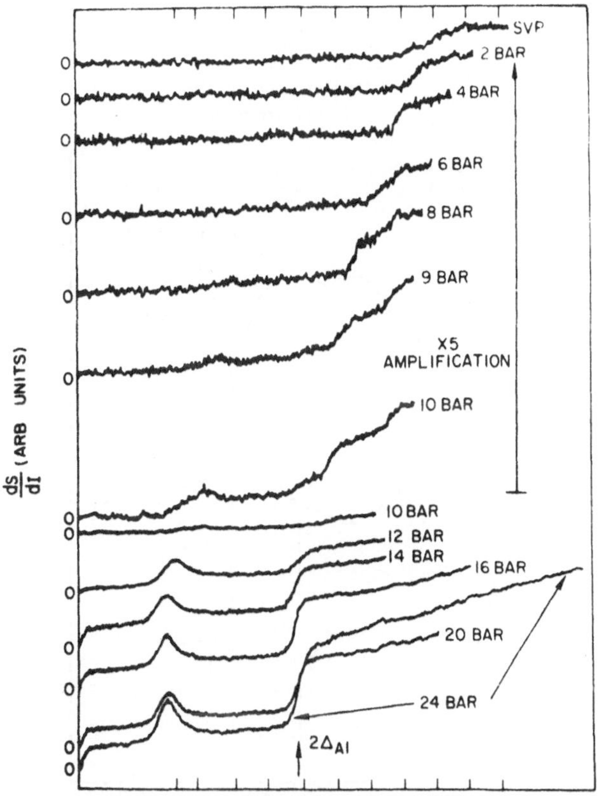

Fig. 18. Transmission of phonons through superfluid helium. Cutoff due to 3-phonon-processes tunes with pressure (Dynes and Narayanamurty 1974).

vectors, there is also a region of upward curvature (greatly exaggerated) in the phonon branch, from zero up to an inflection point. Therefore, there exists a critical wave vector q_c, given by the condition $v(\omega(q_c)) = v(\omega(q_c)/2)$ (resulting from the inequality above) which separates rapidly decaying phonons, with $q<q_c$, from long lived ones with $q>q_c$.

That this is indeed the case was found by Dynes and Narayanamurty (1974) by a "standard" phonon spectroscopy experiment where the sample was superfluid helium under pressure at 0.1K, filling a 0.8 mm gap between two glass supports carrying generator and detector junctions, both made of Al. The results are shown in Fig.18. At low pressures, there is a threshold for phonon propagation at high frequency indeed. With pressure increasing, the threshold shifts to lower and lower frequencies and eventually merges with the $2\Delta_D$-threshold of the detector junction. This demonstrates clearly that the q_c exists, and that it is tunable by pressure.

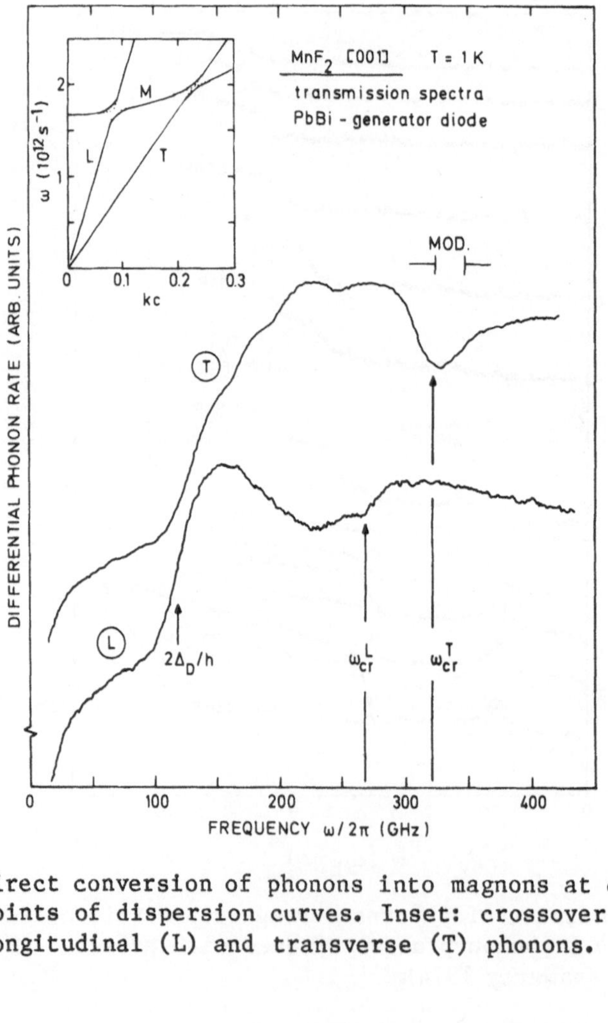

Fig. 19. Direct conversion of phonons into magnons at crossover points of dispersion curves. Inset: crossover differs for longitudinal (L) and transverse (T) phonons.

Phonon-Magnon Interaction

As it is well known, any magnetically ordered substance has excited states where one spin is reversed with respect to that order, and the probability amplitude of finding the reversed spin is a plane wave. These spin waves or magnons are Bose particles. Phonons couple to magnons by magnetostriction and can be converted by direct processes if energies and momenta of both, phonon and magnon are identical, i.e. at crossover points of their respective dispersion curves.

As an example, the dispersion curves for phonons and magnons in (antiferromagnetic) MnF_2 are shown schematically in Fig.19, inset. There is one crossover point for longitudinal phonons, and another one at higher frequencies for transverse phonons. The results of phonon spectroscopy are presented in Fig.19 (Mattes et al. 1978). The strong dip of the transverse phonon trace is clearly absent in that of the longitudinal phonons which has a small dip at lower frequencies instead. Both dips are consistent with the crossover points determined from neutron scattering data. The three weaker dips seen on the transverse trace at lower frequency were attributed to OH^- tunneling states.

Phonon-Electron Interactions in Semiconductors

The interaction of phonons with free electrons in semiconductors was studied in the context of electron-hole droplets in germanium. These are laser-excited regions of a high concentration of electrons and holes, which are bound to each other in much the same way as chlorine and sodium ions in molten rocksalt, although their density is much less.

The conduction band of Ge is well known to have four equivalent valleys in the [111]-directions at the zone boundary. In these valleys, the electrons form Fermi bodies of prolate rotational ellipsoidal shape. Phonons can scatter electrons, by the deformation potential interaction, along the Fermi surface only because their energies are negligible in first approximation. Hence, phonons with wave vectors larger than the large diameter of the Fermi ellipsoids will not be scattered at all. This is the well known "$2k_F$-cutoff".

If the phonon propagation direction is chosen parallel to one of the ellipsoids, this one mainly interacts and produces the cutoff. The number of electrons, and hence $2k_F$, in that valley can be tuned by uniaxial stress. Fig.20 shows the resulting phonon mean free path vs. frequency as it was measured by phonon spectroscopy (Dietsche et al. 1982). The dashed lines represent a theoretical prediction which yields a more gradual cutoff if the finite phonon energy is taken into account. The agreement is quite good and demonstrates clearly the tunability of the Fermi wave vector.

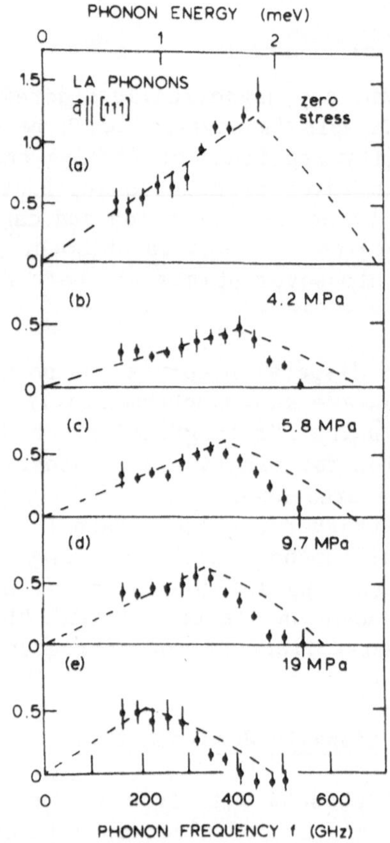

Fig. 20. Mean free path of phonons in electron-hole droplets in Ge. The $2k_F$-cutoff is tuned by stress (Dietsche, Kirch, and Wolfe 1982).

In all of the previous experiments, the attenuation of phonons due to interaction with other excitations was the measured quantity. Another measurable quantity is the velocity of phonons.

Dispersion Measurements in Thin Films

For the measurement of phonon dispersion curves, neutron scattering has been extensively exploited. As it is well known, neutrons couple to the nuclei which have a constant form factor because of their small size and hence, the scattering strength simply goes as $(\vec{q} + \vec{G}).\vec{e}$ where $\vec{q}$ and $\vec{e}$ are wave vector and plarization of the phonon, and $\vec{G}$ is a reciprocal lattice vector. Transverse phonons thus lead to scattering only in higher Brillouin zones, because $\vec{q}.\vec{e} = 0$ in the first zone. And also longitudinal phonons yield better signals in higher zones. From this consideration it becomes clear that neutrons are not well suited for amorphous materials, where higher zones are missing. Phonon spectroscopy can be used in this

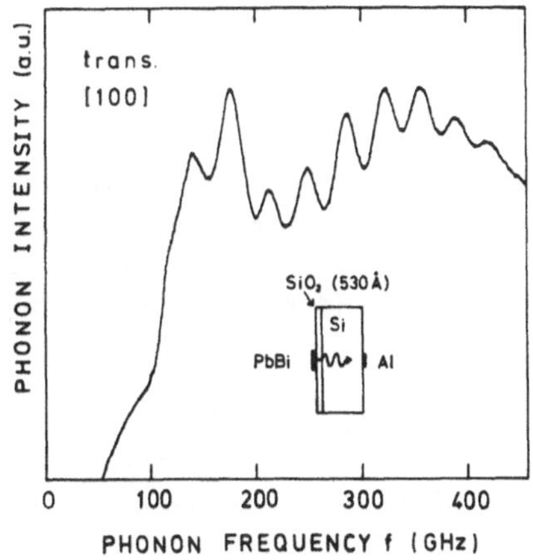

Fig. 21. Thickness resonances in a thin SiO_2 glass film.

case to measure phonon dispersion relations by standing wave resonances in thin films.

The experimental arrangement is shown in the inset of Fig.21. A thin SiO_2 glass film, about 50 nm in thickness was either thermally grown on a silicon crystal or- in other experiments not shown here - evaporated. The PbBi-generator junction transmits phonons through the glass film into the crystal. The transmission becomes large whenever the phonon frequency corresponds to a standing wave resonance, or eigen mode. The wave vectors of the eigen modes at normal incidence are given by

$$q_n = (n+\alpha)\pi/d$$

where n is an integer, α a phase shift due to the boundary condi-

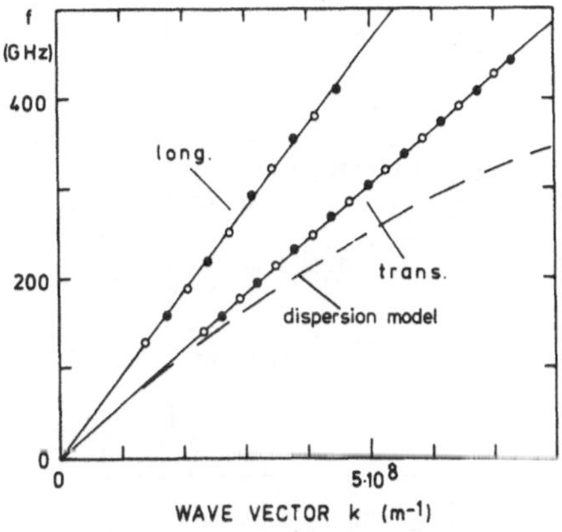

Fig. 22. Dispersion curves derived from measured standing wave resonances, and theoretical prediction.

tions, and d is the film thickness which is measured separately. Fig. 21 shows a sequence of such resonances for transverse waves (Rothenfusser et al. 1983). For each n, then, ω_n is known from phonon spectroscopy and q_n from film thickness. Thus, the dispersion curve is known in discrete points. This was plotted in Fig.22. The experimental dispersion curve was perfectly linear and thus was not consistent with a theory that predicted dispersion as an explanation for the plateau in the thermal conductivity vs. temperature of glasses (Jones et al. 1980).

Other dispersion measurements were done in amorphous Ga, where a softening of the transverse phonons in comparison with the crystalline state, by a factor of three, has been observed (Dietsche et al. 1980).

SPATIALLY RESOLVING DETECTOR JUNCTION

For many purposes, it is inconvenient to have generator and detector at fixed positions. As a mobile generator, a laser spot on a metal film is often used, although the emitted phonons are broad band. As a mobile detector, a large tunnel junction was used recently (Schreyer et al. 1984). This junction was made sensitive on a small spot only, by an incident low power laser beam. The spot was scanned by mirrors (see Fig. 23, inset), and hence the effective detector position was moved.

Due to the laser excitation, the spot has a higher quasiparticle concentration than the remainder of the junction area. As mentioned above, the quasiparticles destroy pair correlations. This

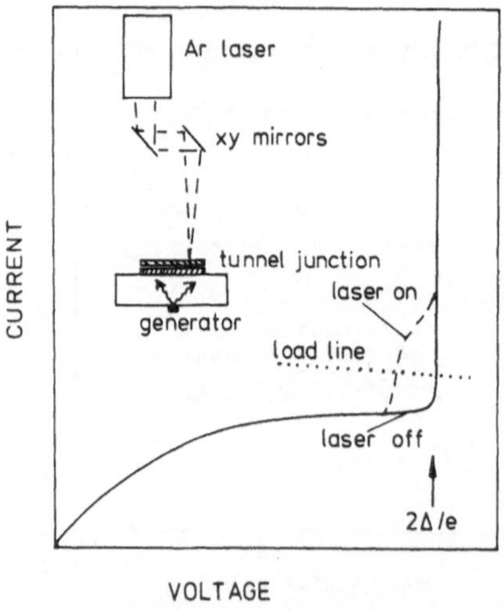

Fig. 23. Principle of operation of a laser scanned junction for spatial resolution.

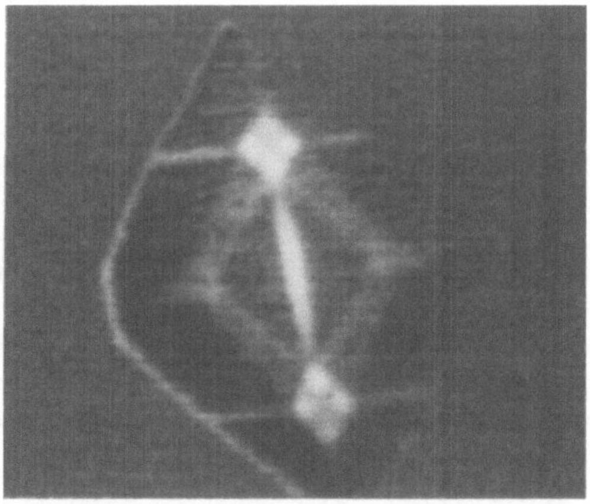

Fig. 24. Phonon focusing pattern of Si-[110] obtained by laser scanning of a tunneling junction (rhomb-shaped outline).

leads also to a reduction of the energy gap. Accordingly, the gap is slightly reduced at the laser spot. This is seen in the tunneling characteristic, exaggerated in Fig. 23, by a "precursor" of the 2Δ-rise. The voltage of the precursor corresponds to the reduced gap, while its height relative to the total rise reflects the spot size in proportion to the total area of the junction. In other words, if the operation point is chosen on the precursor, we can measure the energy gap locally, wherever the laser spot happens to be.

If now phonons are incident into the illuminated junction, they will create additional quasiparticles, i.e. produce an additional gap depression also on the illuminated spot where it is measured by a change of the precursor voltage. The magnitude of this additional depression depends on the local concentration of incident phonons. Biasing the junction to the precursor, and scanning the laser spot across the junction, one can thus monitor the intensity profile of the incident phonons.

As a demonstration, the intensity profile due to "focusing" of phonons emitted by a small heater (see inset of Fig. 23) was measured (Schreyer et al. 1984). The phonon focusing is due to the anisotropic elastic properties of crystals as explained in detail in J. Wolfe's and A. Every's lectures. Fig. 24 shows the intensity profile represented by the brightness of a CRT screen.

While this detector measures the quasiparticle concentration by the gap reducion, the frequency resolving detector to be described next is again "conventional" in the sense that it measures the presence of quasiparticles by their tunneling current, provided they have enough energy.

FREQUENCY RESOLVING DETECTOR

In some cases it is desirable to have a phonon spectrometer which is independent of the sample under study. Then, instead of the stress tuned impurities discussed above one should rather use a heterojunction detector (Dietsche 1978).

A heterojunction is made of two different superconductor films, e.g. Al and Pb. The density of states diagram for this case is presented in Fig.25 (left part) in the form derived above. The resulting tunneling characteristic at finite temperatures is also shown in Fig. 25 (right part). Basically, the current rise by single particle creation is now at $\Delta_1 + \Delta_2$, while the thermal quasiparticle current produces a peak at $\Delta_1 - \Delta_2$, where the singularities in the densities of states are just opposite to each other, so that electron tunneling from left to right is highly probable. At voltages smaller than this peak, most of the thermal quasiparticles can no longer tunnel because of the gap on the right, and only a small current remains due to the Boltzmann tail.

If now phonons of energy $\hbar\omega > 2\Delta_2$ are incident into the junction, they will create two quasiparticles at a time within an energy band ranging from Δ_2 to $\hbar\omega - \Delta_2$, due to energy conservation. This is indicated by dashed lines on the left in Fig.25. Evidently, none of these quasiparticles can tunnel, as long as their maximum energy, $\hbar\omega - \Delta_2$, falls below the gap edge on the right side, $\Delta_1 - eV$. So, they

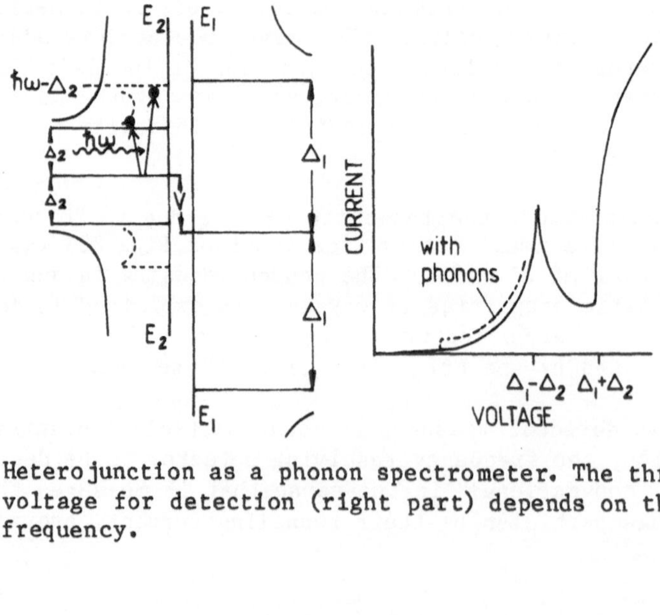

Fig. 25. Heterojunction as a phonon spectrometer. The threshold voltage for detection (right part) depends on the phonon frequency.

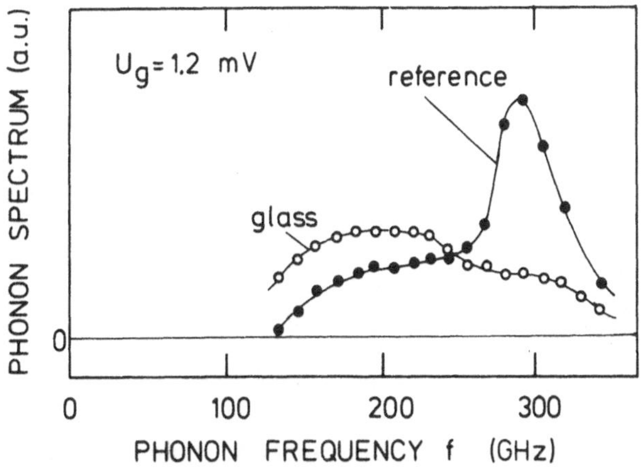

Fig. 26. Inelastic phonon scattering in glass films 1μm in thickness. Spectra were measured by heterojunction.

can lead to a detector signal only if eV is increased such that

$$eV > \Delta_1 + \Delta_2 - \hbar\omega.$$

At this threshold voltage, the detector signal exhibits a step-like rise due to the convolution of singularities (dashed line on right part of Fig. 25). By using a modulation technique to differentiate the signal with respect to the bias voltage, the step is transformed into a peak, so that the incident phonon spectrum is displayed directly. For quantitative measurements, the relaxation processes of the excited quasiparticles must be taken into account in some detail. A numerical deconvolution is then more appropriate than simple differentiation.

The upper fequency limit is given by the zero voltage condition, $\hbar\omega = \Delta_1 + \Delta_2$. Phonons of a higher frequency lead to a negative step at $eV = \hbar\omega - \Delta_1 - \Delta_2$ due to electron tunneling from right to left, as it is evident from the lower half of the energy-of-states diagram.

The heterojunction detector was used to study phonon scattering in SiO_2 glass films (Dietsche and Kinder 1979). These were deposited on Si substrate crystals. Generators were placed on top, and the spectrum of transmitted phonons was monitored on the far Si face by the heterojunction. Fig. 26 shows the spectra for two different thicknesses, 1μm and for reference, 0.1μm. Obviously, the spectrum has radically changed after the transmission through 1μm of thickness. This is interesting for the understanding of the thermal conductivity vs. temperature curves of glasses.

SURFACE PHYSICS WITH PHONONS

Phonon reflection at surfaces and transmission across interfaces to liquid helium has been studied in the context of the Kapitza resistance, for many years (Wyatt 1981). But only recently it turned out that the phonon reflection is extremely sensitive to small amounts of adatoms on the surface.

This was shown in an experiment where a Si surface was cleaned in situ at low temperature by laser annealing, i.e. melting and epitaxial regrowth of a thin surface layer (Basso et al. 1984). After cleaning, a few tenths of a monolayer of gold atoms were deposited onto the fresh surface. After each step, the phonon reflection was measured, once with the surface under vacuum, and once with the surface in contact with helium. The result is presented in Fig. 27: (a) shows the usual "anomalous" Kapitza conductance, i.e. the difference in reflected phonon power between vacuum and helium, of the surface as received; (b) shows the annealed surface where the Kapitza conductance is greatly reduced; (c) shows the fresh surface with 0.4 monolayers of Au, which are obviously enough to induce newly a remarkable conductance; (d) shows the results of the surface with one monolayer.

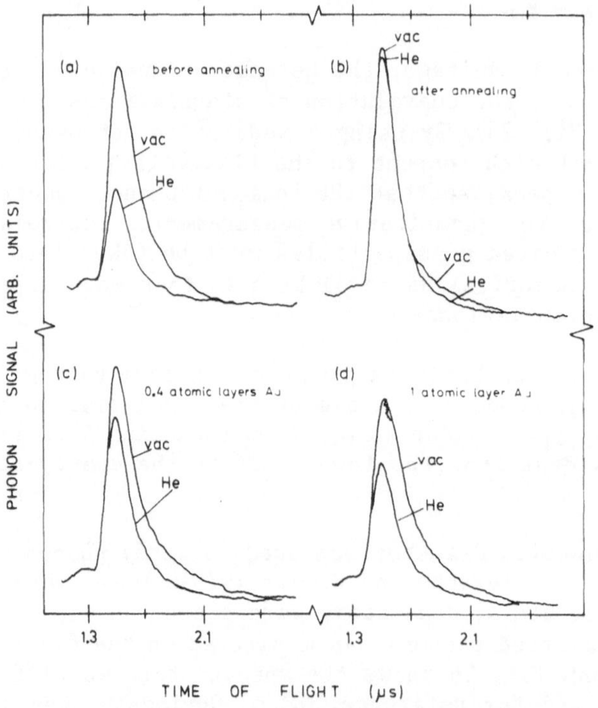

Fig. 27. Reflection of phonons from a surface adjacent to vacuum or to helium. The "anomalous transmission" depends strongly on the state of the surface.

With this high sensitivity to adatoms on previously cleaned surfaces, phonon spectroscopy seems to be well suited as a tool in surface physics at comparatively low energies.

After this presentation of the well established field of monochromatic phonon generation by single particle tunneling, we turn to the more recent method of using pair-tunneling of the same kind of junctions.

JOSEPHSON PHONON GENERATION

Josephson Effect

So far we have discussed tunneling of single electrons only. It was first shown by Josephson that pairs can also tunnel with a comparable probability, between the left and right condensates. Therefore, the junction behaves in many respects like an ordinary superconductor with however reduced electronic density.

In zero magnetic field, the junction sustains a supercurrent at V=0 up to a critical current, as indicated above, in the tunneling characteristic of Fig. 3. In a finite, but small, parallel magnetic field, the junction develops Meissner currents which exhibit a very long penetration depth, of the order of tenths of a millimeter. In somewhat larger fields, flux pentration occurs as in type II superconductors above H_{c1}, and a "mixed state" develops with a linear array of fluxoids parallel to the junction. The fluxoids manifest themselves by vortices of tunneling supercurrents which form the periodic pattern shown in Fig. 28.

Because the currents of the vortex pattern largely compensate each other, the net critical current of the junction is substantially reduced by the magnetic field. For currents larger than critical, the vortices start moving, like in a type II superconductor (without pinning centers) under the action of the Lorentz force. While in type II superconductors the speed of the vortices is limited by the "viscosity" of the vortex core, in junctions it is only limited by the speed of electromagnetic waves, $\bar{c}$, when photons can be excited.

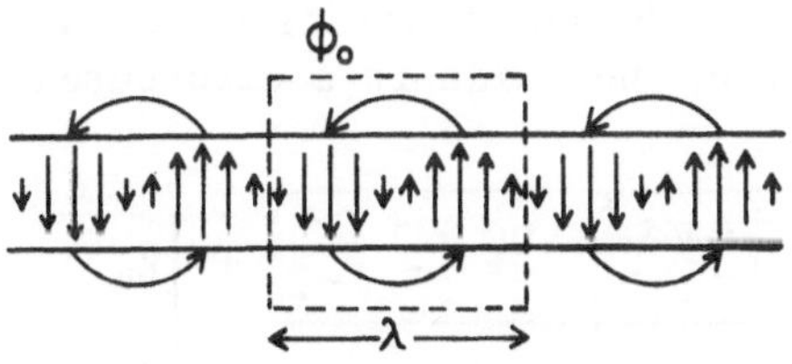

Fig. 28. Pattern of supercurrents in Josephson junction with parallel magnetic field.

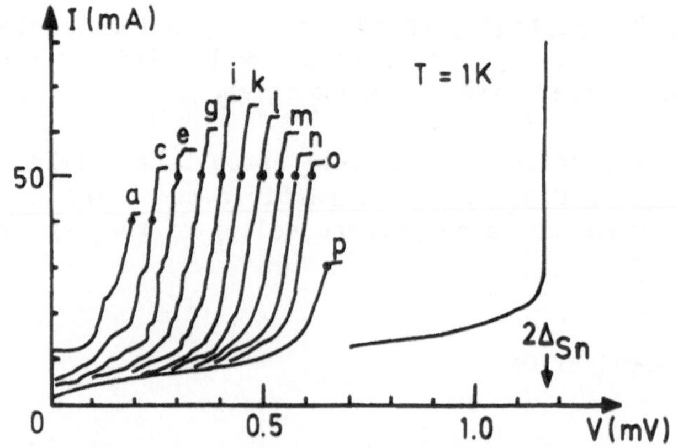

Fig. 29. I-V-characteristic of a Josephson junction at voltages below 2Δ. The voltage drop (and so the Josephson frequency) is tunable by the magnetic field.

The frequency of the photons is $\omega \simeq \bar{c}\, k$, where k is the wave vector of the vortex pattern. The latter is proportional to the magnetic field which can thus be used to tune the frequency. Furthermore, the frequency can be directly read off the voltage drop at the junction, by virtue of the Josephson relation $\omega = 2\ eV/h$. Fig.30 shows a set of I-V-characteristics obtained for a Sn junction in parallel magnetic fields up to 5×10^{-4}T. With increasing magnetic field, the voltage drop moves to the right for a given current. The small steps on the characteristics are due to the finite size of the junction.

Dissipation processes

The junction cannot radiate the photons out to free space, because it is equivalent to a very short antenna. Rather, the photon energy is dissipated entirely inside the junction. Two major channels of dissipation can exist. One is the excitation of quasiparticles, by photon assisted single particle tunneling, photon assisted Josephson tunneling, and direct pair breaking. The excited quasiparticles then relax and recombine by phonon emission of the "bremsstrahlung" and 2Δ-type.

The other channel of photon dissipation is dielectric loss in the oxide of the junction. The oxide is a disordered material with

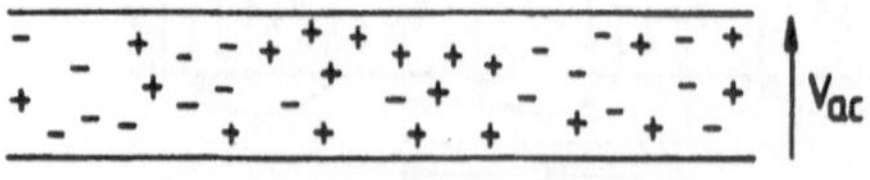

Fig. 30. Random positions of ions lead to local clustering of both, positive and negative charges in the oxide barrier.

ionic bonds. Due to the disorder, the charges of positive and negative ions cannot compensate each other locally. Rather, there will be random deviations from charge neutrality, as indicated schematically in Fig.30. These static charge fluctuations feel the ac voltage of the photons, and hence are forced to oscillate with the same frequency. Phonons of this frequency are thus emitted into random directions due to the changing polarity of the charges. Dielectric loss in the oxide of the junction should therefore result in the generation of monochromatic phonons of Josephson frequency.

Experimental Spectra

The phonons emitted by Josephson junctions were experimentally studied by using the Si:B spectrometer described above in the section on applications. Fig. 31 shows the spectra obtained for the same Sn junction whose characteristics were already shown in Fig.30 with the respective operation points indicated. There is a striking peak in each spectrum whose position follows the voltage drop at the

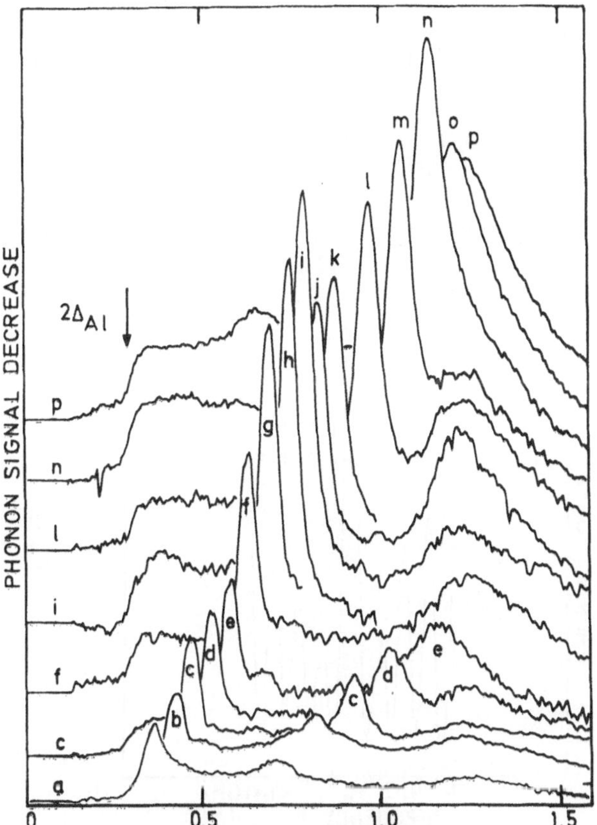

Fig. 31. Phonon spectra emitted by a Josephson junction at the bias voltages indicated in Fig.30.

junction, i.e. the Josephson frequency. This is obviously due to the dielectric loss of the oxide.

The height of the peaks varies from trace to trace. The slow overall increase is due to the spectrometer weight function, i.e. due to the increasing scattering strength of the boron acceptors. The more abrupt changes, most pronounced between traces (i) and (j), or between (n) and (o), are due to changes in the competing dissipation channel of quasiparticle excitation. The first case is the onset of one-photon-assisted single particle tunneling at $3eV = 2\Delta$, and the second one is the onset of direct pair breaking by the photons at $2eV = 2\Delta$.

For the operation points with low bias voltage, there is also a peak of the second harmonic of the Josephson frequency which is generated by photon assisted pair tunneling of the Josephson current by the ac voltage.

Similar spectra were also observed with Pb-oxide-Pb junctions and with Al-oxide-Pb junctions. For the latter junctions, the angular distribution of the emitted phonons was measured and found to be broad, consistent with the origin of the phonons from random charge clusters.

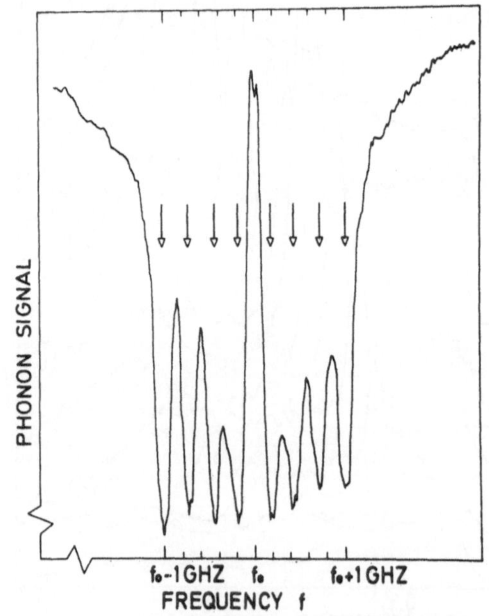

Fig. 32. Hyperfine spectrum of V^{3+} in sapphire, resolved by Josephson phonons.

Josephson Phonon Spectroscopy of $Al_2O_3:V^{3+}$

Because the phonon frequency is tied to the junction voltage by the Josephson relation, one can reach, in principle, a very high frequency resolution by simply stabilizing the dc voltage. However, the junction has always a low electric impedance to yield maximum phonon power, and must therefore always operate in constant current mode. The voltage thus depends on the magnetic field as discussed above. The stability of the magnetic field, and to some extent that of the current, is therefore responsible for the frequency resolution. In the following experiments, the corresponding instrumental resolution was $2x10^{-4}$.

To test the frequency resolution experimentally, and for a first application of Josephson junctions to phonon spectroscopy, we studied once more the system $Al_2O_3:V^{3+}$ which was already discussed in the context of the "bremsstrahlung" method. There, only the effective spin states could be resolved while their 8-fold hyperfine splitting was not seen. The resonant scattering of the Josephson phonons was now studied by sweeping the frequency with the magnetic field (Schick et al. 1984). The result is shown in Fig.32. Eight lines are clearly resolved. In addition, there is a sharp central energy gap due to a small rhombic component of the crystal field near the V^{3+} ions. From the sharpness of these features, a frequency resolution of $2x10^{-4}$ was deduced which is still that of the apparatus used. So, even further progress in frequency resolution should be possible.

CONCLUDING REMARKS

The phonon methods with superconducting tunneling junctions were seen versatile and useful for many applications. Their limitations are mainly set by superconductivity, namely in frequency (2 /h), in tempertature (T_c) and in magnetic field (H_c). Inconvenient features are furthermore the low power (10 W), the fabrication which is never fully reproducible, and their susceptibility to thermal cycling above 77 K. On the other hand, their advantages are mainly the independence from particular host-impurity combinations (Eu^{2+} is not needed), the monochromacy, tunability and the fact that the frequency is exactly known from the bias voltage. The low power can also be an advantage, because of the resultant small perturbation of the system under study.

If someone plans to get started with this technique, I recommend to begin with making aluminium junctions first, because they are easiest, and to use them as detectors, because they are more sensitive than bolometers. After having thus learned how to make junctions, one should naturally also use them as generators. Here, the "bremsstrahlung" technique is easier than the Josephson technique, but the latter gives the highest frequency resolution.

REFERENCES

Basso, H. C., Dietsche, W., and Kinder, H., 1984, Interaction of adsorbed atoms with phonon pulses, in: "LT 17 (Contributed Papers)", U. Eckern, A. Schmid, W. Weber, and H. Wühl, eds., North Holland, Amsterdam, p. 465, and to be publ.

Bates, C. A., and Wardlaw, R. S., 1980, A study of Mn^{3+} ions in Al_2O_3 from uniaxial stress data, J. Phys. C, 13:3609.

Berberich, P., Buemann, R., and Kinder, H., 1982, Monochromatic phonon generation by the Josephson effect, Phys. Rev. Lett., 49:1500.
More detailed accounts: Berberich, P., and Kinder H., 1984, The Josephson junction, a new tunable phonon source with high frequency resolution, in: Phonon Scattering in Condensed Matter, W. Eisenmenger, K. Lassmann, and S. Döttinger, eds., Springer, Berlin, p. 18
Kinder, H., Berberich, P., and Schick, A., 1984, Generation of monochromatic phonons by Josephson junctions, in: "LT 17 (Invited Papers)", U. Eckern, A. Schmid, W. Weber, and H. Wühl, eds., North Holland, Amsterdam, in print.

Berberich, P., and Kinder, H., 1981, The phonon spectrum emitted by superconducting Sn tunnel junctions, J. Physique, C6, 42:374.

Bridges, F., and Zoller, W., 1979, High frequency phonon studies of $KI:OH^-$ under uniaxial stress, Sol. State Comm., 30:717.

Dietsche, W., 1978, Superconducting Al-PbBi tunnel junction as a phonon spectrometer, Phys. Rev. Lett., 40:786.

Dietsche, W., and Kinder, H., 1979, Spectroscopy of phonon scattering in glass, Phys. Rev. Lett., 43:1413.

Dietsche, W., Kinder, H., Mattes, J., and Wühl, H., 1980, Breakdown of shear stiffness in amorphous Ga, Phys. Rev. Lett., 45:1332.

Dietsche, W., Kirch, S. J., and Wolfe, J. P., 1982, Phonon spectroscopy of the electron-hole liquid in germanium, Phys. Rev. B, 26:780.

Dynes, R. C., and Narayanamurti V., 1974, Evidence for upward or "anomalous" dispersion in the excitation spectrum of He II, Phys. Rev. Lett., 33:1195.

Eisenmenger, W., 1976, Superconducting tunneling junctions as phonon generators and detectors, in: "Physical Acoustics", W. P. Mason and R. N. Thurston, eds., Academic, New York, Vol. XII, p. 80.

Eisenmenger, W., 1981, Nonequilibrium phonons, in: "Nonequilibrium Superconductivity, Phonons, and Kapitza Boundaries", K. E. Gray, ed., Plenum, New York, p. 73.

Eisenmenger, W., and Dayem, A. H., 1967, Quantum generation and detection of incoherent phonons in superconductors, Phys. Rev. Lett., 18:125.

Forkel, W., Welte, M., and Eisenmenger, W., 1973, Evidence for 870GHz phonon emission from superconducting Al tunnel diodes through resonant scattering by oxygen in silicon, Phys. Rev. Lett., 31:215.

Jones, D. P., Jäckle, J., and Phillips, W.A., 1980, Dispersion and the thermal conductivity of SiO_2, in: "Phonon Scattering in Condensed Matter", H. J. Maris, ed., Plenum, New York, p. 49.

Kinder, H., 1972, Spectroscopy with phonons on $Al_2O_3:V^{3+}$ using the phonon bremsstrahlung of a superconducting tunnel junction, Phys. Rev. Lett., 28:1564.

Kinder, H., 1973, Spin-phonon coupling of $Al_2O_3:V^{3+}$ by quantitative spectroscopy with phonons, Z. Phys., 262:295.

Kinder, H., and Dietsche, W., 1976, Phonon spectroscopy in Al_2O_3 doped with transition metal impurities, in: "Phonon Scattering in Solids", L. J. Challis, V. W. Rampton, and A. F. G. Wyatt, eds., Plenum, New York, p. 199.

Mattes, J., Berberich, P., and Kinder H., 1978, Resonant scattering of monochromatic phonons by magnons in MnF_2 and in YIG, J. Physique C 6, 39:988.

Rothenfusser, M., Dietsche, W., and Kinder, H., 1983, Linear dispersion of transverse high-frequency phonons in vitreous silica, Phys. Rev. B, 27:5196, and unpublished results.

Schick, A., Berberich, P., Dietsche, W., and Kinder, H., 1984, Zero field hyperfine splitting of $Al_2O_3:V^{3+}$ by Josephson phonon spectroscopy, in: Phonon Scattering in Condensed Matter, W. Eisenmenger, K. Lassmann, and S. Döttinger, eds., Springer, Berlin,, p. 40.

Schreyer, H., Dietsche, W., and Kinder, H., 1984, Laser induced nonequilibrium superconductivity - a spatially resolving phonon detector, in: "LT 17 (Contributed Papers)", U. Eckern, A. Schmid, W. Weber, and H. Wühl, eds., North Holland, Amsterdam, p. 665.

Schwarte, M., and Berberich, P., 1985, Piezo phonon spectroscopy of the ground state of acceptors in Si and Ge, J. Phys. C, in print.

Sussmann, J. A., 1964, Phonon induced tunneling of ions in solids, Phys. Kondens. Materie, 2:146.

Windheim, R., and Kinder, H., 1975, Phonon spectroscopy of OH^- tunneling levels in NaCl, Phys. Lett., 51A:475.

Windheim, R., and Kinder, H., 1976, Phonon spectroscopy of OH^- and Li^+ tunneling states in alkali halides, in: "Phonon Scattering in Solids", L. J. Challis, V. W. Rampton, and A. F. G. Wyatt, eds., Plenum, New York, p. 220, and unpublished results.

Wyatt, A. F. G., 1981, Kapitza conductance of solid-liquid He interfaces, in: "Nonequilibrium Superconductivity, Phonons, and Kapitza Boundaries", K. E. Gray, ed., Plenum, New York, p. 31.

Zoller, W., and Bridges, F., 1981, Phonon spectroscopy of lithium-doped KBr, Phys. Rev. B, 24:4796.

Zoller, W., Dietsche, W., Kinder, H., de Goer, A.-M., and Salce, B., 1980, Phonon scattering in $Al_2O_3:Mn^{3+}$, J. Phys. C, 13:3591.

PHONON IMAGING: THEORY AND APPLICATIONS

G. A. Northrop and J. P. Wolfe

Physics Department and Materials Research Laboratory
University of Illinois at Urbana-Champaign
1110 W. Green Street, Urbana, IL 61801

INTRODUCTION -- HEAT PULSES IN CRYSTALS

The subject of this chapter is the propagation of thermal energy through crystalline solids at low temperatures. A remarkable observation is that thermal energy emanating from a point source of heat is strongly channelled into various directions in the crystal. This effect is simply caused by the elastic anisotropy of the crystal and is known as phonon focusing. Over the past several years phonon imaging techniques have been developed which graphically demonstrate phonon focusing and use it to study the scattering of high frequency phonons in crystals. We will review here the experimental methods and the physics learned from such experiments. We begin with a general discussion of ballistic heat propagation in real crystals.

In non-metallic solids, the principal source of thermal conduction is acoustic phonons -- i.e., lattice waves with the velocity of sound but frequencies in the GHz (10^9 Hz) to THz (10^{12} Hz) range. The maximum frequency of an acoustic phonon in a crystal is the frequency at which the phonon wavelength equals twice the lattice constant, which occurs typically at a few THz. At cryogenic temperatures, the predominant equilibrium phonons have a frequency somewhat below this cutoff frequency; for example, at T = 10 K, the peak in the Planck distribution of phonons is at a frequency ν = 600 GHz. (The peak in the number of phonons per unit frequency is given by $h\nu = 2.8\ k_BT$.) For many years, the propagation and scattering of these thermal phonons has been studied by measuring the thermal conductivity of the solid (Berman, 1976, Parrott and Stukes, 1975) Since the conductivity of an insulating solid is limited by phonon scattering from defects, impurities, or sample boundaries, one can gain valuable information about these scattering processes from a temperature dependence of the thermal conductivity.

At room temperature, the propagation of thermal phonons is highly diffusive, with short mean free paths resulting from phonon scattering. As the temperature is lowered the phonon mean free path in a pure crystal increases dramatically. Measurements of the thermal conductivity κ for a crystal of pure silicon are shown in Fig. 1a. (McCurdy et al., 1970) For a Planck distribution the predominant phonon frequency varies linearly with T, and by measuring $\kappa(T)$ one obtains roughly a phonon transmission spectrum of the crystal. Of course, at a given temperature, the distribution of phonon frequencies is rather broad, as shown by the Planck distributions plotted in Fig. 1b. In its simplest form the thermal conductivity is just $\kappa = (1/3)Cv\ell$, where C is the heat capacity, v is the sound velocity, and ℓ is the mean free path of the phonons. At low temperatures, the heat capacity increases as T^3, so that the results in Fig. 1a imply that the phonon mean free path is independent of temperature below about 8 K. This is the signature of boundary-limited scattering: phonons travel ballistically across the crystal -- i.e., without scattering. If defects are introduced into the crystal, the mean scattering length of the phonons will be shortened. A significant amount of theoretical modelling, however, is required to characterize the scattering processes because thermal conductivity gives us only an average of the scattering lengths over phonon frequency, mode, and direction of propagation.

In 1964, von Gutfeld and Nethercot demonstrated a heat-pulse method which began a revolution in the characterization of heat propagation and scattering in solids (von Gutfeld and Nethercot, 1964, von Gutfeld, 1968). The basic idea was to produce a short burst of non-equilibrium phonons at one surface of the crystal and observe their arrival with a detector at another surface of the crystal. With this method one could choose a particular propagation direction and select the phonon mode by its time-of-

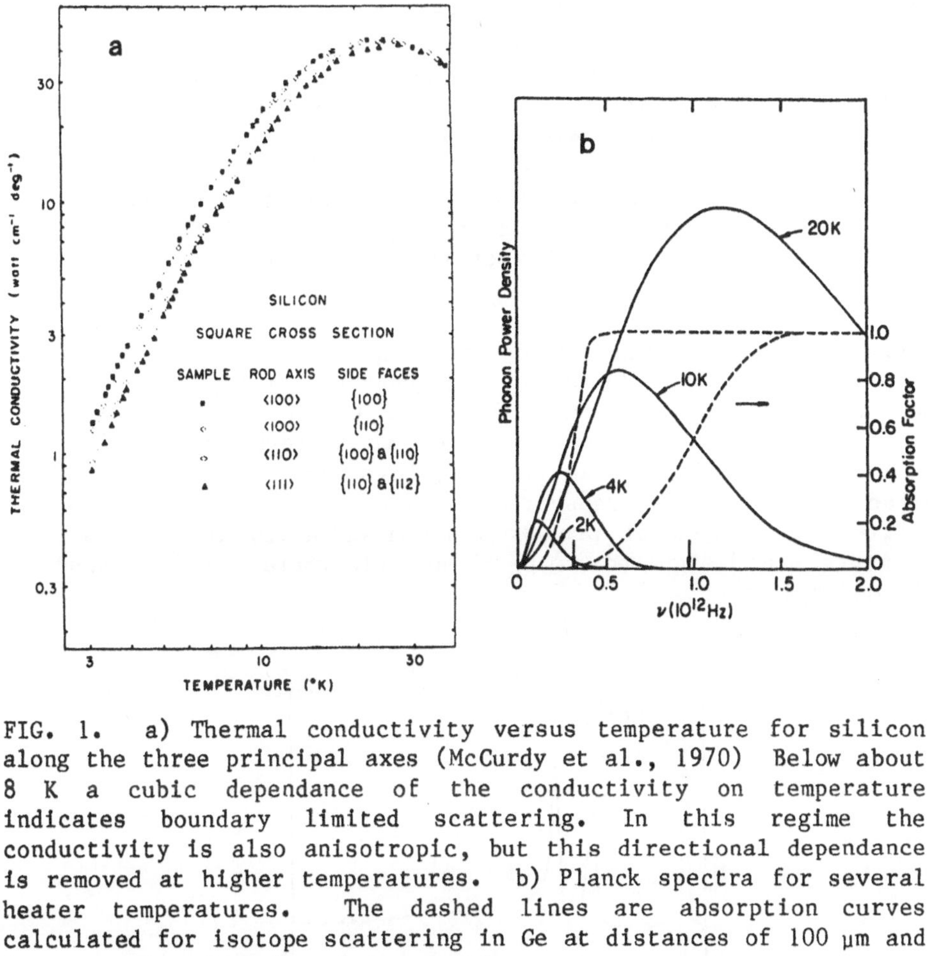

FIG. 1. a) Thermal conductivity versus temperature for silicon along the three principal axes (McCurdy et al., 1970) Below about 8 K a cubic dependance of the conductivity on temperature indicates boundary limited scattering. In this regime the conductivity is also anisotropic, but this directional dependance is removed at higher temperatures. b) Planck spectra for several heater temperatures. The dashed lines are absorption curves calculated for isotope scattering in Ge at distances of 100 μm and 1 cm, as discussed in a later section.

flight across the crystal. A recent version of this experiment for quartz is shown in Fig. 2. In this case, a heat pulse is generated on one face of the crystal by optically exciting a metal film deposited on that surface. A small region of the metal film is heated above the equilibrium temperature of the crystal by excitation with a focused laser beam. On the opposite face of the crystal is evaporated a small strip of aluminum which acts as a bolometer, or heat detector. The temperature of the crystal is adjusted to the point where the Al film is about halfway through its superconducting transition. At this critical temperature, slight alterations in the temperature of the metal film cause a significant change in its resistivity, which can be detected by passing a small current through the film and measuring the voltage drop. Due to its small heat capacity, the metal film is an extremely fast (~10 ns) detector of thermal phonons (Weis, 1969; 1972).

Figure 2 shows the heat-pulse signals arriving at the detector. The arrival times of the phonons correspond to the sound velocities for this crystal, i.e., the heat transport is ballistic. The multiple pulses are due to different propagation modes of the phonons. For an anisotropic solid, one expects one longitudinal and two transverse modes, each with different velocities for a general propagation direction in the crystal. These modes are designated as L, ST and FT, where S and F stand for "slow" and "fast".

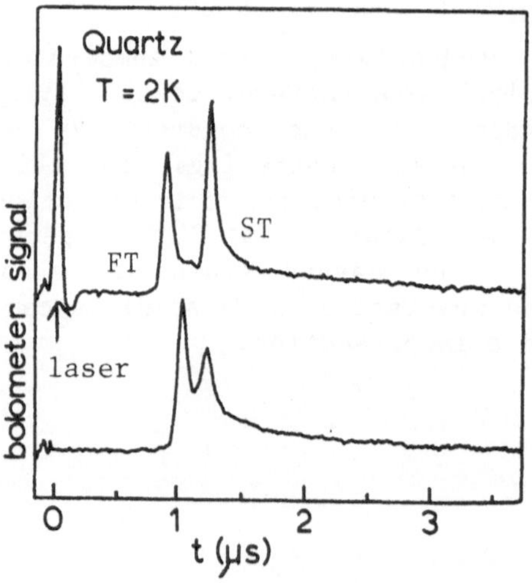

FIG. 2. Typical heat-pulse spectra for two different propagation directions in quartz. The substantial variation in pulse intensities is due to phonon focusing, discussed later. The L mode is too weak to be seen here.

The mean frequency of the phonons in these heat pulses is somewhat above that of the equilibrium phonons in the crystal. Typically, one locally heats the metal film to a few degrees above the crystal temperature, producing a Planck distribution of phonons at that elevated temperature. One can hope to vary the mean frequency of the transmitted phonons by simply varying the excitation level; however, in practice this is not a very effective means of performing "phonon spectroscopy" because the frequency distribution is so broad. However, several forms of phonon spectroscopy which use superconducting tunnel junctions as generators and detectors have been developed. (See Eisenmenger, 1976; Bron, 1980). These techniques have proved useful in the study of phonon scattering from impurities and conduction electrons for frequencies from ~150 GHz to over 1 THz.

The frequencies of the nonequilibrium phonons arriving at the detector cannot be directly determined from the time-of-flights of the heat pulses. In the nondispersive (long-wavelength) limit, the propagation velocity does not depend on frequency. Indeed, the frequencies of the detected heat-pulse phonons are not necessarily distributed as a Planck distribution at the heater temperature. This is because, in any real crystal, phonons scatter, and their mean free paths are strongly dependent on frequency. The scattered phonons do not contribute to the heat-pulse signal at the ballistic time-of-flight. In most materials, there are several atomic isotopes, which are randomly distributed in the crystal. This deviation from periodicity produces a mass-defect type of scattering known as isotope scattering. (Klemens, 1958; Carruthers, 1961; Tamura, 1983a) Theoretically, the scattering rate τ^{-1} of a phonon traveling through an isotopically impure crystal increases as the fourth power of its frequency. For a Ge crystal, we have $\tau^{-1} = A\nu^4$, where $A = 4.5 \times 10^{-41}$ sec^3 from theory (Klemens, 1958) and ν is the phonon frequency. In Fig. 1b, the dashed curves are the attenuation factor for ballistic phonons,

$$\alpha(\nu) = 1-\exp(-A\nu^4 \ell/v), \quad (1)$$

for two path lengths, ℓ = 1 cm and 100 μm, with v the phonon velocity. We see that phonons with frequency greater than about 1 THz scatter within only 100 μm of the source, whereas phonons with frequency less than 250 GHz are transmitted over a distance of 1 cm. In general, phonons emitted from the heater source as a 10 K Planck spectrum do not result in a 10 K Planck spectrum at the detector, since down-conversion of the phonon frequencies will modify the frequency distribution of the detected phonon flux. The process of frequency down-conversion will be discussed shortly.

Phonons scattered in the bulk of the crystal manifest themselves as a delayed signal at the detector, i.e., a heat flux continuing after the ballistic time-of-flight. The amount of this "diffuse" signal depends on both the initial frequency distribution and the scattering processes in the bulk of the crystal. Figure 3 shows a series of time-of-flight spectra (Greenstein et al., 1982) for a photoexcited Ge crystal as the excitation power is increased, thus raising the temperature of the source. All of the time-traces show a ballistic component arriving at the transverse sound velocity for this particular direction in the crystal. (The L signal is not observable due to phonon focusing, as discussed below.) As the excitation power is raised, a tail appears after the ballistic signal, indicating the arrival of scattered phonons. In this particular case, further experiments have shown that this observed delay in propagation time occurs quite near the excitation point. In effect, a hot spot is produced around the excitation point from which low frequency phonons are emitted and propagate ballistically to the detector.

The existence of a heated region inside the crystal is expected from a consideration of Fig. 1b. As the excitation point is heated above 20 K a majority of the phonons created have mean free paths less than 100 μm. These are rather high frequency phonons which can down-convert into lower frequency phonons with longer mean free paths. In addition to these high frequency phonons which are localized near the excitation point, there are phonons with frequencies between 0.3 and 1.0 THz which may propagate diffusely to the detector without down-converting. Clearly the evolution of thermal energy from a hot source is a very complex process involving both elastic and inelastic scattering of phonons, and an understanding of the details of

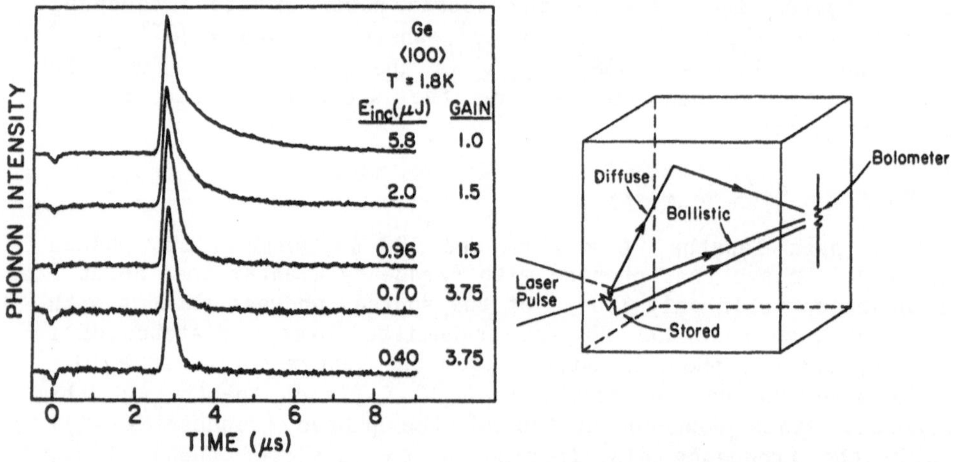

FIG. 3. Heat pulses in Ge for different excitation levels. The tails that occur for the higher powers are due in part to isotope scattering of the higher phonon frequencies present in the hotter source. Schematic drawing: ballistic and scattered phonons.

these processes remain today as one of the important challenges to experimentalists and theorists.

Crystals with impurities and defects commonly show heat pulses with long diffusive tails. If the defect concentration is large, even quite low frequency phonons cannot propagate ballistically across the sample, and it may not be possible to observe a ballistic component at all. Figure 4 shows two time traces of heat pulses in a LiF crystal. One trace corresponds to the as-grown crystal, and the second trace corresonds to a plastically deformed crystal, containing a large number of dislocations. Scattering from these defects is manifested in the delayed signal following the ballistic onset. Except in reasonably pure crystals, long diffusive tails such as these are the rule rather than the exception.

Another interesting case of unusually broad heat pulses has been recently discovered. If one directly photoexcites one surface of a pure GaAs crystal, very broad heat pulses are observed at the opposing surface. It has been hypothesized that the late-arriving phonons in this case are not simply scattered phonons, but are ballistically propagating high-frequency dispersive phonons. (Ulbrich et al., 1980; 1981; Stock et al., 1984) As phonon wavevectors approach the Brillouin zone boundary, the group velocity of the phonon decreases, reaching zero for wavevectors at the zone boundary. The dispersion relation for transverse acoustic phonons in GaAs is shown in Fig. 5a. A large wavevector TA phonon of frequency 1.5 THz, for example, has a group velocity, $d\omega/dk$, about one-half that of the low frequency sound velocity. Thus it

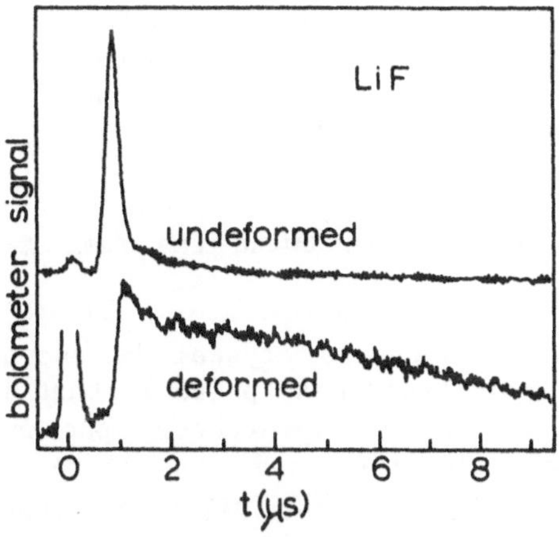

FIG. 4. Phonon time spectra for two different LiF samples. The upper trace is for an unaltered crystal, and the lower trace is for a 10% plastically deformed crystal, showing a long "diffusive" tail due to scattering from dislocations.

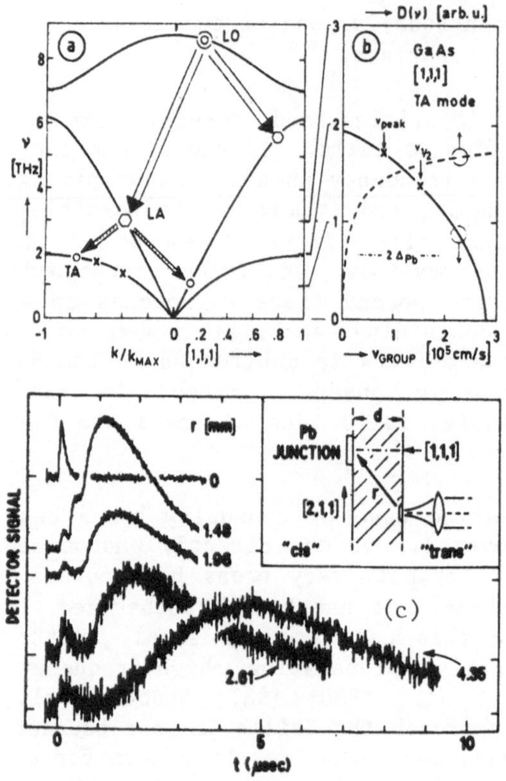

FIG. 5. a) Phonon dispersion curves for GaAs along the [111] direction. (Ulbrich et al., 1980; 1981) The arrows indicate typical three-phonon anharmonic decay channels for optical and longitudinal acoustic phonons. b) The group velocity (solid line) and density of states (dashed line) plotted against frequency for the lowest lying acoustic branch. c) Photoexcited heat-pulse spectra in GaAs for several propagation distances. The use of a Pb tunnel junction as a detector ensures that the detected phonons have a frequency greater than 0.7 THz.

would arrive at the detector in twice the time taken by a low frequency phonon.

Experimental data for a GaAs crystal is shown in Fig. 5c. The interpretation of this broad heat pulse is that it is composed of a distribution of large wavevector phonons, having a distribution of group velocities and, therefore, arrival times (Ulbrich et al., 1980, 1981). The principal support for this hypothesis is that the time of the peak signal and the temporal width of the broad heat pulse scale linearly with source to detector distance, as expected for ballistic propagation. Purely diffusive propagation should yield the arrival time scaling as the square root of the distance. However, this interpretation has been the subject of some controversy. It has been theorized that a "quasi-diffusive" transport mode involving both down conversion and elastic scattering of phonons can produce a linear scaling between distance and time (Guseinov and Levinson, 1983). Phonon imaging experiments by Wolfe and Northrop (1984) support this latter theory for path lengths exceeding about 300 μm.

So far, we have discussed only the temporal features of a heat pulse. Let us now consider the angular distribution of phonon flux emitted from a small heater. From the previous discussion, it is not obvious that the thermal energy flux should be strongly dependent upon the direction of propagation. Certainly, the velocity of the phonons depends on the mode and direction of propagation, due to the elastic anisotropy of the medium. As we will see, the intensity of the heat pulse is even more strongly dependent on propagation direction. This effect -- called "phonon focusing" -- was first discussed by Taylor, Maris and Elbaum in 1969. Looking at data similar to that in Fig. 2, they noted that the intensities of the various phonon pulses (longitudinal and transverse modes) were not the same. The relative intensity of longitudinal vs. transverse phonons depended strongly upon the relative orientation of the source and detector. They could crudely explain this phenomenon not as a differential scattering of the various phonon modes but more generally as a consequence of incoherent wave propagation through an elastically anisotropic medium. (Taylor et al., 1969; 1971; Maris, 1971)

In 1979, we introduced a "phonon imaging" method which produced a global view of the heat-flux anisotropy in a crystal (Northrop and Wolfe, 1979; 1980). An extensive description of the experimental aspects of phonon imaging is given in Appendix A. Intially, such phonon imaging experiments were a demonstration of the elastic properties of well known crystals and provided some useful insights into the details of phonon focusing. (Wolfe, 1980) A knowledge of the anisotropy in heat flux is useful information for any experiments involving high frequency phonons at low temperatures. More recently, the phonon imaging method has been applied to the study of dispersive phonon propagation (Dietsche et al., 1981) and the scattering of high frequency phonons in crystals. Intensity patterns of the thermal energy flux emanating from a point source in a crystal allow the experimentalist to isolate and study the propagation of phonons with a given mode, polarization and propagation direction. Although still in an early stage of development, the phonon imaging method has contributed insights into the problems of phonon scattering from dislocations (Northrop et al., 1982; 1983) and from crystal boundaries (Koos et al., 1983; Northrop and Wolfe, 1984).

This chapter is divided into two main headings. First, we

will consider the ballistic propagation of phonons in anisotropic media. We will examine phonon focusing in several cubic crystals and also non-cubic piezoelectric crystals. Also, the effect of frequency dispersion will be analyzed. The second major topic involves the scattering of these non-equilibrium phonons from defects and boundaries. This will include the scattering of phonons from dislocations in LiF and the reflection and transmission of phonons at a crystal interface. Finally, we will consider the scattering of high frequency phonons from carriers in semiconductors.

BALLISTIC HEAT FLUX AND PHONON IMAGING

Heat flow at ordinary temperatures is a diffusive process. Thermal conduction is described as a second rank tensor, and for cubic crystals, the three eigenvalues of this tensor are equal, implying that diffusive energy flow is isotropic. This is true even though the underlying elastic tensor of the crystal is a fourth rank tensor -- one which relates two second-rank tensors, stress σ_{ij} and strain e_{ij}. In effect, the frequent scattering of phonons washes out the finer details of the elastic anisotropy. At low temperatures, the situation changes. The occupation number of a given phonon mode is much smaller and the anharmonic interaction causing scattering between phonons is greatly reduced. At liquid He temperatures, for example, the mean free path of the equilibrium phonons in the crystal can easily be as large as the crystal itself. Thus, the first key idea that we will deal with is ballistic propagation. Such propagation is clearly anisotropic, as evidenced by the direction-dependent sound velocities of phonons in all crystals. Figure 6 shows schematically the problem we are dealing with. The restoring forces for an elastic wave depends upon both the direction of propagation and the mode of the wave, i.e., whether it is a compression (longitudinal) or shear (transverse) wave. The restoring force for a compression wave is in general larger than that for a shear wave and thus the longitudinal mode propagates with larger velocity than the transverse mode. We will present a mathematical description of this propagation in a moment.

The second essential feature of a heat pulse is that it involves an incoherent distribution of phonons. The difference between coherent and incoherent phonon propagation is illustrated in Fig. 7, which is adapted from a classic paper on phonon focusing by Rösch and Weis (1976). A coherent radiator, such as an rf-driven transducer, produces elastic waves with a unique wavevector. Figure 7a shows the positions of phonons in a quartz crystal 1 μs after emission of a 50 ns long pulse, assuming ballistic propagation. The emission from an incoherent radiator

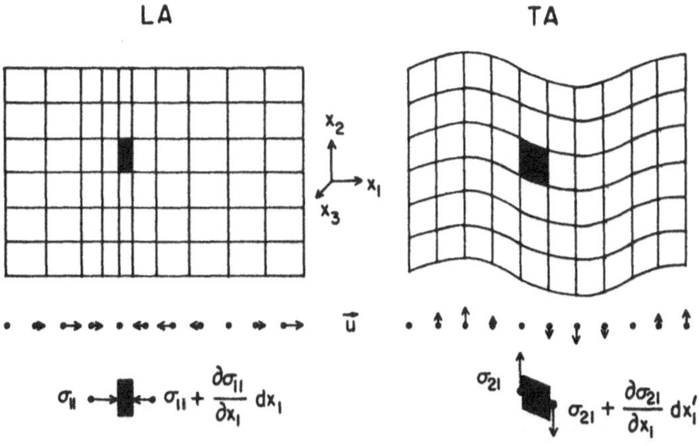

FIG. 6. Schematic representation of longitudinal and transverse waves in a solid medium. An infinitesimal volume element is displaced a distance **u** from its equilibrium position. For the longitudinal wave the net force on a volume element $dx_1dx_2dx_3$ is $(\partial\sigma_{11}/\partial x_1)dx_1dx_2dx_3$, where σ_{11} is the compressive stress. Because the mass of this element is $\rho dx_1dx_2dx_3$, the equation of motion is $\rho\ddot{u}_1 = \partial\sigma_{11}/\partial X_1$. (Wolfe, 1980)

of the same size is quite different, as shown in Fig. 7b. Such a radiator might consist of an ohmically heated metal film which has been deposited on the surface of the quartz crystal. A remarkably complex wave front for each of the different phonon modes is shown in Fig. 7b. (We will see below how such complex wave fronts come about.) This complexity stems from the fact that the incoherent radiator produces wavevectors radiating in all directions in the elastically anisotropic crystal.

Concentrate for a moment on the dashed line connecting the radiator and a detector on the opposite side of the crystal. The wave packets along this direction are similar to those predicted in the case of the coherent radiator, with one important exception: an additional pulse is predicted for one of the transverse modes (FT). This additional wave packet is indeed observed as the "oblique" pulse in the heat-pulse experiment shown in Fig. 7c. The additional packet of thermal energy can only be due to wavevectors with directions other than the line-of-sight direction. In other words, the energy flux for these phonons is not along their wavevector direction! This is one of the essential features of wave propagation in an anisotropic medium. In general, the phonon wavevector $\vec{k}$ and group velocity vector $\vec{V}$ (defining energy flux) are not colinear.

One of the consequences of non-colinear $\vec{k}$ and $\vec{V}$ vectors is the concentration of elastic energy flux along certain directions

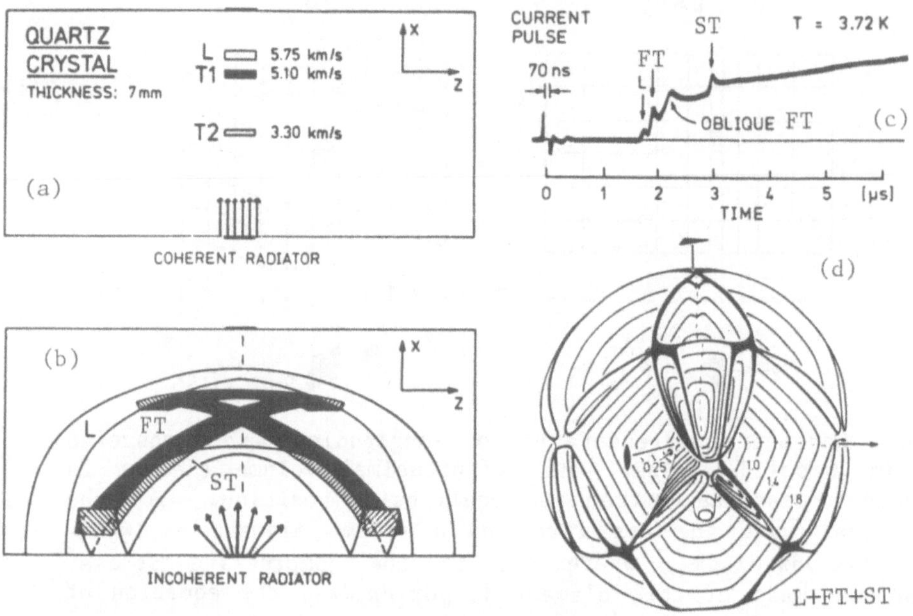

FIG. 7. a) Location of coherent phonon pulse energy emitted into a quartz crystal after 1 μs of propagation. b) Location of incoherent phonon energy after 1 μs. c) Phonon time spectra for X-cut quartz with thermal heat-pulse generation and detection with a tin bolometer. Sharp ballistic pulses and a broad diffusive signal are seen. d) Three dimensional plot of phonon focusing for all three modes of quartz. The radial extent of this surface represents intensity as a function of direction. The lines are contours of constant enhancement, and the surface is truncated at 2.0. (Rosch and Weis, 1976)

of the crystal, which was termed "phonon focusing" by Taylor, Maris and Elbaum. This usage of the term "focusing" does not imply a geometric bending of rays, such as in the case of optics. It is meant to describe a concentration or channeling of thermal energy along certain crystalline directions. Maris (1971) developed a statistical technique for predicting the relative energy flux of phonons along particular crystalline directions. Rösch and Weis (1976) took one step further and calculated the global intensity patterns for several crystals. Figure 7d shows the results of their calculations for quartz. The radius of the "intensity surface" in a given direction is proportional to the thermal flux along that direction. As one can see from this calculation, phonon focusing is hardly a small effect. Indeed, along certain crystalline directions the predicted flux is mathematically infinite, as we will show below.

Theory of Phonon Focusing

To understand phonon focusing, we begin with a description of elastic wave propagation in an anisotropic medium (Musgrave, 1970). To begin, we assume that the elastic waves have wavelengths long compared to the atomic spacings in the lattice; this is known as the long-wavelength or continuum limit. To describe these waves mathematically, one must relate the two second-rank tensors, stress and strain, by the generalization of Hooke's law:

$$\sigma_{ij} = C_{ij\ell m}\, e_{\ell m} \tag{2}$$

where $C_{ij\ell m}$ is the fourth-rank elastic tensor of the crystal, and each subscript represents a Cartesian axis. Summation over repeated indices are assumed in this and the following equations. For a cubic lattice there are only three independent components of the elastic tensor. The definition of the stress tensor (with dimensions force/area) is illustrated in Fig. 6 for compressional (σ_{11}) and shear (σ_{21}) stresses. The strain tensor (which is dimensionless) represents a fractional distortion of the lattice and is defined in terms of the displacement $\vec{u}$ of a volume element from equilibrium, $e_{21} = \frac{1}{2}(\partial u_2/\partial x_1 + \partial u_1/\partial x_2)$. As indicated in the caption of Fig. 6, the equation of motion for a volume element in the crystal is

$$\rho \ddot{u}_i = \frac{\partial \sigma_{ij}}{\partial x_j} = C_{ij\ell m} \frac{\partial^2 u_\ell}{\partial x_m\, \partial x_j} \tag{3}$$

where ρ is the density of the medium. Equation 3 has the wave solution $\vec{u} = \vec{\varepsilon} \exp(i(\vec{k}\cdot\vec{x} - \omega t))$, where $\vec{\varepsilon}$ is the polarization vector and ω is the angular frequency of the wave. Substitution of this solution into Eq. 3 yields a set of linear equations:

$$(C_{ij\ell m} k_j k_m - \rho\omega^2 \delta_{i\ell})\, \varepsilon_\ell = 0. \tag{4}$$

Defining the wave normal $\vec{n} = \vec{k}/|\vec{k}|$ and the phase velocity $v=\omega/|k|$, this equation becomes

$$(D_{i\ell} - v^2 \delta_{i\ell})\, \varepsilon_\ell = 0, \tag{5}$$

with $D_{i\ell} = (1/\rho)\, C_{ij\ell m} n_j n_m$ the Christoffel tensor. (Christoffel, 1877) Since $D_{i\ell}$ depends only on the direction of $\vec{k}$ and not its magnitude, the phase velocity v for a given direction $\vec{n}$ is independent of $|\vec{k}|$. For a given wavevector direction, $\vec{k} = (k, \theta_k, \phi_k)$, the frequency of the wave is

$$\omega = v(\theta_k, \phi_k)\, k. \tag{6}$$

The phase velocity $v(\theta_k, \phi_k)$ can be obtained by solving the characteristic equation of Eq. 5, i.e., setting the determinant of the second-rank tensor within the parentheses equal to zero. This yields a cubic equation in v^2, with the three roots corresponding to one longitudinal and two transverse phonon modes. For each root, v_α, there is a polarization vector $\vec{\varepsilon}_\alpha$, where $\alpha = 0,1,2$. For a given $\vec{k}$ vector, the three polarization vectors $\vec{\varepsilon}_\alpha$ will be mutually orthogonal; however, in general none will be either parallel or normal to $\vec{k}$. The $\alpha = 0$ mode is identified with a predominantly longitudinal mode (L) and $\alpha = 1,2$ are the slow (ST or T2) and fast (FT or T1) transverse modes, respectively.

The energy flux of the elastic wave is parallel to the group velocity. By vector differentiation of Eq. 6, one obtains

$$\vec{V} = \frac{\partial \omega}{\partial \vec{k}} = (v - \vec{n} \cdot \frac{\partial v}{\partial \vec{n}})\vec{n} + \frac{\partial v}{\partial \vec{n}}. \tag{7}$$

The derivative $\partial v / \partial \vec{n}$ is obtained most simply by implicit differentiation of the characteristic equation of Eq. 5, as described in the articles by Northrop and Wolfe (1980) and Every (1980).

A very useful way of displaying these mathematical results is a plot of the constant-frequency surface in wavevector space. This surface is also referred to as the slowness surface, because its radius from the origin is inversely proportional to the phase velocity, i.e., $|\vec{k}| = \omega / v(\theta_k, \phi_k)$. The slowness surface for a given crystal has three sheets, corresponding to the longitudinal and two transverse modes. For brevity, we will refer to individual sheets as the slowness surface of a given mode. In an elastically isotropic medium, there would be two surfaces: an L surface and a (2-fold-degenerate) T surface. All real crystals have non-spherical slowness surfaces. As an example, we show in Fig. 8a a cross-section of the ST surface for Ge for wavevectors in the (001) plane.

As shown in Fig. 8a, the group velocity vector $\vec{V}$ corresponding to a phonon with wavevector $\vec{k}$ is normal to the slowness surface. Immediately we see that an isotropic distribution of wavevectors does not give an isotropic distribution of group velocities; in this particular example, the energy flux seems to be concentrated near to the <100> directions in the crystal. It is very useful, at this point, to construct a second surface consisting of all of the possible group velocities for this mode. This is known as the group-velocity or wave surface, and the section of it corresponding to Fig. 8a is shown in Fig. 8b. Clearly, the wave surface is more complex than the slowness surface.

To get a quantitative idea of the energy flux in a given crystal direction, we assume that the heat pulse contains an isotropic distribution of phonon wavevectors. However, the

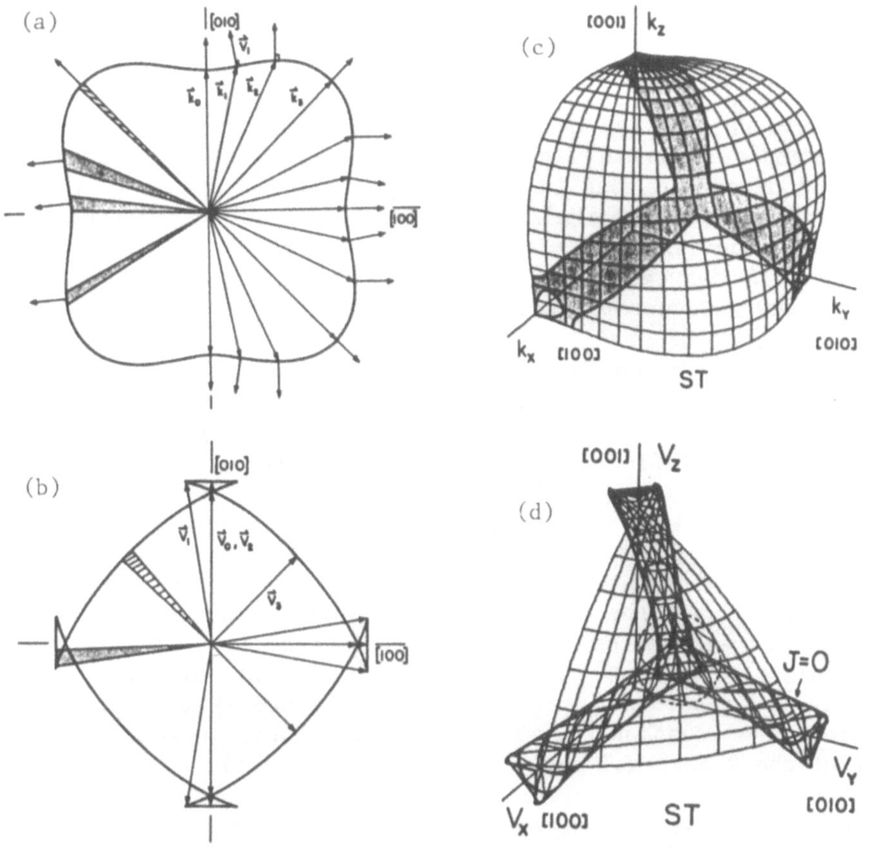

FIG. 8. a) Section of the slowness surface for the ST mode in Ge, plotted in the (100) plane of wavevector space. The propagation direction for each wavevector is along the surface normal. b) Plot of the group velocity vectors corresponding to the wavevectors in a). c) Three dimensional first octant plot for the ST slowness surface of Ge. The grid is formed from lines of constant θ_k and ϕ_k, and the heavy lines mark zero curvature of the surface. d) Group velocity surface formed by plotting the radial magnitude of the group velocity for the grid of lines in c).

detector subtends a certain solid angle in real space, or $\vec{V}$-space. Two equal solid angles in $\vec{V}$-space are shown as shaded regions in Fig. 8b. The corresponding solid angles in wavevector space are shown in Fig. 8a. Note that a given group velocity direction may correspond to more than one $\vec{k}$-vector; this is precisely the origin of the oblique mode shown in Fig. 7b and c. To determine the energy flux along a certain crystalline direction, one can define a phonon flux "enhancement factor" (Maris, 1971) as

$$A \equiv \left| \frac{d\Omega_k}{d\Omega_V} \right|, \tag{8}$$

where $d\Omega_k$ and $d\Omega_V$ are the respective $\vec{k}$- and $\vec{V}$-space solid angles. For the example shown in Fig. 8, we see that the lightly shaded propagation direction contains three pulses of different velocity and corresponding to a relatively large amount of wavevector space; thus phonons are rather strongly focused in this direction. In contrast, the darkly shaded solid angle contains relatively few wavevectors, or a defocusing of heat-pulse energy.

From Fig. 8, one can see that the enhancement factor is large when $\vec{k}$ points to a region on the slowness surface with small curvature. An extreme example of this is the wavevector $\vec{k}_1$ which terminates at an inflection point, or point of zero curvature. The corresponding propagation direction $\vec{V}_1$ defines a fold in the velocity surface. Along $\vec{V}_1$ there is a mathematically infinite (yet integrable) phonon flux. To carry this analysis further, it is necessary to examine the full 3-dimensional slowness surface, as shown in Fig. 8c for the ST mode of Ge. This surface was generated using the calculation of $k = \omega/v(\theta_k,\phi_k)$ described above and a graphics plotting routine on a computer. The local curvature of a surface in 3-dimensions is mathematically defined by the Gaussian curvature. The Gaussian curvature of an elemental surface is the product of the 2 extremal curvatures (inverse radii).

The solid lines drawn on the slowness surface in Fig. 8c define the locus of points with zero Gaussian curvature. The normals to the surface along these "singularity lines" are group velocity directions of mathematically-infinite phonon flux. The singularity lines on the slowness surface divide it into regions of different types of curvature. The shaded segment is a saddle region which lies between the large convex regions around the <110> directions and the small concave regions near the <100> directions. Figure 8d shows the group-velocity surface for this mode plotted with the same net of lines used in Fig. 8c. The folds in this wave surface define the directions of singular flux in real space.

An analytic expression for the enhancement factor A for an arbitrary k-space direction can be obtained as follows. The direction of energy flux (θ_V,ϕ_V) is obtained from the group velocity components, which in turn depend upon (θ_k,ϕ_k). This $\vec{k}$-space to $\vec{V}$-space transformation may be expressed as

$$\begin{aligned} \cos\theta_V &= f\,(\cos\theta_k,\phi_k) \\ \phi_V &= g\,(\cos\theta_k,\phi_k), \end{aligned} \tag{9}$$

where the functions f and g are determined from the components of $\vec{V}$. These equations are a mapping of one 2-dimensional space onto another. The ratio of the product of differentials in these two spaces is the Jacobian of the functions f and g:

$$d\Omega_V = d(\cos\theta_V)d\phi_V = Jd(\cos\theta_k)d\phi_k = Jd\Omega_k, \quad (10)$$

where

$$J = \frac{\partial f}{\partial \eta_k}\frac{\partial g}{\partial \phi_k} - \frac{\partial f}{\partial \phi_k}\frac{\partial g}{\partial \eta_k}, \quad (11)$$

with $\eta_k \equiv \cos\theta_k$. With this choice of coordinates we see that the Jacobian is simply related to the enhancement factor:

$$A = 1/|J|. \quad (12)$$

The Jacobian of the $\vec{k}$-space to $\vec{V}$-space transformation is related to the Gaussian curvature K of the slowness surface by the expression $J = k^2K/\cos\xi$, where ξ is the angle between $\vec{k}$ and $\vec{V}$. (Lax and Narayanamurti, 1980) To obtain the Jacobian for a given real-space direction (θ_V,ϕ_V), one must point-by-point invert the mapping of Eq. 9. As indicated in Fig. 8b, this inverse mapping is generally not unique; there are cases where one $\vec{V}$ direction corresponds to several different $\vec{k}$ directions. This inversion problem makes the direct calculation of flux anisotropy by the enhancement factor difficult.

A practical way of obtaining a global picture of the phonon focusing in a crystal is a Monte Carlo technique. Random k-vectors are chosen with a uniform angular distribution, their corresponding group velocity directions are calculated, and each is mapped onto the desired experimental plane. The technique for doing this in crystals of cubic and other symmetries is described in Appendix B. The result of one such Monte Carlo calculation for Ge is shown in Fig. 9a. The bright regions represent high thermal flux striking the surface of a crystal after propagating from a single point on the opposite surface of the crystal. Even for Ge, which is not an unusually anisotropic cubic crystal, the phonon focusing effect is striking. The borders of the bright regions represent lines of mathematically infinite flux, corresponding to the singularity lines plotted on the slowness surface and the folds in the wave surface as shown in Fig. 8. The square at the center and ramps extending horizontally and vertically correspond to the slow transverse mode in this crystal. The intense X structure corresponds to the fast transverse phonon mode.

A simple method of displaying these phonon focusing patterns is to plot the directions in real space where the enhancement factor is infinite, i.e., where J=0. Once the singularity lines are determined on the slowness surface by a simple root-finding routine, the associated real-space direction (θ_V,ϕ_V) can be found.

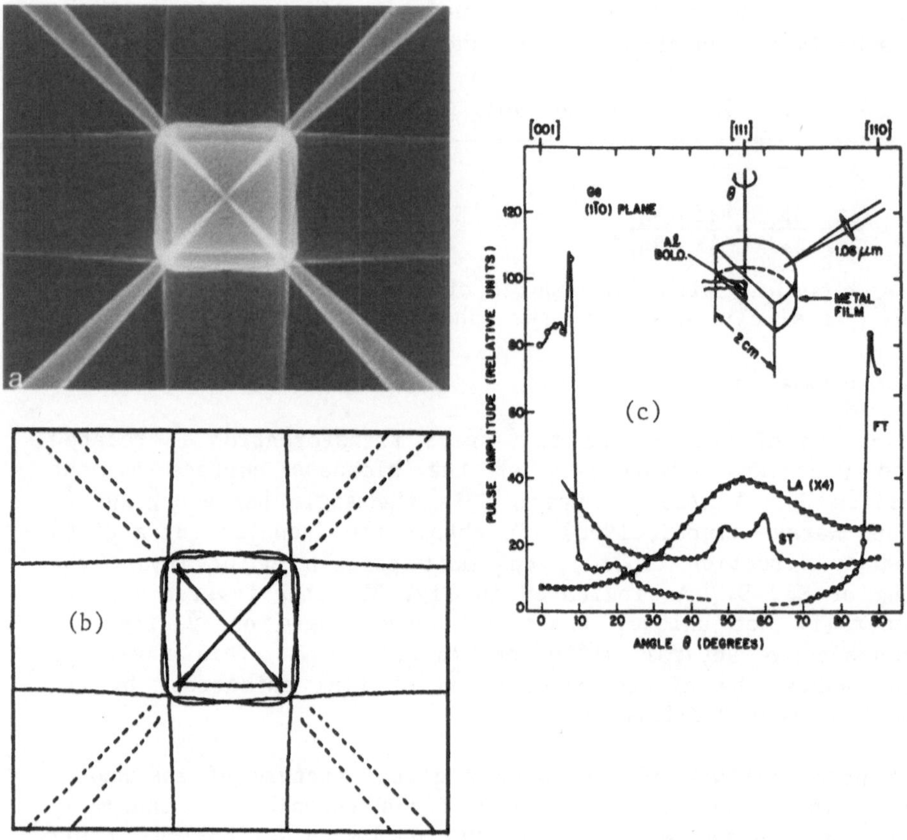

FIG. 9. a) Monte Carlo simulation of phonon focusing for the ST and FT modes of <100> oriented Ge. A horizontal line through the center represents a (110) plane with [100] at the center. The photo spans ~±25° left-to-right in propagation direction. b) Singularity line plot for Ge in the same orientation as a). The solid lines are the ST mode and the dashed lines are the FT mode. The continuation of the dashed lines to the center has been omitted for clarity. (Northrop and Wolfe, 1980). c) Intensity of longitudinal and transverse modes as a function of propagation direction in Ge. (Hensel and Dynes, 1979)

A map of these J=0 directions for the two transverse modes are shown in Fig. 9b. Historically, these complex singularity patterns were observed experimentally before they were predicted by the methods in this section. Let us now turn to a description of the phonon imaging method which makes this possible.

Phonon Imaging

How does one actually go about obtaining a spatial map of the phonon flux emanating from a point source? A first step was taken by Hensel and Dynes (1979). Most previous heat pulse experiments had used current-driven metal films as heat sources which were fixed on the crystal. Instead, Hensel and Dynes used a focused laser beam to heat a small spot on a metal film covering one surface of a crystal. The Ge crystal was cut in the shape of a half-cylinder with the heater film evaporated on the curved surface and an aluminum bolometer strip at the center of the flat surface of the half-cylinder. By rotating the crystal about the cylinder axis and keeping the laser beam fixed, they were able to continuously scan the propagation angle in one plane of the crystal. They obtained a cross section or profile, shown in Fig. 9c, of the theoretical phonon focusing pattern shown in Fig. 9a. Their results agreed well with a statistical calculation based on Maris' technique (Maris, 1971).

A major advance in experimental technique was made by us in that same year. (Northrop and Wolfe, 1979; 1980; Wolfe, 1980) Instead of rotating the sample, we raster-scanned the laser beam in two dimensions over a flat surface of the crystal. Also instead of using a long superconducting strip as a detector, we fabricated a serpentine strip with small dimensions, in order to obtain high angular resolution. An example of one of our first bolometers and its electrical characteristics at low temperature are shown in Fig. 10a. The temperature of the bolometer and crystal was controlled by the vapor pressure of the superfluid helium bath in which the crystal was immersed. A diagram of the crystal geometry is shown in Fig. 10b. The propagation angle of the phonons is varied by scanning the laser beam across the $(00\bar{1})$ surface of the crystal. The bolometer was located at the center of the opposite (001) face. In this first experiment the laser was a Q-switched Nd:YAG laser which produced ~100W pulses of ~300 ns duration at a wavelength of 1.06μm. The laser beam was focused to a 100 μm spot on the sample after reflection from two mirrors which could be rotated by galvanometers, as shown in Fig. 10c. A typical scanning rate of the laser beam was 10 Hz in the horizontal direction and .04 Hz in the vertical direction. The galvanometers were controlled by a microcomputer. The bolometer signals were amplified by an rf preamplifier and fed into a boxcar integrator, which selected the range of observation times after the laser pulse. More details are given in Appendix A.

Typical heat-pulse signals for a fixed laser beam position are shown in Fig. 10d. To produce a phonon image, a boxcar gate as shown in this figure was used to time-integrate the signal of both transverse phonon modes. The output of the boxcar integrator was recorded by the computer for each position of the scanning mirrors. In this way a 256x256 element image of the heat-pulse intensity was collected. This two-dimensional array was viewed by means of a video frame buffer which served as an interface between the microcomputer and a standard video monitor. An image of the heat pulse intensities in Ge is shown in Fig. 11, which is a view along the [001] direction at the center of the photo and spans ±25° in propagation direction from left to right. Upon first observation, these complex focusing structures were quite surprising, to us. This experimental data led us to the topological descriptions of the slowness surfaces as plotted in the previous section. Comparison of the ballistic phonon image in Fig. 11 to the Monte Carlo theory in Fig. 9a shows that theory and experiment are in good accord.

A wide-angle picture of the phonon focusing structures can be gained by tilting a 1 cm cube of Ge in the cryostat and increasing the xy scanning range. Heat pulse signals from three excited surfaces are recorded in the same raster scan, yielding the three-

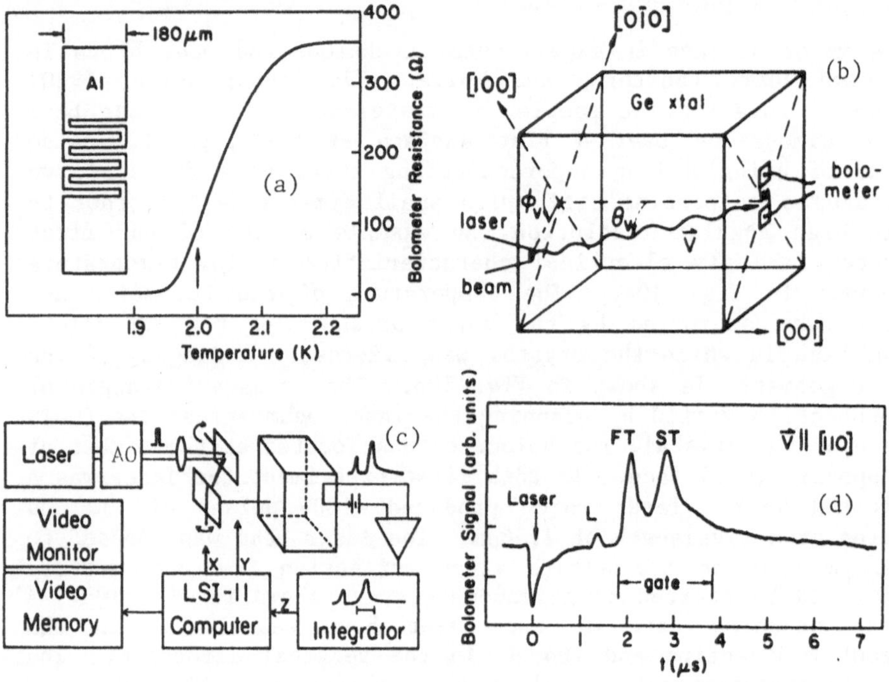

FIG. 10. a) Resistance versus temperature characteristics of the Al bolometer used in an early phonon imaging experiment. The arrow indicates an appropriate operating temperature. The inset shows the detector shape. The sensitivity is confined to the narrow serpentine portion, where the bias-current density is largest. b) Sample geometry used in the first phonon imaging experiment. The propagation direction is explicitly labeled by θ_V and ϕ_V. c) Block diagram of the phonon imaging apparatus. A more detailed description of the experimental methods and equipment may be found in Appendix A. d) Bolometer signal showing the arrival of various phonons for the [110] propagation direction. A typical boxcar gate for the imaging experiment is shown.

FIG. 11. Ballistic phonon image for the sample configuration shown in Fig.10b. In this and all other phonon images in this chapter the degree of brightness at any point represents the heat-pulse intensity sampled for the corresponding point on the scanned surface of the sample. The anisotropic structures in this image correspond well to those in the calculations shown in Fig. 9.

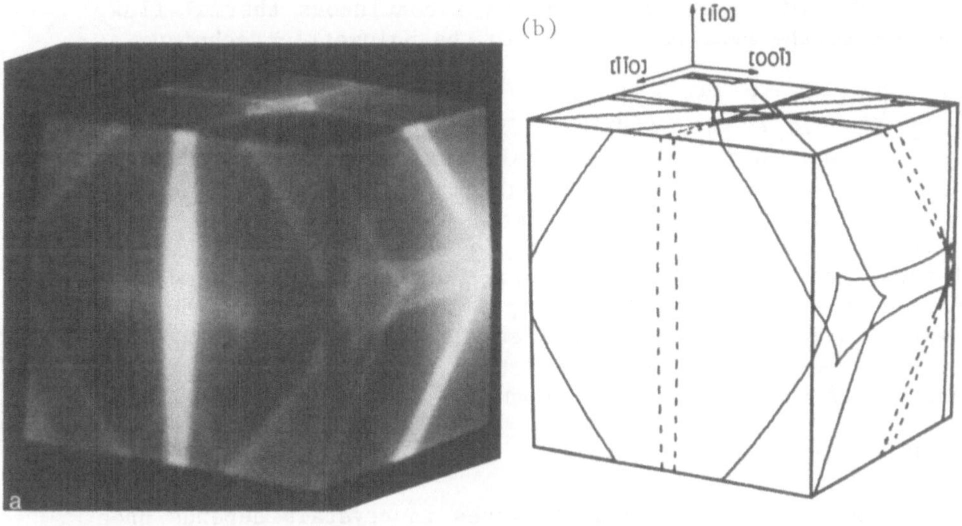

FIG. 12. a) Wide angle phonon image obtained by scanning the laser beam over all three exposed faces of an obliquely oriented sample. The bolometer is in the center of the back right (110) face. b) Theoretical singularities projected onto a cube with the same dimensions as the sample.

dimensional-looking image shown in Fig. 12a. Figure 12b shows the flux singularity lines projected from the bolometer position onto each of the excited surfaces of this crystal. In this particular experiment the bolometer was at the center of the right rear face, which is a (110) face of the crystal, in contrast to the bolometer placement in Fig. 10b. The solid lines shown in Fig. 12b correspond to the singular-flux directions of the slow transverse phonons; a 3-fold cusp structure near the [111] crystal direction is clearly seen. The dashed lines in this figure represent the singular directions for fast transverse phonons in Ge.

A quite different method for observing the anisotropic heat flow in crystals at low temperatures was devised independently by Eisenmenger (1979). His method consisted of partially submersing a Si crystal in a bath of superfluid helium and ohmically heating a small-area film on the center of the submersed side, as shown in the diagram of Fig. 13a. Phonons are channelled as described above, and they locally heat the superfluid film covering the top side of the crystal. Owing to the unusual fountain effect of superfluid helium, the film thickens at points where it is locally heated. This thickening of the superfluid film can be photographed using a light source and a camera adjusted at an oblique viewing angle. The results shown in Fig. 13b and c are consistent with the phonon flux originally calculated by Rösch and Weis (1976) and discussed in the previous section. It is interesting to note that this experiment utilizes a continuous thermal flux, in contrast to the heat pulses used in the bolometric technique.

A variation of the heat-pulse imaging technique was introduced by Eichele et al. (1982a). Instead of using a laser for heat production, they positioned their samples on a cold stage in an electron microscope and excited it with a high intensity focused electron beam. A tiny (2 μm x 2 μm) Al bolometer was used for detection. One possible advantage of this technique is the small focal spot of the electron beam; however, the required high electron energies and intensities significantly increased the size of the excitation region. Eichele et al., (1982b) and Metzger et al. (1984) have demonstrated this method in quartz, sapphire, and Ge, and one of their results is shown in Fig. 18b.

Phonon Focusing in Cubic Crystals

The group velocity of elastic waves in crystals depends upon the crystal density ρ and the elastic tensor $C_{ij\ell m}$. For cubic crystals, only three independent elastic constants are needed: C_{11}, C_{12}, C_{44}, where $1 \equiv xx$, $2 \equiv yy$, $4 \equiv yz$ with x, y and z referring to the (100) axes. The remaining non-equivalent elements of the elastic tensor are zero by symmetry. Actually, the anisotropic flux patterns observed in a phonon image are completely determined by the ratios of elastic constants, $a=C_{11}/C_{44}$ and $b = C_{12}/C_{44}$. Of course, the absolute magnitudes of C_{ij} and ρ are needed to determine the magnitudes of the velocities.

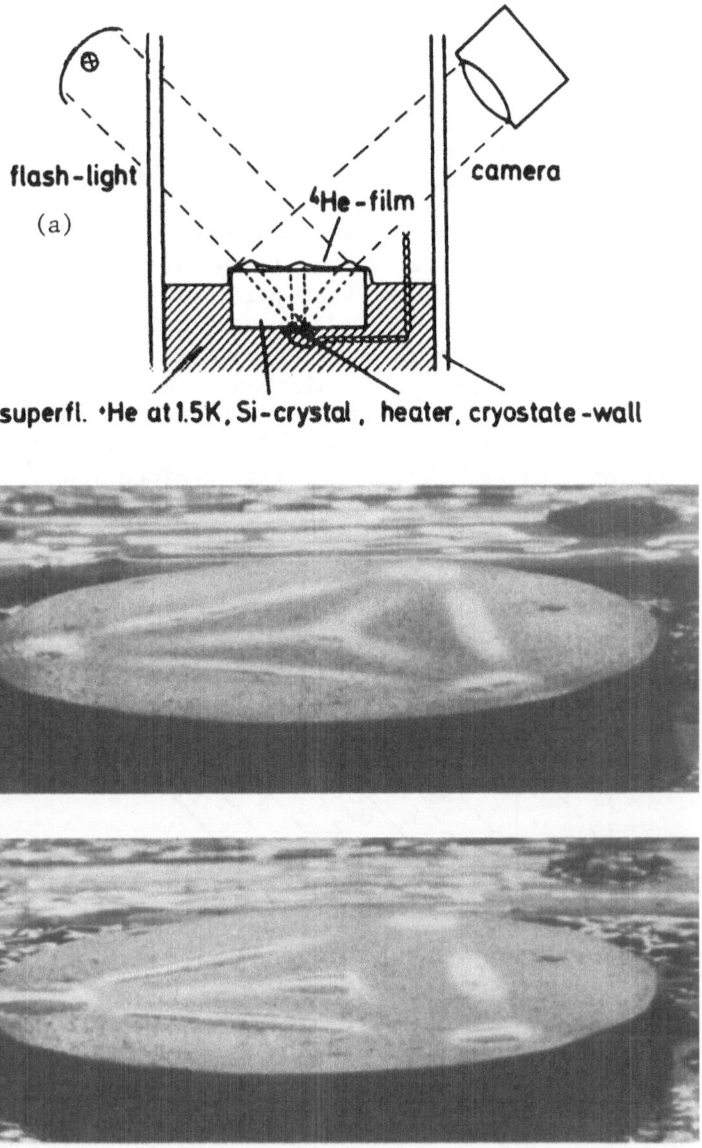

FIG. 13. a) Experimental arrangement for phonon imaging using the ^{4}He fountain presure as a phonon detector (Eisenmenger, 1979). b) Direct photograph of the effects of a continuous heat flow emitted from a heater on one face and striking the other [111] face of a Si crystal. The lighter areas have a thicker film and hence are receiving a larger phonon flux. c) The same as b), but with a lower helium level.

It may seem remarkable that only two independent parameters completely determine the complex phonon focusing patterns we have seen in the last section; the complexity of these patterns are due to the fact that we are dealing with a fourth rank tensor.

Thus, all cubic crystals can be mapped as points in a 2-dimensional "elastic parameter space", and a few examples of common materials are shown in Fig. 14. With regard to phonon focusing, Maris (1971) pointed out that this elastic-parameter space can be divided up into regions with qualitatively different focusing behavior. More recently, Every (1981) conducted a systematic study of the singular flux patterns of cubic crystals with an emphasis on the physical application of catastrophe theory. With the help of more advanced computing techniques -- Monte Carlo flux-intensity simulations and 3-dimensional representations of phonon wave surfaces -- Hurley and Wolfe (1984) have also characterized the systematic changes in phonon focusing structures as one moves throughout the 2-dimensional elastic parameter space. In this section we will show a sampling of their results.

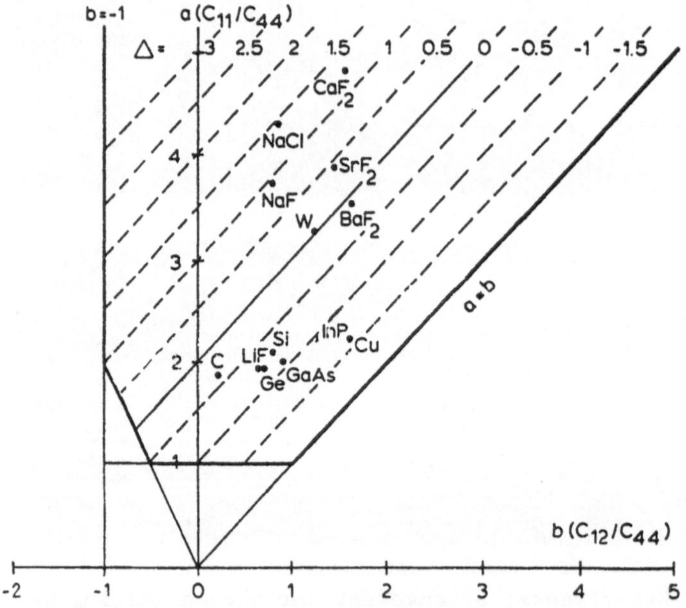

FIG. 14. Normalized elastic constant space for cubic symmetry, with points representing real materials. The phonon focusing is completely determined by the position of a material in this space. The solid diagonal line for which $\Delta = a - b - 2 = 0$ is the isotropic case (exactly spherical slowness surfaces). In general a variation of elastic constants parallel to the isotropy line results in little change in the phonon focusing, while displacement perpendicular to it produces dramatic changes.

As pointed out by Maris and by Every, all real crystals occupy a restricted region in elastic parameter space. The allowed region for real crystals is bounded by the thick lines in Fig. 14. For $a < b$ and $a < -2b$, thermodynamic constraints on the C_{ij} are not met. In addition, for $a < 1$ and $b < -1$, the transverse phase velocity exceeds the longitudinal phase velocity in certain crystalline directions, which is an extremely rare situation (Every, 1981).

There is a locus of points in elastic parameter space corresponding to perfect isotropy, that is, no angular dependence of phase or group velocity and therefore perfectly spherical slowness surfaces. The condition for elastic isotropy is $C_{11}-C_{22}-2C_{44} = 0$. This is shown as the line $\Delta = a-b-2 = 0$ in Fig. 14. This line serves as a natural boundary with which to categorize crystals into two distinct groups: Those for which $\Delta > 0$ ("positive-Δ regime") on the left side of $\Delta = 0$, and those with $\Delta < 0$ ("negative-Δ regime") on the right side. One sees that crystals such as Si, GaAs, and InP fall into the negative-Δ region; hence, they exhibit similar phonon focusing patterns to those shown for Ge in the previous section. In particular, the negative-Δ materials generally exhibit very complex structure near the <100> directions and less complex structures near the <111> directions of the crystal. On the other hand, it is found that crystals falling in the positive-Δ region exhibit complexities near the <111> directions. As an example of a positive-Δ material we will describe the phonon focusing features of CaF_2 from the work of Hurley and Wolfe (1984).

We consider only the transverse acoustic (ST and FT) modes because, in general, the longitudinal mode does not exhibit singular behavior. The slowness and wave surfaces of the two transverse modes of CaF_2 are shown in Fig. 15. As usual, it is useful to divide the slowness surface into regions of different Gaussian curvature. For the FT mode in Fig. 15a, the shaded regions have saddle curvature and the white regions, including the clover leaf structure along <111>, are convex. The locus of all group velocity vectors normal to this slowness surface (i.e., the wave surface) is plotted in Fig. 15b. This is a section of the group velocity surface about the <111> direction, and the labels on this surface indicate the corresponding k-vector directions on the slowness surface of Fig. 15a. The wave surface is very complex, containing multiple folds and cusps. This surface ends on a circle about the <111> axis. It is at this circle that the FT wave surface joins smoothly with the ST wave surface, shown in Fig. 15d. The slowness surface for the ST mode is displayed in Fig. 15c, and again labels are provided to show the correspondence of vectors in k and V spaces. This surface actually contains small concave regions, as defined by the tiny clover leaf structure near the <111> direction.

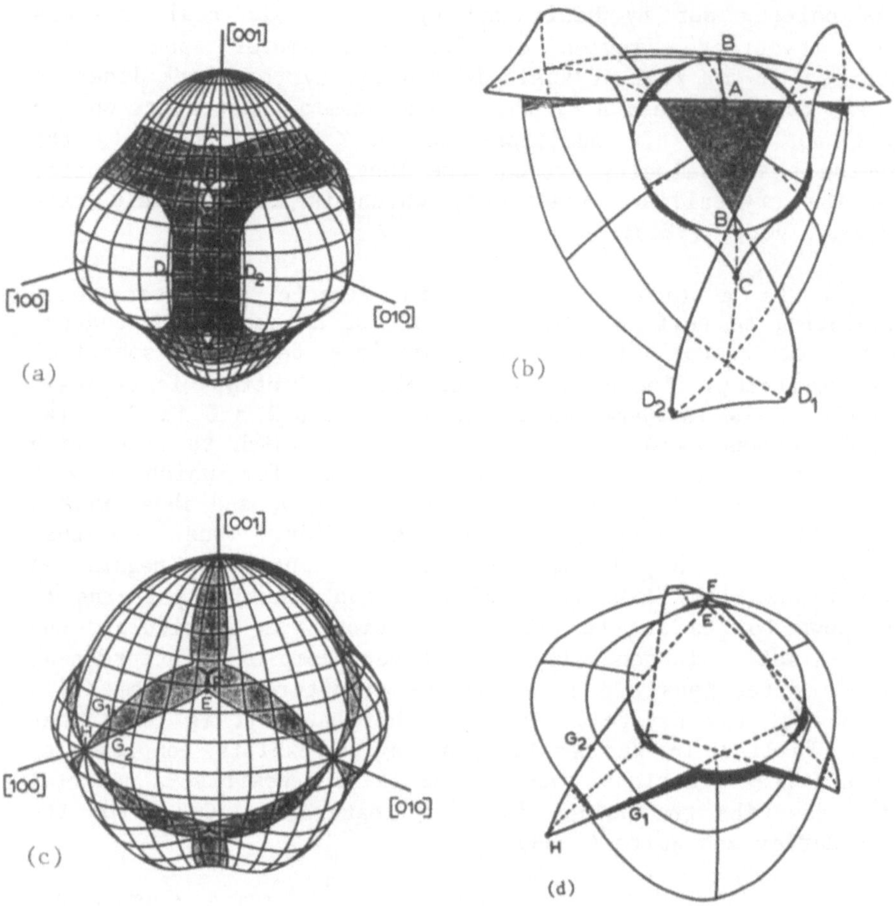

FIG. 15. Slowness surfaces for a) FT and c) ST modes of CaF_2. The grey areas mark negative, or saddle curvature. Corresponding wave surfaces for b) FT and d) ST modes.

As we have seen before, the folds in these wave surfaces produce singularities in phonon flux. Such flux singularities are best exhibited by a Monte Carlo calculation, as shown for Ge in Fig. 9a. Figure 16a is a Monte Carlo calculation of the phonon flux for the FT mode in [111] oriented CaF_2. The fold and cusp features predicted in the wave surface are quite evident here. A similar calculation for the ST mode is shown in Fig. 16b. To connect with reality, we reproduce in Fig. 16c an phonon image of CaF_2. A broad temporal gate was used to collect both FT and ST modes simultaneously. All of the features predicted in the Monte Carlo calculations are accurately represented in the experimental

FIG. 16 Monte Carlo images of a) FT and b) ST modes in CaF_2, with <111> at the center. c) Phonon image for CaF_2 (Hurley, et al., 1985).

data.

Every (1981) has given an interesting analysis of phonon focusing in cubic crystals in terms of elementary catastrophe theory. For a 2d-to-2d transformation such as the k-space to V-space transformation above, only fold and cusp catastrophes occur, as are graphically apparent in the phonon images. However, if one theoretically considers the elastic parameters as variables, a larger variety of elementary catastrophes can occur in this higher dimensional space. Thus, some of the structures seen in phonon focusing patterns can be thought of as cross sections of higher dimensional catastrophes. Catastrophe theory is hardly necessary to understand the physics of phonon focusing, but it provides a means for categorizing the complex topologies encountered in this problem. Further studies of catastrophe theory applied to phonon focusing have been reported by Armbrustor et al. (1984).

Hurley and Wolfe (1985) have examined the changes in the slowness and wave surfaces as the ratios a and b are continuously changed. They have studied the regions in elastic parameter space where certain topological features appear and have given the angular size of the major features. Thus, it is now possible to get a semiquantitative picture of the phonon focusing structures for all realistic values of a and b in the elastic parameter plane. In other words, from their study one can gain a reasonable idea of how the phonon flux is concentrated in any cubic crystal, without resorting to a detailed calculation. Of course, to obtain a quantitative prediction of relative phonon flux, one must resort to a Monte Carlo calculation for the particular crystal in question.

Phonon Focusing in Noncubic and Piezoelectric Crystals

The calculation of phonon flux in non-cubic crystals is somewhat more complex because of the lower symmetry, however computer programs have been written (Northrop, 1982a) to treat the general case. Phonon imaging experiments have so far been reported for quartz (SiO_2) and sapphire (Al_2O_3) (Eichele et al., 1982; Koos et al., 1984; Every et al., 1984) and lithium niobate ($LiNbO_3$) (Koos and Wolfe, 1984), which have trigonal symmetry. Quartz and $LiNbO_3$ belong to an interesting class of crystals having no center-of-inversion symmetry. Such crystals are piezoelectric and require further theoretical consideration in order to predict the focusing of heat flux.

In piezoelectric crystals strain induces an electric polarization and likewise an applied electric field creates a strain in the crystal. Thus, the study of phonons in such crystals requires a consideration of the propagation of both mechanical and electrical fields. It is well known that acoustic wave propagation in piezoelectric crystals can be described by a "stiffened" set of elastic constants (Cady, 1946). In other words, Hooke's law (Eq. 2) must be replaced by a relation which takes into account the stiffening of the lattice due to the piezoelectric effect:

$$\sigma_{ij} = C'_{ij\ell m}\, e_{\ell m}, \tag{13}$$

where $C'_{ij\ell m}$ are the piezoelectrically-stiffened elastic constants for the crystal. The complication is that the stiffened elastic constants depend on the wavevector of the elastic wave. The result can be simply expressed as

$$C'_{ij\ell m} = C^{E}_{ij\ell m} + F_{ij\ell m}(\vec{k}, \overleftrightarrow{e}_p, \overleftrightarrow{\varepsilon}_s), \tag{14}$$

where $C^{E}_{ij\ell m}$ are the elastic stiffness constants at constant electric field and $F_{ij\ell m}(\vec{k}, \overleftrightarrow{e}_p, \overleftrightarrow{\varepsilon}_s)$ is a tensor which depends upon the direction of the phonon wavevector, the piezoelectric coupling tensor $\overleftrightarrow{e}_p$, and the dielectric tensor at constant strain $\overleftrightarrow{\varepsilon}_s$.

By substituting these wave-vector-dependent elastic constants for $C_{ij\ell m}$ in Eq. 4, one can again solve for the constant frequency surfaces and parabolic lines of a given piezoelectric material. The calculation is much more tedious than before because it involves computing the first and second derivatives of the $C'_{ij\ell m}k_\ell k_m$ with respect to the wave-vector components. Such derivatives are required for determining the group-velocity vector and the Gaussian curvature. Nevertheless, a calculation of this sort was carried out by Koos and Wolfe (1984) for the crystals quartz and lithium niobate. They found that for quartz the effect of elastic stiffening on phonon focusing is very small, but for lithium niobate -- which is considerably more piezoelectric than quartz -- the effect of piezoelectric coupling is striking. We summarize their results for these two crystals below.

Quartz has trigonal symmetry (D_3), implying that there are six independent elastic constants. Using the constants determined at low temperature by an rf resonance method, the slowness surfaces for ST and FT modes can be generated as shown in Fig. 17. As usual, the thick lines indicate the locus of points on the slowness surface with zero Gaussian curvature. Figure 17 also shows the wave surfaces for these modes, and the folds in the surface indicate directions of singular flux. A calculation of the singularity pattern for quartz, including both ST and FT modes, is given in Fig. 18a. Various structures in this pattern can be identified with the folds in the wave surfaces of Fig. 17. An inclusion of the appropriate (non-zero) piezoelectric coupling constants has very little effect on the focusing pattern (Koos and Wolfe, 1984).

A phonon image of quartz was first reported by Eichele et al. (1982), who used a scanned electron-beam as a heat-pulse source. Their data is displayed in Fig. 18b as a pseudo-3d representation of the phonon flux. The general features of the experimental phonon image seemed to be explained quite well by the previous calculation. It should be noted however, that time-resolved phonon images obtained by Koos and Wolfe (1984) for this crystal display some anomalies which are not yet understood.

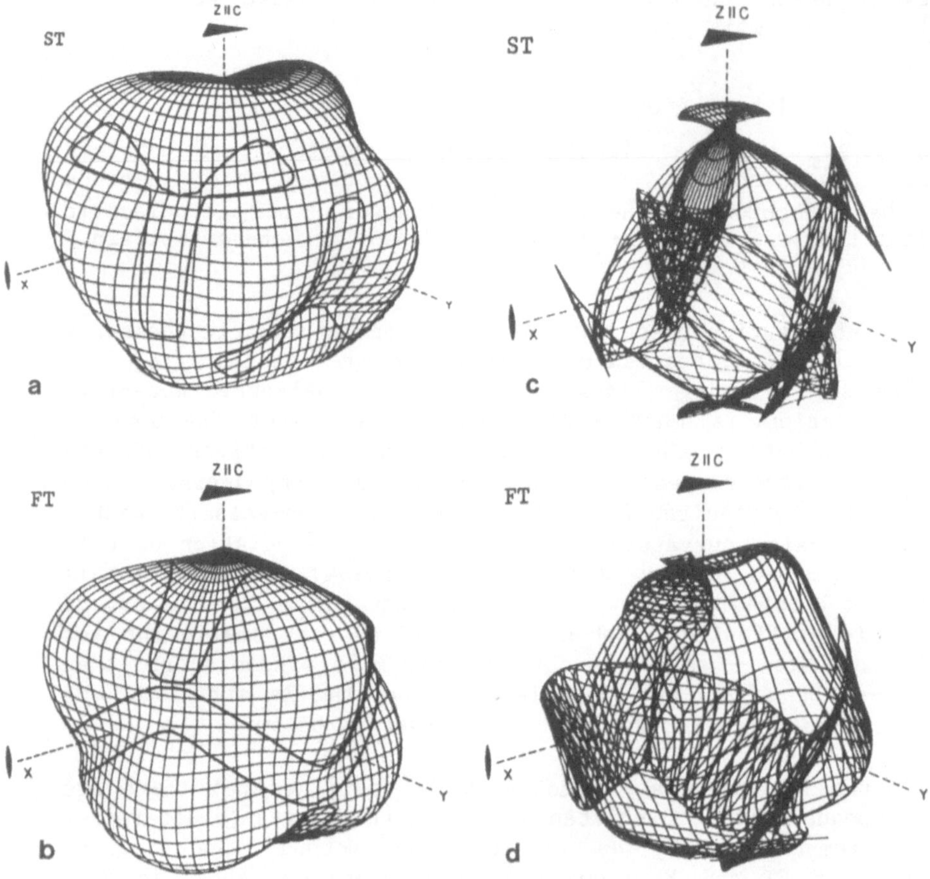

FIG. 17. Slowness surfaces for the a) ST and b) FT modes in quartz. Corresponding wave surfaces for the c) ST and d) FT modes.

More interesting is the case of lithium niobate ($LiNbO_3$), which has been found to be a very useful rf transducer crystal. In this crystal the piezoelectric stress tensor is much larger than for quartz and is found to have a significant effect on the heat-pulse propagation. To demonstrate the effect of piezoelectricity, we plot in Fig. 19 the transverse slowness surfaces in $LiNbO_3$ with and without the piezoelectric coupling. Figure 19a and b are obtained by setting the piezoelectric coupling constants equal to zero, and Fig. 19c and d use the known coupling constants. The effect of piezoelectric coupling is most striking for the FT mode. There is absolutely no singular flux when the coupling constants are set to zero, but parabolic lines do appear on the slowness surface when piezoelectricity is properly included.

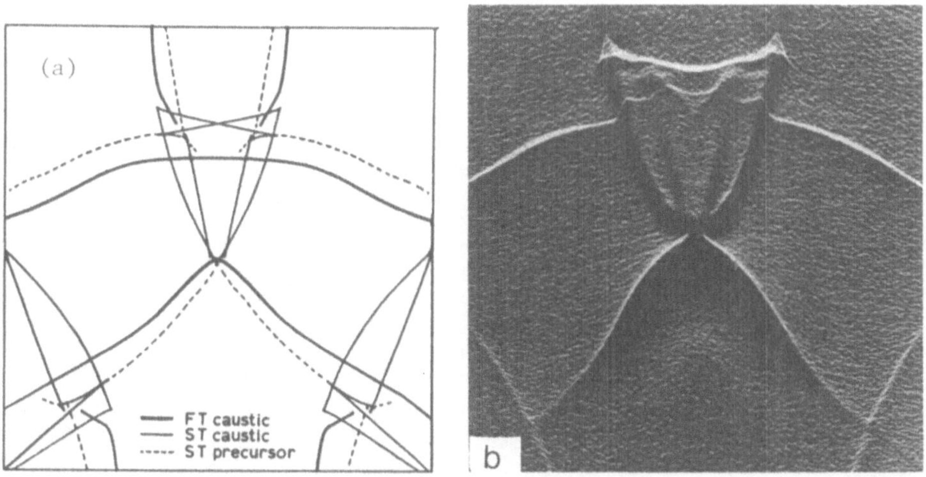

FIG. 18. a) Singularity lines for y-cut Quartz (Koos and Wolfe, 1984). b) Phonon image for Y-cut Quartz taken with a scanned electron beam as the heat-pulse source (Eichele et al., 1982).

It is possible to analyze the effect of these piezoelectric coupling constants by hypothetically multiplying them by a factor f which is varied between 0 and 1. A cross section of the FT and ST sheets in the y-z plane is shown in Fig. 20 for several values of the scale factor f. The FT and ST sheets touch at several points which are known as conic points. As f is increased from 0, the conic point C_1 moves significantly, and at f = 0.95 two new conic points, C_2 and C_3, are generated. Also, the FT (inner) sheet, which would have no saddle regions for f = 0, develops a saddle region which is bordered by singularities S1 and S2 for f>0.9.

This type of analysis provides a useful means for characterizing how the phonon slowness surfaces change when some control parameter is varied. Of course, experimentally we do not have such a control parameter, unless one could change the piezoelectric state of the crystal. Clearly, ballistic heat-pulse imaging could be a very sensitive means of sensing changes in the piezoelectric state of the crystal; however, presently we do not know of any piezoelectric phase changes which occur at the low temperatures accessible to phonon imaging experiments.

To complete the story, we show in Fig. 21a the pattern of flux singularities for a $LiNbO_3$ crystal, with piezoelectricity properly included. This calculation compares favorably to the experimental data shown in Fig. 21b as obtained by Koos and Wolfe (1984). These phonon imaging experiments and corresponding analysis show the profound influence of piezoelectricity on the ballistic propagation of thermal energy in crystals.

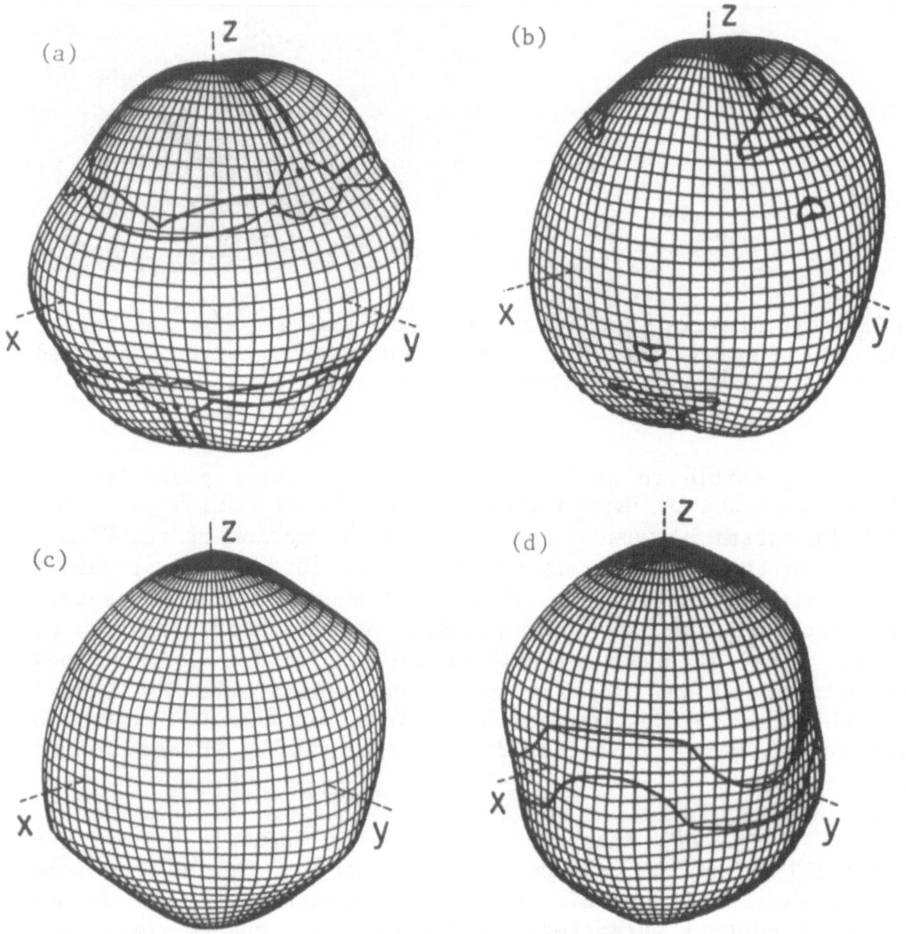

FIG. 19. Calculated slowness surfaces for the ST (a and c) and FT (b and d) modes in $LiNbO_3$. In this calculation a) and b) resulted from simply using the constant-field elastic constants, while c) and d) show the dramatic effect of inclusion of non-zero piezoelectric constants. (Koos and Wolfe, 1984).

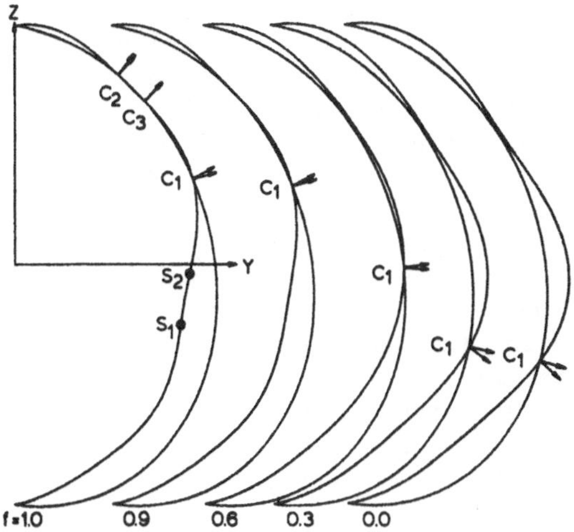

FIG. 20. Cross sections of the ST (outer) and FT slowness surfaces in the yz plane of $LiNbO_3$. The change in the shapes of the surfaces from f=0 to f=1 results from the hypothetical increase of the piezoelectric coupling from zero to the physical value.

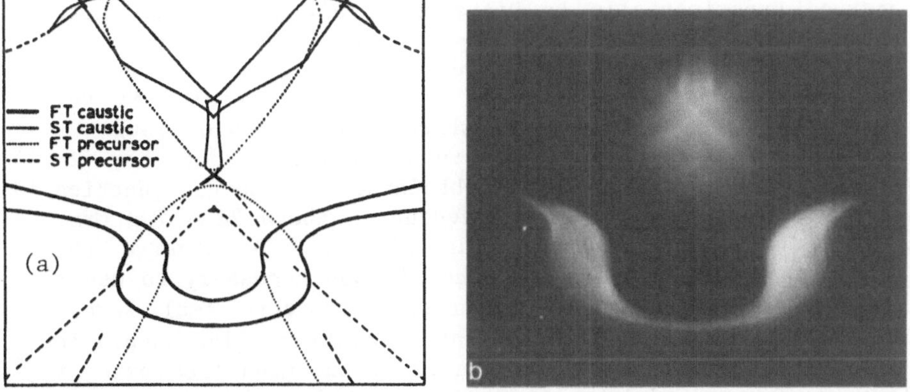

FIG. 21. a) Singularity lines for $LiNbO_3$, including the piezoelectric effect. b) Phonon image taken on y-cut $LiNbO_3$. The normal to the scanned surface is 37° from the z axis in the yz plane.

Focusing of Dispersive Phonons

At progressively higher frequencies the wavelength of a phonon begins to approach the atomic spacing in the crystal, and the continuum approximation used for long-wavelength phonons is no longer valid. The previous calculation of phonon focusing effects was dispersionless -- that is, the group velocity and focusing of ballistic heat flux did not depend on the magnitude of the wavevector. As the wavevector approaches the Brillouin zone boundary of the crystal the situation changes, as can be seen from the dispersion curve of Fig. 5a. Again, it is useful to plot constant-frequency surfaces in wavevector space. The (110) plane for the fast transverse mode in Ge is shown in Fig. 22. The curves are obtained using the lattice dynamics model of Ztetsis et al. (1979) and are labeled by their phonon frequency in THz. As the constant-frequency surfaces approach the Brillouin zone boundary, radical changes occur in their shape. It is useful to examine the evolu-tion of an inflection point, or point of zero curvature, which is represented by the dot in the figure, as the frequency increases in this dispersive regime. In the expanded view of this portion of wavevector space, one can see that as frequency increases the angle of the surface normal (i.e., the group velocity vector) with respect to the <100> direction begins to increase. This angle, denoted α in the figure, represents the half angle of the FT "ridge", displayed as the dashed lines in Fig. 12. A drawing of the dispersive surface in three dimensions has been published by Tamura (1982; 1983b) and is shown in Fig. 22b and c.

This effect of dispersion on the singularity pattern in Ge was observed experimentally by Dietsche et al. (1981) and their results are shown in Fig. 23. The Ge sample is photoexcited directly on its (110) face, and the detector is positioned on the opposite parallel face. The image in Fig. 23a results if the detector is an aluminum bolometer, which detects a wide range of frequencies. If a Pb-oxide-Pb tunnel-junction detector is employed, the image in Fig. 23b is obtained. The tunnel junction selectively detects phonons with frequencies above $2\Delta \sim 700$ GHz, where Δ is the superconducting gap of Pb. To observe the ballistic propagation of these phonons, it was necessary to use a thin crystal and fabricate a detector with very small cross section, 60 x 60 μm^2, for high angular resolution. The sample in this case was only 0.5 mm thick. In fact, the mean free path of the large-wavevector phonons is falling rapidly with increasing frequency due to isotope scattering. When this frequency-dependent attenuation is convolved with the onset of sensitivity in the tunnel-junction, it is expected that only a narrow range of ballistic phonons around 700 to 800 GHz will be detected ballistically. Figure 23b indicates a broadening in the FT ridge by about a factor of 2. This structural change in the phonon

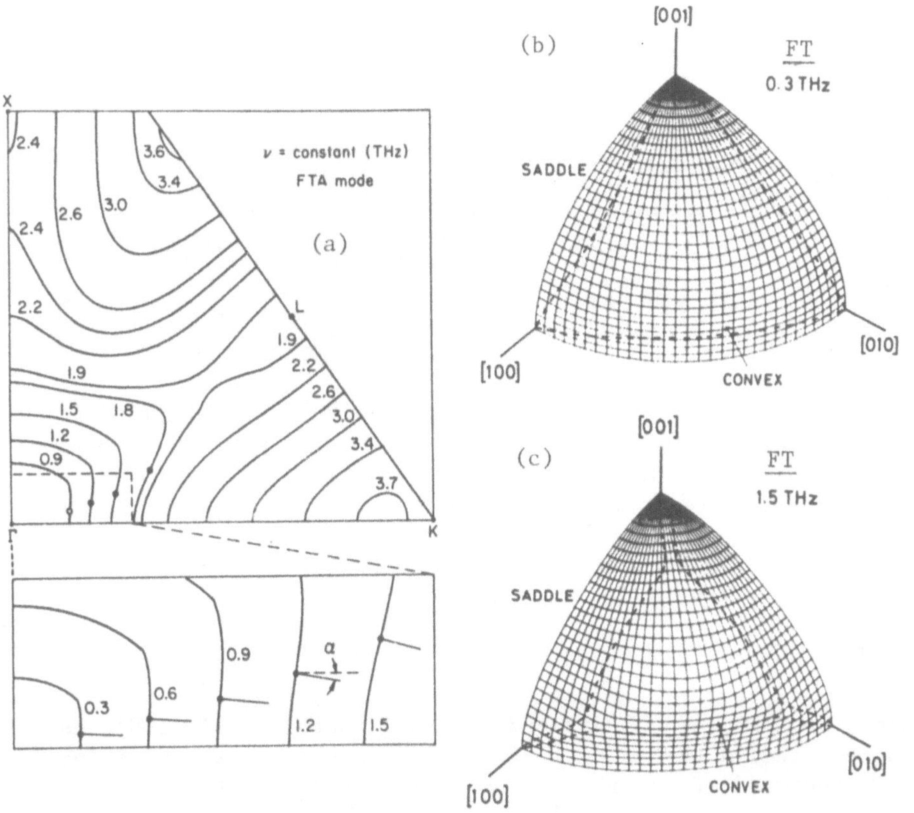

FIG. 22. a) Constant frequency surfaces for the FT mode in Ge as they intersect the (110) plane. Dots mark points of inflection which in turn lead to flux singularities along the surface normals at those points. The enlargement shows how these normals change direction with frequency, as indicated by an increasing angle α. b) and c) Plots of the slowness surface for the FT mode at nondispersive and dispersive frequencies (Tamura, 1983b). The dashed lines mark the singularities, where the curvature of the surfaces changes sign.

focusing pattern is exactly what is expected for phonons with about 700 GHz frequency, as discussed below.

Figure 24 shows the predicted singularity patterns for several phonon frequencies (Northrop, 1982b). These calculations make use of an extended Born-von Karman model for the lattice dynamics of Ge (Zdetsis and Wang, 1979). The positions of the FT singularities at 700 GHz are in reasonable accord with the data. The effect of dispersion at 700 GHz on the slow transverse mode is

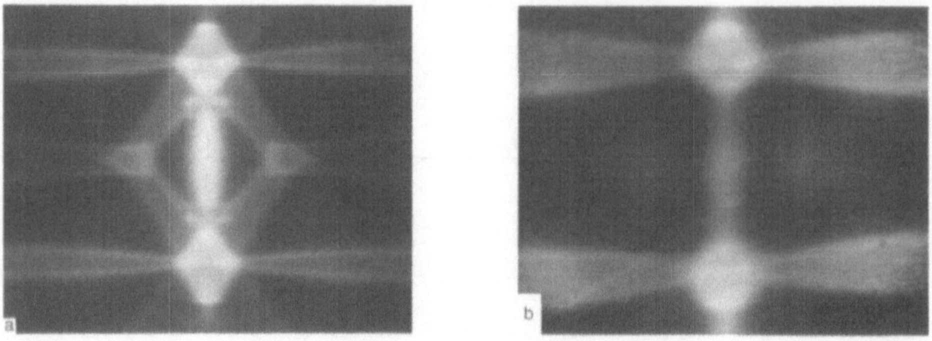

FIG. 23. Phonon images for photoexcitation of a (110) oriented 0.5 mm thick Ge sample. These images cover a large scan range, showing all three principal directions. a) Image obtained using an Al bolometer as detector. b) Image from the same sample, but using a Pb tunnel junction so that only phonons with a frequency greater than 0.7 THz are detected. The changes in the shape of the singularity patterns indicate dispersive phonon focusing.

to cause a rounding of the diamond-shaped structure near [100], as is also observed in the data of Fig. 23b. To observe the striking changes predicted for even higher frequencies, it would be necessary to use smaller path lengths and hence a smaller detector to preserve the angular resolution. Another approach would be to find a material in which the large wavevector phonons propagate ballistically over larger distances.

The experiments illustrated in Fig. 5 of the introduction provided hope that GaAs might be such a crystal (Ulbrich et al., 1980; 1981; Stock et al., 1984). The broad heat pulses observed following photoexcitation of its surface suggested that large-$\vec{k}$ dispersive phonons were propagating millimeter distances in this crystal. However, this claim was not born out by subsequent phonon imaging experiments (Wolfe and Northrop, 1984). No sharp phonon focusing structures are observed at the dispersive time-of-flights corresponding to the broad heat pulse. Our conclusion is that the broad pulses previously reported in GaAs are primarily a result of scattering in the crystal.

However, when the thickness of a GaAs sample was reduced to .25 mm, spatial evidence for dispersive propagation was obtained (Wolfe and Northrop, 1984). The phonon image of Fig. 25b has the distinctly characteristic shape due to dispersive phonons, similar to that predicted for Ge in Fig. 24. Also the FT ridge is broader than observed for longer path lengths. Further experiments using

Pb tunnel-junction detectors show similar dispersive patterns to that in Fig. 25b, but at pathlengths up to ~1 mm (Northrop, Hebboul and Wolfe, 1985). The intensity of the sharp ballistic pattern attenuates rapidly with increasing path length. Velocity selection indicates that the ballistic phonons detected have frequencies close to 700 GHz, which is the lowest detectable frequency of the Pb tunnel junction. We conclude from these experiments that the maximum frequency for ballistic transport in the .25 - 1 mm pathlength range is about 700 GHz.

These results are in reasonable agreement with the standard isotope scattering theory, which predicts that the mean free path of phonons with frequencies above 1 THz is only of order 0.1 mm,

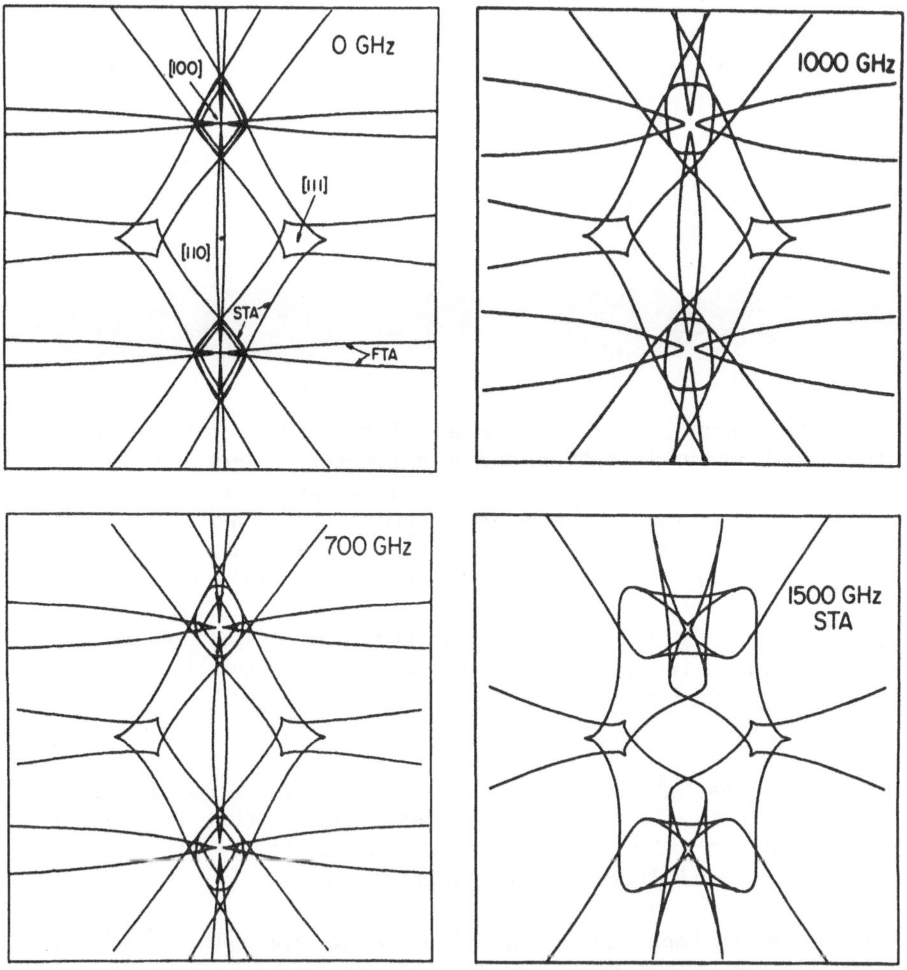

FIG. 24. Calculated phonon focusing singularity patterns for dispersive frequencies in Ge. These were calculated (Northrop, 1982b) using a lattice dynamics model which had been fit to neutron scattering data.

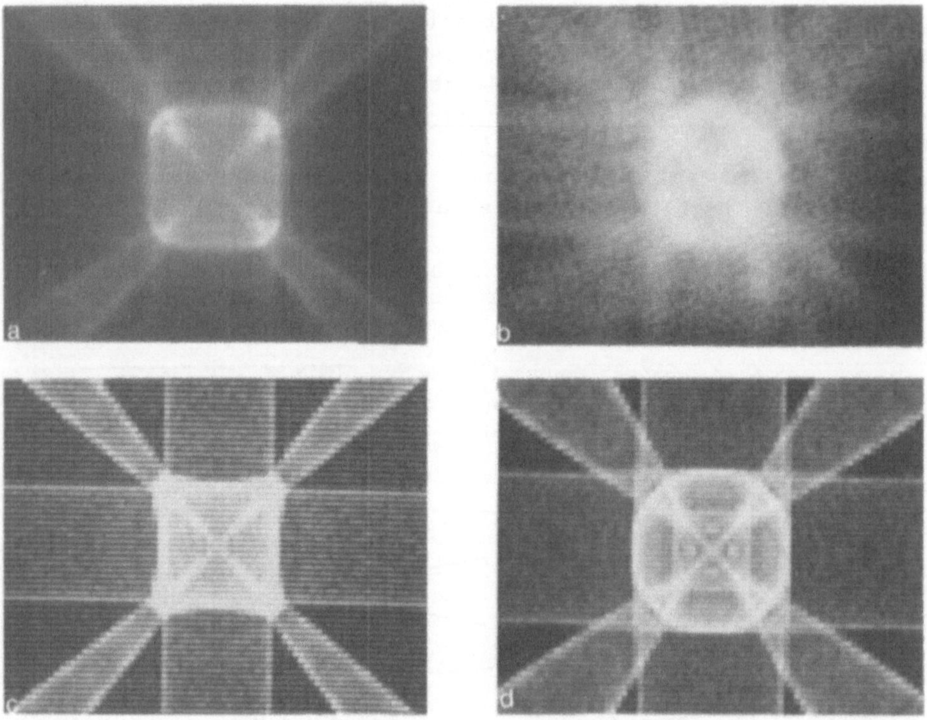

FIG. 25. a) Phonon image of 1.9 mm thick [100] oriented GaAs using low level photoexcitation and a bolometer as detector. The phonon focusing patterns are basically nondispersive. b) Phonon image in GaAs taken using the micro-imaging apparatus for a 0.25 mm thick sample. This image has been enhanced by Fourier transform high-pass filtering to improve on the experimental resolution. The phonon focusing pattern shows a distinct shift from a). Monte Carlo calculations for two GaAs dispersion models are shown for this view in c) (Kunc et al., 1979a) and d) (Kunc et al., 1979b).

not 1-3 mm, as deduced solely from the time-of-flight measurements. It is clear from these measurements that the phonon imaging method provides an important means of distinguishing between ballistic and diffusive propagation modes.

Perhaps more important, the ability to observe dispersive phonon focusing patterns provides an opportunity to test different lattice dynamics models. Models of varying complexity have been reported most of which are reasonably fit to neutron and Raman data, principally along symmetry directions. Monte Carlo images for 700 GHz phonons, using two different lattice dynamics models of Kunc et al. (1979a; 1979b), are shown in Fig. 25c and d. Figure 25c is a dipole model and Fig. 25d is a shell model. Clearly, the phonon imaging data of Fig. 25b favors the shell model. These calculations use the published fit parameters for each model; no attempt has been made to adjust either model to better fit the phonon image data. Such experiments and calculations show that the phonon focusing pattern, which depends on the local curvature of the dispersive slowness surfaces, can be a sensitive probe of phonon dispersion in crystals.

SCATTERING OF NON-EQUILIBRIUM PHONONS

A principal feature of the phonon imaging method is the ability to isolate phonons of a given mode α, wavevector $\vec{k}$ and polarization $\vec{\varepsilon}$. This is especially valuable in the characterization of phonon scattering, where one would like to know how particular phonons are scattered by given defects in the crystal. This ability to select mode, polarization and propagation direction offers a considerable advantage over thermal conductivity measurements, which give the scattering rate averaged over all of the equilibrium phonons. We now turn to a case where the phonon imaging technique has provided valuable new information about scattering from extended defects, namely, dislocations.

Scattering From Dislocations

One of the principal causes for thermal phonon scattering in imperfect crystals is dislocations. A dislocation is a line of imperfection in the lattice structure which extends over many lattice sites. Dislocations have a profound effect on the macroscopic mechanical properties of a crystalline solid. It is observed that most crystals, under the action of an applied shearing stress, deform plastically at stresses far below those predicted for a perfect crystal. The source of mechanical weakness in real crystals is the relative ease of motion of dislocation lines. Thus the characterization of dislocations is extremely important to an understanding of the mechanical strength of materials.

There are two types of dislocations which commonly occur in crystals: edge and screw. Schematic pictures of these dislocations are shown in Fig. 26. Plastic deformation in many crystals occurs by a "slip" of one plane of atoms with respect to an adjacent plane. Such slip motion produced by an applied shear stress produces a slipped area which is bounded by a dislocation loop. Edge dislocations are perpendicular to the slip direction and screw dislocations are parallel to it. Locally, the edge dislocation in Fig. 26a looks like an extra vertical plane of atoms partially inserted into the lattice at line EF. A screw dislocation occurs at the boundary between slipped and unslipped parts of the crystal, as shown in Fig. 26b and c. An example of the slip process for a LiF crystal is shown in Fig. 26d. Here the crystal is compressed along the [100] direction and allowed to expand along the [010] direction, while being constrained to the same dimension along the [001] direction. The favored slip planes for this crystal are (110) planes, as shown in the figure. The larger the plastic deformation, the larger the density of dislocations. Deformations as large as 10% are obtainable in LiF without breaking the crystal.

At first thought, it seems highly unlikely that ballistic

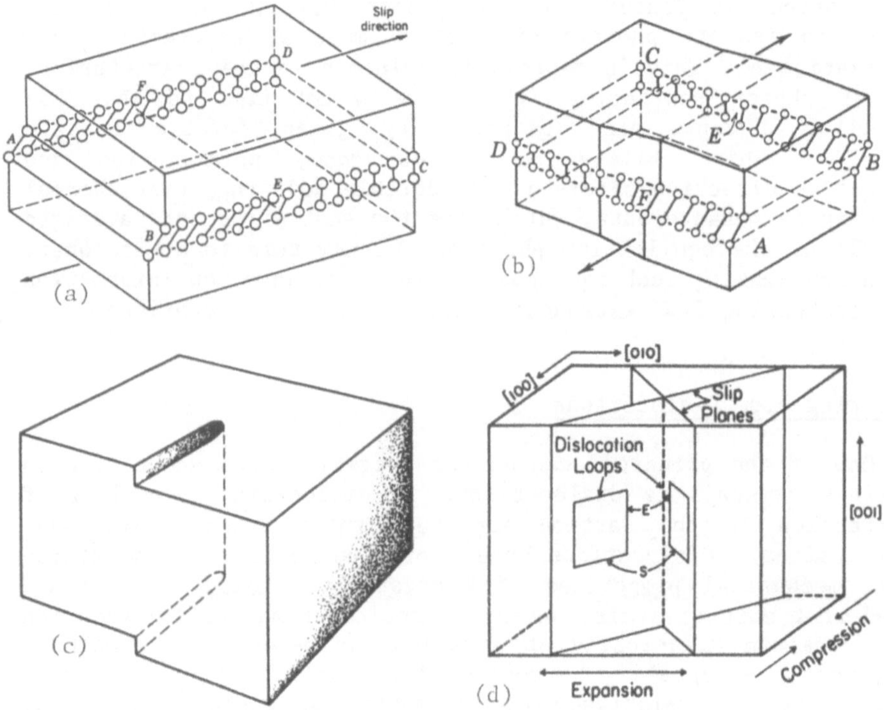

FIG. 26. a) Schematic of an edge dislocation formed along line EF (Kittel, 1976). The area ABEF has slipped, while the area FECD has not. b) A screw dislocation formed along line EF. Note that this dislocation is parallel to the slip direction. c) Schematic of a screw dislocation, showing that a slipped region is bounded by edge and screw dislocations. d) Geometry for the formation of dislocation loops and slip planes in a LiF sample. The deformation geometry shown tends to strongly prefer slip along only two of the six favored (110) glide planes. After deformation the front surface was coated with a metal film and a bolometer was placed in the center of the back face for phonon imaging experiments.

heat pulses could actually traverse a crystal which has been radically deformed and distorted in such a fashion. Nevertheless, thermal conductivity experiments performed on a heavily deformed crystal of LiF still showed a characteristic T^3 dependence at low temperatures, indicating that at least a subset of the thermal phonons had their mean free paths limited only by the boundaries of the crystal (Cotts et al., 1981). The overall thermal conductivity was considerably lower in this case than for the undeformed crystal, and heat pulse experiments with fixed detector and source indicated that the slow transverse phonons were attenuated much more than fast transverse phonons through the deformed crystal (Anderson and Malinowski, 1971). With this motivation, the phonon imaging method was applied to determine the details of the phonon-dislocation scattering process in this system (Northrop et al., 1982; 1983).

At the outset, it should be remarked that there exists some controversy concerning the mechanism of phonon-dislocation scattering in alkali-halide crystals. One school of thought holds that the phonons scatter from the static strain fields surrounding the dislocation lines (Brown, 1981); however, the observed scattering strengths seem to be too large to be explained by this mechanism. On the other hand, a dynamic scattering mechanism has been forwarded by Granato and others (Kneezel and Granato, 1982; Granato, 1958) which views the dislocation line as a "fluttering string" that can absorb a phonon and re-radiate its energy in other directions. Calculations of this dynamic mechanism predict scattering strengths which are in better accord with observations.

Figure 27a shows a ballistic phonon image for an undeformed LiF crystal. A psuedo-3-dimensional representation of this image is given in Fig. 27b. The identification of these structures are the same as for Ge, as previously described in Fig. 9. Again, the (100) face of the crystal is scanned with a laser beam and the detector is on the opposite $(\bar{1}00)$ face. The ballistic phonon image for an identical crystal, except deformed as indicated in Fig. 26d, is shown in Fig. 27d. One can see from this image that the slow transverse phonons are strongly absorbed by the heavily dislocated crystal. It is quite remarkable that this crystal, which has been plastically deformed by 10%, still transmits a subset of fast-transverse phonons, labeled as the horizontal ridge (B) in the photo. The vertical ridge (A) is attenuated much more, but is still visible. Thus the scattering from dislocations is highly mode selective and also highly anisotropic. This is precisely what is predicted for the fluttering dislocation model, as described below. It is important to notice that in this experiment certain slip planes are selected by the deformation geometry, which orients the edge dislocations predominantly along one direction in the crystal. In effect, the crystal has become a "phonon polarizer", in which only one polarization of the fast

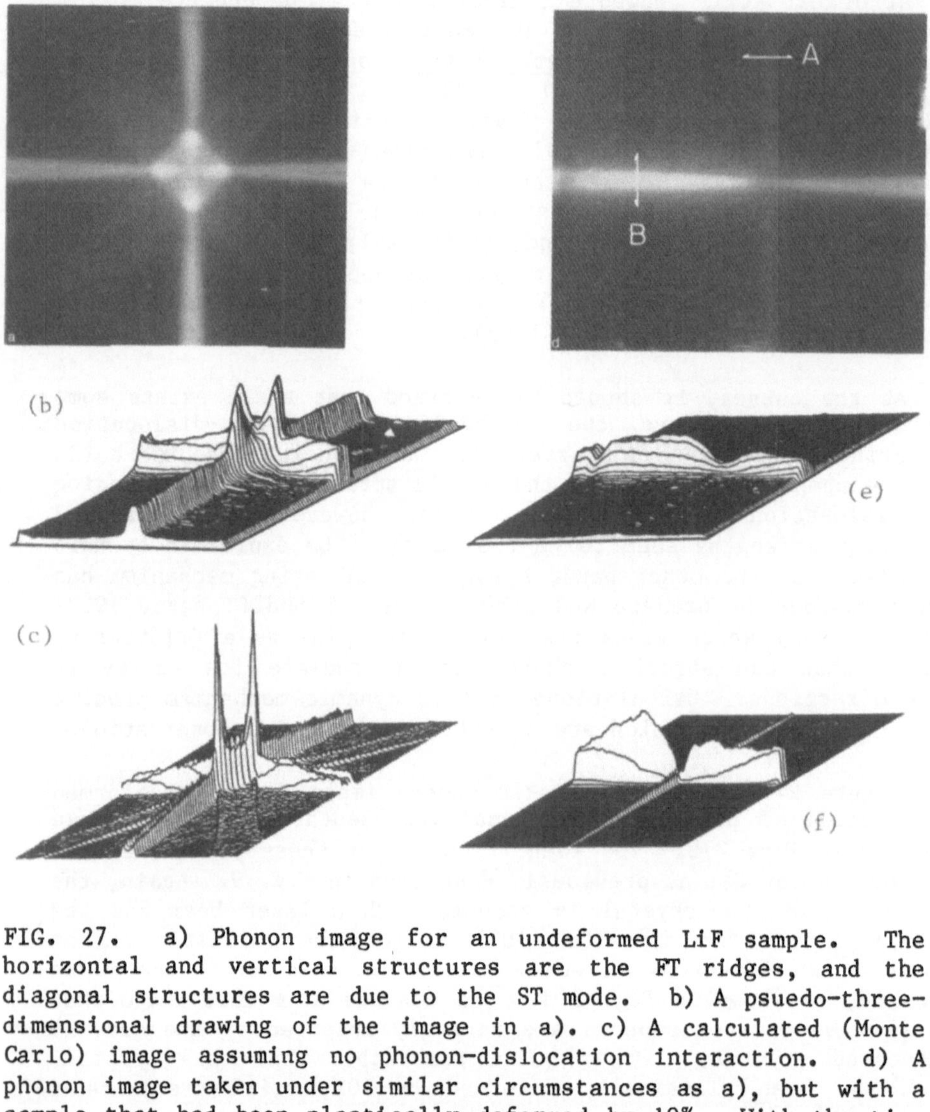

FIG. 27. a) Phonon image for an undeformed LiF sample. The horizontal and vertical structures are the FT ridges, and the diagonal structures are due to the ST mode. b) A psuedo-three-dimensional drawing of the image in a). c) A calculated (Monte Carlo) image assuming no phonon-dislocation interaction. d) A phonon image taken under similar circumstances as a), but with a sample that had been plastically deformed by 10%. With the time gate set to detect only ballistic phonons, the phonon-dislocation scattering is seen as an absorption of the ballistic component. Most of the phonon polarizations are strongly scattered with the exception of the horizontal (H) FT ridge. e) Psuedo- three-dimensional plot of the image in d), showing a weak vertical FT ridge. f) Calculated image for a total scattering strength β=2800.

transverse phonons -- as shown in Fig. 27d -- remains unscattered.

The basic idea of the fluttering dislocation model is that the elastic wave couples to the dislocation through the shear stress component σ_0 which is across the glide plane and in the direction of the Burgers vector $\vec{b}$ of the dislocation. The Burgers vector defines the direction along which one atom plane is slipping with respect to the other during deformation. The shear component of the phonon produces a force per unit length σ_0 on the dislocation, causing it to move in a forced oscillation. The strength of the coupling of the phonon to the dislocation line is proportional to the "resolved shear stress factor", $\Omega = (\sigma_0/\sigma_p)^2$, where σ_p is the total phonon stress. The factor Ω has an anisotropy which is dependent on phonon wavevector $\vec{k} = (k_1, k_2, k_3)$ and polarization $\vec{\varepsilon} = (\varepsilon_1, \varepsilon_2, \varepsilon_3)$, and the orientation of the slip plane. For the (110) and $(\bar{1}10)$ slip systems, Ω takes the form

$$\Omega = (n_1\varepsilon_1 - n_2\varepsilon_2)^2 \tag{13}$$

where $n_i \equiv k_i/k$ are the components of the unit wavevector. Clearly, $\Omega = \Omega(\theta_k, \phi_k)$ is highly anisotropic for a given phonon mode. The ballistic transmission is given by

$$I(\theta_V, \phi_V) = I_0(\theta_V, \phi_V)e^{-\beta\Omega(\theta_V, \phi_V)}, \tag{14}$$

where β is a measure of the total scattering strength, which depends on the density of dislocations and the phonon path length through the crystal. The factor $I_0(\theta_V, \phi_V)$ represents the angular dependence of the phonon flux due to phonon focusing of unscattered phonons.

For heavily dislocated crystals, only phonons with very small dislocation coupling, $\Omega \sim 0$, travel unscattered through the crystal. This is true for FT phonons with $\vec{k}$ exactly in the (001) plane, which have a polarization $\vec{\varepsilon} \parallel [001]$. Thus, $\varepsilon_1 = \varepsilon_2 = 0$, implying $\Omega = 0$ for this subset of phonons. In addition, for $\vec{k}$ exactly in the (010) plane, $\vec{\varepsilon} \parallel [010]$, implying $\varepsilon_1 = 0$ and $k_2 = 0$, leading also to $\Omega = 0$. For $\vec{k}$-vectors slightly out of the (010) plane, the attenuation increases rapidly, so the vertical ridge shown in the data appears much weaker than the horizontal ridge. The polarization vector of the FT phonons in the horizontal ridge does not change rapidly with k-vectors slightly out of the (001) plane, so Ω remains close to zero for all the phonons in this ridge. Contours of constant Ω in wavevector-angle space for LiF are shown in Fig. 28a. The crosshatched areas represent the wavevectors of FT phonons which are transmitted through the 10% deformed sample. Indeed, a comparison of the total transmitted phonon intensity in the horizontal and vertical ridges gives a measure of the total scattering strength β. Figure 28b and c show predicted profiles of the horizontal and vertical ridges for various values of β. The heavy lines correspond to the best fit of the relative intensities for the 10% deformed sample, indicating a total scattering strength β = 2800.

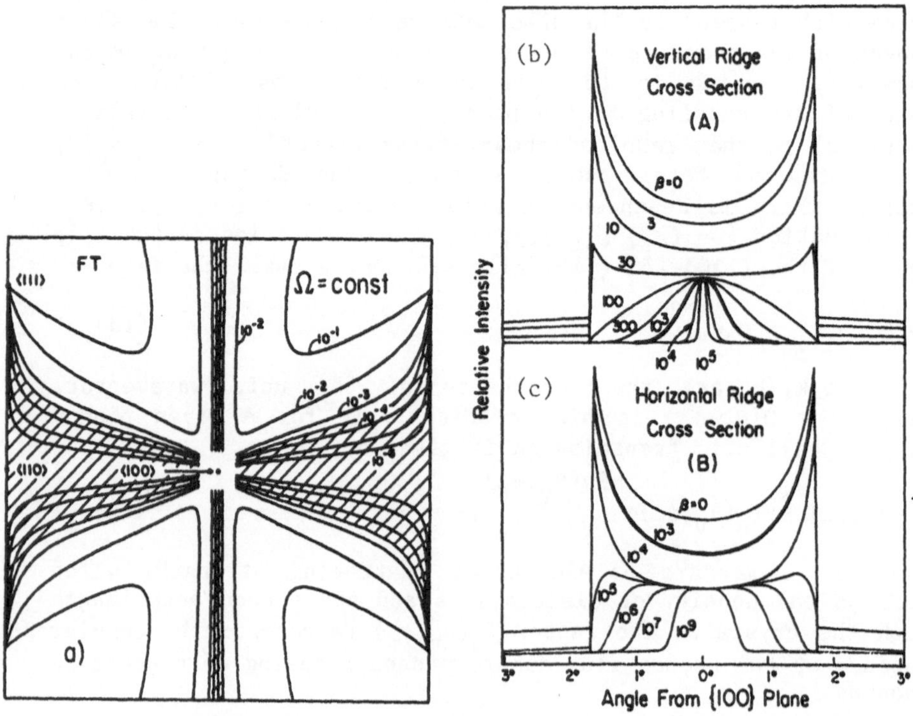

FIG. 28. a) Contours of constant Ω in wavevector space for the FT mode. The cross-hatched area indicates phonons which are transmitted in the 10% deformed sample. b,c) Calculated phonon intensity near the (010) (b) and (001) (c) planes including both phonon focusing and dislocation scattering, for several values of β. The heavy lines indicate the expected vertical and horizontal ridge intensities for $\beta=10^3$.

Monte Carlo calculations have also been performed on this system which show the transmitted intensity as the total scattering strength β is increased. Figures 27c and f show the computed intensities, including both transverse modes, for β=0 and β=2800, respectively. It is clear from a comparison of this Monte Carlo calculation with the data that the fluttering dislocation model explains the highly anisotropic interaction of phonons with dislocations in this crystal.

One interesting anomaly still remains, however. The value of the total scattering strength β = 2800 obtained from the phonon imaging measurements is about a factor of 70 larger than that theoretically estimated using the dislocation density deduced from etch-pit counts on similarly deformed samples. Kneezel and Granato (1982) concluded that a similar discrepancy in thermal conductivity is due to the existence of a high density of dislocation dipoles. They were able to fit the thermal conductivity data by assuming a dipole density of about 30 times the monopole density and an average spacing between the two dislocations of the dipole pair equal to 60 times the Burgers vector. It should be emphasized that this is an effect which occurs at very high dislocation densities. A phonon imaging experiment performed on a 1.6% deformed sample was analyzed to have $\beta \sim 7$, which is in good agreement with the scattering strength expected for the known dislocation density. In conclusion, a measurement of the relative transmission of phonons with various polarizations $\vec{\varepsilon}$ and wavevectors $\vec{k}$ can provide valuable quantitative information about phonon-defect scattering.

Reflection at Crystal Boundaries

The reflection and transmission of thermal phonons at a crystal boundary is one of the foremost problems in phonon physics (Anderson, 1976; Wyatt, 1981). At sub-Kelvin temperatures, the boundary resistance between two carefully bonded solids appears to be satisfactorily described by the acoustic impedance mismatch between the two media (Anderson, 1981). However, recent studies of solid-solid interfaces have indicated a rather universal failure of simple acoustic mismatch theory for higher frequency phonons in the 1-10 K range. The situation is similar to the anomalously low boundary resistance which is observed between a solid and superfluid helium -- commonly known as the Kapitza anomaly. The large difference in acoustic impedance, ρv, between a solid and liquid helium suggests that only about 1% of the crystal phonons striking the boundary will be transmitted into the superfluid helium. The observed transmission factor is closer to 50%. However, in situ cleaved surfaces in superfluid helium do not exhibit this anomaly, suggesting that it arises from surface damage or contamination (Weber et al., 1978). However, for interfaces between a crystal and metal film, the transmission is typically much less than that expected from the impedance mismatch; e.g., 70% transmission instead of 98% transmission (Marx and Eisenmenger, 1982). Again, the question of contamination at the interface and sub-surface damage has been raised.

Recently, it has been suggested that the anomalous transmission (and reflection) coefficients are directly related to whether the thermal phonons are scattered specularly or diffusely from the boundary (Taborek and Goodstein, 1980a; 1980b). Specular reflection is the standard process considered in acoustic mismatch theory and involves an application of Snell's law ($k_{\parallel}$ conservation) to coherent waves of incident, reflected, and transmitted phonons. Diffuse scattering, on the other hand, may come in at least two forms. Roughness which is large on the scale of the phonon wavelength may produce reflection which is locally specular but appears diffuse; or the incident phonon may be absorbed by defects or foreign atoms at the surface and re-radiated in an incoherent fashion. It is postulated that this latter process might increase the total flux of thermal phonons across a largely mismatched boundary.

There have been several recent studies attempting to determine the relative amounts of specular and diffuse phonon scattering at a crystal boundary. Typically, fixed generator and detector films on one surface of the crystal are employed to measure the temporal shape of a heat pulse reflected from another surface. Marx and Eisenmenger (1982; Marx et al., 1978; Marx and Eisenmenger, 1981) have pointed out that both specular and diffuse scattering can produce temporally sharp reflected signals, as were observed in their studies of free Si, Si/metal-film, and Si/rare-gas interfaces. Monte Carlo simulations of the temporal pulses led them to conclude that 96% of the reflected phonons from their free Si surface (i.e., vacuum interface) were diffusely scattered. Using this method in sapphire, Taborek and Goodstein (1980b) observed temporally sharp reflection signals which they attributed to specular reflections. The analysis of specular reflection of phonons at the boundary of an anisotropic solid involves a large number of incident and reflected wave combinations, due to mode conversion at the surface. The observed heat-pulse spectra were often quite complex, but there was generally good identification of the modes involved in each reflected pulse. However, no attempts were made to explain quantitatively the widely differing intensities of the various reflected phonon modes.

Most recently, in an attempt to positively identify diffuse versus specular reflection and understand more quantitatively the reflected intensities associated with mode conversion, we have applied the phonon imaging method to this problem (Northrop and Wolfe, 1984). The technique was first demonstrated on a highly polished sapphire crystal immersed in superfluid helium. The advantage of this "phonon reflection imaging" is the ability to measure the boundary reflection of heat pulses over a wide and continuous range of incident angles. To do this we devised the sample configuration shown in Fig. 29a, which permits continuous scanning from nearly normal incidence (~ 2°) to about 50°. The phonon source is again a scanned laser beam focused onto an evaporated copper film covering most of the generation surface. An aluminum bolometer is evaporated onto the center of this same surface, with insulating gaps in the copper film preventing the bolometer from being shunted. The copper heater film also acts as the contacts for the aluminum bolometer. The sapphire crystal is a standard window with an optical-grade polish on both flat surfaces.

Figure 29b shows schematically the paths of incident and reflected waves for a given displacement $\vec{X}$ between source and detector. For a particular incident-reflected mode pair there is one allowed specular path, but diffuse reflection is allowed for all paths. Generally the diffuse path lengths are equal to or longer than the single specular paths. Thus, for one mode pair, a time-of-flight spectrum should show a single pulse for specular, with a tail indicating diffuse. This picture is greatly complicated by the crystalline anisotropy and the overlap in time of various mode pairs. However, the imaging process should help differentiate specular from diffuse signals by their greatly different angular character.

In order to select a particular phonon velocity, the boxcar gate is continuously adjusted during the raster scan to be proportional to the minimum reflection path length between source and detector. Figure 30a shows heat-pulse signals corresponding to several fixed positions of the laser spot. The broad early-arriving signal corresponds to phonons scattering in the bulk of the crystal between source and detector. The sharp signals are

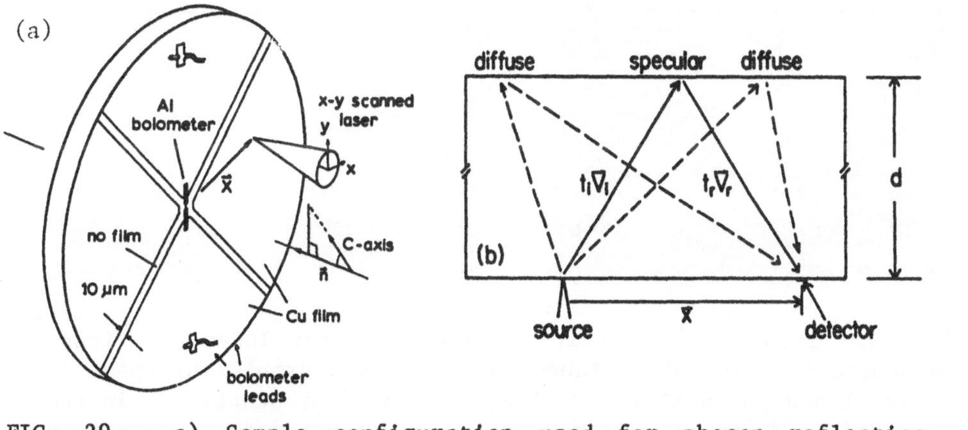

FIG. 29. a) Sample configuration used for phonon reflection imaging. The $(1\bar{1}02)$-oriented sapphire disk has a 3.2 mm thickness and a 30-mm diameter. The phonon detector is a 50 x 50 μm^2 Al bolometer and the 2000 Å Cu heater/contact film has 10-μm insulating gaps. A pulsed Ar^+-ion laser beam is scanned to form the image. b) Schematic of possible diffuse and specular paths for one particular generator/detector position. Normally the diffuse path lengths are greater than or equal to the specular path.

due to phonons reflected from the opposite face of the crystal. The dashed lines indicate times corresponding to three selected phonon velocities. Figures 30b-d show the phonon images corresponding to these three velocities. These velocities are roughly

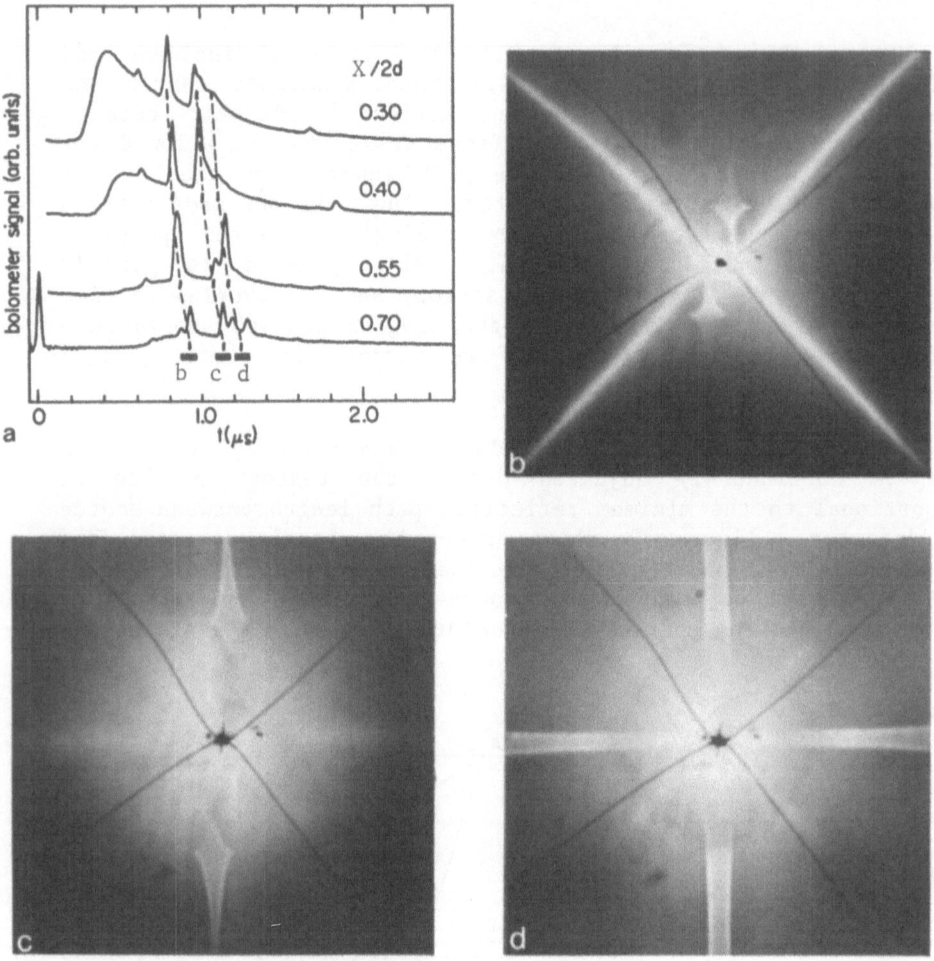

FIG. 30. a) Phonon reflection signals for four laser positions. The source to detector distance is X, and the sample thickness is d. The broad pulse at early times is caused by scattering in the bulk, and it drops off rapidly as the source-to-detector distance is increased. The first sharp pulse at 0.6 μs is an L-to-L reflection, and the weak pulse at 1.8 μs is a double reflection of transverse phonons. The three heavy lines indicate the gates used for constant velocity images. The dashed lines indicate how the gate delays varied with the laser position. The resulting velocity-selected phonon reflection images are for b) $V=8.6 \times 10^5$cm/sec, c) $V=7.0 \times 10^5$, and d) $V=6.4 \times 10^5$. The dark lines correspond to the gaps in the Cu film.

the average velocity for the following incident/reflected mode pairs: (b) L→ST, (c) FT→ST and FT→FT, and (d) ST→ST. Remarkably sharp structures are observed in these images, which we will now show are due to specularly reflected phonons from the crystal boundary.

Figure 29b shows a phonon with initial wavevector $\vec{k}_i$ and mode α_i reflecting into a phonon of wavevector $\vec{k}_r$ and mode α_r, producing a net displacement $\vec{X}$ in the source-detector plane. The displacement is

$$\vec{X} = t_i \vec{V}\,(\vec{k}_i,\alpha_i) + t_r \vec{V}(\vec{k}_r,\alpha_r) \qquad (15)$$

with $t_{i,r} = d/\,\vec{n} \cdot \vec{V}_{i,r}$ the time of flights, d the sample thickness, $\hat{n}$ the surface normal, and $\vec{V}$ the group velocity. Since Snell's law specifies $\vec{k}_r$ as a function of $\vec{k}_i$, we see that for a specified pair of incident and reflected modes $\vec{X}$ is completely defined by the initial wavevector $\vec{k}_i$. That is, there is a unique mapping from a 2-dimensional wavevector space ($\vec{k}_i$) into a 2-dimensional observation space ($\vec{X}$). As in the case of simple phonon focusing, there will be mathematical flux singularities in the observation space. These caustics occur as a locus of propagation directions where the Jacobian of the transformation $\vec{X}\,(\vec{k}_i)$ is equal to zero. Singularity lines for six mode pairs reflecting from a $(1\bar{1}02)$ sapphire surface are plotted in Fig. 31a. The remaining mode pairs (L→L, L→FT, FT→L) show no singular behavior, while the dashed lines indicate near-singular behavior for the FT→FT case.

A Monte Carlo image for all singular mode pairs is shown in Fig. 31b. All of the sharp features in the data of Fig. 30b,c, and d are well explained by this theory. Therefore, this optical-grade surface exhibits remarkably strong specular scattering, implying that the surface is exceptionally good. Recall that phonons in a heat pulse with frequency 300 GHz have a wavelength of only 200 Å. For specular reflection to occur, it seems reasonable to assume that the local surface roughness must be less than the wavelength of the incident phonons.

To see the effect of surface roughness, we degraded the surface by polishing it with a 1 μm diamond paste on a glass slide. The surface was still highly reflecting to the eye. The phonon reflection image obtained from this degraded surface was completely different from the previous images and is shown in Fig. 31c. This image is the sum of phonon images for all the velocities corresponding to Fig. 30b-c. The sharp singularity structures associated with specular reflection are missing.

We now consider the reflection image expected if the scattering off the back surface were <u>diffuse</u>. Diffuse scattering is a random process, with no unique relation between $\vec{k}_i$ and $\vec{X}$. For a specified $\vec{X}$, diffuse scattering from the entire back surface may contribute to the observed flux. Consequently, diffuse scattering does not in general lead to singular behavior. Taborek and Goodstein (1980) qualitatively discussed the reflection intensity as the overlap between a source phonon-focusing pattern and a

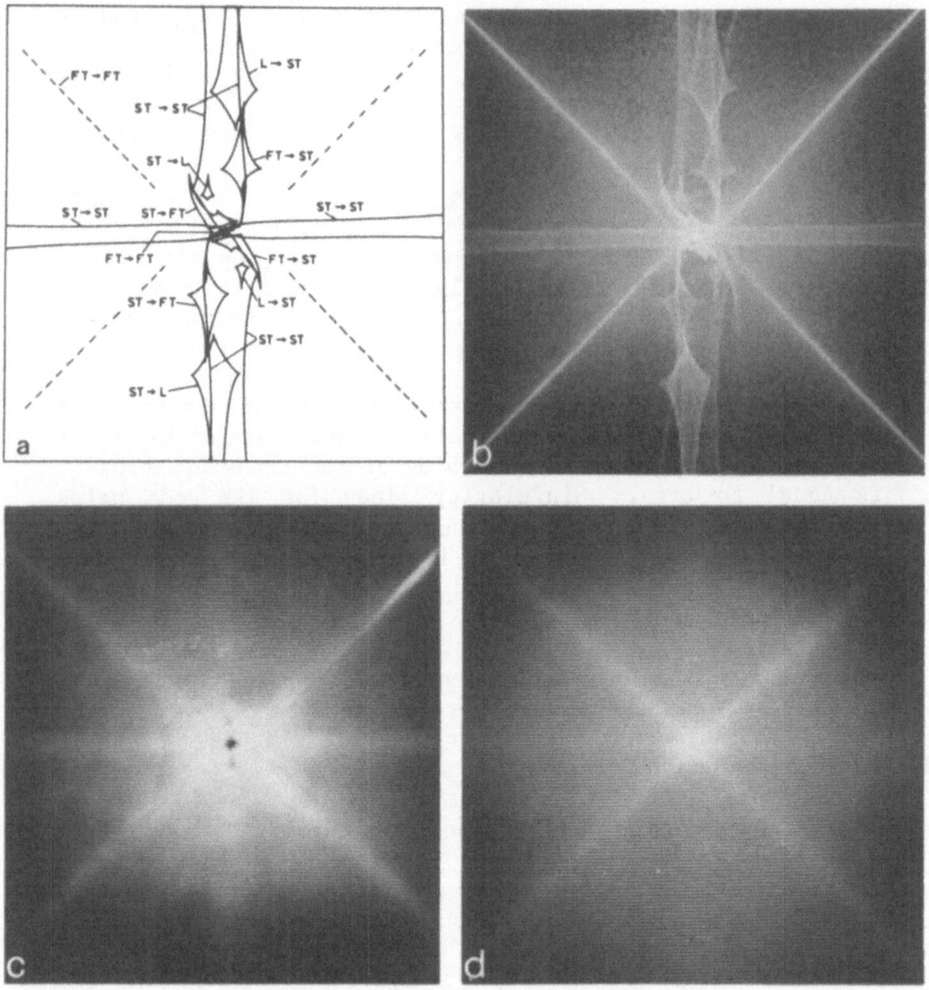

FIG. 31. a) Singularity lines for specular reflection imaging in sapphire. The slight left-right asymmetry is due to a small deviation of the crystal orientation by about 2.5°. b) Monte Carlo calculation of specular reflection image. The mode pairs are weighted by the appropriate branching ratios upon reflection. c) Phonon reflection image for a roughened surface. This image is a composite of images corresponding to the velocities in Fig.30. d) Computed image for diffuse (isotropic) scattering of phonons from the surface.

"detector" phonon-focusing pattern on the reflection surface. We note that, in general, the overlap of two caustic lines (source and detector) occurs at a point, producing no singularity in the reflected flux. In other words, the flux intensity changes smoothly with respect to small changes in X. A Monte Carlo image, obtained by convolving source and detector phonon-focusing patterns is displayed in Fig. 31d. These results are very similar to the phonon reflection image for the degraded surface, indicating that in that case the phonon reflection was primarily diffusive in nature.

In order to get a quantitative measure of the relative amounts of specular and diffuse scattering from the optical-grade surface, we can compare the experimental reflection image (a sum of over all velocities) to a linear combination of the theoretical images for specular and diffuse scattering, Fig. 31b and d. It is found that only (20 $\pm$ 10)% of the phonons are scattered diffusely from this highly polished sapphire surface. A similar comparison applied to the reflection image for the degraded surface shows that the fraction of diffusely scattered phonons has increased to (97 $\pm$ 3)% for this surface.

Such phonon reflection imaging experiments provide a means for isolating and studying specular and diffuse reflection of short wavelength phonons from crystal surfaces. It is clear from the detailed structure in these images that a quantitative understanding of the reflected heat pulse intensities cannot be practically achieved by using fixed phonon sources and detectors. A scanning method would seem to be essential for understanding the phonon mode conversion processes at a surface. So far the method has only been applied to two surface conditions of sapphire. It seems reasonable to assume that phonon reflection imaging will provide new insights into the processes of thermal reflection and transmission at interfaces.

Wavevector Channeling at Solid-Solid Interfaces

In nearly all the above studies thermal energy was introduced into the crystal by heating a metal film evaporated onto its surface. We have implicitly assumed that a more-or-less uniform angular distribution of phonon wavevectors is transmitted through the interface. In fact, the distribution is not isotropic even for a perfect bond, because there is an acoustic mismatch between the two materials. A systematic study of the anisotropic transmission of heat across several crystal-metal interfaces has been conducted by Rösch and Weis (1977). In addition, one expects that the bonding between a crystal and evaporated metal film will not be microscopically perfect and this can affect the transmission of high frequency phonons across the interface. After all, the surface of the crystal substrate is hardly flat on an atomic scale, and generally the surface is not cleaned in an ultra-high vacuum, implying that, with normal procedures, there will almost certainly be contamination by other chemical species at the surface. Also, in the case of some crystals such as Si, oxide layers are rapidly formed at the crystal boundary.

A recent phonon imaging experiment has shown that the transmission of non-equilibrium high frequency phonons through a metal-crystal interface is far from isotropic (Koos et al., 1983; Every et al., 1984). Indeed, there is a preferential transmission of certain phonons, leading to a channeling of certain wavevectors into the crystal. This wavevector channeling gives rise to structures in the phonon images which are not explained by normal phonon focusing theory. The conclusions from these experiments were that the metal-crystal bonding was actually quite weak and that the transmission across this boundary can be enhanced by the existence of "pseudo-surface-waves". A detailed theoretical description of this phenomenon has been given by Every et al. (1984); here we will simply outline the basic ideas and results.

In a first experiment, a 2000-Å copper film was deposited on an optical-grade surface of sapphire and an aluminum bolometer was deposited on the opposite face. The copper film was irradiated by an argon ion laser, as in the usual phonon imaging experiments. The ballistic phonon image for sapphire at 1.6 K with faces cut in the $(1\bar{1}02)$ direction, referred to conventional hexagonal axes, is shown in Fig. 32a. Most of the structure in this image can be understood in terms of the usual phonon focusing theory. In Fig. 33 we show the slowness and wave surfaces for sapphire for both the ST and FT modes. As usual, the thick lines on the slowness surfaces represent the locus of points with zero Gaussian curvature. Structures in the wave surfaces can be readily identified with the ballistic phonon image of Fig. 32a. There is one structure in the image, however, that is not explained by this analysis -- namely the circular-shaped "halo" covering most of this image. When this halo intersects the singularity structures, it seems to reflect off of the folds, an effect which led us to believe that the new structure was associated with an enhancement of phonon wavevectors for a specific angular region in k-space. This is to be contrasted with the usual case of phonon focusing in which a roughly isotropic distribution of wavevectors gives rise to a channeling of group velocity vectors. Figure 32b shows a Monte Carlo image of the ballistic heat flux in sapphire assuming an isotropic distribution of phonon wavevectors. The halo is not present. Clearly, the distribution of wavevectors in the experimental case is not isotropic.

A second clue to the origin of the halo comes from the observation that its position with respect to the normal focusing structures depends upon the orientation of the surface with respect to crystal axes. Indeed, for other crystal orientations, the wavevector channeling structure is more complex and does not even have a near-circular shape. Thus, unlike the bulk phonon

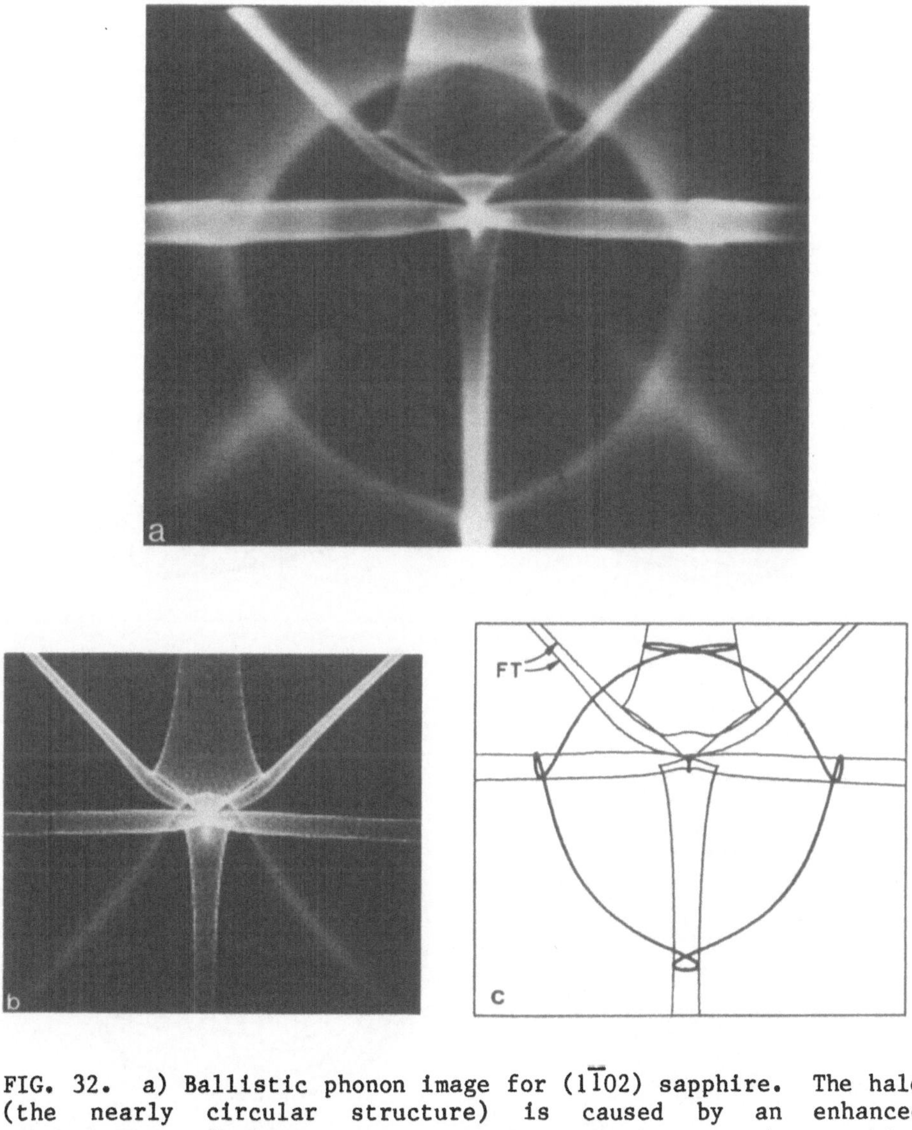

FIG. 32. a) Ballistic phonon image for $(1\bar{1}02)$ sapphire. The halo (the nearly circular structure) is caused by an enhanced transmission of phonons at a certain angle between the sapphire and the metal films. b) Monte Carlo image corresponding to a) for an isotropic distribution of wavevectors. c) Singularity map for this phonon image. The heavy circular line shows the location of the halo predicted by phase matching along the surface between a longitudinal evanescent wave and the bulk transverse wave.

focusing structures, this new structure is related to the surface of the crystal. The halo actually corresponds to a fairly simple distribution in wavevector space; for the data shown, wave vectors are concentrated along a nearly circular cone with axis normal to the surface and half angle of about 33°. These wave vectors seem

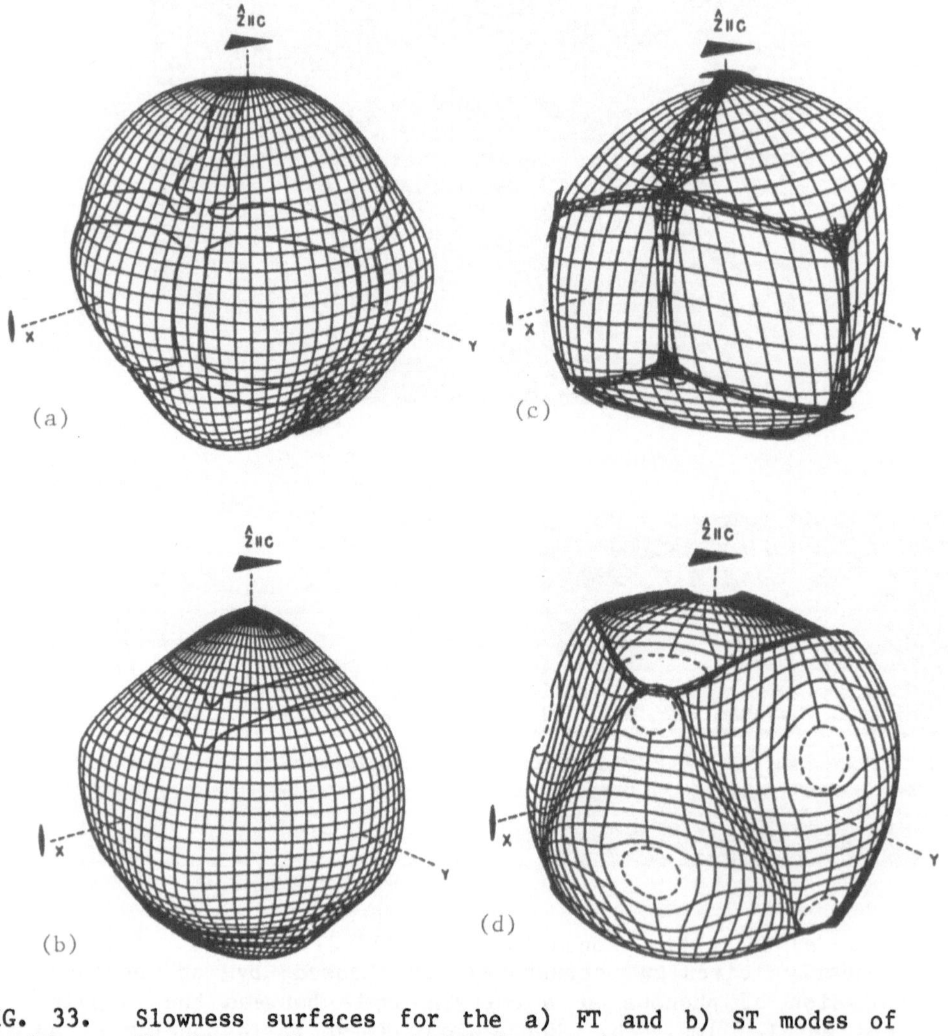

FIG. 33. Slowness surfaces for the a) FT and b) ST modes of sapphire. The heavy lines are singularity lines. Corresponding wave surfaces for the c) FT and d) ST modes. The dashed circles in d) are due to the conic points in the slowness surface.

to be preferentially emitted from the metal heater film.

A semiquantitative explanation of this wavevector-channeling effect is shown in Fig. 34. For simplicity, the crystal is assumed to be isotropic, with spherical L and T surfaces. Imagine first a phonon of wavevector $\vec{k}$ in the sapphire and incident on its surface. If the incident angle θ is less than a certain critical angle, specular reflection allows two reflected bulk waves (L and T) with wavevector component parallel to the crystal surface conserved. As the incident angle is increased, the reflected longitudinal wave propagates more and more parallel to the surface and, above a critical angle, at which the longitudinal wavevector is parallel to the crystal surface, it is no longer possible to create a reflected bulk longitudinal wave with the same $k_\parallel$ as the incident transverse wave. The condition $k_\parallel^2 + k_\perp^2 = k_L^2 = \omega^2/V_L^2$ can, however, be satisfied by taking $k_\perp$ to be imaginary, i.e., corresponding to an evanescent wave with a displacement exponentially decreasing away from the surface. The dashed lines in the inset represent imaginary values of $k_\perp$. This figure shows the squared amplitude $|\Gamma|^2$ of the reflected L wave for a unit-amplitude incident T wave for three different surface conditions. Curve (a) is for a free sapphire surface, curve (b) is for a sapphire surface which is perfectly bonded to copper, and curve (c) is for an intermediate case where the sapphire surface is loosely bonded to the copper film. Thus, for a weakly bonded surface, a rather large amplitude pseudo-surface wave can exist at the surface.

Of course, the "critical cone" is not perfectly circular because of the anisotropy of the slowness surfaces. When mapped into $\vec{V}$-space, the critical cone exhibits folds, due to phonon focusing. A theoretical mapping of the ST-to-L critical cone is shown as the heavy lines in Fig. 32c, which explains well the position of the halo in Fig. 32a.

Every et al. (1984) concluded that the non-equilibrium phonons generated in the copper film couple strongly to the large-amplitude surface wave, thereby creating an enhanced flux of transverse phonons along the critical wavevector cone in sapphire. As shown in the figure, for the enhancement to be significant the bonding between the metal film and the crystal must be relatively weak. A model allowing for weak bonding between the metal and sapphire has been devised by Every (1984). This model explains the critical-cone channeling effect, and a semiquantitative fit to the data indicates that the bonding at the interface is only about 3% of the strength of the bonding of two atoms within the crystal itself! This would be consistent, say, with a boundary layer of 10 atomic spacings and an effective elastic constant about one-third of sapphire.

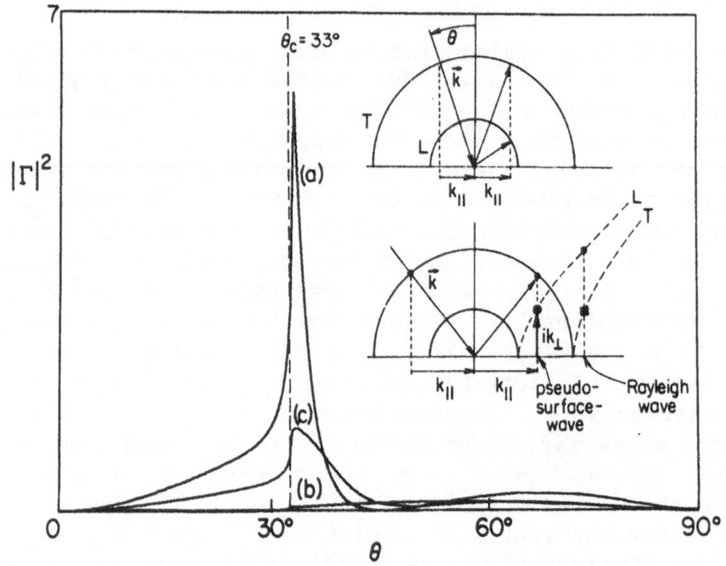

FIG. 34. Mode-conversion strength of transverse to longitudinal waves at a sapphire/copper interface. The dashed line marks the critical angle where an incident transverse wave phase matches to a longitudinal wave travelling parallel to the surface. Just beyond that angle the longitudinal wave becomes evanescent, as shown by the dashed lines in the inset. Curve (a) is for no bonding, (b) is for perfect bonding, and (c) is for loose bonding.

Thus these results indicate that the bonding between sapphire and the copper film is far from perfect. It seems that the critical-cone channeling effect allows us to probe this imperfect bond on a microscopic scale. Since this first experiment, several other systems have been studied. It is found, for example, that when a niobium film is grown epitaxially in ultra high vacuum onto a sapphire substrate, the resulting phonon images show no critical cone channeling effects, indicating that this bond is much better than the copper-sapphire bond described above. Also, phonon imaging experiments have been conducted on a diamond crystal covered with a copper heater film (Hurley, Every and Wolfe, 1984). Very sharp critical-cone channeling effects were observed in this system, both for the heater film and for the aluminum bolometer film. The critical angle in diamond is close to 45°. The sharpness of the critical cone structures in this system arises because diamond has a very small value of C_{12} or effective Poisson's ratio. The sharp critical cone structures are consistent with the weak-bonding model of Every. Such experiments show that phonon imaging can provide a sensitive probe of the bonding between certain solid-solid interfaces.

Scattering From Carriers in Semiconductors

It is interesting to note that the ballistic heat-pulse imaging technique evolved from a study of photoexcited electronic states in semiconductors (Wolfe, 1982). At low temperatures, photoexcitation of a pure semiconductor such as Si or Ge produces electron-hole pairs which interact to form complex excited states of the crystal. The basic excited electronic state of the semiconductor is the exciton, which is simply an electron and hole which are bound together by the Coulomb interaction. The lifetimes of excitons in Ge and Si crystals are several microseconds, long enough for them to interact with each other and form more complex bound states. One of the most interesting discoveries was that a dense gas of excitons can condense into a liquid phase, known as the electron-hole liquid (Hensel et al., 1977). Typically, droplets of electron-hole liquid are formed in a cloud having an average pair density of 1% of the liquid density. The density of pairs in the electron-hole liquid is about 1 millionth the atomic lattice density, yet each droplet of liquid is highly metallic -- having a conductivity comparable to that of copper -- because it is a degenerate Fermi system of particles.

In the past decade there has been considerable interest in the transport of electron-hole droplets (Wolfe, 1982; Bagaev et al., 1983; Wolfe and Jeffries, 1983). Droplets created at the surface of a Ge crystal by laser excitation are found to propagate millimeter distances into the crystal, which is much farther than can be explained by simple diffusion over a time span equal to the droplet's lifetime. In 1976, Keldysh postulated that these electron-hole droplets were pushed into the crystal by a "phonon wind" (Keldysh, 1976). In effect, non-equilibrium ballistic phonons are absorbed by the carriers in the droplet, thereby transferring momentum in the direction of ballistic flux. A luminescent image of the cloud of electron-hole droplets in Ge, as shown in Fig. 35, displays striking anisotropies (Greenstein and Wolfe, 1978; 1981). These data indicate (1) that the phonon wind emanates from a localized source at the excitation point and (2) that the non-equilibrium phonons push much harder in certain directions than in other directions. This experiment and the early heat pulse experiments of Hensel and Dynes (1979, 1977) motivated the present authors to image the phonons themselves, as in Fig. 11.

The principal structures in the cloud of electron-hole droplets in Ge can be understood in terms of phonon focusing. In particular, the sharp flares are due to the intense flux of fast transverse phonons concentrated in the (100) planes of the crystal. The broad lobes are in part due to a large flux of longitudinal phonons around the <111> directions in the crystal. The shape of this cloud has been quite satisfactorily explained from

FIG. 35. Electron-hole liquid cloud in Ge, as imaged by infrared recombination luminescence. This cloud, which is several mm in size, shows several prominent anisotropies: four sharp flares oriented at 45° from vertical, and broad lobes spaced between the flares.

the known ballistic phonon flux and the electron-phonon interaction in this crystal (Greenstein and Wolfe, 1981).

A number of experiments have since been performed by probing electron-hole droplets with ballistic phonons. For example, a phonon spectroscopy experiment has shown that the droplets absorb phonons up to a certain frequency and they are transparent to higher frequency phonons (Dietsche et al., 1982). The high-frequency transmission occurs for phonons with wavevectors larger than twice the Fermi wavevector of the electron-hole liquid. Also recently, phonon imaging has been used to obtain a "shadow" of the electron-hole liquid expanding into the crystal after an intense laser pulse (Kirch and Wolfe, 1984). Figure 36a shows a highly magnified section of the FT ridge in Ge as imaged through a 2.5 cm long crystal. Figure 36b shows this same view shortly after a Q-switched Nd:YAG laser pulse is incident at the right side of the crystal. Electron-hole liquid produced by this pulse absorbs the FT phonons as can be seen from the dark region near the excitation surface. Profiles of the electron-hole liquid expanding into this crystal at successive times after the laser pulse are shown in Fig. 36c.

This "phonon absorption imaging" experiment, as well as previous time-resolved luminescence imaging experiments (Tamor et al., 1983), shows that the droplets continue to expand in the crystal long after the duration of the excitation pulse, indicating that a rather long-lived "hot-spot" is produced near the photo-excitation point. This hot-spot is the same as that observed in the heat pulse experiments shown in Fig. 3. The lifetime of the ballistic heat source depends upon the excitation power. In summary, a combination of luminescence and heat pulse techniques has been used to characterize the interaction of high-frequency phonons with this unique Fermi system.

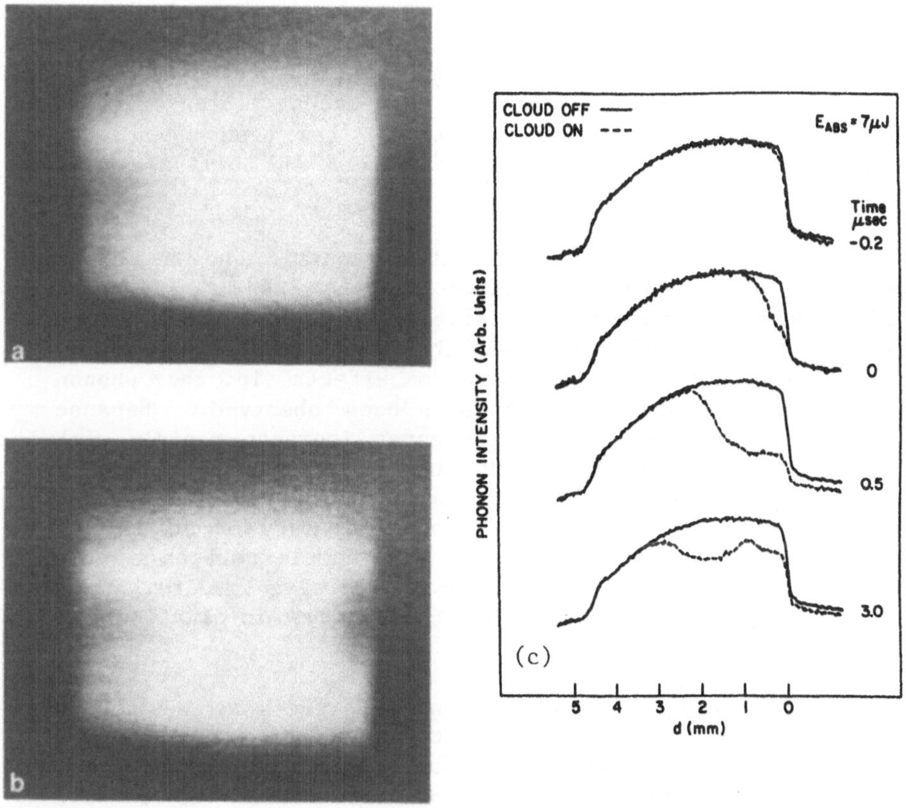

FIG. 36. a) Phonon image along the [110] direction in Ge. Vertical width of the bright ridge is 4°. b) Same image as in a), but with an electron-hole cloud injected into the sample by focusing an intense pulsed laser beam at the right surface. The dark notch on the right shows the absorption of phonons by the EHL. c) Horizontal spatial scans of transmitted phonon intensity as a function of delay after the generation of the cloud. The solid lines are for no cloud, and the dashed lines are with the cloud.

SUMMARY AND PERSPECTIVES

Since its introduction five years ago, the phonon imaging technique has contributed to several major problems in the field of non-equilibrium phonon physics. This review was divided into two main topics: propagation and scattering. Phonon imaging has provided a new and powerful method for observing the anisotropic propagation of ballistic phonons in non-metallic crystals. At low temperatures, any real crystal of sufficient purity is predicted to display highly anisotropic heat flux, due to phonon focusing. Already, a variety of materials have been characterized, including Ge, Si, GaAs, sapphire, quartz, diamond, LiF and CaF_2. With few exceptions, the complex shapes of the observed phonon focusing patterns have been accurately reproduced by calculations based on the continuum elasticity theory and previously determined elastic constants. The imaging method is potentially an accurate means for determining the elastic constants applicable to high frequency phonons. Experiments and calculations have recently been extended to piezoelectric crystals: in $LiNbO_3$, it has been shown that piezoelectric coupling has a profound effect on ballistic heat flow.

The propagation of large-wavevector acoustic phonons is an ongoing pursuit. Based on lattice dynamics models, detailed predictions have been made for the distortion of slowness and wave surfaces in the dispersive regime. By using frequency-sensitive tunnel junction detectors, dispersive effects in the phonon focusing pattern of Ge and GaAs have been observed. Because elastic and inelastic scattering processes increase rapidly with frequency, such experiments require short ballistic path lengths and hence tiny detectors for high angular resolution. Phonon imaging promises to provide a two-dimensional view of the phonon dispersion in a crystal. Considering the complex shapes of the slowness and wave surface in the dispersive regime, such experimental data would produce a wealth of new information for lattice dynamics models.

Phonon scattering from defects and interfaces is the second major topic of our study. Bulk scattering from stress-induced dislocations in LiF was examined in detail by phonon imaging. The principal advantage of this method is the ability to compare scattering strengths of phonons with selected propagation directions and polarizations. The wavevector and polarization of phonons in a particular spatial feature are accurately known from the corresponding phonon focusing calculation. For LiF it was found that the phonon scattering from selectively oriented dislocation lines was strongly dependent on phonon polarization and wavevector. The large anisotropies were well explained by the fluttering string model originally proposed by Granato. The dislocation lines acted as a highly selective phonon polarizer.

The scattering strength of dislocations was understood for moderate plastic deformation (<1.6%), but, as in previous experiments, a highly non-linear increase in scattering strength was observed for larger deformations of the crystal (~10%). Presumably, the increased scattering is due to the effects of dislocation dipoles. So far, the phonon imaging technique has not been applied to the scattering of thermal phonons from point defects in crystals. Again, the angle and polarization dependences of such scattering processes should prove valuable.

The interaction of thermal phonons with interfaces has been a major field of study since Kapitza's first experiments. In recent years experimenters have narrowed in on the two main types of surface scattering: specular and diffuse. It is thought that the diffuse processes contribute to the anomalous Kapitza conductance. By imaging phonons reflected from a crystal surface, one can distinguish specular and diffuse processes by their distinct spatial patterns. Specularly reflected phonons display sharp caustics as a result of phonon focusing and the application of Snell's law at the surface. Diffuse scattering gives rise to a broad non-singular spatial pattern. It was found that internal reflection of thermal phonons from an optical-grade sapphire surface produces sharp, velocity-dependent structures which disappear after the surface has been degraded by mechanical polishing. Monte Carlo calculations specific to this crystal and surface orientation are able to explain the sharp experimental patterns in terms of specularly reflected phonons. The variety of velocity-dependent structures are due to mode conversion of the incident phonons. The ratio of specular-to-diffuse scattering was extremely high (~80%) for the optical-grade sapphire surface, which is quite remarkable in view of the small wavelength (~200Å) of the incident phonons. For future studies, the phonon reflection imaging method will provide a mode- and polarization-sensitive probe of the scattering processes under differing surface conditions.

A related study involves the transmission of thermal phonons across a metal-crystal interface. In the past, it has been assumed that phonons were emitted from a heater film with a rather smooth angular distribution of wavevectors. However, it was discovered by phonon imaging experiments that thermally generated phonons in a metal film were transmitted into a sapphire substrate in a highly anisotropic manner. The phonon images indicated a sharp angular concentration of phonons with wavevectors along the critical cone for conversion of longitudinal to transverse waves. It was concluded that along these critical-cone directions the slow transverse phonons coupled strongly to a longitudinal pseudo-surface-wave, resulting in a larger transmission across the interface. A distinct wavevector-channeling effect was also observed for a metal-diamond interface, and in general the sharpness and intensity of the channeling feature gives a measure of the strength of the metal-crystal bond.

The introduction of the heat pulse method in 1964 greatly expanded the study of thermal transport in crystals by permitting the experimentalist to isolate particular phonon modes by their characteristic time-of-flight. In the ensuing years a variety of phonon spectroscopy techniques have developed which allow in addition a selection of the phonon frequency. The phonon imaging method further permits the experimentalist to select phonons of specific polarization and wave vector as a continuous function of propagation direction. This new ability, which is quite generally applicable to non-metallic crystals, has already proven valuable in the characterization of high-frequency phonon scattering from defects and interfaces.

ACKNOWLEDGEMENTS

The Illinois work was supported in part by the National Science Foundation under the Materials Research Laboratory Grant DMR80-20250 and also by NSF-DMR80-24000.

APPENDIX A -- Phonon Imaging Technique

In this appendix we describe in more detail the phonon imaging technique which has evolved over the past several years. The hardware, shown schematically in Fig. A1, is composed of three principal sections: an excitation source and scanning optics, a thin-film phonon detector and associated amplification and sampling electronics, and a computer used for scanning control and image recording. The rapidly pulsed laser is focused on one surface of the sample, generally covered with a metal film, resulting in the injection of a heat pulse into the sample. The heat-pulse signal is amplified and sampled by a gated integrator (boxcar integrator) which outputs the phonon intensity signal (Z) for a particular laser position (X,Y). The boxcar integrates the heat-pulse signal during a "gate" time Δt centered at a time t after the peak of the laser pulse. This signal may then be recorded as a function of the position (X,Y) of the laser beam on the sample. Data can be recorded in three forms: 1) a "time spectrum" of the detector signal for a fixed laser position, obtained by sweeping the delay of a narrow boxcar gate. 2) A "line scan", giving the Z signal along a single line of propagation directions. 3) an "image", which represents Z as a function of the laser position (X,Y). In the spatial-scanning modes, one often chooses to integrate over a broad time interval Δt centered at a fixed time t. However it is also possible to correlate t with the laser position (X,Y) in order to select a particular velocity for the entire scan range. We refer to this as "constant-velocity scanning". In all, the experimental arrangement of Fig. A1 comprises a flexible and powerful apparatus for spatially and temporally resolving heat pulses.

The samples are cooled by immersion in superfluid helium by means of a cryostat with optical access. A constant temperature is maintained by pumping on the helium and controling its vapor pressure with a manostat. This allows the selection of a fixed temperature from 1.4 K to 4.2 K with a constancy of better than .01 K, which is sufficient for stable biasing of superconducting Al bolometers.

A wide variety of pulsed lasers could be used in this experiment. We frequently use a CW Ar^+ laser which is externally modulated by an acousto-optic chopper (AO) to form pulses. This particular system produces pulses as short as 30 ns, and allows repetition rates as high as several MHz. The phonon time-of-flight sets the limit on practical repetition rate, usually around 100 kHz. The peak power of the pulse at the sample may be varied from 0 to 100 mW. Other pulsing methods may be used to achieve higher powers or shorter time scales. A Nd-YAG laser with an acousto-optic Q-switch produces 150 ns pulses at up to 1 kW peak power, but with a repetition rate of less than 5 KHz. A cavity dumper on the Ar^+ ion laser gives a fixed width pulse of 10 ns with up to 50 W peak power and a high repetition rate. For absolute time calibration or triggering, a glass slide is used to split off about 10% of the modulated laser beam. This portion goes either to a photo-diode or to the detector side of the sample for direct low-level excitation of the bolometer.

The excitation beam, having a nominal diameter of 2 mm, is brought to an initial focus by a 50 mm focal length lens (see Fig. A1). This focal point is then re-imaged 1:1 by a 150 mm "transfer lens", and the 300 mm reconverging beam is reflected from the X-Y scanning mirrors, which are rotated by a pair of galvanometers. This arrangement provides for the scanning of a focused laser beam across the surface of the sample. Alignment and focus of the excitation spot is provided by mounting the first lens on an xyz translation stage. The sharpest focus obtained this way was ~30 μm, and the long distance from the deflection mirrors to the sample (>130mm) allowed distortion-free scanning over lateral distances of at least 10 mm. This arrangement provided sufficient resolution and angular range of phonon propagation direction for samples of thickness 0.5 to 5.0 mm. A nominal angular resolution of 0.3 degrees was achieved for a 5 mm thick quartz crystal.

Samples that are thinner than about 0.5 mm require additional optics to provide sufficient resolution. In this optical scheme the scanned focus is placed about 75 mm outside the cryostat, from

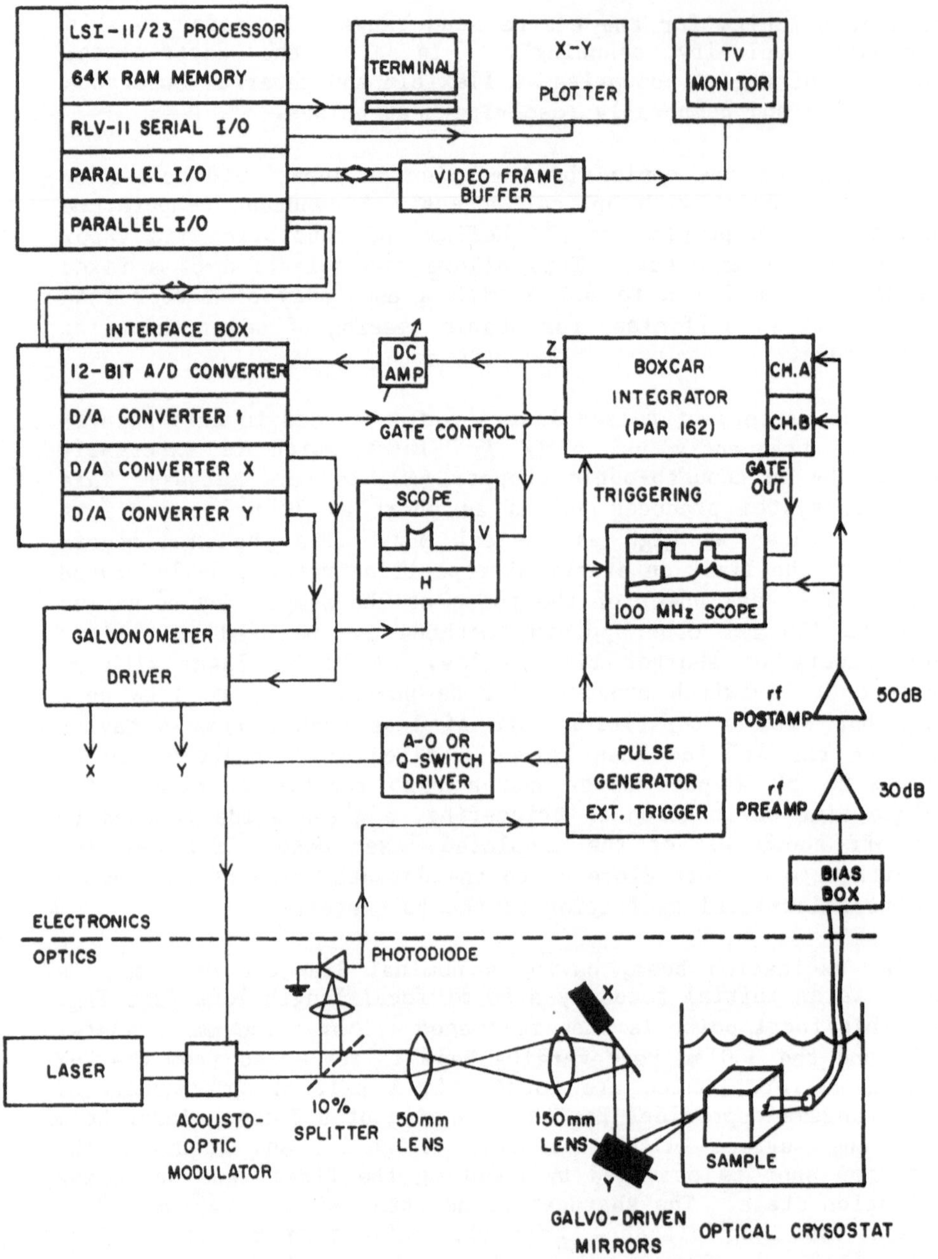

FIG. A1. Block diagram for the phonon imaging system.

which it is transfered once more at a 5:1 reduction to the sample by a fixed 15 mm microscope objective placed inside the cryostat. This results in a focal spot size of about 6 μm. However, since the laser beam is scanned before the final lens its scan range is limited by the size of the lens to about 4 mm, which infers a 0.8 mm scan range of the spot on the sample. This use of a fixed microscope lens in the cryostat allows a potential reduction to propagation lengths of 100 μm. We refer to this optical scheme as "micro-imaging".

Most phonon imaging experiments to date have used a superconducting Al bolometer as a phonon detector, and the size of the bolometer limits the available resolution. Bolometers with a sensitive region as small as 40 μm are fabricated by evaporation through thin stainless steel masks. The higher resolution required by micro-imaging generally requires the use of photo-lithographic techniques directly on the sample. The sample surfaces are prepared by mechanical polishing, often followed by a chemical polish, as dictated by the material under study. Then a pair of Cu pads are evaporated near the center of the detector surface. This is followed by evaporating an Al bolometer strip spanning the gap between the Cu pads. The bolometer is usually about 600 Å thick and evaporated in a partial preassure of O_2 of a few times 10^{-5} Torr at a rate of 10 Å/sec. Two bolometer geometries were used, a serpentine with a 50 x 50 μm^2 sensitive area for the regular imaging, and a straight 5 x 10 μm link for the micro-imaging. To make connection to the bolometer a thin Cu wire is pressed between pieces of In foil, and this sandwich is pressed onto the pads of Cu film.

The wire leads from the bolometer are connected to a small-diameter 50 ohm coaxial cable which runs to the top of the cryostat. There a DC bias current is injected by a high-impedance constant-current source, which is battery-powered and adjustable from 0 to 100 μA. With a bias current of about 1 μA, a μV meter provides a measure of the resistance as the temperature is reduced. In this way the resistance vs. temperature characteristic of the bolometer may be plotted and the optimal biasing temperature determined. The room temperature resistance of the bolometers may range from ~40 ohms to ~300 ohms. When cooled these bolometers have superconducting transition temperatures which range from 1.5 K to 2.2 K. For heat-pulse detection the μV meter is replaced by a low-noise preamp with 50 dB gain and .005-50 MHz band-pass. It is important to use a blocking capacitor before the first amplification stage to prevent the injection of additional DC bias. An additional gain of up to 50 dB is provided by a post amplifier with a DC to 100 MHz bandwidth. This is normally sufficient to bring the signal to the 0.1 to 1.0 volt level required by the gated integrator. The bolometer signal is proportional to any rapid temperature changes of the bolometer, and the detector response time is less than our minimum pulse widths (~30 ns).

The behavior of Al bolometers is strongly affected by both

the ambient temperature and the DC bias current. Initially, the operating temperature is adjusted to where the resistance is half of the normal resistance. Further adjustment is then made after observing the transient response of the bolometer both to heat pulses and to direct illumination by a low power laser pulse. The normal procedure is to maximize the response by adjusting the temperature at a fixed low bias current. Then the bias current is increased, increasing the signal, usually in a strongly non-linear fashion, until operation becomes unstable. An operating point well below that level is chosen. Although the signal is a non-linear function of the bias current, it is usually close to linear with laser power. The detector is then checked for spatial sensitivity by scanning an image. Multiple sensitive areas will produce a sum of spatially offset images (ghosted images). If this ghosting effect occurs it may be reduced or removed by further adjustment of the temperature and bias current.

The amplified bolometer signal is displayed directly on a 100 MHz scope along with a photo-diode signal of the laser pulse. The primary analog processing of the bolometer signal is performed by a two-channel boxcar integrator. It is triggered by the pulse generator used to modulate the laser. One gate (A) is set to sample the arriving phonon signal, while the other (B) is placed before the laser pulse to provide baseline stability and reduce low frequency noise. The delay of the (A) gate may be fixed, or it may be varied by a control voltage from the scanning control system. The difference in the outputs of the two boxcar channels (A-B) becomes the Z intensity. The response time of the Z signal may be controlled both by the input time constants of the integration amplifiers and by a low-pass filter on the output of the difference amplifier. This time constant is set to correspond to the desired spatial resolution divided by the scan speed. Typical values are a response time of 5 ms and a scan speed of 1 second per line. For a repetition rate of 100 kHz, this averages ~500 pulses into each pixel of the image.

The last section of the apparatus concerns the image scan control and recording. The X-Y position of the laser beam is scanned in a raster pattern and the Z signal is recorded as a function of X and Y. We use an LSI-11/23 microcomputer to control the galvos through digital-to-analog converters and to record the integrated bolometer signal through an analog-to-digital converter. Convenient viewing of both real-time and stored image data is provided by a digital frame buffer, and data is permanently stored on magnetic disk. Typically a digital frame buffer is organized as a 256 x 256 array of 6 or 8 bit pixels of solid state memory (RAM). This array is displayed continuously in a video format on a standard video monitor. Random read/write access is provided over the computer bus, usually without interuption of the display. This device is central both to convenient collection and later analysis of image data.

Scanning is done in a raster fashion, with the lines being scanned in one direction only. Increased averaging may be done by summing multiple scans of each line, or by reducing the scan speed and increasing the integration time constants. In addition to imaging, extensive averaging may be done on a single line. Finally the X-Y position may be fixed, and the delay of the A gate swept to obtain a time spectra for a single propagation direction. The three data forms described above (time spectra, line scans, and images) are the principal ones obtainable with this apparatus.

A list of components used in the phonon imaging system is shown in Table A1, along with the commercial sources and a brief description of each component. Virtually all the electronic equipment (or a functional equivalent) in our phonon imaging system is commercially available. The microcomputer should be capable of managing both A-to-D and D-to-A converters, in addition to several serial and parallel I/O ports. Most eight bit processors are probably not fast enough to handle the real-time speed requirements, and the additional computational speed of a 16 bit machine is necessary for constant velocity scanning. A 256x256x6 bit frame buffer should be sufficient for most purposes, but the additional resolution of 512 x 512 x 8 bits is sometimes useful for data analysis. The part of the phonon imaging apparatus most likely to be determined by the particulars of a specific experiment is the laser source. The external modulation of a CW laser described here would be sufficient for many experiments, but a need for more peak power or a shorter pulse may require the use of an intra-cavity device such as a Q-switch, a cavity dumper, or even a mode locker.

As the phonon imaging system described above is heavily dependent upon computer control, it is appropriate to give a brief description of the software that has been developed for this purpose. This software is in the form of a single interactive program which is run under the DEC RT-11 (real time) operating system. This program, known as IMAGER, is written primarily in FORTRAN, with only the low level control routines written in assembly language. IMAGER is divided into three levels: 1) the command level or switchyard, which directs control to a desired function; 2) user routines or functions, which handle interactive I/O and arithmetic processes; and 3) utilities, which handle device control and data I/O. Several examples of functions and utilities are listed in Table A2, along with a description of its purpose. When called, IMAGER presents a prompt and accepts a four letter command. If recognized, the corresponding function is called, and upon completion, control returns to the command level. Each user function exerts control and aquires data through

Table A1 Equipment used in the Illinois phonon imaging system

Component	Manufacturer/model	Remarks
Microcomputer	Digital Equip. Corp. LSI-11/23	Standard Q-bus with serial and parallel I/O boards
Frame buffer	Robot Research 650, or Imaging Tech. IP-512	256x256 pix, 6 bits 512x512 pix, 8 bits
Terminal	Tektronix 4006	Vector graphics
X-Y plotter	Houston Inst HIPLOT	Digital plotter
TV monitor	Panisonic WV-5400	Black & white
Interface box	U. of Illinois	With A/D and D/A cards (similar to CAMAC)
Preamp	Trontech Inc. W50ATC	30 dB gain, 1.3 dB NF
Post-amps	Comlinear E103	35 dB, DC to 100 MHz
Boxcar integrator	PAR 162 and two 164 plug-ins	
Oscilloscope	Tektronix 465	Dual channel, 100 MHz
Pulse gen.	Wavetek 802	Trigger source
Acousto-optic modulator	Intra Action Corp. AOM-125	125 MHz carrier, 15 ns risetime, 50% efficiency
Q-switch (for Nd:YAG)	Intra Action Corp. AQS-25	Produces 150 ns pulses at 1 mJ and 1 kHz
Lasers:	a) Spectra Physics Model 164 Argon ion	1 Watt CW @ 5145 A
	b) Sp. Ph. (Sylvania) Mod 607 Nd:YAG	1 Watt CW @ 1.06 μm
Focusing lens	Nikon 50 mm, f/1.4	Any good imaging lens
Transfer lens	Edmund Scientific 150 mm, f/2	1-to-1 configuration
Microscope lens	Edmund Scientific 10X objective	Mounting threads, etc. machined off
Galvonometers	General Scanning	
	X: GPD-300	With position feedback
	Y: G-100	No feedback
Optical dewar	Janis Research 10DT	Used with forepumps and manostat for control of He vapor pressure

Table A2

Data acquisition software for phonon imaging (IMAGER program.)

Common Fortran Commands (interactive)

Name	Purpose	Utilities called
PSET	Position mirrors at X,Y.	-
LSWP	Continuous scan at line Y.	LSWEP
RSWP	Image sweep to frame buffer.	PHOTO ROBX
OPEN	Open an image file on disk.	-
PLAY	Write disk file to frame buffer.	READIT ROBX
RECD	Record image on disk.	PHOTO WRITIT (ROBX)
RECS	Record constant-velocity image.	PHOTS WRITIT (ROBX)

Main Utility Subroutines (machine language)

Name	Purpose
LSWEP	Set Y mirror and scan X mirror.
READIT	Read a line from a disk file into an array.
WRITIT	Write a line to a disk file into an array.
PHOTO	Set Y, scan X, and record Z into an array.
PHOTS	Same as PHOTO, but adjust boxcar delay during sweep.
ROBX	Transfer an array to the frame buffer.

calls to utility routines. These utilities embody a set of capabilities that may be useful for a variety of functions. As an example the function RSWP raster scans continuously and repetitively through an image, transfering the digitized results to the frame buffer for real-time viewing. This uses two utilities, PHOTO to scan a line and read the 256 intensities into an array, and ROBX to transfer them to the frame buffer. The program structure also includes a large common array for temporary data storage, and makes use of disk overlays to allow transparent access to only the routines in use. This control program is powerful and easy to use, but also may be quickly and easily modified to perform additional tasks without affecting existing functions.

APPENDIX B -- Computational methods

The development of the phonon imaging method has brought about the need for efficient computations of phonon flux anisotropy based upon various models of elastic anisotropy. Such a computational method should emulate the format of the experimental data, while making it simple to change scanning geometries, crystal symmetries, and even elasticity models. The concept of the phonon-focusing enhancement factor (Maris, 1971) gives an analytic expression for the phonon flux anisotropy when the velocity anisotropy may be expressed analytically, which is the case for continuum elasticity. The practical use of such a solution is limited, though, because this enhancement is expressed as a function of wavevector direction, while for comparison to experiment the phonon flux is needed as a function of propagation direction. For even the simplest cases, finding the wavevector for a given propagation direction is a complex process, often resulting in multiple values due to the folding of the velocity surface. Therefore, this analytic approach is tedious and impractical in all but the simplest cases.

A related approach to this problem is to bypass the problem of calculating intensities in all directions and concentrate on the set of lines (ie, a locus of propagation directions) along which flux singularities occur. One may use a root-finding procedure to find lines in wavevector-direction space for which the curvature of the slowness surface is zero. When these singularity lines are mapped into their propagation directions and plotted in the experimental geometry they outline the intense phonon focusing structures. This calculational technique is an efficient means of comparing theory with the experimental phonon focusing patterns, and has even been successful, when used with a dispersive model, in predicting how the singularity pattern shifts with frequency. There are two limitations with this method, though. First, only the shapes of the singularity patterns are found; the intensities of the various structures are not revealed. Second, as one moves from one crystal to another, the topology of the slowness surfaces can change dramatically. The root-finding method requires some preliminary knowledge of the singularity locations in order to initiate the root-finding process. Thus, finding these parabolic (singular flux) lines is not an automatic procedure, and each situation requires individual attention.

We have found a solution to the above problems in what amounts to a more direct simulation of the phonon focusing effect. The physical process of phonon focusing is one of an isotropic (or at least smoothly varying) distribution of phonon wavevectors propagating in directions given by some mapping function $\vec{V}(\vec{k})$. One can simulate that process by generating wavevectors at random (or with a skew proportional to any desired distribution), and then calculating the propagation direction of each according to the mapping and simply tabulating the phonons in equally spaced "bins" in propagation direction. Since most phonon images contain several equivalent symmetry-reduced solid angle sections of propagation direction, phonons need be calculated in only one section and then tabulated in equivalent sections. For cubic symmetry this produces about 8 tabulations for each phonon generated. If one uses a geometry appropriate to the experiment of interest the results should be directly comparable to the data. We refer to this as a Monte Carlo method due to the use of a random number generator. Since there is no randomly selected branching in mid-trajectory, it is not a true Monte Carlo, but rather a "mapping" Monte Carlo. This method is a two-dimensional version of a method commonly used in calculating a single particle density of states.

The Monte Carlo image simulation may be generalized to the case of

$$\begin{aligned} f &= f(x,y) \\ g &= g(x,y), \end{aligned} \tag{B1}$$

where the distribution is observed in the (f,g) space and is generated in the (x,y) space. In the case of continuum elasticity the above mapping is just that given in Equation 9 in the main text. This mapping may be evaluated in a fraction of a second by a VAX 730 computer, making runs of order a few times 10^5 phonons practical. When tabulated in a 256 X 256 array this results in an average of 20 counts per pixel, which is sufficient for most analysis. With the extension of this mapping Monte Carlo technique to dispersive models, however, the computation time increases to several seconds per phonon, making a more efficient algorithm a necessity.

With the exception of points of conic contact between modes (Musgrave, 1970) the mapping expressed by Eq. B1 is usually a well behaved function, with well defined derivatives at each point. If the functions f and g are expanded about $f_0=f(x_0,y_0)$ and $g_0=g(x_0,y_0)$ they may be expressed as

$$\begin{aligned} f &= f_o + f_x dx + f_y dy \\ g &= g_o + g_x dx + g_y dy, \end{aligned} \tag{B2}$$

with the coefficients f_x, f_y, g_x, and g_y being the local first derivatives of the mapping. If the x,y space is divided into small rectangular regions, called neighborhoods, then one may approximate the mapping within a each neighborhood with Eq. B2. The expansion coefficients may be determined once for each neighborhood, and then many points may be efficiently mapped with this expansion. This procedure is then repeated for each of the neighborhoods. This technique, which we refer to as a "neighborhood Monte Carlo", allows the calculation of around 10^6 points in less than one hour on a VAX 730, with results that in almost all cases agree well with the single point procedure.

The concept of the neighborhood Monte Carlo is readily applied to dispersive phonon focusing with one small change. Whereas the phonon propagation direction is a function of only two variables (θ_k, ϕ_k) in the non-dispersive limit, with dispersion it depends also on the magnitude of the wavevector. This means that the neighborhoods are three dimensional, and that the magnitude of the velocities may vary continuously with the dispersion. Since there are a variety of dispersion models that this effort might draw upon, we obtain the frequency as a function of wavevector by interpolation of a rectangular table of $\omega(\vec{k})$ with values filled by the model of interest. For cubic symmetry only 1/48th of the full Brillouin zone need be considered, and wavevectors that fall outside the symmetry-reduced section are mapped into it before interpolation. In this way a three dimensional grid of reasonable density (40 X 40 X 40) may be used. This grid extends only to the largest wavevector of interest, as determined by the frequencies of interest. This look-up-table, once calculated, can be used to interpolate frequency, and from this a double finite difference calculation provides the group velocity and its derivatives with respect to wavevector. The Monte Carlo may then proceed over a three dimensional grid of neighborhoods.

If a dispersive neighborhood Monte Carlo calculation is carried out over a large range of frequencies the singularity patterns caused by dispersion will be smeared out. Frequency resolution is included by tabulating only those neighborhoods that fall within a narrow range of frequencies. This allows one to efficiently calculate the phonon intensity pattern expected for selected frequency bands. Slight program modifications could produce images resolved by magnitude of velocity, etc.

The extensive use of interpolative look-up-tables in the dispersive calculations points out that this method may be generalized to apply to any mapping, even if the mapping procedure is very involved, as long as the mapping is a smooth, well-behaved function. An example of this is the specular reflection image calculation described in the main text. The normal Monte Carlo procedure is to generate a phonon, propagate it to the reflection surface, reflect it specularly into one of three modes, propagate it back to the generation surface, and tabulate it. At a glance, this problem appears not to be a deterministic mapping, since the reflected mode is chosen randomly (but with a weight determined by boundary conditions). It may be made deterministic, however, by forcing a specific mode conversion, and then explicitly including a tabulation weight which is proportional to the branching ratio for that process. Then if the entire generation, propagation, reflection, and return process is expressed as a function or mapping with the initial wavevector direction as input and the final phonon position as output, a look-up table may be calculated, and an appropriately weighted image formed.

With the exception of the dispersive case, none of the above problems absolutely require expansion and interpolation techniques. However, as phonon transport experiments further the study of anisotropic transport in systems that include substantial phonon scattering, there will be a need for true transport Monte Carlo calculations which include anisotropic propagation effects, quite possibly in the dispersive limit. A good example of this is the quasi-diffusive transport mechanism proposed to explain photogenerated heat pulse transport in GaAs (Guseinov and Levinson, 1983). A realistic calculation based upon this idea would involve the true 3-phonon decay (two phonon density-of-states), elastic scattering, and dispersive phonon focusing. Such efforts will clearly require the most efficient computational methods to stand any chance of success.

REFERENCES

Anderson, A.C., and Malinowski, M.E., 1971, Interaction between thermal phonons and dislocations in LiF, Phys. Rev. B5, 3199.

Anderson, A.C., 1976, The thermal boundary resistance, in: "Phonon scattering in solids", L.J. Challis V.W. Rampton, and A.F.G. Wyatt, eds., (Plenum, New York).

Anderson, A.C., 1981, The Kapitza thermal boundary resistance between two solids, in: "Nonequilibrium superconductivity, phonons, and Kapitza boundaries", K.E. Gray, ed., (Plenum, New York).

Armbruster, D., Dangelmayr, G., and Guttinger, W., 1984, Nonlinear phonon focusing, in: "Phonon Scattering in Condensed Matter", W. Eisenmenger, K. Lassmann, and S. Dottinger, eds., (Springer-Verlag, Berlin).

Bagaev, V.S, Galkina, T.I., and Sibeldin, N.N., 1983, Interaction of EHD with deformation field, ultrasound and non-equilibrium phonons, in: "Electron-hole droplets in semiconductors", C.D. Jeffries and L.V. Keldysh, ed., (Elsevier, New York).

Berman, R., 1976, "Thermal Conduction in Solids", (Clarendon, Oxford).

Bron, W.E., 1980, Spectroscopy of High Frequency Phonons, Rep. Prog. Phys. 43, 20.

Brown, R.A., 1981, The effect of dislocations on thermal conductivity, in: "International conference on phonon physics", W.E. Bron, ed., (J. Phys. (Paris) 42, C6).

Cady, W.G., 1946, "Piezoelectricity", (McGraw-Hill, New York).
Carruthers, P., 1961, Theory of thermal conductivity of solids at low temperatures, Rev. Mod. Phys. 33, 92.
Christoffel, E.B., 1877, Ann. Mat. Pura. Appl. 8, 193.
Cotts, E.J., Miliotis, D.M., and Anderson, A.C., 1981, Thermal transport in deformed LiF, Phys. Rev. B24, 7336.
Dietsche, W., Northrop, G.A., and Wolfe, J.P., 1981, Phonon focusing of large-k acoustic phonons in Germanium, Phys. Rev. Lett. 47, 660.
Dietsche, W., Kirch, S.J., and Wolfe, J.P., 1982, Phonon spectroscopy of the electron-hole liquid in Ge, Phys. Rev. B26, 780.
Eichele, R., Huebener, R.P., and Seifert, H., 1982a, Phonon focusing in quartz and sapphire imaged by electron beam scanning, Z. Phys. B48, 89.
Eichele, R., Huebener, R.P., Seifert, H., and Selig, K.P., 1982b, Imaging of ballistic phonon propagation in quartz by electron beam scanning, Phys. Lett. 87A, 469.
Eisenmenger, W., and Dayem, A.H., 1967, Quantum generation and detection of incoherent phonons in superconductors, Phys. Rev. Lett. 18, 125.
Eisenmenger, W., 1976, Superconducting Tunnel Junctions as Phonon Generators and Detectors, Physical Acoustics, Vol. 12, ed. W. P. Mason and R. N. Thurston (New York: Academic), p. 79.
Eisenmenger, W., 1979, Phonon Detection by the Fountain Preassure in Superfluid Helium Films, in: "Phonon Scattering in Condensed Matter", H.J. Maris, ed., (Plenum, New York).
Every, A.G., 1980, General closed-form expressions for acoustic waves in elastically anisotropic solids, Phys. Rev. B22, 1746.
Every, A.G., 1981, Ballistic phonons and the shape of the ray surface in cubic crystals, Phys. Rev. B24, 3456.
Every, A.G., Koos, G.L., and Wolfe, J.P., 1984, Ballistic phonon imaging in sapphire: bulk focusing and critical-cone channeling effects, Phys. Rev. B29, 2190.
Granato, A.V., 1958, Thermal properties of mobile defects, Phys. Rev. 111, 740.
Greenstein, M., and Wolfe, J.P., 1978, Anisotropy in the shape of the electron-hole-drop cloud in germanium, Phys. Rev. Lett. 41, 715.
Greenstein, M., and Wolfe, J.P., 1981, Phonon-wind-induced anisotropy of the electron-hole droplet cloud in Ge, Phys. Rev. B24, 3318.
Greenstein, M., Tamor, M.A., and Wolfe, J.P., 1982, Propagation of laser-generated heat pulses in crystals at low temperature: Spatial filtering of ballistic phonons, Phys. Rev. B26, 5604.
Guseinov, N.M., and Levinson, Y.B., 1983, Diffusion of nondecaying TA phonons, Solid State Commun. 45, 371.
Hensel, J.C., Phillips, T.G., and Thomas, G.A., and Rice, T.M.,

1977, The electron-hole liquid in semiconductors in: "Solid State Physics", Vol. 32, H. Ehrenreich, F. Seitz, and D. Turnbull, eds., (Academic Press, New York).

Hensel, J.C., and Dynes, R.C., 1977, Interactions of electron-hole drops with ballistic phonons in heat pulses: the phonon wind, Phys. Rev. Lett. 39, 969.

Hensel, J.C., and Dynes, R.C., 1979, Observation of singular behavior in the focusing of ballistic phonons in Ge, Phys. Rev. Lett. 43, 1033.

Hurley, D.C., and Wolfe, J.P., 1984, Phonon focusing in cubic crystals (to be published).

Hurley, D.C., Every, A.G., and Wolfe, J.P., 1984, Ballistic Phonon Imaging of Diamond, J. Phys. C.: Solid State Phys. 17, 3157.

Keldysh, L.V., 1976, Phonon wind and dimensions of electron-hole drops in semiconductors, JETP Lett. 23, 86.

Kinder, H., Lassmann, K., and Eisenmenger, W., 1970, Phonon emission by quasiparticle decay in superconducting tunnel junctions, Phys. Lett. 31A, 475.

Kinder, H., 1975, Phonon spectroscopy at ultrahigh frequencies, in: "Proceedings of the 14th international conference on low temperature physics", Vol. 5, M. Krusius and M. Vuorio, eds., (North-Holland, New York).

Kittel, C., 1976, "Introduction to Solid State Physics", Fifth edition (Wiley, New York).

Kirch, S.J., and Wolfe, J.P., 1984, Phonon-absorption imaging of the electron-hole liquid in Ge, Phys. Rev. B29, 3382.

Klemens, P.G., 1958, Thermal conductivity and lattice vibrational modes in: "Solid State Physics", Vol. 7, F. Seitz and D. Turnbull, eds., (Academic Press, New York).

Kneezel, G.A., and Granato, A.V., 1982, Effect of independent and coupled vibrations of dislocations on low-temperature thermal conductivity in alkali halides, Phys. Rev. B25, 2851.

Koos, G.L. and Wolfe, J.P., 1984, Piezoelectricity and ballistic heat flow, Phys. Rev. B29, 6015.

Koos, G.L., Every, A.G., Northrop, G.A. and Wolfe, J.P., 1984, Ballistic Phonon Imaging in Sapphire: Bulk Focusing and Critical-cone Channeling Effects, Phys. Rev. B29, 2190.

Koos, G.L., Every, A.G., Northrop, G.A., and Wolfe, J.P., 1983, Critical-cone channeling of thermal phonons at a sapphire-metal interface, Phys. Rev. Lett. 51, 276.

Koos, G.L., and Wolfe, J.P., 1984a, Phonon focusing in piezoelectric crystals: Quartz and lithium niobate, Phys. Rev. B (September 15).

Kunc, K., Nielsen, O. Holm, 1979a, Lattice dynamics of zincblende structure compounds using deformation-dipole model and rigid ion model, Comp. Phys. Comm. 16, 181.

Kunc, K., Nielsen, O. Holm, 1979b, Lattice dynamics of zincblende structure compounds: II. Shell model, Comp. Phys. Comm. 17, 413.

Lax, M., and Narayanamurti, V., 1980, Phonon magnification and the gaussian curvature of the slowness surface in anisotropic media: detector shape effects with application to GaAs, Phys. Rev. B22, 4876.

Maris, H.J., 1971, Enhancement of heat pulses in crystals due to elastic anisotropy, J. Acoust. Soc. Am. 50, 812.

Marx, D., Buck, J., Lassmann, K., and Eisenmenger, W., 1978, Reflection of high frequency phonons at free silicon surfaces, Journal de Physique 39, C6, 1015.

Marx, D., and Eisenmenger, W., 1981, Reflection of high-frequency phonons at silicon-solid interfaces, Phys. Lett. 82A, 291.

Marx, D., and Eisenmenger, W., 1982, Phonon scattering at siliconcrystal surfaces, Z. Phys. B48, 277.

McCurdy, A.K., Maris, H.J., and Elbaum, C., 1970, Anisotropic heat conduction in cubic crystals in the boundary scattering regime, Phys. Rev. B2, 4077.

Metzger, W., Eichele, R., Seifert, H., and Huebener, R.P., 1984, Phonon focusing in germanium imaged by electron-beam scanning, in: "Phonon Scattering in Condensed Matter", W. Eisenmenger, K. Lassmann, and S. Dottinger, eds., (Springer-Verlag, Berlin).

Musgrave, M.J.P., "Crystal Acoustics", 1970, (Holden-Day, San Francisco).

Northrop, G.A., 1982a, Acoustic phonon anisotropy: phonon focusing, Comp. Phys. Comm. 28, 103.

Northrop, G.A., 1982b, Phonon focusing of dispersive phonons in Ge, Phys. Rev. B26, 903.

Northrop, G.A., and Wolfe, J.P., 1979, Ballistic phonon imaging in solids - a new look at phonon focusing, Phys. Rev. Lett. 43, 1424; 1980, Ballistic phonon imaging in germanium, Phys. Rev. B22, 6196.

Northrop, G.A., Cotts, E.J., Anderson, A.C., and Wolfe, J.P., 1982, Phonon imaging of highly dislocated LiF, Phys. Rev. Lett. 49, 54; 1983, Anisotropic phonon-dislocation scattering in deformed LiF, Phys. Rev. B27, 6395.

Northrop, G.A., and Wolfe, J.P., 1984, Phonon reflection imaging: A determination of specular versus diffuse boundary scattering, Phys. Rev. Lett. 52, 2156.

Northrop, G.A., Hebboul, S., and Wolfe, J.P., 1985, to be published.

Parrott, J.E., and Stukes, Audrey, 1975, "Thermal Conductivity of Solids", (Pion, London).

Rösch, F., and Weis, O., 1976, Geometric propagation of acoustic phonons in monocrystals within anisotropic continuum acoustics, Part I, Z. Physik B25, 101; 1976, Geometric propagation of acoustic phonons in monocrystals within anisotropic continuum acoustics, Part II, Z. Physik B25, 115.

Rösch, F., and Weis, O., 1977, Phonon transmission from incoherent radiators into Quartz, Sapphire, Diamond, Silicon, and Germainium within anisotropic continuum acoustics, Z. Physik B27, 33.

Stock, B., Ulbrich, R.G., and Fieseler, M., 1984, Direct observation of ballistic large-wavevector phonon propagation in gallium arsenide, in: "Phonon Scattering in Condensed Matter", W. Eisenmenger, K. Lassmann, and S. Dottinger, eds., (Springer-Verlag, Berlin).

Taborek, P., and Goodstein, D., 1980a, Phonon focusing catastrophes, Sol. St. Comm. 33, 1191.

Taborek, P., and Goodstein, D., 1980b, Diffuse reflection of phonons and the anomalous Kapitza resistance, Phys. Rev. B22, 1550.

Tamor, M.A., Greenstein, M., and Wolfe, J.P., 1983, Time-resolved studies of electron-hole-droplet transport in Ge, Phys. Rev. B27, 7353.

Tamura, S., 1982, Focusing of high-frequency dispersive phonons, Phys. Rev. B25, 1415.

Tamura, S., 1983a, Isotope scattering of dispersive phonons in Ge, Phys. Rev. B27, 858.

Tamura, S., 1983b, Large-wavevector phonons in highly dispersive crystals: Phonon-focusing effects, Phys. Rev. B28, 897.

Taylor, B., Maris, H.J., and Elbaum, C., 1969, Phonon focusing in solids, Phys. Rev. Lett. 23, 416.

Taylor, B., Maris, H.J., and Elbaum, C., 1971, Focusing of phonons in crystalline solids due to elastic anisotropy, Phys. Rev. B3, 1462.

Ulbrich, R.G., Narayanamurti, V., and Chin, M.A., 1980, Propagation of large wave vector acoustic phonons in semiconductors, Phys. Rev. Lett. 45, 1432; 1981, Ballistic transport and decay of near zone-edge non-thermal phonons in semiconductors, in: "International conference on phonon physics", W.E. Bron, ed., (J. Phys. (Paris) 42, C6).

von Gutfeld, R.J., and Nethercot, A.H., 1964, Heat pulses in Quartz and Sapphire at low temperatures, Phys. Rev. Lett. 12, 641.

von Gutfeld, R.J., 1968, Heat Pulse Transmission, in: "Physical Acoustics", Vol. 5, ed. W.P. Mason (Academic, New York).

Weber, J., Sandemann, W., Dietsche, W., and Kinder, H., 1978, Absence of anomalous Kapitza conductance on freshly cleaved surfaces, Phys. Rev. Lett. 40, 1469.

Weis, O., 1969, Thermal phonon radiation, Z. Angew. Phys. 26, 325.

Weis, O., 1972, The solid-solid interface in thermal phonon radiation, J. Phys. (France) 33, C-4, 48.

Wolfe, J.P., 1980, Ballistic heat pulses in crystals, Phys. Today, 33, 44. (December)

Wolfe, J.P., 1982, Thermodynamics of excitons in semiconductors, Phys. Today, 35, 46. (March)

Wolfe, J.P., Greenstein, M., Northrop, G.A., and Tamor, M.A., 1980, Images of electron-hole droplets and ballistic phonons in Ge, in: "Phonon Scattering in Condensed Matter", H.J. Maris, ed., (Plenum, New York).

Wolfe, J.P., and Jeffries, C.D., 1983, Strain-confined excitons and electron-hole liquid, in: "Electron-hole droplets in semiconductors", C.D. Jeffries and L.V. Keldysh, ed., (Elsevier, New York).

Wolfe, J.P., and Northrop, G.A., 1984, Search for large k-vector phonons in GaAs, in: "Phonon Scattering in Condensed Matter", W. Eisenmenger, K. Lassmann, and S. Dottinger, eds., (Springer-Verlag, Berlin).

Wyatt, A.F.G., 1981, Kapitza conductance of solid-liquid He interfaces, in: "Nonequilibrium superconductivity, phonons, and Kapitza boundaries", K.E. Gray, ed., (Plenum, New York).

Zdetsis, A.D., and Wang, C.S., 1979, Lattice dynamics of Ge and Si using the Born-von Karman model, Phys. Rev. B19, 2999.

AN INTRODUCTION TO CROSSING EFFECTS IN PHONON SCATTERING

L.J. Challis

Department of Physics
University of Nottingham
University Park, Nottingham, NG7 2RD, U.K.

INTRODUCTION

Transitions between two electronic states of a magnetic ion can be induced by a phonon because of the oscillatory electric fields it produces through the motion of the vibrating ligands. The transition probability and hence the phonon scattering rate are both resonant when the phonon frequency ν is equal to the transition frequency ν_o so that phonons can be used for spectroscopy in a similar way to photons. Resonant phonon scattering has now been seen from many different types of ion or impurity although for convenience the discussion here will be largely restricted to magnetic ions. I shall also concentrate on crossing effects seen in recent work where the experimental conditions conformed more closely to those assumed in the theoretical analysis presented here. A more complete description which attempts to give the earlier work its proper role in the development of the subject will be given elsewhere (Challis 1985) although I must mention here the pioneering work of Berman et al (1963) who observed the first frequency crossing signals and Walton (1966, 1967) and Hetzler and Walton (1973) who stressed the use of a known ion as a probe to investigate another.

Crossing effects in phonon scattering is the term used to describe effects that can occur when a phonon beam of broad spectral content is being scattered by two different transitions with similar transition frequencies ν_1, ν_2. The total scattered intensity can change rapidly with $(\nu_1 - \nu_2)$ approaching a turning value as $(\nu_1 - \nu_2) \rightarrow 0$. However, before discussing these effects I should like to review briefly the nature of the resonant phonon scattering process.

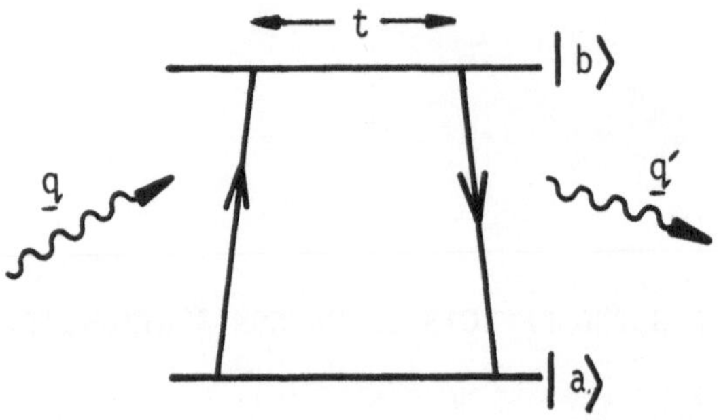

Fig.1 Phonon scattering by the 2nd order resonance fluorescence process.

Let us first consider what happens when the scattering ions are well separated with only weak interactions between them. If the ion is in its ground state, $|a\rangle$ a phonon with $\nu_q \sim \nu_o$ can cause a transition to the upper state $|b\rangle$ of relative energy $h\nu_o$ (Fig.1). After a time t which, by definition, has an average value of T_1, (the spin-lattice relaxation time), the ion will de-excite emitting a phonon $\underline{q}'$ in general in a different direction. So phonon scattering has occurred and this requires a phonon operator $a_q a^+_{q'}$. Now the Hamiltonian coupling a single phonon to an ion describes the change in potential energy eV caused by the instantaneous strain in the lattice it produces. For these small strains, $\mathcal{H}$ is linear in strain and so in phonon operators $(a_q + a^+_{q'})$ and to obtain an operator $a_q a^+_{q'}$ the Hamiltonian describing energy changes due to all phonons must be applied twice. This then is a <u>second-order</u> process. It is also elastic since $\nu_q = \nu_{q'}$ and is equivalent to the resonance fluorescence process by which light is resonantly scattered by atoms. It is variously referred to by all three names either separately or in some combination.

The excited ion may also relax by giving its energy to a neighbour in its ground state (cross-relation) which then gives it to the next and so on. The average time for the ion to relax by this process is T_2, the spin-spin relaxation time, which decreases as the ions get closer together. In this process of phonon scattering, a phonon is only involved in the initial excitation of the ion so this is a <u>first order</u> process requiring a single phonon operator a_q. It can be seen that it is inelastic for the phonon system and is referred to by these names as well as by the <u>direct</u> process.

From this discussion it can be seen that whether the second or the first order process dominates depends on whether the excited ion decays mostly by phonon emission or by cross-relaxation i.e., on whether $T_1 < T_2$ or vice-versa. As a rough guide it appears that for dilute systems, say $\lesssim$100 ppm., the second order process dominates when $\nu_0 \gtrsim$ 100 GHz although this clearly depends on the ions and in particular on the strength of their coupling to the lattice. Detailed discussions of these processes has been given elsewhere (1st order : Orbach (1962), 2nd order : Griffin and Carruthers (1963), Seiden (1963); see also Dreyfus (1972) and Sheard (1976)), and only a simple qualitative description will be given here. We start by considering the first order process. Fermi's Golden rule shows that that the probability of a phonon being absorbed (annihilated) $\propto |\langle n_q -1 | a_q | n_q \rangle|^2 \propto \nu$ and energy conservation to within the requirements of the uncertainty principle gives a Lorentzian lineshape $g(\nu-\nu_0) \propto \Gamma/[(\nu-\nu_0)^2 + \Gamma^2]$where $2\pi\Gamma = 1/2T_2$. So the scattering rate $\tau_R^{-1}(\nu) \alpha\ \nu_o g(\nu-\nu_0)$. This is the rate caused by one ion and for a set, τ_R^{-1} may be somewhat broader and distorted in lineshape since ν_0 will vary from ion to ion in the sample because of random strains etc. Nevertheless, for simplicity, we assume $g(\nu-\nu_0)$ remains Lorentzian so that near to resonance we may write $\tau_R^{-1} = D/[(\nu-\nu_0)^2 + \Gamma^2]$ where the ν in the numerator has been replaced by ν_0 and absorbed into the multiplying constant D and Γ now includes strain broadening. For a second order (elastic) process we have both phonon annihilation and creation giving a combined probability $\propto \nu^2$. An additional factor of ν^2 must be included to allow for the dependence of the probability of emission on the density of phonon states and energy conservation again gives a Lorentzian lineshape with now $2\pi\Gamma=1/2T_1$. For a set of ions this may also be broadened and distorted although we shall ignore any distortion and again subsume any broadening into Γ. So now $\tau_R^{-1} \propto \nu^4 g(\nu-\nu_0) = D/[(\nu-\nu_0)^2 + \Gamma^2]$ near to resonance as before although of course D is different in form, magnitude and even temperature dependence. The effect of this resonant scattering depends on the strength of the background scattering rate τ_B^{-1} (often due to boundaries) so it is convenient to write it in the reduced form

$$\frac{\tau_R^{-1}}{\tau_B^{-1}} = \frac{D\tau_B}{(\nu-\nu_0)^2 + \Gamma^2} = \frac{\Delta^2}{(\nu-\nu_0)^2 + \Gamma^2} \qquad (1)$$

where $\Delta = (D\tau_B)^{1/2}$ has the dimensions of a frequency. We note that $\tau_R^{-1}(\nu_0)/\tau_B^{-1} = \Delta^2/\Gamma^2$ so that $\Delta \gg \Gamma$ for strong resonant scattering $(\tau_R^{-1}(\nu_0) \gg \tau_B^{-1})$ and $\Delta \ll \Gamma$ for weak resonant scattering.

The effect of resonant scattering on thermal conduction

Crossing effects have mostly been seen in thermal conduction although they have also been seen in heat pulse attenuation. In a Debye model at low temperatures, the thermal conductivity of a pure crystal due to phonons in a bandwidth $d\nu$ centred at ν is given by

$$K_o(\nu)d\nu = (1/3)v^2C_s(\nu)\tau_B d\nu = C(\nu)\tau_B d\nu$$

where v is a mean sound velocity, $C_s(\nu)d\nu$, the specific heat of the phonons in the bandwidth and, $C(\nu) = (1/3)v^2C_s(\nu)$. This result is obtained from the Boltzmann equation (e.g. Berman 1976) but is also predictable from simple kinetic theory. The total conductivity $K_o \approx \int_0^\infty K_o(\nu)d\nu$. The addition of a resonant scattering process reduces $K_o(\nu)$ to $K(\nu)$ where

$$K(\nu)d\nu = C(\nu)\ \tau d\nu$$

and $\tau^{-1} = \tau_B^{-1} + \tau_R^{-1}$ and is plotted in fig. 2 in the reduced form τ^{-1}/τ_B^{-1}, assuming τ_B^{-1} to be independent of frequency (normally a good approximation at least in comparison with τ_R^{-1}) and $\Delta = 3\Gamma$. Also shown is $\Delta\tau = \tau_B - \tau$ in the reduced form $\Delta\tau/\tau_B$

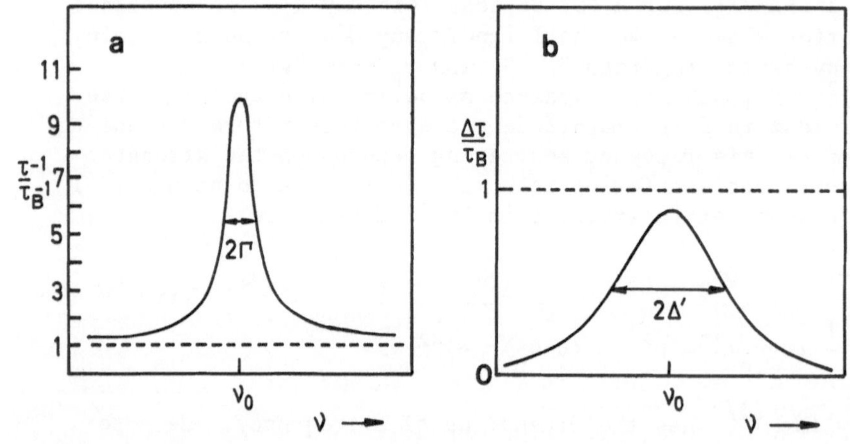

Fig.2(a). The total scattering rate $\tau^{-1} = \tau_R^{-1} + \tau_B^{-1}$ plotted as τ^{-1}/τ_B^{-1} for a system with $\tau_R^{-1}(\nu_o) = 9\tau_B^{-1}$ ($\Delta = 3\Gamma$). The full width of τ_R^{-1} is 2Γ. (b). The decrease, $\Delta\tau$, in total relaxation rate near to resonance plotted as $\Delta\tau/\tau_B$. The full width of $\Delta\tau$ is $2\Delta'$ where $\Delta' = (\Delta^2 + \Gamma^2)^{1/2} = \sqrt{10}\Gamma$.

From equation (1) we can show that

$$\frac{\Delta\tau}{\tau_B} = \frac{\Delta^2}{(\nu-\nu_o)^2 + \Delta^2 + \Gamma^2}$$

which has an area $\pi\Delta^2/(\Delta^2+\Gamma^2)^{1/2}$ and a width $\Delta' = (\Delta^2 + \Gamma^2)^{1/2} = \sqrt{10}\Gamma$ for the case shown in fig. 2. The width approaches limits of Γ and Δ for weak and strong resonant scattering compared with τ_B^{-1}. So for weak scattering, the width of $\Delta\tau$ is independent of the concentration of scattering centres (unless of course this affects Γ) while for strong scattering it $\propto D^{1/2} \propto (\text{conc})^{1/2}$, and the areas $\propto$ (conc) and $(\text{conc})^{1/2}$ respectively. The significance of these parameters is that the fractional decrease in conductivity produced by the resonance is

$$\frac{\Delta K}{K_o} = \int_0^\infty C(\nu)\Delta\tau(\nu)\,d\nu \Big/ \int_0^\infty C(\nu)\tau_B\,d\nu$$

$$\simeq C(\nu_o)\int_0^\infty \frac{\Delta\tau(\nu)\,d\nu}{\tau_B} \Big/ \int_0^\infty C(\nu)\,d\nu \qquad (2)$$

since

$$C(\nu) = \left(\frac{4\pi h^2}{\nu k_B T^2}\right) \frac{\nu^4 \exp(h\nu/kT)}{[\exp(h\nu/kT)-1]^2}$$

has a half width ~ $2k_BT$ ~ 40GHz at 1K and so varies very slowly with frequency compared with $\Delta\tau$ assuming, as we shall throughout, that we are dealing with processes for which Γ and Δ are both very much less than k_BT. Now

$$C(\nu_o) = \left(\frac{4\pi h^2}{\nu k_B T^2}\right) \frac{\nu_o^4 \exp(h\nu_o/kT)}{[\exp(h\nu_o/kT)-1]^2}, \quad \int_0^\infty C(\nu_o) = \frac{16\pi^5 k_B^4 T^3}{15 h^3 v}$$

so that

$$\frac{\Delta K}{K_o} = \frac{15\, x_o f(x_o)}{4\pi^3 \nu_o} \frac{\Delta^2}{(\Delta^2 + \Gamma^2)^{1/2}}$$

where $x_o = h\nu_o/k_BT$ and $f(x) = x^4 e^x/(e^x - 1)^2$. For strong scattering ($\Delta>>\Gamma$) this becomes

$$\frac{\Delta K}{K_o} = \frac{15\, x_o f(x_o)}{4\pi^3} \cdot \frac{\Delta}{\nu_o} = 0.121 x_o f(x_o) \cdot \Delta/\nu_o$$

which has a maximum value at $x_o = 4.9$, when ν_o lies close to the peak of the phonon spectrum, of $2.6(\Delta/\nu_o)$. So, for example, resonance scattering at 200GHz of a strength corresponding to Δ = 1GHz causes a maximum change in thermal conductivity of about 1%. The change is proportional to Δ and so to $(\mathrm{conc})^{1/2}$. The reason for this dependence is, that in this limit, $\tau_R^{-1}(\nu_o) \gg \tau_B^{-1}$, $K(\nu)$ drops to nearly zero at resonance for all values of Δ. So the loss of conductivity, which varies as the area of this hole in $K(\nu)$, increases linearily with its width, Δ.

Similar analysis can be applied to the loss of transmission from a heat pulse or to the modulation window in the phonons from a tunnel junction and again the effective bandwidth is frequently Δ rather than Γ although with τ_B replaced by the transit time of the pulse (Anderson and Challis 1975a).

The simple analysis given above provides an introduction to a consideration of crossing effects. When two resonant scattering processes are present there will be two narrow minima in $K(\nu)$ at ν_1 and ν_2 as shown in fig.3(a). When these are moved towards each other by an external perturbation such as a magnetic field, one might observe changes in $\Delta K/K_o$ due to frequency crossing, level crossing or level anticrossing according to the nature of the systems involved.

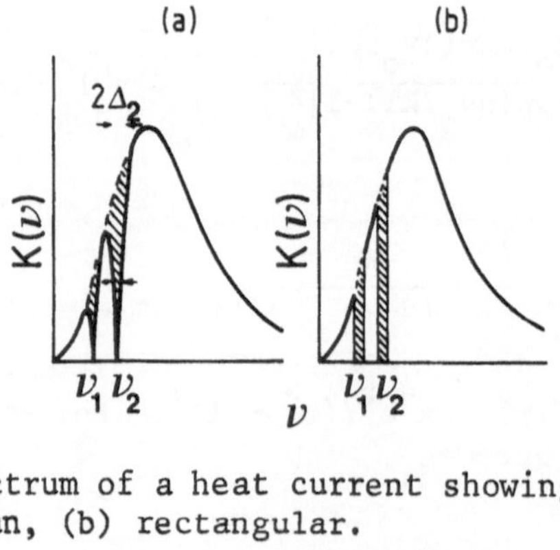

Fig.3 Phonon spectrum of a heat current showing holes at ν_1 and ν_2. (a) Lorentzian, (b) rectangular.

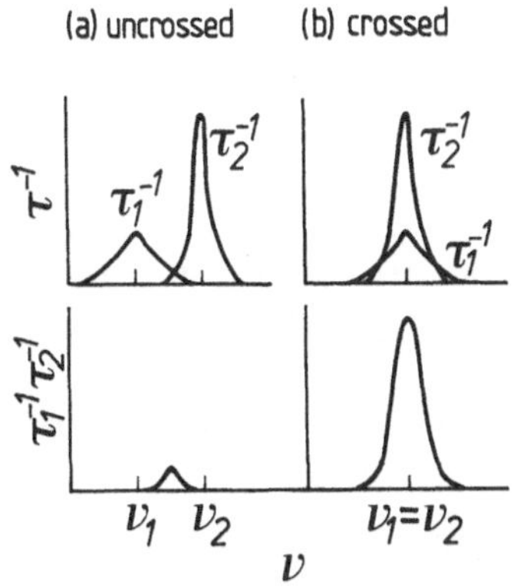

Fig.4 The effect of frequency crossing on τ_1^{-1} τ_2^{-1}.

Frequency crossing

This effect only requires that 2 transition frequencies become equal. It was first observed and accounted for qualitatively by Berman et al (1963) using the heuristic model illustrated in fig 3(b) which assumes that, when resolved, the 2 resonant processes completely block the 2 rectangular conduction channels shown shaded. However when they are crossed, only one channel is blocked so the total conduction rises giving a signal whose size $\delta K_{fc}/K_0$ ~ channel conduction/total conduction and width ~ channel width. In practice, as we have seen, a channel is never entirely blocked. Its conduction $K(\nu)$ depends on $\tau(\nu)$, the phonon relaxation time and since $\tau(\nu) = (\tau_B^{-1} + \tau_1^{-1} + \tau_2^{-1})^{-1}$, where, τ_1^{-1} and τ_2^{-1} are the two resonant scattering rates, $K(\nu)$ depends non-linearly on τ_1^{-1} and τ_2^{-1}. This non-linearity is the origin of the frequency crossing signal. Expansion of $\tau(\nu)$ contains functions of the product $\tau_1^{-1}\tau_2^{-1}$ which is essentially zero at all frequencies when the two processes are resolved but becomes non-zero for $\nu \sim \nu_1 = \nu_2$ when the processes are crossed (fig.4). As a result, as $(\nu_1 - \nu_2)$ is varied, $\tau(\nu)$ and so the thermal conductivity, goes through a maximum at $\nu_1 = \nu_2$. Experimentally, it is usually simpler to work with a fixed heat current and plot the temperature gradient which varies as $W = 1/K$, so that frequency crossing signals usually appear as minima. In general it does not matter whether two transitions are from two different species of ion present in the sample (interdefect crossings) or from the same species (intradefect crossings) although in this latter case, complications associated with level crossing can arise which I shall discuss later.

The size and shape of a frequency crossing signal can be obtained by computing the integral expression for $\Delta K/K_o$ given in equation (2), but with 2 resonant processes (ν_1 and ν_2) included in $\Delta\tau$. As, say, ν_2 is varied while ν_1 is held fixed, $\Delta K/K_o$ will pass through a maximum at the crossing. In some cases though, analytical results can be obtained and as illustration we consider the crossing of two Lorentzian processes for which $\Delta_i >> \Gamma_i$.

In this limit, the reduced resonance scattering rates can be written

$$\frac{\tau_1^{-1}}{\tau_B^{-1}} = \frac{\Delta_1^2}{(\nu-\nu_1)^2} \quad \text{and} \quad \frac{\tau_2^{-1}}{\tau_B^{-1}} = \frac{\Delta_2^2}{(\nu-\nu_2)^2}$$

since the Γ_i^2 terms in the denominator have no significant effect on the reduction in conductivity. (They are only significant in τ when $K(\nu) \simeq 0$). When the processes are resolved, each acts independently, reducing the conductivity by $\Delta K_i/K_o = 0.121 x_i f(x_i) \Delta_i/\nu_i$. Now if $\Delta_i << k_B T$, then, when the processes are just resolved so that $|\nu_1 - \nu_2| > \Delta_1 + \Delta_2$, their difference in frequency is also much less than $k_B T$ so that $x_1 f(x_1)/\nu_1 \simeq x_2 f(x_2)/\nu_2 \simeq x_o f(x_o)/\nu_o$ where ν_o is the mean frequency. So when the frequencies are uncrossed, the two processes reduce the conductivity by

$$\Delta K/K_o = 0.121 x_o f(x_o) \nu_o^{-1} [\Delta_1 + \Delta_2] \qquad (3)$$

When they are crossed with $\nu_1 = \nu_2 = \nu_o$, the total resonance scattering rate, $\tau_R^{-1} = \tau_1^{-1} + \tau_2^{-1}$ is given by

$$\frac{\tau_R^{-1}}{\tau_B^{-1}} = \frac{\Delta_1^2 + \Delta_2^2}{(\nu-\nu_o)^2}$$

and the reduction in conductivity is

$$\Delta K/K_o = 0.121\, x_o f(x_o) \nu_o^{-1} [(\Delta_1^2 + \Delta_2^2)^{1/2})] \qquad (4)$$

which is less than (3) as expected. The maximum change in conductivity caused by the frequency crossing is therefore

$$\delta K_{fc}/K_o = 0.121\, x_o f(x_o) \nu_o^{-1} [\Delta_1 + \Delta_2 + (\Delta_1^2 + \Delta_2^2)^{1/2}] \qquad (5)$$

If $\Delta_1 << \Delta_2$, $\delta K_{fc} \propto \Delta_1$ to a first approximation showing that the size of the crossing signal is determined by the strength of the weaker process.

It does not seem that the shape of the signal can be deduced analytically but from computation it is found that its halfwidth can be represented to an accuracy of better than 1% by the expression

$$\Delta\nu_{1/2}\ (fc) = 0.43\ [\Delta_1 + \Delta_2 + 2\ (\Delta_1{}^2 + \Delta_2{}^2)^{1/2}] \qquad (6)$$

Experimentally, the frequency crossing signal is plotted against magnetic field (or other external perturbation) and $\Delta\nu_{1/2}(fc)$ is determined from the measured width $\Delta B_{1/2}$ (fc) using

$$\Delta\nu_{1/2} = \Delta B_{1/2}(fc)\ (dS/dB) \text{ where } S = |\nu_2 - \nu_1|$$

The results in this section are from Anderson and Challis (1975a) ; in their notation $z_p = \Delta_1/\nu_o$ and $r = \Delta_2/\Delta_1$. The results for other cases have been given by Challis and Wybourne (1979). We stress again that frequency crossing signals can occur for both first or second order resonant scattering.

Experimental techniques for observing crossing signals

An example of a cryostat arrangement for the investigation of frequency crossing signals in thermal resistivity is shown in fig 5. The sample is in contact with a helium bath usually at ~ 1K. The temperature gradient is measured using 2 thermometers placed 20cm from the sample where the magnetic field is much reduced. In the arrangement shown here, this was further reduced by a compensating field and niobium foil. The thermometers are attached to the sample by 0.9mm diameter copper wire supported by thermally insulating 40μm diameter nylon fibres to reduce the effects of vibration. The copper wire is attached to the sample with Stycast 2850FT or indium faced clamps. The thermometers form 2 arms of an inductive bridge, (Automatic Systems Laboratories) with a Brookdeal PSD (9503 SC) balanced in zero field (or at a field just outside the crossing range). Small changes in ΔT are proportional to the out of balance signal and so can be recorded directly on a plotter. The system has a noise level corresponding typically to $\Delta K/K_o \sim 1\text{-}2 \times 10^{-4}$ so that signals ~ 0.1% can readily be measured. An improvement in noise level can be achieved by signal averaging or, better, by field modulation at a few Hz. The modulation has been carried out using small (3mm diameter) superconducting coils on either side of the sample (Fig 6).

The signal from the first PSD now oscillates at the modulation frequency and is detected using a second PSD. The output is now proportional to dK/dB (or d^2K/dB^2 if the reference signal is set at twice the modulation frequency). This moves the noise window due to fluctuation in R, T etc., from 0 to a few H_z and since the noise appears to decrease rapidly with frequency ($\sim 1/f^2$) a significant improvement is achieved. Examples of signals obtained in this way are shown in fig. 7. Further details of these experimental techniques are given by Wybourne et al (1979) and for those used in heat pulse work, by Patel and Wigmore (1977) and references therein.

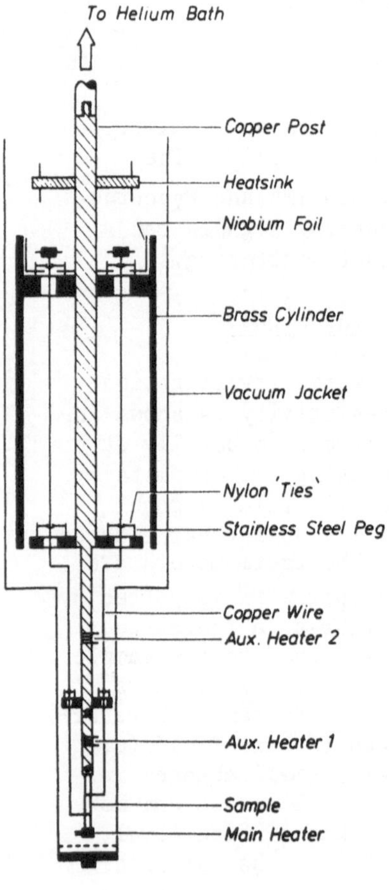

Fig.5 Cryostat for observing crossing signals

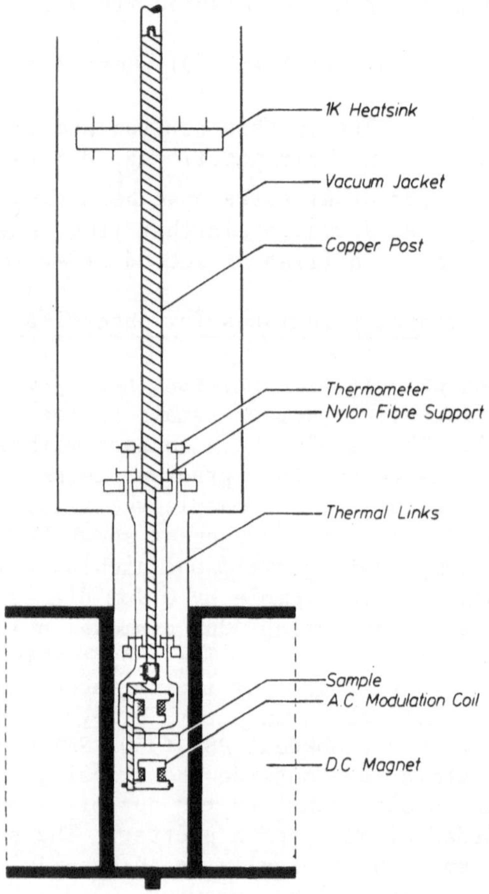

Fig.6 Cryostat fitted with superconducting modulation coils

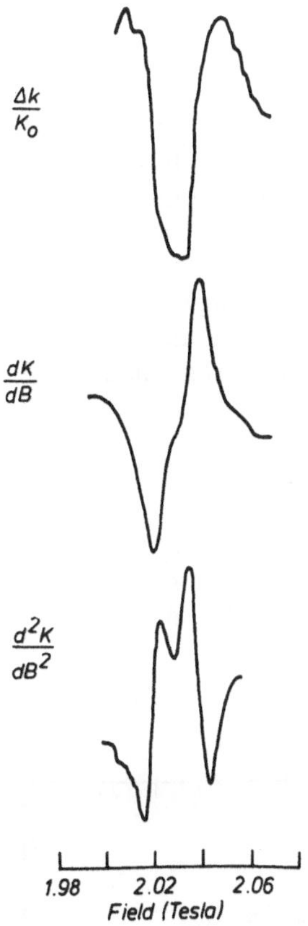

Fig.7 Examples of signals observed in the cryostat shown in Fig.6.

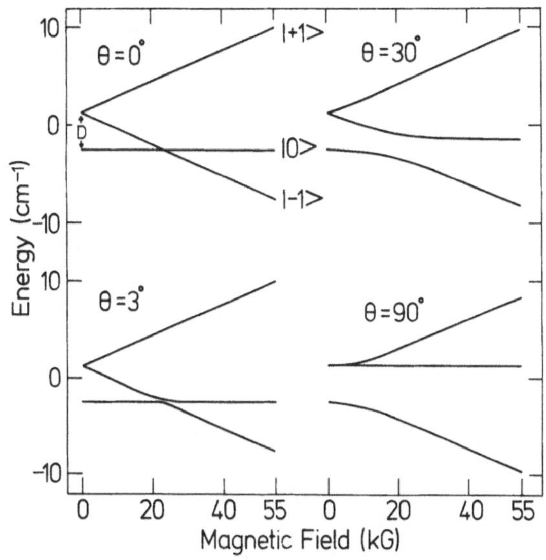

Fig.8 Energy level diagram for the ground state triplet of Fe^{2+} in Al_2O_3 ; θ is the angle between the field and the c-axis

Examples of frequency crossing signals

Many observations of frequency crossing signals have been reported since the first work by Berman et al (1963) but I plan here to concentrate first of all on a system, Fe^{2+} in Al_2O_3, where the results have been compared in some detail with the analysis given previously and then to describe one further system. V^{3+} in Al_2O_3, to demonstrate the sensitivity and resolution that are achievable.

Fe^{2+} in Al_2O_3

The energy level diagram for the ground state triplet is given in fig 8. It can be described by the spin Hamiltonian $\mathscr{H} = D[S_z^2 - 1/3\ S(S+1)] + g\beta \underline{B}.\underline{S}$ with $S = 1$, $g = 3.43$ and $D/h = 112$GHz. Frequency crossing occurs at $B = 0$ and $2B_o^2/(3\cos^2\theta-1)$ where $3g\beta B_o = D$ gives the position at $\theta = 0$ and θ is the angle between the field and the c-axis. The frequency crossing at $B = 0$ also involves a level crossing (or possibly an anticrossing) but, as we shall see, in several respects, the results (Anderson and Challis 1973, 1975b) can be described by the frequency crossing analysis given earlier.

$B = 0$

The 2 transitions are those between 0 and $|\pm1$. The $|\pm1\rangle$ states have time-reversal symmetry so the 2 transition probabilities are equal giving $\Delta(0,1) = \Delta(0,-1) = \Delta$. For this condition the analysis suggests

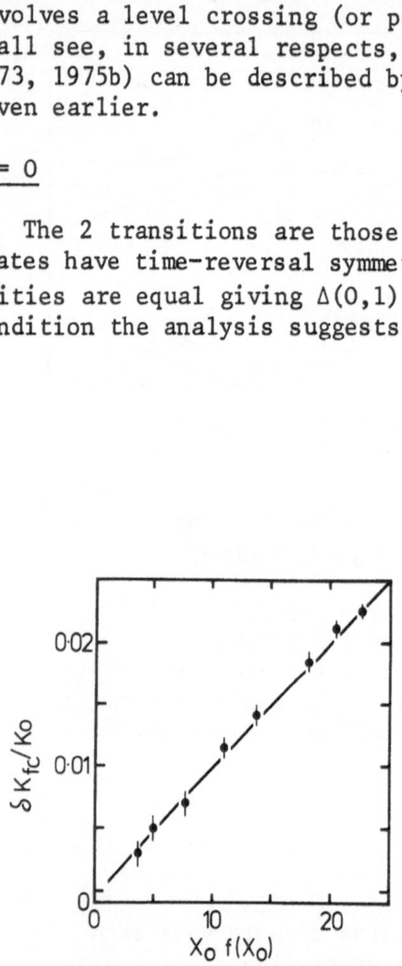

Fig.9 The temperature dependence of the size of the B=0 frequency crossing signal of Fe^{2+} in Al_2O_3. $\delta K_{fc}/K_o$ is plotted against $x_o f(x_o)$ which varies as $T^{-5}\exp(h\nu_o/k_BT)/[\exp(h\nu_o/k_BT)-1]^2$. The theoretical dependence is shown by a solid line whose slope gives a value of $\Delta = 1.6$GHz.

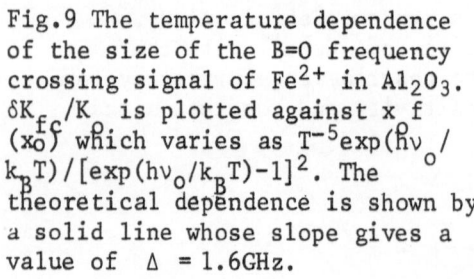

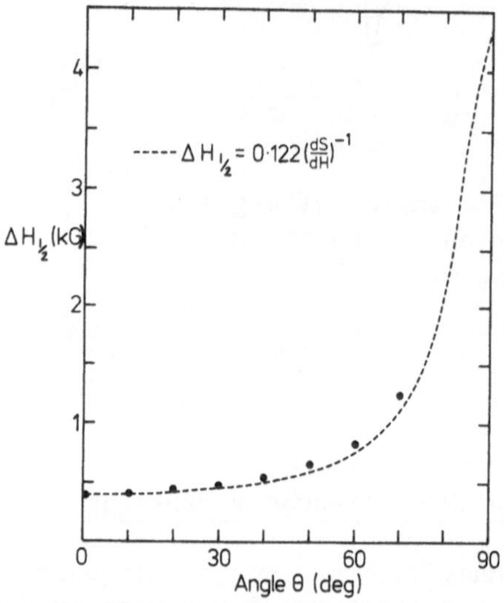

Fig.10 The angular dependence of the width of the B=0 frequency crossing of Fe^{2+} in Al_2O_3. The dashed line shows the theoretical dependence normalised at the value at $\theta = 0$. This gives $\Delta = 0.122/2.08\text{cm}^{-1} = 1.8$GHz

$$\delta K_{fc}/K_o = 0.121(2-\sqrt{2})x_o f(x_o)\nu_o^{-1}\Delta = 0.071\ x_o f(x_o)\nu_o^{-1}\Delta$$

and $$\Delta\nu_{1/2}\ (fc) = 0.86\ (1 + \sqrt{2})\ \Delta = 2.08\Delta$$

so that $\Delta B_{1/2}(fc) = 2.08\Delta(dS/dB)^{-1}$ where $S = |E_1 - E_{-1}|/h$.

Now Δ is independent of T (for a 2nd order process) and of θ so $\delta K_{fc}/K_o$ should be independent of θ and vary with T as $x_o f(x_o)$. $\Delta B_{1/2}$ is independent of T but depends on θ through $(dS/dB)^{-1}$. All these properties were found experimentally and the T-dependence of $\delta K_{fc}/K_o$ and θ-dependence of $\Delta B_{1/2}$ are shown in figs 9 & 10. Values of Δ obtained from $\delta K_{fc}/K_o$ and $\Delta B_{1/2}$ are 1.6 ±0.1 and 1.8 ± 0.2GHz respectively, which are in satisfactory agreement.

$B^2 = 2B_o^2/(3\cos^2\theta - 1)$

This predicted angular dependence of $B(\theta)$ agrees well with experiment as shown in fig.11. Comparison with further theoretical predictions is more complicated in this case because there is no reliable theoretical information on the ratio of the 2 average matrix elements for a phonon transition, f(-1,1)/f(0-1), and so, on $\Delta(-1,1)/\Delta(0,-1)$. Experimental values of $\Delta(-1,1)$ and $\Delta(0,-1)$were determined from the signals using equations 5 and 6 and changed with angle (state mixing) and temperature (population) more or less as expected and provided a value for f(-1,1)/f(0,-1) of ~0.17.

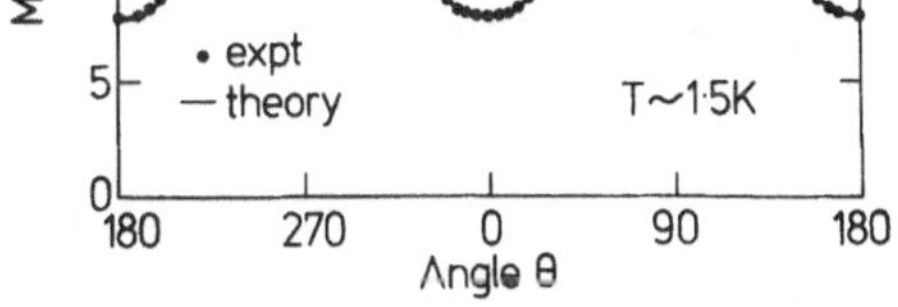

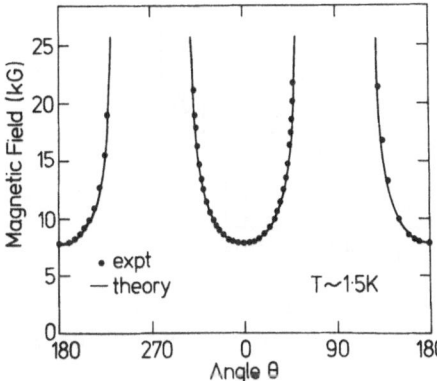

Fig.11. The angular dependence of the frequency crossing of Fe^{2+} in Al_2O_3 that occurs at non-zero field. The line shows the theoretical dependence.

However, although these aspects were satisfactory, the experiment taken with another did reveal a serious deficiency in the theory. Analysis of the width of an anticrossing signal on this sample, which will be discussed later, indicates that 0.5 < f(-1,1) /f(0,-1) <1 in clear disagreement with the value of 0.17 found here. This suggests that this very simple theory in which all quantities are replaced by their angular and polarization averages at an early stage, phonon focussing is ignored etc, is too simple and it would be very desirable to have a more sophisticated theory to test these results against.

V^{3+} in Al_2O_3

This triplet system is quite similar to Fe^{2+} although since the g-value is no longer isotropic the Zeeman term has to be modified. There are two other features though that make it interesting. The first is that most of the V^{3+} ions exhibit hyperfine structure with I = 7/2 so that the doublet levels with non-zero S_z are each split into 8 hyperfine levels separated by 290MHz (fig 12). The $\Delta I_z = 0$ selection rule limits the possible $|-1\rangle \rightarrow |+1\rangle$ transitions to 8 separated by 580MHz and there are also 8 $|0\rangle \rightarrow |-1\rangle$ transitions separated by 290MHz. So as these 2 'combs' of holes in the phonon spectrum pass each other, crossings occur at 22 different but closely separated fields ($\Delta B = h.290 \times 10^6/3g_{\parallel}\beta$=3.6mT). This is shown in fig 13 (Heraud et al 1982).

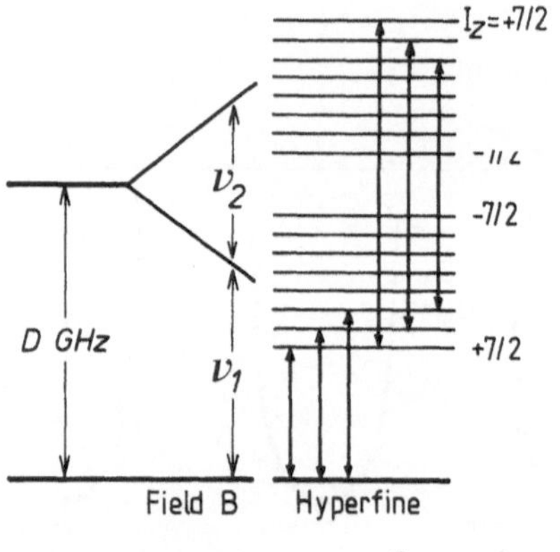

Fig.12 The energy level structure of V^{3+} in Al_2O_3 for B||C showing the allowed transitions between the hyperfine levels.

ΔW does not rise to 0 between the minima showing that the hyperfine holes partially overlap at this concentration (70ppm).

The mean crossing frequency is 170GHz so the resolution required to resolve the structure is 1.7×10^{-3}. The structure can be displayed more clearly using field modulation as described earlier and the results are shown in fig 14; (Wybourne et al, 1980); it would appear that the peaks could still have been resolved for this system even if the splittings had been around 3 times smaller i.e. $\sim 5 \times 10^{-4}$. The loss of resolution for splittings appreciably smaller than this would not, in this case, be due to 'instrument' resolution but to the linewidths $(\Delta^2 + \Gamma^2)^{1/2}$ due to the sample and these presumably, affect all techniques.

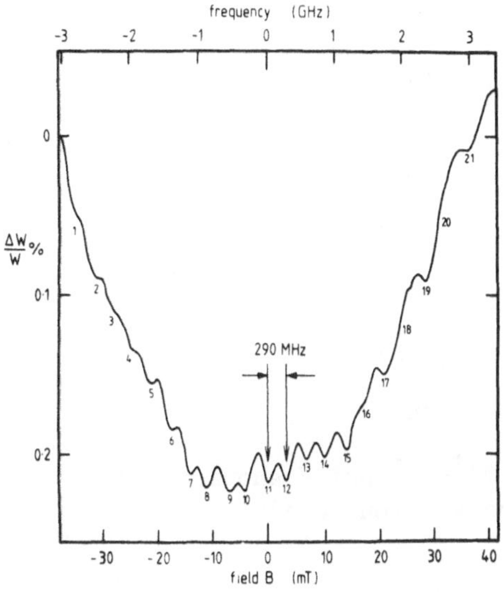

Fig.13 The frequency crossing signal of V^{3+} in Al_2O_3 showing the hyperfine structure.

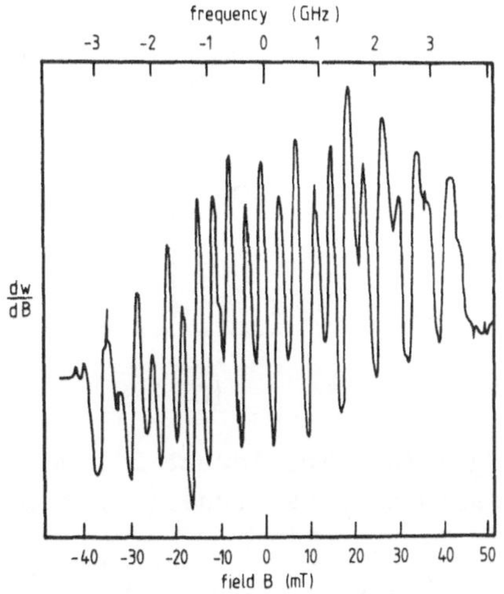

Fig.14 The differential of the crossing signal shown in fig. 13 obtained by field modulation.

The 'instrument' resolution is probably set by $\tau_B^{-1} \sim$ few MHz plus the field inhomogeneity although the effects of this can be smaller than τ_B^{-1} in NMR magnets. Whether in practice though sample linewidths will ever be such that higher resolutions will be needed is not yet clear.

The last example I want to discuss provides an indication of the sensitivity of the technique. This clearly depends on the ion and its coupling to the lattice. As we have seen in fig. 9 a few ppm of Fe^{2+} in Al_2O_3 provides signals ~ 1% in size and since we can observe signals $\lesssim 0.02\%$ we should expect to detect concentrations $<10^{-2}$ppm (we recall that $\Delta K/K_0 \propto (\text{conc})^{1/2}$ provided $\tau_R^{-1}(\nu_0) \gg \tau_B^{-1}$). It is difficult to test this with ions since we should need samples with known concentrations of ions at these very low levels. It can however be tested with ion pairs by using the fact that if ions are distributed randomly the pair concentration ~ (conc)^2 and so can be made very much less than the ion concentration. This approach has been used for V^{3+} - V^{3+} pairs. The ion concentrations varied from 70 to 2000ppm corresponding to pair concentrations from 10^{-2} to 10ppm (Challis et al 1978).

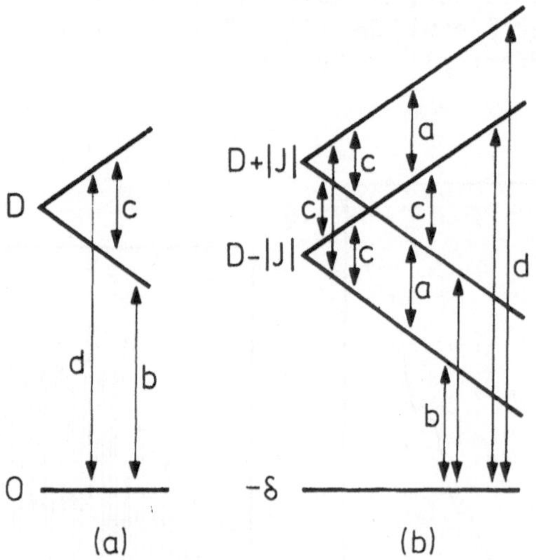

Fig 15(b). The lowest 5 levels of V^{3+}-V^{3+} pairs showing the possible transitions ((a) shows the V^{3+} ion levels)

The lower levels of a V^{3+}-V^{3+} pair are shown in fig 15 for B||c. Many more transitions are possible than in the single ion so the crossing spectrum from transitions within both ions and pairs becomes quite complicated particularly as there are likely to be many different sorts of pairs each with a different J. Here however, I should like to focus on a particular feature, occurring around the ion-ion crossing at $3g_{\parallel}\beta B_o = D$ for B||c. (D/h = 249 GHz, $g_{\parallel}$ = 1.91 so that B_o = 3.11T). 6 equally spaced satellite lines appear, 3 on each side, with a separation ΔB given by $3g_{\parallel}\beta\Delta B = |J|$ which allows us to determine |J|. Data for two samples are shown in fig 16. We can now study the dcerease in the size of these satellites relative to the main line as the concentration is reduced. This was done for a series of samples, and part of the spectrum for an ion concentration of 70ppm is shown in fig 17. The pair lines are clearly visible at a signal level of ~ 0.02% (the noise level was reduced by signal averaging) and as has been noted already, it is estimated that the pair concentration ~ 10^{-2}ppm.

Before leaving this topic, let me point out that the value J/h = 8.7GHz measured in this way at 3.1T differs from the value

J/h = 14 GHz measured recently at B = 0 (Schick et al, 1984). This raises the interesting possibility, as Schick et al suggested, that J is field dependent for these weakly coupled pairs.

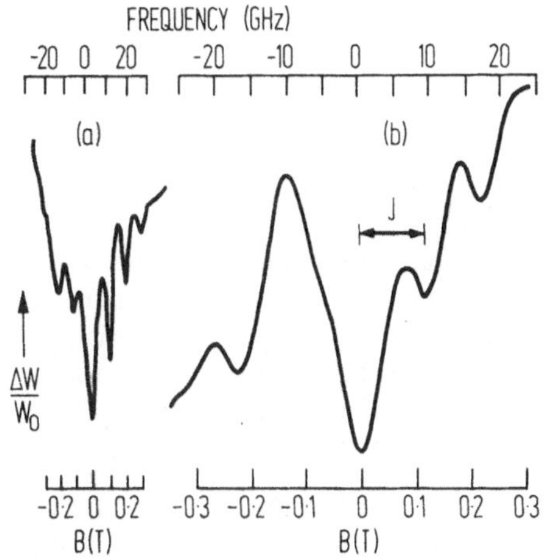

Fig 16 The frequency crossing signal from V^{3+} in Al_2O_3 showing satellite lines due to pairs. In (a), all 6 satellites can be seen although that at the lowest field is only just detectable as a change of slope. In (b) part of the spectrum for another sample is shown at higher resolution. (Ghazi A.A., private communication).

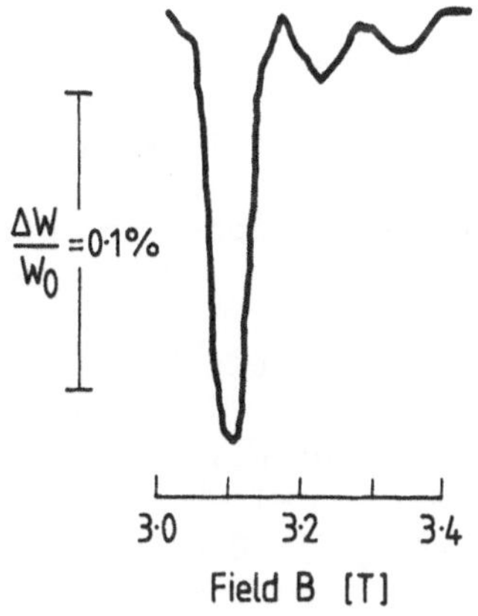

Fig 17 The frequency crossing signal of 70ppm of V^{3+} in Al_2O_3 plus two lines(at higher field) caused by pairs in a concentration 10^{-2}ppm.

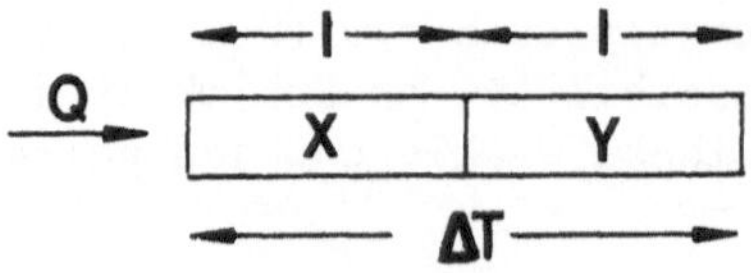

Fig.18 A bicrystal containing different centres in X and Y.

Frequency crossing between spatially separated ions

The discussion so far has assumed that the ions are homogeneously distributed but this is not in fact a necessary condition for the appearance of frequency crossing signals. Consider a sample whose two halves X and Y contain different phonon scattering centres (fig 18). The total phonon relaxation times can be written $\tau_X^{-1} = \tau_B^{-1} + \tau_1^{-1}$ and $\tau_Y^{-1} = \tau_B^{-1} + \tau_2^{-1}$ respectively, with τ_1^{-1} and τ_2^{-1} resonant at frequencies ν_1 and ν_2. If these processes are entirely elastic, we can imagine the heat transport down the rod to take place through a number of independent, parallel channels each covering a narrow bandwidth of frequencies. We suppose however that the channels are kept in thermal equilibrium at each end of the sample through inelastic scattering centres in the material with which they are in contact. Each channel has a total thermal resistance δW equal to the sum of the resistances in X and Y.

$$\delta W = \ell(\delta K_X^{-1} + \delta K_Y^{-1})/A$$

where ℓ is the length of each half, A the cross-sectional area of the sample and δK_X, δK_Y the thermal conductivities of a bandwidth of phonons where, as before, $\delta K_i = C(\nu)\tau_i(\nu)d\nu$.

If the temperature difference across the entire sample is ΔT, the heat flowing down the channel is

$$\delta Q = \frac{\Delta T}{\delta W} = \Delta T \cdot \frac{A}{\ell}(\delta K_X^{-1} + \delta K_Y^{-1})^{-1}$$

So the heat flowing down all the channels is

$$Q = \Delta T \cdot \frac{A}{\ell} \int_0^\infty (\tau_X^{-1} + \tau_Y^{-1})^{-1} C(\nu)d\nu$$

$$= \Delta T \frac{A}{\ell} \int_0^\infty (2\tau_B^{-1} + \tau_1^{-1} + \tau_2^{-1})^{-1} C(\nu)d\nu$$

Expansion of the integrand again leads to terms in $\tau_1^{-1}\tau_2^{-1}$ which peak sharply at the crossing point so giving rise to a crossing signal in Q (or in ΔT if Q is kept fixed). It is also interesting to look at the temperature difference over each half. For a particular channel it is clear that

$$\Delta T_X(\nu) = \frac{\Delta T\,\delta K_X^{-1}}{(\delta K_X^{-1} + \delta K_Y^{-1})} = \frac{\Delta T\ \tau_X^{-1}}{(\tau_X^{-1} + \tau_Y^{-1})}$$

and it can be shown that the average temperature difference of the phonon modes, which is the value measured by two thermometers, is $\Delta T_X(\nu)$ weighted by $C(\nu)$:

$$\Delta T_X = \Delta T \int_0^\infty \tau_X^{-1}(\tau_X^{-1} + \tau_Y^{-1})^{-1}\, C(\nu)\,d\nu \Big/ \int_0^\infty C(\nu)\,d\nu$$

Analysis showing the maximum size of the crossing signal across X is straightforward for 2 Lorentzians with $\Delta_i \gg \Gamma_i$ and gives the result

$$\frac{\delta T_X(fc)}{\frac{1}{2}\Delta T} = \frac{15}{2\sqrt{2}\pi^3}\left(\frac{x_o f(x_o)}{\nu_o}\right)\left(\Delta_1 - \frac{\Delta_1^2}{(\Delta_1^2 + \Delta_2^2)^{\frac{1}{2}}}\right)$$

which differs in form from that of the homogeneous sample discussed earlier. (The temperature difference across X is always $\sim(1/2)\Delta T$, since Δ_1, $\Delta_2 \ll k_BT$. The signals across X and Y can be very different in size. For example if $\Delta_2 \ll \Delta_1$, $\Delta T_X(fc) = \Delta^2_2/\Delta_1$ while $\delta T_Y(fc) \propto \Delta_2$ which can be much greater. Interestingly though, the crossing signal for the whole rod, $[\delta T_X(fc) + \delta T_Y(fc)]/\Delta T$, is identical to that obtained for a homogeneous rod containing the same number of impurity ions: equation (5). This is not evident at first glance since it appears that while the two brackets are the same, the multiplying constant for the inhomogeneous rod is smaller by $\sqrt{2}$ $(0.121 = 15/4\pi^3)$. However, we must recall that if the ions of each species are restricted to half the volume, their concentration in that half is doubled so that Δ_1 and Δ_2 and so the last bracket are increased by $\sqrt{2}$ $(\Delta \propto (\text{conc})^{1/2})$ cancelling out the $1/\sqrt{2}$ in the constant. The analysis in this sectionso far is taken from Challis et al (1982a).

Frequency crossing between spatially separated ions has been demonstrated by Challis et al (1982b) using an Al_2O_3 crystal doped with Fe^{2+} ions in one half and V^{3+} in the other (fig 19). As the magnetic field applied parallel to the c-axis is increased, several crossings occur between the transition frequencies of the two ions and fig 19 shows two of these: lines C and D. The temperature difference ΔT across part of V was measured for heat injected at H_V or H_{Fe}. Signals were observed even when heat was injected at H_V and these were attributed to Fe^{2+} present in V as a trace impurity. However, when the heat was switched to H_{Fe}, the signal increased in size by a factor of 3 demonstrating the increase in size of the Fe^{2+} 'hole' when the heat flows through the Fe doped half.

This result opens up the possibility, for example, of examining resonant centres in diffused or epitaxial layers, electrons in inversion layers etc., by crossing their frequencies with centres in the bulk solid on which they are formed. Of course, if it were possible to bond two crystals together with a glue transparent to phonons, the crossing technique could have wider application in spectroscopy. The problems of doing this though are not insignificant both because of fairly high concentrations of defects normally present in surfaces and the presence of defects - the so-called two level systems - in the usual glues which nearly all contain amorphous regions. So crystalline glues and very carefully prepared surfaces will be needed if this extension is going to be achieved.

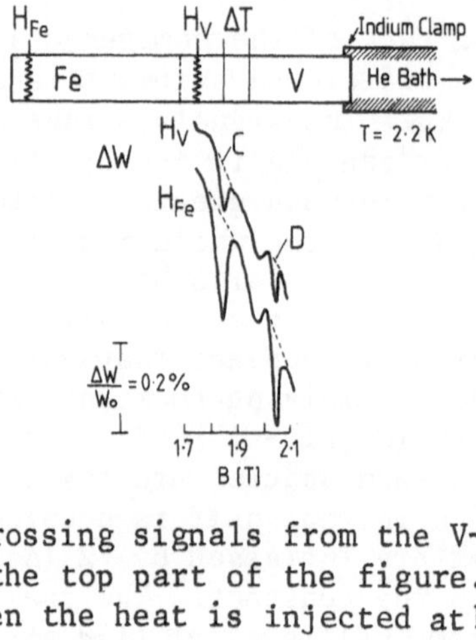

Fig 19 Frequency crossing signals from the V-doped half of a Fe/V bicrystal shown in the top part of the figure. The signals are ~ 3 times larger when the heat is injected at H_{Fe} than when injected at H_V.

Defects should have greatest effect if they provide inelastic scattering. The analysis used so far assumes this to be negligible and so treats the parallel frequency channels as independent. We can see qualitatively the effect of inelastic scattering if we first suppose that there is no resonant scattering in Y. The first half, X, of the composite 'burns a hole' at ν_1 in the phonon spectrum and, in the absence of inelastic scattering, this labelled spectrum passed down Y without distortion. If we now reintroduce the resonant scattering in Y, it will have little effect when it is scattering at the frequency of the hole, i.e. $\nu_1 = \nu_2$, but much more when ν_2 is detuned from ν_1. This is of course the origin of the frequency crossing signals. Inelastic scattering acts to restore thermal equilibrium and we should expect the labelled spectrum entering Y to decay in a distance ~ ℓ_I, the inelastic scattering length, so unless ℓ_I is large compared with the distance from the interface at which we are measuring ΔT, no frequency crossing signal will be seen.

It is evident from fig 19 that the inelastic scattering length must have been bigger than the distance from the interface to the centre of the thermometer contacts, 6mm. To see how much

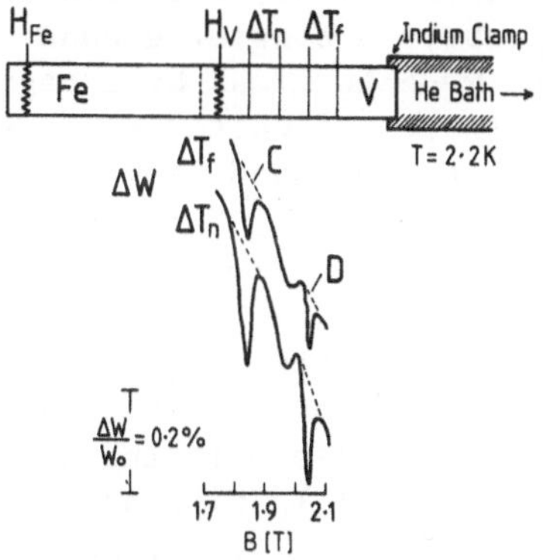

Fig 20. Signals from the V-doped half of a Fe/V bicrystal shown in the top part of the figure. The heat is injected at H_{Fe}. The signals from the nearer pair of contacts, ΔT_n, are ~ 3 times larger than those from the further pair, ΔT_f, showing that the Fe^{2+} holes have effectively decayed by the time they reach ΔT_f.

bigger, measurements were also made with contacts at a mean distance of 12mm, and the results are shown in fig 20 for heat injected at H_{Fe} and compared with those for the original contacts. The signals on ΔT_f are about 3 times smaller than on ΔT_n and about the same as those on both ΔT_n and ΔT_f when heat is injected at H_v. (We recall that these were attributable to Fe^{2+} trace impurities). So the 'hole' has decayed almost completely in a distance from the interface of 12mm and it was estimated that this puts an upper limit to ℓ_I of ~ 4mm at 194GHz and 2K. This is in fact very similar to the diameter of the crystal and it seemed reasonable to suppose that the inelastic scattering was occurring mainly at the surfaces which had been fine ground and then annealed at 1200°C. Work is in progress to extend these measurements in various ways in particular to see if the inelastic scattering can be reduced by surface treatment.

Level Crossing

The level crossing effects that should be observable using phonons are analogous to the Hanle effect seen in light scattering (Hanle 1924). None has been seen so far so this part of the lecture is to encourage you to go and look for them! The effect was first explained by Breit (1933) and since 1959 has been used as a technique in high resolution spectroscopy particularly for measuring radiative lifetimes of excited levels (e.g. Corney 1977).

Suppose that an atom in a gas has 2 levels $|a\rangle$ and $|b\rangle$ with energies $h\nu_a$ and $h\nu_b$ above a singlet ground state $|c\rangle$ and that these 2 levels can be made to cross by a magnetic field to give $\nu_a = \nu_b = \nu_o$ (fig 21(a)). If white light is incident on the atom, a photomultiplier can be used to measure the total intensity I_T of the components at ν_a and ν_b scattered into a particular solid angle by resonance fluorescence. We shall show below that if $|\nu_{ab}| \gg \Gamma$ where ν_{ab} is $\nu_a - \nu_b$ and Γ is the average linewidth of the levels, $I_T = |A_a|^2 + |A_b|^2 = I_a + I_b$ but if $|\nu_{ab}| < \Gamma$, there is coherence between the scattering of the two processes and interference takes place in the emission. So in this limit, $I_T = |A_a + A_b|^2 = I_a + I_b$ plus an interference term. Hence, the intensity in a particular direction changes as the levels cross and the photomuliplier records a signal of width Γ at the crossing point. But there is no change in the total intensity emitted in all directions. This has to equal the intensity absorbed which is not affected by the crossing. There is clearly a close analogy with the phenomenon of interference between two slits. When the slits are close together they produce an interference pattern with a different intensity distribution from that found when they are further apart and can no longer interfere. The same amount of light passes through both slits in the two cases so the total intensity is unchanged.

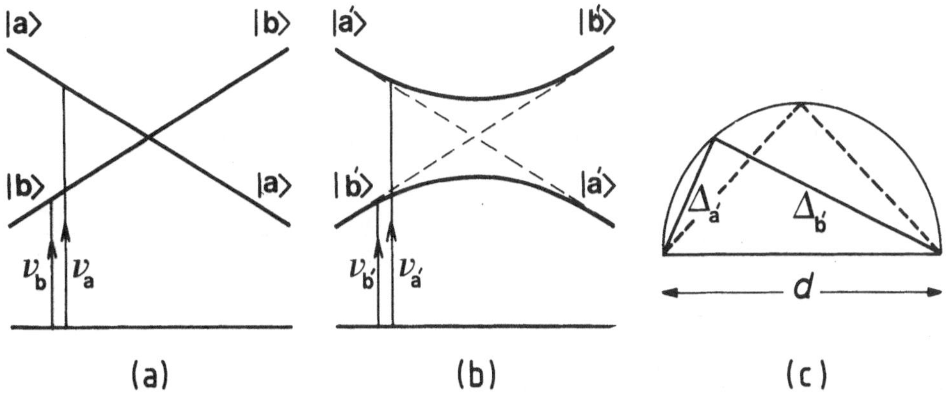

Fig 21 (a) level crossing (b) level anticrossing (c) graphical demonstration that $(\Delta_{a'} + \Delta_{b'}) = d$, when $\Delta_{a'} << \Delta_{b'}$ (or vice-versa) and $= \sqrt{2}d$ when $\Delta_{a'} = \Delta_{b'}$.

The reasons for the coherence effects are not difficult to see. Consider a wave of frequency ν incident on the atom. A photon is absorbed and both states can be excited simultaneously and therefore coherently.† Or, in slightly different language, we can say that the atom is excited into a particular state which is a superposition of the two eigenstates: $\alpha|a> + \beta|b>$. The eigenstates $|a>$ and $|b>$ change with time in the usual way as $\exp(-i2\pi\nu_a t)$ and $\exp(-i2\pi\nu_b t)$ respectively so becoming gradually out of phase by $2\pi\nu_{ab}t$. The significance of this appears when the system decays, spontaneously, after a time τ. There is a phase difference $\Delta\phi = 2\pi\nu_{ab}\tau$ between the amplitudes of the contributions from each state to the amplitude of the emitted photon and so the emitted intensity in a particular direction $\propto|A_a + A_b\exp(i\Delta\phi)|^2$. The intensity in this context means the probability that the photon is emitted in that direction. Now τ is a random quantity with an average value $\tau = (4\pi\Gamma)^{-1}$ where Γ is the level linewidth for spontaneous emission. If τ is sufficiently small that $|\Delta\phi|<<2\pi$ ($|\nu_{ab}|<<4\pi\Gamma$ or $<<\Gamma$), $\exp(i\Delta\phi)\simeq 1$ and these random variations in phase have no significant effect on the angular dependence of the intensity. Every time the atom is excited and decays it does so with the same angular dependent probability $|A_a + A_b|^2$. However if $|\nu_{ab}|>> 4\pi\Gamma$ (or $>>\Gamma$) this probability varies randomly from $|A_a + A_b|^2$ to $|A_a - A_b|^2$ with an average value of $|A_a|^2 + |A_b|^2$. Of course there is no interference between the scattering from different atoms (assuming they are well spaced) since these are randomly placed but each atom behaves in the same average way and the scattering from the gas follows the same behaviour as that described for an atom.

† : This requires $\nu_{ab} \lesssim \Gamma$ which is the condition for coherent emission given below.

This discussion concerned photons and gas atoms but applies equally well to phonons and resonant scattering centres provided the dominant process is the second order resonance fluorescence process discussed earlier. We can now put it on a quantitative basis. In the Born approximation, the scattering amplitude for 2 second-order processes is

$$M = \frac{\langle q', c|\mathcal{H}_{q'}|a\rangle\langle a|\mathcal{H}_q|q,c\rangle}{h\nu - h\nu_a} + \frac{\langle q',c|\mathcal{H}_{q'}|b\rangle\langle b|\mathcal{H}_q|q,c\rangle}{h\nu - h\nu_b}$$

These are processes in which the electronic system makes a transition $c \to a \to c$ (or $c \to b \to c$), a phonon q is absorbed and q', emitted; $h\nu$ is the initial and final energy of the system and $h\nu_a$ ($h\nu_b$) is the intermediate energy. $\mathcal{H}_q$ and $\mathcal{H}_{q'}$, are the Hamiltonians associated with the phonon fields q and q' respectively. The phonon parts of the matrix elements are common and so the scattering rate can be written

$$|M|^2 \propto \left| \frac{f_a g_a^*}{\nu - \nu_a + i\Gamma} + \frac{f_b g_b^*}{\nu - \nu_b + i\Gamma} \right|^2$$

where a damping term $i\Gamma$ has now been included in the denominator and we write f_a for $\langle a|\mathcal{H}_q|c\rangle$ - phonon absorption, and g_a^* for $\langle c|\mathcal{H}_{q'}|a\rangle$ phonon emission. (We note that $\Gamma = \gamma/4\pi$ in the notation of Anderson and Challis (1975c) who worked with ω).

We are interested in the behaviour of $|M|^2$ when $\nu_a \sim \nu_b \sim \nu_0$. There are 4 terms in the expansion of $|M|^2 = M^*M$. Two together vary approximately as $(|f_a g_a|^2 + |f_b g_b|^2)/[(\nu-\nu_0)^2 + \Gamma^2]$ and are clearly not affected by small changes in ν_{ab}. However, the other two terms, the coherent terms, vary as :

$$\frac{f_a f_b^* g_a^* g_b}{(\nu-\nu_a + i\Gamma)(\nu-\nu_b - i\Gamma)} + \text{c.c.}$$

$$= \frac{f_a f_b^* g_a^* g_b}{(\nu-\nu_a)(\nu-\nu_b) + \Gamma^2 + i\Gamma\nu_{ab}} + \text{c.c.}$$

and so clearly do. The total intensity scattered from the beam at all frequencies is found by integration. For the coherent terms, which have poles at $\nu_a - i\Gamma$ and $\nu_b + i\Gamma$ (+ c.c)

$$|M|^2 \propto \frac{f_a f_b^* g_a^* g_b}{2\Gamma - i\nu_{ab}} + c.c$$

by contour integration so for $\nu_{ab} << \Gamma$, $|M|^2 \propto (f_a f_b^* g_a^* g_b + c.c.)/2\Gamma$ while for $\nu_{ab} >> \Gamma$, $|M|^2 \to 0$. Hence, as ν_{ab} increases, the total scattering, $q \to q'$, from all 4 terms changes from $|M|^2 \propto |f_a g_a^* + f_b g_b^*|^2$ to $|f_a g_a^*|^2 + |f_b g_b^*|^2$, that is from $|A_a + A_b|^2$ to $|A_a|^2 + |A_b|^2$. There is no change in the total scattering from q, that is from q into all possible q′, because the angular average of the coherence terms $A_a A_b^* + A_a^* A_b$ is always zero. The angular dependence of A_a, A_b comes through the matrix elements, g, and is due to anisotropy caused by crystal fields, magnetic fields etc.

The principal difficulty in seeing level crossing effects in phonon scattering is that level crossing is inevitably accompanied by frequency crossing. Frequency crossing effects, as we have seen, are caused by non-linearity and vanish at low concentrations when $\tau_R^{-1}(\nu_0) << \tau_B^{-1}$, but all experiments so far have been carried out in the opposite limit where frequency crossing effects would normally seem to be dominant. Indeed calculations suggest that in this limit level crossing effects in thermal conductivity should produce only small changes in the shape of a frequency crossing signal if the crossing is in the excited state and virtually no changes if it is in the ground state unless the ground state is 'prepared' in some way (Anderson and Challis 1975c). The difficulty in the low concentration limit, where in fact the optical work is done, is the small size of the changes expected. The effect of resonant scattering on thermal conduction or heat pulse transmission will be very small so small changes in this will be very difficult to see. A much better approach would be to look at the phonons scattered from a collimated beam. Another problem in both limits is to achieve the rather stringent condition necessary for 2 levels to cross. This can be seen from the secular determinant describing the levels at the crossing point.

$$\begin{vmatrix} E - x & V \\ V & E - x \end{vmatrix} = 0$$

where E is the energy at this point. If there are off-diagonal matrix elements V coupling the two states, the energies are given by $(E-x)^2 = |V|^2$ or $x = E \pm |V|$. So now the levels repel or 'anticross' with a minimum separation of $2|V|$. Off-diagonal matrix elements may result from magnetic field perpendicular to the axis of quantization, random strains etc. So that in reality all level crossings are anticrossings. However, the anticrossing effects are unimportant if $|V|<\Gamma$ when we can continue to speak of level crossings. To achieve this it seems likely that we need to work with zero field crossings (to minimise $B_\perp$) and perhaps with Kramers doublets (to minimise the effects of random strain).

Level anticrossings

Anticrossings are defined by the condition $|V|>\Gamma$ and, interestingly, coherence in the two emission processes is now maintained until $\nu_{ab} \sim |V|$. So in an optical experiment, the signal in the angular distribution of the scattered light as the levels are moved has a width $|V|$ rather than Γ. The feature that leads to this and other effects is the mixing of $|a\rangle$ and $|b\rangle$ that occurs at the anticrossing. Indeed this can result in a change in the total scattered intensity which is very important in the present context.

Anticrossing effects were first seen in light scattering about 20 years ago (Eck et al 1963, Wieder and Eck 1967) and in phonon scattering, 10 years later (Anderson and Challis 1973, 1974, Wigmore and Patel 1974). Only the changes in total scattering rate have been identified in phonon scattering and, as in level crossing, measurements of the angular dependence of the scattering will be needed to demonstrate the coherent effects. We consider two excited states $|a'\rangle$ and $|b'\rangle$ at energies $h\nu_a'$ and $h\nu_b'$ above a ground state $|c\rangle$ (or vice-versa). When the states are well separated, $|a'\rangle \sim |a\rangle$ and $|b'\rangle \sim |b\rangle$ where $|a\rangle$ and $|b\rangle$ are basis states coupled by a matrix element V which, for simplicity, we now assume to be real. As the levels approach

$$|a'\rangle = \cos\alpha|a\rangle + \sin\alpha|b\rangle$$

$$|b'\rangle = -\sin\alpha|a\rangle + \cos\alpha|b\rangle$$

where $\tan 2\alpha = 2V/\Delta_o$. Δ_o is the separation $|a\rangle$ and $|b\rangle$ would have if V were zero. The arrangement of levels is shown in fig 21(b). The matrix elements $f_{a'}$, $f_{b'}$, $g_{a'}$, $g_{b'}$ change rapidly near the anticrossing with

$$f_{a'} = \cos\alpha f_a + \sin\alpha\, f_b, f_{b'} = -\sin\alpha\, f_a + \cos\alpha f_b \qquad (7)$$

with similar expressions for $g_{a'}$ and $g_{b'}$.

In the strong scattering limit $\tau_R^{-1}(\nu_a)$, $\tau_R^{-1}(\nu_b) \gg \tau_B^{-1}$ in which phonon experiments have been done, the analysis is in fact simplest when the anticrossing is in the ground state and we discuss this first. The level scheme of 21(b) is inverted and the total scattering rate of phonons q into q´ occurs through elastic processes such as a→c→a and inelastic processes a→c→b and we note that both these processes are resonant at $\nu = \nu_a$ since it is the process of absorption that is the critical process; spontaneous emission proceeding inevitably.

If we ignore coherent terms of the type $f_a f_b^* g_a^* g_b$, and damping terms, the scattering rate can be written

$$\tau_R^{-1}(q \to q') \propto \frac{(|f_{a'}|^2(|g_{a'}|^2 + |g_{b'}|^2)}{(\nu-\nu_{a'})^2} + \frac{(|f_{b'}|^2(|g_{a'}|^2 + |g_{b'}|^2)}{(\nu-\nu_{b'})^2}$$

$$\propto \frac{|f_{a'}|^2}{(\nu - \nu_{a'})^2} + \frac{|f_{b'}|^2}{(\nu-\nu_{b'})^2}$$

since it is clear from equation (7) that $|g_{a'}|^2 + |g_{b'}|^2 = |g_a|^2 + |g_b|^2$ and so is invariant to state mixing at the anticrossing. The total scattering rate $\tau_R^{-1}(\nu)$ is an average over q and q´ and relative to τ_B^{-1} this can be written

$$\frac{\tau_R^{-1}}{\tau_B^{-1}(\nu)} \doteq \frac{\Delta_{a'}^2}{(\nu-\nu_{a'})} + \frac{\Delta_{b'}^2}{(\nu-\nu_{b'})^2}$$

where $\Delta_{a'}^2 \propto |f_{a'}|^2$ averaged over q and similarly for $\Delta_{b'}^2$ and both are proportional to $(|g_a|^2 + |g_b|^2)$ averaged over q´.

So the analysis reduces to consideration of the effects on thermal conductivity etc., of 2 resonant processes as their frequencies are brought towards each other. There are clear similarities with the analysis of frequency crossing but two important differences. The first is that in anticrossing, the strength of each of the two processes changes as their frequencies approach and the second is that they can only approach to within a closest distance of 2V so that if $2V \gg \Delta_{a'} + \Delta_{b'}$, there can be no overlap of the 2 'holes' and so no frequency crossing effects. We consider this limit first.

$2V >> \Delta_{a'} + \Delta_{b'}$

Since the frequencies are always uncrossed, the two processes act independently and so cause a reduction in conductivity (see equation 3) of

$$\delta K_{fc}/K_o = 0.121\ x_o f(x_o) \nu_o^{-1} (\Delta_{a'} + \Delta_{b'}) \tag{8}$$

$$\propto (\overline{|f_{a'}|} + \overline{|f_{b'}|})$$

where the bar implies that the quantities are averaged over incident directions. With some algebra and using the fact that $(f_a^* f_b + f_a f_b^*)$ averages to zero, we can show that

$$\Delta K \propto (f_a^2 + f_b^2 + 2\ [f_a^2 f_b^2 + \tfrac{1}{4}\ (f_a^2 - f_b^2)^2 \sin^2 2\alpha]^{1/2})^{1/2}$$

where from now on f_a^2 stands for $|\overline{f_a}|^2$ etc. Since $\sin^2 2\alpha = 4V^2/(4V^2+\Delta^2)$, this falls from $[2(f_a^2 + f_b^2)]^{1/2}$ at the anticrossing ($\sin 2\alpha = 1$) to a base-line value ($\sin 2\alpha = 0$) of $[f_a + f_b]$. So the thermal resistivity goes through a maximum at an anticrossing in contrast with the minimum at a frequency crossing. The height of the signal only depends on f_a and f_b (plus of course the other quantities contained in $\Delta_{a'}$, $\Delta_{b'}$ and equation 8) and so is independent of V but this is not the case for the width. This can readily be seen if f_a and f_b are comparable so that $(f_a^2 - f_b^2)^2 << f_a^2 f_b^2$. The brackets can be expanded binomially giving

$$\Delta K = \ell + m \sin^2 2\alpha \quad \text{where } \ell \text{ and m are constants}$$

so that the anticrossing signal is Lorentzian.

$$\Delta K(ac) \propto \sin^2 2\alpha \propto 4V^2/(4V^2 + \Delta^2)$$

with a halfwidth $\Delta_{1/2} = 2V$. In the other limit where f_a and f_b are very different, $(f_a^2 - f_b^2)^2 >> f_a^2 f_b^2$ and expansion leads to

$$\Delta K \propto (1 + \sin 2\alpha)^{1/2}$$

and

$$\Delta K(ac) \propto \left[1 + \frac{2V}{(4V^2 + \Delta^2)^{1/2}}\right]^{1/2} - 1$$

with a halfwidth $\Delta_{1/2} = 3.9V$. So $\Delta_{1/2}$ depends on f_a/f_b and by measuring $\Delta_{1/2}$ we can determine f_a/f_b is V is known. We recall that fields perpendicular to the axis of quantization contribute to V and in principle this can be made the dominant known contribution for anticrossings at non-zero fields. Further spectroscopic information can be obtained from the size of the signal. The existence of a maximum in $\Delta_{a'} + \Delta_{b'}$ and so in thermal resistivity at an anticrossing can be demonstrated graphically (fig 21c). The solid lines have lengths $\Delta_{a'}$, $\Delta_{b'}$ and d. The vertex of the triangle has to lie on a semicircle of diameter d to satisfy the requirement (equation 7) that $\Delta^2_{a'} + \Delta^2_{b'}$ = constant $= d^2$. It is clear that $\Delta_{a'} + \Delta_{b'}$ rises from d when $\Delta_{a'} << \Delta_{b'}$ (or vice-versa) to a maximum of $\sqrt{2}d$ when $\Delta_{a'} = \Delta_{b'}$ at the centre of the anticrossing.

We can also see that if 2V is merely $\gtrsim \Delta_{a'} + \Delta_{b'}$, the edges of the two holes will start to overlap at the very centre of the anticrossing. This should produce a narrow minimum at the peak of the anticrossing maximum.

$2V << \Delta_{a'} + \Delta_{b'}$

Frequency crossing can now occur and indeed will be the dominant effect although the crossing can never be quite complete. In the limit when the two processes are effectively fully crossed throughout the whole of the anticrossing region, the scattering will appear to be from a single process of strength $(\Delta^2_{a'} + \Delta^2_{b'})^{1/2}$ with

$$\Delta K \propto (\Delta^2_{a'} + \Delta^2_{b'})^{1/2} \propto [(f^2_{a'} + f^2_{b'})(g^2_{a'} + g^2_{b'})]^{1/2}$$

which is invariant to state mixing. So there will be a frequency crossing signal but no anticrossing signal in this limit. However, as 2V increases, a small and relatively narrow maximum should appear at the bottom of the frequency crossing minimum.

Anticrossing in the excited state can also produce signals of width ~V, although the analysis (Anderson and Challis 1975 c) suggested these should be very much smaller than the changes from the ground state and only arise through changes in the effects of angular averaging. For $2V >> \Delta_{a'} + \Delta_{b'}$, the signal provides a maximum in thermal resistivity as before but for $2V << \Delta_{a'} + \Delta_{b'}$ the small anticrossing signal at the bottom of the frequency crossing minimum has become a minimum rather than maximum in the ground state case.

This discussion has all been made on the assumption that the phonon scattering is by the second order process and it is clear that first order processes cannot give rise to the coherent effects anticipated in level crossing and level anticrossing experiments. However, we should expect to see the incoherent effects at anticrossings. For 1st order processes, $\Delta_{a'}^2 \propto f_{a'}^2$ so the analysis used for ground state anticrossings can be applied immediately. Patel and Wigmore (1977) have also shown that they lead to anticrossing effects in excited states.

Observation of anticrossing signals

The system that has been studied in most detail so far is Fe^{2+} in Al_2O_3 (Anderson and Challis 1973, 1974). The energy level scheme for an unstrained site was given in fig 8 which shows the development of the anticrossing as the field is moved away from the c-axis. The anticrossing occurs at $g\beta B_0 = D$ and the field component to V is $\langle 1|g\beta B_0 \sin\theta J_x|0\rangle = (\frac{1}{2})^{1/2}\, g\beta B_0 \sin\theta = (\frac{1}{2})^{1/2}$ $g\beta B_0\theta$ for small angles θ to the c-axis. Data for a sample containing a few ppm of Fe^{2+} (200ppm of total Fe) are shown in fig 22.

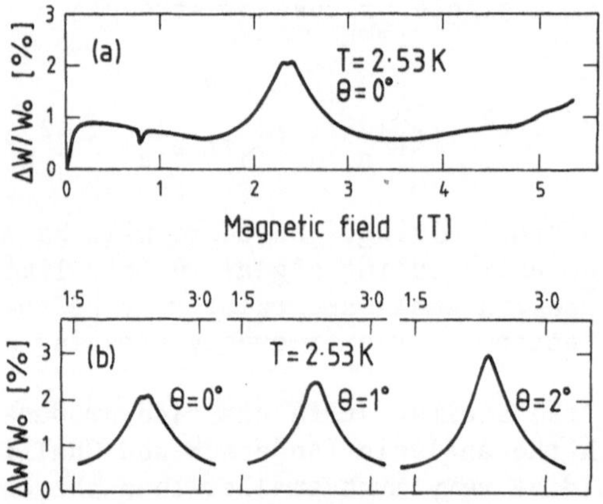

Fig.22 The magnetothermal resistivity of Fe^{2+} in Al_2O_3 (a) the minimum at 0.8T is the frequency crossing discussed earlier and the maximum at 2.4T is the anticrossing.(b) The angular dependence of the anticrossing showing the development of a frequency crossing minimum as $\theta \rightarrow 0^o$.

At $\theta = 2^{\circ}$, the signal has Lorentzian character as expected for $V \gg \Delta_{a}' + \Delta_{b}'$ but as θ and so V is reduced, the expected frequency crossing minimum appears at the centre of the signal. This can be seen more clearly in the raw data (Anderson and Challis 1974) where the Lorentzian shape is also demonstrated quantitatively. The angular dependence of the half-width was consistent with $(\Delta B_{1/2})^2 = AB_{o}^{2}\theta^2 + C^2$ with a slope corresponding to $\Delta_{a}' \sim \Delta_{b}'$ and the value of C gave a constant contribution to the off-diagonal matrix element of ~4GHz. This was consistent with estimates of random strain obtained from the EPR linewidth for this sample.

Anticrossing signals for other systems will be discussed elsewhere (Challis, 1985).

Conclusion

Crossing effects can provide spectroscopic information of rather high resolution in favourable cases. The technique can also be very sensitive : $\lesssim 10^{-2}$ ppm for relatively modestly coupled transitions. The two crossing processes can be in different parts of a sample and it is hoped it may be possible to use the technique for investigating surface layers etc. The theoretical analysis so far has been at a rather idealised level and more sophisticated treatments would be of interest. Anticrossing effects have been seen and can provide information on strain broadening etc., but there have been no reports yet of the predicted coherent effects which should lead to changes in the angular distribution of the scattered phonons. These could provide interesting information on the details of the scattering process and on intrinsic lifetimes etc.

References

Anderson, B.R., and Challis, L.J., 1973, J.Phys.C: Sol State Phys. 6, L266.

Anderson, B.R., and Challis, L.J., 1974, J.Phys.C: Sol State Phys. 7, L440.

Anderson, B.R., and Challis, L.J., 1975a, J.Phys. C: Sol State Phys. 8, 1475.

Anderson, B.R., and Challis, L.J., 1975b, J.Phys. C: Sol State Phys. 8, 1484.

Anderson, B.R., and Challis, L.J., 1975c, J.Phys. C: Sol State Phys. 8, 1495.

Berman, R., 1976, Thermal Conduction in Solids, (Clarendon Press: Oxford).

Berman, R., Brock, J.C.F. and Huntley D.J. 1963, Phys.Letts.3, 310.
Breit, G., 1933, Rev. Mod. Phys. 5, 91.
Challis L.J., 1984, Proc. 4th Int. Conf. on Phonon Scattering in Condensed Matter,(Stuttgart, 1983) Eds. W. Eisenmenger, K. Lassmann and S. Döttinger (Springer, Berlin) p.2
Challis, L.J., 1985, Crossing Effects in Phonon Scattering in Non-equilibrium Phonons in Nonmetallic Crystals, Eds W. Eisenmenger and A.A. Kaplyanskii (North-Holland, Amsterdam).
Challis, L.J., and Wybourne, M.N. 1979, J.Phys. C: Sol State Phys. 12, L711.
Challis, L.J., Ghazi, A.A., Jefferies, D.J., Williams, D.L., and Wybourne, M.N., 1978, Phys. Rev. Letts. 40, 519.
Challis, L.J., Ghazi, A.A., and Wybourne, M.N., 1982a, J.Phys C: Sol. State Phys. 15, 77.
Challis, L.J., Ghazi, A.A., and Wybourne, M.N., 1982b, Phys. Rev Letts 48, 759.
Corney, A., 1977, Atomic and Laser Spectroscopy, (Clarendon Press: Oxford).
Dreyfus, B., 1972, Proc. Int. Conf. on Phonon Scattering in Solids, (Paris, 1972) Ed. H.J. Albany (Centre d'Etudes Nucleaires, Saclay) p.207.
Eck, T.G., Foldy, L.L., and Wieder, H. 1963, Phys.Rev. Letts. 10, C521.
Griffin, A., and Carruthers, P., 1963, Phys. Rev. 131, 1976.
Hanle, W., 1924, Z. Phys. 30, 93.
Hetlzer, M.C. Jr., and Walton, D., 1973, Phys. Rev. B8, 4801.
Heraud, A.P., Ghazi, A.A., and Challis, L.J. 1982, unpublished; see Challis L.J., 1984.
Orbach, R., 1962, Phys. Rev.Letts 8, 393.
Patel, J.L., and Wigmore J.K., 1977, J.Phys. C: Sol State Phys, 10, 1829.
Schick, A., Berberich, P., Dietsche, W., and Kinder, H., 1984, Proc. 4th Int. Conf. on Phonon Scattering in Condensed Matter (Stuttgart, 1983) Eds. W. Eisenmenger, K. Lassman and S. Döttinger (Springer, Berlin) p.40.
Sheard, F.W., 1976, Proc. 2nd Int. Conf. on Phonon Scattering in Solids (Nottingham 1975) eds. L.J. Challis, V.W. Rampton and A.P.G.Wyatt,(Plenum, New York),p154.
Walton, D., 1966, Phys. Rev. 151, 627.
Walton, D., 1967, Phys. Rev. Letts. 19, 305.
Wieder, H., and Eck, T.G., 1967, Phys. Rev. 153, 103.
Wigmore, J.K., and Patel, J.L. 1974, Symposium on Microwave Acoustics,(Lancaster, 1974) eds E.R. Dobbs and J.K. Wigmore, (Institute of Physics, London) p 78.
Wybourne, M.N., Jefferies, D.J., Challis, L.J., and Ghazi, A.A. 1979, Rev.Sci.Inst. 50, 1634.
Wybourne, M.N., Challis, L.J., and Ghazi, A.A., 1980, J.Phys. C: Sol. State Phys. 13, 6495.

Acknowledgements

I am most grateful to all my collaborators whose data and ideas are summarised here, to Dr Ghazi and Mr Heraud for allowing me to show unpublished data, to Dr Sheard for very helpful discussions, to Mrs S Smith for preparing the manuscript and to the Science & Engineering Research Council for their support.

PHONON ECHOES, POLARIZATION ECHOES, AND ACOUSTIC PHASE CONJUGATION IN SOLIDS

Kristian Fossheim

Department of Physics and Mathematics
The Norwegian Institute of Technology
7034 - Trondheim, Norway

INTRODUCTION

In these lecture notes we shall discuss the occurrence of echo phenomena in physical acoustics, the physics of which rest on a wide variety of nonlinear properties in a broad range of substances. These phenomena are, at least formally, related to spin echoes since they all appear through a process of phase reversal. It is therefore quite appropriate to regard them as true echo phenomena.

Here we have already identified two common key factors: The necessity of nonlinearity, and the resulting phase reversal. As it turns out it is only at this level of description, and in the general set-up of the experiments, that the various echoes are similar. As soon as the individual cases are considered a very long list of distinctively different examples are found, varying enormously in material properties exploited, and hence in the physical interpretation of the results.

Already at the outset the terminology provokes a semantic/ pedagogic digression: Unfortunately the term "echo" is already, and for good reasons, commonly used in physical acoustics to describe an acoustic reflection. And a series of reflections is similarly referred to as an "echo train". In these notes, as well as in the whole literature on the present subject, the term echo is used excluselively in the former sense. As will become apparent the echo phenomena,created by phase reversal,are of a far more sophisticated nature than ordinary reflections, and indeed deserve to be thoroughly studied and exploited; and, I should like to add: far more than has so far been done. Echoes have opened up entirely new avenues of research in physical acoustics. And if one looks on a broader scale, including photon echoes, optical phase conjugation, cyclotron echoes, spin echoes, one realizes that a wide area of physics is accessible by related approaches in vastly different fields of fundamental research.

In order to get some idea about the way in which echoes are generated experimentally, let us refer to Fig.1 where a typical trace of receiver output on the time axis is illustrated for echoes (upper trace) and for reflections (lower trace). In this figure all rectangular pulses displayed refer to rf pulses applied to the sample, appearing on the screen essentially without delay. The rounded, sloped pulses are the pulsed responses from the sample. In the upper trace a sequence of echoes (as observed in powders) are displayed following the application of two rf pulses. These are the dynamic echoes. After a long time T when all excitations due to the first pulses have died out, a third pulse is applied. Again an echo appears, now at a time $T+\tau$ where τ was the time between the first two pulses. This is a static echo, or memory echo as explained below.

If we now look at the lower trace we see the familar picture of a series of reflections following the excitation of the sample by a single pulse.

Again, for pedagogic purposes let us spell out some differences between echoes and reflections, using Fig.1 as the illustration.

i) The position of echo pulses on the time axis bear no relationship to the physical size of the specimen studied, but are determined by the observer who sets the time τ .

ii) Multiple, coherent echoes at regularly spaced time intervals are observed, in powders!

iii) Echo signals may be stored by application of two pulses, and then read out by a third puls, even weeks or years (?) later. Note that the stored pulse is a radiofrequency signal. (Holographic storage).

These items, I hope, will suffice to shake off any remaining idea in a newcomer that acoustic echoes are somehow related to reflections, and to make clear the point that entirely new ideas will have to be brought in to describe them.

The subject of acoustic echo phenomena has already been quite

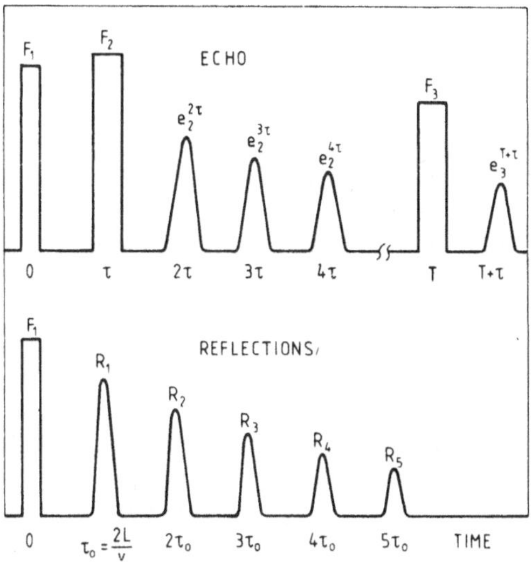

Fig. 1. Comparison of oscilloscope traces in an echo experiment (upper trace) in a powder, and in a pulsed reflection experiment (lower trace) (Fossheim and Holt 1982).

extensively reviewed in the literature. Taken together the following references, from which I will quote below, provide a broad overview: Joffrin and Levelut (1976), Kajimura (1982), Melcher and Shiren (1982), Fossheim and Holt (1982). But the field has had a vivid existence, so, only during the last 4-5 years almost 100 papers have appeared. Here, I have to stay on the main road, and leave the details and the historie development to the curious student.

Some of the papers quoted above dealt rather heavily with the mathematical aspects of the theory. Here, we will proceed at a somewhat more leisurely pace, for the benefit of those unfamilar with the subject. But, first let me discuss briefly why the subject of phonon echoes is treated at this school. The answer has a bearing on the kind of physics we discuss: Firstly, and obviously perhaps, acoustic echoes involve phonons. One may discuss to what extent the phonon concept appiies in ultrasonics at room temperature. But this is mostly a formal issue. In principle, echo experiments may be performed at terahertz frequencies. (That experiment I propose as a challange to the terahertz physicists). Secondly, if one thinks in quantum mechanical terms, and in the spirit of spin echoes, which is often appropriate, a necessary condition for echoes to be generated is the possibility of creating a highly non-equilibrium population of quantum states. So non-equilibrium is indeed a key word in addition to those mentioned before.

Let me now in general terms describe what happens, referred to Fig.1, the upper trace: In a two pulse experiment, which is the simplest case, a primary excitation of the material body is first created at time $t = 0$ by application of an external rf field pulse (F_1). This excitation (which we for now assume to be a forward propagating acoustic wave in a solid) will have a lifetime T_2. Next, at $t = \tau << T_2$ a second pulse (F_2) is applied. Now, if at $t \approx 2\tau$ a coherent pulsed signal is detected, then an echo may be suspected. To be sure, change τ by adjusting the triggering time of the second pulse. Now F_2 moves to a new position on the screen. If the suspected echo pulse moves in such a way on the time axis that it always appears at $t = 2\tau$, then you have an echo. Why should this be so? Because, since echoes arise by phase reversal (this you may take as a definition), the initial phase development which occured in the primary excitation during the time from $t = 0$ to $t = \tau$, must take a time τ to be reversed to its $t = 0$ value. Altogether it therefore must take a time 2τ to create an echo. If τ is varied, the position of the echo on the time axis must vary accordingly, so that the echo appears for any value of $\tau < T_2$.

Again, what mechanism generated this phase reversal? This question must be addressed carefully in each case, and we will revert to it again and again, because this is where the physics is! Assume for the moment that the appropriate mechanism exists, i.e. that the second pulse can mix with the primary excitation in the required manner. In the case of an initial forward propagating wave, the interaction with the second field pulse must have generated a backward acoustic wave which behaves like a time reversed image of the forward one. This is the only logically possible way the echo could arise in this case.

The almost magical quality of this phenomenon is illustrated in Fig.2. Note that no physical boundaries are involved in creating the backward propagating pulse: The forward propagating pulse is "caught in flight", and thrown back by the pump field of the second pulse. This confirms item (i) above.

How strong is the echo signal relative to the forward propagating one? This will vary by many orders of magnitude from one material to the next. In some case you find no echo at all. In other cases like in CdS under illumination we have seen echoes of 100 dB above noise. So, the echo in one sample may be weak, but if you need a strong echo you can have it, and that is very, very important for applications.

Here is another example: When a powder of piezoelectric particles is exposed to a homogeneous electric rf field pulse at $t = 0$ the individual grains are resonantly excited in phase.

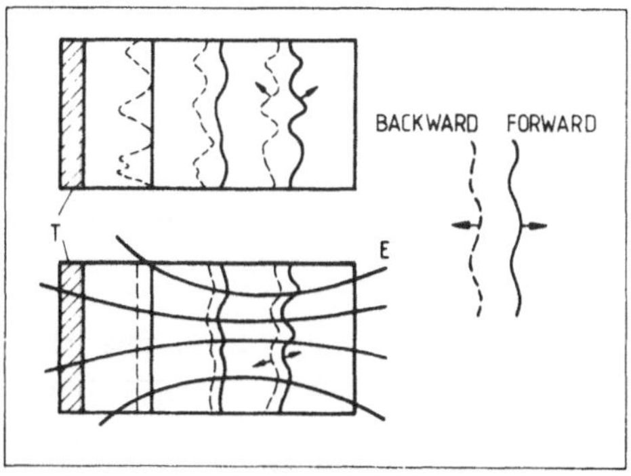

Fig. 2. Comparison of development of wavefronts in inhomogeneous substance for simple reflection (upper figure) and echo (lower figure). T = transducer, E = electric field.

When the field pulse is off again the particles will quickly dephase due to individual variations in size and shape. Now, when a second field pulse is switched on at $t = \tau << T_2$ various non-linearities may cause the excited modes to reverse their phase development. This reversal is of course completed presicely at $t = 2\tau$. Here then the oscillators (and their radiation fields) are in phase, and a coherent radiation is observed. (Between $t = 0$ and $t = 2\tau$ phase cancellation reduces the radiation to a low level). Models developed to describe this phenomenon show that a series of echoes at $m\tau$ $(m = 2,3,4..)$ should be expected. They are indeed observed, as stated in item (ii) above. Of course they have nothing to do with reflections since we are now talking about echoes in powders.

In many cases it turns out that application of the two pulses referred to above causes an echo to be holographically stored in the sample (which may be a powder or a single crystal): Imagine that after you have applied these two pulses (possibly repetitively) at a time interval τ you switch off the whole system and go on vacation. Then imagine that you return at $t = T$, say 4 weeks later, and turn on one, single pulse. On your scope you will see an echo pulse at $t = T + \tau$. (See Fig.1). Now recall that τ was an arbitrarily chosen (and long ago forgotten by you!) pulse separation used 4 weeks ago. Since the equipment has been switched off during that period, obviously the sample must be in possession of a keen memory. It has stored the information for you. In fact it has stored the entire echo which is an rf pulse, and conveniently allowed you to read it out just by applying a single field pulse ("read pulse"). So, the sample memory includes remembering the time τ between applied pulses, their radio frequency ω, and their shapes, since they determined the echo shape just read out. This is one more of the remarkable feats of echo phenomena. (Other astonishing feats will be discussed later). Moreover, the memory is not lost in this process. Repeated reading always gives the same memory echo out any time you like to recall it. This was an example of item (iii) above.

Now it would have been nice to introduce the theory of phonon echoes. But, alas, there is a long list of echo phenomena, and an almost equally long list of possible mechanisms. So, to do the physics correctly a lot of separate cases must be studied. Still, as always in physics, similarities should be exploited, here through a search for the mathematical mapping of various echo problems onto each other. To some extent this has already been carried out, for instance for powder echoes by Kajimura (1982).

Occasionally I receive the impression from colleagues that echoes are considered to be rather "difficult". It is certainly fair to say that they are more demanding to understand than reflection phenomena, but then no doubt they are also much more fun to study! And I happen to think they should be studied much more vigorously than they have been. Although the list of papers on the subject is already frightfully long.

As I shall discuss below, echoes have already for several years been used routinely as a tool in our laboratory in Trondheim, and the use of this method is spreading. But in addition the search for a better understanding of echo mechanisms, and of better echo materials, as well as further applications ought to be carried on. In my own opinion few if any discovery in physical acoustics has been as exciting as echoes.

In what follows the main emphasis will be on travelling wave echoes, like in single crystals, or glasses. This is partly because I want to advocate the use of the echo in single crystals as a practical tool in ultrasonic studies of matter. Echoes in powders are not less interesting, but so far their use in the study of solid state physics has been more restricted. This could change any time though. A brief review of powder echoes is given towards the end.

GENERAL DESCRIPTION OF 2-PULSE ECHOES

Classification

Since the list of observed echoes is very long, both in terms of materials and proposed echo mechanisms, it is important to answer the questions: In what sense are these echoes of similar nature, and in what sense are they different? The latter question is perhaps the most important one, and was basicly answered for two-pulse echoes by Gould (1965), even befor the acoustic echoes were known: There are two general classes of echoes, with characteristically different dependence on pulse separation τ : Either (I) the echo amplitude is decaying like $\exp(-2\tau/T_2)$ from a maximum at $\tau = 0$, as observed for instance in all propagating wave cases (bulk systems); or (II) they have zero amplitude at $\tau = 0$, grow to a maximum at a time τ^* , and finally decay as $\exp(-2\tau/T_2)$ at long times like in powders. In both cases T_2 is a time characteristic of the particular material, depending possibly on intrinsic as well as extrinsic properties. Essentially T_2 is the inverse of the acoustic attenuation of the material.

The former case (I) is referred to as the parametric field-mode case, and the latter case (II) as the anharmonic oscillator case. In case (I) the external rf field applied during the second pulse $\tau \leq t \leq \tau + \Delta_2$ couples parametrically to the modes excited during the first pulse. This interaction generates the phase reversed signal, and an echo is observed at $t = 2\tau$. Several different echo phenomena belong to this class. The main initial observations, taken chronologically are: Spin echoes (Hahn 1950), photon echoes (Kurnit et. al. 1964), ultrasonic spin echoes (Shiren and Kazyaka 1972) forward propagating phonon echoes in glass (Golding and Graebner 1976) and static polarization echoes (memory echoes) in powders, as observed for instance by Melcher and Shiren (1976).

In view of the great variety of systems included here it is clearly necessary to look for a number of different microscopic mechanisms for the parametric interaction. Obviously, several of these cases may be described within the conceptual framework of the original spin echo, even though the details of the experiment are completely different. One way to go about it will obviously be to look for a possible transformation of the equations of motion into a set of spin-dynamical Bloch-type equations. We shall look again at this question at a later point. Although it is not certain that such transformations will always increase the understanding of the physics, it clearly at least saves some work in solving the mathematics.

A qualitative picture

The second class of echoes (II) referred to as nonlinear-oscillator echoes above, are typically those observed in two-pulse powder echo experiments. Again, it appears possible to transform the equations of motion for quite different cases like piezoelectric powders [Kajimura et. al. 1976, Fossheim et. al. 1978] and metal powders [Tsuruoka and Kajimura 1981] into a unified description [Kajimura 1982].

Let us revert now to case (I) and take a forward propagating wave (the probe)

$$A_f\, e^{i(\vec{q}\cdot\vec{x}-\Omega t)} + A_f^*\, e^{-i(\vec{q}\cdot\vec{x}-\Omega t)} \qquad (1)$$

generated in a piezoelectric crystal during the time interval $0 \leq t \leq \Delta_1$ by a transducer at frequency Ω at the $x = 0$ end of the crystal. Next, suppose the other end of the crystal is located in a microwave cavity. Here we can switch on the cavity field at frequency ω during the time interval $\tau \leq t \leq \tau + \Delta_2$ This is the pump:

$$E = E_o\, e^{i\omega(t-\tau)} + E_o^*\, e^{-i\omega(t-\tau)}; \tau \leq t \leq \tau + \Delta_2 \qquad (2)$$

$$\underline{E} = 0 \qquad \text{otherwise}$$

where Δ_2 is the width of the second pulse. If the material is completely linear in its response to these perturbations, then no echo can be generated, since no interaction takes place. (A backward wave can of course only arise by some interaction prosess). In practice no material is completely linear. In fact, higher order terms in the free energy ensure that all materials are in principle candidates for echo formation. (However, not all of these will be good echo materials). In practice, therefore, there will always be mixing, for instance like

$$\gamma\cdot\left[A_f\, e^{i(\vec{q}\cdot\vec{x}-\Omega t)} + A_f^*\, e^{-i(\vec{q}\cdot\vec{x}-\Omega t)}\right]\left[E_o e^{-i\omega(t-\tau)} + E_o^* e^{+i\omega(t-\tau)}\right] \qquad (3)$$

where γ is some coupling constant. Among the resulting terms we now look for a possible backward wave (the signal). The product of the second and third terms gives, after rearrangement, for $\omega = 2\Omega$

$$\gamma A_f^*\, E_o\, e^{-i[\vec{q}\cdot\vec{x}-\Omega(t-2\tau)]} \qquad (4)$$

As long as $\tau \leq t \leq 2\tau$ this represents a wave travelling in the backward direction relative to the initial wave. Furthermore, this wave arrives at $x = 0$ at $t = 2\tau$, as required for an echo signal. Here the phase at all points on the wavefront has the value it had when the initial forward wave was launched at $x = 0$ $t = 0$. These properties of the echo make it clear that it can be described as possessing both phase memory and time reversal . And these it turns out, are the basic common properties of all echo phenomena.

The phenomena discussed here are very closely analogous to

phase conjugation in optics (a fact we shall return to below). Depending on the symmetry properties of the echo material various interactions are possible in a given material, and hence the echo mechanisms may be different from one material to the next. Furthermore, several different mechanisms may be active simultaneously in the same material. However, usually one mechanism will dominate under given conditions (for instance at a given temperature, illumination, doping etc.).

Our discussion above of course said nothing about echo shapes. Since these also hold a clue to what echoes are all about we return to that question below.

Again, to see even the echoes of case (II) (anharmonic oscillator echoes) from the simplest possible viewpoint let us describe these as follows: Take a primary excitation $a_1 e^{-i\Omega_i t}$ of particle i caused by a first field pulse during $0 < t \leq \Delta_1$ and a secondary excitation of the same particle $a_2 e^{-i\Omega_i(t-\tau)}$ produced by the second pulse during $\tau \leq t \leq \tau + \Delta_2$, where a_1 and a_2 are the respective amplitudes, and Ω_i is again the resonance frequency of particle i. In a linear system the total result is a simple superposition, summed over all N particles. Neglecting acoustic damping the radiation from the system is proportional to

$$\sum_{i=1}^{N}\left[a_1 e^{-i\Omega_i t} + a_2 e^{-i\Omega_i(t-\tau)}\right] \tag{5}$$

Since we assume N to be a large number, and since Ω_i is different in different particles, the net result is zero due to phase cancellation, i.e. no radiation; hence no echo.

Now allow for the nonlinearities, which we know are always present, as pointed out above. The coupling between the modes of (5) will produce terms like

$$\sum_{i=1}^{N}\left[a_1^* e^{i\Omega_i t} \cdot a_2^2 e^{-i2\Omega_i(t-\tau)}\right], \tag{6}$$

where the power two in the second term is due to the higher order nature of the nonlinearity. Rearranging the terms in (6) gives the radiation proportional to

$$\sum_{i=1}^{N} a_1^* a_2^2 e^{-i\Omega_i(t-2\tau)} \tag{7}$$

Here we see that at $t = 2\tau$ all phases are the same, and the radiation from all particles add to give strong coherent radiofrequency signal, although the different particles have different eigenfrequencies. This again is the amazing result of phase reversal, or phase memory and time reversal.

To pursue this point the analogy between phase coherence in case (I) and case (II) is the following: The phase coherence at $t = 2\tau$ of particles with different eigenfrequencies is analogous to a situation where, in the travelling wave case the primary excitation was generated from a rough surface. The only way in which a signal resulting from such initial conditions can ever become coherent is precisely by the reversal of the signal as described above, since then the backward wavefront will travel to the generating surface, and the wavefront will precisely match the shape of the surface from which the forward wave was generated. Thus case (I) is no less amazing than case (II).

It is to be expected that such phenomena may be exploited for many practical purposes in the future, and even more so when the static, or memory effects are brought into the picture.

Briefly, the memory phenomena are possible due to the creation of a "frozen state" in which the material retains a holographic image containing the full phase information about the two first pulses. This halogram may be read by exciting it with a third pulse. The resulting echo signal will have properties quite similar to the dynamic echo resulting from only two pulses. Detailed models for such echoes have been worked out for both class I and class II echoes. We will discuss these also briefly in what follows.

The study of acoustic echoes has naturally generated its own jargon and terminology. As an aid to the reader I have listed some of their vocabulary in Appendix A.

Some experimental facts

To substantiate the qualitative discussion given above on 2-pulse (dynamic) echoes and their separation into two classes we look briefly at some experimental results, illustrated in Figs 3-5.

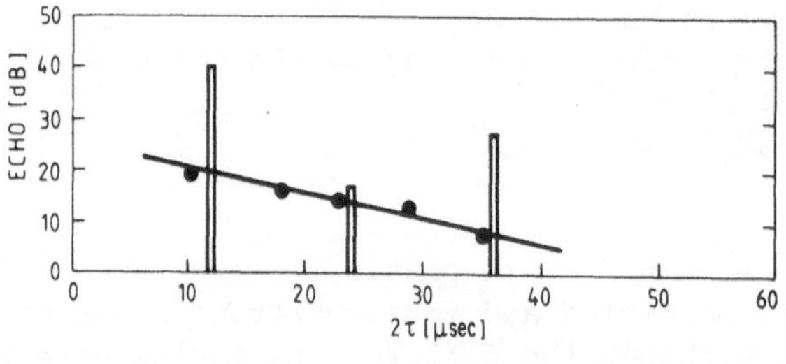

Fig. 3. Comparison of decay of reflections (bars) and echoes (dots) observed in $Bi_{12}GeO_{20}$ (BGO). Echo decays exponentially (Fossheim and Holt 1982).

Fig. 3 shows experimental observations of the decay of 2-pulse echoes in BGO ($Bi_{12}GeO_{20}$), compared to reflections in the same material. The main point here is to observe that the echo decays exponentially, even though the reflections die out in an irregular manner due to nonparallel end faces. These aspects illustrate both phase reversal, and the fact that the decay is simple exponential in the field-mode parametric interaction.

Fig. 4 is a further manifestation of these results showing that the echo may even be substantially stronger than reflection under conditions where phase cancellation of reflections occurs due to nonparallel end faces.

Fig. 5 shows experimental results on 2-pulse echoes in a powder, where as indicated previously, the decay behaviour is more complex: First a rise to a maximum, and then an approximately experimental decay at large τ.

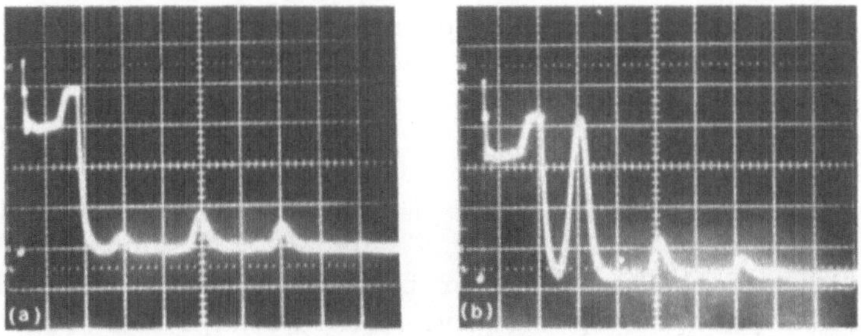

Fig. 4. Comparison of reflection signals in BGO (left) and reflections plus echo (right). Here the echo appears stronger than the first reflection.

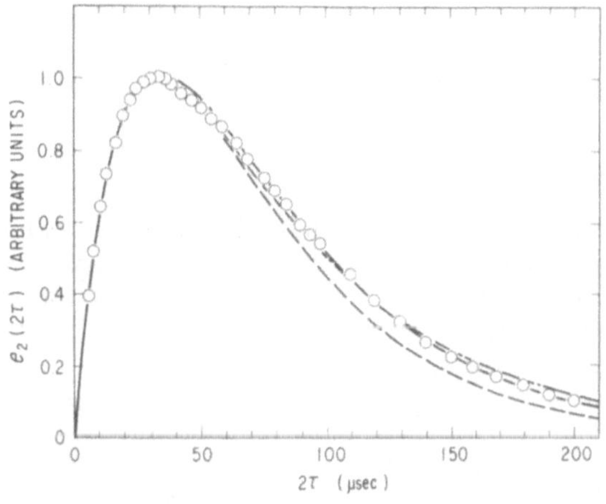

Fig. 5. Typical "rise-and-decay" of two-pulse powder echoes. Data from Nb metal powder (Tsuruoka and Kajimura (1980).

SELECTION RULES IN BULK DIELECTRICS

The echo generation process of course has to obey energy and momentum conservation. This places definite restrictions on the frequencies and momenta of participating phonons and photons for electroacoustic echoes,as was in fact already implied in the arguments leading to the expressions (4) and (7) above. Let us now discuss two specific examples: Two-pulse Ω, 2Ω echoes, and two-pulse Ω, Ω echoes. Note here that although the phonon symbol Ω is used for both frequencies the meaning is that the first Ω refers to a phonon frequency, while the second one simply implies that the photon frequency in $\omega = 2\Omega$ in the first case, and $\omega = \Omega$ in the second case. We discuss Ω, 2Ω echoes first. Here the implication is then that both the forward and backward acoustic wave is at Ω, while the rf field is at 2Ω. We will show that such a case is possible due to socalled second order piezoelectricity, i.e. an interaction $f \cdot ES^2$, where f is some constant, E is the electric field, and S is the acoustic strain. Equating the momentum and energy before and after the process, as required, we have:

$$\vec{q}_f + \vec{k}_p = \vec{q}_f + \vec{q}_b + \vec{q}_f \qquad (8a)$$

$$\Omega_f + \omega_p = \Omega_f + \Omega_b + \Omega_f \qquad (8b)$$

where subscripts f, b, and p refer to forward wave, backward wave (echo), and photon respectively. Here we have included the primary phonon $(\Omega_f, \vec{q}_f)$. Since this quantum persists on both sides we may remove it to obtain the net balance. Furthermore, at experimental frequencies the momentum $\vec{k}_p$ of the photon is so small that we may safely set it equal to zero. Now (8a) becomes

$$\vec{k}_p = 0 = \vec{q}_f + \vec{q}_b$$

or

$$\vec{q}_b = - \vec{q}_f \qquad (9)$$

Here we see the remarkable result that the backward phonon has a momentum presicely opposite of the forward one, implying wave-vector reversal. Assuming we stay on the same acoustic branch before and after this also implies $\Omega_b = \Omega_f$.
Hence the frequency condition

$$\omega_p = \Omega_f + \Omega_b$$

reduces to

$$\omega_p = 2\Omega_f \equiv 2\Omega \qquad (10)$$

The required electric field frequency is therefore twice that of the forward propagating phonon. We conclude that the interaction $f \cdot E S^2$ allows generation of Ω, 2Ω echoes.

Furthermore the analysis has shown this to be a 3-wave mixing (TWM) since 3 waves participated in the net process, one photon and two phonons. It is a nondegenerate process in the sense that not all three waves are of the same frequency. The size of f determines the strength of the echo. Here the material properties enter in a desicive way.

The fact which we observe above that the original forward phonon $(\Omega_f, \vec{q}_f)$ did not directly participate means that it plays only a stimulating role. The generation of the backward wave is due to the photon breaking up into phonons travelling in opposite directions.

In engineering language the waves referred to above are: The forward wave is the "idle" or "probe", the electric field is the "pump", and the backward wave (echo) is the "signal".

Next we look at the case of Ω, Ω echo, and inquire whether it may arise through an interaction term of the type $g E^2 S^2$, where g is a constant. The conservation laws now require two photons because of the E^2- dependence. Again we start by including the primary, forward wave.

$$\vec{q}_f + \vec{k}_{p1} + \vec{k}_{p2} = \vec{q}_f + \vec{q}_f + \vec{q}_b \tag{11a}$$

$$\Omega_f \quad \omega_{p1} + \omega_{p2} = \Omega_f + \Omega_f + \Omega_f \tag{11b}$$

or, removing the "redundant" forward wave:

$$\vec{k}_{p1} + \vec{k}_{p1} = \vec{q}_f + \vec{q}_b \tag{12a}$$

$$\omega_{p1} + \omega_{p2} = \Omega_f + \Omega_b \tag{12b}$$

This type of process is often referred to as 4-wave mixing, similar to that used in optical phase conjugation. (See also Appendix A regarding this terminology).

Since the photons come from the same low frequency monochromatic source they are identical, i.e. $\omega_{p1} = \omega_{p2} = \omega_p$; $\vec{k}_{p1} = \vec{k}_{p2} = \vec{k}_p \approx 0$. The latter condition means that

$$\vec{q}_f + \vec{q}_b = 0 \Rightarrow \vec{q}_b = -\vec{q}_f \tag{13}$$

Again we have found wavevector reversal. And, again assuming the phonons belong to the same branch (13) implies that the phonon frequencies are equal, i.e.:

$$\Omega_b = \Omega_f$$

Using this in (12b) we get

$$\omega_p = \Omega_f \equiv \Omega \tag{14}$$

So, in this case, when the interaction is gE^2S^2 the electric field should be at the same frequency as the forward wave. This we called the Ω, Ω echo, and it is indeed allowed.

As it turns out the pump need not be the directly applied cavity field, but may for instance be the superposition of the cavity field and the acoustoelectric field of the forward wave.

Further analysis shows that other interactions may exist. We refer to the literature on that subject, except for certain cases to be discussed under CdS.

Symmetry considerations

We have seen that echoes may be generated through different types of interaction. Whether both of these, or only one of those discussed are possible in a given material depends on its crystal symmetry. Since f is a fifth rank tensor it can only exist in crystals without a center of symmetry. It will therefore exist in piezoelectrics, but not for example in glasses. Hence Ω, 2Ω echoes are possible in piezoelectrics, but not in glasses, or in any centrosymmetric crystals.

The gE^2S^2 coupling on the other hand exists in all matter, including also glass, and Ω, Ω backward echoes may therefore be produced here. Experimentally this was confirmed by Shiren et.al. (1977). Again, by the same argument all centrosymmetric crystals are also possible candidates. Still, the fact that this is a fourth order term may reduce the intensity and hence also the chances of finding the effect. Piezoelectrics are especially interesting echo materials because they allow both the Ω, Ω and the Ω, 2Ω echo to be produced, as has been confirmed in for instance CdS. (Joffrin and Levelut) 1972, Fossum et.al. 1981). In glass on the other hand no Ω, 2Ω backward wave echo is found. We emphasize the fact that we have discussed backward echoes only. In glasses it turns out that even forward wave echoes may be generated at sufficiently low temperature (Golding and Graebner 1976). This is a different problem, analogous to acoustic spin echo.

In addition to the specific interactions mentioned so far other terms E^nS^n are possible i.e. interactions of order $n+m$. Since the principles remain the same we omit further discussion here.

DYNAMIC ECHOES IN BULK DIELECTRICS: THEORY

General considerations

We now study the conditions for dynamic echo generation in dielectrics. We mainly discuss development which have taken place during the last 6-8 years. The earliest work in the field has been thoroughly described in reviews already referred to above. Since echoes in dielectrics have been so much studied we emphasize these.

Following Fossheim and Holt (1982) the thermodynamic potential is taken as

$$G = \frac{1}{2}CS^2 + \frac{1}{3!}C^{NL}S^3 - \frac{1}{2}\varepsilon E^2 - \frac{1}{3!}\chi^{NL}E^3 - eES$$
$$- qE^2S + \frac{1}{2}fES^2 + \frac{1}{2}gE^2S^2 + \ldots \quad . \tag{15}$$

Here S and E are strain and electric field respectively, and terms up to fourth order have been included. (The expansion is complete only to third order). The mechanical stress and the electric displacement are derived by differentiation:

$$T = \partial G/\partial S \tag{16}$$

$$D = - \partial G/\partial E \ . \tag{17}$$

Using (15) we find

$$T = CS + \frac{1}{2}C^{NL}S^2 - eE + fES + gE^2S \tag{18}$$

$$D = \varepsilon E + \frac{1}{2}\chi^{NL}E^2 + eS + 2qES - \frac{1}{2}fS^2 - gES^2 \tag{19}$$

This one-dimensional exposition will be sufficient to bring about echo generation. An important distinction must be made between materials with and without space charge. We discuss first the case of zero space charge.

No space charge : $\nabla\vec{D} = 0$

The first task is to set up the equation of notion. Due to the higher order terms in G we obtain source terms which may be of arbitrary order in E reflecting different possible mechanisms. For simplicity we suppress the explicit S-dependense by introducing an effective elastic constant

$$C_{eff} = \sum_{n=o}^{\infty} C_n E^n \tag{20}$$

where the strain-dependence is hidden in C_n. Using this expansion Newton's second law leads to

$$\rho \frac{\partial^2 u}{\partial t^2} = C_{eff} \frac{\partial^2 u}{\partial z^2} \tag{21}$$

The first few terms C_n in (20), assuming $e >> qE$ and $\varepsilon >> \chi^{NL}E$ are, when the explicitly strain dependent terms are neglected:

$$C_o = C + e^2/\varepsilon \tag{22}$$

$$C_1 = f + 4eq/\varepsilon - e^2\chi^{NL}/\varepsilon^2 \tag{23}$$

$$C_2 = g + 4q^2/\varepsilon - 4eq\chi^{NL}/\varepsilon^2 + e^2(\chi^{NL})^2/\varepsilon^3 \tag{24}$$

Writing now $C = C_o + C_n E^n$ the equation of motion takes the form

$$\frac{1}{v_o^2} \frac{\partial^2 u}{\partial t^2} - \frac{\partial^2 u}{\partial z^2} = \frac{C_n E^n}{C_o} \frac{\partial^2 u}{\partial z^2} \tag{25}$$

where $v_o = (C_o/\rho)^{\frac{1}{2}}$. Here, ρ is the density of the material.

The echo generation occurs through the interaction ("mixing") of the acoustic forward wave and the electric field of the cavity, as explained before. This process, as seen in a space-time representation, is shown in Fig.6.

Mathematically a forward porpagating acoustic pulse u_1 of amplitude u_o, frequency ω, wavevector q, width Δ_1 and velocity v_o may be written as

$$u_1 = u_o \cos(\omega t - qz) f_1 \left(\frac{t - z/v_o}{\Delta_1}\right) \tag{26}$$

where f_1 is a symbolic way of stating that the pulse envelope is centered at z/v_o and is of width Δ_1.

Similarly then E_2 is written as

$$E_2 = E_o \cos(2\omega t/n) f_2 [(t-\tau)/\Delta_2] \tag{27}$$

Here we have chosen the electric field to have a frequency $2\omega/n$. This ensures that the product

$$u_1 E_2^n \propto \cos(\omega t - qz) \cos^n(2\omega t/n) \tag{28}$$

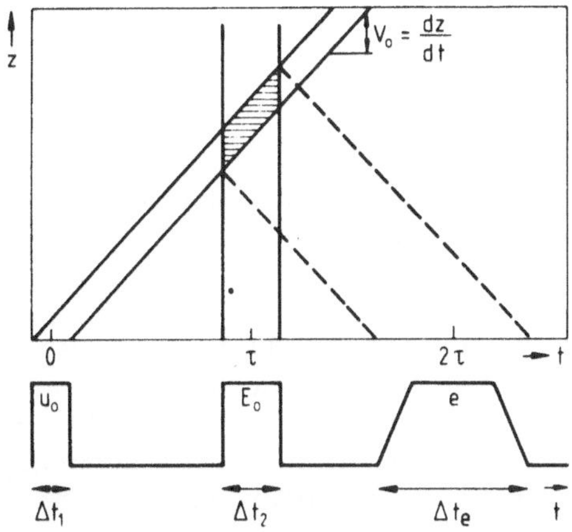

Fig. 6. Space-time representation of echo generation for travelling waves in bulk. Full lines with positive slopes delineate forward acoustic pulse. Hatched region indicates interaction with field at $t = 2\tau \pm \frac{\Delta}{2}$. Dotted lines of negative slope indicate echo propagation. Pulses are shown on time axis below. (Fossheim and Holt 1982).

contains a backward component of wavevector $-q$, precisely what we want for an echo. The procedure could be expanded to 3 dimensions to obtain reversal of the entire wavevector $\vec{q}$. We choose a trial solution

$$u(z,t) = u_1(z,t) + e(z,t) \tag{29}$$

and assume that the second term on the right hand side is weak. Insertion into (25) gives

$$\left(\frac{1}{v_o^2}\frac{\partial^2}{\partial t^2} - \frac{\partial^2}{\partial z^2}\right)e(z,t) = \frac{C_n}{C_o}\frac{\partial^2 u_1(z,t)}{\partial z^2}E_2^n(t) \tag{30}$$

We are here only looking for backward propagating terms, i.e. terms like $\sin(\omega t+qz)$. The n'th order echo, i.e. the echo derived from the term containing E_2^n has the form

$$e(z,t) = \frac{1}{2^{n+2}}\frac{C_n}{C_o}u_o E_o^n\,\omega f_e\left(\frac{t+z/v_o-2\tau}{\Delta_1+2\Delta_2}\right)\sin(\omega t+qz) \tag{31}$$

Here f_e is the echo shape function which is a convolution of the shapes f_1 and f_2 according to:

$$f_e\left(\frac{t+z/v_o-2\tau}{\Delta_1+2\Delta_2}\right) = f_1(t/\Delta_1) * f_2^n(t/2\Delta_2) \tag{32}$$

Here the widths Δ_1 and Δ_2 are arbitrary, contrary to the situation in spin echoes (Hahn 1950) and other echoes describable by a spin echo formalism.

Equation (32) expresses the fact that the echo will be found at a position z which depends on τ through the condition

$$t - (2\tau - z/v_o) = 0 .$$

For instance, at $t = 2\tau$, the echo will be observed at $z = 0$, as it should since the generating transducer was placed there.

Furthermore Eq.(32) states that the total echo width will be $\Delta_1 + 2\Delta_2$. Due to the nature of convolution one finds also that input pulses with $\Delta_1 = 2\Delta_2$ produce triangular shaped echoes, while if $\Delta_1 \neq 2\Delta_2$ the shape is trapezoidal. A further convolution property is that the echo amplitude depends on echo width, being proportional to the lesser of the widths Δ_1 or $2\Delta_2$. The echo will of course be subject to the same inelastic losses as an ordinary acoustic wave, and this will be the same as the loss of the forward wave if the two are of the same frequency, and of the same acoustic mode. (An exception will be discussed for $\nabla\cdot\vec{D}\neq 0$).

The formalism used here, although classical, gives some clue also to the quantum picture (Bajak 1978): One may regard the n'th order echo generation as a process where first n photons produce one photon $n\omega_2$, subsequently decaying into two phonons, one forward and one backward. This leads again to the condition $\vec{q}_b = - \vec{q}_f$ discussed before.

Echo generation when $\nabla\cdot\vec{D} \neq 0$

Since CdS has been one of the most interesting echo materials to study in the past, and since most echoes here involve space charge it is obviously necessary to consider situations when $\nabla\cdot\vec{D} \neq 0$, i.e. echo generation under conditions where space charge is present. For this the previous analysis must be expanded by the condition

$$\nabla\cdot\vec{D} = - Qn_s \qquad (33)$$

where Q is the elementary charge, and n_s is the charge number density. While previously the divergence of $\vec{D}$ was given by use of Eq.(19), we now have to apply the condition (33) in combination with (19). When this is followed through new source terms are found in the equation of motion, terms which are proportional to Qn_s . Previously we found only source terms which arose from the nonlinearities of C .

The equation for a backward propagating wave is now

$$\rho \frac{\partial^2 u}{\partial t^2} - C_o \frac{\partial^2 u}{\partial z^2} = \frac{eQn_s^B}{\varepsilon} \tag{34}$$

where Qn_s^B is a backward propagating Fourier component of space charge density. The equation of charge continuity is

$$Q \frac{\partial n_s}{\partial t} = - \frac{\partial J}{\partial z} \tag{35}$$

where J contains contributions due to both drift and diffusion. We may write it as

$$J = Q\mu n_c E + QD_n \frac{\partial n_c}{\partial z} \tag{36}$$

where D_n is the diffusion constant and μ is the electron mobility. The density n_c here refers to the total number density of electrons in the conduction band. (We have an n-type semiconductor in mind). This density may be split up into an equilibrium part n_0 and a term $f'n_s$ where f' gives the fraction of space charge in the conduction band. The rest of course will be that residing in traps. For a discussion of f' we refer to Greebe (1966). Using then the definition

$$n_c = n_0 + f'n_s \tag{37}$$

we may combine Eqs.(35), (36) and (37) into a differential equation for n_s. Using also $\partial D^B/\partial z = - Qn_s^B$ n_s^B may be calculated as (B = backward, F = forward)

$$n_s^B = \frac{e}{Q} \frac{i\omega_c/\omega}{\Gamma^B} \frac{\partial S^B}{\partial z} + \frac{\mu f'}{v_s} \frac{n_s^{2\omega}E^F + E^{2\omega}n_s^F}{\Gamma^B} \tag{38}$$

The new quantities introduced here are: $\omega_c = \sigma/\varepsilon$, where the conductivity σ is $Q\mu n_0$; $\omega_D = v_s^2/f'D_n$ is the diffusion frequency, and $v_s = \omega/q$ is the sound velocity. Furthermore

$$\Gamma^B = \gamma^B - i(\frac{\omega_c}{\omega} + \frac{\omega}{\omega_D})$$

with $\gamma^B = 1 - v_d/v_s$; $v_d = f'\mu E_0$ is the drift velocity of the electrons, E_0 is the applied electric field, while E^2 and $n_s^{2\omega}$ are the non-propagating 2ω-Fourier components of the electric field and space charge density.

The source n_s^B is given by the products $n_s^{2\omega}E^F + n_s^F E^{2\omega}$, as may easily be verified.

When this source is inserted into the wave equation (34) the following is obtained

$$\rho\frac{\partial^2 u^B}{\partial t^2} - C^B \frac{\partial^2 u^B}{\partial x^2} = \frac{eQ\mu f'}{\varepsilon v_s} \frac{1}{\Gamma^B}(n_s^{2\omega}E^F + E^{2\omega}n_s^F) \tag{39}$$

with

$$C^B = C(1+K^2 \frac{\gamma^B - i\omega/\omega_D}{\Gamma^B}) \tag{40}$$

This result was found by Fossum and Holt (1982). Here we have identified the complex elastic constant which may be analyzed further with respect to sound velocity and attenuation in line with White (1962). The attenuation may be expressed as

$$\alpha = \frac{k^2}{2v_s}\omega\frac{\gamma\omega_c}{\omega}[\gamma^2 + (\frac{\omega_c}{\omega} + \frac{\omega}{\omega_D})^2]^{-1} \tag{41}$$

Here we have introduced the electromechanical coupling constant $K^2 = e^2/\varepsilon C$ and γ represents either γ^B or γ^F. Since $\gamma^B \neq \gamma^F$ with a DC field present the forward and backward waves are not equally damped.

In the early papers of Shiren and Melcher (1975) and Melcher and Shiren (1975) echo generation was discussed on the basis of the term $n_s^{2\omega}E^F$, as well as the effects of an electric field E_0. They found that in the case of a homogeneous AC-field of frequency $2\omega/n$ with $E_0 = 0$ only $n = 2,4,6$ ... would give echoes. So an $\Omega, 2\Omega$ cannot be generated from these terms.

However, as pointed out by Fossum and Holt (1982) the term $E^{2\omega}n_s^F$ gives an $\Omega, 2\Omega$ echo even when $E_0 = 0$. Assuming also that $\mathrm{Im}C^B \ll \mathrm{Re}C^B$, and deriving the forward propagator

$$n_s^F = -\frac{e}{Q}\frac{i\omega_c/\omega}{\Gamma^F}\frac{\partial S^F}{\partial x} \tag{42}$$

with

$$\Gamma^F = \gamma^F + i(\frac{\omega_c}{\omega} + \frac{\omega}{\omega_D})$$

where

$$\gamma^F = 1 + v_d/v_s$$

they find S^F to be the main source of n_s^F.

Using these results as well as the expression for the attenuation the wave equation becomes

$$\frac{1}{v_s^2}\frac{\partial^2 u^B}{\partial t^2} - \frac{\partial^2 u^B}{\partial z^2} = i\alpha \frac{2\mu f'}{\omega} E^{2\omega} \frac{\partial s^F}{\partial z} \qquad (43)$$

Without a DC-field an $\Omega, 2\Omega$ echo is obtained:

$$u^B = \frac{1}{8}\frac{C_1}{C_0} u_0^F E_0^{2\omega} \omega \tilde{f}^B \sin(\omega t + qz)$$

$$+ \frac{1}{4}\mu f'\alpha u_0^F E_0^{2\omega} \tilde{f}^B \cos(\omega t + qz) \qquad (44)$$

Here the contribution from $E^{2\omega} n_s^B$ was neglected.

The most interesting part of this result is the second term on the right hand side since here the electronic properties enter through α and f' . Fossum and Holt give the final expression for this part as

$$u_0^B = \frac{1}{4}\mu f'\alpha u_0^F E_0^{2\omega} f_e \cdot \exp[-(\alpha_0 + \alpha)2v_s\tau] \qquad (45)$$

where f_e is the convolved echo evelope, and the attenuation α_0 is of non-electronic origin. Under conditions where $\omega_C/\omega \ll 1$ and $\omega/\omega_D \ll 1$ the electric attenuation is $\alpha = (K^2/2v_s)\omega_C$.

These theoretical results were confirmed experimentally as will be shown below.

Intrinsic versus extrinsic echo mechanisms

A very important question which arises in all echo work in dielectrics is about the role of defects. In other words: Are the observed echoes of intrinsic or of extrinsic origin? This is an area where much more work is needed. The answer so far is that in a lot of cases defects are known to be the main source of echoes in dielectrics, both in insulators and semiconductors. In other cases the nature of the nonlinearity is not known in detail. This is true in most cases. Examples:

Equation (45) already carries implications about the importance of trapped charge through the factor f' . Furthermore since α is in function of the conductivity one immediately recognizes the possibility of having the electronic properties of a semiconductor directly mirrored in the echo properties. Such behaviour was studied particularly in memory echoes in a number of papers by Shiren, Melcher and coworkers in the early 70's, and later in dynamic echoes by Fossum and Holt (1982).

The possibility of observing echoes caused by internal strains was pointed out by Holt, Fossum and Fossheim (1984), and verified by Miyasato, Wigmore and Meredith (1984). Experiments on quartz showed that the echo amplitude increased on quenching and decreased on annealing the sample.

Shiren, Arnold and Kazyaka (1977) discovered backward-wave phonon echoes in several glasses and attributed the effect to the presence of OH^- groups forming a distribution of inhomogeneously broadened resonances which were sensitive both to ultrasonic and electromagnetic fields. The echo amplitude was found to be proportional to the concentration of the OH^- groups.

Golding and Graebner (1976) discovered forward propagating phonon echoes in Suprasil W glass at mK temperatures. Two-level tunnelling systems were found to cause the echoes, and a pseudo-spin formalism was used to describe the observations. The exact nature of these two level systems is not known, it may of course be intrinsic.

Melcher (1979) developed a backward phonon spectroscopy in indium doped silicon. The backward wave phonon spectrum as a function of magnetic field was used to identify the source of the nonlinearity as resonance transitions within the neutral acceptor ground state.

For certain memory echoes observed in powders a deformation model has been developed to explain the long lifetimes of the "frozen state", by Laikhtman (1977).

This listing should not be taken as evidence that echo phenomena in dielectrics are always of extrinsic origin. Rather I would say that such a possibility always exists, and should be considered in every case.

The situation is different in spin systems. Here the echo will normally be of strictly intrinsic origin. Acoustic spin echoes were discovered by Shiren and Kazyaka (1972) in MgO doped with Ni^{2+}, Fe^{3+} and Mn^{2+}. Here both forward and backward echoes were observed, as well as stimulated echoes.

DYNAMIC EXPERIMENTS IN DIELECTRICS

Historically the first observation of echoes in bulk dielectrics were made independently by Thompson and Quate (1970) in $LiNbO_3$, and by Popov and Krainik (1971) in SbSI. Before this, however, Svaasand (1969) had observed the electric field resulting from two counterpropagating waves, a result closely related to echoes. Chaban (1967) had predicted the existence of parametric echoes well in advance of these observations.

At first, since $LiNbO_3$ and SbSI were both examined in the ferroelectric phase, it was thought that ferroelectricity was necessary for echo generation. Since Joffrin and Levelut (1972) found echoes in CdS this notion had to be discarded.

The experimental set-up used to study echoes varies considerably, depending on frequency range and material under investigation. Fig.7 gives a rather detailed block diagram of apparatus used in our group at ~100 - 500 MHz (Fossheim and Holt 1982). In addition, when studying CdS, illumination and temperature control were added.

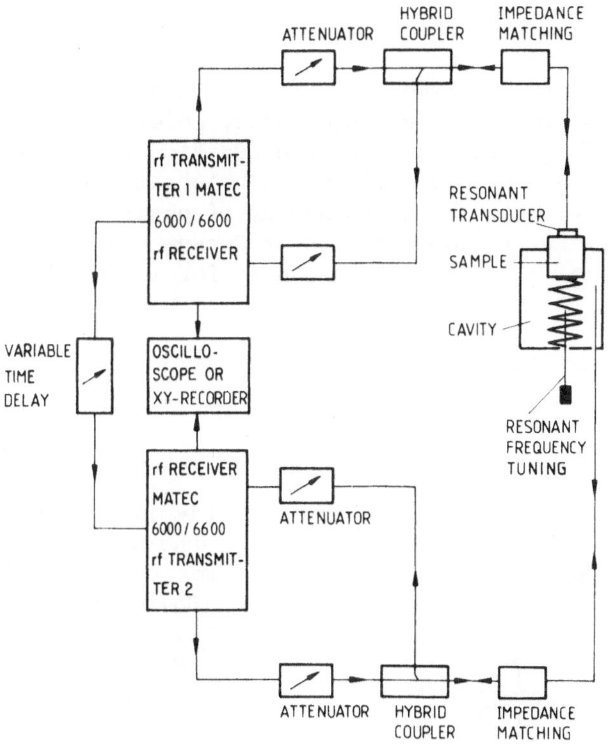

Fig. 7. Block diagram of experimental set-up for echo experiments using the broadband tunable spiral cavity developed by Fossheim and Holt (1978). (After Fossheim and Holt 1982).

The corresponding set-up for experiments in the microwave region is illustrated in Fig.8, from a review by Shiren and Melcher (1977).

The predictions made in the theoretical section above may be checked one by one both with regard to the echo shapes, dependence on applied signal strength and frequency, decay behaviour, dependence on temperature and illumination etc. In BGO ($Bi_{12}GeO_{20}$) we found (Fossheim and Holt 1982) all predictions for insulators to be corroborated by experiments. Specificly the echo amplitude varied linearly with the amplitudes of the primary acoustic wave and the electric cavity field (Fig.9), and with the lesser of the two pulse widths Δ_1 and $2\,\Delta_2$ (Fig.10). The work of Joffrin and Levelut (1972) and later, and of Shiren and Melcher and coworkers in a series of papers from the same period gave insight into echo generation in the presence of space charge. A wealth of physics was found in CdS in particular. Echoes, both dynamic and static, were studied with respect to selection rules, temperature influence, illumination, power dependence etc. A specific mechanism was proposed and developed for CdS-type systems involving charge transfer from filled donors both to the conduction band and to empty traps. Most predictions were borne out by experiment. Of particular importance was the dynamic model involving conduction electrons, and the memory model involving trapped charge. The latter leads to the creation of charge gratings storing the phase information in holograms with lifetimes of months to years. From the great number of observations in CdS we show in Fig.11, 12, 13 some results of Fossum and Holt (1982) on the Ω, 2Ω echo which was not accounted for in the original

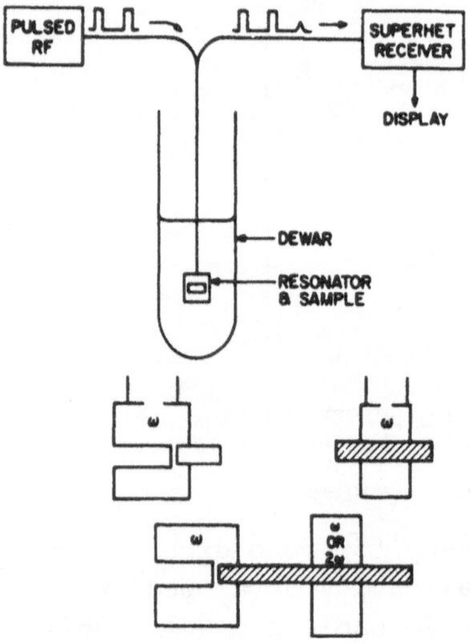

Fig. 8. Simplified view of echo set-up in the micro-wave region (Shiren and Melcher 1977) showing various sample-cavity configurations.

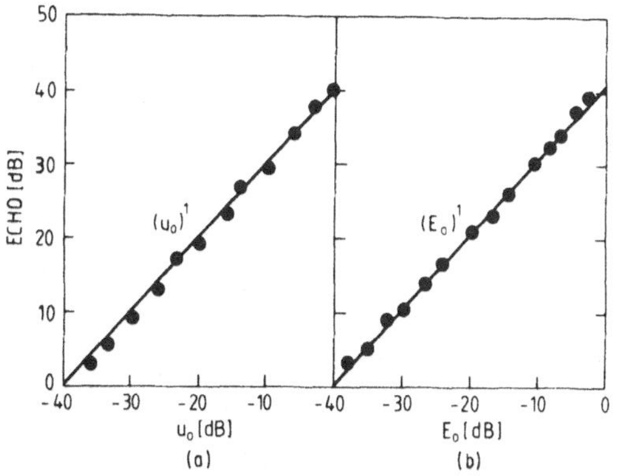

Fig. 9. Measurements of echo amplitude versus amplitude of primary (forward) wave u_o (a) and electric field of second pulse E_o (b) in BGO. (Fossheim and Holt 1982).

model of Shiren and Melcher. The main point here is a linear function of the attenuation, which again is proportional to the conductivity. Hence the variation of the echo is explained as being of electronic origin as proposed in the analysis of Fossum and Holt. As valuable references to the work by the IBM group we point to a few papers: "Polarization echoes in piezoelectric semiconductors" by Shiren and Melcher (1975). "Acoustically induced charge transfer and storage in piezoelectric semiconductors" by Shiren and Melcher (1974). A brief review was also given by the same authors, Shiren and Melcher (1977).

Applications of echoes

The pioneering work of Shiren and Melcher referred to above demonstrated several applications of echoes, as did also the work of Joffrin and Levelut (1976). One particularly promising application is provided by the exploitation of wave-vector reversal already explained earlier in this paper. While the work referred to here took advantage of the echo properties in studying the echo material itself, in our group we have gone a step further and used it as a tool in studying other materials (Fossheim and Holt (1980)). In this case the material we want to study is inserted between the acoustic transducer and the echo material placed in a cavity. When wavefront distortions, even of very severe nature, occur in the sample, these are "repaired" by the echo, and accurate measurements may be made. The method and results were discussed extensively in the review by Fossheim and Holt (1982). Fig.14 shows how the experiments are made, and an example of signal improvement.

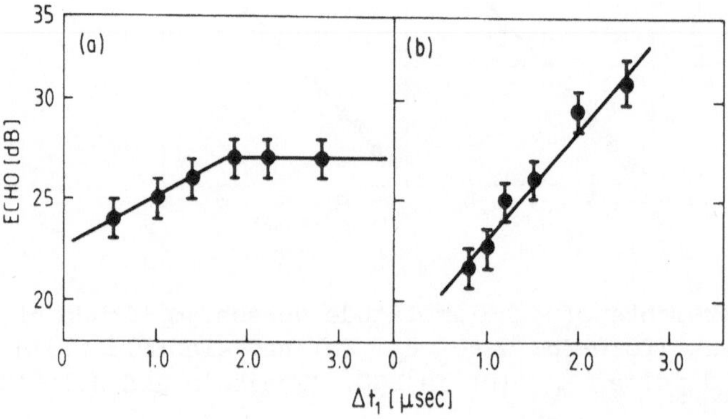

Fig. 10. Data from BGO confirming the proportionality of echo amplitude with the lesser of the two pulse widths, with $\Delta t_2 = 0,75\ \mu s$ kept constant in (a) and with $\Delta t_1 = 2\Delta t_2$ in (b). (Fossheim and Holt 1982).

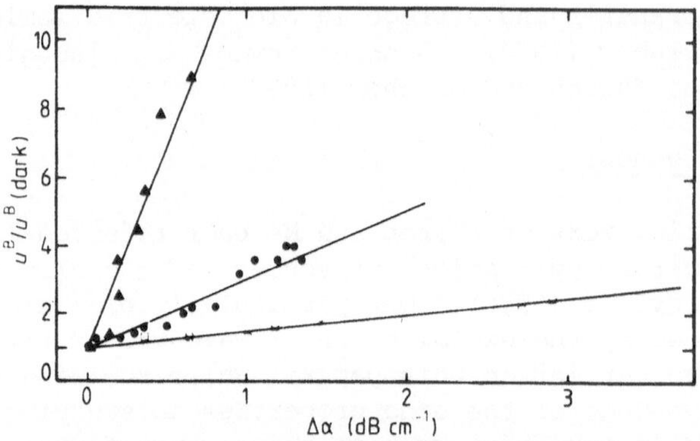

Fig. 11. Experiments on three different CdS samples with different impurity concentrations demonstrating $u^B \sim \Delta\alpha$, i.e. echo proportional to electronic attenuation. All are $\Omega, 2\Omega$ echoes at room temperature. $(\Omega/2\pi = 90\ \text{MHz})$. $\Delta\alpha$ was induced by illumination (Fossum and Holt 1982).

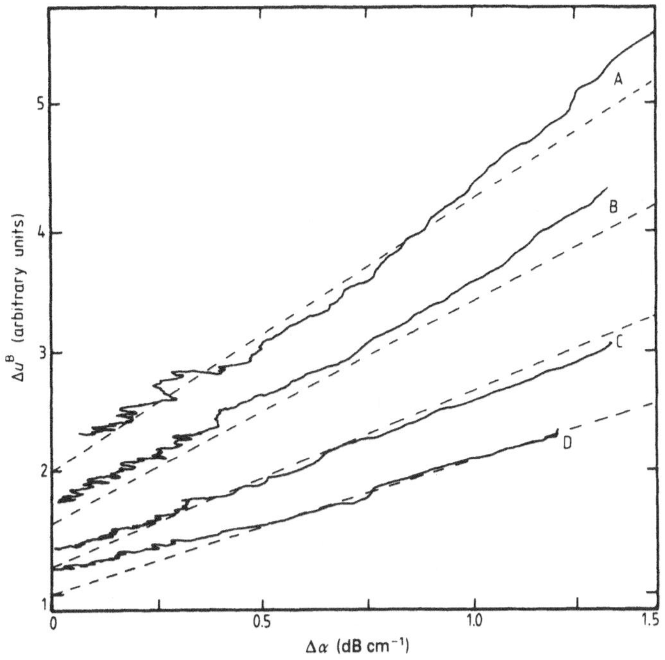

Fig. 12. $\Omega, 2\Omega$ echo measurements, u^B in CdS versus illumination induced change of attenuation $\Delta\alpha$ at four different levels of 2Ω-field, each 2dB apart (Fossum and Holt 1982).

Shiren and Melcher also developed the storage of RF signals to high sophistication in CdS by inducing space charge grating as mentioned before.

Also implied by the analysis made above, convolution and correlation may be achieved. This aspect has been stressed also by the French group, Joffrin and Levelut and coworkers. A brief review has been given by Levelut (1983).

Shiren and Melcher (1977) have demonstrated pulse compression by a factor of 50 resulting from the correlation of an input chirped 9 GHz acoustic pulse with a stored chirped charge grating in CdS.

Recently a lot of interest has been shown in applications of powder echoes which may be used for similar signal manipulation. We cannot go into these problems any further here, but refer particularly to recent Russian literature on this subject.

Further work on acoustic phase conjugation must be done in the future. Of particular importance would be the development of a technique directly analogous to phase conjugation in optics.

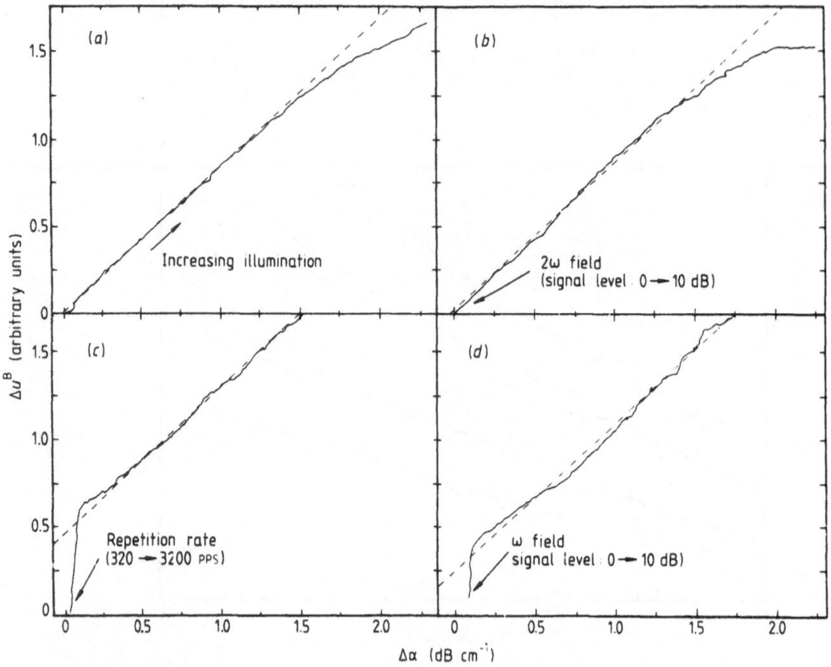

Fig. 13. Ω, 2Ω echo, u^B versus ultrasonic attenuation in CdS. Change of α created by white light illumination (a), by sudden increase of 2Ω field (b), by increase of RF pulse repetition (c) and by increase of 2Ω field (d).

This requires that the pump field can be omitted entirely, and that momentum and energy conservation is taken care of by the phonons in the reflecting substance itself which would act as a phase conjugation mirror. Some work in this field has already been done (Bunkin, Vlasov and Kravtsov 1981).

Here we have pointed out a few examples of applications. There are many more which could have deserved to be mentioned.

ECHOES IN POWDERS

Although echoes in powders were discovered before those in bulk, by Rubinstein and Stauss (1965) this subject came to the forefront of research only during the last 10 years. We have already pointed to the basic characteristics of powder echoes, or polarization echoes in powders as they are often called. Detailed models were developed, based on the idea of nonlinear oscillators, by Kajimura et.al. (1976) and by Fossheim et. al. (1978). The subject was recently reviewed by Fossheim and Holt

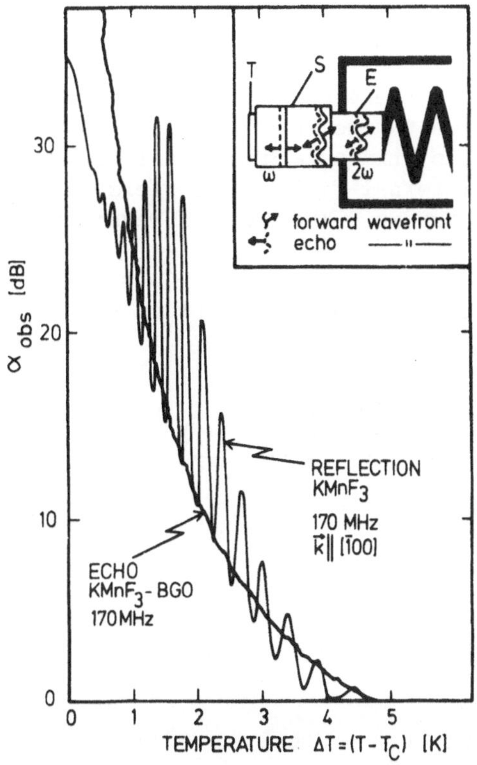

Fig. 14. Echo technique for measurements of attenuation in samples which do not generate echoes. Mounting illustrated in inset. The main figure shows simultaneous recording of attenuation as seen by reflections (strong interference influence is seen) and by the echo (interferences removed). (Fossheim and Holt 1980).

(1982) on the small signal behaviour, by Kajimura (1982) on large signal behaviour, and by Melcher and Shiren (1982) on memory echoes.

Since the subject is so wide, and mathematically quite demanding we can here only give a simple outline of the necessary theoretical apparatus.

Dynamic echoes

Following the work just quoted, as formulated by Kajimura (1982), we take the internal energy density to be

$$U = \frac{1}{2}C_2S^2 + \frac{1}{24}C_4S^4 + \alpha SF_i - \frac{1}{2}\kappa F_i^2 \qquad (46)$$

where C_2 and C_4 are the second and fourth-order elastic constants. In addition to the pure strain terms we have included the interaction between strains and internal field F_i, and the field energy of the internal field. The field F_i is not limited to electric field of course. The case of magnetic field is equally interesting.

The wave equation obtained from (46) in the elastic displacement $\xi(z,t)$ is

$$\frac{\partial^2\xi}{\partial t^2} + 2\Gamma\frac{\partial\xi}{\partial t} = \frac{C_2}{\rho}\frac{\partial\xi}{\partial z}\frac{\partial^2\xi}{\partial z^2} - \frac{1}{\rho}\frac{\partial}{\partial z}(\alpha F_i) \qquad (47)$$

where $\Gamma = \frac{1}{T_2}$ is the damping constant, and ρ is the density of the material.

The applied RF-field will be taken to be retangular pulses of frequency ω, amplitude F, and width Δ, which may be written as

$$F(t) = F\sin\omega t[U(t-t_o) - U(t - T_o - \Delta)] \qquad (48)$$

where U is the unit step function. For the displacement $\xi(z,t)$ we take a trial function

$$\xi(z,t) = a(t)\, m(z)\exp[-i\omega(t-t_o)] + c.c. \qquad (49)$$

where $m(z)$ is a normal-mode function normalized as

$$\frac{1}{V}\int m^2 dV = bd^2$$

integrated over the volume V, in a plate of thickness d ; b is a normalization constant.

In the slowly varying envelope approximation the 2. order time derivatives of $\xi(z,t)$ in Eq.(47) are reduced to first order derivatives of $a(t)$ by inserting (49) in Eq.(47). The simplified equation is

$$\frac{\partial a}{\partial t} + [\Gamma - i(\omega - \Omega)]a - i\gamma a|a|^2$$

$$= -\frac{\beta F\cos\theta}{4\omega}[U(t-t_o) - U(t - t_o - \Delta)] \qquad (50)$$

where Ω is the normal mode frequency of the particle, and γ a nonlinearity constant as defined in the literature referred to above. The constant β depends on boundary conditions, and has been discussed in the papers by Fossheim et. al (1978) and by Tsuruoka and Kajimura (1980).

Equation (50) may be solved in two cases: Either for $\gamma = 0$, or for $F = 0$. This is sufficient to carry out the analysis, assuming that the applied pulses are short. The strategy which may be used then is the following: Allow the field $F \neq 0$ during pulses but neglect the nonlinearity, i.e. take $\gamma = 0$ during this interval. Next, solve the equation of motion with $\gamma \neq 0$ when $F = 0$. Since the decay constant T_2 is normally orders of magnitude longer than the pulse width Δ this is a safe procedure.

The solution for $a(t)$ after the first pulse is

$$a(t) = [a(t_o)+a_o\cos\Theta G(\xi)]\exp\{-[\Gamma-i(\omega-\Omega)](t-t_o)\} \tag{51}$$

for $t > t_o + \Delta$ with intial amplitude $a(t_o)$. Here $a_o = -\beta F\Delta/4\omega$; $G(\xi) = (\sin\xi/\xi)\exp(-i\xi)$, $\xi = (\omega'+i\Gamma)\Delta/2$; $\omega' = \omega - \Omega$. After exposing the particles to a second pulse similar to the first, i.e. of the same frequency ω, and of width Δ_2, after averaging over all orientations and summing the radiated field as detected in an external circuit one arrives at the rather complicated expression for the observed echo:

$$e_2(t) = \frac{\gamma\beta^2}{40\Gamma\omega^3}(F_1\Delta_1)(F_2\Delta_2)^2\exp[-\Gamma(t+2\Delta_2)]$$

$$\times\{1 - \exp[-2\Gamma(t-\tau-\Delta_2)]\}f_2(t)\sin\omega(t-2\tau) \tag{52}$$

where $f_2(t)$ is the shape function

$$f_2(t) = \int_{-\infty}^{\infty} G^*(\xi_1)G^2(\xi_2)\exp[i\omega'(t-2\tau)]d\omega' \tag{53}$$

Eq.(52) gives a number of predictions which can be checked experimentally. Detailed verification was carried out by Fossheim et. al (1978) in piezoelectric powders, and by Tsuruoka and Kajimura (1980) in metals. The theory outlined here being limited to low signals turns out to be quite adequate in that regime (see Fig.5). But at intermediate or high signals improvements are necessary. The echo subject at this stage becomes very complex, and is not suitable for discussion in a brief review. We mention here the calculations by Fossheim et.al (1978), Stokka and Fossheim (1978), Asadullin (1980), and Kajimura (1982). The last reference here is by far the most extensive study of the subject, including also detailed analysis of echo shapes under various circumstances.

As further references to work in the area of powder echoes we mention that of Laikhtman (1976), Kessel' et. al (1976) Smolensky et. al (1976); Berezov and Romanov (1977); Kupca and Searle (1975).

Memory echoes in powders

This is again a subject which has received much attention during the last decade. Memory phenomena have been studied in a large number of powder substances. Here we recommend the review by Melcher and Shiren (1982) on this extensive subject. There has been some apparent controversy over the possible mechanisms. It is by now quite clear that the torque-rotation model proposed by Melcher and Shiren (1976) is correct in a large number of cases. In some instances it appears necessary to supplement it with additional mechanisms, like for instance the plastic deformation model described by Kessel et. al (1976) and Kessel (1978), or the spectral hole burning model of Berezov et. al (1977), or by the charge transfer model of Shiren and Melcher (1974).

Acknowledgements

The author is indebted to several coworkers for invaluable cooperation in the field of echo research, in particular to Drs. K. Kajimura, R.L. Melcher, N.S. Shiren, at IBM Yorktown Heights and to my former students, Drs. R.M. Holt, S. Stokka, and J.O. Fossum here in Trondheim.

Appendix A

Echo terminology

Echo:	Coherent pulsed signal observed at time $t = 2\tau$, generated by phase reversal of an initial pulsed excitation created at $t = 0$, through interaction with a second pulsed field or excitation applied at $t = \tau$ ("Two-pulse echo"). A third pulse applied at $t = T$ may create "three-pulse echo" at $t = T + \tau$. Multiple echo may also occur.
Dynamic echo:	Echo generated during $t < T_2$ where T_2 is the lifetime of excitations involved.
Memory echo:	Echo observed at times $t >> T_2$, i.e. after all dynamics due to primary excitation have died out. = static echo or holographic echo.
Stimulated echo:	Echo observed after third applied pulse. May be dynamic or static.
Aftereffect:	Analog of free induction decay, i.e. coherent radiation emitted after application of single pulse ("t = 0 echo").
Phonon echo:	Ill-defined term. May be any echo associated with acoustic excitation.
Polarization echo:	Used when polarization of substance is involved in echo generation.
Ω, Ω echo:	Echo produced when exciting pulses at $t = 0$ and $t = \tau$ are at the same frequency. The first Ω refers to pulse creating phonons, the second Ω refers to the condition that applied radiation is of frequency $\omega = \Omega$.
Ω, 2Ω echo:	First exciting pulse ($t = 0$) is at Ω Second pulse ($t = \tau$) is at $\omega = 2\Omega$.
Phase conjugation:	Refers particularly to photon echoes and phonon echoes where analysis shows that backward propagating wave is phase conjugated with respect to forward wave.

Parametric echo:	Echo generated through "parametric interaction". Engineering term which loosely applies to most two-pulse echoes where a forward acoustic mode couples to an applied homogeneous field to create backward wave (echo).
Spin echo:	Original discovery of echo by Hahn 1950, in spin system.
4-wave mixing (FWM):	Echo process may be viewed as interaction of 4 fields, like gS^2E^2, i.e. involving two phonons and two photons: $$gS^2E^2=gE(\omega=\Omega,0)E(\omega=\Omega,0)S(\Omega,\vec{q})S(\Omega,-\vec{q})$$ This process is called degenerate (DFWM) because all waves are at the same frequency. Analogous to that used in optical phase conjugation, i.e.: $$\chi^{NL}E(\omega,\vec{K})E(\omega,-\vec{K})E(\omega,\vec{q})E(\omega,-\vec{q})$$
3-wave mixing (TWM):	Three wave participate, i.e. net process involves one photon and two phonons: $$fES^2=fE(\omega=2\Omega,0)S(\Omega,\vec{q})S(\Omega,-\vec{q})$$
Idle, probe, pump, signal:	Engineering terms for parametric processes corresponding to echo generation.

References

Asadullin, Ya., Ya., 1980, Pis'ma Zh. Exp. Teor. Fiz. 32, 405.

Bajak, I. L., 1978, Phys. Rev. B 18, 2405.

Berezov, V. M., Asadullin, Ya. Ya., Korepanov, V.D., and Ramanov, V. S., 1975, Sov. Phys. JETP (Engl. transl.) 42, 851.

Berezov, V. M. and Romanov, V. S., 1977, JETP Lett. (Engl. transl.) 25, 151.

Bunkin, F. V., Vlasov, D. V., and Kravtsov, Yu. A., 1981, Sov. Tech. Phys. Lett. (Engl. transl.) 7 (3), 138.

Chaban, A. A., 1967, JETP Lett. (Engl. transl.) 6, 381.

Fossheim, K., and Holt, R. M., 1980, Phys. Rev. Letters 45, 730.

Fossheim, K., and Holt, R. M., 1982, in Physical Acoustics (W.P. Mason and R.N. Thurston, eds.) Vol. 16, p 217 (Academic Press, New York).

Fossheim, K., Kajimura, K., Kazyaka, T.G., Melcher, R.L., and Shiren, N.S., 1978, Phys. Rev. B 17, 964.

Fossum, J. O., Fossheim, K., Aune, H. J., and Holt, R. M., 1981, J. Physique C6, 867.

Fossum, J. O. and Holt, R. M., 1982, J. Phys. C: Solid State, 15, 6613.
Golding, B. and Graebner, J. E., 1976, Phys. Rev. Letters 37,852.
Gould, R. W., 1965, Phys. Lett. 19, 477.
Greebe, C. A. A. J., 1966, IEEE Trans. Sonics and Ultrasonics SU-13, 54.
Hahn, E. L., 1950, Phys. Rev., 80, 580.
Holt, R. M., Fossum, J. O., and Fossheim, K., Proc. of "Ultrasonics International 1983" p 142, Butterworth & Co. 1983.
Joffrin, J., and Levelut, A., 1972, Phys. Rev. Letters 29, 1325.
Joffrin, J., and Levelut, A., 1976, in New Direction in Physical Acoustics, LXIII Corso (Soc. Italiana de Fisica, Bologna, Ed.) p 291.
Kajimura, K., 1982, in Physical Acoustics (Mason, W. P. and Thurston R. N. eds.) vol. 16, p 295, Academic Press, New York.
Kajimura, K., Fossheim, K., Kazyaka, T. G., Melcher, R. L., and Shiren, N. S., 1976, Phys. Rev. Letters 37, 1151.
Kessel', A. R., 1978, Ferroelectrics 22, 759.
Kessel',A. R., Zel'dovich, S. A., and Gurevich, I. L., 1976, Sov. Phys. Solid State (Engl. transl.) 18, 473.
Kupca, S., and Searle, C. W., 1975, Can. J. Phys. 53, 2622.
Kurnit, N. A., Abella, I. D., and Hartmann, S. R., 1964, Phys. Rev. Letters, 13, 567.
Laikhtman, B. D., 1976, Sov. Phys. Solid State (Engl. transl.) 17, 2154 ; 18 357.
Levelut, A., 1983, J. Physique 44, C2-61.
Melcher, R. L., 1979, Phys. Rev. Letters 43, 939.
Melcher, R. L., and Shiren, N. S., 1975, Phys. Rev. Letters 34,731.
Melcher, R. L., and Shiren, N. S., 1976, Phys. Rev. Letters 36, 888.
Melcher, R. L., and Shiren, N. S., 1982, in Physical Acoustics (Mason, W. P., and Thurston, R. N., eds.) Vol. 16 p 341, Academic Press, New York.
Miyasato, T., Wigmore, J. K., Meredith, D. J., 1984, Phys. Rev. Letters 52, 843.
Popov, S. N. and Krainik, N.N., 1971, Sov. Phys. Solid State (Engl. transl.) 12, 2440.
Rubinstein, M. and Stauss, G. H., 1965, Phys. Lett. 14, 277.
Shiren, N. S. and Kazyaka, T. G., 1972, Phys. Rev. Letters 20, 1304.
Shiren, N. S. and Melcher, R. L., 1974, Proceedings IEEE Symposium on Sonics and Ultrasonics (IEEE, New York 1974) p 558.
Shiren, N. S. and Melcher, R. L., 1975, Journal of Electronic Materials 4, 1143.
Shiren, N. S. and Melcher, R. L., 1977, in Internal Friction and Ultrasonic Attenuation in Solids (Hasiguti, R. R. and Mikoshiba, N. eds.) University of Tokyo Press p 11.
Shiren, N. S., Arnold, W., and Kazyaka, T. G., 1977, Phys. Rev. Letters 39, 239.
Smolensky, G. A., Krainik, N. N., Popov, S. N., and Laikhtman, B. D., 1976, Ferroelectrics 14, 571.

Svaasand, L., 1969, Appl. Phys. Lett. 15, 300.
Stokka, S. and Fossheim, K., 1978, J. Phys. C 11, 3948.
Thompson, R. B. and Quate, C. F., 1970, Appl. Phys. Lett. 16, 295.
Tsuruoka, F., and Kajimura, K., 1980, Phys. Rev. B 22, 5092.
White, D. L., 1962, J. Appl. Phys. 33, 2547.

INTRODUCTION TO PHONON HYDRODYNAMICS

Rudolf Klein

Fakultät für Physik
Universität Konstanz
D-7750 Konstanz, Federal Republic of Germany

I. INTRODUCTION

The subject of phonon hydrodynamics is the study of the collective behavior of the interacting phonon system at low frequencies and long wavelengths. For most problems in lattice dynamics the anharmonic interactions among the phonons of a pure dielectric are weak perturbations, which have the effect of a slight temperature dependent frequency shift of the phonons in the harmonic approximation and of introducing a finite lifetime. These quantities were calculated by Maradudin and Fein (1962) by applying many-body techniques to the phonon system. Lorentzian line shapes for the response functions, which are measured in various scattering experiments, are very often satisfactory descriptions of the result of these measurements. At low frequencies, however, in the first sound region (Cowley 1967), the frequency Ω is much less than the inverse lifetime Γ of most of the phonons in the crystal. Therefore, the latter are able to follow the low-frequency excitation to a certain extent. If the low-frequency excitation is connected with a slowly varying temperature or density, the distribution of thermal phonons will adjust to the local values of these thermodynamic variables, since many anharmonic processes among the thermal phonons take place during one period of the excitation. Such phenomena, which are connected to space and time dependent phonon density fluctuations and are therefore similar to the situation in gases or liquids, are the subject of interest in phonon hydrodynamics.

There are two routes to gain insight into the collective behavior of the interacting phonon system. The first one is to use the phenomenological Boltzmann equation for phonons (Peierls 1929,1955). The occurrence of hydrodynamic phenomena such as second sound and

Poiseulle flow was predicted on this basis by Sussmann and Thellung (1963) and by Gurzhi (1964). A more systematic investigation was performed by Prohofsky and Krumhansl (1964), Guyer and Krumhansl (1964, 1966a, 1966b), Krumhansl (1965), Guyer (1966) and Kwok (1967). Comprehensive reviews have been given by Beck et al.(1974) and Beck (1975).

The second approach is a microscopic one using many-body theory, which starts from the Hamiltonian of the anharmonic lattice and derives transport equations. If $\Omega < \Gamma$, low-order perturbation theory breaks down, but the summation of certain interaction processes to infinite order leads to a generalized Boltzmann equation. Various Green function methods were used for these derivations. The extensive literature on this subject is compiled in the above mentioned reviews.

In the following we will first give a short outline of the phenomenological approach and list a few results. Then, a few introductory remarks are made concerning the microscopic theory. The results of this approach will be used to demonstrate for a few cases the importance of phonon transport. It will be shown that the singular behavior of the self-energy of a long-wavelength phonon as $\underline{Q}$ and Ω go to zero is important to understand the thermodynamics of the interacting phonon system, such as the difference between adiabatic and isothermal elastic constants. The response function, which governs the scattering of photons from a dielectric at very low energy transfer, is modified by phonon density fluctuations, which consist partly of entropy fluctuations and of additional "dielectric" fluctuations away from local thermodynamic equilibrium. The light scattering spectrum will be calculated and it is shown that two types of central peaks can exist, one due to entropy fluctuations and a much broader one arising from non-thermodynamic phonon density fluctuations. Finally, there is the possibility of coupling light directly to the varying phonon density, which can lead to a strong modification of the total scattered intensity and to interference phenomena in the spectrum of scattered light. The possible importance of these results to experiments will also be discussed.

II. PHENOMENOLOGICAL APPROACH

II.1. The Peierls-Boltzmann Equation

The basis of the phenomenological description of phonon transport phenomena is the concept of a space and time dependent phonon distribution function $n_q(\underline{r},t)$ (Peierls 1955). The implicite assumption behind this concept is a sufficiently slowly varying situation, in which the mean free path of phonons $\underline{q}$ is small compared to the distance over which $n_q(\underline{r},t)$ changes appreaciably and the assumption of many phonon-phonon interactions on the time scale of change of

$n_q(\underline{r},t)$. The equation of motion of $n_q(\underline{r},t)$ is then given by the Peierls-Boltzmann equation

$$\frac{\partial n_q(\underline{r},t)}{\partial t} + \underline{v}_q \cdot \frac{\partial n_q(\underline{r},t)}{\partial \underline{r}} = \mathcal{L}[n_q] \; . \qquad \text{(II.1)}$$

Here, $\underline{v}_q = \partial\omega_q/\partial\underline{q}$ is the group velocity of phonons of frequency ω_q and $\mathcal{L}[n_q]$ denotes the three-phonon collision operator. The explicit form of $\mathcal{L}$ is not of primary interest at the moment, since general solutions of (II.1) are not available. It suffices to remark that $\mathcal{L}$ consists of a part $\mathcal{L}_N$ due to three-phonon Normal processes, which conserve quasi-momentum, and a part $\mathcal{L}_R$, which describes all resistive phonon collisions, such as phonon Umklapp processes, boundary scattering and impurity scattering. Restricting the analysis to the infinite ideal crystal, $\mathcal{L}_R = \mathcal{L}_U$. Because of energy conservation, the phonon energy ω_q is an eigenfunction of $\mathcal{L}$, $\mathcal{L}_N$ and $\mathcal{L}_U$ with eigenvalue zero, and the conservation of quasi-momentum in N-processes finds its expression in $\underline{q}$ being an eigenfunction of $\mathcal{L}_N$ with eigenvalue zero. From these properties it follows that the local equilibrium distribution

$$n_q^{LE}(\underline{r},t) = \{\exp[\beta(\underline{r},t)\,\omega_q] - 1\}^{-1} \qquad \text{(II.2)}$$

is unaffected by $\mathcal{L}$ and that the drifting local equilibrium distribution

$$n_q^{LE,d}(\underline{r},t) = \{\exp[\beta(\underline{r},t)(\omega_q - \underline{u}(\underline{r},t)\cdot\underline{q})] - 1\}^{-1} \qquad \text{(II.3)}$$

is unaffected by $\mathcal{L}_N$. Here, $\beta = (kT)^{-1}$, so that $\beta(\underline{r},t)$ describes the temperature field varying slowly in space and time; $\underline{u}(\underline{r},t)$ is the phonon drift velocity. From (II.3) it follows that N-processes can relax an arbitrary non-equilibrium phonon distribution $n_q(\underline{r},t)$ only to a drifting local equilibrium $n_q^{LE,d}(\underline{r},t)$, whereas U-processes reduce it to $n_q^{LE}(\underline{r},t)$.

From the Boltzmann equation (II.1) one can easily derive macroscopic conservation laws by multiplying it by ω_q and $\underline{q}$, respectively, and summing over $\underline{q}$. This procedure introduces the following quantities:

Energy density

$$E(\underline{r},t) = \sum_q \omega_q\, n_q(\underline{r},t) \qquad \text{(II.4)}$$

Energy current density

$$S^\alpha(\underline{r},t) = \sum_q \omega_q\, v_q^\alpha\, n_q(\underline{r},t) \qquad \text{(II.5)}$$

Quasi-momentum density

$$P^{\alpha}(\underline{r},t) = \sum_q q^{\alpha}\, n_q(\underline{r},t) \tag{II.6}$$

Quasi-momentum flux denxity

$$T^{\alpha\beta}(\underline{r},t) = \sum_q q^{\alpha}\, v_q^{\beta}\, n_q(\underline{r},t) \ . \tag{II.7}$$

On the right-hand side expressions like

$$\sum_q \omega_q\, \mathcal{L}[n_q] \quad \text{and} \quad \sum_q \underline{q}\, \mathcal{L}[n_q]$$

appear. Due to energy conservation the first one vanishes, whereas the second one is finite due to U-processes. The resulting conservation laws are

$$\frac{\partial E(\underline{r},t)}{\partial t} + \frac{\partial}{\partial r^{\alpha}} S^{\alpha}(\underline{r},t) = 0 \tag{II.8}$$

$$\frac{\partial P^{\alpha}(\underline{r},t)}{\partial t} + \frac{\partial}{\partial r^{\beta}} T^{\alpha\beta}(\underline{r},t) = \sum_q q^{\alpha}\, \mathcal{L}_U[n_q] \ . \tag{II.9}$$

As remarked before, the general solution of (II.1) is impossible with the full expression for $\mathcal{L}$. A first insight into the phenomena described by the transport equation can be obtained by using a relaxation time approximation, which according to the comments in connection with eqs. (II.2) and (II.3) has the following form

$$\frac{\partial n_q(\underline{r},t)}{\partial t} + \underline{v}_q \cdot \frac{\partial n_q(\underline{r},t)}{\partial \underline{r}} =$$
$$= - \frac{1}{\tau_N(q)} \left(n_q(\underline{r},t) - n_q^{LE,d}(\underline{r},t)\right) - \frac{1}{\tau_U(q)} \left(n_q(\underline{r},t) - n_q^{LE}(\underline{r},t)\right) \tag{II.10}$$

$\tau_{N,U}(q)$ are the relaxation times due to N- and U-processes, respectively. This equation can be solved in an approximation, which treats the deviations from equilibrium in a linear fashion. One expands

$$n_q(\underline{r},t) = n_q + m_q\, g_q(\underline{r},t) \tag{II.11}$$

where $n_q = \left(\exp(\beta_0\, \omega_q) - 1\right)^{-1}$ is the Bose-Einstein distribution and

$$m_q = - dn_q/d\omega_q = \beta_o\, n_q(n_q + 1) \; ; \tag{II.12}$$

$\beta_o = (kT_o)^{-1}$ and T_o is the equilibrium temperature. Similarly,

$$n_q^{LE}(\underline{r},t) = n_q - m_q\, \omega_q \frac{\delta\beta(\underline{r},t)}{\beta_o} \tag{II.13}$$

$$n_q^{LE,d}(\underline{r},t) = n_q - m_q\left[\omega_q \frac{\partial\beta(\underline{r},t)}{\beta_o} - \underline{u}(\underline{r},t)\cdot\underline{q}\right] \quad , \tag{II.14}$$

where $\delta\beta(\underline{r},t) = \beta(\underline{r},t) - \beta_o$.

Using (II.11), (II.13) and (II.14) in (II.10) gives the following equation of motion for the unknown function $g_q(\underline{r},t)$:

$$\left(\frac{\partial}{\partial t} + \underline{v}_q \cdot \frac{\partial}{\partial \underline{r}} + \frac{1}{\tau(q)}\right) g_q(\underline{r},t) = - \frac{\omega_q}{\tau(q)} \frac{\delta\beta(\underline{r},t)}{\beta_o} + \frac{1}{\tau_N(q)}\, \underline{u}(\underline{r},t)\cdot\underline{q} \tag{II.15}$$

where

$$\tau^{-1}(q) = \tau_N^{-1}(q) + \tau_U^{-1}(q) \; . \tag{II.16}$$

Introducing Fourier transforms eq. (II.15) becomes

$$g_q(\underline{Q},\Omega) = \frac{1}{1-i\Omega\tau(q)+i\underline{v}_q\cdot\underline{Q}\tau(q)} \left[-\omega_q \frac{\delta\beta(\underline{Q},\Omega)}{\beta_o} + \frac{\tau(q)}{\tau_N(q)}\, \underline{u}(\underline{Q},\Omega)\cdot\underline{q}\right] . \tag{II.17}$$

The simplest approximation consists in assuming that U-processes are unimportant, since they vanish exponentially with decreasing temperature. Therefore, we are in the regime $\Omega\tau_U >> 1$, the lifetime due to U-processes becomes very large. With this assumption

$$g_q(\underline{Q},\Omega) = - \omega_q \frac{\delta\beta(\underline{Q},\Omega)}{\beta_o} + \underline{u}(\underline{Q},\Omega)\cdot\underline{q} \; . \tag{II 18}$$

With this approximation and (II.11) the conserved quantities (II.4) to (II.7) are easily calculated. For the energy density one obtains

$$\begin{aligned} E(\underline{r},t) &= \sum_q \omega_q\, n_q + \sum_q \omega_q\, m_q\, g_q(\underline{r},t) \\ &= E_o - 4\, E_o \frac{\delta\beta(\underline{r},t)}{\beta_o} \; . \end{aligned} \tag{II.19}$$

The second term in (II.18) does not contribute. The quantity

$$E_o = \frac{\pi^2}{30} (k_B T_o)^4 \sum_{j=1}^{3} c_j^{-3} \tag{II.20}$$

is the thermal energy of the equilibrium phonon density, and the dispersion curves have been approximated by a Debye-model of three acoustic branches $\omega_\lambda \equiv \omega_{qj} = c_j q$. In a similar manner,

$$S^\alpha(\underline{r},t) = \frac{4}{3} E_0\, u^\alpha(\underline{r},t) \tag{II.21}$$

$$P^\alpha(\underline{r},t) = \frac{4}{9c_2^2} E_0\, u^\alpha(\underline{r},t) \tag{II.22}$$

$$T^{\alpha\beta}(\underline{r},t) = \frac{1}{3}\,\delta^{\alpha\beta}\, E(\underline{r},t)\ . \tag{II.23}$$

Here, the velocity c_2 has been introduced by

$$c_2^2 = \frac{1}{3}\frac{\sum_j c_j^{-3}}{\sum_j c_j^{-5}}\ . \tag{II.24}$$

Introducing the results (II.19) and (II.21) to (II.23) in the conservation laws (II.8) and (II.9), leads to macroscopic hydrodynamic equations for the coupled temperature and phonon drift velocity fields, $\beta(\underline{r},t)$ and $\underline{u}(\underline{r},t)$, respectively. In the spirit of the present approximation one neglects the Umklapp contribution on the right-hand side of (II.9) and finds

$$\frac{\partial}{\partial t}\,\frac{\delta\beta(\underline{r},t)}{\beta_0} = \frac{1}{3}\,\frac{\partial u^\alpha(\underline{r},t)}{\partial r^\alpha} \tag{II.25}$$

$$\frac{\partial u^\alpha(\underline{r},t)}{\partial t} = 3\,c_2^2\,\frac{\partial}{\partial r^\alpha}\,\frac{\delta\beta(\underline{r},t)}{\beta_0}\ . \tag{II.26}$$

Eliminating the drift velocity $\underline{u}(\underline{r},t)$ leads to

$$\left(\frac{\partial^2}{\partial t^2} + c_2^2\,\frac{\partial^2}{\partial r^\alpha \partial r^\alpha}\right)\,\delta\beta(\underline{r},t) = 0\ . \tag{II.27}$$

It is seen that under the present assumption of vanishing U-processes the temperature deviation $\delta\beta(\underline{r},t)$ propagates in an undamped manner with the velocity c_2, eq. (II.24), which in a one-branch model reduces to one third of the sound velocity.

After this demonstration of the origin of phonon hydrodynamic equations within a rather simplified approximation (undamped second sound), one can continue in the same manner and improve the solution by using (II.17) instead of (II.18). This procedure was used by Sussmann and Thellung (1963), who derived in this way the complex dispersion relation for second sound and also considered Poiseuille flow. The dispersion relation thus obtained is (Beck 1975)

$$\Omega \simeq c_2 Q \left[1 - \frac{1}{4}\left(\frac{\overline{\tau_U^{-1}}}{Qc_2} + \bar{\tau}_N Q c_2\right)^2\right]^{1/2} - i(\overline{\tau_U^{-1}} + Q^2 c_2^2 \bar{\tau}_N) \qquad \text{(II.28)}$$

where $\bar{\tau}_N$ and $\overline{\tau_U^{-1}}$ are rather complicated averages of the q dependent relaxation times. Eq. (II.28) shows that the damping of second sound depends on N- and U-processes. The characteristic features of this result are exactly what one expects on physical grounds: (a) There should be frequent N-processes ($\Omega\bar{\tau}_N << 1$), in order to establish quickly the drifting local equilibrium distribution and (b) U-processes should be rare ($\Omega\tau_U >> 1$), since they are momentum destroying and reducing the drift. These two conditions taken together lead to the famous window condition for the existence of propagating second sound:

$$\Omega\, \tau_N << 1 << \Omega\, \tau_U , \qquad \text{(II.29)}$$

where we have replaced the averaged relaxation times by (q independent) numbers τ_N and τ_U.

The results given so far have in many ways been generalized. For instance, the velocity c_2 of second sound (II.24) can easily be derived without the restrictions of the Debye model. Kwok (1967) has shown that for a cubic crystal

$$c_2^{\,2} = \frac{\left(\sum_\lambda \frac{c_\lambda}{\omega_\lambda}\, \underline{q}\cdot \underline{v}_q\right)^2}{3\left(\sum_\lambda c_\lambda\right)\left(\sum_\lambda c_\lambda\, \underline{q}^2/\omega_\lambda^2\right)} , \qquad \text{(II.30)}$$

where $\lambda = (\underline{q}\, j)$ stands for wavevector and branch index j and c_λ is the contribution of the mode λ to the specific heat. By a numerical evaluation of (II.30) Hardy and Jaswal (1971) found for NaF that c_2 decreases by 24 % as the temperature is increased from 10 to 30K because of the dispersion of the phonon frequency spectrum.

The treatment by Kwok was further extended by Maris (1981) who calculated c_2 for anisotropic crystals. It is shown that c_2 is often significantly smaller than (II.24) and that this calculation is in good agreement with measurements of the angle dependence of the second sound velocity in hcp ^4He.

II.2. The Formal Solution of the Peierls-Boltzmann Equation

A more systematic solution of the Boltzmann equation was developed and used by Guyer and Krumhansl (1966a). The solution is expanded in terms of the eigenfunctions of the part $\mathcal{L}_N$ of the collision operator. Because of energy and quasi-momentum conservation ω_q and $\underline{q}$ are eigenfunctions of $\mathcal{L}_N$ with eigenvalues zero:

$$\mathcal{L}_N \chi_0(q) = 0 \quad ; \quad \chi_0(q) = \mu_0 \omega_q \tag{II.31}$$

$$\mathcal{L}_N \chi_1^\alpha(q) = 0 \quad ; \quad \chi_1^\alpha(q) = \mu^\alpha q^\alpha \quad ; \alpha = 1,2,3. \tag{II.32}$$

The coefficients μ_0, μ^α are normalizing factors so that χ_0, χ_1^α; $\alpha = 1,2,3$ are four mutually perpendicular and normalized vectors in a Hilbert space of functions with scalar product

$$\sum_q (f_q^{(1)})^* m_q f_q^{(2)} \equiv < f_q^{(1)} | f_q^{(2)} > . \tag{II.33}$$

Besides the four eigenfunctions (II.31), (II.32), which are the known collision invariants, there are of course finite eigenvalues of $\mathcal{L}_N$, which are however not known. One expands the deviation of the phonon density from its equilibrium value as

$$g_q(\underline{Q},\Omega) = a_0(\underline{Q},\Omega)\chi_0(q) + \sum_{\alpha=1}^{3} b^\alpha(\underline{Q},\Omega)\chi_1^\alpha(q) + c_A(\underline{Q},\Omega)\chi_A(q) \tag{II.34}$$

where all the unknown eigenfunctions belonging to finite eigenvalues are lumped together in the third term.

One can relate $a_0(\underline{Q},\Omega)$ to the temperature fluctuations $\delta T(\underline{Q},\Omega)$ and $b^\alpha(\underline{Q},\Omega)$ to the momentum density fluctuations (or the local drift velocity). This means that the third term in (II.34) represents fluctuations away from drifting local equilibrium. These can therefore be interpreted as truely non-hydrodynamic fluctuations; they were called dielectric fluctuations by Cowley and Coombs (1973).

To demonstrate the meaning of $a_0(\underline{Q},\Omega)$ we observe that the energy density fluctuations $\delta E(\underline{Q},\Omega)$ are given from (II.4), (II.11) and (II.33) as

$$\delta E(\underline{Q},\Omega) = < \omega_q | g_q(\underline{Q},\Omega) > = \frac{1}{\mu_0} < \chi_0 | g_q > . \tag{II.35}$$

Using now the expansion (II.34),

$$\delta E(\underline{Q},\Omega) = \frac{1}{\mu_0} a_0(\underline{Q},\Omega) . \tag{II.36}$$

On the other hand, the energy fluctuations are directly related to temperature fluctuations by

$$\delta E(\underline{Q},\Omega) = C_V \, \delta T(\underline{Q},\Omega) . \tag{II.37}$$

The specific heat C_V of the equilibrium system is given by

$$C_V = k_B \sum_q \left(\frac{\omega_q}{k_B T_0}\right)^2 n_q(n_q + 1) = \frac{1}{T_0} < \omega_q | \omega_q >, \tag{II.38}$$

whereas the normalizing constant μ_0 is $\mu_0 = <\omega_q|\omega_q>^{-1/2}$. Putting the last three equations together gives

$$a_0(\underline{Q},\Omega) = < \omega_q | \omega_q >^{1/2} \frac{\delta T(\underline{Q},\Omega)}{T_0} . \tag{II.39}$$

The macroscopic energy conservation (II.8) can therefore also be expressed as

$$- i \Omega C_V \; \delta T(\underline{Q},\Omega) + i \; \underline{Q} \cdot \underline{S}(\underline{Q},\Omega) = 0 \quad . \tag{II.40}$$

In a similar manner the quasi-momentum density fluctuations are given by (II.6), (II.11) and (II.33) as

$$\delta P^\alpha(\underline{Q},\Omega) = <q^\alpha|g_q(\underline{Q},\Omega)> = \frac{1}{\mu^\alpha} <\chi_1^\alpha|g_q> \; ; \; (\text{no sum over } \alpha)$$

$$= <q^\alpha|q^\alpha>^{1/2} \; b^\alpha(\underline{Q},\Omega) \quad . \tag{II.41}$$

In a Debye model, where $v_q^\alpha = \partial\omega_q/\partial q^\alpha = cq^\alpha/|q|$, the energy current density is

$$\delta S^\alpha(\underline{Q},\Omega) = c^2 \; \delta P^\alpha(\underline{Q},\Omega) \tag{II.42}$$

and is therefore also simply related to $b^\alpha(\underline{Q},\Omega)$.

With the expansion (II.34) for $g_q(\underline{Q},\Omega)$ the Boltzmann equation

$$(- i\Omega + i\underline{Q}\cdot\underline{v}_q) \; g_q(\underline{Q},\Omega) = (\mathcal{L}_N + \mathcal{L}_U) \; g_q(\underline{Q},\Omega) \tag{II.43}$$

becomes a homogeneous set of algebraic equations by multiplying it with $m_q\chi_0$, $m_q\chi_1^\alpha$ and $m_q\chi_A$, respectively, and subsequent summation over $\underline{q}$: (II.44)

$$\begin{pmatrix} - i\Omega & i\underline{Q}\cdot\underline{v}_{o1} & i\underline{Q}\cdot\underline{v}_{oA} \\ i\underline{Q}\cdot\underline{v}_{10} & -i\Omega+i\underline{Q}\cdot\underline{v}_{11}-(\mathcal{L}_U)_{11} & i\underline{Q}\cdot\underline{v}_{1A}-(\mathcal{L}_U)_{1A} \\ i\underline{Q}\cdot\underline{v}_{1A} & i\underline{Q}\cdot\underline{v}_{1A}-(\mathcal{L}_U)_{A1} & -i\Omega+i\underline{Q}\cdot\underline{v}_{AA}-(\mathcal{L}_U)_{AA}-(\mathcal{L}_N)_{AA} \end{pmatrix} \begin{pmatrix} a_0(\underline{Q},\Omega) \\ \underline{b}(\underline{Q},\Omega) \\ c_A(\underline{Q},\Omega) \end{pmatrix} = 0$$

Here the abbriviations

$$\underline{v}_{o1} = < \chi_0 | \underline{v}_q | \chi_1 > \quad ; \quad (\mathcal{L}_U)_{1A} = < \chi_1 | \mathcal{L}_U | \chi_A > \text{, etc.}$$

have been introduced. On the basis of (II.44) Guyer and Krumhansl (1966a, 1966b) have discussed all the phonon hydrodynamic phenomena including their static limits. For instance, one can eliminate c_A from (II.44) to get an equation relating a_0 to $\underline{b}$. Since a_0 is proportional to δT and $\underline{b}$ to $\delta \underline{S}$ one has Fourier's heat conduction law. Therefore, an expression for the $\underline{Q}$ and Ω dependent thermal conductivity $\kappa(\underline{Q},\Omega)$ in terms of the matrix elements appearing in (II.44) can be obtained. Although these results are exact, rather little has been done to evaluate the complicated matrix elements of the collision operator. One exception is the static thermal conductivity κ_0 in the socalled Ziman limit, where U-processes are very infrequent compared to N-processes. Here, κ_0 is essentially given by the matrix elements $(\mathcal{L}_U)_{11}$, which were evaluated numerically by Werthamer and Chiu (1972a, 1972b) for the case of hcp ^4He. By using the realistic phonon spectrum in the calculation these authors were able to describe the rather large anisotropy of the thermal conductivity in this substance.

In most cases the exact matrix elements in (II.44) have to be approximated by relaxation times. This has been done for second sound and heat pulse experiments (Beck and Beck 1973). From a comparison of the theoretical results, which depend on the various relaxation times (including those for impurity and boundary scattering), with the experimentally obtained temperature dependence of the second sound velocity it is possible to estimate the temperature dependence of the relaxation times. From fitting to second sound data in NaF, Beck (1976) finds $\tau_N \sim T^3$, $\tau_U \sim T^4 \exp(-\Theta_D/T)$ and $\tau_I \sim T^4$. But it has to be mentioned that slightly different exponents in τ_N and τ_U had to be used by other authors. For the analysis of heat pulse experiments the appropriate equation of motion for the heat current has been solved in finite geometries (Overton 1980).

III. OUTLINE OF PHONON TRANSPORT THEORY

III.1. Response and Correlation Functions

The results obtained so far are all based on the Peierls-Boltzmann equation which is essentially a phenomenological transport equation for the distribution functions of phonons. One may ask how this equation arises from a more basic or microscopic description of an anharmonic lattice. At the same time, a derivation which starts from a microscopic level, can give further insight into the validity and the possible limitations of a phenomenological equation. Besides this somewhat formal task phonon transport theory is of interest in understanding the low-frequency or hydrodynamic excitations of dielectric materials and their response to external probes.

Microscopically the anharmonic lattice is described by the Hamiltonian ($\hbar=1$)

$$H = H_0 + H_A \tag{III.1}$$

$$H_0 = \sum_\lambda \omega_\lambda \left(a^+(\lambda)\, a(\lambda) + \frac{1}{2}\right) \tag{III.1a}$$

$$H_A = \sum_{n\geq 3} \sum_{\lambda_1 \ldots \lambda_n} V_n(\lambda_1 \ldots \lambda_n) A(\lambda_1) \ldots A(\lambda_n) . \tag{III.1b}$$

The $a(\lambda)$, $a^+(\lambda)$ are phonon annihilation and creation operators which are related to the quantized normal coordinates $A(\lambda)$ by

$$A(\lambda) = a(\lambda) + a^+(\bar{\lambda}) \tag{III.2a}$$

$$\dot{A}(\lambda) = -i\, \omega_\lambda \left[a(\lambda) - a^+(\bar{\lambda})\right] \quad ; \tag{III.2b}$$

λ denotes wavevector $\underline{q}$ and branch index j, $\lambda = (\underline{q}\, j)$ and $\bar{\lambda} = (-\underline{q}\, j)$. The phonon operators satisfy the algebra

$$\left[a(\lambda), a(\lambda')\right] = 0 = \left[a^+(\lambda), a^+(\lambda')\right] \; ; \; \left[a(\lambda), a^+(\lambda')\right] = \delta_{\lambda\lambda'}, \tag{III.3a}$$

whereas

$$\left[A(\lambda), A(\lambda')\right] = 0 = \left[\dot{A}(\lambda), \dot{A}(\lambda')\right] \; ; \; \left[A(\lambda), \dot{A}(\lambda')\right] = 2\omega_\lambda\, i\, \delta_{\lambda\lambda'} . \tag{III.3b}$$

$V_n(\lambda_1 \ldots \lambda_n)$ denotes the anharmonic couplings of order n. Restricting ourselves to cubic and quartic anharmonicities only ($n=3$ and 4 in eq. (III.1b)), the basic equation of motion follows from $\dot{A}(\lambda) = -i\, [A(\lambda), H]$ and is given by

$$\begin{aligned} \ddot{A}(\lambda) + \omega_\lambda^2 A(\lambda) = -2\omega_\lambda \Big\{ 3 \sum_{\lambda_1\lambda_2} & V(\bar{\lambda}\lambda_1\lambda_2) A(\lambda_1) A(\lambda_2) \\ + 4 \sum_{\lambda_1\lambda_2\lambda_3} & V(\bar{\lambda}\lambda_1\lambda_2\lambda_3) A(\lambda_1) A(\lambda_2) A(\lambda_3) \Big\} . \end{aligned} \tag{III.4}$$

In order to investigate non-equilibrium phenomena the system of interacting phonons is subjected to an external perturbation $J(\lambda,t)$ which couples to the normal coordinates by adding to the Hamiltonian H a further term

$$H_t' = \sum_\lambda A(\bar{\lambda})\, J(\lambda)\, e^{-i\omega t}\, e^{\varepsilon t} . \tag{III.5}$$

The simplest response of the system is the normal coordinate itself, being given by linear response theory as

$$\langle A(\lambda)\rangle_t = \langle A(\lambda)\rangle - i\sum_{\lambda'} \int_{-\infty}^{\infty} \theta(t-t')\langle[A(\lambda,t),A(\bar{\lambda}',t')]\rangle J(\lambda',t')dt'. \tag{III.6}$$

This describes the motion of the elastic lattice in response to the external field. The expectation values <...> are taken in equilibrium as long as the perturbation is small, which we will assume to be the case. The retarded commutator in (III.6) is the retarded one-phonon Green function

$$G_t^r(\lambda|\lambda') = -i\,\theta(t)\,\langle[A(\lambda,t),A(\lambda',0)]\rangle \tag{III.7}$$

so that (III.6) can also be written as

$$\langle A(\lambda)\rangle_t = \langle A(\lambda)\rangle + \sum_{\lambda'} G_\omega^r(\lambda|\bar{\lambda}')\, J(\lambda',t)\ . \tag{III.8}$$

The operators inside the commutator are in the interaction picture with respect to $H + H_t'$:

$$A(\lambda,t) = e^{iHt}\, A(\lambda)\, e^{-iHt}\ . \tag{III.9}$$

It is clear from the form of (III.6) that the response function can also be interpreted as the functional derivative of the induced (J-dependent) thermal expectation value of $A(\lambda,t)$ with respect to the external field:

$$\frac{\delta\langle A(\lambda)\rangle_t}{\delta J(\lambda',t)} = -iG_t^r(\lambda|\lambda')\ . \tag{III.10}$$

In the harmonic approximation we have for the Fourier transform

$$G_\omega^{r,(o)}(\lambda|\lambda') = \frac{2\,\omega_\lambda}{(\omega+i\varepsilon)^2 - \omega_\lambda^2}\,\Delta(\underline{q}+\underline{q}')\,\delta_{jj'}\quad ;\ \varepsilon\to 0^+$$

so that in this case a response arises only for that mode ω_λ which has the same frequency as the external source. In the anharmonic case the situation is of course more complicated and more interesting.

So far we have considered the response of the normal coordinate to the external field. A similar treatment is possible also for other quantities; the external field will create deviations from the corresponding equilibrium values. This applies also to the occupation numbers of the phonon modes, which in equilibrium are given by

$$n_\lambda \equiv n(\omega_\lambda) = \left(\exp(\beta\, \omega_\lambda) - 1\right)^{-1} . \tag{III.11}$$

If the perturbation is of sufficiently long wavelength and low frequency, the anharmonic interactions among the different modes will be able to change the n_λ's so that they can adjust to a certain extent to the varying displacement field, temperature, etc. which is introduced by the perturbation. According to linear response theory we may write in complete analogy to (III.6) for the phonon density of mode λ' varying with a wavevector $\underline{q}$

$$\langle n(\lambda'|\underline{q})\rangle_t = n_{\lambda'}\, \delta_{\underline{q},o} - i \int_{-\infty}^{\infty} \theta(t-t')\langle [n(\lambda'|\underline{q},t-t'), H'_{t'}]\rangle dt'. \tag{III.12}$$

In order to evaluate the commutator on the right hand side an expression in terms of phonon operators for the non-equilibrium situation is needed. In quantum statistical mechanics of particle systems the Wigner operator is used and it can be shown (Klein and Wehner 1968) that a similar construction is possible for phonons by writing

$$n(\lambda'|\underline{q}) = a^+(\underline{q}' - \tfrac{1}{2}\,\underline{q},j')\; a(\underline{q}' + \tfrac{1}{2}\,\underline{q},j') . \tag{III.13}$$

Using (III.13) and (III.5) in (III.12) results in (III.14)

$$\langle n(\lambda'|\underline{q})\rangle_t = n_{\lambda'}\, \delta_{\underline{q},o} + \sum_j G^r_\omega \left(a^+(\underline{q}'- \tfrac{1}{2}\,\underline{q},j')a(\underline{q}'+ \tfrac{1}{2}\,\underline{q},j')\,|A(-\underline{q}j)\cdot \cdot J(\underline{q}j,t)\right).$$

The retarded Green function appearing in this expression depends on three operators.

III.2. Some Results from the Functional Derivative Method

For a systematic calculation of the Green functions it is convenient to consider instead of the retarded functions the more general time-ordered functions

$$G_n(1,\ldots,n) = \langle T\, A(\lambda_1,t_1)\ldots\, A(\lambda_n,t_n)\rangle \tag{III.15}$$

where T is the time-ordering operator. With the help of the equation of motion (III.4), coupled equations for Green functions can be derived. In order to have a convenient method of generating higher order Green functions and decoupling their equations of motion systematically, one uses functional methods, which generalize the $G_n(1,\ldots,n)$ in (III.15) to non-equilibrium functions $G_{n,J}(1,\ldots,n)$, which depend on the external field.

Whereas equilibrium expectation values are calculated with the help of the density matrix $\rho = Z^{-1}\exp(-\beta H)$, where $Z = \mathrm{Tr}\,\exp(-\beta H)$ is the partition function of the interacting phonon system in equilibrium, the presence of the external field modifies the partition function to a functional of J

$$\begin{aligned} Z[J] &= \mathrm{Tr}\,\{\exp[-\beta(H + H')]\} \\ &\equiv \mathrm{Tr}\,\{[\exp(-\beta H)]S\}\ . \end{aligned} \tag{III.16}$$

Here the operator S is a functional of J :

$$S = T\exp\{-\int_0^\beta dt'\ H'(t')\}\quad . \tag{III.17}$$

Its derivative with respect to J is

$$\frac{\delta S}{\delta J(\lambda,t)} = -\,T\,(S\,A(\lambda,t))\ . \tag{III.18}$$

Taking higher order derivatives of S generates higher order products of time-ordered operators A. Taking the expectation values of such products leads directly to Green functions. If the expectation values are calculated with $J \neq 0$ one gets non-equilibrium functions $G_{n,J}(1,\ldots,n)$, which reduce to those in eq. (III.15) if one sets $J=0$.

In the extensive literature on phonon transport theory (Horie and Krumhansl 1964, Kwok and Martin 1966, Sham 1967, Götze and Michel 1967a, 1967b, 1969, Klein and Wehner 1968, 1969, Niklasson and Sjölander 1968, Ranninger 1968, 1969, 1972, Enz 1969, 1974, Niklasson 1969, 1970, 1972, Meier 1969, Beck and Meier 1970, Beck 1971) a number of different routes have been developed in handling the source field $J(\lambda,t)$. One can use in (III.17) imaginary times in the intervall $[0,i\beta]$, which makes an analytical continuation to real times (or frequencies) necessary at the end of the calculation. One can avoid this and take $J(\lambda,t)$ as a double-valued quantity along the real time axis. Again different choices have been made in the literature. All these cases are discussed in detail by Bilz, Strauch and Wehner (1984) who give a thorough representation of the functional approach to the Green function method for phonon systems. The application of these procedures to the problem of phonon transport in reviewed by Beck, Meier and Thellung (1974) and by Beck (1975).

Here, we will only indicate some steps in the derivation following Klein and Wehner (1968, 1969). As indicated after eq. (III.18), $Z[J]$ or $\langle S\rangle_J$ are generating functions for non-equilibrium Green functions according to

$$G_{n,J}(1,\ldots,n) = \frac{i^n}{\langle S\rangle_J} \frac{\delta^n}{\delta J(1)\ldots\delta J(n)} \langle S\rangle_J \ . \tag{III.19}$$

In addition, n-point functions $f_{n,J}(1,\ldots,n)$ are introduced

$$f_{n,J}(1,\ldots,n) = i^n \frac{\delta^n}{\delta J(1)\ldots\delta J(n)} \log \langle S\rangle_J \ . \tag{III.20}$$

There are, obviously, relations between the G_n and f_n ; the first few are

$$\begin{aligned} G_1(1) &= f_1(1) \\ G_2(1,2) &= f_2(1,2) + f_1(1)\, f_1(2) \\ G_3(1,2,3) &= f_3(1,2,3) + f_1(1)f_2(2,3) + f_1(2)f_2(1,3) + f_1(3)f_2(1,2) \\ &\quad + f_1(1)f_1(2)f_1(3) \ . \end{aligned} \tag{III.21}$$

The reason for introducing the functions f_n is to approximate the higher-order Green functions G_n in a systematic way in terms of a few phonon correlations f_n of low order.

Since G_1 is the time-independent thermodynamic expectation value (for J=0) of the operator $A(\lambda)$, we may assume $G_1(1)=f_1(1)=0$ as long as static rigid displacements can be neglected. In this case, the most important function is f_2. Its equation of motion can be written in the form of a Dyson equation

$$f_2(\lambda t,\lambda' t') = f_2^{(o)}(\lambda t,\lambda' t') - i\, f_2^{(o)}(\lambda t,1)\Sigma(1,2)f_2(2,\lambda' t') \tag{III.22}$$

where $f_2^{(o)}$ is the harmonic function satisfying

$$\left(\frac{\partial^2}{\partial t^2} + \omega_\lambda^2\right) f_2^{(o)}(\lambda t,\lambda' t') = -\, 2\, i\, \omega_\lambda\, \Delta(\underline{q}+\underline{q}')\delta_{jj'}\, \delta(t-t'). \tag{III.23}$$

In eq. (III.22) the repeated variables denote $k = (\lambda_k t_k)$ and are understood to include a summation over λ_k and a time integration from $-\infty$ to $+\infty$. $\Sigma(1,2)$ is the phonon self-energy, which arises from all anharmonicities. Restricting ourselves to three-phonon processes only, the self-energy is given by

$$\Sigma(1,2) = 3\, V_3(1,3,4)f_2(3,3')f_2(4,4')\Gamma_3(3',4',2) \ . \tag{III.24}$$

This can be represented by the following diagram

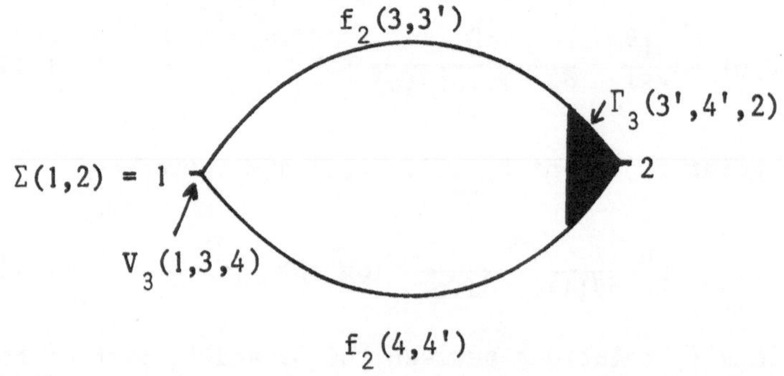

(III.24a)

The quantity $\Gamma_3(1,2,3)$ is called a vertex correction (or vertex part) which in lowest order is just proportional to the instantaneous cubic anharmonicity $V_3(1,2,3)$, but it includes higher order cubic interactions between the intermediate phonons described by the two f_2 functions in (III.24). These processes are of importance for phonon transport, as will be demonstrated lateron. Here it suffices to state that $\Gamma_3(1,2,3)$ satisfies the following integral equation

$$\Gamma_3(1,2,3) = \Gamma_3^{(0)}(1,2,3) + \Gamma_3^{(0)}(1,4',5)f_2(4',4'')\Gamma_3^{(0)}(2,4'',6)\cdot \\ \cdot f_2(5,5')f_2(6,6')\Gamma_3(5',6',3) \ , \tag{III.25}$$

where

$$\Gamma_3^{(0)}(1,2,3) = -6\,i\,V_3(\lambda_1\lambda_2\lambda_3)\delta(t_1-t_3)\delta(t_2-t_3) \tag{III.26}$$

In diagrams

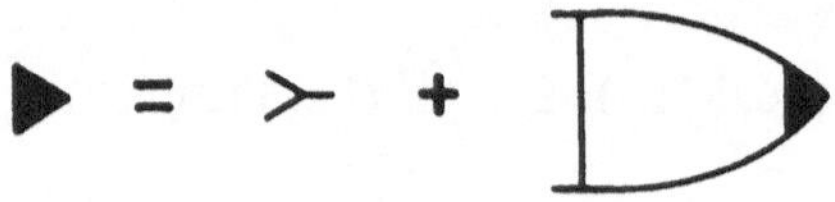

(III.25a)

Using the iteration of this integral equation in the diagram for $\Sigma(1,2)$ gives the infinite series of bubble diagrams with an increasing number of ladder rungs:

(III.27)

In eq. (III.14) the response of the phonon density was represented by a Green function depending on three operators. It is connected with the function $G_3(1,2,3)$ for which the general formalism gives

$$G_3(1,2,3) = f_2(1,1')f_2(2,2')f_2(3,3')\Gamma_3(1',2',3') \qquad \text{(III.28)}$$

where again the vertex corrections enter.

As shown in the examples given above, for the response of the phonon normal coordinate eq. (III.6) and for the response of the phonon density eq. (III.14), the retarded Green functions are needed. They can be obtained as boundary values of the time-ordered ones, for example

$$G^r_\omega(A|B) = -\, i \lim_{\eta\to 0^+} G_\omega(A|B)\,. \qquad \text{(III.29)}$$

From (III.22) one obtains by this procedure a Dyson equation for the retarded one-phonon Green function:

$$G^r_\omega(\lambda|\lambda') = G^{r,(0)}_\omega(\lambda|\lambda') - \sum_{\lambda_1\lambda_2} G^{r,(0)}_\omega(\lambda|\lambda_1)\Pi_{\lambda_1\lambda_2}(\omega)G^r_\omega(\lambda_2|\lambda')\,, \qquad \text{(III.30)}$$

where $\Pi_{\lambda_1\lambda_2}(\omega)$ is the retarded self-energy corresponding to the time-ordered self-energy $\Sigma(1,2)$ in eq. (III.24).

If one considers only the first diagram in the series (III.27), which neglects all vertex corrections, one obtains a simple approximation for the self-energy. Neglecting also polarization mixing this contribution is given by

$$\Pi'_\lambda(\omega) = 18 \sum_{\lambda_1\lambda_2} |V(\bar{\lambda}\lambda_1\lambda_2)|^2\, G^r_{\lambda_1\lambda_2}(\omega)\,. \qquad \text{(III.31)}$$

It describes summation and difference processes of phonons $(\underline{q}j)$ into phonons λ_1, λ_2, which are described by the two-phonon retarded Green function $G^r_{\lambda_1\lambda_2}(\omega)$. This expression was evaluated by Maradudin and Fein (1962) to give through eq. (III.30) the shift of the phonon frequency from its harmonic value ω_λ to a temperature dependent value $\bar{\omega}_\lambda$ and to provide for a damping of the phonons due to the presence of third-order anharmonicities. Writing (III.30) as

$$G^r_\omega(\lambda|\lambda') = \frac{2\,\omega_\lambda}{(\omega+i\varepsilon)^2 - \omega_\lambda^2 - 2\omega_\lambda\,\Pi'_\lambda(\omega)}\,\delta_{\lambda\lambda'} \qquad \text{(III.32)}$$

and

$$\Pi'_\lambda(\omega) = \Delta_\lambda(\omega) - i\,\Gamma_\lambda(\omega)$$

the real part $\Delta_\lambda(\omega)$ gives the shift and $\Gamma_\lambda(\omega)$ the width of the phonon λ. Since we are here mainly interested in lifetime effects, the real part $\Delta_\lambda(\omega)$ of the self-energy is always assumed to be included in the phonon frequency. Considering three-phonon processes only, $\Gamma_\lambda(\omega)$ is of the order of $|V_3|^2$.

This contribution to the self-energy, which is obtained from the lowest-order term in (III.27), is usually sufficient for all lifetime effects of thermal phonons, since practically all phonon frequencies ω_λ, which are involved in the lowest-order diagram, are large compared to the widths Γ_λ. Suppose, however, that we are interested in the self-energy of a low frequency and long-wavelength phonon. If $\underline{Q}$ denotes the wavevector of this phonon, its interactions are with thermal phonons of wavevectors $\underline{q}_1$ and $\underline{q}_2$ such that $\underline{q}_1+\underline{q}_2=\underline{Q}$. Since $|\underline{Q}| \ll |\underline{q}_{1,2}|$, the two thermal phonons have wavevectors of nearly the same lengths but of essentially opposite direction. If in addition they are from the same branch, we may write

$$\lambda_1 = (\underline{q}' + \underline{Q}/2,j') \; ; \; \lambda_2 = (-\underline{q}' + \underline{Q}/2,j') \; ; \; |\underline{Q}| \ll |\underline{q}'| \qquad \text{(III.33)}$$

and introduce a mean frequency $\omega_{\lambda'}$ corresponding to $\underline{q}'$, which is given by

$$\omega_{\lambda_1} = \omega_{\lambda'} + \frac{1}{2}\underline{Q}\cdot\underline{v}_{\lambda'} \; ; \quad \omega_{\lambda_2} = \omega_{\lambda'} - \frac{1}{2}\underline{Q}\cdot\underline{v}_{\lambda'} \qquad \text{(III.34)}$$

where $\underline{v}_{\lambda'} = \partial\omega_{\lambda'}/\partial\underline{q}'$ is the group velocity. Since the two thermal phonons λ_1, λ_2 in (III.31) are nearly equivalent, it is reasonable to assume $\Gamma_{\lambda_1} \approx \Gamma_{\lambda_2} \approx \Gamma_{\lambda'}$. Then the two-phonon Green function in the self-energy of the long-wavelength phonon $(\underline{Q},\Omega)$ is given by

$$G^r_{\lambda_1\lambda_2}(\Omega) = -\frac{2n_{\lambda'}+1}{\omega_{\lambda'}} - 2\beta n_{\lambda'}(n_{\lambda'}+1) + \beta n_{\lambda'}(n_{\lambda'}+1)\Omega\left(\frac{1}{\Omega-\underline{Q}\cdot\underline{v}_{\lambda'}+2i\Gamma_{\lambda'}} + \frac{1}{\Omega+\underline{Q}\cdot\underline{v}_{\lambda'}+2i\Gamma_{\lambda'}}\right). \qquad \text{(III.35)}$$

When the broadening of the thermal phonons, $\Gamma_{\lambda'}$, is comparable with Ω and $\underline{Q}\cdot\underline{v}_{\lambda'}$, or even larger than the latter quantities, the two-phonon Green function in eq. (III.35) contributes a factor $O(V_3^{-2})$ to the self-energy of the phonon $\underline{Q}$. This means that according to (III.31) the self-energy due to the bubble diagram is of $O(1)$ although the anharmonic coupling coefficient V_3 might be small. This situation was first pointed out by Sham (1967), who also noticed that the ladder-type diagrams in (III.27) all contribute in $O(1)$ to the self-energy at low frequencies and long wavelength. Each rung of the ladder introduces a factor V_3^2 from the two cubic vertices, but they are canceled by the contributions from the additional pair of one-phonon Green functions. This shows that for the case of the low frequency response it is not possible to get reliable results from low-order perturbation theory, but that one has to sum the series of ladder-diagrams to all orders. This is accomplished by working with eq. (III.24) instead of (III.31). Therefore, the vertex correction $\Gamma_3(1,2,3)$ is the quantity of central importance for the low-frequency response of dielectrics. This is also evident for the phonon density fluctuations (III.14), since the response function appearing there is related to G_3 of (III.28), which in turn is again determined by $\Gamma_3(1,2,3)$.

A detailed study of the integral equation for Γ (1,2,3) shows (Wehner and Klein 1972) that a function $Y_{\lambda\lambda'}(\underline{Q}\Omega)$, which is closely connected to the retarded boundary value of Γ_3, satisfies an equation very similar to a phonon Boltzmann equation. This derivation is technically somewhat involved and will not be reproduced here. The result is

$$(\Omega-\underline{Q}\cdot\underline{v}_\lambda)Y_{\lambda'\lambda}(\underline{Q}\Omega) = \delta_{\lambda'\lambda} - \frac{i}{n_{\lambda'}(n_{\lambda'}+1)} \sum_{\lambda''} \mathcal{L}_{\lambda'\lambda''} Y_{\lambda''\lambda}(\underline{Q}\Omega) \qquad \text{(III.36)}$$

where $\mathcal{L}$ is the linearized collision operator for three-phonon processes given by Peierls:

$$\sum_{\lambda''} \mathcal{L}_{\lambda'\lambda''} Y_{\lambda''\lambda} = 72 \sum_{\lambda''\bar{\lambda}'''} |V_3(\lambda'\lambda''\bar{\lambda}''')|^2 (n_{\lambda'}+1)(n_{\lambda''}+1)n_{\lambda'''} \cdot$$

$$\cdot\ \delta(\omega_{\lambda'}+\omega_{\lambda''}-\omega_{\lambda'''})(Y_{\lambda'\lambda}+Y_{\lambda''\lambda}-Y_{\lambda'''\lambda})$$

$$+ 36 \sum_{\lambda''\bar{\lambda}'''} |V(\lambda'\bar{\lambda}''\bar{\lambda}''')|^2 (n_{\lambda'}+1)n_\lambda\, n_{\lambda'''} \cdot$$

$$\cdot\ \delta(\omega_{\lambda'}-\omega_{\lambda''}-\omega_{\lambda'''})(Y_{\lambda'\lambda}-Y_{\lambda''\lambda}-Y_{\lambda'''\lambda}).$$

Taking also four-phonon processes into account adds further terms, but they will not be considered here, since the qualitative aspects are not changed.

The solution $Y_{\lambda'\lambda}$ of the transport equation (III.36) determines the self-energy

$$\Pi_{jj'}(\underline{Q}\Omega) = -\,36\,i \sum_{\lambda\lambda'} V_3(\bar{\Lambda}\lambda_1\lambda_2)N_{\lambda\lambda'}(\underline{Q}\Omega)V_3(\lambda_3\lambda_4\Lambda'), \qquad \text{(III.37)}$$

where $\Lambda = (\underline{Q}j)$, $\Lambda' = (\underline{Q}j')$, $\lambda_{1,2} = (\pm\underline{q} + \frac{1}{2}\underline{Q},j)$, $\lambda_{3,4} = (\mp\underline{q}' - \frac{1}{2}\underline{Q},j')$,

and the fluctuating phonon density of mode λ

$$\delta n_\lambda(\underline{Q}\Omega) = -\,6i \sum_{j\lambda'} N_{\lambda\lambda'}(\underline{Q}\Omega)\; V_3(\lambda_3\lambda_4\Lambda)\; A_j(\underline{Q}\Omega) \quad . \qquad \text{(III.38)}$$

In these expressions, $N_{\lambda\lambda'}(\underline{Q}\Omega)$ is given in terms of $Y_{\lambda\lambda'}(\underline{Q}\Omega)$:

$$N_{\lambda\lambda'}(\underline{Q}\Omega) = i\,\beta\, n_\lambda (n_\lambda+1) \left[-\delta_{\lambda\lambda'} + \Omega\, Y_{\lambda\lambda'}(\underline{Q}\Omega)\right] \quad . \qquad \text{(III.39)}$$

$A_j(\underline{Q}\Omega)$ denotes the strain $\langle A(\underline{q}j)\rangle_{Q\Omega}$ introduced by the external perturbation $J(\underline{Q}j)\exp(-i\,\Omega\, t)$. For a long-wavelength longitudinal phonon traveling in the x direction, $A(\underline{Q}\Omega)$ is connected with a deformation tensor component $u_{xx}(\underline{Q}\Omega) \equiv u(\underline{Q}\Omega)$ by

$$u(\underline{Q}\Omega) = -\,(2\,\omega_Q\,\rho\, V)^{-1/2}\, A(\underline{Q}\Omega) \quad . \qquad \text{(III.40)}$$

In the same long wavelength approximation the anharmonic coupling constants in (III.37) and (III.38) are related to Grüneisen parameters $\gamma_{xx}(q) \equiv \gamma_q$ of the mean thermal phonon q through

$$V_3(-\underline{Q},\underline{q}_1,\underline{q}_2) = \frac{1}{6}\,(2\,\omega_Q\,\rho\, V)^{-1/2}\, Q\,\omega_Q\,\gamma_q \quad , \qquad \text{(III.41)}$$

where γ_q is defined by writing the frequency shift $\delta\omega_q$ of a thermal phonon due to the deformation $u(\underline{Q}\Omega)$ as

$$\delta\omega_q(\underline{Q}\Omega) = -\,\omega_q\,\gamma_q\, u(\underline{Q}\Omega) \quad . \qquad \text{(III.42)}$$

IV. SOME CONSEQUENCES OF PHONON TRANSPORT

IV.1. The Self-energy of a Long Wavelength and Low Frequency Phonon

In order to calculate the response at low frequencies Ω and long wavelength $\underline{Q}$ the transport equation (III.36) is now solved by the methods introduced in section II. The function $Y_{\lambda\lambda'}$ is expanded in terms of the eigenfunctions $\omega_{\lambda'}$ and q_i', i = 1,2,3 of $\mathcal{L}^N_{\lambda\lambda'}$ belonging to the eigenvalue zero and a rest, which belongs to all other eigenvalues. In order to simplify the notation and the calculation we also specialize to a one-branch Debye-model. One writes

$$\begin{aligned} \sqrt{n_{q'}(n_{q'}+1)}\; Y_{q'q}(\underline{Q}\Omega) &\equiv Z_{q'q}(\underline{Q}\Omega) \\ &= a^o_{\underline{q}}(\underline{Q}\Omega)\,\eta^o_{q'} + \sum_{i=1}^{3} a^i_{\underline{q}}(\underline{Q}\Omega)\,\eta^i_{\underline{q}'} + \tilde{Z}_{q'q}(\underline{Q}\Omega) \end{aligned} \qquad \text{(IV.1)}$$

where

$$\begin{aligned} \eta^o_{q'} &= \mu_o\,\beta\sqrt{n_{q'}(n_{q'}+1)}\;\omega_{q'} \\ \eta^i_{q'} &= \mu_i\,\beta\sqrt{n_{q'}(n_{q'}+1)}\; q'_i \quad ; \quad i=1,2,3 \end{aligned} \qquad \text{(IV.2)}$$

and $\mu_0 = \beta(k_B/VC_V)^{1/2}$, $\mu_i = \beta v(3k_B/VC_V)^{1/2}$. Since $\tilde{Z}$ describes the deviation of the phonon distribution function from a drifting local equilibrium distribution one approximates the action of $\mathcal{L}$ on $\tilde{Z}$ in the spirit of eq. (II.10) by relaxation times τ_N and τ_U respectively:

$$\sum_{q''} [n_q(n_q+1)n_{q''}(n_{q''}+1)]^{-1/2} \mathcal{L}^N_{qq''} \tilde{Z}_{q''q'} = \frac{1}{\tau_N} \tilde{Z}_{qq'}(\underline{Q}\Omega) \tag{IV.3a}$$

$$\sum_{q''} [n_q(n_q+1)n_{q''}(n_{q''}+1)]^{-1/2} \mathcal{L}^U_{qq''}[\sum_{i=1}^{3} a^i_{q'} \eta^i_{q''} + \tilde{Z}_{q''q'}] = \frac{1}{\tau_U} [\sum_{i=1}^{3} a^i_{q'} \eta^i_q + \tilde{Z}_{qq'}]. \tag{IV.3b}$$

The various factors $[n_q(n_q+1)]^{1/2}$ in (IV.1) and (IV.3) are introduced in order to have a Hermitean collision operator. Using (IV.1) and the above definitions of the relaxation times in (III.36) leads to the following solution of the transport equation

$$Z_{q'q}(\underline{Q}\Omega) = \frac{1}{\Omega-\underline{Q}\cdot\underline{v}_{q'}+i/\tau} \{\sqrt{n_{q'}(n_{q'}+1)}\, \delta_{q'q} + \frac{i}{\tau} a^0_q \eta^0_{q'} + \frac{i}{\tau_N} \sum_i a^i_q \eta^i_{q'}\} \tag{IV.4}$$

where $\tau^{-1} = \tau_U^{-1} + \tau_N^{-1}$. The functions $a^0_q(\underline{Q}\Omega)$ and $a^i_q(\underline{Q}\Omega)$, i = 1,2,3, are still unknown. They are determined from a system of linear algebraic equations, which are obtained from (III.36) by multiplication with $\eta^0_{q'}$ and $\eta^i_{q'}$, respectively, and a subsequent summation over q'. Using further the orthonormality of the eigenfunctions in the expansion (IV.1), one obtains

$$\Omega a^0_q - \underline{c}\cdot\underline{Q}\, a^i_q = [n_q(n_q+1)]^{1/2} \eta^0_q \tag{IV.5a}$$

$$-\underline{c}\cdot\underline{Q}\, a^0_q + (\Omega+i/\tau_U)a^i_q - \sum_{q'} \eta^i_{q'} \underline{Q}\cdot\underline{v}_{q'} \tilde{Z}_{q'q} = [n_q(n_q+1)]^{1/2}\eta^i_q \; ; \quad i = 1,2,3 \tag{IV.5b}$$

Here

$$\underline{c}\cdot\underline{Q} = \underline{Q}\cdot\sum_{q'} \eta^0_{q'} \underline{v}_{q'} \eta^i_{q'} = v\, Q/\sqrt{3} \equiv \omega_2 \tag{IV.6}$$

for a Debye model with sound velocity v . $\tilde{Z}$ in (IV.5b) can be expressed in terms of η^0_q and η^i_q by using (III.36), (IV.1) and (IV.3). The solution of the system of equations (IV.5) for the coefficients $a^0_q(\underline{Q}\Omega)$ and a^i_q is

$$\frac{a_q^{\,0}}{[n_q(n_q+1)]^{1/2}} = \frac{\eta_q^{\,0}\,\{\Omega+i\gamma(\underline{Q}\Omega)\} + \omega_2\,\eta_q^{\,1}\left(1+\frac{\underline{Q}\cdot\underline{v}_q}{\Omega-\underline{Q}\cdot\underline{v}_q+i/\tau}\right)}{\Omega^2 - \frac{i}{\tau}M_{11}(\underline{Q}\Omega)\,\omega_2^{\,2} + i\Omega\,\gamma(\underline{Q}\Omega)} \tag{IV.7a}$$

$$\frac{a_q^{\,i}}{[n_q(n_q+1)]^{1/2}} = \frac{\frac{i}{\tau}M_{11}\,\omega_2\,\eta_q^{\,0} + \Omega\,\eta_q^{\,i}\left(1+\frac{\underline{Q}\cdot\underline{v}_q}{\Omega-\underline{Q}\cdot\underline{v}_q+i/\tau}\right)}{\Omega^2 - \frac{i}{\tau}M_{11}(\underline{Q}\Omega)\,\omega_2^{\,2} + i\Omega\,\gamma(\underline{Q}\Omega)} \tag{IV.7b}$$

with

$$\gamma(\underline{Q}\Omega) = \tau^{-1} - \tau_N^{-1}(\Omega + i\,\tau^{-1})\,M_{11}(\underline{Q}\Omega) \tag{IV.8}$$

$$M_{11}(\underline{Q}\Omega) = \sum_{q'} \eta_{q'}^{\,1}\,\eta_{q'}^{\,1}\,(\Omega - \underline{Q}\cdot\underline{v}_{q'} + i\,\tau^{-1})^{-1}. \tag{IV.9}$$

In the limits where $\Omega\tau \ll 1$ and $\underline{Q}\cdot\underline{v}_q\tau \ll 1$, we have $iM_{11}\tau^{-1} \to 1$ and $(\Omega+i\tau^{-1})M_{11} \to 1$, so that $\gamma(\underline{Q}\Omega) \to \tau_U^{-1}$.

Now we are in the position to calculate the self-energy $\Pi(\underline{Q}\Omega)$ of a long-wavelength phonon from (III.37), (III.39), (III.41), (IV.4) and (IV.7). The result can be written in different forms, either by separating the thermodynamic fluctuations completely from those away from drifting local equilibrium, or in a form which makes the transition from high frequency to low frequency response particularly transparent. Choosing the latter, one has

$$\Pi(\underline{Q}\Omega) = \frac{Q^2\,T\,C_V}{2\,\omega_Q\,\rho}\Big\{-\,(\overline{\gamma^2}-\bar{\gamma}^2) + \frac{1}{VC_V}\sum_q c_q\,\gamma_q(\gamma_q-\bar{\gamma})\,\frac{\Omega}{\Omega-\underline{v}_q\cdot\underline{Q}+i\tau^{-1}} - \bar{\gamma}^2\,\frac{\omega_2}{2}\,\tilde{G}_{22}(\underline{Q}\Omega)\Big\} \tag{IV.10}$$

where

$$\tilde{G}_{22}(\underline{Q}\Omega) = \frac{2\omega_2(\Omega + i\tau^{-1})\,M_{11}(\underline{Q}\Omega)}{\omega_2^{\,2}\;i\tau^{-1}\,M_{11}(\underline{Q}\Omega) - \Omega^2 - i\,\Omega\,\gamma(\underline{Q}\Omega)}\;. \tag{IV.11}$$

Furthermore, $\bar{\gamma}$ and $\overline{\gamma^2}$ are defined by

$$C_V\,V\,\overline{\gamma^n} = \sum_q c_q\,\gamma_q^{\,n} \quad ; \quad n = 1,2 \tag{IV.12}$$

where c_q is the contribution of mode q to the specific heat.

IV.2. Adiabatic and Isothermal Elastic Constants

As a first application of the result (IV.10) one can discuss the difference between adiabatic and isothermal elastic constants. A $(\underline{Q}\Omega)$-dependent elastic constant $c(\underline{Q}\Omega)$ may be introduced by (Cowley 1967)

$$\frac{Q^2}{\rho} c(\underline{Q}\Omega) = \omega_Q^2 + 2\,\omega_Q\,\Pi(\underline{Q}\Omega). \tag{IV.13}$$

Because of the terms in (IV.10), which originate from phonon transport, $c(\underline{Q}\Omega)$ depends on the order in which $\underline{Q}$ and Ω go to zero. If first $\Omega \to 0$, there will be sufficient time for the thermal phonons to adjust via anharmonicities to the external perturbation, so that all temperature gradients can disappear. Therefore, this limit corresponds to the isothermal case. Taking the limit in the opposite order, the propagation of the perturbation is fast and no heat conduction can take place. Therefore, the isothermal and the adiabatic elastic constants are given from (IV.13) and (IV.10) by (Wehner and Klein 1971)

$$c_T = \lim_{Q\to 0}\lim_{\Omega\to 0} c(Q\Omega) = c^{(o)} - T\,C_V\,\overline{\gamma^2} \tag{IV.14a}$$

$$c_S = \lim_{\Omega\to 0}\lim_{Q\to 0} c(Q\Omega) = c^{(o)} - T\,C_V(\overline{\gamma^2} - \bar{\gamma}^2) \tag{IV.14b}$$

where $c^{(o)} = \rho\,v^2$ is the harmonic elastic constant. The difference between c_S and c_T depends on $\bar{\gamma}$. But all terms in the self-energy (IV.10), which are proportional to $\bar{\gamma}$, originate from phonon transport, meaning that they arise from diagrams other than the simple bubble diagram in (III.27), whose contribution is proportional to $\bar{\gamma}^2$. The difference $c_S - c_T = T\,C_V\bar{\gamma}^2$ is in agreement with a well-known result from thermoelasticity and demonstrates the essential role of phonon transport in understanding such results on the basis of anharmonic lattice dynamics.

The Green function of the long wavelength phonon is

$$G(\underline{Q}\Omega) = \frac{2\,\omega_Q}{\omega_Q^2 - \Omega^2 + 2\,\omega_Q\,\Pi(\underline{Q}\Omega)} \tag{IV.15}$$

with $\Pi(\underline{Q}\Omega)$ given by (IV.10) and ω_Q the harmonic frequency. Using the first term in (IV.10) renormalizes according to (IV.14b) ω_Q^2 to the adiabatic frequency

$$\omega_S^2 = \omega_Q^2 - \frac{T\,C_V\,Q^2}{\rho}\,(\overline{\gamma^2} - \bar{\gamma}^2)\ . \tag{IV.16}$$

This shows that part of the anharmonicities described by $\Pi(Q\Omega)$ renormalize the phonon to an adiabatic sound wave, which is usually called first sound. But it is also evident from (IV.15) and (IV.10) that first sound is coupled to two further modes which are represented by the second and third term in (IV.10). The second term in (IV.10) represents the dielectric fluctuations or fluctuations away from local equilibrium. They may be interpreted as a second viscosity. The third term originates from thermodynamic fluctuations and it will be shown to represent entropy fluctuations. The dielectric fluctuations do not contribute to the difference $c_T - c_S$, since this term vanishes in both limits. In contrast, the third term in (IV.10) gives $-\bar{\gamma}^2$ in the isothermal limit, so that the renormalized frequency in the static limit is the isothermal one.

In the opposite limit of high frequencies Ω the thermodynamic fluctuations die out, as expected, but from the dielectric fluctuations one obtains as the renormalized frequency for zero sound

$$\tilde{\omega}^2(Q\Omega) = \omega_T^2\,\Big(1 - \frac{Q^2\,T\,C_V}{\rho\,\omega_Q^2}\,\overline{\gamma^2}\Big) \quad ; \qquad \Omega \to \infty \tag{IV.17}$$

which agrees with a result obtained by Maris (1967).

IV.3. Coupling of First and Second Sound

We will now investigate the coupling of the adiabatic (first) sound wave to the entity $\tilde{G}_{22}(\underline{Q}\Omega)$ in (IV.10). It will be shown that $\tilde{G}_{22}(\underline{Q}\Omega)$ is the propagator for entropy fluctuations. The entropy of a Bose system is

$$S(\underline{r},t) = -\frac{k}{V}\sum_q \{n_q(\underline{r},t)\ln n_q(\underline{r},t) - [n_q(\underline{r},t)+1]\ln[n_q(\underline{r},t)+1]\}. \tag{IV.18}$$

The fluctuating part of S is given by the fluctuations of the occupation numbers

$$\delta S(Q\Omega) = \frac{k}{V}\sum_q \beta\,\omega_q\,\delta n_q(Q\Omega)\ . \tag{IV.19}$$

But by (III.38) and (III.39) $\delta n_q(Q\Omega)$ is expressed in terms of the solution of the transport equation. Using this solution, one obtains

$$\delta n_q(\underline{Q}\Omega) = \delta n_q^{(1)}(\underline{Q}\Omega) + \delta n_q^{(2)}(\underline{Q}\Omega) \tag{IV.20}$$

$$\delta n_q^{(1)}(\underline{Q}\Omega) = \beta\, n_q(n_q+1)(\bar{\gamma}-\gamma_q)\omega_q \frac{\underline{v}_q\cdot\underline{Q} - i\tau^{-1}}{\Omega-\underline{v}_q\cdot\underline{Q}+i\tau^{-1}}\, u(Q\Omega) \tag{IV.20a}$$

$$\delta n_q^{(2)}(\underline{Q}\Omega) = \beta\, n_q(n_q+1)\bar{\gamma}\,\omega_q\Big\{ \frac{i\,\tau^{-1}}{\Omega-\underline{v}_q\cdot\underline{Q}+i\tau^{-1}}\frac{\omega_2}{2}\tilde{G}_{22}(\underline{Q}\Omega)$$

$$- \frac{\underline{v}_q\cdot\underline{Q}}{\Omega-\underline{v}_q\cdot\underline{Q}+i\tau^{-1}}\,\frac{i\,\omega_2^2\,\tau^{-1}\,M_{11}(\underline{Q}\Omega)-\Omega(\Omega+i\tau^{-1})}{i\,\omega_2^2\,\tau^{-1}\,M_{11}(\underline{Q}\Omega)-\Omega\{\Omega+i\gamma(\underline{Q}\Omega)\}}\Big\}u(\underline{Q}\Omega). \tag{IV.20b}$$

Only the first part of $\delta n_q^{(2)}$ contributes to entropy fluctuations with the result

$$\delta S(\underline{Q}\Omega) = C_V\,\frac{\omega_2}{2}\,\tilde{G}_{22}(\underline{Q}\Omega)\bar{\gamma}\; u(\underline{Q}\Omega)\ . \tag{IV.21}$$

This represents the response of the entropy to the deformation $u(Q\Omega)$, induced by the external perturbation. In the limit of very fast Normal processes ($\Omega\tau_N\to 0$), the relaxation of the phonon distribution to a local equilibrium is practically instantaneous. In this case the dielectric fluctuations described by the second term in (IV.10) vanish and $\tilde{G}_{22}$ simplifies to

$$G_{22}^{(1)}(Q\Omega) = \frac{2\,\omega_2}{\omega_2^2 - \Omega^2 - i\,\Omega\,\tau_U^{-1}}\ . \tag{IV.22}$$

Since $\omega_2 = v\,Q/\sqrt{3}$, this is just the propagator for (uncoupled) second sound and the damping arises only from Umklapp processes. Denoting the self-energy in the present approximation of very fast Normal processes by $\Pi_N(Q\Omega)$, we have from (IV.10), (IV.16) and (IV.22)

$$\omega_Q^2 + 2\omega_Q\,\Pi_N(Q\Omega) = \omega_S^2 - \frac{\bar{\gamma}^2\,T\,C_V\,Q^2\,\omega_2}{2\rho}\,G_{22}^{(1)}(Q\Omega) \tag{IV.23}$$

so that the Green function of the long wavelength phonon (IV.15) becomes

$$G(Q\Omega) = \frac{\omega_Q}{\omega_S}\;G_{11}(Q\Omega) \tag{IV.24}$$

where

$$G_{11}(Q\Omega) = \frac{2\omega_S}{\omega_S^2 - \Omega^2 - 2\omega_S \Pi_{12}^2 G_{22}^{(1)}(Q\Omega)}$$

$$= \left[G_{11}^{(o)-1} - \Pi_{12}^2 G_{22}^{(1)}(Q\Omega)\right]^{-1} \qquad \text{(IV.25)}$$

is the first sound propagator (adiabatic sound wave), which is coupled through

$$\Pi_{12} = \Pi_{21} = \left(\frac{T\, C_V\, \omega_2}{\rho\, \omega_S}\right)^{1/2} Q\, \bar{\gamma}/2 \qquad \text{(IV.26)}$$

to the second sound propagator (entropy wave). The function

$$G_{11}^{(o)} = \frac{2\,\omega_S}{\omega_S^2 - (\Omega + i\varepsilon)^2} \qquad \text{(IV.27)}$$

denotes the uncoupled first sound. Using (IV.22) in (IV.25) shows that second sound appears as a second pole in the one-phonon Green function (Kwok and Martin 1966).

It is also easy to find the response of the temperature to the fluctuating deformation $u(Q\Omega)$ in the present case of local equilibrium. Comparing the macroscopic expression

$$\delta n_q(Q\Omega) = \beta\, n_q(n_q+1)\omega_q\left\{ -\frac{\delta\omega_q(Q\Omega)}{\omega_q} + \frac{\delta T(Q\Omega)}{T}\right\}$$

with the microscopic result (III.38), (III.39), it is evident that $\delta T(Q\Omega)$ is directly related to the function $Y_{qq'}(Q\Omega)$ which is the solution of the Boltzmann equation. One obtains

$$\frac{\delta T(Q\Omega)}{T} = \frac{\Omega^2 + i\,\Omega\,\tau_U^{-1}}{2\omega_2} G_{22}^{(1)}(Q\Omega)\, \bar{\gamma}\, u(Q\Omega). \qquad \text{(IV.28)}$$

It is interesting to note that the expressions (IV.21) and (IV.28) for the fluctuations of the entropy and the temperature behave in the adiabatic and isothermal limits as they should. First putting $\Omega \to 0$ and then $Q \to 0$ results in $\delta T = 0$, and letting first $Q \to 0$ and then $\Omega \to 0$ gives $\delta S = 0$.

The form (IV.25) of the first sound propagator, which contains in the denominator the coupling to second sound suggests to describe the low frequency response of a dielectric in terms of two coupled oscillators (Klein and Wehner 1972). The first one is the long wavelength phonon mode ($Q\Omega$) and the other one the collective phonon density mode which we have interpreted as the second sound or entropy fluctuation. The external perturbation $J(Q\Omega)$ couples to the first sound mode, whose normal coordinate is denoted by A_1, and results in the response

$$\langle A_1(Q\Omega)\rangle = G_{11}(A_1|A_1)\; J(Q\Omega) \tag{IV.29}$$

where G_{11} is given by (IV.25). But A_1 is coupled through anharmonicities to second sound, whose normal coordinate is denoted by A_2. The response of A_2 to $J(Q\Omega)$ is written as

$$\langle A_2(Q\Omega)\rangle = G_{21}(A_2|A_1)\; J(Q\Omega) \quad . \tag{IV.30}$$

In the same way as $G_{11}(Q\Omega)$, eq. (IV.25), represents first sound being coupled to second sound, the propagator for second sound being coupled to first sound is

$$G_{22}(Q\Omega) = \frac{2\,\omega_2}{\omega_2^2-\Omega^2-i\Omega\tau_U^{-1}-2\omega_2\,\Pi_{12}^2\,G_{11}^{(o)}(Q\Omega)} = \tag{IV.31}$$

$$= \left[G_{22}^{(1)^{-1}} - \Pi_{12}^2\,G_{11}^{(o)}\right]^{-1}$$

(For $\Pi_{12}=0$, G_{22} reduces to $G_{22}^{(1)}$, eq. (IV.22)).

If there are two coupled modes A_1,A_2, the Dyson equation (III.30) takes the form

$$G_{ik} = G_{ii}^{(o)}\,\delta_{ik} - \sum_{\ell} G_{ii}^{(o)}\,\Pi_{i\ell}\,G_{\ell k} \quad . \tag{IV.32}$$

In our case, $\Pi_{11}=0$, $\Pi_{22}=-i\Omega\tau_U^{-1}$ and $\Pi_{12}=\Pi_{21}$ is given by (IV.26); $G_{11}^{(o)}$ and $G_{22}^{(o)}$ are the uncoupled and undamped propagators. Then one finds from (IV.32)

$$G_{12} = G_{21} = -\,G_{22}^{(1)}\,\Pi_{21}\,G_{11} = -\,G_{11}^{(o)}\,\Pi_{12}\,G_{22} \tag{IV.33}$$

and (IV.30) becomes

$$\langle A_2(Q\Omega)\rangle = -\,G_{22}^{(1)}(Q\Omega)\,\Pi_{21}\,\langle A_1(Q\Omega)\rangle \quad . \tag{IV.34}$$

Using (IV.22), (III.40) and (IV.26) one finds that the expectation value of A_2 is related to the entropy fluctuations by

$$< A_2(Q\Omega) > = \left(\frac{2VT}{\omega_2 C_V}\right)^{1/2} \delta S(Q\Omega) \ . \tag{IV.35}$$

The importance of the coupling of the two modes can be nicely illustrated by calculating the mean square fluctuations of the entropy. Denoting the autocorrelation function of the entropy by

$$J_S(Q\Omega) = < \delta S(Q\Omega)\ \delta S(-Q\Omega) > \ , \tag{IV.36}$$

the mean square fluctuations are

$$< |\delta S(Q)|^2 > = \frac{1}{2\pi} \int_{-\infty}^{\infty} d\Omega\ J_S(Q\Omega) \tag{IV.37}$$

$$= \frac{1}{2\pi} \frac{\omega_2 C_V}{2VT} \int_{-\infty}^{\infty} d\Omega\ J_{A_2}(Q\Omega) \ ;$$

here, (IV.35) was used and J_{A_2} is the correlation function of A_2. Since in the classical limit $(\hbar\Omega << kT)$

$$J_{A_2}(Q\Omega) = (2kT/\Omega)\ \mathrm{Im}\ G_{22}(Q\Omega) \ , \tag{IV.38}$$

one obtains from the Kramers-Kronig relation

$$< |\delta S(Q)|^2 > = (kC_V\omega_2/2V)\ G_{22}(Q,\Omega=0) \ . \tag{IV.39}$$

If the coupling to first sound in (IV.31) is neglected ($\Pi_{12}=0$), eq. (IV.39) yields

$$< |\delta S(Q)|^2 > = k\ C_V\ /\ V \ . \tag{IV.40}$$

In order to obtain the correct thermodynamic result, one has to include the coupling to first sound. From (IV.31), (IV.26) and (IV.27),

$$G_{22}(Q,\Omega=0) = \frac{2}{\omega_2} \frac{\omega_S^2}{\omega_S^2 - T\ C_V\ \bar{\gamma}^2\ Q^2/\rho} \ . \tag{IV.41}$$

Using the results (IV.14) for the thermodynamic elastic constants, (IV.41) and (IV.39) give, instead of (IV.40),

$$< |\delta S(Q)|^2 > = \frac{k\ C_V\ c_S}{V\ c_T} = \frac{k\ C_P}{V} \ , \tag{IV.42}$$

where C_p is the heat capacity at constant pressure. This result is in accordance with thermodynamics.

IV.4. Dielectric Fluctuations and the Second Viscosity

The general expression (IV.10) for the self-energy contains terms describing the dielectric fluctuations, which we have neglected in the model of the two coupled oscillators for first and second sound. If one considers more general non-equilibrium situations, for which $\Omega\tau_N$ is not vanishing small, the local equilibrium phonon distribution will not be able to adjust instantaneously to the external perturbation. The state of the system is in this case not completely described by the two variables $u(Q\Omega)$ and $\delta S(Q\Omega)$. Instead, additional internal degrees of freedom are necessary to describe the fact that the phonon distribution lags behind the external perturbation. This situation is similar to a sound wave in a gas of molecules with internal degrees of freedom. If the sound frequency is such that the corresponding local density and local temperature change faster than the internal molecular degrees of freedom can adjust to the varying external thermodynamic conditions, there is an additional dissipation, sometime called the second viscosity (Landau and Lifschitz 1971).

This situation has been discussed and experimentally investigated by light scattering from molecular liquids such as CCl_4 (Gornall et al. 1966, Carome et al. 1968). The spectra show a prominent continuous background, centered at zero frequency transfer and extending out to the Brillouin doublets in addition to the central peak arising from entropy fluctuations. It was shown by Zwanzig (1965) that the bulk viscosity of a fluid composed of molecules with internal degrees of freedom, which are weakly coupled to the translational motion, gets an additional frequency dependent contribution. The latter is determined by the characteristic frequency of the correlation function of the fluctuating energy of the internal degrees of freedom. The corresponding light scattering spectrum for such molecular liquids was calculated by Mountain (1966, 1968) and by Desai and Kapral (1972).

Let us consider the generalized elastic constant $c(Q\Omega)$ introduced in eq. (IV.13). It is the response function of the stress $\sigma(Q\Omega)$ with respect to the deformation field $u(Q\Omega)$, so that

$$\delta\sigma(Q\Omega) = c(Q\Omega)\, u(Q\Omega) . \qquad \text{(IV.43)}$$

There will be two contributions to $\delta\sigma$; the first one is a direct part, which is the harmonic contribution, and the second one arises from the fluctuations of the phonon density due to the presence of $u(Q\Omega)$:

$$\delta\sigma(Q\Omega) = \left(\frac{\partial\sigma}{\partial u}\right)_{\delta n_q} u(Q\Omega) + \sum_q \left(\frac{\partial\sigma}{\partial\delta n_q}\right)_u \delta n_q(Q\Omega) \quad . \tag{IV.44}$$

According to (IV.43) the elastic "constant" $c(Q\Omega)$ is therefore

$$c(Q\Omega) = \left(\frac{\partial\sigma}{\partial u}\right)_{\delta n_q} + \sum_q \left(\frac{\partial\sigma}{\partial\delta n_q}\right)_u \frac{\partial\delta n_q(Q\Omega)}{\partial u(Q\Omega)} \quad . \tag{IV.45}$$

In order to calculate the last factor in the second term, one uses (III.38), (III.40) and (III.41), which reads in the present isotropic model

$$\delta n_q(Q\Omega) = i \sum_{q'} \omega_{q'} \gamma_{q'} N_{qq'}(Q\Omega)\, u(Q\Omega) \quad . \tag{IV.46}$$

Similarly, the self-energy becomes from (III.37)

$$\Pi(Q\Omega) = - \frac{Q^2}{2\omega_Q \rho V} \sum_q \gamma_q \omega_q \, i \sum_{q'} \gamma_{q'} \omega_{q'} N_{qq'}(Q\Omega) \quad . \tag{IV.47}$$

Using (IV.47) and (IV.46) in (IV.13) and comparing with (IV.45),

$$\left(\frac{\partial\sigma}{\partial\delta n_q}\right)_u = - \frac{1}{V} \omega_q \gamma_q = \frac{1}{V} \frac{\delta\omega_q(Q\Omega)}{u(Q\Omega)} \quad , \tag{IV.48}$$

where in the last step the Grüneisen approximation (III.42) was used. This result shows that the coupling of the stress to the phonon density arises from the change of the phonon frequencies due to fluctuations in the volume.

To evaluate (IV.45) further, the solution (IV.20) of the Boltzmann equation is used. Taking first that part of the total phonon density fluctuation which gives a non-vanishing contribution to the entropy fluctuations leads in (IV.45) to the contribution

$$\sum_q \left(\frac{\partial\sigma}{\partial\delta n_q}\right)_u \frac{\partial\delta n_q^{(2)}(Q\Omega)}{\partial u(Q\Omega)} = - T \bar{\gamma} \frac{\delta S(Q\Omega)}{u(Q\Omega)} \quad . \tag{IV.49}$$

But from thermoelasticity, $-T\bar{\gamma} = (\partial\sigma/\partial S)_u$, so that finally

$$c(Q\Omega) = c^{(o)} + \left(\frac{\partial\sigma}{\partial S}\right)_u \frac{\delta S(Q\Omega)}{u(Q\Omega)} + \sum_q \left(\frac{\partial\sigma}{\partial\delta n_q}\right)_u \frac{\partial\delta n_q^{(1)}(Q\Omega)}{\partial u(Q\Omega)} \quad . \tag{IV.50}$$

This shows that it is sufficient to treat the part $\delta n_q^{(1)}(Q\Omega)$ as internal degrees of freedom, since $\delta n_q^{(2)}$ can be expressed by the macroscopic entropy degree of freedom.

We now introduce the viscous stress $\delta\sigma^V$ as the difference between the stress induced by an adiabatic sound wave in the general non-equilibrium state and the stress induced by an adiabatic sound wave in local equilibrium:

$$\begin{aligned} \delta\sigma^V(Q\Omega) &= \delta\sigma_S(Q\Omega) - \delta\sigma_S^{LE}(Q\Omega) \\ &\equiv \eta(Q\Omega)\; u(Q\Omega) \quad . \end{aligned} \tag{IV.51}$$

The subscript S indicates that $\delta S=0$. For a plane wave like deformation, $u(Q\Omega) = - i\,\Omega\, u(Q\Omega)$, so that the second viscosity $\eta(Q\Omega)$ is given by

$$- i\,\Omega\,\eta(Q\Omega) = \frac{\partial\delta\sigma^V(Q\Omega)}{\partial u(Q\Omega)} \quad . \tag{IV.52}$$

Putting $\delta S=0$ in (IV.50), the generalized adiabatic elastic constant is

$$c_S(Q\Omega) = c^{(o)} + \sum_q \left(\frac{\partial\sigma}{\partial\delta n_q}\right)_u \frac{\partial\delta n_q^{(1)}(Q\Omega)}{\partial u(Q\Omega)} \quad . \tag{IV.53}$$

It represents the adiabatic response of the stress to $u(Q\Omega)$:

$$\delta\sigma_S(Q\Omega) = c_S(Q\Omega)\; u(Q\Omega) \quad . \tag{IV.54}$$

Using (IV.51) to (IV.53), the viscosity becomes

$$\begin{aligned} -i\,\Omega\,\eta(Q\Omega) &= c_S(Q\Omega) - c_S^{LE}(Q\Omega) \\ &= \sum_q \left(\frac{\partial\sigma}{\partial\delta n_q}\right)_u \frac{\partial}{\partial u}\left(\delta n_q^{(1)}(Q\Omega) - \delta n_q^{(1)LE}(Q\Omega)\right). \end{aligned} \tag{IV.55}$$

Here, the local equilibrium limit $\delta n_q^{(1)LE}$ of $\delta n_q^{(1)}$ is obtained by letting $\Omega\tau_N \to 0$ in $\delta n_q^{(1)}$. Therefore, the complex viscosity is

$$\eta(Q\Omega) = \frac{iT}{V}\sum_q c_q\, \gamma_q(\gamma_q - \bar{\gamma})\, \frac{1}{\Omega - Q\, v_q + i\,\tau^{-1}} \quad . \tag{IV.56}$$

The self-energy (IV.10) can finally be written as

$$\Pi(Q\Omega) = \frac{Q^2}{2\omega_Q\rho} \{- T\, C_V(\overline{\gamma^2}-\bar{\gamma}^2) - T\, \bar{\gamma}\, \frac{\delta S(Q\Omega)}{u(Q\Omega)} - i\,\Omega\,\eta(Q\Omega)\} \quad . \qquad (IV.57)$$

If this is used in the one-phonon Green function of a long wavelength phonon, the first term renormalizes the harmonic frequency to the adiabatic first sound frequency, the second term is the coupling to the entropy fluctuations, which we treated above in detail (in a local equilibrium approximation) and the third term is the coupling to the viscosity.

IV.5. Light Scattering at Low Frequency Transfer

The hydrodynamic properties of the interacting phonon gas, as they have been developed above, should modify a number of properties of a dielectric at sufficiently low frequencies and long wavelength. One example is the spectrum of scattered light (Griffin 1968, 1969, Wehner and Klein 1972), which was of particular interest in order to investigate second sound by other methods than phonon pulse experiments. The light scattering spectrum is given by the susceptibility autocorrelation function

$$J_\chi(\underline{Q}\Omega) = \int_{-\infty}^{\infty} dt\; e^{i\Omega t} \langle \delta\chi(\underline{Q},t)\; \delta\chi(-\underline{Q},0)\rangle \quad , \qquad (IV.58)$$

where $\underline{Q}$ is the transfered wavevector and Ω the frequency shift. Since the expectation value is simply related to the imaginary part of a retarded Green function, the spectrum can also be expressed as

$$J_\chi(Q\Omega) = \frac{2kT}{\Omega}\, \mathrm{Im}\; G_\Omega^r\, (\delta\chi|\delta\chi) \;, \qquad (IV.59)$$

where $\Omega \ll kT$ has been used. The fluctuations of the susceptibility arise from the atomic motions of the lattice and can therefore be represented as an expansion in terms of phonon normal coordinates

$$\delta\chi(\underline{Q}) = \frac{1}{V}\, \{P_1(\underline{Q})\; A(\underline{Q}) + \sum_q P_2(\underline{q}_1,\underline{q}_2)\; A(\underline{q}_1)\; A(\underline{q}_2)\} \qquad (IV.60)$$

$$\underline{q}_{1,2} = \pm\underline{q} + \underline{Q}/2 \quad .$$

The coefficient $P_1(\underline{Q})$ describes ordinary one-phonon scattering giving rise to Brillouin lines. Its magnitude is determined by the elasto-optical coefficient p_{12}

$$P_1(Q) = \left(\frac{V}{2\omega_Q\rho}\right)^{1/2} \frac{Q\varepsilon^2}{4\pi}\, p_{12} \;;\quad p_{12} = -\frac{4\pi}{\varepsilon^2}\frac{\partial\chi_{zz}}{\partial u_{xx}} \quad . \qquad (IV.61)$$

ε is the optical dielectric constant. Here, the scattering geometry is such that incomming and scattered wave vectors are in the xy-plane, the wavevector transfer lies in the x direction and the light wave is polarized in the z direction.

The magnitude of the coefficient P_2 is related to the temperature derivative of χ. Taking the thermal average of $\delta\chi(\underline{Q}=0)$ in (IV.60) one obtains

$$\delta\chi = \langle\delta\chi(\underline{Q}=0)\rangle = \frac{1}{V}\sum_{q} P_2(\underline{q},-\underline{q})\ (2n_q + 1)\ , \tag{IV.62}$$

showing that this part represents a temperature dependent contribution to the susceptibility. The derivative of it with respect to temperature at constant volume is

$$\left(\frac{\partial\chi}{\partial T}\right)_V = \frac{2}{C_V}\sum_q c_q \frac{P_2(\underline{q},-\underline{q})}{\omega_q} \equiv \frac{2}{C_V}\sum_q c_q\, \mu_q \equiv 2\,\bar{\mu}\ . \tag{IV.63}$$

Let us first consider the one-phonon spectrum only. Using (IV.60) in (IV.59), this spectrum is

$$J_\chi^{(1)}(\underline{Q}\Omega) = \frac{2kT}{V^2} P_1^{\,2}(Q)\,\frac{1}{\Omega}\,\mathrm{Im}\, G_\Omega^r \left(A(\underline{Q})\,|\,A(\underline{Q})\right)\ , \tag{IV.64}$$

where G_Ω^r is the one-phonon Green function (IV.15), so that

$$J_\chi^{(1)}(\underline{Q}\Omega) = \frac{2kT\, P_1^{\,2}(Q)}{V^2}\,\frac{1}{\Omega}\,\mathrm{Im}\,\frac{2\omega_Q}{\omega_Q^{\,2}-\Omega^2+2\omega_Q\,\Pi(Q\Omega)}\ . \tag{IV.65}$$

With the self-energy $\Pi(\underline{Q}\Omega)$ as given by (IV.10) or (IV.57) this spectrum will contain more structure than just Lorentzian Brillouin lines. In the coupled oscillator model für vanishing $\Omega\tau_N$, the Green function in (IV.64) is given by (IV.24) and (IV.25), which means that the light couples to the adiabatic first sound, giving rise to Brillouin lines at $\pm\omega_S$. But in addition there is anharmonic coupling to the damped second sound propagator $G_{22}^{(1)}(\underline{Q}\Omega)$ which introduces additional structure at frequencies below ω_S. For $\Omega\tau_U >> 1$, the window condition for propagating second sound is satisfied, and there is a secondary peak at $vQ/\sqrt{3}$, the second sound frequency. For $\Omega\tau_U=1$ this peak vanishes and for smaller $\Omega\tau_U$ second sound is overdamped so that a central peak in the spectrum arises.

This situation is in complete analogy to the scattering from simple liquids. There one treats the susceptibility as a function of density ρ and temperature T, so that the fluctuations are given by

$$\delta\chi(\underline{r},t) = \left(\frac{\partial\chi}{\partial\rho}\right)_T \delta\rho(\underline{r},t) + \left(\frac{\partial\chi}{\partial T}\right)_\rho \delta T(\underline{r},t) \quad . \tag{IV.66}$$

This decomposition is equivalent to (IV.60). Neglecting the temperature dependence of the susceptibility, which is a very good approximation in most liquids, one is left with the first term in (IV.66). Then one can decompose the density fluctuations into adiabatic pressure fluctuations and isobaric entropy fluctuations to get

$$\delta\chi(\underline{r},t) \simeq \left(\frac{\partial\chi}{\partial\rho}\right)_T \left\{ \left(\frac{\partial\rho}{\partial p}\right)_S \delta p(\underline{r}t) + \left(\frac{\partial\rho}{\partial S}\right)_p \delta S(\underline{r},t)\right\} . \tag{IV.67}$$

Note, that the coupling of light is via the density dependence of χ only, but the entropy fluctuations are nevertheless present through their coupling (proportional to thermal expansion α) to first sound. If one uses (IV.67) in (IV.58), several correlation functions appear, which can be calculated from the dynamic equations for the liquid. These are the Navier-Stokes equation, the heat flow equation and the continuity equation. The result is the well-known three-peak spectrum consisting of emission and absorption of propagating adiabatic sound waves and of a central peak of overdamped entropy fluctuations. The relative intensities of scattering from second sound and first sound are given by the Landau-Placzek relation

$$R_{LP} = \frac{I_2}{2I_1} = \frac{C_P}{C_V} - 1 = \frac{c_S}{c_T} - 1 \quad . \tag{IV.68}$$

The solution (IV.10) for the self-energy lifts the restriction of very fast Normal processes and allows for an investigation of the influence of non-equilibrium fluctuations of the phonon density on the scattering spectrum. To this end one has to estimate the difference $\gamma^2-\bar{\gamma}^2$. Since this is related to the difference between the adiabatic elastic constants c_S and the elastic constant c_Z under zero sound conditions, it was assumed that c_Z is 1 % larger than c_S. Fig. 1 shows the spectrum for $\omega_Q\tau_U = 100$ and different values of $\omega_Q\tau_N$. Choosing $\omega_Q\tau_N = 10^{-10}$ results in well-developed second sound and the coupled oscillator model is perfectly sufficient. At $\omega_Q\tau_N = 10^{-1}$ the second sound peak is comewhat rounded off and an increasing background appears for $0 < \Omega < \omega_Q$. For $\omega_Q\tau_N = 1$, the second peak has disappeared, since the window condition is no longer satisfied. The increasing background is due to the dielectric fluctuations and extends from zero to the Brillouin line in the isotropic Debye model used here. Although all the derivations were made in the one-branch Debye model, let us tentatively assume that the thermal phonons are from another branch. Then the group velocity v_q in the second term in (IV.10) belongs to this other branch. If this v_q is smaller than the sound velocity of the long-wavelength phonon to which the light couples, this broad central component will be

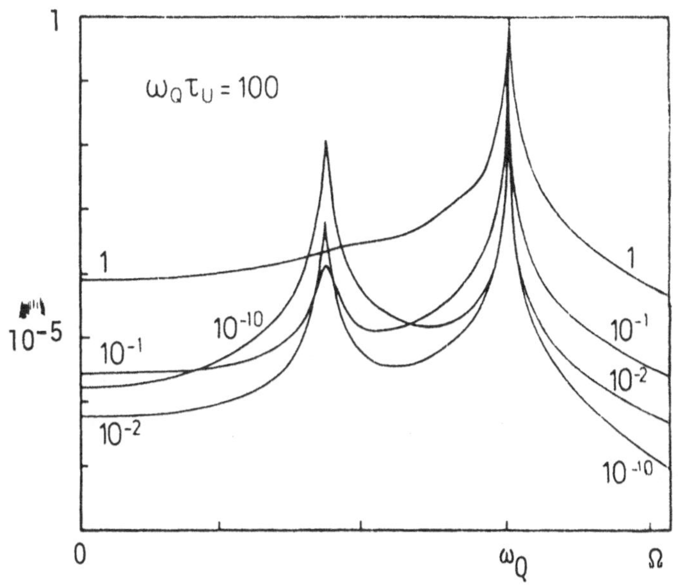

Fig. 1. Light scattering spectrum for $\omega_Q\tau_U = 100$ and $\omega_Q\tau_N = 10^{-10}$; 10^{-2}; 10^{-1}; 1.

narrower and a separation of it from the phonon peaks is found. It must also be mentioned that under this condition the dependence of this central peak on $\underline{Q}$ is different from the one of the entropy peak; in the limit $v_q=0$, the width is independent of $\underline{Q}$, whereas the entropy peak has a width proportional to Q^2.

In Fig. 2 the spectrum is plotted for $\omega_Q\tau_U = 10^{-2}$. Here the window condition is never satisfied and the second sound gives only rise to a central peak. The broad background appears at $\omega_Q\tau_N = 10^{-3}$. For $\tau_N > \tau_U$ there are no additional changes, since the viscosity term depends on $\tau = \tau_N\tau_U/(\tau_N+\tau_U)$. For $\tau_N > \tau_U$, τ is determined by τ_U.

IV.6. Direct Coupling of Light to Phonon Density Fluctuations

Going back to eq. (IV.60) we now consider $P_2 \neq 0$ which describes the coupling of light to a two-phonon state, where the two phonons have nearly the same magnitude of wavevectors but of opposite direction. Using (IV.60) in (IV.59) gives three additional terms besides $J_\chi^{(1)}$, eq. (IV.64). Two of the resulting Green functions depend on three phonon coordinates and one of them depends on four A's. It was shown in eq. (III.28) that the three-point function involves the vertex corrections Γ_3 and therefore phonon transport. The four-point function is somewhat more complicated and we refer to the paper by Wehner and Klein (1972) where it is shown that G_4 can be expressed

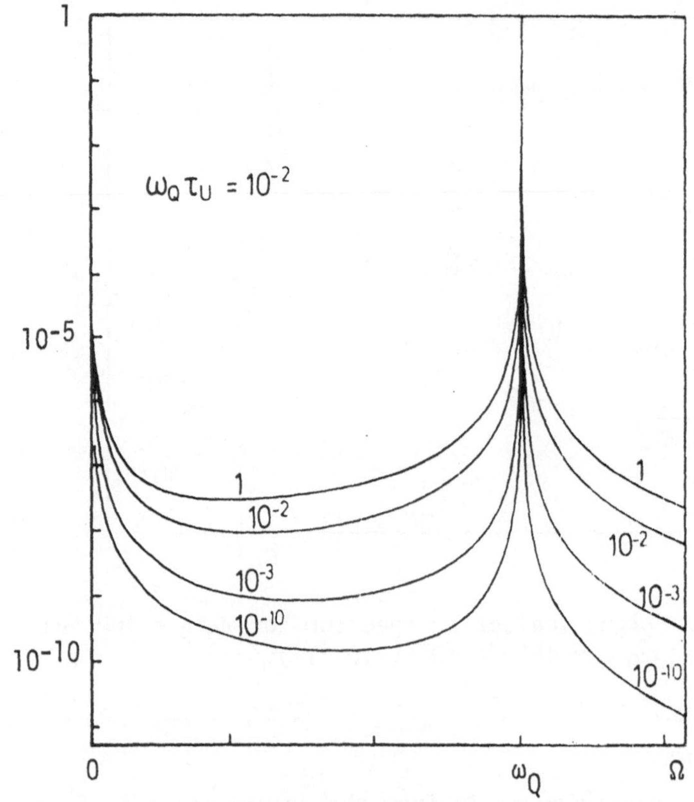

Fig. 2. Light scattering spectrum for $\omega_Q\tau_U = 10^{-2}$ and $\omega_Q\tau_N = 10^{-10}$; 10^{-3}; 10^{-2}; 1.

by Γ_3 and a vertex correction Γ_4. The latter quantity is different from zero even if quartic anharmonicities are neglected. Γ_4 can be related in a similar manner as Γ_3 to the function $N_{qq'}(\underline{Q})$ which is in turn given in terms of the solution of the transport equation. The result of this procedure for the spectrum is

$$J_\chi(\underline{Q}\Omega) = \frac{2kT}{V^2\Omega} \operatorname{Im} \{\tilde{P}_1^2(\underline{Q}\Omega) \frac{2\,\omega_Q}{\omega_Q^2 - \Omega^2 + 2\omega_Q \Pi(\underline{Q}\Omega)} + \tag{IV.68}$$

$$+ 4i \sum_{qq'} P_2(\underline{q}_1,\underline{q}_2)\, N_{qq'}(\underline{Q}\Omega)\, P_2(\underline{q}_3,\underline{q}_4)\} .$$

Here, $\underline{q}_1,\underline{q}_2$ are as in (IV.60) and

$$\underline{q}_{3,4} = \mp \underline{q} - \underline{Q}/2 . \tag{IV.69}$$

Furthermore, $\tilde{P}_1(\underline{Q}\Omega)$ is a (dynamically) renormalized one-phonon coupling coefficient given by

$$\tilde{P}_1(\underline{Q}\Omega) = P_1(\underline{Q}) - 12\, i \sum_{qq'} P_2(\underline{q}_1,\underline{q}_2) N_{qq'}(Q\Omega) V_3(\underline{q}_3,\underline{q}_4,\underline{Q}). \quad (IV.70)$$

Instead of reproducing the derivation of (IV.68) and (IV.70), we will give a simple physical argument, which illustrates the structure of these results. The central peak, which results from one-phonon scattering only (P_2=0), appears because of anharmonic coupling (through Π_{12}) to the collective excitations described by the second and third terms in $\Pi(Q\Omega)$. Denoting for the moment the phonon propagator to which the light couples by G_1 and the propagator for the collective excitations by G_2, we draw the following diagram

(A)

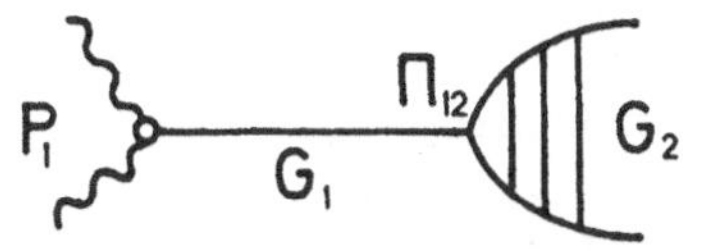

In the somewhat simplified picture of the two coupled oscillators G_1 is given in eq. (IV.24) and G_2 is the second sound propagator $G_{22}^{(1)}$, appearing in the denominator of eq. (IV.25). This mechanism to detect the collective excitations might be called indirect, since the light "sees" G_2 only through the anharmonic coupling of G_2 to G_1 . The spectrum is in this case given by $J_\chi^{(1)}$, eq. (IV.65).

If in addition the second term in (IV.60) is included, one also takes the direct coupling of light to phonon density fluctuations into account, which is illustrated by the following diagram

(B)

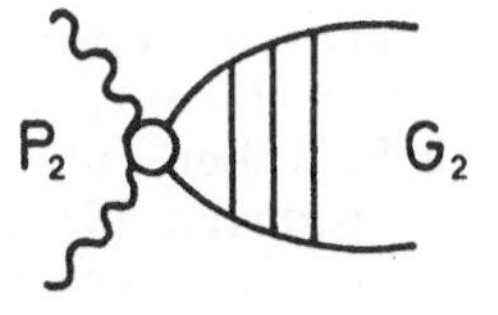

Since the spectrum (IV.58) contains the susceptibility fluctuations squared there are, however, interference effects between direct and indirect coupling processes, which are proportional to P_1P_2. These are represented by

(C)

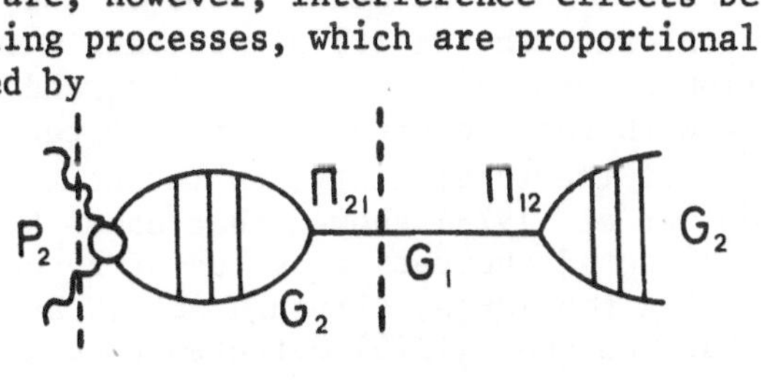

That part of this diagram which is between the two vertical broken lines is essentially $P_2G_2\Pi_{21}$, so that processes (A) and (C) together are of the form of diagram (A), but with a renormalized coupling

$$\tilde{P}_1 = P_1 + P_2G_2\Pi_{21}.$$

This is of the form of (IV.70), where G_2 is represented by the phonon density fluctuations described in $N_{qq'}(Q\Omega)$, and the anharmonic coupling Π_{21} is proportional to V_3.

Since the full spectrum (IV.68) is proportional to $\tilde{P}_1^2$, there is a term $P_1P_2G_2\Pi_{21}$, which describes interference effects between the two coupling mechanisms of light to phonon density fluctuations. The sign of this term can be positive or negative, and therefore constructive and destructive interferences between the two coupling mechanisms are possible.

The analogy of the direct coupling mechanism in the case of the simple liquid consists of expanding $\delta\chi(\underline{r},t)$ in (IV.66) in terms of pressure and entropy fluctuations and keeping both terms. This leads to a modification of the Landau-Placzek ratio (IV.68) by a term depending on the temperature dependence of the susceptibility (Fabelinski 1968)

$$R = R_{LP}\left[1 + \frac{C_P}{\alpha C_V}\;\frac{(\partial\chi/\partial T)_\rho}{(\partial\chi/\partial\rho)_S}\right]^2 \qquad \text{(IV.71)}$$

where α is the thermal expansion coefficient.

Whereas the correction term to R_{LP} is very small for most liquids, Wehner and Klein (1972) have estimated the importance of the direct coupling mechanism from the thermal-expansion coefficient and the elasto-optic and thermo-optic constants of a number of materials. Significant deviations were found for several substances (diamond, Si, GaP, MgO, Ge, $SrTiO_3$ and ZnS), indicating that the direct coupling mechanism should be taken into account. It was also found that for some materials the process (C) leads to destructive interference. A particularly interesting case from the point of view of propagating second sound is NaF. Heat pulse experiments (McNelly et al. 1970, Jackson and Walker 1971) had clearly established the existence of second sound in NaF. The measurement of the thermo-optic coefficient by Pohl and Schwarz (1973) for this material between 300K and 10K allowed for a determination of P_1 and P_2 with the result that $R < R_{LP}$ with total destructive interference at 34K. Subsequent measurements of the total scattered intensity as a function of temperature (Pohl et al. 1973) showed that above 80K good agreement exists with the corrected theoretical value of R, but below 80K the measured signal is too large, although at 34K an indication of a minimum exists. Lateron Pohl (1976) detected second sound by forced Rayleigh scattering between 17K and 25K.

Going back to the general expression (IV.68) for the spectrum of scattered light it should be noted that the collective phonon density fluctuations appear in three different places, in $\tilde{P}_1^2(Q)$ through the second term in (IV.70), in $\Pi(Q\Omega)$ in the denominator of the phonon propagator and in the term which is proportional to P_2^2 in (IV.68). Using the results from the solution of the Boltzmann equation, one obtains

$$\tilde{P}_1(\underline{Q}\Omega) = P_1(Q) + 2TC_VQ\left[\frac{V}{2\omega_Q\rho}\right]^{1/2}\{-(\overline{\mu\gamma}-\bar{\mu}\,\bar{\gamma}) + \frac{\Omega}{VC_V}\sum_q c_q\,\mu_q(\gamma_q-\bar{\gamma})\frac{1}{\Omega-\underline{v}_q\cdot\underline{Q}+i\tau^{-1}} - \bar{\mu}\,\bar{\gamma}\,\frac{\omega_2}{2}\,\tilde{G}_{22}(\underline{Q},\Omega)\} \tag{IV.72}$$

Here, μ_q and $\bar{\mu}$ are defined in (IV.63) and $\overline{\mu\gamma}$ is the average of $\mu_q\gamma_q$ weighted with the specific heat c_q. The second term in (IV.70) becomes

$$4i\sum_{qq'}\mu_q\,\mu_{q'}\,\omega_q\,\omega_{q'}\,N_{qq'}(\underline{Q},\Omega) = -4TC_VV\{-(\overline{\mu^2}-\bar{\mu}^2) + \frac{\Omega}{VC_V}\sum_q c_q\,\mu_q(\mu_q-\bar{\mu})\frac{1}{\Omega-\underline{v}_q\cdot\underline{Q}+i\tau^{-1}} - \bar{\mu}^2\,\frac{\omega_2}{2}\,\tilde{G}_{22}(\underline{Q},\Omega)\}\ . \tag{IV.73}$$

In both expressions terms arise which are similar to those describing the generalized viscosity and the entropy, respectively.

The spectrum has been calculated from (IV.68), (IV.72) and (IV.73) with the assumptions

$$\frac{\overline{\gamma^2}}{\bar{\gamma}^2} - 1 = \frac{\overline{\mu\gamma}}{\bar{\mu}\,\bar{\gamma}} - 1 = \frac{\overline{\mu^2}}{\bar{\mu}^2} - 1 = 10\ . \tag{IV.74}$$

Where the coupled oscillator model ($\Omega\tau_N$=0) was used (Wehner and Klein 1972), it was found that the spectrum is very sensitive to a variation of $\bar{\mu}$. If $\Omega\tau_U$=50 and $\bar{\mu}$=0 (no direct coupling) there is a well-defined second sound peak at $\omega_Q/\sqrt{3}$. A reasonable value for $\bar{\mu}$ is of the order of 10^{-13} in cgs-units, although the sign may be positive or negative. Taking $\bar{\mu} = 1.0\cdot 10^{-13}$ in this example, the second sound peak has not only completely disappeared, but at its position an antiresonance dip is found. Changing $\bar{\mu}$ by ±10 % results in a reduced second sound peak and a shift of the antiresonance to

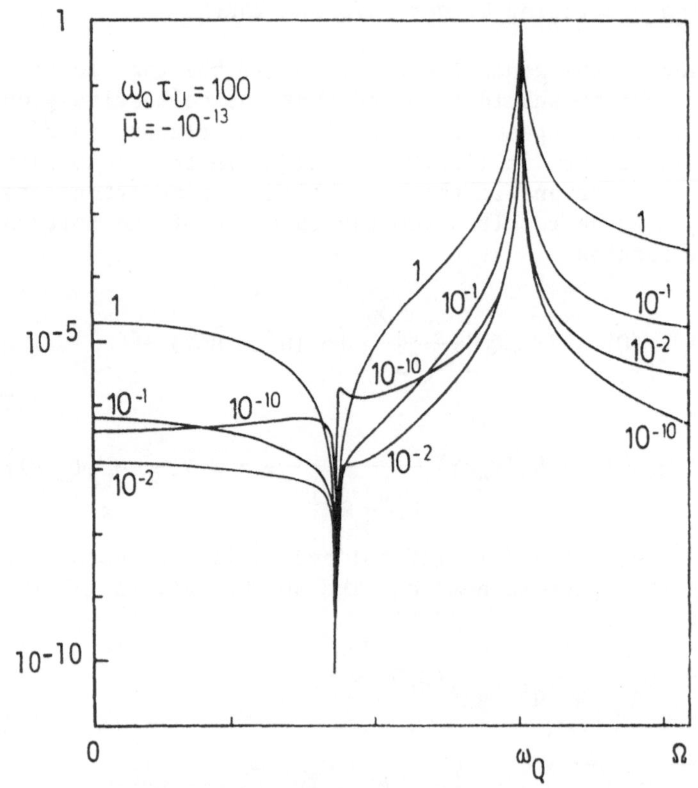

Fig. 3. Light scattering spectrum due to direct and indirect coupling for $\omega_Q\tau_U = 10^2$, $\bar{\mu} = -10^{-13}$ and $\omega_Q\tau_N = 10^{-10}$; 10^{-2}; 10^{-1}; 1.

lower (higher) frequencies.

In Fig. 3 the effects of finite values of $\omega_Q\tau_N$ are shown for a fixed value of $\bar{\mu}$ which corresponds to a disappearing second sound mode due to interference. Increasing $\omega_Q\tau_N$ does not change the dip, but the additional dielectric fluctuations lead again to a broad increase of the spectrum in qualitative agreement with Fig. 1.

The theory developed so far is in many respects oversimplified in order to be applicable to real solids. The assumption of an isotropic one-branch model is rather crude, in particular it seems important to consider the pair of thermal phonons as belonging to all existing branches and to take the dispersion of $\omega_j(q)$ into account. Furthermore, it is necessary to investigate the symmetry properties and the magnitudes of the various anharmonic coupling parameters $V_3(-\underline{Q}j,\underline{q}_1j',\underline{q}_2j')$, which enter the couplings between the long-wavelength mode (Qj) and the thermal phonons $(\underline{q}_1, j')$, $(\underline{q}_2, j')$ for $j=j'$ and $j \neq j'$ (Cowley and Coombs 1973). Nevertheless, it is pointed out that there are experimental results which show, in qualitative agreement with the predictions of the theory, two central peaks (Lyons and Fleury 1976, Sandercock 1978, 1982; Fleury and Lyons 1979), a narrow one and a rather broad one which extends far out, up to the one-phonon Brillouin lines. Among the materials, which were investigated, are $KTaO_3$ and Si. In both cases, the Landau-Placzek ratio for the integrated intensities is very small. Assuming $KTaO_3$ to be similar to $SrTiO_3$, both experiments were performed on materials, for which the direct coupling mechanism is very important (Wehner and Klein 1972). The narrow central peak is due to entropy fluctuations, its width is given by $\Gamma_S = 2 D_t Q^2$, where $D_t = \kappa/\rho C_p$ is the thermal diffusity and κ the thermal conductivity. Using the parameters for $KTaO_3$, Lyons and Fleury (1976) find a calculated value for Γ_S of 2.8 GHz for $90°$ scattering which is in good agreement with their experimental result of 2.3 GHz. Furthermore, they find the Q^2 dependence of Γ_S. The broad central peak is found to have a Q independent width, which has a linear temperature dependence. The Q independence results, if phonons $\underline{q}_{1,2} = \pm\underline{q} + \underline{Q}/2$ are mainly from a flat portion of a dispersion curve. It should be emphasized that there is no evidence of divergent behavior of the central peaks in approaching the structural transitions in $KTaO_3$ and $SrTiO_3$. This together with the observation of similar behavior in Si suggests a possible explanation of the broad central peak in terms of phonon density fluctuations, although further theoretical work and more systematic experimental results are definitely necessary.

REFERENCES

Beck, H., 1971, Phys. Condens. Matter 12, 330 - 369.
Beck, H., 1975, Second Sound and Related Thermal Conduction Phenomena, in: Dynamical Properties of Solids, G.H. Horton and A.A. Maradudin, eds., vol.2, 205 - 284 (Amsterdam, North-Holland).
Beck, H., in: Phonon Scattering in Solids, L.J. Challis, V.W. Rampton, and A.F.G. Wyall, eds. (New York, Plenum), 96 - 98.
Beck, H., and Beck, R., 1973, Phys. Rev. B8, 1669 - 1679.
Beck, H., and Meier, P.F., 1970, Phys. Condens. Matter 12, 16 - 46.
Beck, H., Meier, P.F., and Thellung, A., 1974, phys. stat. sol. (a) 24, 11 - 63.
Bilz, H., Strauch, D., and Wehner, R.K., 1984, in: Handbuch für Physik, Band XXV/2d, L. Genzel, ed. (Berlin, Springer).
Carome, E.F., Nichols, W.H., Kunsitis-Swyt, C.R. and Singal, S.P., 1968, J. Chem. Phys. 49, 1013 - 1017.
Cowley, R.A., 1967, Proc. Phys. Soc.90, 1127 - 1147.
Cowley, R.A., and Coombs, G.J., 1973, J. Phys. C6, 121 - 142.
Desai, R.C., and Kapral, R., 1972, Phys. Rev. A6, 2377 - 2390.
Enz, C., 1968, Ann. Phys. 46, 114 - 173.

Enz, C., 1974, Rev. Mod. Phys. 46, 705 - 753.
Fabelinski, I.L., 1968, Molecular Scattering of Light, (New York, Plenum).
Fleury, P.A., and Lyons, K.B., 1979, Solid State Comm. 32, 103 - 109.
Götze, W., and Michel, K.H., 1967a, Phys. Rev. 156, 963-975.
Götze, W., and Michel, K.H., 1967b, Phys. Rev. 157, 738-743.
Götze, W., and Michel, K.H., 1969, Z. Phys. 223, 199-256.
Gornall, W.S., Stegeman, G.I.A., Stoicheff, B.P., Stolen, R.H., and Volterra, V., 1966, Phys. Rev. Letters 17, 297-299.
Griffin, A., 1968, Rev. Mod. Phys. 40, 167-205.
Griffin, A., 1969, Rev. Mod. Phys. 41, 274-274.
Gurzhi, R.N., 1964, Soviet Phys. -JETP 19, 490-493.
Guyer, R.A., 1966, Phys. Rev. 148, 789-797.
Guyer, R.A., and Krumhansl, J.A., 1964, Phys. Rev. 133, A1411-1417.
Guyer, R.A., and Krumhansl, J.A., 1966a, Phys. Rev. 148, 766-778.
Guyer, R.A., and Krumhansl, J.A., 1966b, Phys. Rev. 148, 778-788.
Hardy, R.J., and Jaswel, S.S., 1971, Phys. Rev. B3, 4385-4387.
Horie, C., and Krumhansl, J.A., 1964, Phys. Rev. 136, A1397-1407.
Jackson, H.E., and Walker, C.T., 1971, Phys. Rev. B3, 1428-1439.
Klein, R., and Wehner, R.K., 1968, Phys. Condens. Matter 8, 141-166.
Klein, R., and Wehner, R.K., 1969, Phys. Condens. Matter 10, 1-20.
Klein, R., and Wehner, R.K., 1972, in: Intern. Conference on Phonon Scattering in Solids, H.J. Albany, ed., 18-23 (Paris, Service de Documentation du CEN Saclay).
Krumhansl, J.A., 1965, Proc. Phys. Soc. 85, 921-930.
Kwok, P.C., 1967, Physics 3, 221-229.
Kwok, P.C., and Martin, P.C., 1966, Phys. Rev. 142, 495-504.
Landau, L.D., and Lifschitz, E.M., 1971, Hydrodynamik, Lehrbuch der theoretischen Physik, Band VI (Berlin, Akademie-Verlag).
Lyons, K., and Fleury, P.A., 1976, Phys. Rev. Letters 37, 161-164.
Maradudin, A.A., and Fein, A.E., 1962, Phys. Rev. 128, 2589-2608.
Maris, H.J., 1967, Phil. Mag. 16, 331-340.
Maris, H.J., 1981, Phys. Rev. B24, 1205-1208.
McNelly, T.F., Rogers, S.J., Channin, D.J., Rollefson, R.J., Goubau, W.M., Schmidt, G.E., Krumhansl, J.A., and Pohl, R.O., 1970, Phys. Rev. Letters 24, 100-102.
Meier, P.F., 1969, Phys. Condens. Matter 8, 241-267.
Mountain, R.D., 1966, J. Res. Nat. Bur. Stand. 70A, 207.
Mountain, R.D., 1968, J. Res. Nat. Bur. Stand. 72A, 75.
Niklasson, G., 1969, Fortschr. Physik 17, 235-275.
Niklasson, G., 1970, Ann. Phys. 59, 263-322.
Niklasson, G., 1972, Phys. Condens. Matter 14, 138-184.
Niklasson, G., and Sjölander, A., 1968, Ann. Phys. 49, 249-295.
Overton, W.C., 1980, in: Phonon Scattering in Condensed Matter, H.J. Maris, ed., (New York, Plenum) 133-135.
Peierls, R.E., 1929, Ann. Physik, 3, 1055-1101.
Peierls, R.E., 1955, Quantum Theory of Solids, (Oxford, Oxford University Press).
Pohl, D.W., 1976, in: Phonon Scattering in Solids, L.J. Challis, V.W. Rampton, and A.F.G. Wyatt, eds. (New York, Plenum).

Pohl, D.W., and Schwarz, S.E., 1973, Phys. Rev. B 7, 2735-2739.
Pohl, D.W., Schwarz, S.E., and Irniger, V., 1973, Phys. Rev. Letters 31, 32-35.
Prohofsky, E.W., and Krumhansl, J.A., 1964, Phys. Rev. 133, A1403-1410.
Ranninger, J., 1968, Ann. Phys. 49, 297-308.
Ranninger, J., 1969, J. Phys. C 2, 929-940.
Ranninger, J., 1972, Phys. Rev. B 5, 3315-3321.
Sandercock, J.R., 1978, Solid State Comm. 26, 547-551.
Sandercock, J.R., 1982, in: Light Scattering in Solids II, M. Cardona and G. Güntherodt, eds., Topics in Applied Physics, vol. 51 (Berlin, Springer).
Sham, L.J., 1967, Phys. Rev. 156, 494-500.
Sussmann, J.A., and Thellung, A., 1963, Proc. Phys. Soc. 81, 1122-1130.
Wehner, R.K., and Klein, R., 1971, Physica 52, 92-108.
Wehner, R.K., and Klein, R., 1972, Physica 62, 161-197.
Werthamer, N.R., and Chui, S.T., 1972a, Phys. Letters 41A, 157-158.
Werthamer, N.R., and Chui, S.T., 1972b, Solid State Comm. 10, 843-846.
Zwanzig, R., 1965, J. Chem. Phys. 43, 714-720.

ELECTRON-PHONON INTERACTION, SCREENING AND PHONON-GENERATION

J.T. Devreese°

Physics Department, University of Antwerpen (U.I.A.), Universiteitsplein 1, B-2610 Wilrijk-Antwerpen, Belgium

CHAPTER I: ELECTRON-PHONON SCATTERING (One-Electron Approximation)

1. INTRODUCTION

The understanding of electron-phonon scattering is crucial to the analysis of phenomena like phonon-generation or phonon optics. In the present section I will discuss the Hamiltonian governing electron-phonon scattering and review some techniques available at present to calculate the influence of electron-phonon scattering on dynamical processes occurring in solids.

2. THE HAMILTONIAN FOR AN ELECTRON INTERACTING WITH PHONONS

2.1. Electron-LO Phonon Interaction

The Fröhlich Hamiltonian for an electron interacting with the phonon field has been derived and re-derived many times. Maybe the most appealing derivation remains the one given by Fröhlich [1]. In ref. [1] Fröhlich also reviews the history of the problem. The introduction of field theoretical concepts to study the electron-phonon interaction goes back to the 1950 paper of Fröhlich et

---

°Also at: University of Antwerpen (R.U.C.A.), Groenenborgerlaan 171, B-2020 Antwerpen, Belgium, and University of Technology, Eindhoven, the Netherlands.

al. [2]. Rather than repeat the derivation by Fröhlich here, I will present a short derivation of the structure of the Fröhlich Hamiltonian due to Feynman. Consider the case of the interaction between the electron and LO phonons ("polaron problem"). As an electron moves through a polar crystal it generates a distortion which acts back on the electron ("the polaron is an electron digging its own grave"). The distortion induced by the electron has a wave number $\vec{k}$. What is the interaction between the electron and the induced distortion? The assumption - typical for the Fröhlich Hamiltonian - is that the crystal can be regarded as a continuous dielectric, in which polarization waves propagate. $\vec{P}$ be the polarization; only longitudinal polarization is considered here because in the continuum approximation, $\vec{\nabla}.\vec{P} = o$ and therefore the interaction with the transverse modes is negligible. The polarization at $\vec{r}$ can be written as:

$$\vec{P} = \frac{\vec{k}}{k} a_k e^{i\vec{k}.\vec{r}} \tag{1}$$

The charge density, from the ions, is then given by Poisson's equation:

$$\rho = \vec{\nabla}.\vec{P} = k a_k e^{i\vec{k}.\vec{r}} \tag{2}$$

The potential V is related to the charge density by

$$\rho = \nabla^2 V \tag{3}$$

It then follows that the interaction between the polarization wave and the electron is proportional to the sum over all $\vec{k}$-vectors of $\frac{q_k}{k} e^{i\vec{k}.\vec{r}}$, where q_k is the amplitude of the longitudinal wave characterized by $\vec{k}$ ($q_k \sim a_k$).

The preceding argument results in the following structure of the standard Fröhlich Hamiltonian:

$$H = \frac{p^2}{2m} + \sum_k \hbar\omega_{LO} a_k^+ a_k + \sum_k (V_k a_k e^{i\vec{k}.\vec{r}} + h.c.) \tag{4}$$

with $V_k \sim \frac{1}{k}$.

It is customary to write V_k as follows:

$$V_k = i \frac{\hbar\omega_{LO}}{k} \sqrt[4]{\frac{\hbar}{2m\omega_{LO}}} \sqrt{\frac{4\pi\alpha}{\Omega}} \tag{5}$$

where m is the electron band-mass, ω_{LO} the long wavelength LO phonon frequency, Ω the crystal volume and

$$\alpha = \frac{e^2}{\hbar} \sqrt{\frac{m}{2\hbar\omega_{LO}}} \left(\frac{1}{\varepsilon_\infty} - \frac{1}{\varepsilon_0}\right) \qquad (6)$$

with ε_∞, ε_0 respectively the electronic and static dielectric constant of the polar solid.

Because the continuum approximation is made, the polaron radius is considerably larger than the lattice parameter and consequently only long wavelength phonons interact with the electrons ($k < \sqrt{\frac{2m}{\hbar}}$). For a derivation of the form of $V_k \sim \frac{1}{k}$ and α, I refer to ref. [1].

A simple intuitive "order-of magnitude" argument can be given to see how the various physical parameters appear in α. First note that to second order in perturbation theory ("weak coupling") the groundstate energy for the Hamiltonian (4) is $E_0 = -\alpha\hbar\omega_{LO}$. This form is pleasing to field theorists who like to see the self energy of the particle interacting with a field as equal to the energy of the quantum of the field times some coupling constant". What expression do we expect for $\frac{E_0}{\hbar\omega_{LO}}$ at weak coupling? For sufficiently weak coupling the polaron radius is given approximately by $\Delta x \sim \sqrt{\frac{\hbar}{m\omega_{LO}}}$ (see footnote (a)). The potential energy of a (classical) charge, extended over a sphere of radius Δx, in a medium of dielectric constant $\bar{\varepsilon}$, is

$$-\frac{e^2}{\bar{\varepsilon}\Delta x} = \frac{-e^2}{\bar{\varepsilon}} \sqrt{\frac{m\omega_{LO}}{\hbar}} = \frac{-e^2}{\bar{\varepsilon}\hbar} \sqrt{\frac{m}{\hbar\omega_{LO}}}\, \hbar\omega_{LO} \qquad (7)$$

From this expression we expect $\alpha \sim \frac{e^2}{\hbar} \sqrt{\frac{m}{\hbar\omega_{LO}}} \frac{1}{\bar{\varepsilon}}$ which is similar to the rigorous definition (eq. (6)). A more refined argument shows that $\frac{1}{\bar{\varepsilon}} = \frac{1}{\varepsilon_\infty} - \frac{1}{\varepsilon_0}$.

(a) Indeed during a time ω_{LO}^{-1} an electron can travel a distance $\Delta x \cong (\Delta v)\, \omega_{LO}^{-1}$, where Δv is the electron's velocity M.S.D. From the uncertainty relations: $m\, \Delta v . \Delta x \sim \hbar$ or $\frac{m}{\omega_{LO}} \Delta v\, \Delta v \sim \hbar$. It then follows that

$$\Delta v \sim \sqrt{\frac{\hbar\omega_{LO}}{m}} \text{ and } \Delta x \sim \sqrt{\frac{\hbar}{m\omega_{LO}}}\,.$$

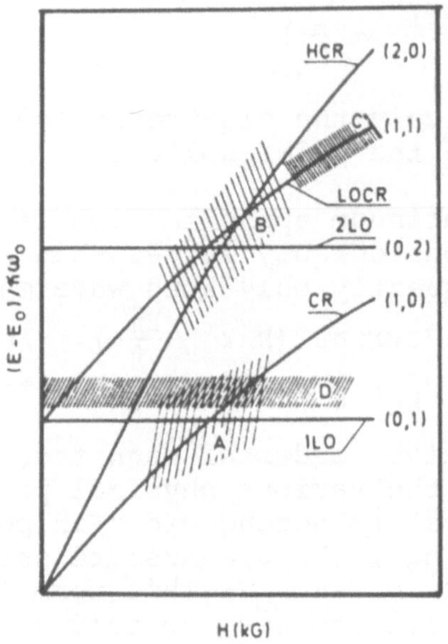

Fig. 1. Schematic plot of the hypothetical magneto-optical spectrum in the absence of polaron interaction.

The Hamiltonian (4) is a cornerstone for the analysis of electron-phonon interaction phenomena. For many polar semiconductors and ionic crystals this Hamiltonian describes the magneto-optical, transport and other properties with surprising accuracy. As an example, I show in fig. 1 the two-phonon magneto-optical spectrum for polarons in InSb; in InSb the coupling constant α is very weak ($\alpha \approx 0.2$) and yet the electron-LO phonon coupling effects are clearly seen and accurately described by the Fröhlich Hamiltonian.

Of course when looking e.g. at the generation rate of phonons by electrons one should include other interactions, in particular the piezoelectric interactions and the deformation potential interactions.

2.2. Deformation Potential

The deformation potential concept goes back to Bardeen [3]. The basic idea is that conduction electrons feel a potential due to the variation in the band gap resulting from the vibrations of the atoms; as a phonon wave travels by, the interatomic distances are modulated locally. This results in a change of the band gap and therefore leads to a potential for the electron. The deformation potential coupling to optical phonons turns out to be weak compared to the deformation potential coupling to acoustical phonons (Ehrenreich, Overhauser, 1958).

The Hamiltonian describing the deformation potential coupling to acoustical phonons is of the same structure as the Fröhlich Hamiltonian (4) with a modified form for the Fourier component V_k.

The interaction energy takes the form

$$H_{e-ph} = D \sum_q \left(\frac{\hbar}{2\rho\nu\omega_q}\right)^{1/2} |\vec{q}| \; (a_q + a^+_{-q}) \tag{8}$$

ρ is the density of the solid (grams per cm^3), $\nu = \frac{MN}{\rho}$; M is the ion mass, N is the total number of ions. D is called the deformation constant. ω_q is the relevant phonon frequency and a_q, a^+_{-q} are the field operators for the phonon field. The expression (8) arises if the $q \to o$ limit is taken in a more general expression for the electron-phonon interaction which is derived e.g. in Mahan's book (ref. [5], p. 36, eq. 1.3.1). Under quite general circumstances the electron interacts only with the long wavelength optical phonons. The expression (8) contains only interaction with longitudinal modes which is sufficient if the bands are nondegenerate.

Mahan has also considered the case of degenerate valence bands; in that case electrons or hole excitations couple to transverse phonons also [5].

In principle deformation constants can be determined experimentally by measuring the pressure dependence of the energy bands. In practice, however, several complications can arise, e.g. the difficulty to distinguish the conduction band and valence band contributions.

The deformation potential scattering is of importance in studying electronic transport phenomena in semiconductors e.g. in the hot electron regime. The dependence of the electron mobility on the deformation potential D is relatively weak; therefore accurate calculations (and measurements) of the Hall mobility are necessary. It also turns out to be difficult to disentangle the different contributions (electron-LO phonon scattering;

deformation potential scattering; impurity scattering) from each other in electrical transport measurements.

2.3. Piezoelectric Scattering

Again the interaction Hamiltonian is of the Fröhlich type, the only difference being in the form of V_k. The original derivation was given by Hutson [6]. Mahan has given a very transparant derivation which we follow here [5].

Piezoelectric crystals are characterized by the absence of an inversion center. The physical origin of the piezoelectric effect is the generation of an electric field when a crystal is subjected to pressure (and the inverse effect). The acoustical phonons induce periodic electric fields as they correspond to local pressure variations.

The electric field, induced by a stress S_{ij} on the crystal, is given by (in Mahan's notation)

$$E_k = \sum_{ij} M_{ijk} S_{ij} \tag{9}$$

The stress S_{ij} can be expressed as follows:

$$S_{ij} = \frac{1}{2} \sum_q \left(\frac{\hbar}{2\rho\nu\omega_q}\right)^{1/2} (\xi_j q_i + \xi_i q_j)(a_q + a^+_{-q}) e^{i\vec{q}.\vec{r}} \tag{10}$$

All symbols in eq. (10) have been explained previously, except ξ_j. The ξ_j are components of the polarization vectors. The index j refers to the $\vec{k}$ vector and the polarization λ of the phonons: $j \equiv (k,\lambda)$ ($\xi_{-k} = -\xi_k$ etc. ...). The index λ runs over all possible normal modes. The electric field E_k can be written as:

$$E_k = -\nabla_k\, \phi(\vec{r}) = -\frac{1}{\nu^{1/2}} \sum_{\vec{q}} i q_k\, \phi_q\, e^{i\vec{q}.\vec{r}} \tag{11}$$

Indeed, it can be shown that the electric field is longitudinal and is in the direction $\vec{q}$ of the phonon and therefore can be written as the gradient of a potential $\phi(\vec{r})$.

It then follows:

$$\phi(\vec{r}) = i \sum_{\vec{q}\lambda} \left(\frac{\hbar}{2\rho V \omega_{\vec{q}\lambda}}\right)^{1/2} M_\lambda(\vec{q}) e^{i\vec{q}\cdot\vec{r}} (a_q + a^+_{-q}) \tag{12}$$

(the potential $\phi(\vec{r})$ is proportional to the displacement) and the piezoelectric interaction leads to an electron-phonon interaction of the form:

$$H_{ep} = i \sum_{\vec{q}\lambda} \left(\frac{\hbar}{2\rho V \omega_{\vec{q}\lambda}}\right)^{1/2} M_\lambda(\vec{q}) \; \rho(\vec{q}) \; (a_q + a^+_{-q}) \tag{13}$$

It should be pointed out that $M_\lambda(\vec{q})$ does not depend on the magnitude of $\vec{q}$, but is very direction-dependent, reflecting the anisotropic character of the piezoelectric interaction. M_λ is different for LA and TA modes. In many applications, nevertheless, $M_\lambda(\vec{q})$ is considered to be a constant.

The II-VI compound semiconductors (CdS, ZnO) are very piezoelectric, III-V compounds are weakly piezoelectric.

An interesting feature of the piezoelectric interaction (and similarly for the acoustic deformation interaction) consists of the fact that simultaneously weak and strong interaction between phonons and electrons is present; this is because phonons with energy much smaller than the average kinetic energy of the electron ("strong coupling") as well as phonons with energies considerably larger than the electron kinetic energy are present ("weak coupling").

2.4. Note on the Mathematical Techniques for Study of the Electron-Phonon Interaction

Although the Fröhlich Hamiltonian looks quite simple in structure it has not been possible to diagonalize it rigorously. Several approximate methods have been devised and are, by now, well known (ref. [1], [7]). These methods include weak coupling perturbation theory, strong coupling adiabative theory, Green's function theory and - above all - the Feynman path integral theory.

Weak coupling theory is easy to treat and provides the relevant physical quantities characterizing the polaron in many semiconductors; e.g. for the effective mass ($m^\star$), the self energy (ΔE), the mobility μ, the optical absorption $\varepsilon_2(\nu)$, the following expressions are obtained: (we limit ourselves to the interaction between the electron and the LO phonons here)

$$\frac{m_e}{m^{\star}} = 1 - \frac{\alpha}{6} + 0.02263\ \alpha^2 + O(\alpha^3) \tag{14}$$

From ref. [8]

$$\frac{\Delta E}{\omega_{LO}} = -\ \alpha - 0.0159\ \alpha^2 - 0.008765\ \alpha^3 + O(\alpha^4) \tag{15}$$

from ref. [8]

$$\mu = \frac{e}{2m\alpha\omega_{LO}}\ e^{\hbar\omega_{LO}/kT} \tag{16}$$

from ref. [2], and from ref. [9], [10]

$$\varepsilon_2(\nu) \propto \alpha\ \frac{\sqrt{\nu-\omega_{LO}}}{\nu^{3/2}} \tag{17}$$

Some difficulties remain, however, even at weak coupling: e.g. no proof exists that the expansions of $\frac{m_e}{m^\star}$ or $\frac{\Delta E}{\hbar\omega_{LO}}$ are convergent (they might be asymptotic). The expression (16) for the mobility is only valid for low temperature. In ref. [11] the Boltzmann equation has been solved for all temperatures in the limit $\alpha\to o$. The results are shown in Table 1.

Table 1: Mobility for various temperatures as obtained from the present calculation (μ_{pr}), from the relaxation time approximation (μ_{REL}), and from the path-integral formulation (μ_{PI}) in units such that $\hbar = m = \omega_{LO} = 1$.

T/θ	μ_{pr}	μ_{REL}	μ_{PI}
0.1	6.28×10^5	5.51×10^5	7.71×10^4
0.2	4.30×10^3	3.69×10^3	9.74×10^2
0.5	151	160	91.8
1.0	46.2	43.0	41.8

As an illustration of the type of calculations involved in weak coupling theory I refer to ref. [12], for a calculation of $\varepsilon_2(\nu)$ in which a simplifying trick is used based on the consideration of the force-force correlation function: the operator expressing the force is proportional to $\sqrt{\alpha}$, the correlation function is therefore proportional to α, and for calculations to order α, it is therefore sufficient to consider the vacuum for the wave functions in the matrix elements; this simplifies the calculations considerably.

A quantity of great interest for the study of phonon generation and phonon optics is the probability for the emission Π^e respectively absorption Π^a of a phonon by an electron. At weak coupling (i.e. to order α) these expressions are:

$$\Pi^a(\vec{p},\vec{p}\,') = N \frac{\alpha}{\pi\sqrt{2}} \frac{1}{|\vec{p}-\vec{p}\,'|^2} \delta\left(\frac{p^2}{2} + 1 - \frac{p'^2}{2}\right) \tag{18}$$

$$\Pi^e(\vec{p},\vec{p}\,') = (N+1) \frac{\alpha}{\pi\sqrt{2}} \frac{1}{|\vec{p}-\vec{p}\,'|^2} \delta\left(\frac{p^2}{2} + 1 - \frac{p'^2}{2}\right) \tag{19}$$

(with $\hbar = m_e = \omega_{LO} = 1$). N is the occupation number for real phonons. Such expressions are the cornerstone for all weak coupling calculations of transport properties of electrons, phonon generation phenomena and phonon optics in general. Note that the expressions for Π^a and Π^e can be equally well obtained for deformation scattering and piezoelectric scattering: it is sufficient to modify α and the argument of the energy conserving δ-function in (18-19).

It is necessary to have at one's disposal expressions for phonon-probabilities, self-energies, effective mass, etc. ... at all coupling. Indeed, although many of the semiconductors of interest have a relatively small α, the interaction with acoustical phonons leads to the simultaneous action of effectively "weak", "strong", and also intermediate coupling (as stated above).

The best model to treat all coupling is the celebrated Feynman-polaron model [13]. The key idea of Feynman is the rigorous elimination of the phonon variables which leads to a one-particle problem when the electron interacts with itself at all previous times. Using a quadratic approximation for this retarded self-interaction Feynman obtains the lowest variational results for all

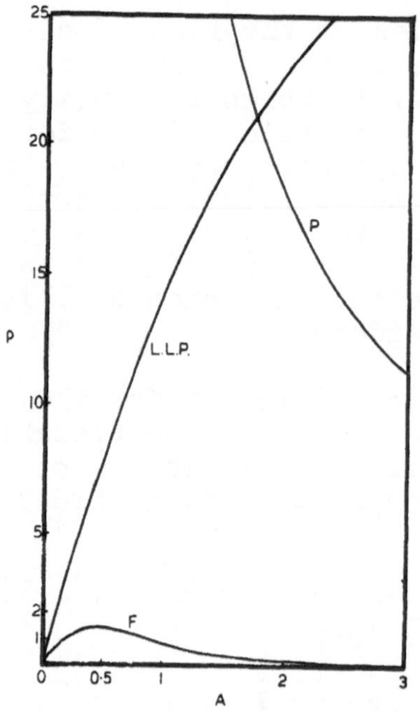

Fig. 2. Relative error of different polaron theories on the self-energy of the symmetrical model (P = Pekar; L.L.P. = Lee, Low, Pines; F = Feynman).

coupling for the polar groundstate energy¶. Fig. 2 illustrates the adequacy of the Feynman model shown by applying it to a model which I solved rigorously, and comparing its accuracy to that of other methods [14].

As is well known the Feynman formulation is based on the expression of the quantum mechanical amplitudes as sums over all trajectories or paths:

$$K(\vec{r}''t'', \vec{r}'t') = \int D\vec{r}(\tau)\ e^{i/\hbar\ S} \tag{20}$$

with S the classical action.
Feynman et al. [15] applied his model also to the study of dynamical response of interacting electron-phonon systems. The present author and coworkers have calculated the optical absorption and magneto-optical absorption of the polaron with the Feynman model [16] . Peeters and Devreese have shown how the dynamical properties of

¶ It must be conceded that as $\alpha \to \infty$ an upper bound for ΔE can be obtained which is about one percent lower than the Feynman bound but this is not very relevant.

polarons (in the case of interaction with LO-phonons) can be obtained with simple operator techniques [17], [18]. From the work of Peeters and Devreese also some of the deficiencies of the study of the dynamics with the Feynman model became apparent [19]. E.g. it was shown that the transport properties calculated in ref. [15] correspond to a choice of a drifted-Maxwellian distribution for the electron, a choice which is not adequate for certain temperatures and values of the coupling constant. A full solution to this interesting problem has not been presented as yet.

The strong coupling polaron theory which (for LO-phonons) leads e.g. to

$$\frac{\Delta E}{\hbar\omega_{LO}} = - 0.108 \ \alpha^2 \tag{21}$$

a variational upper bound (see e.g. ref. [1]) or

$$\frac{m^*}{m_e} = 0.0208 \ \alpha^4 \tag{22}$$

is based on a "Product-Ansatz" for the polaron wave function and is very easy to handle.

The strong coupling theory (combined with weak coupling theory) can be useful for applications if piezo-coupling is considered. However, it is better to use at once the Feynman model, which covers all coupling strengths. Again, the simple operator techniques used in refs. [17] and [18] simplify the calculations considerably.

It is relative simple to apply all the above mentioned techniques to the case where electrons and phonons are treated in one dimension (1D) or two dimensions (2D) rather than in three dimensions (3D) (see ref. [19]).

CHAPTER II

1. INTRODUCTION

For the study of phonon optics and phonon generation the screening of the electron-phonon interaction is of great importance if degenerate semiconductors are treated. This point was stressed and examined e.g. in the work of Lax and Narayamurti [19].

Before discussing the screening of the electron-phonon interaction it is necessary to understand the electron-electron interaction. Therefore this chapter is devoted to a brief discussion of aspects of the many electron problem. Apart from the "classical literature" (Pines, Nozières-Pines, Fetter-Walecka), I use in this chapter also elements from a review paper by J. Devreese and F. Brosens [20].

1.1. Basic Concepts. The Groundstate Energy of Jellium (from ref. [20])

Consider a system of N electrons in an arbitrarily large box of volume Ω, and imagine the total charge neutralized by a uniform positive background. The single parameter characterizing the system is then the averaged electron density n : N/Ω. In practice, one often uses the Wigner-Seitz radius r_s, which is the radius (measured in units of the Bohr radius a_0) of a sphere which has a volume equal to the averaged volume per electron: $\frac{1}{n} = \frac{4\pi}{3} r_s^3 a_0^3$. Some values of r_s for real metals, taking the valence electrons into account, are listed in table II.

If one assumes a uniform electron density, the simplest possible approximation would be to neglect the Coulomb forces. Indeed, if the electrons were not fermions, the Coulomb forces from the electrons and the uniform background would exactly cancel each other. The individual electrons are then described by plane waves, on which periodic boundary conditions are imposed in order to eliminate surface effects.

Table II: Values of r_s and valence s.

Metal	s	r_s	Metal	s	r_s	Metal	s	r_s
Li	1	3.24	Be	2	1.87	Cd	2	2.60
Na	1	3.96	Mg	2	2.66	Ra	2	3.70
K	1	4.96	Cs	2	3.26	Al	3	2.07
Rb	1	5.23	Zn	2	2.30	Ca	3	2.18
Cs	1	5.63	Sr	2	3.54	In	3	2.41

However, the electrons are fermions, and the exclusion principle prevents two electrons to occupy the same state. Therefore, a given wave vector $\vec{k}$ cannot be occupied by more than two electrons of opposite spin. The maximum occupancy of the states with lowest possible kinetic energy defines the Fermi sphere in wave vector space. Its radius k_F is given by

$$k_F^3 = 3\pi^2 n = \frac{9\pi}{4} \frac{1}{r_s^3 a_o^3} \tag{23}$$

This condition follows by imposing that the number of occupied states equals the number of electrons.

However, due to the exclusion principle this free-particle description breaks down because in this approximation the total electron wave function is assumed to be a product of single-particle states, which is not antisymmetric for interchanging two particles. The exclusion principle is automatically built in if one considers a Slater determinant of single-particle wave functions as a trial wave function. On applying the variational principle of quantum mechanics, one then obtains the Hartree-Fock equation for the single-particle wave functions. In a uniform compensating background, plane waves still are a solution of the Hartree-Fock equations. For the total energy however one now obtains an exchange contribution, apart from the kinetic terms. The single-particle energy for an electron with wave vector $\vec{k}$ then becomes:

$$E_k = \frac{\hbar^2 k^2}{2m} - \frac{e^2 k_F}{\pi} \left(1 + \frac{1-\eta^2}{2\eta} \ln \left|\frac{1+\eta}{1-\eta}\right|\right) \tag{24a}$$

with

$$\eta = k/k_F \tag{24b}$$

and the total Hartree-Fock energy of the system equals:

$$E_{HF} = N \left\{\frac{3}{10} \frac{\hbar^2 k_F^2}{m} - \frac{3}{4\pi} e^2 k_F\right\}$$

$$= N \left\{\frac{2.2099}{r_s^2} - \frac{0.9163}{r_s}\right\} \frac{e^2}{2a_o} \tag{25}$$

Instead of considering a Slater determinant of plane waves, where each wave vector is occupied by two electrons of opposite spins, one can examine whether a different spin occupation produces a lower Hartree-Fock energy. This question was already raised in 1929 by Bloch [21]. He found that depending on the density either the above mentioned paramagnetic state, or the ferromagnetic state (i.e. all electrons with parallel spin) has the lowest Hartree-Fock energy. For the ferromagnetic case, one obtains:

$$(E_{HF})_{ferro} = N \left\{\frac{3.508}{r_s^2} - \frac{1.155}{r_s}\right\} \frac{e^2}{2a_o} \tag{26}$$

which is smaller than the paramagnetic Hartree-Fock energy for $r_s \gtrsim 5.47$. As was shown in Table II, most simple metals have a smaller r_s.

Already some 50 years ago [22], attempts have been made to estimate the correlation energy, i.e. the difference between the exact groundstate energy and the Hartree-Fock energy. A well known result was obtained by Gell' Mann and Brückner [23]:

$$E = \left\{\frac{2.2099}{r_s^2} - \frac{0.9163}{r_s} - 0.094 + 0.0622 \ln(r_s) + \ldots\right\} . \frac{e^2}{2a_o} \tag{27}$$

which was derived by a diagrammatic expansion. A discussion of this correction and subsequent alternative approaches, lies beyond the scope of the present paper. A concise review can e.g. be found in [24].

Apart from these studies of the homogeneous electron distribution, one also has examined the possibility of a non-uniform electron gas (in a uniform background). At extremely low density, Wigner [25] has shown that a Hartree-Fock state with electrons localized near lattice positions, has lower energy than the paramagnetic Hartree-Fock state. The result which he obtained served as a basis for interpolating between the low- and high-density regime.

The existence of the Wigner lattice in three-dimensional solids remained a rather academic problem. However, recently experimental evidence was obtained, showing the occurrence of Wigner crystallization in a two-dimensional electron gas [26]. In contrast to a three-dimensional electron gas, two-dimensional electron systems form a Wigner lattice at high areal densities [27].

A second important deviation from the paramagnetic Hartree-Fock groundstate has been found by Overhauser [28,29], who finds that spin- and/or charge-density waves lower the total energy as compared to the homogeneous electron distribution.

Thus in a simple model like jellium, with well-defined interactions, even the problem of the groundstate energy is far from solved. The dynamical properties of this system are still more complex, and no exact solution has been found. We will try to show some of the more recent developments under the assumption of a homogeneous paramagnetic groundstate. But first of all we will briefly sketch what can be learned about the response properties of the jellium from a simple perturbative treatment.

1.2. The Random Phase Approximation (Screening and Single-Particle Excitations) (from ref. [20])

Although a detailed calculation of the dielectric response properties of the electron gas requires a many-body treatment, the single-particle picture is physically very transparent in discussing some of the important concepts in linear-response theory. This treatment has also the advantage to be closely connected to the standard pseudopotential theory of simple metals (see e.g. [30], [31]).

Starting from a free-electron gas, with the electrons occupying free-electron states $|\vec{k}\rangle$ with energy $E_{\vec{k}} = \hbar^2k^2/2m$, the interaction between the electrons, with the background charges, and with possible external perturbations, is described by some time-dependent operator W, which in general is non-local. This means that the interaction potential depends on the electron state under consideration. For simplicity, we will consider a local interaction, and treat the non-locality in the many-body treatment of part II. Also the spin dependence is not taken into account in the present simple illustrative treatment.

The potential W, seen by the electrons, can be expanded in a Fourier series:

$$W(\vec{r},t) = \sum_{\vec{q},\omega} W_{\vec{q}\omega}\, e^{i(\vec{q}.\vec{r}-\omega^+ t)} \tag{28}$$

where $\omega^+ = \omega + i\delta$ accounts for switching on the interaction adiabatically at $t = -\infty$. Considering then the time evolution of the electron wave function:

$$|\psi_{\vec{k}}(t)\rangle = |\vec{k}\rangle e^{-iE_{\vec{k}}t/\hbar} + \sum_{\vec{q}} a_{\vec{q}}(\vec{k},t)\,|\vec{k}+\vec{q}\rangle e^{-iE_{\vec{k}+\vec{q}}t/\hbar} \tag{29}$$

the expansion coefficients $a_{\vec{q}}(\vec{k},t)$ can easily be determined to first order in the perturbation W, from the Schrödinger equation:

$$i\hbar \frac{\partial}{\partial t} |\psi_{\vec{k}}(t)\rangle = (T+W)\,|\psi_{\vec{k}}(t)\rangle \tag{30}$$

where T is the kinetic energy operator, whose eigenstates are the initial plane waves $|\vec{k}\rangle$:

$$T|\vec{k}\rangle = E_{\vec{k}}|\vec{k}\rangle \;; \qquad E_{\vec{k}} = \frac{\hbar^2 k^2}{2m} \tag{31}$$

One readily obtains:

$$\sum_{\vec{q}} [i\hbar \frac{\partial}{\partial t} a_{\vec{q}}(\vec{k},t)]\,|\vec{k}+\vec{q}\rangle e^{-iE_{\vec{k}+\vec{q}}t/\hbar} = W|k\rangle e^{-iE_{\vec{k}}t/\hbar} + \sum_{q} a_{\vec{q}}(\vec{k},t) W|\vec{k}+\vec{q}\rangle e^{-iE_{\vec{k}+\vec{q}}t/\hbar} \tag{32}$$

Multiplying from the left with $|\vec{k}+\vec{q}\,'|$, and omitting the last term in (32) which is of second order, one finds

$$i\hbar [\frac{\partial}{\partial t} a_{\vec{q}}(\vec{k},t)] e^{-iE_{\vec{k}+\vec{q}}t/\hbar} = \langle\vec{k}+\vec{q}|W|\vec{k}\rangle e^{-iE_{\vec{k}}t/\hbar} \tag{33}$$

Given the boundary condition $a_{\vec{q}}(\vec{k},t = -\infty) = 0$, one obtains

$$i\hbar\, a_{\vec{q}}(\vec{k},t) = \int_{-\infty}^{t} dt\, \langle\vec{k}+\vec{q}|W|\vec{k}\rangle e^{-i(E_{\vec{k}} - E_{\vec{k}+\vec{q}})t/\hbar} \tag{34}$$

With the local approximation (28), the matrix element $\langle\vec{k}+\vec{q}|W|\vec{k}\rangle$ becomes

$$\langle\vec{k}+\vec{q}|W|\vec{k}\rangle = \sum_{\omega} W_{\vec{q}\omega}\, e^{-i\omega^{+}t} \tag{35}$$

and the expansion coefficient $a_{\vec{q}}(\vec{k},t)$ of the wave function to first order in $W_{\vec{q}\omega}$ is given by

$$a_{\vec{q}}(\vec{k},t) = \sum_{\omega} W_{\vec{q}\omega} \frac{e^{-it(\hbar\omega^{+}+E_{\vec{k}}-E_{\vec{k}+\vec{q}})/\hbar}}{\omega^{+}+E_{\vec{k}}-E_{\vec{k}+\vec{q}}} \tag{36}$$

The electron density $n(\vec{r},t)$ to first order is then readily obtained:

$$n(\vec{r},t) = \sum_{|\vec{k}|<k_F} |\psi_{\vec{k}}(\vec{r},t)|^2$$

$$= n + \frac{1}{\Omega}\sum_{\vec{q}}{}'\,(e^{i\vec{q}.\vec{r}} \sum_{|\vec{k}|<k_F} a_{\vec{q}}(\vec{k},t)\,e^{-i(E_{\vec{k}+\vec{q}}-E_{\vec{k}})t/\hbar} + c.c.) \tag{37}$$

where n is the averaged density N/Ω. Inserting (36), and performing the Fourier expansion of the density:

$$n(\vec{r},t) = n + \sum_{\vec{q}\omega}{}'\, n_{\vec{q}\omega} e^{i(\vec{q}.\vec{r}-\omega^{+}t)} \tag{38}$$

the density fluctuations $n_{\vec{q}\omega}$ are given by

$$n_{\vec{q}\omega} = \frac{1}{\Omega} \sum_{|\vec{k}|<k_F} \left[\frac{W_{\vec{q}\omega}}{\hbar\omega^{+}+E_{\vec{k}}-E_{\vec{k}+\vec{q}}} + \frac{W^{\star}_{-\vec{q},-\omega}}{-\hbar\omega^{+}+E_{\vec{k}}-E_{\vec{k}+\vec{q}}} \right] \tag{39}$$

Because the interaction potential has to be hermitian, it follows that $W^{\star}_{-\vec{q},-\omega} = W_{\vec{q}\omega}$. If furthermore in the last term $\vec{k}$ is replaced by $-\vec{k}$, (39) becomes:

$$\frac{4\pi e^2}{q^2} n_{\vec{q}\omega} = -Q_0(q,\omega)\, W_{\vec{q}\omega} \tag{40}$$

where

$$Q_0(q,\omega) = -\frac{4\pi e^2}{q^2} \sum_{|\vec{k}|<k_F} \left\{ \frac{1}{\hbar\omega^{+}+E_{\vec{k}}-E_{\vec{k}+\vec{q}}} - \frac{1}{\hbar\omega^{+}-E_{\vec{k}}+E_{\vec{k}+\vec{q}}} \right\} \tag{41}$$

The response function $Q_o(q,\omega)$ thus relates the induced Coulomb potential to the averaged potential W felt by the electrons. Replacing the summation by an integral ($\sum_{|\vec{k}|} \rightarrow \frac{2\Omega}{(2\pi)^3}\int d^3k$ where a factor of 2 accounts for the spins) the integral can be done analytically, giving

$$\mathrm{Re}\, Q_o(q,\omega) = \frac{1}{2}\frac{k_{TF}^2}{q^2}\; 1+\frac{m^2}{2k_F q^3\hbar^2}[4E_F E_q-(E_q+\hbar\omega)^2]\ln\left|\frac{E_q+\hbar q v_F+\hbar\omega}{E_q-\hbar q v_F+\hbar\omega}\right|$$

$$+\frac{m^2}{2k_F q^3\hbar^2}[4E_F E_q-(E_q-\hbar\omega)^2]\ln\left|\frac{E_q+\hbar q v_F-\hbar\omega}{E_q-\hbar q v_F-\hbar\omega}\right| \tag{42}$$

$$\mathrm{Im}\, Q_o(q,\omega) = \frac{\pi}{2}\frac{\omega}{qv_F}\frac{k_{FT}^2}{q^2} \qquad \text{if } |E_q-\hbar q v_F|>\hbar\omega>0$$

$$= \frac{\pi}{4}\frac{k_F}{q}\left[1-\left(\frac{\hbar\omega-E_q}{qv_F}\right)^2\right]\frac{k_{FT}^2}{q^2} \qquad \text{if } E_q+ qv_F>\hbar\omega>|E_q-\hbar q v_F|$$

$$= 0 \qquad \text{elsewhere}$$

and

$$\mathrm{Im}\, Q_o(q,-|\omega|) = -\mathrm{Im}\, Q_o(q,|\omega|) \tag{43}$$

In these expressions, k_{FT} is the Thomas-Fermi wave vector, defined by

$$k_{FT}^2 = \frac{4mk_F e^2}{\pi\hbar^2} \tag{44}$$

The imaginary part of this response function $Q_o(q,\omega)$ results from the resonances which are apparent in (41), and which are due to the excitation of electron-hole pairs if the exciting energy $\hbar\omega$ equals the difference in energy between states inside and outside the Fermi sphere. Two possibilities for resonance are possible. The first term in (41) gives rise to the absorption process:

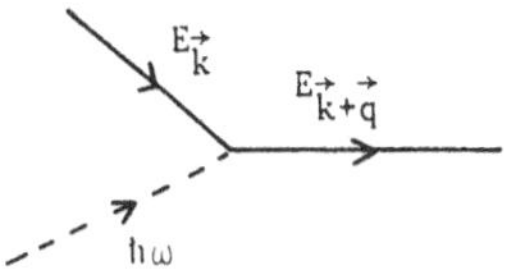

(a)

which implies the condition $\hbar\omega = E_{\vec{k}+\vec{q}} - E_{\vec{k}}$. For $|k| < k_F$ this condition implies

$$-qk_F + \frac{q^2}{2} \leqslant \frac{m\omega}{\hbar} \leqslant qk_F + \frac{q^2}{2}$$

The second term involves the emission process

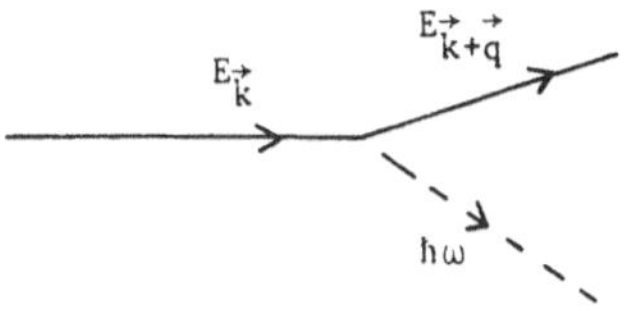

(e)

with the condition $E_{\vec{k}} = \hbar\omega + E_{\vec{k}+\vec{q}}$, or:

$$-qk_F - \frac{q^2}{2} \leqslant \frac{m\omega}{\hbar} \leqslant qk_F - \frac{q^2}{2}$$

These resonances determine the continuum of single particle excitations in the (ω,q) plane.

The function $Q_0(q,\omega)$ forms the basic ingredient of the Lindhard dielectric function [32], which is commonly called the RPA dielectric function, from a famous attempt to describe the electron gas in terms of collecting coordinates [33,34].

This dielectric function is given by

$$\varepsilon^{RPA}(q,\omega) = 1 + Q_0(q,\omega) \tag{45}$$

and is readily derived if one makes the Hartree approximation for the effective potential, i.e. if one assumes that the Coulomb interactions between the electrons governs their dynamical behaviour. In this approximation, the potential W felt by the electrons consists of the induced Coulomb interaction $\frac{4\pi e^2}{q^2} n_{\vec{q}\omega}$, and the applied external potential $W^{ext}_{\vec{q}\omega}$:

$$W_{\vec{q}\omega} \overset{\cong}{\text{Hartree}} \frac{4\pi e^2}{q^2} n_{\vec{q}\omega} + W^{ext}_{\vec{q}\omega} \tag{46}$$

Inserting this approximation (46) into the response equation (40), and using the definition of the dielectric function

$$V^{tot}_{\vec{q}\omega} = W^{ext}_{\vec{q}\omega} + \frac{4\pi e^2}{q^2} n_{\vec{q}\omega} = \frac{W^{ext}_{\vec{q}\omega}}{\varepsilon(q,\omega)} \tag{47}$$

one obtains the RPA approximation (45) for the dielectric function. Incidentally, the RPA approximation (46) assumes that the potential W felt by the electrons equals the potential $V^{tot}_{\vec{q}\omega}$ which a test probe outside the medium would see.

However, because the electrons are fermions, also an exchange and correlation interaction has to be introduced to account for the exclusion principle and correlation effects. If we formally represent the exchange and correlation potential to first order in the induced electron density by:

$$V^{xc}_{\vec{q}} = -\frac{4\pi e^2}{q^2} G(q,\omega)\, n_{\vec{q}\omega} \tag{48a}$$

the potential acting upon the electrons becomes:

$$W_{\vec{q}\omega} = \frac{4\pi e^2}{q^2} (1-G(q,\omega))\, n_{\vec{q}\omega} + W^{ext}_{\vec{q}\omega} \tag{48b}$$

Solving then the response equation (40) combined with (48) for the density, one obtains:

$$\frac{4\pi e^2}{q^2} n_{\vec{q}\omega} = -\frac{Q_o(q,\omega)}{1+Q_o(q,\omega)\,[1-G(q,\omega)]} W^{ext}_{\vec{q}\omega} \tag{49}$$

whereas the potential W becomes:

$$W_{\vec{q}\omega} = \frac{1}{1+Q_o(q,\omega)\,[1-G(q,\omega)]} W^{ext}_{\vec{q}\omega} \tag{50}$$

Inserting (49) into the defining equation (47) for the dielectric function, an expression for the dielectric function is obtained which is commonly used in the electron gas theory:

$$\varepsilon(q,\omega) = 1 + \frac{Q_o(q,\omega)}{1-G(q,\omega)Q_o(q,\omega)} \tag{51}$$

The function $G(q,\omega)$ in (51) describes the exchange and correlation effects in the dielectric function. Its calculation is one of the fundamental problems in many-body theory.

It should be noted that the potential $V^{tot}_{\vec{q}\omega}$, which defines the dielectric function, is different from the potential acting upon the electrons. The dielectric function $\varepsilon(q,\omega)$ in fact is a test-charge - test-charge dielectric function. In analogy with (47), one could consider (50) as the defining equation for the electron - test-charge dielectric function

$$W_{\vec{q}\omega} = \frac{W^{ext}_{\vec{q}\omega}}{\varepsilon^{et}(q,\omega)} \tag{52}$$

which describes how the potential from an external test-charge upon an electron is screened by the surrounding electrons, and which is given by

$$\varepsilon^{et}(q,\omega) = 1 + Q_o(q,\omega)\,[1-G(q,\omega)] \tag{53}$$

The distinction between both dielectric functions has been emphasized in the past [35-38]. Recently, a further development has been presented [39] to account for electron-electron screening (taking also the spin dependence into account).

In the RPA, the function $G(q,\omega)$, describing the exchange and correlation hole, is supposed to be zero, so that no distinction is made between the electron - test-charge and the test-charge - test-charge dielectric functions (53) and (51). Despite the neglect of exchange and correlation, the RPA dielectric function has been widely used because of its simple conceptual basis, and also because of some remarkable results, especially in the long wave length-limit, where the long-range Coulomb forces are the dominant interactions.

1.3. Relation to the Density Functional Formalism (see ref. [20])

For a quasi uniform electron gas, the single particle treatment with the local exchange and correlation potential (eq. (48)) is consistent with the Hohenberg-Kohn formulation for the quasi-homogeneous many-electron system. For the details the reader is referred to p. 159-162 of ref. [20].

1.4. The Dielectric Function and Related Physical Quantities

For reference I include some well known properties of the electron gas expressed in terms of the dielectric function. First define the pair correlation function at $\vec{r}$:

$$g(\vec{r}) = \frac{1}{nN} \sum_{i \neq j}^{N} \delta(\vec{r}_i - \vec{r}_j - \vec{r}) \tag{54}$$

where $\vec{r}_i, \vec{r}_j$ denote electron positions.

The pair correlation function is related to the static structure factor $S(q)$:

$$g(r) - 1 = \frac{1}{n} \frac{1}{(2\pi)^3} \int d^3q \, [S(q) - 1] \, e^{i\vec{q}.\vec{r}} \tag{55}$$

The static structure factor $S(q)$ is related to the energy $\text{Im} \frac{1}{\varepsilon(q,\omega)}$ via the fluctuation-dissipation theorem:

$$S(q) = \frac{q^2}{4\pi e^2} \frac{\hbar}{nN} \int_0^{\infty} d\omega \, \text{Im} \frac{1}{\varepsilon(q,\omega)} \tag{56}$$

In fig. 3 a measurement of $S(q)$ for Al is shown [40].

2. TIME-DEPENDENT HARTREE-FOCK FOR THE ELECTRON GAS

In this section a brief review is given on a variational treatment of the time-dependent Hartree-Fock theory for the electron gas. This work was started by F.

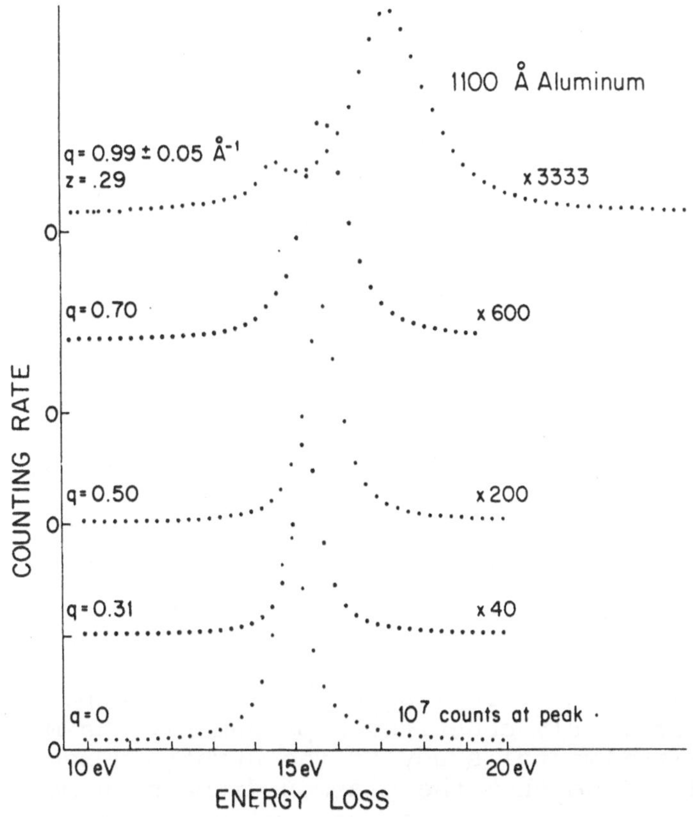

Fig. 3. Measurement of energy loss of e^- in Al.

Brosens, L. Lemmens, J. Devreese (ref. [41]) and developed further by Brosens and Devreese [42] also in collaboration with H. Nachtegaele [43]. The variational treatment of the electron gas, introduced in [41] constitutes a considerable improvement over the R.P.A.

This work is based on a decoupling of the equations of motion (Liouville equations) for the Wigner distribution function $f(p,q,\omega)$ of the electron gas. This decoupling is consistent with the Pauli principle. It results in an integro-differential equation for $f(p,q,\omega)$ which we have solved with a variational approximation. This method leads to the following expression for $\varepsilon(q,\omega)$:

$$\varepsilon^{TDHF}(q,\omega) = 1 + \frac{Q_o(q,\omega)}{1-G(q,\omega)Q_o(q,\omega)} \quad (57)$$

$G(q,\omega)$ is a highly intricate integral which we have been the first to calculate after a two-year effort, using two different methods and finalized in 1977 [44] . The same integral was rederived with still another method in 1979 in ref. [45]. Although the calculation and derivation of $G(q,\omega)$ in the TDHF variational approximation is tedious, the result is quite simple and was tabulated in extenso in ref. [41,42] . With eq. (57) it is then easy to calculate the time-dependent Hartree-Fock expression for $\varepsilon(q,\omega)$.

2.1. Discussion of the Time-Dependent Hartree-Fock Dielectric Function

We now have several pieces of evidence indicating that $\varepsilon^{TDHF}(q,\omega)$ is a considerable improvement over the R.P.A. This evidence is briefly summarized here.

i) the plasmon dispersion in Al. The experimental work of Schnatterly constitutes presumably the most profound experimental study of the structure factor in Al. In fig. 4 we show the plasmon dispersion measured by Schnatterly et al. and compare with our theoretical result determined from the maxima in $- \mathrm{Im}\, \frac{1}{\varepsilon^{TDHF}(q,\omega)}$. It is apparent that the TDHF considerably improves the agreement with experiment, in particular at short wavelength, i.e. in the Landau damping regime, where the R.P.A. fails.

ii) scaling of $G^{TDHF}(\frac{q}{k_F}, \frac{\hbar\omega}{E_F})$. We have given a proof that $G^{TDHF}(\frac{q}{k_F}, \frac{\hbar\omega}{E_F})$ scales with r_s (see ref. [41]). This scaling prediction is compared with experiment in fig. 5 from ref. [42]. This figure suggests that the scaling property predicted from variational TDHF is satisfied up to $r_s \sim 4$.

iii) energy loss of ions in plasmas. As can be seen from fig. 6 Sayanov [46] has shown that our expression for G^{TDHF}, reflected in ε^{TDHF}, leads to a substantially better prediction for the energy loss of ions in plasmas than the R.P.A.

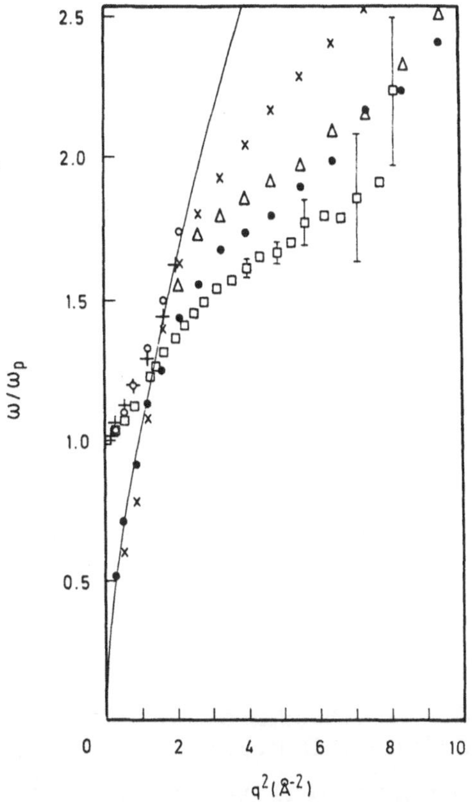

Fig. 4. Plasmon dispersion in Al versus q^2.
Squares: experiment.
Triangles: theory of ref. 45 (Holas et al.).
Crosses: R.P.A.
Circles: T.D.H.F.

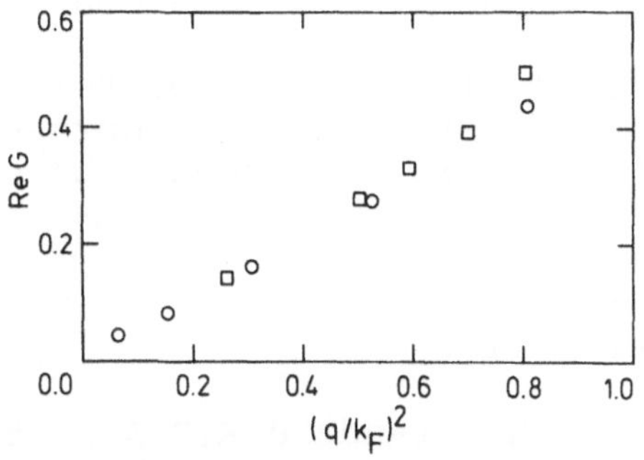

Fig. 5. Fitted values of the real part of G(q,ω) for different wave vectors.
Circles: Na (r_s = 3.96).
Squares: Al (r_s = 2.07).

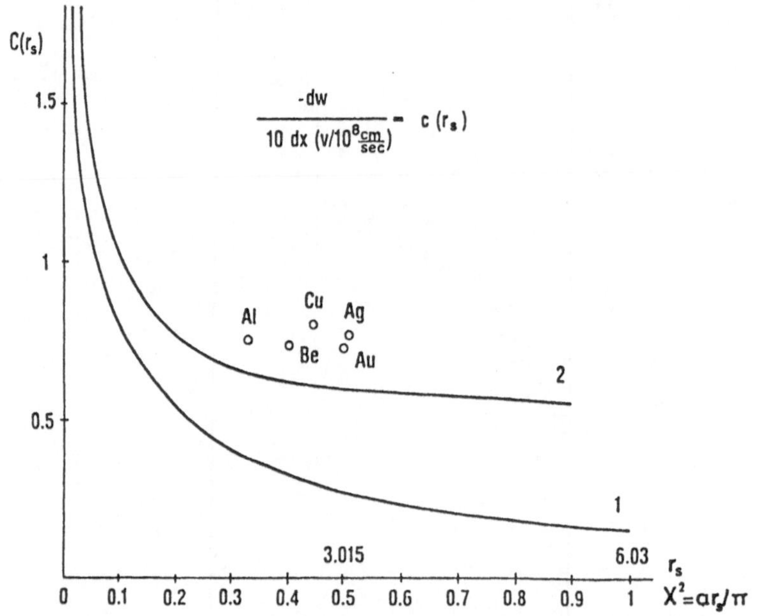

Fig. 6. Energy loss of ions in plasmas: dW/dx (in units eV/Å).
Curve 1: G = 0 (Lindhard)
Curve 2: G from T.D.H.F.
Circles: experimental results.

3. COMPARISON TO OTHER DECOUPLING SCHEMES

I refer to ref. [47] and [48] for this topic. It may be pointed out however, that the representation of our results in ref. [48] is different from our view of these results.

4. SUMMARY OF CHAPTER II

In this chapter the standard R.P.A. for the electron gas has been described. Subsequently the variational solution to the TDHF equations [41] was briefly reviewed. The key feature is in the behaviour of $G(q,\omega)$. This function $G(q,\omega)$ can also be tested in phonon-optics experiments as described in Chapter III.

CHAPTER III: ELECTRON-PHONON INTERACTION, SCREENING AND PHONON GENERATION

1. INTRODUCTION

The present chapter gives a brief account of the description of phonon generation. Electron-phonon interaction and screening, as discussed in chapters I and II, are necessary ingredients for this chapter. My discussion provides an introduction to some of the work of M. Lax and V. Narayanamurti [19] which it reviews. Of particular importance is the fact that the influence of local field corrections in the many-body problem, as described by the function G(q) of chapter II, in principle can be observed in the case of degenerate semiconductors. The case of acoustic phonon generation in n-GaAs epilayers is used by Lax and Narayanamurti as an example. From the point of view of the many-electron theory and e.g. the work of ref. [41,42,43] my main interest is in possible experimental information on G(q).

2. AN EXAMPLE OF PHONON GENERATION (see ref. [19])

M. Lax, V. Narayanamurti and co-workers studied the generation of phonons by the application of current pulses to thin n-layers, p-layers or p-n junctions of GaAs. With a current pulse in an n-layer at low temperature, e.g. (~ 2 K) the donors are ionized. The resulting electrons then emit phonons nonradiatively and return towards equilibrium. The emitted phonons travel through a thick crystal (~ 2 mm) of insulating GaAs and are detected using a superconducting bolometer or tunnel junction.

In fig. 7, taken from ref. [19], the resulting ballistic signals observed in this way by Narayanamurti and Lax are shown. The beauty of the experiment of Narayanamurti and Lax is in that longitudinal and transverse phonons are clearly separated because their ballistic flight times are different. Lax and Narayanamurti discuss LA and TA phonon generation only but I refer the interested reader to ref. [49] for work on the LO phonon cascades. The dominant mechanisms at work in the Lax-Narayanamurti experiment are GaAs deformation potential scattering (longitudinal phonons) and piezoelectric scattering (transverse phonons).

The rate of phonon production turns out to depend on the phonon type, the direction of propagation of the

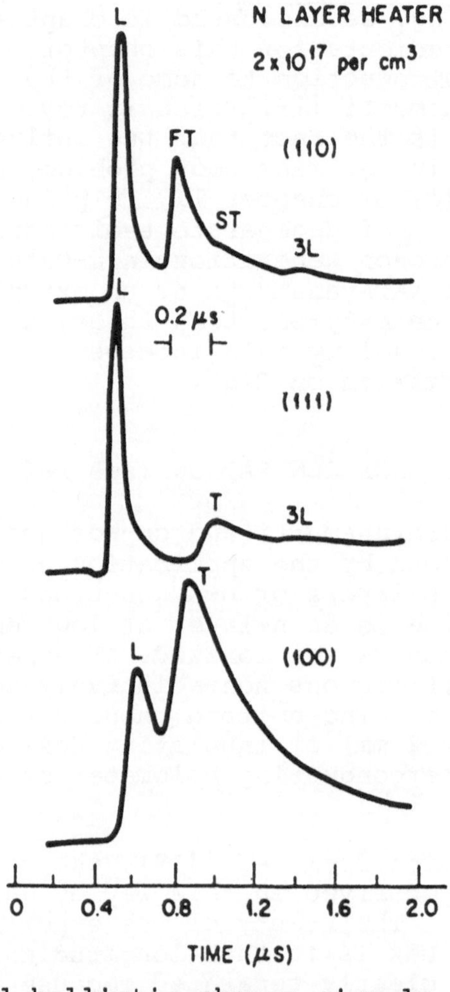

Fig. 7. Typical ballistic phonon signals observed with a superconducting bolometer as a function of orientation. N layer heater. Carrier concentration ~ $2\times10^{17}cm^3$ (from ref. [19])

phonons, lattice temperature, and, importantly, the direct effect of carrier concentration and screening. The sensitive dependence on these various parameters makes phonon generation experiments a very useful tool in solid state physics (see also the work of Bron [50]).

The samples of Narayamurti and Lax contain typically 10^{16} electrons cm^{-3}, therefore considerable screening occurs and piezoelectric and deformation potential scattering rates become comparable.

2.1. Screened Piezoelectric and Deformation Potential Interaction

A significant point in the work of Narayanamurti and Lax is that they analyze the effect of screening on the piezoelectric and deformation interaction and the resulting expressions. The importance of screening is obvious if one realizes that GaAs at a few degrees K and with 10^{17} electrons cm^{-3} "behaves as an excellent metal".

Lax and Narayamurti then examine what the predictions are for the screening of different many-body theories (e.g. Thomas-Fermi, Hedin-Lundqvist, Lindhard (R.P.A.); Hubbard, Geldart-Vosko-Taylor, Toigo-Woodruff, Singwi et al., Brosens-Devreese-Lemmens).

Lax and Narayamurti show that the effective potential seen by an electron can be represented in the form:

$$V_{eff}(\vec{q}) = f_j^{dp} u_j(q) + f_j^{pe} u_j(q) \tag{58}$$

where the u_j describe the momentum (q)-dependent displacement field (dp refers to deformation potential, pe to piezoelectric). The f_j with vertex corrections are:

$$f_j^{dp} = \Lambda_V(q)\ i\ E_{jk}\ q_k\ \left(1 + \frac{\lambda^2}{q^2}\right)^{-1} \tag{59a}$$

$$f_j^{pe} = \Lambda_V(q)\ e\ s_i\ e_{ijk}\ s_k\ \left[\varepsilon_b\left(1 + \frac{\lambda^2}{q^2}\right)\right]^{-1} \tag{59b}$$

$\vec{s} = \frac{\vec{q}}{q}$ is a unit vector in the direction of propagation, E_{jk} occurs in the definition of the deformation potential, $V_{dp}(r) = E_{jk} \frac{\partial u_j}{\partial x_k}$, e_{ijk} are the Levi-Civita symbols, e is the electron charge, ε_b is the background permittivity, $\lambda(q)$ is the inverse screening length ($\lambda^2(q) = q^2\Phi(q)\,\Pi(q)$ with $\Pi(q)$ the proper polarization and $\Phi(q)$ the Coulomb propagator). Furthermore:

$$\Lambda_V = \frac{1}{1-\Phi(q)\Pi_o(q)G(q)} \qquad (60)$$

and

$$\Pi(q) = \Pi_o(q)\Lambda(q) \quad ; \; Q_o = \Phi(q)\;\Pi_o(q) \qquad (61)$$

The relevant point here is that the screened electron-phonon interaction contains G(q), the local field correction.

2.2. Transition Rates and Phonon Transport

The Fröhlich type Hamiltonians, discussed in chapter I, are the basis for the calculation of the phonon emission rate (see eqs. (18,19)).

As shown by Lax and Narayamurti the net rate of phonon production is given by

$$\left.\frac{\partial N(q)}{\partial t}\right|_k = \sum_k{}' \; T_{k,k'}\; \delta_{\vec{k}'+\vec{q},\vec{k}}\; \delta(E(k')-E(k)) \qquad (62)$$

All the information on electron-phonon coupling of Chapter I and on screening of chapter II is contained in the expression for $T_{k,k'}$ which contains G(q), the local field correction. In (62), N(q) is the number of phonons with wave number q; the meaning of the other symbols is obvious.

2.3. Relative Importance of Piezoelectric and Deformation Potential Phonon Production

Using the Fröhlich type interaction for piezoelectric and deformation potential, the general expression for the dielectric function $\varepsilon(q,\omega) = 1 + \frac{Q_o(q,\omega)}{1-G(q,\omega)Q_o(q,\omega)}$ and the equation (62) for $\frac{\partial N(q)}{\partial t}$, the total power radiated into the phonons can be calculated. For details I refer to [19].

For the rate of production of phonon energy per carrier $\frac{P}{N}$, Lax and Narayamurti find, for screened deformation potential interaction,

$$[\frac{P}{N}]_{def} = 4.4\ 10^{-16}\ (T_e - T_p)\ y\ f(y)\ J/sec \qquad (63)$$

where $y = \frac{6.0292}{r_s}$. T_p is the lattice temperature, T_e is electron temperature.

It is the function f(y) which contains the information on the details of the electron screening and in particular on G(q), the local field correction.

Lax and Narayamurti have calculated f(y) for different many-electron theories; it is seen that this function is rather sensitive to the type of many-body approximation as seen in fig. 8, taken from ref. [19]. Similar expansions to (63) were calculated in ref. [19] for

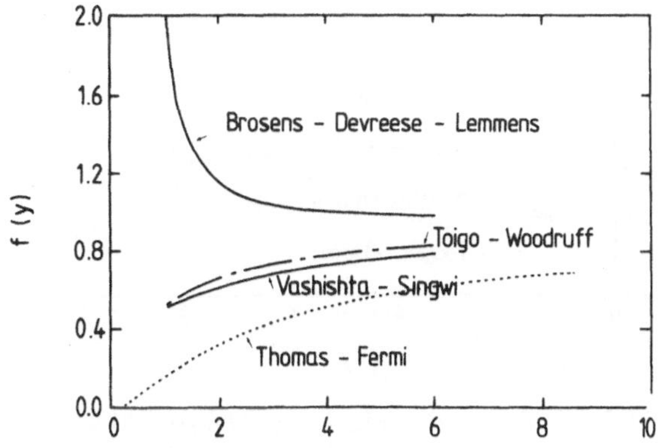

Fig. 8. Screening factor f(y) (appropriate to deformation-potential scattering) plotted against $y = \pi k_F a_B = 6.03/r_s$ a variable proportional to the cube root of the concentration.

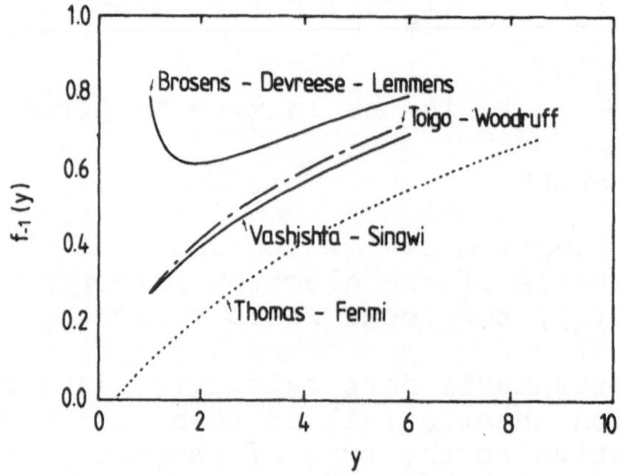

Fig. 9. Screening factor $f_{-1}(y)$ appropriate to ionized impurity scattering is plotted vs $y = \pi k_F a_B$.

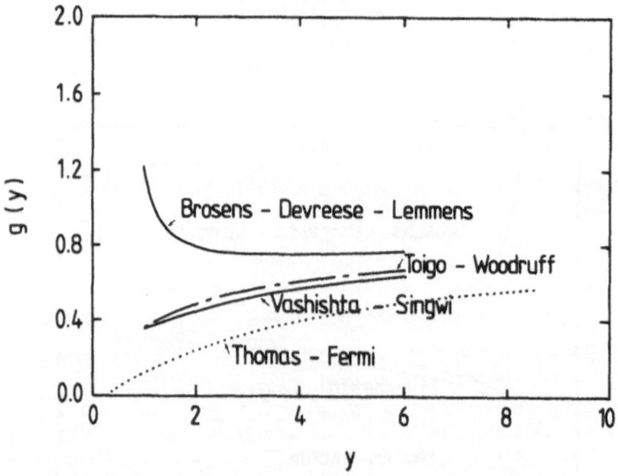

Fig. 10. Screening factor $g(y)$ appropriate to piezo-electric scattering is plotted vs $y = \pi k_F a_B$.

piezoelectric scattering and ionized impurity scattering. Rather than present further formulas, I include fig. 9 and fig. 10 which show the functions g(y) and $f_{-1}(y)$, the analogs of f(y) for the case of piezoelectric scattering (g(y)) and ionized impurity scattering ($f_{-1}(y)$).

Finally fig. 11 shows the relative importance of piezoelectric screening to deformation potential screening expressed in $\frac{g(y)}{f(y)}$ from the above discussion.

The authors of ref. [19] also discuss phonon selection rules, finite aperture focussing and infinitesimal aperture results but for these topics I refer both to ref. [19] (in fact I had originally conceived the present lectures as an introduction to the then planned lectures of Prof. Melvin Lax) and to the lectures of Prof. Bron, Prof. Challis, Prof. Kinder, Prof. Renk at the present Advanced Study Institute.

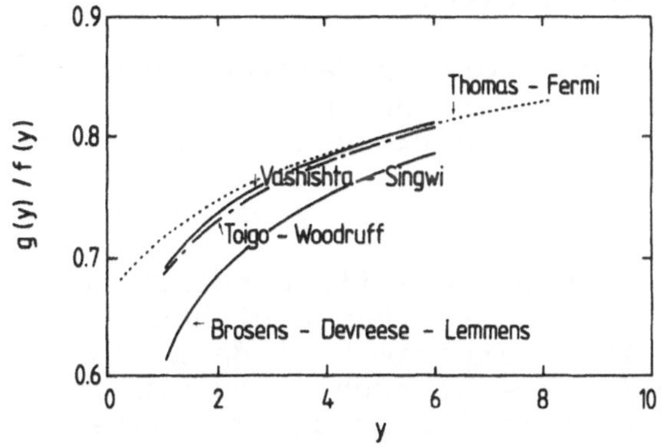

Fig. 11. Ratio g(y)/f(y) of piezoelectric screening to deformation-potential screening is plotted vs $y = \pi k_F a_B$.

My main interest in the work of Lax and Narayamurti is in the possibility to obtain experimental information on G(q) from yet another source than that discussed in chapter II: plasma dispersion and energy loss of charged particles moving through metals. Of great importance is the tunability of r_s in the GaAs systems studied in ref. [19]. The results obtained in figs. 8 to 11 clearly show the sensitivity of the phonon generation to the details of G(q), which is the central ingredient in modern many-electron theories. Experimental data on phonon generation in degenerate GaAs as a function of density, could greatly help us to improve existing screening theories and to test, and compare with each other, the existing theories, including the one developed by us [41].

It is therefore our hope that the paper announced in ref. [19] on the experimental data to be compared with figs. 8 to 11 will appear soon.

2.4. Discussion

The present series of lectures introduces the student to the theoretical description of electron-phonon interaction and screening in the many-electron system in a bath of phonons. With these ingredients it is then possible to calculate phonon generation rates, taking into account many-body effects.

The theoretical work of Narayanamurti and Lax suggests that experiments on the generation of phonons might provide direct experimental information on the so-called "local field correction function" G(q), a fundamental ingredient in the theory of the electron gas.

Our work on this local field correction (see ref. [41,42,43,44,47]) suggests that G(q) has a pronounced maximum at $q = 2k_F$. Experimental data confirming or refuting the existence of such a maximum of G(q) would be of great importance for an understanding of the theory of the electron gas.

From the experimental point of view it seems less clear whether the studies of Narayanamurti and Lax have indeed revealed information capable of differentiating between different theories for the dielectric function of the electron gas. Difficulties in this respect concern i.a. the role of scattering of electrons at impurities, LA-TA phonon "mixing" etc. ... I refer the interested reader to the lectures of Prof. R.G. Ulbrich and Prof.

W. Bron at the present Advanced Study Institute. Also ref. [51] contains several articles relevant to this question.

Acknowledgment: It is my pleasure to thank Profs. W. Bron, M. Lax, V. Narayamurti, R. Ulbrich for illuminating discussions.

REFERENCES

1. H. Fröhlich, in "Polarons and Excitons", C. Kuper and G.D. Whitfield, eds., Oliver and Boyd, Edinburgh and London (1963).
2. H. Fröhlich, H. Pelzer and S. Zienau, Phil. Mag. 41, 221 (1950).
3. J. Bardeen, Phys. Rev. 52, 688 (1937).
4. H. Ehrenreich and A. Overhauser, Phys. Rev. 104, 331 (1958).
5. G.D. Mahan, "Many-Particle Physics", Plenum Press, New York and London (1981).
6. A.R. Hutson, J. Appl. Phys. 32, 2287 (1961).
7. "Polarons in Ionic Crystals and Polar Semiconductors", J.T. Devreese, ed., North Holland, Amsterdam (1972).
8. P. Sheng and J. Dow, Phys. Rev. B4, 1343 (1971).
9. V. Gurevich, I. Lang and Y. Firsov, Sov. Phys. - Solid State 4, 918 (1962).
10. J. Devreese, W. Huybrechts and L. Lemmens, Phys. Stat. Sol. (b) 48, 77 (1971).
11. F. Brosens and J.T. Devreese, Phys. Stat. Sol. (b) 108, K29 (1981).
12. W. Huybrechts and J. Devreese, Phys. Rev. B8, 5754 (1973).
13. R.P. Feynman, Phys. Rev. 97, 660 (1955).
14. J. Devreese and R. Evrard, Phys. Letters 23, 196 (1966).
15. R. Feynman, R. Hellwarth, C. Iddings and P. Platzman, Phys. Rev. 128, 1599 (1962).
16. J.T. Devreese, J. De Sitter and M. Goovaerts, Phys. Rev. B5, 2367 (1972).
17. F.M. Peeters and J.T. Devreese, Phys. Rev. B23, 1936 (1981).
18. F.M. Peeters and J.T. Devreese, Phys. Rev. B28, 6051 (1983).
19. M. Lax and V. Narayamurti, Phys. Rev. B24, 4692 (1981).
20. J.T. Devreese and F. Brosens, in "Electron Correlations in Solids, Molecules and Atoms", Plenum

Press, New York and London (1983). Edited by the same authors.
21. F. Bloch, Z. Phys. 55, 545 (1929).
22. E.P. Wigner and F. Seitz, Phys. Rev. 43, 804 (1933); 46, 509 (1934).
23. M. Gell'Mann and K. Brueckner, Phys. Rev. 106, 364 (1957).
24. See ref. 5.
25. E.P. Wigner, Trans. Faraday Soc. 34, 678 (1938).
26. C.C. Grimes and G. Adams, Phys. Rev. Lett. 42, 795 (1979).
27. R.S. Crandall and R. Williams, Phys. Letters 34A, 404 (1971).
28. A.W. Overhauser, Phys. Rev. Lett. 4, 462 (1960).
29. A.W. Overhauser, Adv. Phys. 27, 343 (1978) and references cited therein.
30. W.A. Harrison, "Pseudopotentials in the Theory of Simple Metals", Benjamin Inc., New York (1966).
31. V. Heine, M.L. Cohen and D. Weaire, Solid State Phys. 24, F. Seitz and D. Turnbull, eds., Academic Press, New York (1964).
32. J. Lindhard, Kgl. Dan. Videnskab. Selsk. Mat. Fys. Medd. 28, (8) (1954).
33. D. Pines and D. Bohm, Phys. Rev. 85, 338 (1952); 92, 609 (1953).
34. P. Nozières and D. Pines, Phys. Rev. 111, 442 (1958).
35. L. Kleinman, Phys. Rev. 160, 585 (1967); 172, 383 (1968).
36. R.W. Shaw, Jr., J. Phys. C: Solid State Phys. 3, 1140 (1970).
37. R. Lobo, Phys. Rev. B8, 5348 (1973).
38. L. Hedin and B.I. Lundqvist, J. Phys. C: Solid State Phys. 4, 2064 (1971).
39. C.A. Kukkonen and A.W. Overhauser, Phys. Rev. B20, 550 (1979).
40. See e.g. S. Schnatterly in ref. 20.
41. J.T. Devreese, F. Brosens and L.F. Lemmens, Phys. Rev. B21, 1349 (1980).
F. Brosens, J.T. Devreese and L.F. Lemmens, Phys. Rev. B21, 1363 (1980).
42. See e.g. F. Brosens and J.T. Devreese, Helv. Phys. Acta 56, 223 (1983).
43. H. Nachtegaele, F. Brosens and J.T. Devreese, Phys. Rev. B28, 6064 (1983).
44. F. Brosens, L.F. Lemmens and J.T. Devreese, Phys. Stat. Sol. (b) 80, 99 (1977).
45. A. Holas, P. Aravind and K. Singwi, Phys. Rev. B20, 4912 (1979).
46. Yu. Sayasov, Journ. de Phys. 44, C8-1 (1983).

47. F. Brosens, L.F. Lemmens and J.T. Devreese, Phys. Stat. Sol. (a) 59, 447 (1980).
48. K. Singwi and M. Tosi, Solid State Physics 36, 177 (1981).
49. R. Evrard, in "Polarons and Excitons", J.T. Devreese, F. Peeters, eds., Plenum Press, New York and London (1984).
50. W. Bron, e.g. in lectures at the present Advanced Study Institute.
51. "Phonon Scattering in Condensed Matter", Springer Series in Solid State Sciences, vol. 51 (W. Eisenmenger et al., eds.), in particular part III.

SURFACE ACOUSTIC WAVES

Alexei A. Maradudin

Department of Physics
University of California
Irvine, CA 92717, U.S.A.

1. Introduction

In 1887 Lord Rayleigh[1] showed that a semi-infinite, isotropic, elastic medium, bounded by a single, stress-free, planar surface, can support surface vibration modes that are wavelike in directions parallel to the surface of the solid, but whose amplitudes decay exponentially with increasing distance into the solid from the surface, with a decay length that is of the order of the wavelength of the wave along the surface. The displacement vector of these waves lies in the sagittal plane, i.e. in the plane defined by the direction of propagation of the wave and the normal to the surface. These waves are acoustic waves in that their frequencies are linear in the magnitude of the two-dimensional wave vector characterizing their propagation along the surface. They are consequently nondispersive, i.e. their speed of propagation is independent of their wavelength parallel to the surface, which is due to the absence of a characteristic length in the system under consideration. Their frequencies also lie below the continuum of frequencies allowed the normal vibration modes of an infinite elastic medium for the same value of the two-dimensional wave vector. Such surface acoustic waves are now known as Rayleigh Waves.

In the years following the pioneering work of Lord Rayleigh surface acoustic waves continued to be investigated theoretically on the basis of elasticity theory. They were initially studied at very low frequencies (< 100 Hz) in the context of the study of the propagation of seismic shocks in the earth's crust. Because the radius of the earth is much larger than the wavelength of seismic disturbances, this problem was simplified by neglecting the curvature of the earth and considering the surface of the earth to be the top of an infinite half-space. It was on the basis of this model that Lord Rayleigh first predicted the existence of the surface acoustic waves that now bear his name. Until approximately the 1950's progress in the study of surface acoustic waves was associated primarily with the relaxation of the several assumptions underlying Rayleigh's work, but still within the context of seismology or seismic testing. Thus, for example, the propagation of surface acoustic waves on inhomogeneous (stratified) elastic media was investigated when it was recognized that the elastic properties of the earth's crust differ from those of the material supporting it. The assumption of a planar stress-free surface was also subsequently relaxed, and the properties of surface acoustic waves on curved (viz. cylindrical and spherical) surfaces determined. The extension of the preceding analysis to anisotropic media was also accomplished in certain special cases, and it was found that the decay of the amplitudes of surface acoustic waves with increasing distance into the medium from the surface could have a more complex form than the simple exponential law followed by Rayleigh waves, viz. a product of an exponential and a trigonometric function. Indeed, the contributions made by geophysicists to the theory of surface acoustic waves have been many and diverse, but often unknown to physicists who have, in some cases, rediscovered them many years later.

In the mid 1940's surface acoustic waves with frequencies of the order of 10^6 Hz began to be used in nondestructive testing for cracks in castings,(2,3) and subsequently have become an important tool for the nondestructive evaluation of sample surfaces and surface layers. They have more recently been used in ultrasonic delay lines.(4)

The last ten to fifteen years have seen an enormous increase in the interest shown in surface acoustic waves by physicists, electrical engineers, and materials scientists. In contrast with the low frequencies of the surface acoustic waves of interest to geophysicists, the frequencies of the surface acoustic waves of current interest are very high, in the range from 10^6 to 10^{11} Hz. The reason for this interest in surface acoustic waves in this frequency range is the recognition that they can provide a new approach to signal processing, with important applications for radar and communications(5-7). At the same time surface acoustic waves possess an intrinsic scientific interest because of the basic physics that plays a role in their formation, and because of the physical properties of solids they can probe.

The properties of surface acoustic waves that make them useful for such applications are several, and serve as an introduction to the discussion of these waves that follows(7).

The most important property of these waves is their extremely low speed, about 10^{-5} that of electromagnetic waves. Because of their low speeds acoustic waves also have very small wavelengths compared with the wavelengths of electromagnetic waves of the same frequency. The reduction in size is also of the order of 10^{-5}. Acoustic wave devices thus offer the possibility of dramatic reductions in size and weight when compared with electromagnetic devices. In addition, because surface acoustic wave devices are fabricated on the surface of a crystal they tend to be rugged and reliable. Surface acoustic waves travel along the surface of a solid and are confined to its vicinity. They are therefore easier to modify or tap into than are bulk acoustic waves. The use of surface waves also permits surface wave devices to be compatible with integrated circuit technology and to allow their fabrication by lithographic techniques that produce surface acoustic wave devices of comparatively low cost with precise and reproducible characteristics. Surface acoustic waves on semi-infinite media are dispersionless, i.e. their speed is independent of their wavelength. This is a useful property for certain types of applications, and has its origin in the absence of any characteristic length in the problem. If, however, some dispersion in a surface wave is desired, this can be achieved by adding a thin film of one material onto a substrate of a different material. The presence of a characteristic length in such a structure, viz. the thickness of the film, renders surface acoustic waves in it dispersive, in a manner than can be varied by changing the thickness of the film and the material properties of the film and substrate. Finally, surface acoustic waves can be guided[8,9] and amplified[10,11], and these properties provide the possibility of analogues of electromagnetic wave guide devices constructed on the basis of elastic media, with the reduced dimensions mentioned above.

With such applications as a motivation the range of materials supporting surface acoustic waves has also been expanded from the purely elastic media that were considered exclusively until the 1960's. Surface acoustic waves on piezoelectric media have been investigated intensively in recent years, because they have an accompanying quasistatic electric field. This feature of such waves makes possible their excitation by means of metal electrodes deposited on the surface. In this way very high frequency ($\sim 10^{10}$ Hz) surface acoustic waves can be generated. At the same time, the presence of this electric field enables surface acoustic waves on piezoelectric media to couple to external electrical circuits, to other surface waves, or to charge carriers in semiconductors. More recently surface acoustic waves on various types of magnetic media have also begun to be studied also for possible use in technological applications.

Properties of surface acoustic waves, such as the power flow in them, their attenuation due to a variety of mechanisms, their focusing on surfaces of anisotropic media, and their interactions due to elastic and electrical nonlinearities in the media supporting such waves have also been investigated.

It is not coincidental that the increase in the interest shown in surface acoustic waves coincided with the advent of high speed computers. It is the availability of the latter that has made possible the modern work on surface acoustic waves in realistic systems, such as on anisotropic elastic media, piezoelectric media, magnetoelastic media, and inhomogeneous media; at solid-solid and solid-liquid interfaces; on curved and other nonplanar surfaces, and in spatially restricted geometries. We will see many examples of this in the present chapter.

As a consequence of the generalizations of Rayleigh's pioneering work described above, many new types of surface acoustic waves have been discovered. In the present chapter we present the basic elements of the theory of a wide variety of different types of surface acoustic waves, we describe methods for their excitation and detection, and survey various properties possessed by these waves. The emphasis in the discussion is on the theoretical aspects of the subject, but here and there relevant experimental results will be presented as well. The possible technological applications of surface acoustic waves will not be discussed here, however. Such discussions should be written by those who know something of the subject, and are well presented in Refs. 3, 10, 12-17. In view of the limitations of space and time no attempt at completeness of coverage of the subject has been made. Nevertheless, I hope that the material presented provides an adequate introduction to the continuum theory of vibrational surface waves that will enable the reader to pursue by himself the topics that have not been covered here.

2. Surface Acoustic Waves on Various Media

Surface acoustic waves can propagate along the surfaces of, and interfaces between, several different kinds of media, elastic, piezoelectric, or magnetic. In this section we present the theory of such waves, and discuss the distinctive features imparted to them by the nature of the medium over which they propagate, or the media between which they propagate.

2.1. Elastic Media

Many materials of physical interest are purely elastic, i.e. are neither piezoelectric nor magnetic. Among these are the non-transition metals and homopolar semiconductors. The propagation of surface acoustic waves over their surfaces can be of technological utility, e.g. in nondestructive testing for cracks[2,3]. At the same time, the theory of surface acoustic waves on such media is somewhat simpler than the theory of surface acoustic waves on piezoelectric and magnetic media, and serves as a good introduction to the latter, as well as to the several types of guided waves associated with an overlayer, or with spatially varying material properties, that will be discussed below.

We begin this section, then, with the theory of surface acoustic waves on a semi-infinite elastic medium bounded by a stress-free planar surface, the Rayleigh waves. We consider first the special, and simple, case of an isotropic elastic medium, and then turn to the case of a general anisotropic elastic medium. We point out other types of surface acoustic waves, besides Rayleigh waves, that can exist on such media, viz. generalized Rayleigh waves and pseudosurface (or leaky) surface waves. We then examine how these waves are modified when an overlayer of a different elastic material is present on a semi-infinite elastic substrate. Depending on the thickness of the overlayer, and the polarization of the waves being studied, one can obtain in this way new types of surface acoustic waves called Love waves, Stoneley waves, and Sezawa waves. Of interest originally in seismological contexts, these waves are acquiring a new interest, for electronic signal processing applications. We conclude this subsection with an investigation of the propagation of surface acoustic waves on elastic media whose material properties vary continuously with distance from the surface.

2.1.1. Homogeneous Media

Surface acoustic waves can exist on the planar, stress-free surface of either a homogeneous or an inhomogeneous elastic medium. Historically, the former case was studied first, and is the simpler of the two cases to consider. If the elastic medium is inhomogeneous, however, i.e. if its material properties vary from point to point in the medium, surface acoustic waves become possible that have no counterpart in a homogeneous medium. In this subsection we consider the kinds of surface acoustic waves that can exist on a homogeneous medium. Some of the surface acoustic waves that can exist on an inhomogeneous medium whose material properties are functions of the distance into the medium from the surface are considered in subsection 2.1.2.

We consider an elastic medium occupying the upper half-space $x_3 > 0$. The surface $x_3 = 0$ is assumed to be stress-free. Within the framework of the linear theory of elasticity the equations of motion of the medium are

$$\rho \frac{\partial^2}{\partial t^2} u_\alpha = \sum_\beta \frac{\partial T_{\alpha\beta}}{\partial x_\beta}, \qquad \alpha = 1,2,3, \tag{2.1}$$

where ρ is the mass density of the medium, $u_\alpha(\vec{x},t)$ is the α Cartesian component of the displacement of the medium at the point $\vec{x}$ and time t, and $T_{\alpha\beta}(\vec{x},t)$ is the stress tensor. The latter is given by (Hooke's law)

$$T_{\alpha\beta} = \sum_{\mu\nu} C_{\alpha\beta\mu\nu} \frac{\partial u_\mu}{\partial x_\nu} \qquad \alpha,\beta = 1,2,3, \tag{2.2}$$

where the $\{C_{\alpha\beta\mu\nu}\}$ are the elements of the elastic modulus tensor. The elements of the tensor $C_{\alpha\beta\mu\nu}$ are symmetric in α and β; in μ and ν; and in the pairs $\alpha\beta$ and $\mu\nu$. When the elastic medium is subjected to an arbitrary real, orthogonal transformation S that sends a point $\vec{x}$ in it into a point $\vec{x}'$, where both $\vec{x}$ and $\vec{x}'$ are defined with respect to a set of Cartesian axes fixed in space (active convention), according to

$$\vec{x}' = \overleftrightarrow{S}\,\vec{x}, \tag{2.3}$$

where $\overleftrightarrow{S}$ is the 3 × 3 real, orthogonal matrix that represents the transformation S, the elastic modulus tensor of the transformed medium is given by

$$C'_{\alpha\beta\mu\nu} = \sum_{\alpha'\beta'\mu'\nu'} S_{\alpha\alpha'} S_{\beta\beta'} S_{\mu\mu'} S_{\nu\nu'} C_{\alpha'\beta'\mu'\nu'}. \tag{2.4}$$

If the transformation S is one of the elements R of the crystal class G to which the elastic medium belongs, the medium is sent into itself, and we can remove the prime from the right hand side of Eq. (2.4). The transformation law for the elements of the elastic modulus tensor under the operations of G is therefore

$$C_{\alpha\beta\mu\nu} = \sum_{\alpha'\beta'\mu'\beta'} R_{\alpha\alpha'} R_{\beta\beta'} R_{\mu\mu'} R_{\nu\nu'} C_{\alpha'\beta'\mu'\nu'}. \tag{2.5}$$

Although the elastic modulus tensor $C_{\alpha\beta\mu\nu}$ is a fourth rank tensor, the fact that it is symmetric in α and β, in μ and ν, and in $\alpha\beta$ and $\mu\nu$ makes it possible to express it equivalently in a two subscript notation, $C_{\alpha\beta\mu\nu} \to c_{ij}$, according to the scheme

$$\begin{array}{cccccc} 11 & 22 & 33 & 23 = 32 & 31 = 13 & 12 = 21 \\ 1 & 2 & 3 & 4 & 5 & 6 \end{array} \tag{2.6}$$

The resulting 6 × 6 matrix representation of the elastic modulus tensor is referred to as the contracted, or Voigt, representation of this tensor. It should be emphasized, however, that the $\{c_{ij}\}$ are not the elements of a second rank tensor, since they do not transform as one under real, orthogonal transformations of the medium.

When we substitute Eq. (2.2) into Eq. (2.1) we obtain the equations of motion of the medium in the form

$$\rho \frac{\partial^2 u_\alpha}{\partial t^2} = \sum_{\beta\mu\nu} C_{\alpha\beta\mu\nu} \frac{\partial^2 u_\mu}{\partial x_\beta \partial x_\nu} \qquad \alpha = 1,2,3. \tag{2.7}$$

For future reference we note that the Lagrangian density $\mathcal{L}$ from which Eqs. (2.7) can be obtained through the use of the Lagrange equations of motion[18],

$$- \frac{\partial F}{\partial \dot{\phi}_j} + \frac{\partial \mathcal{L}}{\partial \phi_j} - \frac{d}{dt} \frac{\partial \mathcal{L}}{\partial \dot{\phi}_j} - \sum_\alpha \frac{\partial}{\partial x_\alpha} \frac{\partial \mathcal{L}}{\partial(\partial \phi_j / \partial x_a)} = 0, \tag{2.8}$$

where the $\{\phi_j\}$ are the fields that appear in the Lagrangian density and F is the dissipation function that is non-zero if dissipative processes are present in the system, is given by

$$\mathcal{L} = \frac{1}{2} \rho \sum_\alpha \dot{u}_\alpha^2 - \frac{1}{2} \sum_{\alpha\beta\mu\nu} C_{\alpha\beta\mu\nu} \frac{\partial u_\alpha}{\partial x_\beta} \frac{\partial u_\mu}{\partial x_\nu} . \tag{2.9}$$

The equations of motion (2.7) have to be supplemented by the conditions expressing the fact that the stresses acting on the surface $x_3 = 0$ vanish:

$$T_{\alpha 3}\Big|_{x_3 = 0} = 0 \qquad \alpha = 1, 2, 3, \tag{2.10}$$

or

$$\sum_{\mu\nu} C_{\alpha 3\mu\nu} \frac{\partial u_\mu}{\partial x_\nu}\Big|_{x_3 = 0} = 0 \qquad \alpha = 1, 2, 3. \tag{2.11}$$

To solve Eqs. (2.7) we make the <u>Ansatz</u>

$$u_\alpha(\vec{x},t) = U_\alpha e^{i\vec{k}_\parallel \cdot \vec{x}_\parallel - k_\parallel \beta x_3 - i\omega t} \tag{2.12}$$

where $\vec{x}_\parallel = \hat{x}_1 x_1 + \hat{x}_2 x_2$ is a position vector on the surface $x_3 = 0$, while $\vec{k}_\parallel = \hat{x}_1 k_1 + \hat{x}_2 k_2$ is the two-dimensional wave vector characterizing the propagation of the wave along this surface. Here $\hat{x}_1$ and $\hat{x}_2$ are unit vectors along the x_1- and x_2-directions, respectively. The form of the solution given by Eq. (2.12) manifestly satisfies vanishing boundary conditions at $x_3 = \infty$, provided Re $\beta > 0$. The substitution of Eq. (2.12) into Eqs. (2.7) yields three homogeneous equations for the determination of the amplitudes $\{U_\alpha\}$, that we write as

$$\sum_\mu [\Gamma_{\alpha\mu}(\hat{k}_\parallel;\beta) - \delta_{\alpha\mu} v^2] U_\mu = 0, \tag{2.13}$$

where $\hat{k}_\parallel = \vec{k}_\parallel / k_\parallel$, v is the phase velocity of the wave, $v = \omega/k_\parallel$, and

$$\Gamma_{\alpha\mu}(\hat{k}_\parallel;\beta) = \frac{1}{\rho} \sum_{\beta\nu} C_{\alpha\beta\mu\nu} [(1 - \delta_{\beta 3})\hat{k}_\beta + i\delta_{\beta 3}\beta] \times$$
$$\times [(1 - \delta_{\nu 3})\hat{k}_\nu + i\delta_{\nu 3}\beta]$$
$$= \Gamma_{\mu\alpha}(\hat{k}_\parallel;\beta) . \tag{2.14}$$

In order that Eq. (2.13) have nontrivial solutions for the $\{U_\alpha\}$, the determinant of their coefficients in this equation must vanish. When this determinant is expanded it results in a sixth order equation for the decay constant β. There are therefore six values of β for each $\hat{k}_\parallel$ and v. The nature of these six roots can be determined as follows. If we replace $i\beta$ by γ, we see from Eqs. (2.13)-(2.14) that the sixth order equation for γ has real coefficients (for real values of k_α.) This means that the six values of γ are real or occur in complex conjugate pairs. This means in turn that the six values of β are purely imaginary, $\beta = -i\beta_1$, or have the forms $\beta = \pm \beta_2 - i\beta_3$, where β_1, β_2, β_3 are real quantities. Since we are concerned here with surface waves, it is only in the values of β whose real parts are positive that we are interested. In the general case there are three such roots, and we will denote them by $\beta_j(\hat{k}_\parallel v)$ $(j = 1,2,3)$. (Any other number of $\beta_j(\hat{k}_\parallel v)$ with positive real parts would make it impossible to satisfy the boundary conditions (2.11), and surface acoustic waves would not exist for the corresponding values of $\vec{k}_\parallel$ and ω. This does not appear to happen[19,20].) The amplitudes $\{U_\alpha^{(j)}\}$ corresponding to $\beta_j(\hat{k}_\parallel v)$ are related by

$$\frac{U_1^{(j)}}{C_1^{(j)}} = \frac{U_2^{(j)}}{C_2^{(j)}} = \frac{U_3^{(j)}}{C_3^{(j)}} = K_j \qquad j = 1,\ 2,\ 3, \tag{2.15}$$

where the $\{K_j\}$ are constants, and the $\{C_\alpha^{(j)}(\hat{k}_\parallel v)\}$ are the cofactors of the elements in the first row of the determinant of the coefficients of the $\{U_\alpha\}$ in Eq. (2.12), evaluated for $\beta = \beta_j(\hat{k}_\parallel v)$.

To satisfy the boundary conditions (2.11) we superpose the solutions corresponding to the three $\beta_j(\hat{k}_\parallel v)$ to obtain

$$u_\alpha(\vec{x},t) = \sum_{j=1}^{3} C_\alpha^{(j)}(\hat{k}_\parallel v) K_j\, e^{i\vec{k}_\parallel\cdot\vec{x}_\parallel - k_\parallel \beta_j(\hat{k}_\parallel v)x_3 - ik_\parallel vt} \tag{2.16}$$

Substitution of this form for the displacement field into the boundary conditions (2.11) leads to a set of homogeneous equations for the $\{K_j\}$,

$$\sum_{j=1}^{3} M_{\alpha j}(\hat{k}_\parallel v)K_j = 0 \qquad \alpha = 1,\ 2,\ 3, \tag{2.17}$$

where

$$M_{\alpha j}(\hat{k}_\parallel v) = \frac{1}{\rho}\sum_{\mu\nu} c_{\alpha 3\mu\nu} C_\mu^{(j)}(\hat{k}_\parallel v)\,\left[(1-\delta_{\nu 3})i\hat{k}_\nu - \delta_{\nu 3}\,\beta_j(\hat{k}_\parallel v)\right]. \tag{2.18}$$

The conditions for the existence of a nontrivial solution of Eqs. (2.17) is that the determinant of the coefficients of the $\{K_j\}$ vanish,

$$\left| M_{\alpha j}(\hat{k}_{\parallel} v) \right| = 0. \tag{2.19}$$

Equation (2.19) is the dispersion relation for surface acoustic waves: its solution gives the phase velocity $v(\hat{k}_{\parallel})$ of these waves as a function of their direction of propagation $\hat{k}_{\parallel}$. We have thus obtained the important result that a surface acoustic wave on a semi-infinite elastic substrate is nondispersive, i.e. its phase velocity is independent of the magnitude of its wave vector $k_{\parallel}$, and hence of the wavelength of the wave parallel to the surface $\lambda = 2\pi/k_{\parallel}$, although it does depend on the direction of propagation of the wave. As we have observed above, this is due to the absence of any characteristic length in a semi-infinite system against which the wavelength of the surface wave can be compared.

Before proceeding to a discussion of the solutions of Eq. (2.19) in the general case, which can only be obtained numerically, we consider the particular case of an isotropic elastic medium, for which all the results can be obtained essentially analytically.

2.1.1.1. Isotropic Media

The elastic modulus tensor for an isotropic elastic medium has only two, independent, nonzero components[21], and we write it in the form

$$C_{\alpha\beta\mu\nu} = \rho(c_{\ell}^2 - 2c_t^2)\delta_{\alpha\beta}\delta_{\mu\nu} + \rho c_t^2(\delta_{\alpha\mu}\delta_{\beta\nu} + \delta_{\alpha\nu}\delta_{\beta\mu}) , \tag{2.20}$$

where c_{ℓ} and c_t are the speeds of sound for longitudinal and transverse waves in an isotropic medium, respectively. They are given in terms of the Lamé constants λ and μ by[22]

$$c_{\ell}^2 = (\lambda + 2\mu)/\rho, \quad c_t^2 = \mu/\rho. \tag{2.21}$$

It is readily verified that the expression given by Eq. (2.20) satisfies Eq. (2.5) identically for any 3×3 real, orthogonal matrix, not just for those that correspond to the elements of a particular crystal class.

Since all directions are equivalent in an isotropic medium, with no loss of generality we can restrict our attention to waves propagating in the x_1-direction. We write the displacement field in this case as

$$u_\alpha(\vec{x},t) = U_\alpha e^{ikx_1-\alpha x_3-i\omega t}. \qquad (2.22)$$

The equation for the amplitudes $\{U_\alpha\}$ that results from the substitution of Eq. (2.21) into Eqs. (2.7) is (see Eqs. (2.13)-(2.14))

$$\begin{pmatrix} c_\ell^2k^2-c_t^2\alpha^2-\omega^2 & 0 & i(c_\ell^2-c_t^2)k\alpha \\ 0 & c_t^2k^2 - c_t^2\alpha^2 - \omega^2 & 0 \\ i(c_\ell^2-c_t^2)k\alpha & 0 & c_t^2k^2-c_\ell^2\alpha^2-\omega^2 \end{pmatrix} \begin{pmatrix} U_1 \\ U_2 \\ U_3 \end{pmatrix} = 0. \qquad (2.23)$$

We see from this equation that the 2-component of the displacement is decoupled from the 1- and 3-components, and we consider it first. From Eqs. (2.22) and (2.23) we see that this component is given by

$$u_2(\vec{x},t) = U_2 e^{ikx_1- \alpha_t(k\omega)x_3-i\omega t}, \qquad (2.24)$$

where

$$\alpha_t(k\omega) = \left(k^2 - \frac{\omega^2}{c_t^2}\right)^{1/2}. \qquad (2.25)$$

In order that Eq. (2.24) describe an exponentially decaying function α_t must be real and positive (since k^2, ω^2, c_t^2 are themselves real and positive, α_t can only be purely real or purely imaginary). This requires that

$$k > \omega/c_t. \qquad (2.26)$$

The only boundary condition (2.11) involving $u_2(\vec{x},t)$ can be written in the form

$$\left.\frac{\partial u_2(\vec{x},t)}{\partial x_3}\right|_{x_3=0} = 0 \ . \tag{2.27}$$

When we substitute Eq. (2.24) into Eq. (2.27), the latter becomes

$$\alpha_t U_2 = 0. \tag{2.28}$$

This equation requires that $\alpha_t = 0$, $U_2 \neq 0$; $\alpha_t \neq 0$, $U_2 = 0$; or $\alpha_t = U_2 = 0$. None of these possibilities yields an elastic wave localized at the surface. We conclude, therefore, that in an isotropic medium there is no surface acoustic wave that is polarized perpendicular to the sagittal plane.

The only nontrivial case, viz. $\alpha_t = 0$, $U_2 \neq 0$, corresponds to a bulk transverse acoustic wave skimming along the surface of the elastic medium. Although it is not a surface acoustic wave it has nevertheless been employed successfully in the construction of devices that have been configured into bandpass filters, delay lines, oscillators, and resonators.(23)

This surface skimming transverse bulk wave has been called "unstable" by Viktorov,(24) in the sense that a slight variation of the boundary conditions or properties of the medium convert it into a surface wave.In succeeding sections of this chapter we will see that surface acoustic waves polarized normal to the sagittal plane (shear horizontal surface acoustic waves) can exist when the boundary condition (2.27) is altered by inhomogeneities in the material properties of the elastic medium near the surface (subsection 2.1.2), by piezoelectricity (subsection 2.2.2), by magnetoelasticity (subsection 2.3.2), by curvature of the surface (Section 5), or by spatial dispersion (Section 10).

Turning now to the pair of coupled equations for the amplitudes U_1 and U_3, we find that the vanishing of the determinant of their coefficients yields two values of the decay constants α, viz.

$$\alpha = \alpha_\ell(k\omega), \quad \alpha = \alpha_t(k\omega) \ . \tag{2.29}$$

The decay constant $\alpha_t(k\omega)$ has been defined in Eq. (2.25), $\alpha_\ell(k\omega)$ is defined by

$$\alpha_\ell(k\omega) = \left(k^2 - \frac{\omega^2}{c_\ell^2}\right)^{1/2} . \tag{2.30}$$

The requirement that α_ℓ be real and positive restricts the allowed values of the wave vector k to the range

$$k > \omega/c_\ell , \tag{2.31}$$

for a given value of ω.

The physically permissible values of c_t^2/c_ℓ^2 range from 0, corresponding to an incompressible solid, to 3/4, which is the largest value consistent with the stability of the elastic medium (i.e. consistent with a positive definite strain energy)[25]. Consequently, since c_ℓ is always larger than c_t, the more restrictive condition of the two given by Eqs. (2.26) and (2.31) is the former, which thus defines the region of (ω,k)-space in which a Rayleigh wave can exist.

If we superpose the solutions corresponding to the two values of α with the aid of Eqs. (2.15), we obtain

$$u_1(\vec{x},t) = -(c_\ell^2 - c_t^2)[k^2 K_\ell e^{-\alpha_\ell x_3} + \alpha_t^2 K_t e^{-\alpha_t x_3}] e^{ikx_1 - i\omega t} \tag{2.32a}$$

$$u_3(\vec{x},t) = -i(c_\ell^2 - c_t^2)k[\alpha_\ell K_\ell e^{-\alpha_\ell x_3} + \alpha_t K_t e^{-\alpha_t x_3}] e^{ikx_1 - i\omega t} \tag{2.32b}$$

The boundary conditions (2.11) that involve u_1 and u_3 can be written in the forms

$$c_t^2\left[\frac{\partial u_1}{\partial x_3} + \frac{\partial u_3}{\partial x_1}\right]_{x_3=0} = 0; \quad \left[(c_\ell^2 - 2c_t^2)\frac{\partial u_1}{\partial x_1} + c_\ell^2 \frac{\partial u_3}{\partial x_3}\right]_{x_3=0} = 0. \tag{2.33}$$

When Eqs. (2.32) are substituted into Eqs. (2.3) the following pair of homogeneous equations for K_ℓ and K_t is obtained,

$$\begin{pmatrix} 2\alpha_\ell k^2 & \alpha_t(k^2+\alpha_t^2) \\ k^2+\alpha_t^2 & 2\alpha_t^2 \end{pmatrix} \begin{pmatrix} K_\ell \\ K_t \end{pmatrix} = 0 \ . \tag{2.34}$$

The solvability condition for these equations yields the equation

$$4\alpha_t\alpha_\ell k^2 - (k^2+\alpha_t^2)^2 = 0, \tag{2.35}$$

that relates the frequency of the Rayleigh wave to its wave vector k.

To obtain the Rayleigh wave solution of Eq. (2.35) we note first of all that from the form of the equation ω must be a linear function of k. It is also the case, from Eq.(2.26), that $\omega < c_t k$. Thus set

$$\omega = c_R k \ , \tag{2.36a}$$

where c_R is the speed of Rayleigh waves, and in addition set

$$c_R = \xi c_t \ , \tag{2.36b}$$

where $\xi < 1$. When we substitute Eqs. (2.36) into Eq. (2.35), square the latter, and rearrange the terms, the equation for ξ takes the form of a cubic equation in ξ^2,

$$\xi^6 - 8\xi^4 + 8\left(3-2\frac{c_t^2}{c_\ell^2}\right)\xi^2 - 16\left(1-\frac{c_t^2}{c_\ell^2}\right) = 0 \ . \tag{2.37}$$

Of the three solutions of this equation for ξ^2 only one, the smallest, satisfies the condition $\xi < 1$, that is the condition for the existence of a localized wave. An approximate expression for this root, that is in error by less than 0.5%, is[26]

$$\xi = \frac{0.87 + 1.12\sigma}{1+\sigma} \ , \tag{2.38a}$$

where

$$\sigma = \frac{\lambda}{2(\lambda+\mu)} = \frac{1 - 2(c_t^2/c_\ell^2)}{2[1 - (c_t^2/c_\ell^2)]} \tag{2.38b}$$

is called Poisson's ratio, and decreases from 1/2 to -1 as c_t^2/c_ℓ^2 increases from 0 to 3/4. (For actual materials σ ranges from 0 to 1/2.) A plot of c_R/c_t as a function of σ is given in Fig. 2.1.[27]

The expressions for the displacement components in a Rayleigh wave, Eqs. (2.32) take much simpler forms if we make the replacements

$$K_\ell = -\frac{A_\ell}{(c_\ell^2 - c_t^2)k^2}, \quad K_t = -\frac{A_t}{(c_\ell^2 - c_t^2)\alpha_t^2} \tag{2.39}$$

and use the fact that from Eqs. (2.34), (2.35), and (2.39) the new amplitudes A_t and A_ℓ are related by

$$\frac{A_t}{A_\ell} = -\frac{2\alpha_t\alpha_\ell}{2k^2 - \omega^2/c_t^2} = -\frac{2\alpha_t\alpha_\ell}{k^2 + \alpha_t^2} = -\frac{(\alpha_t\alpha_\ell)^{1/2}}{k}$$

$$= -\left(1 - \frac{c_t^2}{c_\ell^2}\xi^2\right)^{1/4}\left(1 - \xi^2\right)^{1/4} = -\left(1 - \tfrac{1}{2}\xi^2\right). \tag{2.40}$$

In this way we find that the displacement field of a Rayleigh wave is given by

$$u_1(\vec{x},t) = A_\ell \left\{e^{-\alpha_\ell x_3} - \frac{(\alpha_t\alpha_\ell)^{1/2}}{k} e^{-\alpha_t x_3}\right\} e^{ikx_1 - i\omega t} \tag{2.41a}$$

$$u_3(\vec{x},t) = i\frac{\alpha_\ell}{k} A_\ell \left\{e^{-\alpha_\ell x_3} - \frac{k}{(\alpha_t\alpha_\ell)^{1/2}} e^{-\alpha_t x_3}\right\} e^{ikx_1 - i\omega t}, \tag{2.41b}$$

where now

$$\alpha_\ell = k\left(1 - \frac{c_t^2}{c_\ell^2}\xi^2\right)^{1/2}, \qquad \alpha_t = k\left(1 - \xi^2\right)^{1/2}. \tag{2.42}$$

On the assumption that A_ℓ is real, the physical displacement components, given by the real part of Eqs. (2.41), are

$$u_1(\vec{x},t) = A_\ell \left\{e^{-\alpha_\ell x_3} - \frac{(\alpha_t\alpha_\ell)^{1/2}}{k} e^{-\alpha_t x_3}\right\}\cos(kx_1 - \omega t) \tag{2.43a}$$

$$u_3(\vec{x},t) = -\frac{\alpha_\ell}{k} A_\ell \left\{e^{-\alpha_\ell x_3} - \frac{k}{(\alpha_t\alpha_\ell)^{1/2}} e^{-\alpha_t x_3}\right\} \sin(kx_1 - \omega t) \ . \tag{2.43b}$$

From Eqs. (2.42) we find that the attenuation lengths of a Rayleigh wave into the elastic medium, α_ℓ^{-1} and α_t^{-1}, are of the order of the wavelength $\lambda_R = (2\pi/k)$ of the wave along the surface. For example, in the typical case in which $c_\ell^2 = 3c_t^2$ (the so-called Poisson case, for which the two Lamé constants are equal), we find that $\xi = (c_R/c_t) = [2 - \frac{2}{3}\sqrt{3}\,]^{1/2} = 0.9194$, whence it follows that

$$\alpha_\ell^{-1} = \frac{1.18}{2\pi}\lambda_R = 0.188\lambda_R, \qquad \alpha_t^{-1} = \frac{2.54}{2\pi}\lambda_R = 0.404\lambda_R. \tag{2.44}$$

We also see, from Eqs. (2.43), that the particle displacements in a Rayleigh wave execute ellipses in the sagittal plane. Near the surface the Rayleigh wave motion is retrograde, and it reverses its sense at depths greater than approximately one-fifth of a wavelength. The major axes of the ellipses are normal to the surface, and the aspect ratio varies with depth. The components of the displacement field as functions of the distance into the medium from the surface are shown in Fig. 2.2.

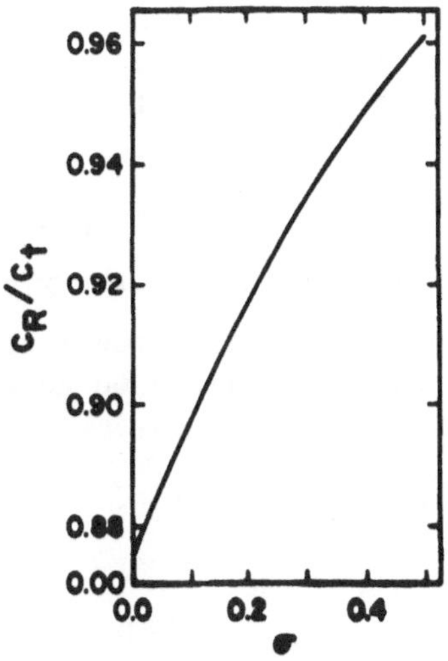

Fig. 2.1. The speed of Rayleigh waves, c_R/c_t, as a function of Poisson's ratio σ [Ref. 27].

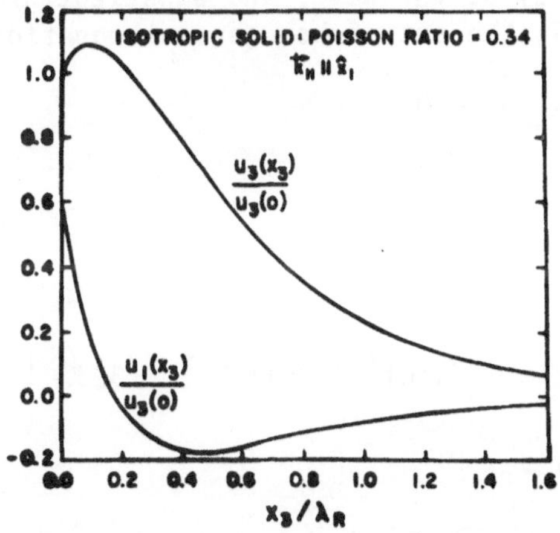

Fig. 2.2. Normalized displacements in a Rayleigh wave on an isotropic medium as functions of the distance into the medium from the surface [Ref. 3].

2.1.1.2. Anisotropic Media

When we turn our attention to surface acoustic waves on anisotropic elastic media, we are required to solve Eqs. (2.13) and (2.19) simultaneously. In general this requires the use of a computer. However, there are situations in which essentially analytic solutions are still possible.

To explore them we rotate the crystal in such a way that the wave vector $\vec{k}_{\parallel}$ in Eq. (2.12) is always in the x_1-direction. The necessary transformation of the elastic moduli is carried out with the use of Eq. (2.4), and in what follows we assume this has been done.

A displacement field of the form of

$$u_\alpha(\vec{x},t) = U_\alpha \, e^{ikx_1 - k\beta x_3 - i\omega t} \qquad \mathrm{Re}\,\beta > 0, \tag{2.45}$$

describes a straight crested wave that propagates in the x_1-direction and whose amplitude decays exponentially with increasing distance into the medium from the surface. Setting $\omega = vk$, where v is the phase velocity of the wave, and substituting Eq. (2.45) into Eqs. (2.7) yields the set of homogeneous equations

$$\sum_\mu [\Gamma_{\alpha\mu}(\beta) - \delta_{\alpha\mu} v^2] U_\mu = 0 \tag{2.46a}$$

for the amplitudes $\{U_\alpha\}$, where

$$\Gamma_{\alpha\mu}(\beta) = \frac{1}{\rho}\,[c_{\alpha 1\mu 1} + (c_{\alpha 1\mu 1} + c_{\alpha 3\mu 1})\, i\beta - c_{\alpha 3\mu 3}\beta^2].$$

$$= \Gamma_{\mu\alpha}(\beta) \; . \tag{2.46b}$$

In order that Eqs. (2.46) have nontrivial solutions the determinant of the coefficients must vanish. For each value of the phase velocity v, setting this determinant equal to zero yields a sixth order equation for the decay constant β. There are therefore six values of β for each value of v. These six roots are either real or members of complex conjugate pairs. Since we are concerned here with surface waves, it is only in the values of β whose real parts are positive that we are interested. In the general case there are three such roots, and we will denote them by $\beta_j(v)$ ($j = 1, 2, 3$). The amplitudes $\{U_\alpha\}$ for any β_j are related by

$$\frac{U_1^{(j)}}{C_1^{(j)}} = \frac{U_2^{(j)}}{C_2^{(j)}} = \frac{U_3^{(j)}}{C_3^{(j)}} = K_j \qquad j = 1, 2, 3, \tag{2.47}$$

where the $\{K_j\}$ are constants, and $C_\alpha^{(j)}(v)$ ($\alpha = 1, 2, 3$) are the cofactors of the elements in the first row of the matrix $\Gamma_{\alpha\mu}(\beta_j(v)) - \delta_{\alpha\mu}v^2$. $C_\alpha^{(j)}(v)$ is therefore a function of $\beta_j(v)$ and of v:

$$C_1^{(j)}(v) = [\Gamma_{22}(\beta_j(v))-v^2][\Gamma_{33}(\beta_j(v))-v^2] - \Gamma_{23}^2(\beta_j(v)) \tag{2.48a}$$

$$C_2^{(j)}(v) = -\Gamma_{21}(\beta_j(v))[\Gamma_{33}(\beta_j(v))-v^2] + \Gamma_{23}(\beta_j(v))\Gamma_{31}(\beta_j(v)) \tag{2.48b}$$

$$C_3^{(j)}(v) = -\Gamma_{31}(\beta_j(v))[\Gamma_{22}(\beta_j(v))-v^2] + \Gamma_{32}(\beta_j(v))\Gamma_{21}(\beta_j(v)) \, . \tag{2.48c}$$

To satisfy the boundary conditions (2.11) we superpose the solutions corresponding to the three $\beta_j(v)$ to obtain

$$u_\alpha(\vec{x},t) = \sum_{j=1}^{3} C_\alpha^{(j)}(v)K_j e^{ikx_1 - k\beta_j(v)x_3 - ivkt} \, . \tag{2.49}$$

Substitution of this form for the displacement field into Eqs. (2.10) leads to a set of homogeneous linear equations for the $\{K_j\}$. The condition for a nontrivial solution is that the determinant of the coefficients of the $\{K_j\}$ vanish,

$$|M_{\alpha j}(v)| = 0, \tag{2.50}$$

where

$$M_{\alpha j}(v) = \frac{1}{\rho}\sum_\mu [C_{\alpha 3\mu 1} + iC_{\alpha 3\mu 3}\beta_j(v)]C_\mu^{(j)}(v) \tag{2.51}$$

$$\alpha = 1, 2, 3; \; j = 1, 2, 3.$$

Equation (2.50) is the equation for the phase velocity v of the surface acoustic wave. In general it has to be solved numerically.

The preceding analysis simplifies considerably for propagation of a surface acoustic wave along a direction of high symmetry on a surface of high symmetry[28]. Thus, for example if $\Gamma_{12}(\beta) = \Gamma_{21}(\beta)$ and $\Gamma_{23}(\beta) = \Gamma_{32}(\beta)$ vanish in Eq. (2.46a), the latter system of equations breaks up into the pair

$$\begin{pmatrix} \Gamma_{11}(q) - v^2 & \Gamma_{13}(q) \\ \Gamma_{31}(q) & \Gamma_{33}(q) - v^2 \end{pmatrix} \begin{pmatrix} U_1 \\ U_3 \end{pmatrix} = 0 \tag{2.52a}$$

and

$$(\Gamma_{22}(\beta) - v^2)U_2 = 0. \tag{2.53a}$$

From Eq. (2.46b) this requires the vanishing of six elements of the elastic modulus tensor, viz. those that contain the index 2 only once:

$$C_{1121} \equiv c_{16} = 0, \; C_{1123} \equiv c_{14} = 0; \; C_{1321} \equiv c_{56} = 0;$$

$$C_{1323} \equiv c_{54} = 0; \; C_{2133} \equiv c_{63} = 0; \; C_{2333} \equiv c_{43} = 0. \tag{2.54}$$

This will be the case, for example, when x_2 is parallel to a two-fold rotation axis of the crystal or is perpendicular to a reflection plane. This result can be obtained from the transformation law for the elastic modulus tensor under an operation from the point group of the crystal, Eq. (2.5), with the use of

$$\overleftrightarrow{R} = \begin{vmatrix} -1 & 0 & 0 \\ 0 & 1 & 0 \\ 0 & 0 & -1 \end{vmatrix} \tag{2.55a}$$

in the case that R is a rotation through 180° about x_2, and

$$\overleftrightarrow{R} = \begin{vmatrix} 1 & 0 & 0 \\ 0 & -1 & 0 \\ 0 & 0 & 1 \end{vmatrix} \tag{2.55b}$$

in the case that R is reflection in the x_1x_3-plane.

The boundary conditions also decouple in this case:

$$[(c_{15}\frac{\partial}{\partial x_1} + c_{55}\frac{\partial}{\partial x_3})u_1 + (c_{55}\frac{\partial}{\partial x_1} + c_{35}\frac{\partial}{\partial x_3})u_3]_{x_3=0} = 0 \quad (2.52b)$$

$$[(c_{13}\frac{\partial}{\partial x_1} + c_{35}\frac{\partial}{\partial x_3})u_1 \ (c_{35}\frac{\partial}{\partial x_1} + c_{33}\frac{\partial}{\partial x_3})u_3]_{x_3} = 0 = 0 \quad (2.52c)$$

and

$$[(c_{46}\frac{\partial}{\partial x_1} + c_{44}\frac{\partial}{\partial x_3})u_2]_{x_3= 0} = 0. \quad (2.53b)$$

Let us focus our attention for the moment on the equations for $u_2(\vec{x},t) = 0$. From Eq. (2.53a) and Eqs. (2.46b) and (2.54), we find that for U_2 not to be identically zero we must have

$$\frac{1}{\rho}[c_{66} + 2c_{46}i\beta - c_{44}\beta^2] - v^2 = 0. \quad (2.56)$$

The solution of this equation for that has a positive real part is

$$\beta = \frac{1}{c_{44}}[c_{44}(c_{66} - \rho v^2) - c_{46}^2]^{1/2} + i\frac{c_{46}}{c_{44}}$$

$$\equiv \beta_R(v) + i\beta_I \quad (2.57)$$

provided that

$$c_{44}(c_{66} - \rho v^2) - c_{46}^2 > 0. \quad (2.58)$$

It is only when Eq. (2.58) is satisfied that a surface wave can exist. The 2-component of the displacement field therefore has the form

$$u_2(\vec{x},t) = U_2 e^{ikx_1 - k(\beta_R(v)+i\beta_I)x_3 - ivkt} . \quad (2.59)$$

The boundary condition (2.53b) yields the equation for determining v:

$$[c_{46} + c_{44}(i\beta_R(v) - \beta_I)]U_2 = 0 \tag{2.60a}$$

Or, in view of Eq. (2.57),

$$\beta_R(v)U_2 = 0. \tag{2.60b}$$

Thus, the satisfaction of this equation requires that (a) $U_2 \neq 0$, $\beta_R(v) = 0$; (b) $U_2 = 0$, $\beta_R(v) = 0$; or (c) $U_2 = 0$, $\beta_R(v) = 0$. None of these alternatives corresponds to a surface acoustic wave. (See, however, the discussion following Eq. (2.28).)

We have therefore reached an important conclusion: A surface acoustic wave propagating in the x_1-direction in the plane $x_3 = 0$, when the x_2-axis is parallel to a two-fold rotation axis or perpendicular to a reflection plane, must be polarized in the sagittal plane.

The solution of Eqs. (2.52) to obtain the dispersion relation and displacement field of a surface acoustic wave polarized in the sagittal plane now proceeds in exactly the same fashion as in the case of an elastically isotropic medium discussed in the preceding subsection, and will not be repeated here.

In the case of propagation in an arbitrary direction on an anisotropic surface (i.e. in a direction other than one perpendicular to a two-fold reflection axis lying the the surface or in a direction for which the sagittal plane is a plane of reflection symmetry) we are forced to use Eqs. (2.13) and (2.19) to obtain the dispersion relation for surface acoustic waves and the associated displacement field. The resulting surface acoustic waves have the following general properties.

1. The surface acoustic wave generally has three nonzero displacement components and is the superposition of three partial waves.
2. The partial waves are characterized by the corresponding decay constants $\{\beta_j(\hat{k}_\parallel v)\}$. Three different situations are possible.
   a) All three $\{\beta_j(\hat{k}_\parallel v)\}$ are real and positive. The resulting surface acoustic wave is called an ordinary Rayleigh wave.
   b) It can happen that the sixth degree equation for the $\{\beta_j(\hat{k}_\parallel v)\}$ degenerates into a cubic equation in β^2 with real coefficients. Thus not only is β^* a solution if β is, but so is $-\beta$. It is then possible for one of the

$\{\beta_j(\hat{k}_\parallel v)\}$ to be real and the other two to be complex conjugates of each other, with Re $\beta_j(\hat{k}_\parallel v)$ positive in each case. The amplitudes of the latter two waves are such that the corresponding contribution to the displacement field has the form $\exp(-\hat{k}_\parallel \beta_R(\hat{k}_\parallel v)x_3)\sin k_\parallel (\beta_I(\hat{k}_\parallel v)x_3 + \delta)$. The decay of the amplitude of this part of the displacement field thus has a more complicated dependence on x_3 than the simple exponential decay characteristic of Rayleigh waves on isotropic media. The resulting surface acoustic wave is called a <u>generalized Rayleigh wave</u>.(29,30) These waves appear to have been discovered by Stoneley(30) who, however, discarded them, and was therefore led to the erroneous conclusion that surface acoustic waves were possible on the (001) face of only a very restricted class of cubic crystals. A similar suggestion was made by Synge(31), viz. that Rayleigh waves can travel unattenuated only in certain discrete directions on the free surface of an anisotropic elastic half-space. The basis for this suggestion was the observation that the dispersion relation determining the phase velocity of the surface wave is apparently complex, so that any solution of this equation must be complex, which corresponds to a wave that is attenuated. Only for certain directions of propagation would the dispersion relation have a real solution for the phase velocity. However, Stroh(33) showed that Synge's suggestion is incorrect by showing that the dispersion relation can always be reduced to one purely real equation. An improved proof of Stroh's result was subsequently given by Currie.(34) Thus, there is no difference in this respect between isotropic and anisotropic media: a Rayleigh wave can travel in any direction provided the single real dispersion relation has a real root corresponding to a surface wave. In addition, Burridge(35) showed indirectly that Synge's suggestion is incorrect by using an energy argument. Subsequently, Lim and Farnell(19,20), apparently unaware of Stroh's earlier work, sought numerically to find directions on free crystal surfaces in which it is impossible to propagate unattenuated Rayleigh waves, and found none.

c) One or more of the $\beta_j(\hat{k}_\parallel v)$ is complex in such a way that the corresponding partial wave radiates energy into the interior of the solid from the surface. By conservation of energy the wave is therefore attenuated in the direction of propagation, and the components of $\vec{k}_{\parallel}$ become complex. Such surface acoustic waves are called pseudo-surface or leaky surface acoustic waves. They will be discussed in greater detail in the following subsection.

3. The group velocity vector $\vec{v}_R(\vec{k}_{\parallel}) = \nabla_{\vec{k}_{\parallel}} \omega_R(\vec{k}_{\parallel})$ for a surface acoustic wave is not parallel to the wave vector $\vec{k}_{\parallel}$ in general. This gives rise to the possibility of the focusing of surface acoustic waves that will be discussed in Section 8.

2.1.1.3. Leaky Surface Acoustic Waves

It was pointed out in subsection 2.1.1.2 that one of the consequences of elastic anisotropy for surface acoustic waves is the possibility of the existence of a pseudosurface, or leaky surface acoustic wave. This is a wave that depends on the presence of a surface for its existence, and whose phase velocity is greater than that of the slowest bulk transverse wave. It can therefore decay into the continuum of bulk modes, and acquires a finite lifetime thereby, through what amounts to Landau damping.[36] Another way of describing the finite lifetime is that the pseudosurface wave is attenuated as it propagates along the surface of a solid. As a consequence the components of $\vec{k}_{\parallel}$ acquire imaginary parts whose signs are such that $\mathrm{Im}\vec{k}_{\parallel} \cdot \vec{x}_{\parallel}$ is positive when $\mathrm{Re}\vec{k}_{\parallel} \cdot \vec{x}_{\parallel}$ is positive so that the wave is attenuated essentially in the direction of propagation. The requirement that a leaky surface acoustic wave decay in the direction of propagation serves as a kind of boundary condition in the study of these waves. The decay constants $\alpha_j(\vec{k}_{\parallel}\omega)$ for such a wave acquire imaginary parts as a consequence of the complex nature of $\vec{k}_{\parallel}$, and the sign of at least one of them is such that energy is radiated away from the surface into the interior of the solid. Since the solid is not a lossy medium, i.e. the elastic moduli are assumed to be real, it is this radiation of energy away from the surface that, by energy conservation, leads to the attenuation of a pseudosurface wave as it propagates across the surface, and gives rise to the alternative, and more descriptive, name of leaky surface wave that will be used here. In many cases the attenuation is small enough that the leaky surface acoustic wave is readily observable. Such waves on the surfaces of semi-infinite anisotropic elastic media were discovered by Engan *et al*[37] in 1967.

Leaky surface acoustic waves on anisotropic media were subsequently studied extensively, both theoretically and experimentally, by several authors.[38-45] A typical result[40], which shows theoretical and experimental phase velocities for surface acoustic waves propagating along the (100) face of copper as a function of the angle θ between the sagittal plane and the [010] direction, is shown in Fig. 2.3. The velocities are normalized to the shear wave velocity $c_s = (c_{44}/\rho)^{1/2}$. The curve is symmetric about the value of $\theta = 45^o$ (the [011] direction). It is seen that in addition to a generalized Rayleigh wave whose phase velocity lies below that of the slowest bulk wave for all values of the angle θ in the range 0-45°, there is a leaky surface wave whose phase velocity is greater than that of the slowest bulk wave, and which exists for only a limited range of angles greater than about 22°. If the displacement field in this leaky surface wave is written in the form

$$u_\alpha(x,t) = \sum_{j=1}^{3} C_\alpha^{(j)} K_j e^{ik(\ell_1 x_1 + \ell_2 x_2) - k\beta_j x_3 - ikvt} \tag{2.61}$$

on the assumption that the medium occupies the half-space $x_3 > 0$, then a numerical calculation(44) for $\theta = 30^o$ yields the following values for the parameters appearing in it:

$$v = 2268\text{m/sec}$$

$$\ell_1 = \frac{\sqrt{3}}{2}(1 + i9.1 \times 10^{-5})$$

$$\ell_2 = \frac{1}{2}(1 + i9.1 \times 10^{-5})$$

$$\beta_1 = 0.297 + i0.609 \qquad |K_1| \approx |K_2| \approx \frac{2}{\sqrt{2}} \tag{2.62}$$

$$\beta_2 = 0.297 - i0.609$$

$$\beta_3 = -3.1 \times 10^{-4} - i0.544 \qquad |K_3| = 0.125 \ .$$

From these results we see that this leaky surface wave is primarily the superposition of two waves that are localized to the surface ($\mathrm{Re}\beta_{1,2} > 0$), because K_1 and K_2 are much larger in magnitude than K_3. The ratio of the imaginary to the real part is

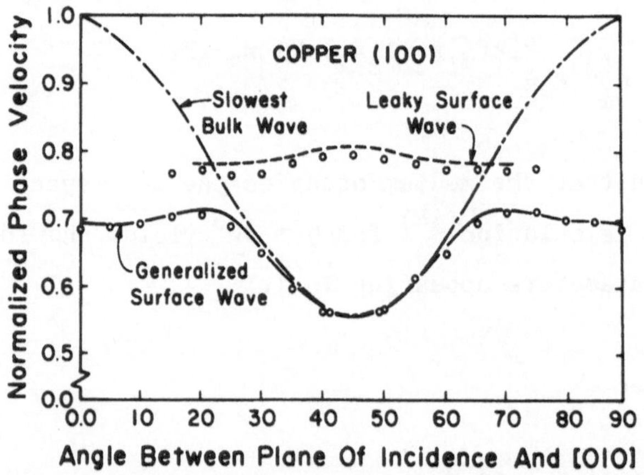

Fig. 2.3. Theoretical and experimental phase velocities for surface acoustic waves on the (100) face of copper as functions of the angle between the sagittal plane and the [010] direction [Ref. 40].

equal for ℓ_1 and ℓ_2, so that the attenuation of the wave is in a direction normal to the surface wavefront and is small. However, from the signs of the real and imaginary parts of β_3 we see that the leaky surface acoustic wave not only has a component that radiates energy into the bulk but one whose amplitude <u>grows</u> exponentially with increasing distance into the elastic medium from the surface. Thus a leaky surface acoustic wave does not satisfy the vanishing boundary condition at $x_3 = \infty$ that is required of ordinary Rayleigh and generalized Rayleigh waves. We will argue below that this property of leaky surface waves is an unavoidable consequence of the divergence of the displacement field as $x_1 \to -\infty$, $x_2 \to -\infty$, caused by the imaginary parts of ℓ_1 and ℓ_2.

It is now known that the presence of anisotropy is not necessary for the existence of leaky surface acoustic waves. In fact, the earliest example of a leaky surface acoustic wave was a wave at the plane interface between an isotropic elastic medium and a liquid(46-49). In this case, the wave is localized in the elastic medium, but as it propagates along the interface it radiates energy into the liquid, and is attenuated in the direction of propagation as a result. Other types of leaky surface acoustic waves at an interface in which attenuation takes place due to the radiation of energy into the medium adjacent to the one in which the wave is localized are Lamb waves in plates immersed in a liquid(50,51), waves at the boundary of a liquid with a solid layer(52), and waves at the plane interface between two solid media(53,54).

A more interesting type of leaky surface acoustic wave is the class of waves at a solid-vacuum interface for which energy is radiated not into an adjacent medium but back into the same medium in which the wave is localized. We have already seen an example of this in the case of leaky surface acoustic waves in anisotropic elastic media. An earlier example is the case of Rayleigh waves propagating circumferentially around a cylindrical isotropic solid surface that is concave toward the vacuum.(55-57). Yet another example is provided by a semi-infinite isotropic elastic medium in contact with a thin liquid layer(58-60).

Perhaps on the basis of the preceding examples the belief has arisen that anistropy, or a solid-liquid or solid-solid interface, or a curved surface are <u>necessary</u> for the existence of leaky surface acoustic waves. In fact, it has been stated explicitly(61) that "solutions of the pseudosurface-wave type do not occur in isotropic media..." In fact, as will be shown below, this statement is incorrect. Leaky surface acoustic waves are known to exist on the planar, stress-free surface of a semi-infinite, isotropic elastic medium(62-65), and their existence has led to the recent discovery of a new type of leaky surface acoustic wave in anisotropic media(66). A discussion of the isotropic case is useful, because its simplicity allows many of the necessary calculations to be carried out essentially analytically, rather than numerically, and thus to reveal features of the solution that may be obscured by a purely numerical calculation.

We have already seen (Eq. (2.35)) that the dispersion relation for sagitally polarized surface acoustic waves propagating in the x_1-direction on the planar, stress-free surface of a semi-infinite, isotropic elastic medium occupying the region $x_3 > 0$ is

$$4k^2\alpha_\ell(k\omega)\alpha_t(k\omega) - (k^2 + \alpha_t^2(k\omega))^2 = 0. \qquad (2.35)$$

In analyzing this equation and its solutions it is convenient to make the replacements

$$k^2 = u, \quad a = \omega^2/c_\ell^2, \quad b = \omega^2/c_t^2, \quad a < b, \qquad (2.63)$$

after which it becomes

$$S(u) = 4u\sqrt{u-a}\,\sqrt{u-b} - (2u-b)^2 = 0. \qquad (2.64)$$

The problem of obtaining k from a solution u of Eq. (2.64) is simplified by the fact that we seek a displacement field that propagates in the x_1-direction and attenuates in the direction of propagation. Thus, if we assume ω real and k complex,

$$k = k_R + ik_I$$

then we must have that both k_R and k_I are positive. These requirements tell us which sign of the square root of u gives the physically acceptable solution.

The square roots in the expression on the left hand side of this equation give rise to two branch points in the complex u-plane at u = a and u = b. We take as branch cuts segments of the real u-axis, $(-\infty,a)$ for $\sqrt{u-a}$ and $(-\infty,b)$ for $\sqrt{u-b}$ (Fig. 2.4). The Riemann surface on which S is single-valued consists of four sheets, each of which is defined by a combination of sheets of $\sqrt{u-a}$ and $\sqrt{u-b}$, and which intersect along certain segments of the real u-axis. These can be denoted by $S_1 = (+\,+)$, $S_2 = (+$

-), $S_{-1} = (- -)$, and $S_{-2} = (- +)$, according to

$$(\text{sgn Re}\sqrt{u - a},\ \text{sgn Re}\sqrt{u - b}). \tag{2.65}$$

In fact, we can confine our attention to the sheets S_1 and S_2 because the remaining sheets S_{-1} and S_{-2} yield identical solutions.

The physically acceptable solution of Eq. (2.64) on the sheet $S_1 = (+ +)$ is just the solution corresponding to the Rayleigh wave, and will not be discussed further here.

Solutions on the sheet $S_2 = (+ -)$ have been obtained by a complex root search method. For example, in the case that $c_t/c_\ell = 0.5$ it is found that

$$u = \frac{\omega^2}{c_t^2}(0.26016 + i0.07676) \tag{2.66}$$

which corresponds to

$$k = k_R + ik_I \equiv \frac{\omega}{c_1} + i\frac{\omega}{c_2} \tag{2.67}$$

with

$$\frac{c_1}{c_t} = 1.94\ c_t,\quad \frac{c_2}{c_t} = 13.43\ c_t. \tag{2.68}$$

In this case

$$\begin{aligned} \alpha_\ell(k\omega) &= \sqrt{u - a} = \frac{\omega}{c_t}(0.2097 + i0.1831) \\ \alpha_t(k\omega) &= \sqrt{u - b} = \frac{\omega}{c_t}(-0.04475 - i0.8611)\ . \end{aligned} \tag{2.69}$$

Since in this case $k_I/k_R = c_1/c_2 = 0.144$, we see that this is a highly damped wave. From Eqs. (2.41) or (2.43) we see that it is a superposition of a longitudinal wave that is bound to the surface ($\text{Re}\alpha_\ell(k\omega) > 0$) and propagating toward it ($\text{Im}\alpha_\ell(k\omega) > 0$), and a transverse wave that radiates energy into the bulk of the medium ($\mathrm{Im}\alpha_t(k\omega) < 0$) and at the same time has an amplitude that grows exponentially with increasing distance into the medium ($\mathrm{Re}\alpha_t(k\omega) < 0$). The component of the time averaged Poynting vector normal to the surface for this wave, $\langle P_3 \rangle$, obtained from Eqs. (3.9) and (3.11) is positive far from the surface, which shows that there is a net flux of energy away from the surface into the interior of the medium, that is the reason for the attenuation of the wave as it propagates along the surface.

The phase velocities c_1/c_t and c_2/c_t defined by Eq. (2.67) are plotted as functions of c_ℓ/c_t in Figs. (2.5a) and (2.5b), respectively.

We now argue that the exponential growth of the amplitude of the leaky component of the displacement field is a physically necessary consequence of the leaky nature of this component.[65]

At time t, at some point of observation $(0, x_3)$ above the surface we are seeing the wave that left the surface from the point $(x_1, 0)$ such that $(x_1^2 + x_3^2)^{1/2} = \tilde{c}t$, where $\tilde{c}$ is the propagation speed of the wave. At the same time t (same $\tilde{c}t$) at a higher point $(0, x_3)$ the point of origin $(x_1, 0)$ of the wave must correspond to a smaller value of x_1. Consequently, the amplitude of the source on the surface in the latter case must be larger than in the former, according to the damping law $\exp(-k_I x_1)$. The wave on the surface that is attenuated as $\exp(-k_I x_1)$ has an unavoidable divergence at $x_1 = -\infty$, that is the source of the apparent divergence at $x_3 = +\infty$. In a real experiment the wave is launched from some finite value of x_1, and there is no wave beyond some value of x_3 for each value of x_1.

Glass and Maradudin[65] have shown that this leaky surface acoustic wave contributes a damped pole to the imaginary part, $\mathrm{Im}\, D_{ii}(\vec{k}_\parallel \omega | x_3 x_3)$, of the dynamical Green's tensor for a semi-infinite, isotropic elastic medium bounded by a planar, stress-free surface, considered as a function of ω. This function, which will be discussed further in Section 8, is proportional to the

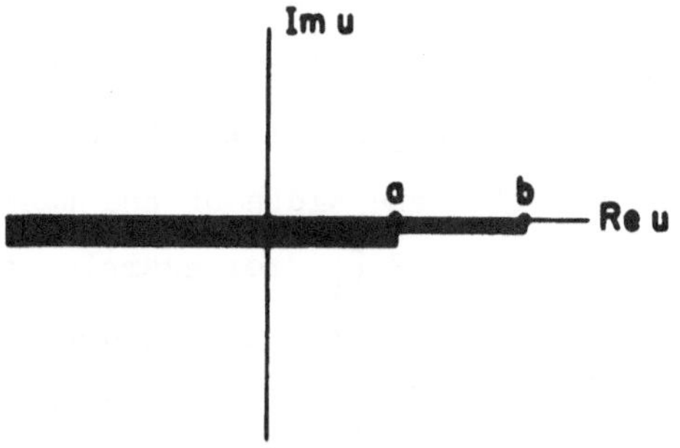

Fig. 2.4. The branch cuts defining the Riemann surface on which the Rayleigh dispersion relation, Eq. (2.64), is single-valued. In this figure $a = \omega^2/c_\ell^2$, $b = \omega^2/c_t^2$.

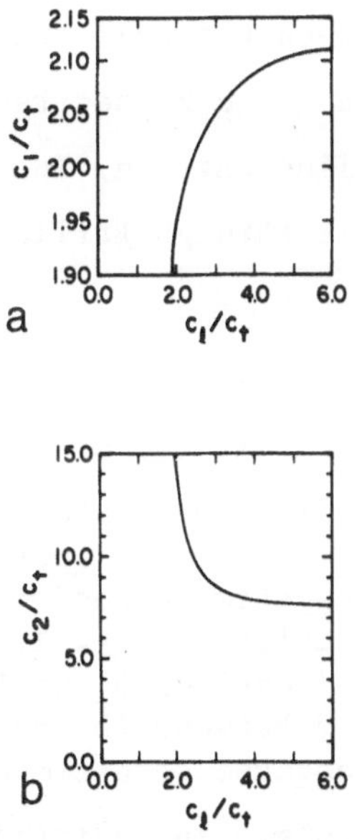

Fig. 2.5. (a) The phase velocity c_1/c_t (Eq. (2.67)) for a leaky surface acoustic wave on a semi-infinite, isotropic elastic medium as a function of c_ℓ/c_t; (b) the phase velocity c_2/c_t (Eq. (2.67)) for a leaky surface acoustic wave on a semi-infinite, isotropic elastic medium as a function of c_ℓ/c_t.

partial density of vibrational modes in the medium at a distance x_3 into the bulk. It has no structure for $i = 3$, corresponding to transverse waves; however, for $i = 1$ -- longitudinal waves -- there is a broad, but well defined, peak for $c_t k < \chi < c_\ell k$ when $x_3 \lesssim k^{-1}$ (see Fig. 2.6). As x_3 increases this peak eventually vanishes, while a new peak at the bulk frequency $\omega = c_t k$ forms. If the position and width of the peak in $\mathrm{Im}\, D_{11}(\vec{k}_\parallel \omega | x_3 x_3)$, as a function of c_ℓ, for example, for some fixed, small, value of x_3, the same functional dependence is found as when the solutions of Eqs. (2.35) for the real and imaginary parts of k are similarly plotted. In fact, the results of the two approaches can be made to overlap for the proper choice of x_3 in the Green's tensor (between $\frac{1}{4} k^{-1}$ and $\frac{1}{2} k^{-1}$).

The existence of short-lived surface states of longitudinal character with frequencies between the bulk transverse and longitudinal thresholds, $\omega = c_t k$ and $\omega = c_\ell k$, has been demonstrated experimentally for polycrystalline (Aℓ, Cu, Ta, Ni, Fe) as well as crystalline (Cr, GaAs) materials through Brillouin scattering by Sandercock(67) (see Fig. 2.7).

Recently, such high frequency leaky surface acoustic waves have been found by Camley and Nizzoli(66) on the (100) surface of a cubic crystal. Their results for the phase velocities of the several waves that can propagate on such a surface as functions of the angle the direction of propagation makes with the [001] direction are depicted schematically in Fig. 2.8. The leaky surface acoustic wave discovered earlier by Rollins, Lim, and Farnell(40) has a phase velocity between the phase velocities of the slower and faster bulk transverse acoustic waves, that are denoted by T_1 and T_2, respectively. In addition to this leaky surface acoustic wave there is a second, high frequency, leaky surface acoustic wave whose phase velocity is above that of the T_2 bulk transverse acoustic wave, but below that of the bulk longitudinal wave (L) propagating along the surface. This wave is the analogue of the one discussed by Glass and Maradudin(65), and like that one is strongly attenuated. This can be understood as

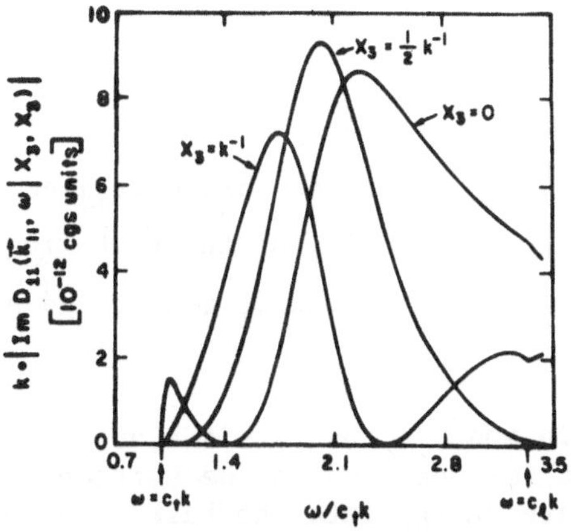

Fig. 2.6. Imaginary part of the 11 component of the dynamic Green's tensor for a semi-infinite, isotropic elastic medium ($c_t/c_\ell = 0.3$) bounded by planar, stress-free surface, for $\vec{k}_\parallel = \hat{x}_1 k$, i.e. for longitudinal vibrations, and for three distances into the medium [Ref. 65].

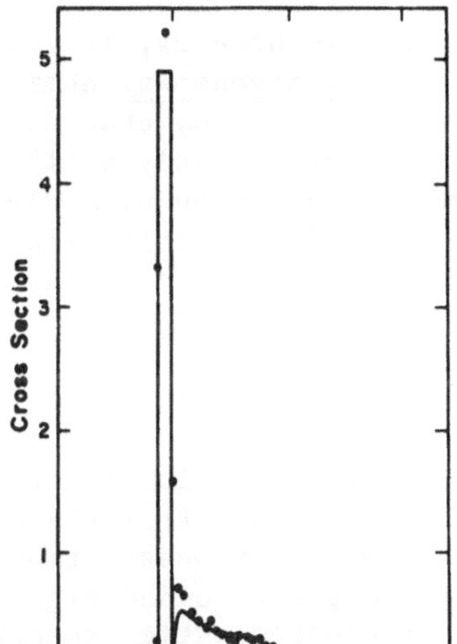

Fig. 2.7. Experimental Brillouin spectrum for scattering from the surface of a polycrystalline sample of Aℓ(•) [Ref. 67], together with a theoretical calculation of it in the approximation in which polycrstalline aluminum is replaced by an isotropic elastic medium with $c_\ell = 2c_t$ (---) [Ref. 68]. The shoulder on the spectrum in the vicinity of $\omega < c_\ell k_\parallel$ is evidence for a high frequency, intrinsically leaky surface acoustic wave.

follows. As one approaches the isotropic limit, the phase velocity curves labeled T_1 and T_2 collapse into one curve with a phase velocity c_t. The low frequency leaky surface acoustic wave is essentially squeezed out of existence, since it lies between T_1 and T_2. However, the phase velocity of the high frequency leaky surface acoustic wave remains between T_1 and L, and so remains in the isotropic limit.

The nature of this mode is that it is a superposition of a longitudinal wave that is localized to the surface, and two transverse waves that radiate energy into the bulk, and whose amplitudes grow exponentially with increasing distance into the bulk of the solid. Camley and Nizzoli argue that this new leaky surface acoustic wave should be observable in Brillouin scattering experiments.

2.1.2. Inhomogeneous Media

We have already noted that Rayleigh surface acoustic waves were first studied in the context of the propagation of seismic shocks through the earth's crust. However, it was soon recognized that Rayleigh's assumption of a homogeneous, semi-infinite elastic medium to model the earth was too simple, even if the wavelengths of the seismic shock waves are sufficiently small compared with the earth's radius that the approximation of a flat, stress-free surface was justified. The fact is that the elastic properties of the earth's crust are different from those of the underlying material.

In the years following Rayleigh's pioneering work many papers have been written on the propagation of surface acoustic waves on inhomogeneous elastic media, whose elastic properties vary with distance into the interior of the media from the surface. The simplest model of such a system, on the assumption of a planar surface, consists of an elastic plate bonded rigidly to a half space (or substrate) with different elastic properties. Even more complicated structures involving multiple layers of different materials of different thicknesses are often required in the context of seismological problems.

In parallel with studies of the propagation of surface acoustic waves on such layered media, similar calculations have been carried out for inhomogeneous media in which the elastic properties vary continuously with distance into the medium from the surface.

In the remainder of this subsection we will consider the propagation of surface acoustic waves on both types of inhomogeneous media. We will find how the properties of surface acoustic waves that exist in the absence of these inhomogeneities are modified by the latter, and we will see that under appropriate conditions these inhomogeneities can give rise to types of surface acoustic waves that cannot exist on homogeneous media.

2.1.2.1. Layered Media

In this subsection we will discuss three different types of elastic guided waves that can exist in the case of a single elastic layer on a different elastic substrate. Originally of interest only to seismologists, such waves have taken on a new importance in recent years in the context of high frequency acoustic surface wave devices for electronic signal processing.

In the structure we will consider, the region $0 < x_3 < d$ is occupied by an elastic medium characterized by a mass density ρ', and elastic moduli $C'_{\alpha\beta\mu\nu}$. The semi-infinite region $x_3 > d$ is filled by an elastic medium whose density is ρ and whose elastic moduli are $C_{\alpha\beta\mu\nu}$ (Fig. 2.9). The equations of motion of this system are

$$\rho' \ddot{u}_\alpha = \sum_{\beta\mu\nu} C'_{\alpha\beta\mu\nu} \frac{\partial^2 u_\mu}{\partial x_\beta \partial x_\nu} \qquad 0 < x_3 < d \tag{2.70a}$$

$$\rho \ddot{u}_\alpha = \sum_{\beta\mu\nu} C_{\alpha\beta\mu\nu} \frac{\partial^2 u_\mu}{\partial x_\beta \partial x_\nu} \qquad x_3 > d \,. \tag{2.70b}$$

At the interface $x_3 = d$ between the two media the displacement components must be continuous,

$$u_\alpha \Big|_{x_3 = d+} = u_\alpha \Big|_{x_3 = d-} , \tag{2.71}$$

and the stresses acting in this plane must also be continuous,

$$T_{\alpha 3} \Big|_{x_3 = d+} = T_{\alpha 3} \Big|_{x_3 = d-} , \tag{2.72}$$

where

$$T_{\alpha\beta} = \sum_{\mu\nu} C'_{\alpha\beta\mu\nu} \frac{\partial u_\mu}{\partial x_\nu} \qquad 0 < x_3 < d \tag{2.73a}$$

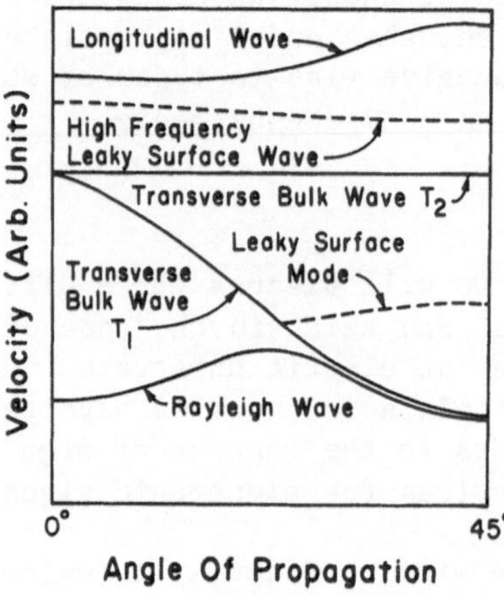

Fig. 2.8. Schematic illustration of the various bulk and surface modes that propagate on a (001) surface of a cubic crystal. The angle of propagation is the angle from the [100] direction [Ref. 66].

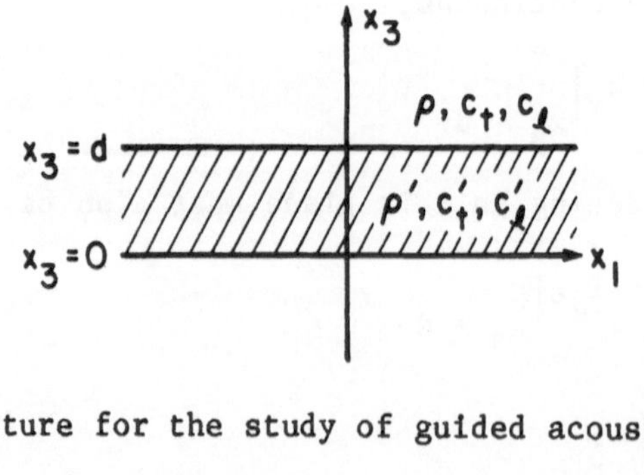

Fig. 2.9. Structure for the study of guided acoustic waves.

$$= \sum_{\mu\nu} C_{\alpha\beta\mu\nu} \frac{\partial u_\mu}{\partial x_\nu} \qquad x_3 > d. \tag{2.73b}$$

The surface $x_3 = 0$ is assumed to be stress-free,

$$T_{\alpha 3}\Big|_{x_3 = 0} = 0. \tag{2.74}$$

In the discussion that follows we will assume that the elastic media constituting the layer and the substrate are isotropic, so that from Eq. (2.19)

$$C'_{\alpha\beta\mu\nu} = \rho'(c'^2_\ell - 2c'^2_t)\delta_{\alpha\beta}\delta_{\mu\nu} + \rho' c'^2_t(\delta_{\alpha\mu}\delta_{\beta\nu} + \delta_{\alpha\nu}\delta_{\beta\mu}) \tag{2.75a}$$

$$C_{\alpha\beta\mu\nu} = \rho(c^2_\ell - 2c^2_t)\delta_{\alpha\beta}\delta_{\mu\nu} + \rho c^2_t(\delta_{\alpha\mu}\delta_{\beta\nu} + \delta_{\alpha\nu}\delta_{\beta\mu}). \tag{2.75b}$$

The introduction of anisotropy into the problem yields no qualitatively different results, but the analysis is complicated significantly thereby.

We now turn to a discussion of the propagation of elastic waves in this structure in directions parallel to the plane $x_3 = 0$. Because of the elastic isotropy of each medium, we can assume that these waves propagate in the x_1-direction.

2.1.2.1.1. Love Waves

The simplest solutions to the problem of obtaining the vibrations of a plate on a semi-infinite medium are Love waves.(69) These are waves that are polarized perpendicular to the sagittal plane (shear horizontal surface acoustic waves), have the character of standing waves in the plate, and whose amplitudes decay exponentially with increasing distance into the substrate from the interface. In this section we obtain the dispersion relation for these waves. We recall that waves of this polarization cannot exist on the surface of a homogeneous medium.

The displacement field in each medium consists of only a component parallel to the x_2-direction that is a function of x_1 and x_3 only. For such a field the equations of motion of the material in the plate and in the substrate are

$$\ddot{u}_2(x_1x_3;t) = c_t'^{-2}\left(\frac{\partial^2}{\partial x_1^2} + \frac{\partial^2}{\partial x_3^2}\right)u_2(x_1x_3;t) \qquad 0 < x_3 < d \tag{2.76a}$$

$$\ddot{u}_2(x_1x_3;t) = c_t^2\left(\frac{\partial^2}{\partial x_1^2} + \frac{\partial^2}{\partial x_3^2}\right)u_2(x_1x_3;t) , \qquad x_3 > d, \tag{2.76b}$$

respectively. We assume solutions of the form

$$u_2(x_1x_3;t) = A(k\omega|x_3)e^{ikx_1-i\omega t} \qquad 0 < x_3 < d \tag{2.77a}$$

$$= B(k\omega|x_3)e^{ikx_1-i\omega t} \qquad x_3 > d, \tag{2.77b}$$

in the two media, where $A(x_3)$ and $B(x_3)$ satisfy the equations

$$\left[\frac{d^2}{dx_3^2} + \hat{\alpha}^2\right] A(x_3) = 0 \qquad 0 < x_3 < d \tag{2.78a}$$

$$\left[\frac{d^2}{dx_3^2} - \alpha^2\right] B(x_3) = 0 \qquad x_3 > d. \tag{2.78b}$$

The coefficients $\hat{\alpha}$ and α entering these equations are defined by

$$\hat{\alpha} = \left(\frac{\omega^2}{c_t'^2} - k^2\right)^{1/2} \tag{2.79a}$$

$$\alpha = \left(k^2 - \frac{\omega^2}{c_t^2}\right)^{1/2} , \tag{2.79b}$$

and are required to be real and positive. This means that we are restricting our attention to that portion of the (ω,k) plane in which

$$\frac{\omega}{c_t} < k < \frac{\omega}{c_t'} , \tag{2.80}$$

and this requires that $c_t' < c_t$. The solutions of Eqs. (2.78) that decay to zero as $x_3 \to -\infty$ are

$$A(x_3) = A^{(+)} e^{i\hat{\alpha} x_3} + A^{(-)} e^{-i\hat{\alpha} x_3} \qquad 0 \leq x_3 \leq d \tag{2.81a}$$

$$B(x_3) = B e^{-\alpha x_3} . \qquad x_3 \geq d. \tag{2.81b}$$

The boundary conditions satisfied by the displacement field are

$$u_2(x_1 x_3;t)\Big|_{x_3 = d-} = u_2(x_1 x_3;t)\Big|_{x_3 = d+} \tag{2.82a}$$

$$T_{23}\Big|_{x_3 = d-} = T_{23}\Big|_{x_3 = d+} ; \; T_{23}\Big|_{x_3 = 0} = 0, \tag{2.82b}$$

where

$$T_{23}(x_1 x_3;t) = \rho' c_t'^2 \frac{\partial u_2}{\partial x_3} \qquad 0 \leq x_3 \leq d \tag{2.83a}$$

$$= \rho c_t^2 \frac{\partial u_2}{\partial x_3} \qquad x_3 \geq d. \tag{2.83b}$$

When we substitute Eqs. (2.77) and (2.81) into these boundary conditions, we obtain a set of three homogeneous linear equations for the coefficients $A^{(+)}$, $A^{(-)}$, and B:

$$A^{(+)} e^{i\hat{\alpha} d} + A^{(-)} e^{-i\hat{\alpha} d} = B e^{-\alpha d} \tag{2.84a}$$

$$A^{(+)} e^{i\hat{\alpha} d} - A^{(-)} e^{-i\hat{\alpha} d} = - \frac{\alpha}{i\hat{\alpha}} \frac{\rho c_t^2}{\rho' c_t'^2} B e^{-\alpha d} \tag{2.84b}$$

$$A^{(+)} - A^{(-)} = 0. \tag{2.84c}$$

The conditions that this set of equations have a nontrivial solution yields the dispersion relation for Love waves,

$$\frac{\rho c_t^2 \alpha}{\rho' c_t'^2 \hat{\alpha}} = \tan \hat{\alpha} d \ . \tag{2.85}$$

This dispersion relation can be solved graphically. We first eliminate k between Eqs. (2.79a) and (2.79b) to obtain

$$\alpha^2 = \omega^2\left(\frac{1}{c_t'^2} - \frac{1}{c_t^2}\right) - \hat{\alpha}^2 \ . \tag{2.86}$$

Graphical plots are then made of this equation and of Eq. (2.85), rewritten as

$$\alpha = \frac{\rho' c_t'^2}{\rho c_t^2} \hat{\alpha} \tan \hat{\alpha} d \ , \tag{2.87}$$

in the coordinates α^2 and $\hat{\alpha}^2$. Solutions for $\hat{\alpha}(\omega)$ are obtained from the intersection points. Since α must be positive, branches with $\hat{\alpha}\tan\hat{\alpha}d$ negative are excluded in plotting Eq. (2.87). The dispersion curve is then obtained by solving for $k(\omega)$ from Eq. (2.79a),

$$k^2(\omega) = \frac{\omega^2}{c_t'^2} - \hat{\alpha}^2(\omega) \ , \tag{2.88}$$

using the solutions obtained from the above graphical construction. Each positive branch of $\hat{\alpha}\tan\hat{\alpha}d$ corresponds to a separate Love wave. The dispersion curves for the three lowest Love waves are plotted in Fig.2.10. At high frequencies these curves all approach the curve $\omega = c_t k$ from above. We note that these waves are dispersive. This is due to the presence of a characteristic length in this problem, the thickness of the slab, d.

The displacement field in a Love wave is finally found to be

$$u_2(x_1 x_3;t) = e^{ikx_1 - i\omega t} B e^{-\alpha d} \frac{\cos\hat{\alpha}x_3}{\cos\hat{\alpha}d} \qquad 0 \leq x_3 \leq d \tag{2.89a}$$

$$= e^{ikx_1 - i\omega t} B e^{-\alpha x_3} \qquad x_3 > d. \tag{2.89b}$$

The dispersion relation for shear horizontal surface acoustic waves for an hexagonal plate on an isotropic substrate has been obtained in Ref. 71.

2.1.2.1.2 Stoneley Waves

The Love Waves discussed in the preceding subsection have their displacement vector perpendicular the sagittal plane. They are localized to the interface between the plate and the substrate only from the substrate side. In this and the next subsection we study waves in the layered system we are considering that are polarized in the sagittal plane, and that are localized to the interface from both the substrate and plate sides, and from only the substrate side, respectively.

The simpler of these modes are the former, the so-called Stoneley waves[72]. The theory of Stoneley waves is simplest when the thickness of the slab d is much larger than their wavelength. In this case the amplitude of the wave in the slab has decayed essentially to zero before the free surface $x_3 = 0$ is reached, so that the boundary condition (2.74) can be ignored. What we have in effect then is two different, semi-infinite elastic media in contact across a planar interface. The region $x_3 > 0$ is filled by a medium characterized by the mass density ρ and the speeds of transverse and longitudinal sound c_t and c_ℓ, respectively; the region $x_3 < 0$ is filled by a medium characterized by the material properties ρ', c_t', and c_ℓ'. This is the model we will use in our discussion of Stoneley waves.

The displacement field in a Stoneley wave has the form

$$\vec{u}(x,t) = (\bar{u}_1(k\omega|x_3),\ 0,\ \bar{u}_3(k\omega|x_3))e^{ikx_1 - i\omega t} \qquad (2.90)$$

where, with no loss of generality, we have assumed a wave propagating in the x_1-direction. The amplitudes $\bar{u}_1(x_3)$ and $\bar{u}_3(x_3)$ satisfy the differential equations (with $D \equiv d/dx_3$)

$$\begin{pmatrix} c_t^2D^2-c_\ell^2k^2+\omega^2 & ik(c_\ell^2-c_t^2)D \\ ik(c_\ell^2-c_t^2)D & c_\ell^2D^2-c_t^2k^2+\omega^2 \end{pmatrix} \begin{pmatrix} \bar{u}_1(x_3) \\ \bar{u}_3(x_3) \end{pmatrix} = 0 \quad x_3 > 0 \tag{2.91a}$$

$$\begin{pmatrix} c_t'^2D^2-c_\ell'^2k^2+\omega & ik(c_\ell'^2-c_t'^2)D \\ ik(c_\ell'^2-c_t'^2)D & c_\ell'^2D^2-c_t'^2k^2+\omega^2 \end{pmatrix} \begin{pmatrix} \bar{u}_1(x_3) \\ \bar{u}_3(x_3) \end{pmatrix} = 0 \quad x_3 < 0. \tag{2.91b}$$

The solutions of Eqs. (2.91a) that vanish as $x_3 \to \infty$ are

$$\bar{u}_1(x_3) = A^{(\ell)}e^{-\alpha_\ell x_3} + A^{(t)}e^{-\alpha_t x_3} \tag{2.92a}$$

$$\bar{u}_3(x_3) = i\left(\frac{\alpha_\ell}{k} A^{(\ell)}e^{-\alpha_\ell x_3} + \frac{k}{\alpha_t} A^{(t)}e^{-\alpha_t x_3}\right) \tag{2.92b}$$

with

$$\alpha_\ell = \left(k^2 - \frac{\omega^2}{c_\ell^2}\right)^{1/2} \qquad k^2 > \frac{\omega^2}{c_\ell^2} \tag{2.93a}$$

$$\alpha_t = \left(k^2 - \frac{\omega^2}{c_t^2}\right)^{1/2} \qquad k^2 > \frac{\omega^2}{c_t^2} . \tag{2.93b}$$

The solutions of Eqs. (2.91b) that vanish as $x_3 \to -\infty$ are

$$\bar{u}_1(x_3) = B^{(\ell)}e^{\alpha'_\ell x_3} + B^{(t)}e^{\alpha'_t x_3} \qquad (2.94a)$$

$$\bar{u}_3(x_3) = -i\left(\frac{\alpha'_\ell}{k} B^{(\ell)}e^{\alpha'_\ell x_3} + \frac{k}{\alpha'_t} B^{(t)}e^{\alpha'_t x_3}\right) \qquad (2.94b)$$

where

$$\alpha'_\ell = \left(k^2 - \frac{\omega^2}{c'^2_\ell}\right)^{1/2} \qquad k^2 > \frac{\omega^2}{c^2_\ell} \qquad (2.95a)$$

$$\alpha'_t = \left(k^2 - \frac{\omega^2}{c'^2_t}\right)^{1/2} \qquad k^2 > \frac{\omega^2}{c^2_t} \quad . \qquad (2.95b)$$

The boundary conditions that have to be satisfied at the plane $x_3 = 0$ are the continuity of the displacement components across it,

$$u_1(\vec{x},t)\Big|_{x_3 = 0-} = u_1(\vec{x},t)\Big|_{x_3 = 0+} \qquad (2.96a)$$

$$u_3(\vec{x},t)\Big|_{x_3 = 0} = u_3(\vec{x},t)\Big|_{x_3 = 0+} \quad , \qquad (2.96b)$$

and the continuity of the stress components T_{13} and T_{33} across it,

$$\rho' c'^2_t\left(\frac{\partial u_1}{\partial x_3} + \frac{\partial u_3}{\partial x_1}\right)\Big|_{x_3= 0-} = \rho c^2_t\left(\frac{\partial u_1}{\partial x_3} + \frac{\partial u_3}{\partial x_1}\right)\Big|_{x_3= 0+}$$

$$\left[\rho' c'^2_\ell \frac{\partial u_3}{\partial x_3} + \rho'(c'^2_\ell - 2c'^2_t) \frac{\partial u_1}{\partial x_1}\right]\Big|_{x_3 = 0-} \qquad (2.97a)$$

$$= [[\rho c_\ell^2 \frac{\partial u_3}{\partial x_3} + \rho(c_\ell^2 - 2c_t^2) \frac{\partial u_1}{\partial x_1}]|_{x_3 = 0+} \quad . \tag{2.97b}$$

When the solutions given by Eqs. (2.90), (2.92), and (2.94) are substituted into Eqs. (2.97), we obtain a set of homogeneous linear equations for the coefficients, $A^{(\ell)}$, $A^{(t)}$, $B^{(\ell)}$, $B^{(t)}$,

$$\begin{pmatrix} 1 & 1 & -1 & -1 \\ \frac{\alpha_\ell}{k} & \frac{k}{\alpha_t} & \frac{\alpha_\ell'}{k} & \frac{k}{\alpha_\ell} \\ 1 & \frac{k^2+\alpha_t^2}{2\alpha_t\alpha_\ell} & \frac{\rho' c_t'^2 \alpha_\ell'}{\rho c_t^2 \alpha_\ell} & \frac{\rho' c'_t \alpha_\ell^2}{\rho c_t^2 \alpha_\ell} \frac{k^2 + \alpha_t'^2}{2\alpha_t'\alpha_\ell'} \\ \frac{k^2+\alpha_t^2}{2k^2} & 1 & -\frac{\rho' c_t'^2}{\rho c_t^2} \frac{k^2+\alpha_t'^2}{2k^2} & -\frac{\rho' c_t'^2}{\rho c_t^2} \end{pmatrix} \begin{pmatrix} A^{(\ell)} \\ A^{(t)} \\ B^{(\ell)} \\ B^{(t)} \end{pmatrix} = 0 \tag{2.98}$$

If we equate to zero the determinant of the coefficients in this equation, we obtain the dispersion relation for the waves that can propagate in the structure under consideration. Although the resulting dispersion relation can be simplified somewhat, its solution has to be obtained numerically in general.

The presence of the functions $\alpha_t(k\omega)$, $\alpha_\ell(k\omega)$, $\alpha_t'(k\omega)$, and $\alpha_\ell'(k\omega)$ in the dispersion relation has the consequence that the latter, regarded as a function of k for a fixed, real value of ω, is single-valued on a sixteen-sheeted Riemann surface in the complex $u = k^2$ plane. These sixteen sheets are defined by the sixteen values

$$(\text{sgnRe}\alpha_t(k\omega), \ \text{sgnRe}\alpha_\ell(k\omega), \ \text{sgnRe}\alpha_t'(k\omega), \ \text{sgnRe}\alpha_\ell'(k\omega)) \tag{2.99}$$

can assume. (In fact, only eight of the sheets yield independent solutions, since the dispersion relation obtained from Eq. (2.98) is invariant against the simultaneous change of sign of all four decay constants.)

It has been found that a solution of the dispersion relation on the sheet (+,+,+,+) can exist for real values of k. The amplitude of the resulting wave thus decays exponentially with increasing distance into each of the media in contact. This is the Stoneley wave. We can see from Eq. (2.98) that the frequency of Stoneley waves is a linear function of the wave vector k,

$$\omega = c_S k. \tag{2.100}$$

Like Rayleigh waves, Stoneley waves are nondispersive due to the absence of any characteristic length in the problem.

The conditions under which Stoneley waves can exist have been investigated by several authors(73,80). It has been shown(75) that the speed of Stoneley waves, c_S, must lie between the speed of Rayleigh waves and transverse waves in the denser medium, i.e.

$$c_R' < c_S' < c_t' \quad , \tag{2.101}$$

where $\rho' > \rho$. Limits on the values of $\rho c_t^2/\rho' c_t'^2$ and ρ/ρ' for which Stoneley waves can exist are given in Fig. 2.11 for the Poisson case: $c_\ell/c_t = c_\ell'/c_t' = \sqrt{3}$. The solutions are very sensitive to the ratio c_ℓ/c_t in the more dense of the two media, but are only slightly influenced by c_ℓ/c_t in the less dense medium.

The solutions of the dispersion relation obtained from Eq. (2.98) for $u = k^2$ on the remaining fifteen sheets of the Riemann surface on which it is single-valued have been traced out by Pilant(81), to whose paper the reader is referred for the details of the calculation and of the results obtained. These solutions correspond to leaky interface modes that are attenuated as they propagate along the interface due to the radiation of energy into one or the other of the two media in contact, or both. Each contribution to the displacement field that is radiative also grows exponentially in amplitude with increasing distance into the medium into which the radiation occurs.

2.1.2.1.3. Sezawa Waves

The third type of guided acoustic wave that can exist in the structure depicted in Fig. 2.9 is the generalized Lamb wave(82). This is a wave that is polarized in the sagittal plane, like a Rayleigh wave or a Stoneley wave. However, it has the character

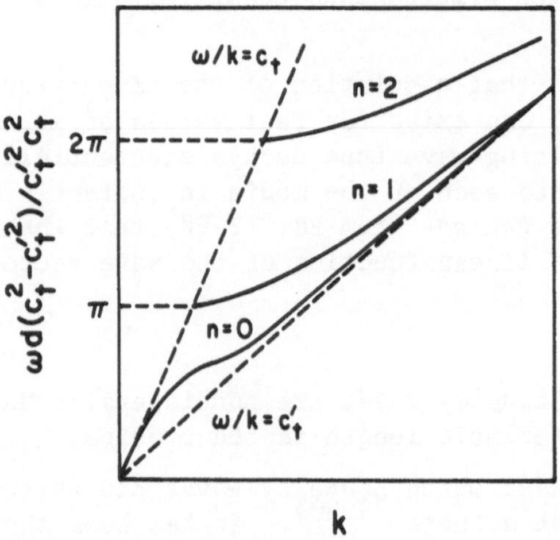

Fig. 2.10. Dispersion curves for the three lowest frequency Love waves [Ref. 70].

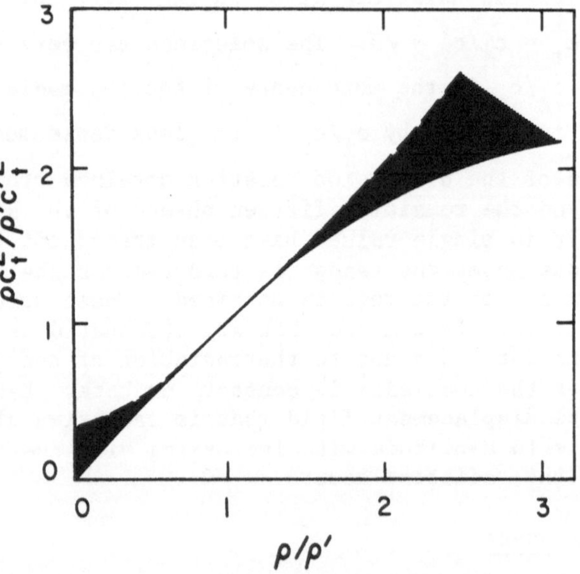

Fig. 2.11. Region of existence of Stoneley waves for the Poisson case $c_\ell/c_t = c'_\ell/c'_t = \sqrt{3}$. [Ref. 75].

of a standing wave in the plate, and its amplitude decays exponentially with increasing distance into the substrate from the interface. In the latter respect it is reminiscent of a Love wave.

The displacement field in a generalized Lamb wave has the form given by Eq. (2.90). The amplitudes $\bar{u}_1(k\omega|x_3)$ and $\bar{u}_3(k\omega|x_3)$ have the following forms in the regions $0 < x_3 < d$ and $x_3 > d$:

$0 < x_3 < d$

$$\bar{u}_1(k\omega|x_3) = A_\ell^{(+)} e^{i\hat{\alpha}_\ell(k\omega)x_3} + A_\ell^{(-)} e^{-i\hat{\alpha}_\ell(k\omega)x_3} + A_t^{(+)} e^{i\hat{\alpha}_t(k\omega)x_3} + A_t^{(-)} e^{-i\hat{\alpha}_t(k\omega)x_3} \tag{2.102a}$$

$$\bar{u}_3(k\omega|x_3) = \frac{\hat{\alpha}_\ell(k\omega)}{\kappa} A_\ell^{(+)} e^{i\hat{\alpha}_\ell(k\omega)x_3} - \frac{\hat{\alpha}_\ell(k\omega)}{\kappa} A_\ell^{(-)} e^{-i\hat{\alpha}_\ell(k\omega)x_3} - \frac{\kappa}{\hat{\alpha}_\ell(k\omega)} A_t^{(+)} e^{i\hat{\alpha}_t(k\omega)x_3} + \frac{\kappa}{\hat{\alpha}_t(k\omega)} A_t^{(-)} e^{-i\alpha_t(k\omega)x_3} \tag{2.102b}$$

where

$$\hat{\alpha}_\ell(k\omega) = \left(\frac{\omega^2}{c_\ell^2} - \kappa^2\right)^{1/2}, \quad \hat{\alpha}_t(k\omega) = \left(\frac{\omega^2}{c_t^2} - \kappa^2\right)^{1/2}. \tag{2.102c}$$

$x_3 > d$

$$\bar{u}_1(k\omega|x_3) = B_\ell e^{-\alpha_\ell(k\omega)x_3} + B_t e^{-\alpha_t(k\omega)x_3} \tag{2.103a}$$

$$\bar{u}_3(k\omega|x_3) = i\left[\frac{\alpha_\ell(k\omega)}{\kappa} B_\ell e^{-\alpha_\ell(k\omega)x_3} + \frac{\kappa}{\alpha_t(k\omega)} B_t e^{-\alpha_t(k\omega)x_3}\right] \tag{2.103b}$$

where

$$\alpha_\ell(k\omega) = \left(k^2 - \frac{\omega^2}{c_\ell^2}\right)^{1/2}, \quad \alpha_t(k\omega) = \left(k^2 - \frac{\omega^2}{c_t^2}\right)^{1/2}. \tag{2.103c}$$

We consider first the boundary conditions that must be satisfied at the interface $x_3 = d$. These require, first, that the displacement components u_1 and u_3 be continuous across this plane, and yield the pair of equations

$$e^{-\alpha_\ell d} B_\ell + e^{-\alpha_t d} B_t - e^{i\hat{\alpha}_\ell d} A_\ell^{(+)} - e^{-i\hat{\alpha}_\ell d} A_\ell^{(-)} -$$

$$- e^{i\hat{\alpha}_t d} A_t^{(+)} - e^{-i\hat{\alpha}_t d} A_t^{(-)} = 0 \tag{2.104a}$$

$$i\frac{\alpha_\ell}{\kappa} e^{-\alpha_\ell d} B_\ell + i\frac{k}{\alpha_t} e^{-\alpha_t d} B_t - \frac{\hat{\alpha}_\ell}{\kappa} e^{i\hat{\alpha}_\ell d} A_\ell^{(+)} + \frac{\hat{\alpha}_\ell}{\kappa} e^{-i\hat{\alpha}_\ell d} A_\ell^{(-)} +$$

$$+ \frac{\kappa}{\hat{\alpha}_t} e^{i\hat{\alpha}_t d} A_t^{(+)} - \frac{\kappa}{\hat{\alpha}_t} e^{-i\hat{\alpha}_t d} A_t^{(-)} = 0. \tag{2.104b}$$

The boundary conditions at the interface $x_3 = d$ also require that the stresses T_{13} and T_{33} be continuous across this plane. These stresses are given by

$$T_{13} = \rho c_t^2\left(\frac{\partial u_1}{\partial x_3} + \frac{\partial u_3}{\partial x_1}\right) \qquad x_3 > d \tag{2.105a}$$

$$= \rho' c_t'^2\left(\frac{\partial u_1}{\partial x_3} + \frac{\partial u_3}{\partial x_1}\right) \qquad 0 < x_3 < d \tag{2.105b}$$

$$T_{33} = \rho(c_\ell^2 - 2c_t^2)\frac{\partial u_1}{\partial x_1} + \rho c_\ell^2 \frac{\partial u_3}{\partial x_3} \qquad x_3 > d \tag{2.106a}$$

$$= \rho'(c_\ell'^2 - 2c_t'^2)\frac{\partial u_1}{\partial x_1} + \rho' c_\ell'^2 \frac{\partial u_3}{\partial x_3} \quad 0 < x_3 < d. \tag{2.106b}$$

The following two equations result from these conditions:

$$-\rho c_t^2\left[2\alpha_\ell e^{-\alpha_\ell d} B_\ell + \frac{\alpha_t^2+k^2}{\alpha_t} e^{-\alpha_t d} B_t\right] -$$

$$-\rho' c_t'^2\left[2i\hat{\alpha}_\ell e^{i\hat{\alpha}_\ell d} A_\ell^{(+)} - 2i\hat{\alpha}_\ell e^{-i\hat{\alpha}_\ell d} A_\ell^{(-)} + \right.$$

$$\left. + i\frac{\hat{\alpha}_t^2-k^2}{\hat{\alpha}_t} e^{i\hat{\alpha}_t d} A_t^{(+)} - i\frac{\hat{\alpha}_t^2-k^2}{\hat{\alpha}_t} e^{-i\hat{\alpha}_t d} A_t^{(-)}\right] = 0 \tag{2.107a}$$

$$- i\rho c_t^2\left[\frac{\alpha_t^2+k^2}{k} e^{-\alpha_\ell d} B_\ell + 2ke^{-\alpha_t d} B_t\right] +$$

$$+ i\rho' c_t'^2\left[\frac{k^2-\hat{\alpha}_t^2}{k} d^{i\hat{\alpha}_\ell d} A_\ell^{(+)} + \frac{k^2-\hat{\alpha}_\ell d}{k} e^{-i\hat{\alpha}_\ell d} A_\ell^{(-)} + \right.$$

$$\left. + 2ke^{i\hat{\alpha}_t d} A_t^{(+)} + 2ke^{-i\hat{\alpha}_t d} A_t^{(-)}\right] = 0. \tag{2.107b}$$

Turning now to the boundary conditions at the free surface $x_3 = 0$, we find from Eqs. (2.74) that they require the vanishing of the stresses T_{13} and T_{33} on this surface. This yields the pair of equations

$$i\rho' c_t'^{2}\left[2\hat{\alpha}_\ell A_\ell^{(+)} - 2\hat{\alpha}_\ell A_\ell^{(-)} + \frac{\hat{\alpha}_t^2-k^2}{\hat{\alpha}_t} A_t^{(+)} - \frac{\hat{\alpha}_t^2-k^2}{\hat{\alpha}_t} A_t^{(-)}\right] \approx 0 \tag{2.108a}$$

$$i\rho' c_t'^{2}\left[\frac{\hat{\alpha}_t^2-k^2}{k} A_\ell^{(+)} + \frac{\hat{\alpha}_t^2-k^2}{k} A_\ell^{(-)} - 2kA_t^{(+)} - 2kA_t^{(-)}\right] \approx 0. \tag{2.108b}$$

The dispersion relation for generalized Lamb waves in the structure under consideration is obtained by equating to zero the determinant of the coefficients in Eqs. (2.104), (2.107), and (2.108). The roots of this determinantal equation must be found numerically. The resulting modes are dispersive, because the thickness of the slab enters the theory through Eqs. (2.104) and (2.107).

As in the case of Love waves, the characteristics of generalized Lamb waves depend strongly on the ratio of the speed of transverse waves in the substrate c_t, to the speed of transverse waves in the plate, c_t' (83). If the two speeds are appreciably different, and $c_t' \gg c_t$, there is only one generalized Lamb wave solution. In this case both $\hat{\alpha}_\ell$ and $\hat{\alpha}_t$ are pure imaginary, and this solution reduces to a Rayleigh wave on the substrate when $kd \to 0$, and exists only over the range of kd for which

$$\frac{\omega}{k} < c_t \ . \tag{2.109}$$

When $c_t' \ll c_t$ there is an infinite number of solutions that fall into two families of modes, often called the M_1 series and the M_2 series. These reduce to the symmetric and antisymmetric Lamb waves, respectively, when the density or elastic moduli of the substrate go to zero. Lamb waves(84-87) are waves, polarized in the sagittal plane, in a plate of finite thickness, both of whose surfaces, at $x_3 = 0$ and $x_3 = d$, are stress-free. The dispersion relation for these modes is obtained by imposing the boundary condition that the stresses T_{13} and T_{33} vanish at the surface $x_3 = d$ on the displacements given by Eqs. (2.90) and (2.102), i.e. by setting B_ℓ and B_t equal to zero in Eqs. (2.107):

$$i\rho' c_t'^{-2}\Big[2\hat{\alpha}_\ell e^{i\hat{\alpha}_\ell d} A_\ell^{(+)} - 2\hat{\alpha}_\ell e^{-i\hat{\alpha}_\ell d} A_\ell^{(-)} +$$

$$+ \frac{\hat{\alpha}_t^2 - k^2}{\hat{\alpha}_t} e^{i\hat{\alpha}_t d} A_t^{(+)} - \frac{\hat{\alpha}_t^2 - k^2}{\hat{\alpha}_t} e^{-i\hat{\alpha}_t d} A_t^{(-)}\Big] = 0 \tag{2.110a}$$

$$i\rho' c_t'^{-2}\Big[\frac{\hat{\alpha}_t^2 - k^2}{k} e^{i\hat{\alpha}_\ell d} A_\ell^{(+)} + \frac{\hat{\alpha}_t^2 - k^2}{k} e^{-i\hat{\alpha}_\ell d} A_\ell^{(-)} -$$

$$-2k e^{i\hat{\alpha}_t d} A_t^{(+)} - 2k e^{-i\hat{\alpha}_t d} A_t^{(-)}\Big] = 0. \tag{2.110b}$$

The solvability condition for the system given by Eqs. (2.108) and (2.110) yields the dispersion relation for the symmetric and antisymmetric Lamb waves:

$$\frac{\tan \frac{1}{2}\hat{\alpha}_\ell d}{\tan \frac{1}{2}\hat{\alpha}_t d} = -\frac{(\hat{\alpha}_t^2 - k^2)^2}{4k^2 \hat{\alpha}_t \hat{\alpha}_\ell} \qquad \text{(symmetric)} \tag{2.111a}$$

$$\frac{\tan \frac{1}{2}\hat{\alpha}_\ell d}{\tan \frac{1}{2}\hat{\alpha}_t d} = -\frac{4k^2 \hat{\alpha}_t \hat{\alpha}_\ell}{(\hat{\alpha}_t^2 - k^2)^2} \qquad \text{(antisymmetric)} \tag{2.111b}$$

These modes are called symmetric and antisymmetric because the 3-component of the displacement field for these waves is odd and even in the variable $x_3 - \frac{1}{2} d$, respectively. The displacements of free-plate surfaces corresponding to symmetric (M_1) and antisymmetric (M_2) modes of propagation for Lamb waves are shown schematically in Fig. 2.12.

The fundamental modes of the two series of generalized Lamb waves, that are denoted by M_{11} and M_{21}, have particularly interesting properties. For plate thickness approaching zero ($kd \to 0$) the M_{11} mode approaches a Rayleigh type of surface wave on the substrate, while the higher order M_1 modes and all the M_2 modes are leaky waves, i.e. they radiate energy into the substrate. As the plate thickness is increased, the first additional mode to become trapped is the M_{21} mode, which is called the Sezawa wave[88] in this region. For very high frequencies or thick plates ($kd \to \infty$) the M_{11} mode approaches a Rayleigh type of surface wave on the upper boundary of the plate. All higher modes (M_{1i}, M_{2i}, $i > 1$) behave like the Lamb waves of a free thick plate as $kd \to \infty$. The behavior of the mode M_{21} depends critically on the relative material properties of the plate and the substrate. For certain special combinations of material parameters with $c_t' \approx c_t$ it becomes a bound, or surface-type, wave at the interface between the plate and the substrate when $kd \to \infty$. These are the Stoneley waves discussed in the preceding subsection. Waves of this type can also exist when $c_t \approx c_t'$ and $c_t' > c_t$. In this case the Rayleigh wave solution for the unplated substrate ($kd = 0$) becomes a bound state at the interface (with $\omega/k < c_t$) in the limit as $kd \to \infty$. If c_t' and c_t are appreciably different the mode M_{21} approaches a vertically polarized shear wave in the limit as $kd \to \infty$, i.e. it approaches a Lamb wave of a free thick plate.

Viktorov[89] has shown that in the limit $\rho/\rho' \ll 1$ the dispersion relation obtained from Eqs. (2.104), (2.107), and (2.108) has a solution

$$k = k_0'(1 + \delta) \tag{2.112a}$$

where

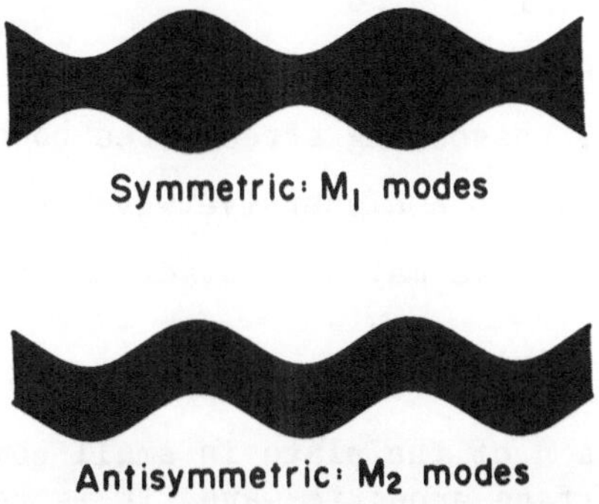

Fig. 2.12. Displacements of the free surfaces of plates corresponding to symmetric (M_1) and antisymmetric (M_2) Lamb modes.

$$k_o' = \left(\frac{\omega}{c_t'}\right)^{1/2}\left(\frac{6(1-\delta')}{d}\right)^{1/2} \tag{2.112b}$$

$$\delta = \frac{(\xi')^2(2-(\xi')^2)}{2(1-(\xi')^2)}\,\frac{\xi^2}{1+\xi^2}\left(\frac{c_t}{\omega d}\right)^2\frac{\rho}{\rho'} \tag{2.112c}$$

where $\xi = c_t/c_\ell$ $\varepsilon' = c_t'/c_\ell'$, and σ' is Poisson's ratio $\sigma' = [1-2(c_t'^{-2}/c_\ell'^{-2})]/2[1-(c_t'^{-2}/c_\ell'^{-2})]$. k_o' is the wave number for the flexural (M_2) vibrations of a very thin plate with material properties ρ', c_t', c_ℓ', possessing stress-free boundaries.[90] This solution corresponds to a dispersive wave that is slow. It is suggested that such a wave may have practical applications.

2.1.2.1.4. Thin Plate Limit

When the thickness d of the plate is small compared with the wavelength λ of the surface acoustic wave, it is possible to eliminate explicit consideration of the plate and to supplement the equations of motion of the substrate with effective boundary conditions at its surface that reproduce the effects of the plate exactly to $O(d/\lambda)$.[91-95]

We consider the geometry depicted in Fig. 2.9. In this subsection it will be convenient to label the region $x_3 > d$ as region 1; the region $0 < x_3 < d$ will be labeled region 2. With this labeling the boundary conditions (2.71)-(2.74) can be written in the forms

$$T_{\alpha 3}^{(2)}\Big|_{x_3=0} = 0 \tag{2.113a}$$

$$u_\alpha^{(1)}\Big|_{x_3=d} = u_\alpha^{(2)}\Big|_{x_3=d} \tag{2.113b}$$

$$T_{\alpha 3}^{(1)}\Big|_{x_3 = d} = T_{\alpha 3}^{(2)}\Big|_{x_3 = d} . \tag{2.113c}$$

We will combine them into a set of boundary conditions at $x_3 = 0$ alone.

We begin by rewriting $T_{\alpha 3}^{(2)}\Big|_{x_3 = d}$ as

$$T_{\alpha 3}^{(2)}\Big|_{x_3 = d} = T_{\alpha 3}^{(2)}\Big|_{x_3 = 0} + d \frac{\partial}{\partial x_3} T_{\alpha 3}^{(2)}\Big|_{x_3 = 0} + 0(d^2) \tag{2.114a}$$

$$= d \frac{\partial}{\partial x_3} T_{\alpha 3}^{(2)}\Big|_{x_3 = 0} + 0(d^2) , \tag{2.114b}$$

where we have used Eq. (2.113a) in going from Eq. (2.114a) to Eq. (2.114b). At the same time we have that

$$T_{\alpha 3}^{(1)}\Big|_{x_3 = d} = T_{\alpha 3}^{(1)}\Big|_{x_3 = 0} + d \frac{\partial}{\partial x_3} T_{\alpha 3}^{(1)}\Big|_{x_3 = 0} + 0(d^2). \tag{2.115}$$

On substituting Eqs. (2.114b) and (2.115) into Eq. (2.113c), we obtain

$$T_{\alpha 3}^{(1)}\Big|_{x_3 = 0} = d\Big[\frac{\partial}{\partial x_3} T_{\alpha 3}^{(2)} - \frac{\partial}{\partial x_3} T_{\alpha 3}^{(1)}\Big]_{x_3 = 0} + 0(d^2)$$

$$= d\Big[C'_{\alpha 3\mu\nu} \frac{\partial^2 u_\mu^{(2)}}{\partial x_3 \partial x_\nu} - C_{\alpha 3\mu\nu} \frac{\partial^2 u_\mu^{(1)}}{\partial x_3 \partial x_\nu}\Big]_{x_3 = 0} + 0(d^2), \tag{2.116}$$

where summation over a repeated index is assumed. We must now eliminate $u_\mu^{(2)}$ from the right hand side of this equation.

To do this we note that

$$C'_{\alpha 3\mu\nu} \frac{\partial^2 u_\mu^{(2)}}{\partial x_3 \partial x_\nu}\bigg|_{x_3 = 0} = C'_{\alpha 3\mu\nu} \frac{\partial^2 u_\mu^{(2)}}{\partial x_3 \partial x_\nu}\bigg|_{x_3 = d} + O(d)$$

$$= \left[C'_{\alpha 3\mu 1} \frac{\partial^2 u_\mu^{(2)}}{\partial x_3 \partial x_1} + C'_{\alpha 3\mu 2} \frac{\partial^2 u_\mu^{(2)}}{\partial x_3 \partial x_2} + C'_{\alpha 3\mu 3} \frac{\partial^2 u_\mu^{(2)}}{\partial x_3^2} \right]_{x_3 = d} + O(d). \tag{2.117}$$

Then with the aid of the equations of motion of medium 2,

$$\rho' \ddot{u}_\alpha^{(2)} = C'_{\alpha\beta\mu\nu} \frac{\partial^2 u_\mu^{(2)}}{\partial x_\beta \partial x_\nu}$$

$$= C'_{\alpha 3\mu 3} \frac{\partial^2 u_\mu^{(2)}}{\partial x_3^2} + C'_{\alpha 3\mu\delta} \frac{\partial^2 u_\mu^{(2)}}{\partial x_3 \partial x_\delta} +$$

$$+ C'_{\alpha\delta\mu 3} \frac{\partial^2 u_\mu^{(2)}}{\partial x_\delta \partial x_3} + C'_{\alpha\delta\mu\delta'} \frac{\partial^2 u_\mu^{(2)}}{\partial x_\delta \partial x_{\delta'}} , \tag{2.118}$$

where here and in what follows we use the convention that $\delta, \delta' = 1, 2$, Eq. (2.117) becomes

$$C'_{\alpha 3\mu\nu} \frac{\partial^2 u_\mu^{(2)}}{\partial x_3 \partial x_\nu}\bigg|_{x_3 = 0} = \left[\rho' \ddot{u}_\alpha^{(2)} - C'_{\alpha\delta\mu 3} \frac{\partial^2 u_\mu^{(2)}}{\partial x_\delta \partial x_3} - \right.$$

$$\left. - C'_{\alpha\delta\mu\delta'} \frac{\partial^2 u_\mu^{(2)}}{\partial x_\delta \partial x_{\delta'}} \right]_{x_3 = d} + O(d) . \tag{2.119}$$

Since, according to Eqs. (2.113b) u_μ (and $\ddot{u}_\mu$) is continuous across the plane $x_3 = d$, so are its tangential derivatives, and Eq. (2.119) becomes

$$C'_{\alpha 3\mu\nu} \frac{\partial^2 u_\mu^{(2)}}{\partial x_3 \partial x_\nu}\bigg|_{x_3 = 0} = [\rho' \ddot{u}_\alpha^{(1)} - C'_{\alpha\delta\mu\delta'} \frac{\partial^2 u_\mu^{(1)}}{\partial x_\delta \partial x_{\delta'}}]_{x_3 = d} -$$

$$- C'_{\alpha\delta\mu 3} \frac{\partial^2 u_\mu^{(2)}}{\partial x_\delta \partial x_3}\bigg|_{x_3 = d} + O(d). \tag{2.120}$$

Now, from Eq. (2.113c)

$$C_{\alpha 3\mu\nu} \frac{\partial u_\mu^{(1)}}{\partial x_\nu}\bigg|_{x_3 = d} = C'_{\alpha 3\mu\nu} \frac{\partial u_\mu^{(2)}}{\partial x_\nu}\bigg|_{x_3 = d}$$

or

$$[C_{\alpha 3\mu\delta} \frac{\partial u_\mu^{(1)}}{\partial x_\delta} + C_{\alpha 3\mu 3} \frac{\partial u_\mu^{(1)}}{\partial x_3}]_{x_3 = d}$$

$$= [C'_{\alpha 3\mu\delta} \frac{\partial u_\mu^{(2)}}{\partial x_\delta} + C'_{\alpha 3\mu 3} \frac{\partial u_\mu^{(2)}}{\partial x_3}]_{x_3 = d} . \tag{2.121}$$

If we assume that

$$u_\alpha(\vec{x}, t) = \hat{u}_\alpha(\vec{k}_\parallel \omega | x_3) e^{i\vec{k}_\parallel \cdot \vec{x}_\parallel - i\omega t} , \tag{2.122}$$

Eq. (2.121) becomes

$$C'_{\alpha 3\mu 3} \frac{\partial \hat{u}_\mu^{(2)}}{\partial x_3}\bigg|_{x_3 = d} = C_{\alpha 3\mu 3} \frac{\partial \hat{u}_\mu^{(1)}}{\partial x_3}\bigg|_{x_3 = d} +$$

$$+ ik_\delta (C_{\alpha 3\mu\delta} - C'_{\alpha 3\mu\delta}) \hat{u}_\mu^{(1)}\bigg|_{x_3 = d}, \tag{2.123}$$

where we have again used the result that the tangential derivatives of u_μ are continuous across $x_3 = d$. We next define the symmetric matrices

$$\Gamma'_{\alpha\mu} \equiv C'_{\alpha 3\mu 3} , \quad \Gamma_{\alpha\mu} = C_{\alpha 3\mu 3} , \tag{2.124}$$

and rewrite Eq. (2.123) with their aid as

$$\left(\frac{\partial \hat{u}_\mu^{(2)}}{\partial x_3}\right)_{x_3 = d} = \Gamma'^{-1}_{\mu\nu}\Gamma_{\nu\rho}\left(\frac{\partial \hat{u}_\rho^{(1)}}{\partial x_3}\right)_{x_3 = d} +$$

$$+ \Gamma'^{-1}_{\mu\nu} ik_\delta (C_{\nu 3\rho\delta} - C'_{\nu 3\rho\delta})\hat{u}_\rho^{(1)})\hat{u}_\rho^{(1)}\Big|_{x_3 = d}$$

$$= \Gamma'^{-1}_{\mu\nu}\Gamma_{\nu\rho}\left(\frac{\partial \hat{u}_\rho^{(1)}}{\partial x_3}\right)_{x_3 = 0} + \Gamma'^{-1}_{\mu\nu} ik_\delta (C_{\nu 3\rho\delta} - C'_{\nu 3\rho\delta})\hat{u}_\rho^{(1)}\Big|_{x_3 = 0} +$$

$$+ O(d) . \tag{2.125}$$

Then, from Eqs. (2.120), (2.122), and (2.125) we obtain

$$C'_{\alpha 3\mu\nu} \frac{\partial^2 u_\mu^{(2)}}{\partial x_3 \partial x_\nu}\Big|_{x_3 = 0} = [\rho' \ddot{u}_\alpha^{(1)} - C'_{\alpha\delta\mu\delta'} \frac{\partial^2 u_\mu^{(1)}}{\partial x_\delta \partial x_{\delta'}}]_{x_3 = 0} -$$

$$- C'_{\alpha\delta\mu 3}\, \Gamma'^{-1}_{\mu\nu}\Gamma_{\nu\rho} \left(\frac{\partial^2 u_\rho^{(1)}}{\partial x_\delta \partial x_3}\right)_{x_3 = 0} -$$

$$- C'_{\alpha\delta\mu 3}\, \Gamma'^{-1}_{\mu\nu}(C_{\nu 3\rho\delta'} - C'_{\nu 3\rho\delta'})\left(\frac{\partial^2 u_\rho^{(1)}}{\partial x_\delta \partial x_{\delta'}}\right)_{x_3 = 0} . \tag{2.126}$$

When Eq. (2.126) is substituted into Eq. (2.116) the boundary conditions at $x_3 = 0$ are expressed in terms of the displacement field in medium 1 alone:

$$T_{\alpha 3}^{(1)}\Big|_{x_3=0} = d\Big[\rho'\ddot{u}_\alpha^{(1)} - C'_{\alpha\delta\mu\delta'}\frac{\partial^2 u_\mu^{(1)}}{\partial x_\delta \partial x_{\delta'}} - C'_{\alpha\delta\mu 3}\Gamma'^{-1}_{\mu\nu} \times$$

$$\times\, \Gamma_{\nu\rho}\frac{\partial^2 u_\rho^{(1)}}{\partial x_\delta \partial x_3} - C'_{\alpha\delta\mu 3}\,\Gamma'^{-1}_{\mu\nu}\left(C_{\nu 3\rho\delta'} - C'_{\nu 3\rho\delta'}\right)\frac{\partial^2 u_\rho^{(1)}}{\partial x_\delta \partial x_{\delta'}} -$$

$$- C_{\alpha 3\mu\delta}\frac{\partial^2 u_\mu^{(1)}}{\partial x_3 \partial x_\delta} - C_{\alpha 3\mu 3}\frac{\partial^2 u^{(1)}}{\partial x_3^2}\Big]_{x_3=0} + O(d^2). \tag{2.127}$$

Equation (2.127) can be simplified somewhat. From the equations of motion of medium 1,

$$\rho\ddot{u}_\alpha^{(1)} = C_{\alpha\beta\mu\nu}\frac{\partial^2 u_\mu^{(1)}}{\partial x_\beta \partial x_\nu}\ , \tag{2.128}$$

we find that

$$C_{\alpha 3\mu 3}\frac{\partial^2 u_\alpha^{(1)}}{\partial x_3^2} = \rho\ddot{u}_\alpha^{(1)} - \left(C_{\alpha 3\mu\delta} + C_{\alpha\delta\mu 3}\right)\frac{\partial^2 u_\mu^{(1)}}{\partial x_\delta \partial x_3} -$$

$$- C_{\alpha\delta\mu\delta'}\frac{\partial^2 u_\mu^{(1)}}{\partial x_\delta \partial x_{\delta'}}\ . \tag{2.129}$$

If we substitute this result into the last term on the right hand side of Eq. (2.127), and write out explicitly the elements of the stress tensor appearing on the left hand side of the latter equation, we obtain

$$C_{\alpha 3\mu\nu} \frac{\partial u_\mu^{(1)}}{\partial x_\nu}\Big|_{x_3=0} = d\{(\rho'-\rho)\ddot{u}_\alpha^{(1)} + \left(C_{\alpha\delta\rho 3} - C'_{\alpha\delta\mu 3}\Gamma'^{-1}_{\mu\nu}\Gamma_{\nu\rho}\right) \times$$

$$\times \frac{\partial^2 u_\rho^{(1)}}{\partial x_\delta \partial x_3} - [C'_{\alpha\delta\rho\delta'} - C_{\alpha\delta\rho\delta'} +$$

$$+ C'_{\alpha\delta\mu 3}\,\Gamma'^{-1}_{\mu\nu}(C_{\nu 3\rho\delta'} - C'_{\nu 3\rho\delta'})]\,\frac{\partial^2 u_\rho^{(1)}}{\partial x_\delta \partial x_{\delta'}}\}_{x_3=0} + O(d^2)\ . \qquad (2.130)$$

These are the boundary conditions sought.

The physical content of Eqs. (2.130) is that in the presence of the overlayer the surface of the substrate is no longer stress-free, but is acted on by external stresses that, to first order in d, are given by the terms on the right hand side of Eq. (2.130).

To illustrate their use, and at the same time to point out their limitations, we apply them to obtain the dispersion curve for Love waves propagating in the x_1-direction in the structure depicted in Fig. 2.9. It is assumed that each of the media 1 and 2 possesses cubic symmetry, with the cube axes coinciding with the coordinate axes. The displacement field in this case is given by

$$\vec{u}(\vec{x},t) = A(0,1,0)e^{ikx_1-\alpha(k\omega)x_3-i\omega t} \qquad (2.131)$$

where, from Eq. (2.128) it is found that

$$\alpha(k\omega) = \left(k^2 - \frac{\omega^2}{c_t^2}\right)^{1/2}\ , \qquad (2.132)$$

with $c_t^2 = c_{44}/\rho$. The only nonzero component of the stress tensor in the present case is $T_{23}^{(1)}$. When Eq. (2.131) is substituted into Eq. (2.130) with $\alpha = 2$, we obtain the following equation:

$$-\,c_{44}\alpha(k\omega) = d[-(\rho'-\rho)\omega^2 + (c'_{44} - c_{44})\,k^2]\ ,$$

or

$$\left(k^2 - \frac{\omega^2}{c_t^2}\right)^{1/2} = \frac{d}{c_{44}}\,[(\rho' - \rho)\omega^2 - (c'_{44} - c_{44})k^2] > 0. \qquad (2.133)$$

Since we see from this equation that $\omega^2 = c_t^2k^2 + O(d^2)$, we can replace ω^2 on the right hand side of this equation by $c_t^2k^2$, and obtain

$$\left(k^2 - \frac{\omega^2}{c_t^2}\right)^{1/2} = dk^2\,\frac{\rho'}{\rho}\left(1 - \frac{c_t'^2}{c_t^2}\right) > 0, \qquad (2.134)$$

where $c_t'^2 = c'_{44}/\rho'$. The requirement that the right hand side of this equation be positive, which follows from the necessity of $\alpha(k\omega)$ being real for a surface localized mode to exist, has the consequence that the inequality

$$c_t'^2 < c_t^2 \qquad (2.135)$$

must be satisfied. If we now solve Eq. (2.134) for $\omega(k)$, we obtain finally

$$\omega(k) = c_tk\left[1 - \frac{1}{2}\left(\frac{\rho'}{\rho}\right)^2\left(1 - \frac{c_t'^2}{c_t^2}\right)^2 (kd)^2\right]\ . \qquad (2.136)$$

The dispersion curve given by Eq. (2.136) coincides to the order in (kd) indicated with the solution of Eq. (2.85) for the frequency of the lowest branch (n = 0) of the spectrum of Love waves. This branch corresponds to a displacement field that possesses no nodes (as a function of x_3) within the plate, i.e. is the most smoothly varying branch within the plate. Thus, the results obtained in this subsection, in addition to yielding results that are exact only to the leading nonzero order in kd, are further limited to yielding the dispersion curve of only the mode that is the smoothest function of x_3 within the plate. This is reasonable because only the monotonically decaying displacement field $\vec{u}^{(1)}(\vec{x},t)$ of the substrate is used in satisfying the boundary conditions (2.130).

The approach taken in this section, viz. the replacement of an overlayer by modified boundary conditions on the surface of the substrate can be applied to other situations as well, e.g. to a piezoelectric overlayer on a nonpiezoelectric or piezoelectric substrate. The resulting boundary conditions are clearly more complicated than those given by Eqs. (2.130).

2.1.2.2. Media With Continuously Varying Material Properties

Although systems consisting of a slab or overlayer of one material on a substrate of a second material are encountered frequently in applications, and provide models for systems in which the material properties vary with distance from the surface, situations arise in which it is desirable to take into account a continuous, rather than a discrete, variation of these properties. A sheet of metal that has been cold rolled, for example, might be expected to have a different mass density, and different elastic moduli, in the vicinity of its surfaces from those in the interior, with an essentially continuous variation of these properties across the thickness of the sheet.

In this subsection we sketch a method(96) for obtaining the dispersion curve and associated displacement field of a surface acoustic wave propagating across the surface of a semi-infinite elastic medium, occupying the region $x_3 > 0$, in which the mass density $\rho(x_3)$ and the elastic moduli $\{C_{\alpha\beta\mu\nu}(x_3)\}$ are continuous functions of the distance into the medium from the stress-free surface $x_3 = 0$.

In the presence of spatially varying elastic moduli, the equations of motion of an elastic medium take the form

$$\rho(x_3)\ddot{u}_\alpha(\vec{x},t) = \sum_{\beta\mu\nu} \frac{\partial}{\partial x_\beta}\left(C_{\alpha\beta\mu\nu}(x_3)\frac{\partial u_\mu(\vec{x},t)}{\partial x_\nu}\right)$$

$$= \sum_{\beta\mu\nu}\left\{\frac{\partial C_{\alpha\beta\mu\nu}(x_3)}{\partial x_\beta}\frac{\partial u_\mu(\vec{x},t)}{\partial x_\nu} + C_{\alpha\beta\mu\nu}(x_3)\frac{\partial^2 u_\mu(\vec{x},t)}{\partial x_\beta \partial x_\nu}\right\}. \qquad (2.137)$$

In the present case we have that

$$\rho(x_3) = \theta(x_3)\hat{\rho}(x_3) \tag{2.138a}$$

$$C_{\alpha\beta\mu\nu}(x_3) = \theta(x_3)c_{\alpha\beta\mu\nu}(x_3) , \tag{2.138b}$$

where $\theta(x_3)$ is the Heaviside unit step function

$$\begin{aligned} \theta(x_3) &= 1 \quad x_3 > 0 \\ &= 0 \quad x_3 < 0. \end{aligned} \tag{2.139}$$

Equations (2.137) thus take the form

$$\begin{aligned} \theta(x_3)\hat{\rho}(x_3)\ddot{u}_\alpha &= \delta(x_3) \sum_{\mu\nu} c_{\alpha 3\mu\nu}(o) \frac{\partial u_\mu}{\partial x_\nu} + \\ &+ \theta(x_3) \sum_{\mu\nu} \left[\frac{\partial c_{\alpha 3\mu\nu}(x_3)}{\partial x_3} \frac{\partial u_\mu}{\partial x_\nu} + \sum_\beta c_{\alpha\beta\mu\nu}(x_3) \frac{\partial^2 u_\mu}{\partial x_\beta \partial x_\nu}\right]. \end{aligned} \tag{2.140}$$

The vanishing of the coefficient of $\delta(x_3)$ on the right hand side of this equation at $x_3 = 0$ (since $\delta(x_3)f(x_3) = \delta(x_3)f(o)$) expresses the stress-free boundary conditions in this system. Thus these boundary conditions are now included in the equations of motion of the medium.

If we now direct our attention to the case of a shear horizontal wave propagating along the x_1-direction ([100]) on the (001) surface of a cubic medium whose cube axes are oriented along the coordinate axes, the displacement field has the form

$$\vec{u}(\vec{x},t) = (0,\hat{u}_2(k\omega|x_3),0)e^{ikx_1 - i\omega t} . \tag{2.141}$$

The equations of motion (2.140) in this case reduce to a single equation

$$-\theta(x_3)\hat{\rho}(x_3)\omega^2\hat{u}_2(k\omega|x_3) = c_{44}(o)\delta(x_3)\,\frac{d\hat{u}_2(k\omega|x_3)}{dx_3} +$$

$$+ \theta(x_3)\; c'_{44}(x_3)\,\frac{d\hat{u}_2(k\omega|x_3)}{dx_3} +$$

$$+ \theta(x_3)c_{44}(x_3)[-k^2\hat{u}_2(k\omega|x_3) + \frac{d^2\hat{u}_2(k\omega|x_3)}{dx_3^2}] \;, \tag{2.142}$$

in the Voigt notation, where $c'_{44}(x_3) \equiv dc_{44}(x_3)/dx_3$.

To solve Eq. (2.142) we expand $\hat{u}_2(k\omega|x_3)$ according to

$$\hat{u}_2(k\omega|x_3) = \sum_{n=0}^{\infty} a_n(k\omega)\phi_n(x_3;\alpha) \;, \tag{2.143}$$

where

$$\phi_n(x_3;\alpha) = \alpha^{1/2}\, e^{-1/2\,\alpha x_3}\, \frac{L_n(\alpha x_3)}{n!}, \tag{2.144}$$

with $L_n(x)$ the nth Laguerre polynomial. The functions $\{\phi_n(x_3;\alpha)\}$ are complete and orthonormal in the interval $0 < x_3 < \infty$:

$$\int_0^\infty dx_3\; \phi_m(x_3;\alpha)\phi_n(x_3;\alpha) = \delta_{mn} \;. \tag{2.145}$$

The parameter α is arbitrary, and can be varied to improve the rate of convergence of the expansion (2.143). If we substitute Eq. (2.143) into Eq. (2.143), multiply the resulting equation from the left by $\phi_m(x_3;\alpha)$, and integrate the product over x_3 from 0 to ∞, we obtain the matrix equation

$$\omega^2 \sum_{n=0}^{\infty} N_{mn}\, a_n(k\omega) = \sum_{n=0}^{\infty} M_{mn}(k)a_n(k\omega), \quad m = 0, 1, 2,... \tag{2.146}$$

where

$$N_{mn} = \langle m|\hat{\rho}(x_3)|n\rangle \tag{2.147}$$

$$M_{mn}(k) = -c_{44}(0)\ \langle m|\delta(x_3)\frac{d}{dx_3}\ |n\rangle - \langle m|c_{44}'(x_3)\frac{d}{dx_3}|n\rangle +$$

$$+ k^2\langle m|c_{44}(x_3)|n\rangle - \langle m|c_{44}(x_3)\frac{d^2}{dx_3^2}|n\rangle, \tag{2.148}$$

with

$$\langle m|f(x_3)|n\rangle = \int_0^\infty dx_3\phi_m(x_3;\alpha)f(x_3)\phi_n(x_3;\alpha). \tag{2.149}$$

Equation (2.146) can be transformed into an eigenvalue equation of the standard type,

$$\omega^2 a_m(k\omega) = \sum_{n=0}^{\infty}(\overleftrightarrow{N}^{-1}\overleftrightarrow{M}(k))_{mn}\ a_n(k\omega), \tag{2.150}$$

so that the frequencies of the surface (and/or guided) waves supported by the system being studied are the eigenvalues of the matrix $\overleftrightarrow{N}^{-1}\overleftrightarrow{M}(k)$.

Although the elements of the matrices $\overleftrightarrow{N}$ and $\overleftrightarrow{M}(k)$ can be obtained by numerical integration for arbitrary dependences of $\hat{\rho}(x_3)$ and $\hat{c}_{44}(x_3)$ on x_3, the situation is considerably simplified if these functions can be represented as sums of a finite number of exponentials,

$$\hat{\rho}(x_3) = \sum_i R_i e^{-\beta_i x_3} \tag{2.151a}$$

$$\hat{c}_{44}(x_3) = \sum_i C_i\, e^{-\gamma_i x_3} , \tag{2.151b}$$

because in this case the matrix elements required can be obtained analytically with the aid of the generating function.

$$\alpha^{1/2} \frac{e^{-\frac{1}{2}\alpha x_3 \frac{1+s}{1-s}}}{1 - s} = \sum_{m=0}^{\infty} s^m \Phi_m(x_3;\alpha). \tag{2.152}$$

Thus, we have that[96]

$$\langle m|n\rangle = \delta_{mn} \tag{2.153a}$$

$$\langle m|e^{-\beta x_3}|n\rangle = \sum_{p=0}^{\min(m,n)} \frac{(m+n-p)!}{(m-p)!(n-p)!p!} \frac{\alpha(\alpha-\beta)^p \beta^{m+n-2p}}{(\alpha+\beta)^{m+n+1-p}} \tag{2.153b}$$

$$\langle m|\delta(x_3) \frac{d}{dx_3}|n\rangle = -\alpha^2(n + \tfrac{1}{2}) \tag{2.153c}$$

$$\langle m|e^{-\beta x_3} \frac{d}{dx_3}|n\rangle = -\frac{\alpha^2}{2} \sum_{p=0}^{\min(m,n)} \frac{(m+n-p)!}{(m-p)!(n-p)!p!} \frac{(\alpha-\beta)^p \beta^{m+n-2p}}{(\alpha+\beta)^{m+n+1-p}} -$$

$$-\alpha^2 \sum_{r=0}^{n-1} \sum_{p=0}^{\min(m,r)} \frac{(m+r-p)!}{(m-p)!(r-p)!p!} \frac{(\alpha-\beta)^p \beta^{m+r-2p}}{(\alpha+\beta)^{m+r+1-p}} \tag{2.153d}$$

$$\langle m| \frac{d^2}{dx_3^2} |n\rangle = \frac{\alpha^2}{4} \delta_{mn} + \alpha^2(n-m)\, \theta(n-m-1) \tag{2.153e}$$

$$\langle m|e^{-\beta x_3} \frac{d^2}{dx_3^2} |n\rangle = a_{mn} + 4 \sum_{p=0}^{n} a_{mp}(n-p)\theta(n-p-1) \tag{2.153f}$$

where

$$a_{mn} = \frac{\alpha^3}{4} \sum_{r=0}^{\min(m,n)} \frac{(m+n-r)!}{(m-r)!(n-r)!r!} \frac{\beta^{m+n-2r}(\alpha-\beta)^r}{(\alpha+\beta)^{m+n+1-r}}, \tag{2.154a}$$

and

$$\theta(n) = 1 \qquad n = 0, 1, 2, 3, \ldots$$

$$= 0 \qquad n = -1, -2, -3, \ldots . \tag{2.154b}$$

The eigenvalue problem (2.150) has been solved for the following forms for $\hat{\rho}(x_3)$ and $c_{44}(x_3)$:

$$\hat{\rho}(x_3) = \rho(1 + e^{-ax_3}) \tag{2.155a}$$

$$c_{44}(x_3) = c_{44}\left(1 + \frac{1}{2} e^{-\frac{3}{2} ax_3}\right). \tag{2.155b}$$

The position dependent speed of transverse sound waves in this medium is given by

$$c_t(x_3) = \left(\frac{c_{44}(x_3)}{\hat{\rho}(x_3)}\right)^{1/2} = c_t(\infty)\left(\frac{1 + \frac{1}{2} e^{-\frac{3}{2} ax_3}}{1 + e^{-ax_3}}\right)^{1/2}, \tag{2.156}$$

where $c_t(\infty) = (c_{44}/\rho)^{1/2}$. We see from Eq. (2.156) that $c_t(0) < c_t(\infty)$, which is a necessary condition for the existence of Love waves in this medium (see subsection 2.1.2.1.1).

The dispersion curves of the guided waves supported by the structure defined by Eqs. (2.155) are shown in Fig. 2.13. Only modes in the region of the (ω,k)-plane for which $k^2 > \omega^2/c_t^2(\infty)$ can be localized. This condition follows from the fact that in the limit as $x_3 \to \infty$ Eq. (2.142) takes the form

$$-\rho\omega^2 \hat{u}_2 = c_{44}\left(-k^2 + \frac{d^2}{dx_3^2}\right)\hat{u}_2 , \tag{2.157}$$

whose solution that vanishes at infinity is

$$\hat{u}_2(k\omega|x_2) = \text{const.}\ \exp(-(k^2 - (\rho\omega^2/c_{44}))^{1/2} x_3), \tag{2.158}$$

provided that $k^2 > \rho\omega^2/c_{44} \equiv \omega^2/c_t^2(\infty)$.

Typically only the first ten terms in the expansion (2.143) were required to obtain the lowest eigenfrequency to four-figure accuracy. More terms are needed to obtain the higher frequency modes with comparable accuracy, but the method described here is simple to implement nonetheless.

We conclude this subsection by pointing out that, despite some efforts in this direction[(97)], the inverse problem, viz. that of determining $c_t(x_3)$ from given results for $\omega(k)$, remains unsolved. It is a challenging problem that merits attention in view of the potential the use of surface acoustic waves possesses for the nondestructive characterization of solid surfaces.

2.2 Piezoelectric Media.

Deforming a piezoelectric medium produces an electric field in it; applying an electric field to it deforms it. Acoustic surface waves can exist on such media. Piezoelectricity causes only second order effects on the dominantly mechanical nature of acoustic surface waves. However, the elastic deformation of the medium in such waves is accompanied by an electric field. It is this electric field that makes piezoelectric surface waves interesting from a technological standpoint, for it represents the way in which a surface wave is usually coupled to external electrical circuits, to other surface waves, or to charge carriers in semiconductors.

The equations of motion of a piezoelectric medium are still

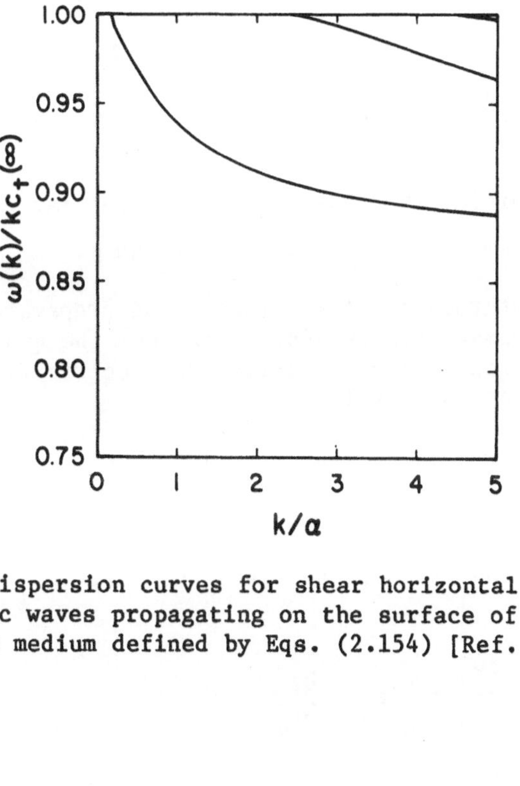

Fig. 2.13. Dispersion curves for shear horizontal surface acoustic waves propagating on the surface of an inhomogeneous medium defined by Eqs. (2.154) [Ref. 96].

given by Eqs. (2.1), except that the stress tensor now possesses an electrical contribution, in addition to the usual mechanical contribution,

$$T_{\alpha\beta} = \sum_{\mu\nu} C_{\alpha\beta\mu\nu} \frac{\partial u_\mu}{\partial x_\nu} - \sum_{\mu} e_{\mu\alpha\beta} E_\mu , \tag{2.159}$$

where $\vec{E}$ is the macroscopic field in the medium, the $\{C_{\alpha\beta\mu\nu}\}$ are the elastic moduli (at constant macroscopic field), and $e_{\mu\alpha\beta}$ is the piezoelectric tensor. It is symmetric in the second pair of indices. The equation of motion has to be supplemented by the constituitive relation connecting the electric displacement $\vec{D}$ with the displacement gradients and the macroscopic electric field,

$$D_\alpha = \sum_{\mu\nu} e_{\alpha\mu\nu} \frac{\partial u_\mu}{\partial x_\nu} + \sum_{\beta} \varepsilon_{\alpha\beta} E_\beta , \tag{2.160}$$

where $\varepsilon_{\alpha\beta}$ is the dielectric tensor of the medium (at constant strain). It is symmetric in the indices α and β.

Since the disturbances of interest to us propagate with sonic rather that electromagnetic speeds, we can use the quasistatic, or electrostatic, approximation and write the electric field as the gradient of a scalar potential,

$$E_\alpha = - \frac{\partial \phi}{\partial x_\alpha} . \tag{2.161}$$

Equations (2.159) and (2.160) then take the forms

$$T_{\alpha\beta} = \sum_{\mu\nu} C_{\alpha\beta\mu\nu} \frac{\partial u_\mu}{\partial x_\nu} + \sum_{\mu} e_{\mu\alpha\beta} \frac{\partial \phi}{\partial x_\mu} \tag{2.162}$$

$$D_\alpha = \sum_{\mu\nu} e_{\alpha\mu\nu} \frac{\partial u_\mu}{\partial x_\nu} - \sum_{\beta} \varepsilon_{\alpha\beta} \frac{\partial \phi}{\partial x_\beta} . \tag{2.163}$$

The equations of motion now become

$$\rho \frac{\partial^2}{\partial t^2} u_\alpha = \sum_{\beta\mu\nu} c_{\alpha\beta\mu\nu} \frac{\partial^2 u_\mu}{\partial x_\beta \partial x_\mu} + \sum_{\beta\mu} e_{\mu\alpha\beta} \frac{\partial^2 \phi}{\partial x_\beta \partial x_\mu} . \qquad (2.164)$$

A closed system of equations is obtained when we include the Maxwell equation

$$\nabla \vec{D} = \sum_{\alpha\mu\nu} e_{\alpha\mu\nu} \frac{\partial^2 u_\mu}{\partial x_\alpha \partial x_\nu} - \sum_{\alpha\beta} \varepsilon_{\alpha\beta} \frac{\partial^2 \phi}{\partial x_\alpha \partial x_\beta} = 0. \qquad (2.165)$$

The Lagrangian density $\mathcal{L}$ from which the equations of motion (2.164) and (2.165) can be obtained with the use of Eq. (2.8) is

$$\mathcal{L} = \frac{1}{2} \rho \sum_\alpha \dot{u}_\alpha^2 - \frac{1}{2} \sum_{\alpha\beta\mu\nu} c_{\alpha\beta\mu\nu} \frac{\partial u_\alpha}{\partial x_\beta} \frac{\partial u_\mu}{\partial x_\nu} -$$

$$- \sum_{\alpha\mu\nu} e_{\alpha\mu\nu} \frac{\partial \phi}{\partial x_\alpha} \frac{\partial u_\mu}{\partial x_\nu} + \frac{1}{2} \sum_{\alpha\beta} \varepsilon_{\alpha\beta} \frac{\partial \phi}{\partial x_\alpha} \frac{\partial \phi}{\partial x_\beta} . \qquad (2.166)$$

In what follows we use Eqs. (2.164) and (2.165) as the basis for a study of surface acoustic waves on semi-infinite piezoelectric media. We first formulate the problem of obtaining their dispersion curves in some generality, and then specialize the general treatment to the particular case of a shear horizontal surface acoustic wave that has no counterpart for an elastic medium.

Analogues of Love, Stoneley, and Sezawa waves can exist in piezoelectric films on piezoelectric substrates. They will not be considered here. A very comprehensive survey of the properties of such waves has been given by Farnell and Adler[80], to which the interested reader is referred.

2.2.1. Piezoelectric Surface Acoustic Waves

In seeking solutions of Eqs. (2.164) and (2.165) that describe surface waves, we assume forms for the displacement amplitudes and for the potential ϕ that describe a wave propagating in the x_1 direction that decays exponentially in the x_3-direction, and is independent of the coordinate x_2:

$$u_\beta(\vec{x},t) = A_\beta e^{ikx_1 - \alpha x_3 - i\omega t} \qquad \beta = 1, 2, 3 \qquad (2.167a)$$

$$\phi(\vec{x},t) = A_4 e^{ikx_1 - \alpha x_3 - i\omega t} . \tag{2.167b}$$

When we substitute these expressions into Eqs. (2.164) and (2.165), we obtain the following set of equations for the amplitudes $\{A_\beta\}$,

$$\begin{pmatrix} \Gamma_{11} - \rho v^2 & \Gamma_{12} & \Gamma_{13} & \Gamma_{14} \\ \Gamma_{12} & \Gamma_{22} - \rho v^2 & \Gamma_{23} & \Gamma_{24} \\ \Gamma_{13} & \Gamma_{23} & \Gamma_{33} - \rho v^2 & \Gamma_{34} \\ \Gamma_{14} & \Gamma_{24} & \Gamma_{34} & \Gamma_{44} \end{pmatrix} \begin{pmatrix} A_1 \\ A_2 \\ A_3 \\ A_4 \end{pmatrix} = 0, \tag{2.168}$$

where the Γ coefficients that couple the elastic displacements are

$$\begin{aligned} \Gamma_{11} &= c_{11} + 2c_{15} i\beta - c_{55}\beta^2 \\ \Gamma_{22} &= c_{66} + 2c_{46} i\beta - c_{44}\beta^2 \\ \Gamma_{33} &= c_{55} + 2c_{35} i\beta - c_{33}\beta^2 \\ \Gamma_{12} &= c_{16} + (c_{14} + c_{56})\, i\beta - c_{45}\beta^2 \\ \Gamma_{13} &= c_{15} + (c_{13} + c_{55}) i\beta - c_{55}\beta^2 \\ \Gamma_{23} &= c_{56} + (c_{36} + c_{45}) i\beta - c_{34}\beta^2 . \end{aligned} \tag{2.169}$$

The Γ coefficients that couple the potential to the elastic displacements are

$$\Gamma_{14} = e_{11} + (e_{15} + e_{31})i\beta - e_{35}\beta^2$$

$$\Gamma_{24} = e_{16} + (e_{14} + e_{36})i\beta - e_{34}\beta^2$$

$$\Gamma_{34} = e_{15} + (e_{13} + e_{35})i\beta - e_{33}\beta^2. \qquad (2.170)$$

The purely electrical term is

$$\Gamma_{44} = - [\varepsilon_{11} + 2\varepsilon_{13}i\beta - \varepsilon_{33}\beta^2]. \qquad (2.171)$$

In writing these equations we have set

$$\beta = \alpha/k, \ v = \omega/k, \qquad (2.172)$$

and have used the Voigt, or contracted, notation in writing the elastic moduli and the elements of the piezoelectric tensor.

In order that Eq. (2.168) have a nontrivial solution the determinant of the coefficients must be equated to zero. This leads to an eighth degree equation for β. For real values of v only four of these roots have a positive real part and so can describe a wave localized at the surface. We denote these by β_j. They are functions of v. For each of these roots there is a four-component eigenvector $\left(A_1^{(j)}, A_2^{(j)}, A_3^{(j)}, A_4^{(j)}\right)$, for which we have the relations

$$\frac{A_1^{(j)}}{C_1^{(j)}} = \frac{A_2^{(j)}}{C_2^{(j)}} = \frac{A_3^{(j)}}{C_3^{(j)}} = \frac{A_4^{(j)}}{C_4^{(j)}} = K_j , \qquad (2.173)$$

where $C_\alpha^{(j)}$ is the cofactor of the element in the αth column of the first row of the matrix appearing on the left hand side of Eq. (2.168), evaluated at $\beta = \beta_j$, and the $\{K_j\}$ are new amplitudes.

We superpose these four solutions to satisfy the boundary conditions at the surface x_3:

$$u_\alpha(\vec{x},t) = \sum_{j=1}^{4} K_j C_\alpha^{(j)} e^{ik(x_1-\beta_j x_3-vt)} \qquad \alpha = 1,\ 2,\ 3, \qquad x_3 > 0 \tag{2.174a}$$

$$\phi(\vec{x},t) = \sum_{j=1}^{4} K_j C_4^{(j)} e^{ik(x_1-\beta_j x_3-vt)} \qquad x_3 > 0 \tag{2.174b}$$

In the vacuum region $x_3 < 0$ only the scalar potential is nonzero. It satisfies Laplace's equation

$$\left(\frac{\partial^2}{\partial x_1^2} + \frac{\partial^2}{\partial x_3^2}\right)\hat{\phi} = 0. \tag{2.175}$$

The solution of this equation that vanishes as $x_3 \to -\infty$ is

$$\hat{\phi}(\vec{x},t) = K_5 e^{ikx_1 + |k|x_3 - i\omega t} \qquad x_3 < 0\ , \tag{2.176}$$

where C_5 is the potential in the medium evaluated at the surface $x_3 = 0$. According to Eq. (2.174b) this is given by

$$K_5 = \sum_{j=1}^{4} K_j C_4^{(j)}\ . \tag{2.177}$$

The satisfaction of Eq. (2.177) ensures the continuity of the tangential components of the electric field across the surface $x_3 = 0$.

The boundary conditions that must be satisfied at the surface $x_3 = 0$ are now four in number, which equals the number of unknowns in the problem, viz. K_1, K_2, K_3, K_4. These fall into three categories: the mechanical boundary conditions involving transverse surface stresses; the remaining electrical boundary conditions; and the sagittal plane mechanical boundary conditions. They are:

(a) Mechanical Transverse

(1) Vanishing of the transverse shear stress

$$T_{23} = 0; \tag{2.178}$$

(b) Electrical

(2) Continuity of the normal component of the electric displacement

$$D_3 = \hat{D}_3 = -\frac{\partial}{\partial x_3}\hat{\phi}\ ; \tag{2.179}$$

(c) Mechanical Sagittal

(3) Vanishing of the sagittal shear stress

$$T_{13} = 0; \tag{2.180}$$

(4) Vanishing of the vertical compressional stress

$$T_{33} = 0\ . \tag{2.181}$$

When the solutions given by Eqs. (2.174a) and (2.174b) are substituted into these homogeneous boundary conditions, and the stress components and D_3 are obtained from Eqs. (2.162) and (2.163), a set of four homogeneous linear equations for the coefficients is obtained. This equation can be written in the form

$$\sum_j M_{ij}K_j = 0, \tag{2.182}$$

where the elements of the matrix are given by

$$M_{1j} = \sum_\mu (C_{32\mu 1} + C_{32\mu 3}\beta_j)C_\mu^{(j)} + \\ + (e_{132} + ie_{332}\beta_j)C_4^{(j)} \tag{2.183a}$$

$$M_{2j} = \sum_\mu (e_{3\mu 1} + ie_{3\mu}\beta_j)C_\mu^{(j)} + \\ + (\varepsilon_{31} + i\varepsilon_{33}\beta_j)C_4^{(j)} \tag{2.183b}$$

$$M_{3j} = \sum_\mu (C_{31\mu 1} + iC_{31\mu 3}\beta_j)C_\mu^{(j)} + \\ + (e_{131} + ie_{331}\beta_j)C_4^{(j)} = 0 \tag{2.183c}$$

$$M_{4j} = \sum_{\mu} (C_{33\mu 1} + iC_{33\mu 3}\beta_j)C_{\mu}^{(j)} +$$

$$+ (e_{133} + ie_{333}\beta_j)C_4^{(j)} . \qquad (2.183d)$$

The indexing of the rows in this matrix follows the order of the four boundary conditions above.

The condition that the set of equations (2.182) have a non-trivial solution is that the determinant of the matrix $\overleftrightarrow{M}$ vanish,

$$|M_{ij}| = 0 \qquad (2.184)$$

The resulting equation determines the value of v, the speed of surface piezoelectric waves.

Equation (2.184) has to be solved numerically, in general. Even without such a solution, however, we see from the form of Eqs. (2.182) and (2.183) that piezoelectric surface acoustic waves, like Rayleigh waves, are nondispersive, due to the absence of a characteristic length in the problem. Also, like Rayleigh waves in anisotropic media, the particle motion in a piezoelectric surface wave is not confined to the sagittal plane: although nothing varies with x_2, there is in general a nonvanishing 2-component of the displacement field.

Numerical solutions of Eq. (2.184) have been carried out by several authors(98-100) for several piezoelectric surfaces. It is found that the displacement field in a piezoelectric surface acoustic wave is little changed by the piezoelectricity from what it would be for the corresponding purely elastic medium ($e_{\mu\alpha\beta} \equiv 0$, $\varepsilon_{\alpha\beta} \equiv 0$). The new feature is the electric field associated with the surface acoustic wave that is localized to the surface. In Fig. 2.14 we have plotted the results of Tseng and White(98) for the normalized piezoelectric field components versus distance from the surface, both into the medium and into the vacuum outside it, for a surface acoustic wave propagating on the basal plane of CdS. Both the piezoelectric fields and the particle displacement vector are elliptically polarized in the sagittal plane. The variation of the electric field outside the crystal with distance from the surface has been measured(98). It is found to decrease exponentially and to have a decay length that is approximately equal to the wavelength of the wave parallel to the surface divided by 2π. Both of these experimental results are in agreement with the predictions of theory(98).

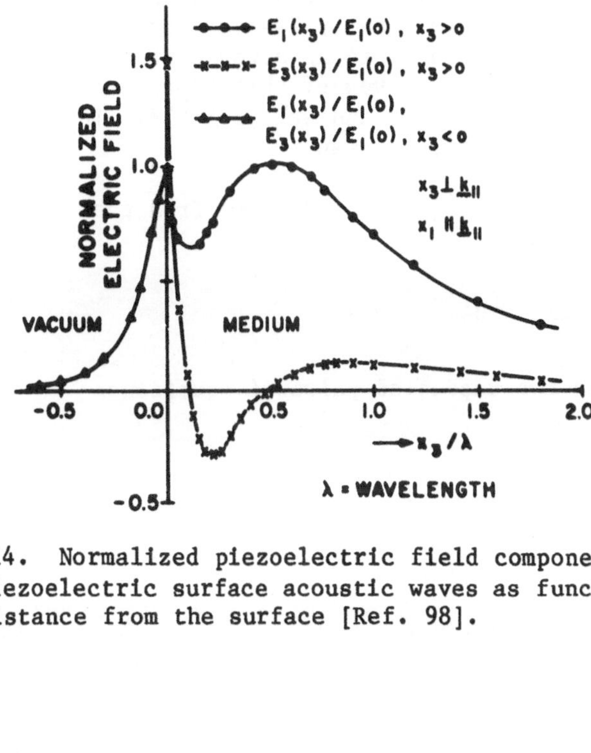

Fig. 2.14. Normalized piezoelectric field components in a piezoelectric surface acoustic waves as functions of distance from the surface [Ref. 98].

2.2.2. Bleustein-Gulyaev Surface Acoustic Waves

Just as the equations of motion, and the associated boundary conditions, governing surface acoustic waves on an anisotropic elastic medium simplify greatly for directions of propagation of such a wave along a symmetry direction on a surface that is itself a plane of symmetry, so do the equations of motion and the associated boundary conditions governing surface acoustic waves on a piezoelectric medium. Moreover, under such conditions a new type of surface acoustic wave becomes possible that has no counterpart for purely elastic media. This is a shear horizontal surface acoustic wave.

The earliest suggestion that surface acoustic waves polarized perpendicular to the sagittal plane can exist on a piezoelectric surface appears to have been made by Kaganov and Sklovskaya.(101) Unfortunately, that work contained an error that invalidated their results.(102) Subsequently, waves of this kind were discovered theoretically by Bleustein(103) and Gulayev(104). In what follows we outline the theory of these waves.

Let us return to Eqs. (2.168). We see from them that if

$$\Gamma_{12} = \Gamma_{14} = \Gamma_{23} = \Gamma_{34} \equiv 0 \qquad (2.185)$$

they separate into two uncoupled pairs of equations:

$$(\Gamma_{11} - \rho v^2)A_1 + \Gamma_{13}A_3 = 0 \qquad (2.186a)$$

$$\Gamma_{13}A_1 + (\Gamma_{33} - \rho v^2)A_3 = 0 \qquad (2.186b)$$

and

$$(\Gamma_{22} - \rho v^2)A_2 + \Gamma_{24}A_4 = 0 \qquad (2.187a)$$

$$\Gamma_{24}A_2 + \Gamma_{44}A_4 = 0. \qquad (2.187b)$$

From Eqs. (2.169)-(2.171) we see that the satisfaction of Eqs.

(2.185) requires that

$$c_{14} = c_{1123} = 0 \qquad e_{11} = e_{111} = 0$$

$$c_{16} = c_{1112} = 0 \qquad e_{13} = e_{133} = 0$$

$$c_{34} = c_{3323} = 0 \qquad e_{15} = e_{113} = 0$$

$$c_{36} = c_{3312} = 0 \qquad e_{31} = e_{311} = 0$$

$$c_{54} = c_{1323} = 0 \qquad e_{33} = e_{333} = 0$$

$$c_{56} = c_{1312} = 0 \qquad e_{35} = e_{313} = 0 \qquad (2.188)$$

The vanishing of the indicated elements of the elastic modulus tensor, in which the index 2 appears only once, will occur if the x_1x_3-plane is a plane of reflection symmetry, or if the x_2-axis is a two-fold rotation axis. The vanishing of the indicated elements of the piezoelectric tensor will occur if the x_2-axis is a two-fold rotation axis. Thus we focus our attention on the case that x_2 is a two-fold rotation axis.

The boundary conditions at the surface $x_3 = 0$ also decouple in this case:

$$\left(c_{15}\frac{\partial}{\partial x_1} + c_{55}\frac{\partial}{\partial x_3}\right)u_1 + \left(c_{55}\frac{\partial}{\partial x_1} + c_{35}\frac{\partial}{\partial x_3}\right)u_3 = 0 \qquad (2.189a)$$

$$\left(c_{13}\frac{\partial}{\partial x_1} + c_{35}\frac{\partial}{\partial x_3}\right)u_1 + \left(c_{35}\frac{\partial}{\partial x_1} + c_{33}\frac{\partial}{\partial x_3}\right)u_3 = 0 \qquad (2.189b)$$

and

$$\left(c_{46}\frac{\partial}{\partial x_1} + c_{44}\frac{\partial}{\partial x_3}\right)u_2 + \left(e_{14}\frac{\partial}{\partial x_1} + e_{34}\frac{\partial}{\partial x_3}\right)\phi = 0 \qquad (2.190a)$$

$$\left(e_{36}\frac{\partial}{\partial x_1} + e_{34}\frac{\partial}{\partial x_3}\right)u_2 - \left(\varepsilon_{31}\frac{\partial}{\partial x_1} + \varepsilon_{33}\frac{\partial}{\partial x_3}\right)\phi = -\frac{\partial\hat{\phi}}{\partial x_3}\ . \qquad (2.190b)$$

$$\phi = \hat{\phi} \tag{2.190c}$$

If solutions to Eqs. (2.186) and (2.189) exist, they correspond to a surface acoustic wave polarized in the sagittal plane, from which piezoelectric effects are absent. We will not consider them further here, since they are equivalent to the similarly polarized surface acoustic waves that propagate on a purely elastic medium.

The solutions of Eqs. (2.187) and (2.190), if they exist, describe a surface acoustic wave polarized perpendicular to the sagittal plane. Since such a wave cannot exist on the surface of a purely elastic medium, they owe their existence in the present case to the presence of piezoelectricity.

Rather than maintaining complete generality let us assume the symmetry of our system is so high that

$$c_{46} = 0, \ c_{66} = c_{44} \tag{2.191a}$$

$$e_{14} = e_{36} = 0, \quad e_{34} = e_{16} \tag{2.191b}$$

$$\varepsilon_{13} = 0, \ \varepsilon_{33} = \varepsilon_{11} \ . \tag{2.191c}$$

These conditions will be satisfied, for example, for a crystal, such as CdS, belonging to the crystal class C_{6v}, whose axis of six-fold symmetry is parallel to the x_2-direction. In this case Eqs. (2.187) take the form

$$\begin{pmatrix} c_{44}(1-\beta^2)-\rho v^2 & e_{16}(1-\beta^2) \\ e_{16}(1-\beta^2) & -\varepsilon_{11}(1-\beta^2) \end{pmatrix} \begin{pmatrix} A_2 \\ A_4 \end{pmatrix} = 0. \tag{2.192}$$

From the solvability condition for this system of equations, and Eqs. (2.167), we obtain the result that inside the medium

$$u_2(x_1 x_3;t) = e^{ik(x_1-vt)} \frac{\varepsilon_{11}}{e_{16}} A_4^{(2)} e^{-k\beta x_3} \tag{2.193}$$

$$\phi(x_1 x_3;t) = e^{ik(x_1-vt)}[A_4^{(1)}e^{-kx_3} + A_4^{(2)}e^{-k\beta x_3}]. \tag{2.194}$$

In these expressions

$$\beta = \left(1 - \frac{\rho v^2}{\bar{c}_{44}}\right)^{1/2} \tag{2.195}$$

is real and positive, where

$$\bar{c}_4 = c_{44} + \frac{e_{16}^2}{\varepsilon_{11}} \tag{2.196}$$

is a piezoelectrically stiffened elastic modulus. The satisfaction of the condition that β be real and positive requires that

$$v^2 < \frac{\bar{c}_{44}}{\rho} . \tag{2.197}$$

The scalar potential in the vacuum is

$$\hat{\phi}(x_1 x_3,t) = Be^{ik(x_1-vt)} e^{kx_3} \tag{2.198}$$

The boundary conditions (2.190) at $x_3 = 0$ now take the form

$$c_{44} \frac{\partial}{\partial x_3} u_2 + e_{16} \frac{\partial}{\partial x_3} \phi = 0 \tag{2.199a}$$

$$e_{16} \frac{\partial}{\partial x_3} u_2 - \varepsilon_{11} \frac{\partial}{\partial x_3} \phi = - \frac{\partial \hat{\phi}}{\partial x_3} \tag{2.199b}$$

$$\phi = \hat{\phi} \tag{2.199c}$$

When the solutions given by Eqs. (2.193), (2.194), and (2.198) are substituted into Eqs. (2.199), the solvability condition for the resulting set of homogeneous equations for $A_4^{(1)}$, $A_4^{(2)}$, and B yields the phase velocity for Bleustein-Gulyaev surface acoustic waves,

$$v^2 = \frac{\bar{c}_{44}}{\rho}\left[1 - \frac{e_{16}^4}{\bar{c}_{44}^2\varepsilon_{11}^2(1+\varepsilon_{11})}\right]. \tag{2.200}$$

The corresponding value of β is

$$\beta = \frac{e_{16}^2}{\varepsilon_{11}\bar{c}_{44}}\,\frac{1}{1+\varepsilon_{11}}\,. \tag{2.201}$$

The continuity of ϕ and of D_3 are not the only boundary conditions we can assume in this problem. An alternative set of boundary conditions are the short circuit boundary conditions. These apply when the surface $x_3 = 0$ is lightly metallized so that the mechanical boundary conditions are not affected but the electrical boundary condition becomes that $\hat{\phi}$ vanishes identically. It follows that ϕ must vanish at $x_3 = 0$, and so must D_3:

$$\phi = 0 \tag{2.202a}$$

$$\varepsilon_{16}\frac{\partial}{\partial x_3}u_2 - \varepsilon_{11}\frac{\partial}{\partial x_3}\phi = 0. \tag{2.202b}$$

When Eqs. (2.193) and (2.194) are substituted into Eqs. (2.199a) and (2.202), we find that

$$v^2 = \frac{\bar{c}_{44}}{\rho}\left[1 - \frac{e_{16}^4}{\bar{c}_{44}^2\varepsilon_{11}^2}\right]. \tag{2.203}$$

The corresponding value of β is

$$\beta = \frac{e_{16}^2}{\varepsilon_{11}\bar{c}_{44}} . \tag{2.204}$$

From a comparison of the expressions for β given by Eqs. (2.201) and (2.204) we see that Bleustein-Gulayev waves on a metallized surface are more useful in device applications than such waves on a surface with free electrical boundary conditions, because β in the former case is larger by a factor of $1 + \varepsilon_{11} \sim 10$ than β in the latter case. This means that the fields both mechanical and electrical, and hence the energy carried by the wave (Section 3), are more tightly bound to the surface in the case of a metallized surface.[105]

We emphasize that a shear horizontal surface acoustic wave does not exist on non-piezoelectric substrates. That such a wave exists on a piezoelectric substrate is due directly to the piezoelectricity of the medium. Indeed the reason that the condition (2.197) is satisfied by the solutions given by Eqs. (2.200) and (2.203) is the presence of a nonzero piezoelectric constant e_{16}. For CdS $v^2 = 0.9999875\ \bar{c}_{44}/\rho$, for the free boundary case.

The first experimental observation of Bleustein-Gulyaev surface acoustic waves was by Maerfeld et al.[106] Subsequently, Koerber and Vogel[107] described the criteria for the existence of these waves in crystal classes other than C_{6v}, and they have now been studied theoretically in rhombic and monoclinic crystals.[108-110].

2.3. Magnetic Media

Surface waves have now been studied theoretically and experimentally on ferromagnetic, antiferromagnetic, and paramagnetic surfaces. Due to limitations on length, in this section we discuss only surface waves on ferromagnetic media, following primarily the work of Scott and Mills[111] and Camley and Scott[112].

A ferromagnetic medium has several formal similarities to a piezoelectric medium. Straining such a medium, for example, gives rise to a net magnetization in it due to the magnetoelastic coupling between the strain and the spins in the system; conversely, magnetizing the medium gives rise to a strain in it. Surface acoustic waves, called magnetoelastic surface waves, can exist on ferromagnetic media; the mechanical displacements in such a wave are accompanied by a magnetic field in the medium and outside it. A magnetoelastic analogue of the Bleustein-Gulyaev surface wave can also be found. However, there is one large difference between magnetoelastic surface acoustic waves and those on the surface of a piezoelectric medium. Surface waves on a ferromagnetic medium are nonreciprocal, i.e. the frequency of a surface wave with wave vector $\vec{k}_\parallel$ may be different from that of a surface wave with wave vector $-\vec{k}_\parallel$, while those on a piezoelectric medium are reciprocal, $\omega(-\vec{k}_\parallel) = \omega(\vec{k}_\parallel)$. The reason for this difference is ultimately the difference between the axial vector nature of a magnetic field and the polar vector nature of an electric field. In what follows we will see examples of all of these kinds of magnetoelastic surface acoustic waves.

Surveys of recent work on other kinds of surface waves on ferromagnetic media and of surface waves on other types of magnetic media can be found in Refs. 113 and 114.

2.3.1. Magnetoelastic Surface Acoustic Waves

We consider a semi-infinite ferromagnetic material that is taken to be cubic, with the cube axes along the coordinate axes. The crystal occupies the half-space $x_2 > 0$, with an externally applied dc magnetic field $\vec{H}_o$ and the saturation magnetization $\vec{M}_s$ directed along the x_3-axis (see Fig. 2.15). As usual we linearize the dynamical problem by writing the total magnetic field $\vec{H}$ and the total magnetization $\vec{M}$ in our system in the forms

$$\vec{H} = \hat{x}_3 H_o + \vec{h} \tag{2.205a}$$

$$\vec{M} = \hat{x}_3 M_s + \hat{x}_1 m_1 + \hat{x}_2 m_2 , \tag{2.205b}$$

and neglecting terms of higher than first order in the components of $\vec{h}$ and $\vec{m}$ in the equations of motion of the system. In Eqs. (2.205) $\vec{h}$ is the demagnetizing field associated with spin motion, and $\vec{m}$ is the transverse magnetization vector.

The frequencies of the surface acoustic waves of interest to us lie below a few GHz, and correspond to wave vectors of the order of, or smaller than, $10^5 cm^{-1}$. Under these conditions the effects of retardation can be neglected, and the quasistatic (magnetostatic) approximation used. Maxwell's equations in this

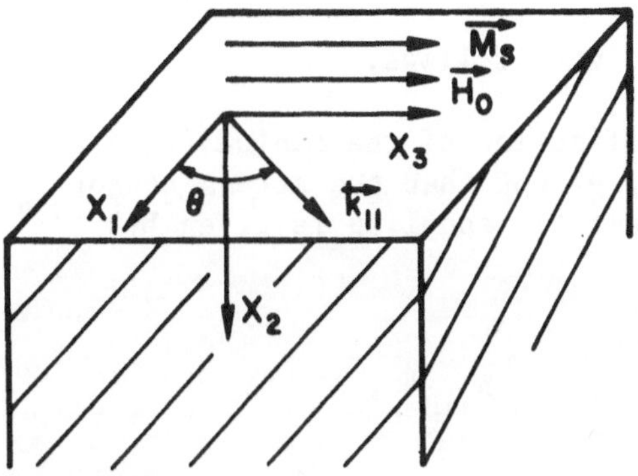

Fig. 2.15. Geometry used in the study of magnetoelastic surface acoustic waves.

approximation become

$$\nabla \times \vec{h} = 0 \quad (2.206a)$$

$$\nabla \cdot \vec{b} = 0 \ , \quad (2.206b)$$

where

$$\vec{b} = \vec{h} + 4\pi\vec{m} \quad (2.206c)$$

is the dynamic magnetic induction in our system. As a consequence of Eq. (2.206a) we can express $\vec{h}$ as the gradient of a magnetic scalar potential,

$$h_\alpha = -\frac{\partial\phi}{\partial x_\alpha} \qquad \alpha = 1, 2, 3, \quad (2.207)$$

and we will do so in what follows.

The equations of motion of the medium ($x_2 > 0$) are again given by Eqs. (2.1), except that the stress tensor $T_{\alpha\beta}$ in the presence of magnetoelastic coupling is given by[115]

$$T_{\alpha\beta} = \sum_{\mu\nu} C_{\alpha\beta\mu\nu} \frac{\partial u_\mu}{\partial x_\nu} + \sum_\mu b_{\mu\alpha\beta} m_\mu , \quad (2.208)$$

where the $\{C_{\alpha\beta\mu\nu}\}$ are the elements of the elastic modulus tensor (at constant magnetization), and the coefficients $\{b_{\mu\alpha\beta}\}$ are the elements of the magnetoelastic coupling tensor. They are symmetric in the second and third indices. In the case of a cubic medium, with the cube axes along the coordinate axes, the only elements of this tensor that enter the present discussion can be written in the form

$$b_{113} = b_{131} = b_{223} = b_{232} = \frac{b_2}{M_s} \quad (2.209)$$

in a conventional notation.[115] The equations of motion of the medium are thus

$$\rho\ddot{u}_\alpha = \sum_{\beta\mu\nu} C_{\alpha\beta\mu\nu} \frac{\partial^2 u_\mu}{\partial x_\beta \partial x_\nu} + \sum_{\beta\mu} b_{\mu\alpha\beta} \frac{\partial m_\mu}{\partial x_\beta} \qquad \alpha = 1, 2, 3. \quad (2.210)$$

Equations (2.210) have to be supplemented by the equations of motion of the components of the transverse magnetization vector $\vec{m}$. These are

$$\left(\frac{d}{dt} + \frac{1}{\tau}\right)m_1 = \omega_H m_2 + \omega_M \frac{\partial\phi}{\partial x_2} - D\nabla^2 m_2 + \omega_M \sum_{\mu\nu} b_{2\mu\nu} \frac{\partial u_\mu}{\partial x_\nu} \tag{2.211a}$$

$$\left(\frac{d}{dt} + \frac{1}{\tau}\right)m_2 = - \omega_H m_1 - \omega_M \frac{\partial\phi}{\partial x_1} + D\nabla^2 m_2 - \omega_M \sum_{\mu\nu} b_{1\mu\nu} \frac{\partial u_\mu}{\partial x_\nu} , \tag{2.211b}$$

and are just the Landau-Lifshitz[116,117] equations for the magnetization in the presence of magnetoelastic coupling.[115] In these equations τ is the transverse relaxation time for the spins in the magnetic subsystem; we have set $\omega_H = \gamma H_o$ and $\omega_M = \gamma M_s$, where γ is the gyromagnetic ratio (it is useful to recall that γ is negative so that ω_H and ω_M are also negative quantities); and D is the exchange constant[115-117].

Together with Eqs. (2.206b), which can be rewritten in the form

$$-\nabla^2\phi + 4\pi\left(\frac{\partial m_1}{\partial x_1} + \frac{\partial m_2}{\partial x_2}\right) = 0 , \tag{2.212}$$

Eqs. (2.210) and (2.211) constitute the full set of equations of motion of the magnetoelastic medium. They can be derived from Eqs. (2.8) if the Lagrangian density $\mathcal{L}$ and the dissipation function F are chosen in the following forms:

$$\mathcal{L} = \frac{1}{\omega_M} m_2 \dot{m}_1 - \frac{\omega_H}{2\omega_M}(m_1^2 + m_2^2) - \frac{D}{2\omega_M}\left[(\nabla m_1)^2 + (\nabla m_2)^2\right] +$$

$$+ \frac{1}{8\pi}(\nabla\phi)^2 - \sum_\alpha m_\alpha \frac{\partial\phi}{\partial x_\alpha} + \frac{1}{2}\rho \sum_\alpha \dot{u}_\alpha^2 -$$

$$- \frac{1}{2}\sum_{\alpha\beta\mu\nu} c_{\alpha\beta\mu\nu} \frac{\partial u_\mu}{\partial x_\beta}\frac{\partial u_\mu}{\partial x_\nu} - \sum_{\alpha\mu\nu} b_{\alpha\mu\nu} m_\alpha \frac{\partial u_\mu}{\partial x_\nu} \tag{2.213}$$

$$F = \frac{1}{\omega_M \tau}(m_2\dot{m}_1 - m_1\dot{m}_2) \ . \qquad (2.214)$$

In the vacuum filling the region $x_3 < 0$ only the magnetic scalar potential $\hat{\phi}(\vec{x},t)$ is nonzero, and it satisfies the equation

$$\nabla^2\hat{\phi} = 0. \qquad (2.215)$$

The boundary conditions at the vacuum-solid interface $x_2 = 0$ are seven in number in the general case. There are three conditions that ensure that the surface $x_2 = 0$ be stress free:

$$T_{\alpha 2}\Big|_{x_2=0+} = 0 \qquad \alpha = 1, 2, 3. \qquad (2.216)$$

There are two more conditions that express the requirement that the tangential components of $\vec{h}$ and the normal component of $\vec{b}$ be continuous across this surface:

$$\phi\Big|_{x_2 = 0+} = \hat{\phi}\Big|_{x_2 = 0-} \qquad (2.217a)$$

$$\left[\frac{\partial\phi}{\partial x_2} - 4\pi m_2\right]_{x_2 = 0+} = \frac{\partial\hat{\phi}}{\partial x_2}\Big|_{x_2=0-} . \qquad (2.217b)$$

Finally, in the presence of short-range exchange interactions between spins in a semi-infinite ferromagnet, the boundary conditions on the magnetization at the surface become $[\partial m_1/\partial n]_{x_2=0+} = 0$ and $[\partial m_2/\partial n]_{x_2=0+}=0$, where $\partial/\partial n$ denotes the derivative along the normal to the surface (in the present case $\partial/\partial n = \partial/\partial x_2$).(115) In practice this turns out to be an oversimplified description of these boundary conditions. The reason is that the spins at the surface often feel effective pinning fields, arising from a variety of intrinsic and extrinsic sources, that inhibit their motion. To take these effective pinning fields into account phenomenologically, the boundary conditions on the transverse components of the magnetization at the surface can be taken to have the form

$$\left[\frac{\partial m_1}{\partial x_2} - \lambda m_1\right]_{x_2=0+} = 0 \tag{2.218a}$$

$$\left[\frac{\partial m_2}{\partial x_2} - \lambda m_2\right]_{x_2=0+} = 0, \tag{2.218b}$$

where λ is a phenomenological parameter that is a measure of the strength of the surface pinning field. In the limit $\lambda = 0$ we recover the boundary conditions given above, that are appropriate to slowly varying disturbances in a semi-infinite Heisenberg ferromagnet. In the opposite limit $\lambda = \infty$ the boundary conditions become $m_1\big|_{x_2=0+} = m_2\big|_{x_2=0+} = 0$, which corresponds to the limit of very strong surface pinning, with the motion of the surface spins completely inhibited by the pinning. It should be pointed out, moreover, that the phenomenological constant λ that appears in the boundary conditions (2.218) need not be the same for m_1 and m_2. These boundary conditions could be generalized by replacing λ by λ_1 in the equation for m_1 and by λ_2 in the equation for m_2. Such a modification would correspond to the assumption that the surface spins precess with an ellipticity that differs from that in the bulk. While the inclusion of this effect in the present discussion is straightforward, I will not consider it further here.

In the absence of the magnetoelastic terms in its Hamiltonian ($b_{\mu\alpha\beta} = 0$) the system under consideration consists of two non-interacting subsystems, a purely elastic system that can support Rayleigh surface acoustic waves, and a purely magnetic system that can support magnetic surface excitations. In the presence of the magnetoelastic coupling the Rayleigh wave can excite the magnetic subsystem as it propagates along the surface of the medium. It thus carries with it a disturbance of the spin system and becomes a magnetoelastic surface acoustic wave. Its dispersion relation is modified by the coupling to the spins, and it is attenuated by this coupling.

The origin of this attenuation is the following. The normal modes of the bulk spin system (spin waves) have frequencies that lie in a restricted range[118]. For example, for a wave vector in the x_2x_3-plane (i.e. in the plane defined by the saturation magnetization and the normal to the surface), the frequencies of the bulk spin waves lie in the range $\omega_H < \omega < (\omega_H\omega_B)^{1/2}$, where $\omega_B = \omega_H + 4\pi\omega_M$. If the frequency of the Rayleigh waves lies in this region, the surface wave becomes a leaky wave that is attenuated by its magnetoelastic coupling to the bulk spin waves. This coupling also leads to structure in the Rayleigh wave dispersion, i.e. the surface wave becomes dispersive. We now turn to the solution of the coupled equations of motion for the elastic and magnetic subsystems.

In the general case, the solutions of the equations of motion in the medium ($x_2 > 0$) can be written in the form

$$v_i(\vec{x},t) = \sum_{j=1}^{6} C_i^{(j)}(\vec{k}_\parallel\omega)K_j e^{i\vec{k}_\parallel\cdot\vec{x}_\parallel-\alpha_j(\vec{k}_\parallel\omega)x_2-i\omega t} \tag{2.219}$$

where now $\vec{k}_\parallel = \hat{x}_1 k_1 + \hat{x}_3 k_3$, and we have adopted a notation in which

$$v_1 = u_1,\ v_2 = u_2,\ v_3 = u_3\ ,$$

$$v_4 = m_1,\ v_5 = m_2,\ v_6 = \phi. \tag{2.220}$$

The $\{\alpha_j(k_\parallel\omega)\}$ are obtained from the solvability condition for the set of homogeneous equations that result from substituting a solution of the form

$$v_i(\vec{x},t) = V_i e^{i\vec{k}_\parallel\cdot\vec{x}_\parallel - \alpha x_2 - i\omega t} \qquad i = 1, 2,\ldots, 6 \tag{2.221}$$

into the equations of motion (2.210), (2.211), and (2.212). The $\{C_i^{(j)}(\vec{k}_\parallel\omega)\}$ are the cofactors of the elements in the first row of the determinant of the coefficients of the $\{V_i\}$ in this system of homogeneous equations, and the $\{K_j\}$ are new amplitudes.

In the vacuum filling the region $x_2 < 0$, we have the solution

$$\hat{\phi}(\vec{x},t) \equiv v_7(\vec{x},t) = K_7\ e^{i\vec{k}_\parallel\cdot\vec{x}_\parallel + k_\parallel x_2 - i\omega t} \tag{2.222}$$

Thus, there are as many unknown amplitudes ($K_1, K_2,\ldots,K_7$) as there are boundary conditions, Eqs. (2.216), (2.217), and (2.218), to be satisfied.

Substitution of Eqs. (2.219) and (2.222) into these boundary conditions yields a set of seven homogeneous equations for the seven unknown coefficients K_j. The solvability condition for this sytem of equations is the dispersion relation for magnetoelastic surface waves.

The program described here, which requires the use of a computer, has been carried out by Camley and Scott[(112)]. In earlier work Scott and Mills[(111)] solved the simpler problem that is obtained in the absence of the exchange interaction between spins ($D = 0$), and we consider their results here. Because of the leaky nature of these modes when their frequency overlaps the frequency range in which bulk spin waves exist the wave number $k_\parallel$ is complex for real ω. Scott and Mills[(111(a))] therefore calculated $c_R \mathrm{Im} k_\parallel/\omega_H$ and $c_R \mathrm{Re}(k_\parallel - \omega/c_R)/\omega_H$ as functions of ω/ω_H, where c_R is the speed of Rayleigh waves in the absence of magnetoelastic coupling. The results are depicted in Figs. 2.16(a) and 2.16(b), respectively, for the case of YIG in an external magnetic field $H_0 = 50G$. The direction of propagation of the surface wave is parallel to the saturation magnetization (i.e. along the x_3-axis). The width of the peak in the frequency dependence of $\mathrm{Im} k_\parallel$ is insensitive to the parameter $\Gamma = (\omega_H\tau)^{-1}$, which is a measure of the spin relaxation rate. This fact identifies the origin of the attenuation peak as leakage of surface wave energy into the bulk normal modes of the system, since the density of the bulk modes is insensitive to τ^{-1}. The sensitivity of the height of the attenuation peak to Γ is presumably due to the fact that the attenuation peak occurs near $\omega = \omega_H$, where the amplitude of the response of the spin system is sensitive to τ. The feature in Fig. 2.16(a) at $\omega \lesssim (\omega_H\omega_B)^{1/2}$ arises from the fact that in this frequency range there is a peak in the density (per unit frequency) of bulk spin waves with wave vectors having a component parallel to the surface, which is a measure of the number of bulk modes available to carry the surface wave energy off into the bulk. The smallness of the amplitude of this feature is due to the fact that one of the $\alpha_j(\vec{k}_\parallel\omega)$ in this frequency range is pure imaginary and large. The resulting rapid oscillations of the magnetization induced by

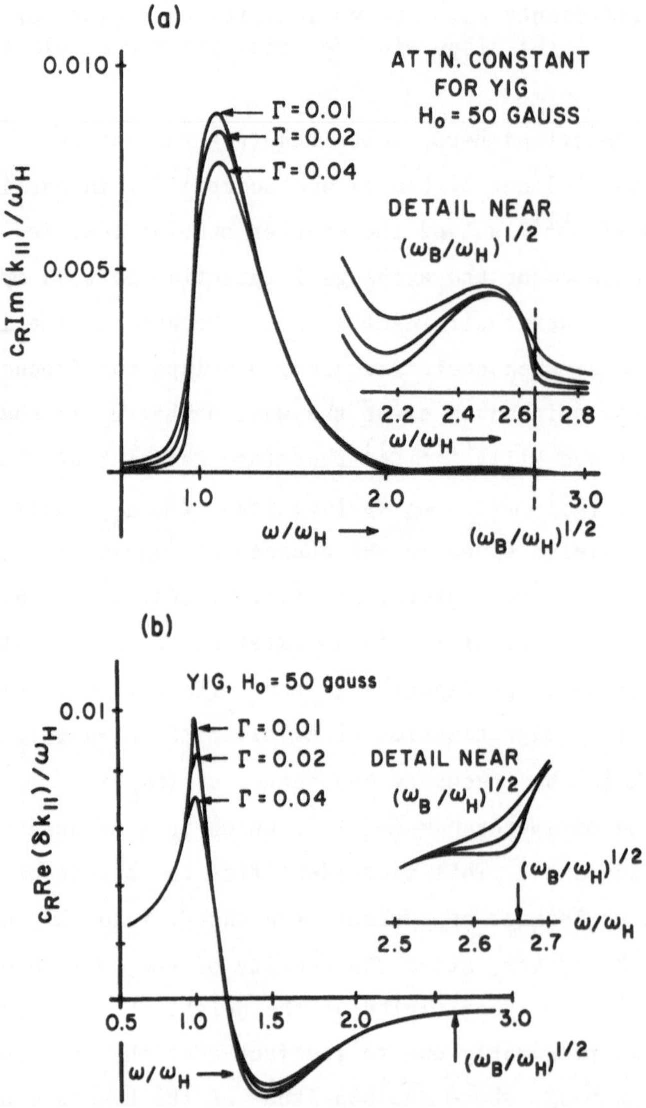

Fig. 2.16. (a) The attenuation constant $c_R \text{ Im } k_{\|}/\omega_H$ for a magnetoelastic surface acoustic wave propagating parallel to the saturation magnetization, as a function of ω/ω_H; (b) the wave number shift due to the magnetoelastic coupling, $\delta k_{\|} = \text{Re}(k_{\|} - \omega/c_R)$, of a magnetoelastic surface acoustic wave propagating parallel to the saturation magnetization [Ref. 111(a)].

the Rayleigh wave as a function of x_3 reduces the coupling between the Rayleigh wave and bulk spin waves. A peak in the same density of bulk spin waves for ω near ω_H is responsible for the asymmetric peak in $\mathrm{Im}k_{\parallel}$ for $\omega \approx \omega_H$.

Similar results are obtained for propagation in other directions[111(b)]. The nonreciprocity of these waves manifests itself through the fact that plots of $\mathrm{Im}k_{\parallel}$ vs. ω for propagation angles θ (Fig.2.17) in the range $\pi/2 < \theta < \pi$ are different from the plots of this function for θ in the range $0 < \theta < \pi/2$. A more explicit manifestation of this nonreciprocity will be presented in the following subsection.

2.3.2. Shear Horizontal Magnetoelastic Surface Acoustic Waves

For propagation perpendicular to the saturation magnetization (i.e. along the $-x_1$-direction) a magnetoelastic surface acoustic wave polarized perpendicular to the sagittal plane (the x_1x_2-plane) can exist,[112,119,120] in addition to a Rayleigh wave that is uncoupled from the spin system. The situation is analogous to that encountered when a piezoelectric surface acoustic wave propagates in a direction that is perpendicular to an axis of two-fold symmetry lying in the surface (subsection 2.2.2.). This wave owes its existence to the presence of the magnetoelastic coupling, and is a magnetic analogue of the Bleustein-Gulyaev wave on the surface of a piezoelectric medium[103,104]. The situation in the magnetoelastic case, however, is considerably more interesting than in the piezoelectric case, in that the spin system can exhibit a resonance response to the strain field that drives it.

To bring out the features of this surface acoustic wave in the clearest fashion,[121] we set the applied field H_0, and thus ω_H, equal to zero. We also neglect the effects of exchange (this is valid provided the wave vector of the mode is smaller than $\sim 10^6\,\mathrm{cm}^{-1}$), and do not consider the effects associated with a relaxation time ($1/\tau = 0$). We assume a displacement field polarized in the x_3-direction, and that it, the scalar potentials ϕ and $\hat{\phi}$, and the components of the magnetization vector, m_1 and m_2, are all independent of the coordinate x_3. The equations of motion of the system in this case are four in number:

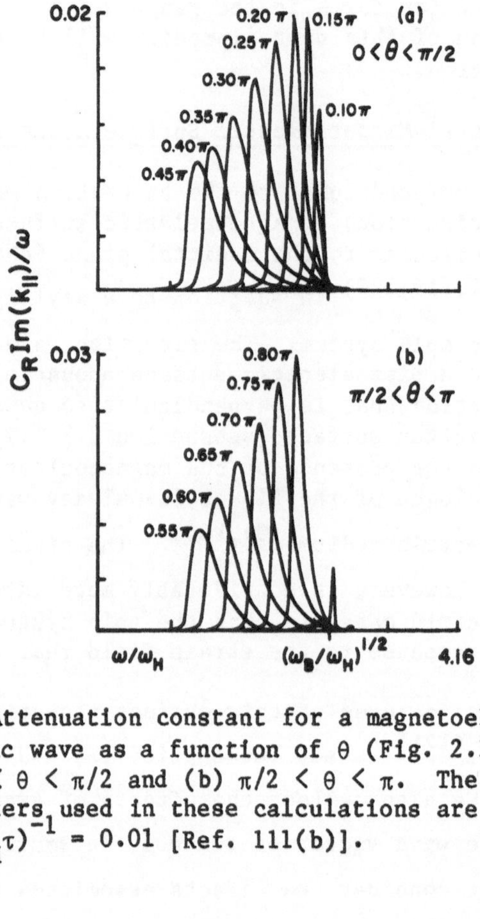

Fig. 2.17. Attenuation constant for a magnetoelastic surface acoustic wave as a function of θ (Fig. 2.15) for (a) $0 < \theta < \pi/2$ and (b) $\pi/2 < \theta < \pi$. The values of the parameters used in these calculations are $H_0 = 50G$, $\Gamma = (\omega_H\tau)^{-1} = 0.01$ [Ref. 111(b)].

$$\ddot{u}_3 = c_t^2\left(\frac{\partial^2 u_3}{\partial x_1^2} + \frac{\partial^2 u_3}{\partial x_2^2}\right) + \frac{b_2}{\rho M_s}\left(\frac{\partial m_1}{\partial x_1} + \frac{\partial m_2}{\partial x_2}\right) \tag{2.223a}$$

$$\frac{dm_1}{dt} - \omega_M \frac{\partial \phi}{\partial x_2} - \gamma b_2 \frac{\partial u_3}{\partial x_2} = 0 \tag{2.223b}$$

$$\frac{dm_2}{dt} + \omega_M \frac{\partial \phi}{\partial x_1} + \gamma b_2 \frac{\partial u_3}{\partial x_1} = 0 \tag{2.223c}$$

$$-\frac{\partial^2 \phi}{\partial x_1^2} - \frac{\partial^2 \phi}{\partial x_2^2} + 4\pi\left(\frac{\partial m_1}{\partial x_1} + \frac{\partial m_2}{\partial x_2}\right) = 0. \tag{2.223d}$$

where we have set $\rho c_t^2 = C_{2323} = c_{44}$. The equation for the scalar potential in the vacuum region ($x_2 < 0$) is

$$\frac{\partial^2 \hat{\phi}}{\partial x_1^2} + \frac{\partial^2 \hat{\phi}}{\partial x_2^2} = 0 \ . \tag{2.224}$$

The boundary conditions are:

$$\left[\rho c_t^2 \frac{\partial u_3}{\partial x_2} + \frac{b_2}{M_s} m_2\right]_{x_2=0} = 0 \tag{2.225a}$$

$$\phi\big|_{x_2=0} = \hat{\phi}\big|_{x_2=0} \tag{2.225b}$$

$$\left[\frac{\partial \phi}{\partial x_2} - 4\pi m_2\right]_{x_2=0} = \frac{\partial \hat{\phi}}{\partial x_2}\bigg|_{x_2=0} \ . \tag{2.225c}$$

If we assume expressions of the form

$$u_3(x_1 x_2;t) = u(k\omega|x_3)e^{ikx_1 - i\omega t} \tag{2.226}$$

for the several field components, the solutions of Eqs. (2.223) -

(2.224) that satisfy the boundary conditions at infinity are

$$u_3 = e^{ikx_1 - i\omega t} \frac{i\omega}{\gamma b_2 \alpha} C e^{-\alpha x_2} \tag{2.227a}$$

$$m_1 = e^{ikx_1 - i\omega t} \left[Be^{-|k|x_2} + Ce^{-\alpha x_2} \right] \tag{2.227b}$$

$$m_2 = e^{ikx_1 - i\omega t} \left[i\sigma Be^{-|k|x_2} + \frac{ik}{\alpha} Ce^{-\beta x_2} \right] \tag{2.227c}$$

$$\phi = e^{ikx_1 - i\omega t} \frac{i\omega}{\omega_M |k|} Be^{-|k|x_2} \tag{2.227d}$$

$$\hat{\phi} = e^{ikx_1 - i\omega t} De^{|k|x_2} \tag{2.227e}$$

where σ = sgn k, and

$$\alpha = \left(k^2 - \frac{\omega^2}{c_t^2}\right)^{1/2} . \tag{2.228}$$

Substitution of these expressions into the boundary conditions (2.225) yields the equations

$$\sigma \frac{b_2}{M_s} B - \left(\omega \frac{\rho c_t^2}{\gamma b_2} - \frac{b_2}{M_s} \frac{k}{\alpha}\right) C = 0 \tag{2.229a}$$

$$\frac{i\omega}{\omega_M |k|} B = D \tag{2.229b}$$

$$\left(\frac{\omega}{\omega_M} + 4\pi\sigma\right) B + \frac{4\pi k}{\alpha} C = iD|k| . \tag{2.229c}$$

The solvability condition for this system of equations yields the dispersion relation for these waves in the form

$$\alpha = \frac{k\delta}{\omega+\sigma\omega_s}\ , \qquad \delta = \frac{\gamma b_2^2}{\rho c_t^2 M_s}\ , \tag{2.230}$$

where $\omega_s = 2\pi m_M$ is the frequency of the surface spin wave in the absence of magnetoelastic coupling, or equivalently

$$k^2 = \frac{\omega^2}{c_t^2} \frac{(\omega+\sigma\omega_s)^2}{(\omega+\sigma\omega_s)^2-\delta^2}\ . \tag{2.231}$$

The nonreciprocal nature of the dispersion curve in this case is evident from the presence of σ in the dispersion relation (2.231).

The results expressed by Eq. (2.231) are plotted in Figs. (2.18a) and (2.18b). The values ω_s/γ = 190G and δ/γ = 7.9G appropriate to DyIG were used in obtaining these plots. In Fig. (2.18a) is plotted the dispersion relation given by Eq. (2.231) for propagation in the $+x_1$-direction ($\sigma = +1$). We see that as the frequency ω approaches $\omega_s-\delta$ the dispersion curve bends. This bending indicates that the mode is changing from a primarily elastic wave to a primarily magnetic wave.

When propagation is in the $-x_1$-direction ($\sigma = -1$) the results are significantly different. The corresponding dispersion curve is depicted in Fig. (2.18b). Here we see that the dispersion curve is nearly linear. It lies close to the curve $\omega = c_s k$, where $c_s^2 = c_t^2[1 - (\delta^2/\omega_s^2)]$, for low frequencies, and as ω increases past $\omega_s - \delta$ the dispersion curve approaches the line $\omega = c_t k$.

The reason that the dispersion curves for both $\sigma = +1$ and $\sigma = -1$ are dispersive in the present case is that, unlike the situation with purely elastic and piezoelectric media, there is a characteristic length in the present problem, not connected with the structure of the system being studied. It is the wavelength of the magnetic excitation in the medium that is responsible for the resonant response of the spin system to the strain field that drives it. The latter is most readily seen if we use Eqs. (2.229) to rewrite Eqs. (2.227) in the following forms:

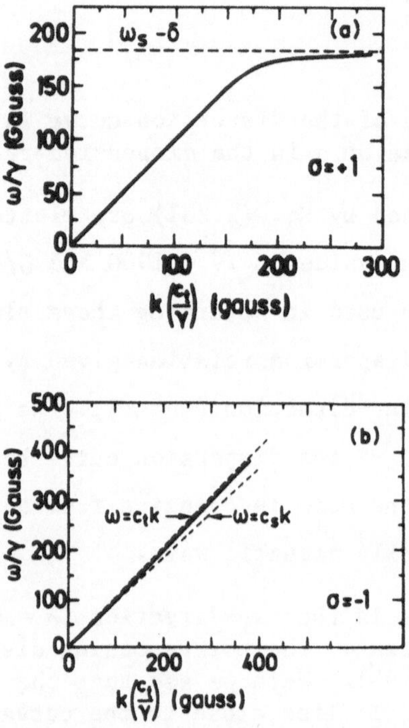

Fig. 2.18. (a) Dispersion curve obtained from Eq. (2.231) for a shear horizontal magnetoelastic surface acoustic wave propagating in the $+x_1$-direction ($\sigma = +1$). (b) Dispersion curve obtained from Eq. (2.231) for a shear horizontal magnetoelastic surface acoustic wave propagating in the $-x_1$ direction ($\sigma = -1$) [Ref. 121].

$$u_3 = u_o e^{ikx_1 - i\omega t} e^{-\alpha x_2} \tag{2.232a}$$

$$m_1 = u_o e^{ikx_1 - i\omega t} \frac{\gamma b_2}{i\omega} \left[- \frac{\omega_s k}{\omega + \sigma\omega_s} e^{-|k| x_2} + \alpha e^{-\alpha x_2} \right] \tag{2.232b}$$

$$m_2 = u_o e^{ikx_1 - i\omega t} \frac{\gamma b_2}{\omega} \left[- \frac{\omega_s |k|}{\omega + \sigma\omega_s} e^{-|k| x_2} + k e^{-\alpha x_2} \right] \tag{2.232c}$$

$$\phi = u_o e^{ikx_1 - i\omega t} \left(- \frac{2\pi\sigma}{\omega + \sigma\omega_s} \right) \gamma b_2 e^{-|k| x_2} \tag{2.232d}$$

$$\hat{\phi} = u_o e^{ikx_1 - i\omega t} \left(\frac{-2\pi\sigma}{\omega + \sigma\omega_s} \right) \gamma b_2 e^{|k| x_2} , \tag{2.232e}$$

where ω is now a function of k through Eq. (2.231). It is the presence of the factor $(\omega+\sigma\omega_s)^{-1}$ in the expressions for m_1, m_2, ϕ, and $\hat{\phi}$ that demonstrates this resonant excitation of the spin system by the strain field of the wave.

We conclude this discussion by pointing out that although the surface wave dispersion curve is nonreciprocal, the dispersion curve of the corresponding bulk waves is described by a function that is even in the components of the wave vector. There is a simple symmetry argument that indicates the reason why this is so[III(b)].

Consider a cubic material, infinite in extent, with magnetic field parallel to the x_3-direction (Fig. 2.15). Let $\omega_v(k_1,k_3)$ be the dispersion relation for a bulk excitation (a polariton, a magnetoelastic wave,...). We show that $\omega_v(k_1,k_3)$ is an even function of both k_1 and k_3. First reflect the whole system through the x_1x_2-plane. This changes k_3 to $-k_3$, but does not change the direction of the magnetic field, since the latter transforms as an axial vector under reflection. Since the operation reverses the sign of k_3 and leaves both the crystal and magnetic field invariant, we have $\omega_v(k_1,k_3) = \omega_v(k_1,-k_3)$. A reflection in the x_2x_3-plane changes the sign of k_1. But since the magnetic field is an axial vector, reflection in a plane parallel to the magnetic field changes its sign. The operation is thus not a symmetry operation of the system. Now if we next reflect in the x_1x_3-plane, k_1 and k_3 are unaffected, while the magnetic field is restored to its original direction. Thus, the combination of the two operations is a symmetry operation, and leads us to conclude that $\omega_v(k_1,k_3) = \omega_v(-k_1,k_3)$, i.e., the dispersion relation is reciprocal for the bulk mode.

Now consider a surface mode with dispersion relation $\omega_s(k_1,k_3)$. Again a magnetic field is parallel to the x_3-axis, as in Fig. 2.15. By precisely the argument given above, we may prove $\omega_s(k_1,k_3) = \omega_s(k_1,-k_3)$. However, we cannot prove $\omega_s(k_1,k_3)$ is even in k_1, because the reflection through the x_1x_3-plane is no longer a symmetry operation. It takes a crystal initially in the half-space $x_2 > 0$ and flips it into the half-space $x_2 < 0$. Thus, $\omega_s(k_1,k_3) \neq \omega_s(-k_1,k_3)$ because of the surface.

For additional work on various aspects of magnetoelastic surface waves the reader is referred to Refs. 122 and 123.

3. Power Flow in Surface Acoustic Waves

A practically important characteristic of surface acoustic waves is the power flow associated with them. This is given in terms of the energy flux vector $\vec{P}$, that has the dimensions of energy per unit area per unit time. In electromagnetic theory this vector is called the Poynting vector. In terms of this vector the differential form of the law of conservation of energy is given by

$$\frac{dH}{dt} + \nabla\cdot\vec{P} = - f, \tag{3.1}$$

where H is the energy density in the wave and f is the rate at which energy is absorbed by dissipative processes in the system supporting the wave. The vector $\vec{P}$ gives the energy crossing unit area normal to the direction of propagation of the wave per unit time, as can be seen by integrating Eq. (3.1) throughout some volume and using the divergence theorem. Knowledge of $\vec{P}$ is useful in calculating cross sections for the scattering of surface elastic waves from point defects in solids[(124)], in the calculation of the efficiencies of second harmonic generation of Rayleigh waves[(125)], and in obtaining reflection and transmission coefficients for Rayleigh waves incident on material and/or geometrical discontinuities[(126)].

In this section we present a simple derivation of the expression for $\vec{P}$, and apply it to the particular cases of elastic, piezoelectric, and magnetoelastic media. Our starting point is the expression for the energy density (Hamiltonian density) for a system described by a Lagrangian density $\mathcal{L}$,

$$H = \sum_j \dot{\psi}_j \frac{\partial \mathcal{L}}{\partial \dot{\psi}_j} - \mathcal{L} \quad , \tag{3.2}$$

where the $\{\psi_j\}$ are the fields appearing in the Lagrangian density. The time derivative of this quantity is

$$\frac{dH}{dt} = \sum_j \left(\ddot{\psi}_j \frac{\partial \mathcal{L}}{\partial \dot{\psi}_j} + \dot{\psi}_j \frac{d}{dt} \frac{\partial \mathcal{L}}{\partial \dot{\psi}_j} \right) - \frac{d\mathcal{L}}{dt} \quad . \tag{3.3}$$

The time derivative of the Lagrangian density is given by

$$\frac{d\mathcal{L}}{dt} = \sum_j \left(\frac{\partial \mathcal{L}}{\partial \psi_j} \dot{\psi}_j + \frac{\partial \mathcal{L}}{\partial \dot{\psi}_j} \ddot{\psi}_j + \sum_\alpha \frac{\partial \mathcal{L}}{\partial (\partial \psi_j / \partial x_\alpha)} \frac{\partial \dot{\psi}_j}{\partial x_\alpha} \right) \quad . \tag{3.4}$$

When we use Eqs. (2.8) and (3.4) in Eq. (3.3), together with the identity

$$\frac{\partial}{\partial(\partial \psi_j / \partial x_\alpha)} \frac{\partial \dot{\psi}_j}{\partial x_\alpha} = \frac{\partial}{\partial x_\alpha} \left(\dot{\psi}_j \frac{\partial}{\partial(\partial \psi_j / \partial x_\alpha)} \right) - \dot{\psi}_j \frac{\partial}{\partial x_\alpha} \left(\frac{\partial}{\partial(\partial \psi_j / \partial x_\alpha)} \right) , \tag{3.5}$$

we obtain the result that

$$\frac{dH}{dt} = -\sum_\alpha \frac{\partial}{\partial x_\alpha} P_\alpha - \sum_j \dot{\psi}_j \frac{\partial F}{\partial \dot{\psi}_j} \quad . \tag{3.6}$$

The α Cartesian component of the vector $\vec{P}$ on the right hand side of this equation is given by

$$P_\alpha = \sum_j \dot{\psi}_j \frac{\partial \mathcal{L}}{\partial(\partial \psi_j / \partial x_\alpha)} \quad . \tag{3.7}$$

The second term on the right hand side of Eq. (3.6) gives the rate at which energy is absorbed by the dissipative processes in our system. Comparison of Eq. (3.6) with Eq. (3.1) thus yields the result that the quantity f appearing in the latter is given by

$$f = \sum_j \dot{\phi}_j \frac{\partial F}{\partial \dot{\phi}_j} . \tag{3.8}$$

The time averaged power flow is given by

$$\langle P_\alpha \rangle = \text{Re}\, \mathcal{P}^c_\alpha , \tag{3.9}$$

where the complex power flow vector $\vec{\mathcal{P}}^c$ has the components

$$\mathcal{P}^c_\alpha = \frac{1}{2} \sum_j \dot{\phi}^*_j \frac{\partial \mathcal{L}}{\partial(\partial\phi_j/\partial x_\alpha)} . \tag{3.10}$$

For a purely elastic medium we obtain from Eqs. (3.10) and (2.9) the result that

$$\mathcal{P}^c_\alpha = -\frac{1}{2} \sum_\beta \dot{u}^*_\beta T_{\beta\alpha} , \tag{3.11}$$

where $T_{\alpha\beta}$ is given by Eq. (2.2). For a piezoelectric medium, Eqs. (3.11) and (2.166) yield the result that

$$\mathcal{P}^c_\alpha = -\frac{1}{2} \{\sum_\beta \dot{u}^*_\beta T_{\beta\alpha} + \dot{\phi}^* D_\alpha\} , \tag{3.12}$$

where $T_{\alpha\beta}$ and D_α are given by Eqs. (2.162) and (2.163), respectively. Finally, for a magnetoelastic medium we find from Eqs. (3.10) and (2.113) that

$$\mathcal{P}^c_\alpha = -\frac{1}{2} \{\frac{1}{4\pi} \dot{\phi}^* b_\alpha + \frac{D}{\omega_M} \sum_\beta \dot{m}^*_\beta \frac{\partial m_\beta}{\partial x_\alpha} + \sum_\beta \dot{u}^*_\beta T_{\beta\alpha}\} , \tag{3.13}$$

where b_α is given by Eqs. (2.206c) and (2.207), m_α is given by Eqs. (2.211), and $T_{\alpha\beta}$ is given by Eq. (2.208).

Since a surface wave has its fields localized in the vicinity of the surface, the vector $\vec{\mathcal{P}}^c$ is also localized in the vicinity of the surface. More useful than $\langle P_\alpha \rangle$, therefore, is the time-averaged power flow per unit width of the surface wave. If the surface is the plane $x_n = 0$, and the surface wave propagates in the α direction

$$W_\alpha = \int_{-\infty}^{\infty} dx_n \langle P_\alpha \rangle . \tag{3.14}$$

There is no power flow normal to the surface for non-leaky surface waves. Therefore the vector $\vec{W}$ has no component in this direction.

To illustrate the application of Eq. (3.14) to a concrete situation, we calculate the total power carried parallel to the surface in a Rayleigh wave on an isotropic substrate. If the solid occupies the region $x_3 > 0$, the vector $\vec{W}$ is given by Eqs. (3.9), (3.11), and (3.14) as

$$W_\alpha = -\frac{1}{2} \mathrm{Re} \int_0^\infty i\omega \sum_\beta T_{\alpha\beta} u_\beta^* \, dx_3 . \tag{3.15}$$

From the results of subsection 2.1.1.1 we find in this case that

$$u_1 = A[e^{-\alpha_\ell x_3} - (1 - (c_R^2/2c_t^2))e^{-\alpha_t x_3}]e^{ikx_1 - i\omega t} \tag{3.16a}$$

$$u_3 = iA(1-(c_R^2/c_\ell^2))^{1/2}\left[e^{-\alpha_\ell x_3} - \frac{e^{-\alpha_t x_3}}{(1-(c_R^2/2c_t^2)}\right]e^{ikx_1 - i\omega t} \tag{3.16b}$$

$$T_{11} = 2ik\rho c_t^2 A\{[1+(1 - 2\frac{c_t^2}{c_\ell^2})\frac{c_R^2}{2c_t^2}]e^{-\alpha_\ell x_3} -$$

$$- (1 - \frac{c_R^2}{2c_t^2})\, e^{-\alpha_t x_3}\}e^{ikx_1 - i\omega t} \tag{3.17a}$$

$$T_{13} = -2k\rho c_t^2 A(1 - \frac{c_R^2}{c_\ell^2})^{1/2}\{e^{-\alpha_\ell x_3} - e^{-\alpha_t x_3}\}e^{ikx_1 - i\omega t} , \tag{3.17b}$$

where

$$\alpha_\ell = k(1-(c_R^2/c_\ell^2))^{1/2} , \ \alpha_t = k(1-(c_R^2/c_t^2)^{1/2} . \tag{3.18}$$

It follows from these results that the only nonzero component of the vector $\vec{W}$ is W_1. If we substitute into Eq. (3.11) the results given by Eqs. (3.16)-(3.17), we obtain the result that

$$\wp_1^c = |A|^2 \omega\rho c_t^2 k\{\zeta_{\ell\ell} e^{-2\alpha_\ell x_3} + \zeta_{\ell t} e^{-(\alpha_\ell+\alpha_t)x_3} + \zeta_{tt} e^{-2\alpha_t x_3}\}, \tag{3.19}$$

where

$$\zeta_{\ell\ell} = 2 + \frac{c_R^2}{2c_t^2}\left(1 - 4\frac{c_t^2}{c_\ell^2}\right) \tag{3.20a}$$

$$\zeta_{\ell t} = -\left(1 - \frac{c_R^2}{2c_t^2}\right)\left(2 + \frac{c_R^2}{2c_t^2} - \frac{c_R^2}{c_\ell^2}\right) - \frac{(1 - (c_R^2/c_\ell^2))(2-(c_R^2/2c_t^2))}{(1 - (c_R^2/2c_t^2)} \tag{3.20b}$$

$$\zeta_{tt} = \frac{1}{(1-(c_R^2/2c_t^2))}\left\{\left(1 - \frac{c_R^2}{c_\ell^2}\right) + \left(1 - \frac{c_R^2}{2c_t^2}\right)^3\right\}. \tag{3.20c}$$

Because $\wp_1^c$ is real, the time average of the 1-component of the power flow vector of a Rayleigh wave is given simply by

$$\langle P_1\rangle = \wp_1^c \,. \tag{3.21}$$

The time averaged power flow in the x_1-direction per unit width of the Rayleigh wave (in the x_2-direction) is obtained from Eqs. (3.14), (3.19), and (3.21) as

$$W_1 = \int_0^\infty dx_3 \,\langle P_1\rangle$$

$$= |A|^2\omega^2\rho \frac{c_t^2}{c_R}\left[\frac{\zeta_{\ell\ell}}{2\alpha_\ell} + \frac{\zeta_{\ell t}}{\alpha_\ell+\alpha_t} + \frac{\zeta_{tt}}{2\alpha_t}\right]. \tag{3.22}$$

4. Generation and Detection of Surface Acoustic Waves

A variety of experimental techniques has been used over the past twenty-five years for the excitation and detection of surface acoustic waves in the laboratory. Some of the early methods for this are described in the book by Viktorov(3). More recent, and more extensive, surveys of techniques for the generation and detection of surface acoustic waves can be found in the review articles by White[16] and Nizzoli and Stegeman[17]. In this section we describe briefly several of these techniques. For a fuller account the reader is referred to the articles cited above.

One of the earliest methods used for ultrasonic surface acoustic wave generation used wedge shaped transducers[127-129]. In this method (Fig. 4.1) a prism or wedge, usually made of plastic, is acoustically coupled through one of its faces to the surface of the solid. A piezoelectric plate is bonded to the sloping surface of the wedge, and emits a bulk longitudinal wave whose speed is c_W that impinges on the interface between the solid and the wedge at an angle of incidence θ. The optimum value of θ is given by $\sin\theta_R = c_W/c_R$, where c_R is the speed of Rayleigh waves on the sample. The requirement that $c_W < c_R$ explains why the wedge is usually made of plastic. A periodic perturbation is thereby created at the interface between the solid and the wedge, whose spatial period equals the wavelength of the Rayleigh wave on the solid. Since θ_R is greater than the critical for total internal reflection of bulk longitudinal and transverse waves, the waves transmitted into the solid are evanescent. This perturbation thus excites a Rayleigh wave that propagates over the surface of the solid in the positive x_1-direction. The optimum excitation occurs when the projection of the edge of the wedge onto the sloping surface coincides with the leading edge of the piezoelectric plate, as shown in Fig. 4.1. The inverse process can serve as a way of detecting a Rayleigh wave.

Wedge-shaped transducers that convert bulk transverse waves into Rayleigh waves have also been used[130,131]. The slower speed of bulk transverse waves relative to bulk longitudinal waves has the consequence that the wedge need not be made of plastic, but can be made of metal. This has the advantage of simpler fabrication, less wear, and a better capability for matching the wedge and sample materials.

Subsequently, deposited interdigital comb structures photoetched onto piezoelectric substrates[132,133] have made possible the generation of surface acoustic waves in the gigahertz frequency range[134]. The wavelength of the surface acoustic waves generated in this way is determined by the distance between the consecutive "fingers" of the interdigital transducer. Such interdigital comb structures can also serve as detectors of surface acoustic waves (Fig. 4.2). Theories of the generation and detection of piezoelectric surface acoustic waves have been constructed on the assumption that the total piezoelectric coupling is weak, i.e. that the electric field produced by the surface wave on a piezoelectric medium is small in comparison with the field used to excite them[135-137]. A theory of this mode of generation and detection of piezoelectric surface acoustic waves when the total piezoelectric coupling is not weak has been developed by Emtage[138].

A variant of these approaches uses a light beam instead of a transducer for detection[139,140].

The optical probing of surface acoustic waves has been used for the measurement of the amplitude of the surface acoustic wave and for the measurement of their attenuation in nonpiezoelectric materials[141]. In this method an optical wave receives a spatial phase modulation when it is reflected from the surface or passes through the substrate layer perturbed by the surface wave. This phase modulation results from two effects induced by a surface acoustic wave (Rayleigh wave): a sinusoidal corrugation of the surface, and a periodic variation of the refractive index of the medium supporting the surface wave in the region closer to the surface than the penetration depth of the surface wave, due to the elasto-optic effect. This phase modulation can be related to the amplitude of the surface corrugation,[141] and can be obtained experimentally from the intensities of the diffraction maxima of the $\pm 1^{st}$ orders. In a recent variant of this method[142,143] its sensitivity is enhanced by propagating the surface acoustic wave across a stationary reference grating whose period coincides with the wavelength of the surface acoustic wave. By this method it has been possible to detect surface acoustic wave amplitudes of the order of 10^{-3} Å. [143]

A novel method for the generation of large amplitude, nonthermal surface acoustic waves at gigahertz frequencies on semiconductor surfaces has been developed by Schmidt and Dransfeld[144]. They subjected the surface of GaAs crystal to intense illumination with 5145Å radiation from an Ar^+ laser at room temperature. Since the frequency of the incident light is larger than that of the band gap in GaAs, electron-hole pairs are created that can recombine via two processes: (1) optical luminescence; and (2) nonradiative transitions in which phonons are emitted instead of photons. In their work Schmidt and Dransfeld found that part of the absorbed optical energy was converted into surface acoustic waves via nonradiative transitions. The surface

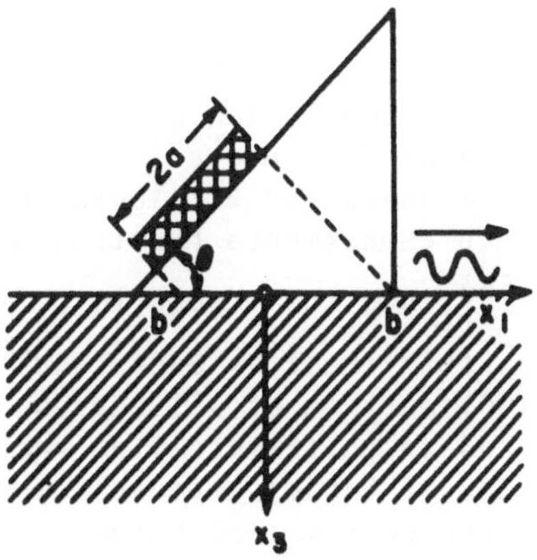

Fig. 4.1. A wedge transducer for the generation of Rayleigh surface acoustic waves [Ref. 127].

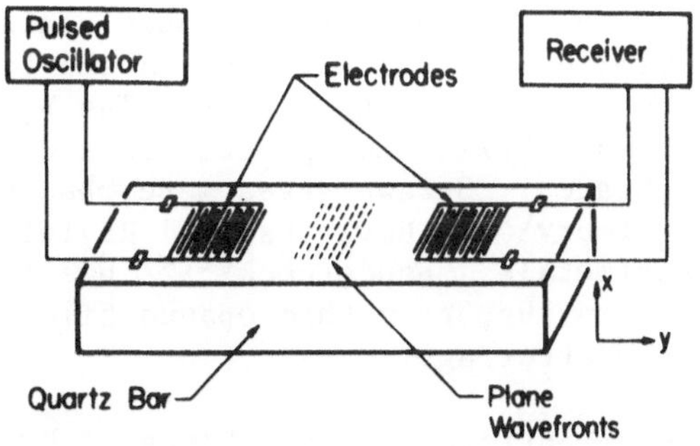

Fig. 4.2. Arrangement for surface acoustic wave transduction by interdigital comb structures on a crystalline quartz bar.

acoustic waves thus produced were detected by high resolution Brillouin spectroscopy, in which the radiation from the same Ar^+ laser used for exciting the electron-hole pairs was used to probe the surface acoustic wave, whose frequency was 7 GHz. The mechanism by which the absorbed optical energy of 2.4 eV (5145Å) is transformed into surface acoustic waves at gigahertz frequencies is not yet understood.

All of the preceding detection methods possess the common feature of requiring external generation of intense surface waves. Recently, thermally excited surface acoustic waves in the 1-20 GHz frequency range have been measured by Brillouin scattering techniques. The theory of such experiments has been worked out by several authors(68,145-168), and the reader is referred to their papers for the details of this theory. Sandercock(67) has studied Rayleigh waves on the (100) surface of silicon, on crystalline samples of GaAs and Cr, and on polycrystalline samples of Aℓ, Cu, Ta, Ni, and Fe. The speeds of Rayleigh waves measured by Sandercock in this work were consistently 1-5% below the values calculated from the bulk elastic moduli. This, however, appears to be the consequence of a calibration error in the spectrometer.(169) Stoneley waves have been observed in Brillouin scattering from a glass-mercury interface by Dil et al.(170), the only experimental observation of these thermally excited interface modes to date. Several Brillouin scattering studies have been carried out of surface acoustic waves in thin films on substrates. Thus Rowell and Stegeman(159) observed both Rayleigh waves and Sezawa waves in a thin film of Corning 7059 glass supported on a fused silica substrate. Bortolani et al.(171) studied thermally excited surface acoustic waves on GaAs and Si substrates coated with a thin Aℓ film. A Rayleigh wave was observed in the former case, and the first two Sezawa waves in the latter. The same authors(172) also obtained Brillouin spectra for aluminum and gold films on silicon substrates. Sezawa waves were observed in the results. Harley and Fleury(173) have observed Rayleigh waves on layered metals and semimetals. Sandercock(174) has shown that generalized Lamb modes in supported thin opaque films can also be observed by Brillouin scattering.

5. Some Applications of Surface Acoustic Waves as Probes of Physical and Electronic Properties of Solids

In addition to being objects of study in themselves, Rayleigh waves have been used recently to study some physical and electronic properties of solids. The temperature dependence of the speed of Rayleigh waves in the (100) direction on the (001) surface of $SrTiO_3$ has been measured at several different frequencies at temperatures just above the temperature of the structural phase transition at 105 K(175) (Fig. 5.1). The aim of the investigation

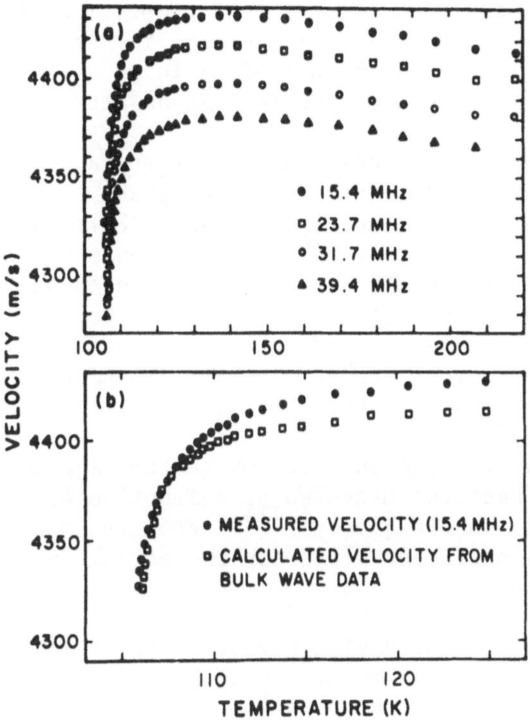

Fig. 5.1. (a) Measured critical speeds of Rayleigh waves propagating in the [100] direction on a (001) surface of $SrTiO_3$ at different frequencies as functions of temperature in the vicinity of the structural phase transition at 105K. (b) Measured Rayleigh wave speeds compared with speeds calculated from bulk wave measurements [Ref. 175].

was to determine how the material properties of the crystal differ from their bulk values within a distance from the surface of the order of the correlation length over which the order parameter has a surface-induced spatial dependence. Since this correlation length diverges as the transition temperature is approached from either side, surface waves will be affected more and more by the surface layer as the critical temperature is approached. Such effects were not observed in the experiments of Bjerkan and Fossheim[175], who concluded that the critical properties of $SrTiO_3$ are dominated by the bulk critical properties in the frequency range studied. Nevertheless, the use of Rayleigh waves for the study of surface critical properties remains an attractive one.

Experimental and theoretical results for surface acoustic wave propagation near the spin reorientation phase transition in $ErFeO_3$ in zero field and various nonzero magnetic fields have been reported by Gorodetsky and Shaft[176]. The purpose of the study was an investigation of the effect of the surface on the nature of this phase transition. The differences observed between the results of the surface wave measurements and those of bulk wave measurements in the critical region were attributed to the effects of the surface on the magnetic anisotropy.

The attenuation of Rayleigh waves on Cu/Pb-, Ag/Pb-, and Aℓ/Pb-double layers has been measured as a function of temperature, magnetic field, and frequency[177]. From the results the electron interaction parameter N(0)V for the normal metals Cu and Ag has been determined.

Changes in the speed of Rayleigh waves with as little as a few hundredths of an adsorbed monolayer of gas atoms can be measured[178], which offers the possibility of monitoring atom-surface interaction phenomena in this way.

Additional applications of Rayleigh waves to the study of physical and electronic properties of solids have recently been described by Borisov and Konstantinov[179].

6. Surface Acoustic Waves on Curved Surfaces

Up to now we have considered the propagation of surface acoustic waves only along planar surfaces or interfaces. Interesting results are obtained when the restriction to planar surfaces is lifted, and the surface is allowed to be curved. In particular, the frequencies of Rayleigh waves propagating over curved surfaces become dispersive. However, in addition to Rayleigh waves, a curved (convex) boundary can support other types of waves, localized to the curved boundary. Such surface waves were first studied, in the context of a problem in acoustics, by Lord Rayleigh[180], and today are sometimes called "whispering gallery" waves.

Subsequently, Rayleigh waves on the surface of a cylinder and of a sphere have been studied by several authors(181), since these are the only structures, apart from a semi-infinite medium, for which the equations of motion of elasticity, and the appropriate boundary conditions can be solved exactly. To obtain results for other surfaces one has to resort to approximate methods of solution, but perhaps more interesting results are obtained. In this section I consider surface acoustic wave propagation for three types of curved surfaces: cylindrical surfaces; surfaces with periodic, one-dimensional gratings ruled on them; and randomly rough surfaces.

6.1. Cylindrical Surfaces

In this subsection we consider the simplest case of a surface acoustic wave propagating across a curved surface, viz. a shear horizontal surface wave propagating circumferentially around a portion of a cylindrical surface of radius R. Although simple, this case displays several of the interesting features of surface acoustic wave propagation on a curved surface that are absent when the surface is flat.

In cylindrical coordinates (r, θ, z), the elements of the stress tensor τ_{ij} for an isotropic elastic medium are given by(182)

$$\tau_{ii} = \lambda\theta + 2\mu e_{ii} \tag{6.1a}$$

$$\tau_{ij} = 2\mu e_{ij} \quad (i \neq j) \tag{6.1b}$$

where λ and μ are the Lamé constants,

$$\theta = e_{11} + e_{22} + e_{33} , \tag{6.1c}$$

and the elements of the strain tensor e_{ij} are given by

$$e_{rr} = \frac{\partial u_r}{\partial r} , \quad e_{\theta\theta} = \frac{1}{r}\frac{\partial u_\theta}{\partial \theta} + \frac{u_r}{r} , \quad e_{zz} = \frac{\partial u_z}{\partial z}, \tag{6.2a,b,c}$$

$$e_{r\theta} = \frac{1}{2}\left(\frac{1}{r}\frac{\partial u_r}{\partial \theta} + \frac{\partial u_\theta}{\partial r} - \frac{u_\theta}{r}\right) = e_{\theta r} \tag{6.2d}$$

$$e_{rz} = \frac{1}{2}\left(\frac{\partial u_z}{\partial r} + \frac{\partial u_r}{\partial z}\right) = e_{zr} \tag{6.2e}$$

$$e_{\theta z} = \frac{1}{2}\left(\frac{\partial u_\theta}{\partial z} + \frac{1}{r}\frac{\partial u_z}{\partial \theta}\right) = e_{z\theta} \ . \tag{6.2f}$$

In these expressions $u_r(r,\theta,z)$, $u_\theta(r,\theta,z)$, $u_z(r,\theta,z)$ are the components of the displacement vector.

The equations of motion of the medium take the forms

$$\rho\ddot{u}_r = \frac{\partial \tau_{rr}}{\partial r} + \frac{1}{r}\frac{\partial \tau_{r\theta}}{\partial \theta} + \frac{\partial \tau_{rz}}{\partial z} + \frac{\tau_{rr} - \tau_{\theta\theta}}{r} \tag{6.3a}$$

$$\rho\ddot{u}_\theta = \frac{\partial \tau_{r\theta}}{\partial r} + \frac{1}{r}\frac{\partial \tau_{\theta\theta}}{\partial \theta} + \frac{\partial \tau_{\theta z}}{\partial z} + \frac{2}{r}\tau_{r\theta} \tag{6.3b}$$

$$\rho\ddot{u}_z = \frac{\partial \tau_{rz}}{\partial r} + \frac{1}{r}\frac{\partial \tau_{\theta z}}{\partial \theta} + \frac{\partial \tau_{zz}}{\partial \theta} + \frac{1}{r}\tau_{rz} \ . \tag{6.3c}$$

In the case of a shear horizontal wave propagating circumferentially around a cylinder whose axis is along the z-axis the displacement field has the form

$$\vec{u}(r,\theta,z;t) = \hat{z}u_z(r,\theta|\omega)e^{-i\omega t} . \tag{6.4}$$

The nonzero elements of the strain tensor are

$$e_{rz} = \frac{1}{2}\frac{\partial u_z}{\partial r} , \quad e_{\theta z} = \frac{1}{2}\frac{1}{r}\frac{\partial u_z}{\partial \theta} , \tag{6.5}$$

so that the nonzero elements of the stress tensor are

$$\tau_{rz} = \mu\frac{\partial u_z}{\partial r} , \quad \tau_{\theta z} = \frac{\mu}{r}\frac{\partial u_z}{\partial \theta} . \tag{6.6}$$

The equation for $u_z(r,\theta)$ thus becomes

$$\rho\ddot{u}_z = \mu\left(\frac{\partial^2 u_z}{\partial r^2} + \frac{1}{r}\frac{\partial u_z}{\partial r} + \frac{1}{r^2}\frac{\partial^2 u_z}{\partial \theta^2}\right) . \tag{6.7}$$

With the aid of Eq. (6.4) this equation can be rewritten as

$$\left(\frac{\partial^2}{\partial r^2} + \frac{1}{r}\frac{\partial}{\partial r} + \frac{1}{r^2}\frac{\partial^2}{\partial \theta^2} + \frac{\omega^2}{c_t^2}\right)u_z(r,\theta|\omega) = 0. \tag{6.8}$$

where we have used the relation $\mu/\rho = c_t^2$. This equation must be supplemented by the condition that the stress acting on the surface $r = R$ vanish:

$$\tau_{rz}\Big|_{r=R} = \mu\left.\frac{\partial u_z}{\partial r}\right|_{r=R} = 0 . \tag{6.9}$$

We solve equation (5.8) by separating the variables. We write

$$u_z(r,\theta) = R(r)\Theta(\theta), \tag{6.10}$$

and find that $\Theta(\theta)$ and $R(r)$ satisfy the equations

$$\Theta'' + \nu^2\Theta = 0 \tag{6.11a}$$

$$R'' + \frac{1}{r}R' + \left(\frac{\omega^2}{c_t^2} - \frac{\nu^2}{r^2}\right)R = 0, \tag{6.11b}$$

respectively, where ν^2 is the separation constant. The sign of ν^2 has been chosen in such a way that the solution of Eq. (6.11a) corresponds to a wave propagating circumferentially around the cylinder, viz.

$$\Theta(\theta) = e^{i\nu\theta}. \tag{6.12}$$

Since we are considering propagation over only a portion of a cylindrical boundary, it is not necessary to impose a single-valuedness requirement on the displacement component $u_z(r,\theta)$, and hence on $\Theta(\theta)$. Thus, ν need not be an integer. If we rewrite Eq. (6.12) in the form

$$\Theta(\theta) = e^{i \frac{\nu}{R} (R\theta)} , \tag{6.13}$$

and recall that $R\theta$ is the path length measured along the cylindrical surface, we see that

$$\frac{\nu}{R} \equiv k \tag{6.14}$$

can be regarded as the wavenumber characterizing the propagation of this circumferential wave.

In solving Eq. (6.11b) there are two cases to consider: (i) the solid is convex toward the vacuum (Fig. 6.1(a)); and (ii) the solid is concave toward the vacuum (Fig. 6.1(b)). We consider them in turn.

(i) <u>solid convex to vacuum</u>

The solutions of Eq. (6.11b) are Bessel functions. In the present case, for a reason discussed below, we choose for the solution of Eq. (6.11b)

$$R(r) = J_\nu\left(\frac{\omega}{c_t} r\right) , \tag{6.15}$$

where $J_\nu(x)$ is a Bessel function of order ν. The displacement $u_z(r,\theta|\omega)$ thus takes the form

$$u_z(r,\theta|\omega) = \text{const.}\ J_\nu\left(\frac{\omega}{c_t} r\right) e^{i\nu\theta} . \tag{6.16}$$

Application of the boundary condition (6.9) yields the condition

$$J_\nu'\left(\frac{\omega}{c_t} R\right) = 0 \tag{6.17}$$

that relates the frequency ω of the wave to the wave number $k = \nu/R$. In Eq. (6.17) the prime denotes differentiation with respect to argument.

The Bessel function $J_\nu(x)$ has the following property that is of central importance here, and that dictates the choice of

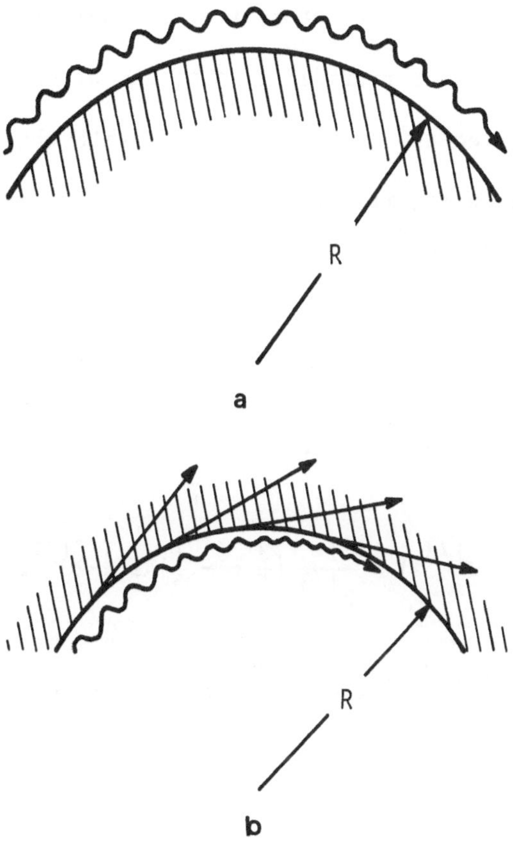

Fig. 6.1. Shear horizontal surface acoustic waves propagating circumferentially around a portion of a cylindrical surface on an isotropic elastic medium that (a) is convex toward the vacuum, and (b) is concave toward the vacuum.

solution given by Eq. (6.15).[183] For a fixed value of (real, nonzero) ν, $J_\nu(x)$ increases exponentially with increasing x, until a value of $x \sim \nu$ is reached, at which point $J_\nu(x)$ acquires an oscillatory dependence on x that continues for $x > \nu$. Since we are concerned with a solution for r in the range $0 \leqslant r \leqslant R$ that is localized for r in the vicinity of R, i.e. is small for $r \sim 0$, $J_\nu(\frac{\omega}{c_t} r)$ has this behavior provided ν is of the order of $(\omega/c_t)R$. This requirement on ν is consistent with the boundary condition (6.9), which requires that $r = R$ be in the range of r where $J_\nu(\frac{\omega}{c_t} r)$ has an oscillatory rather than an exponentially increasing nature.

To solve Eq. (6.17) we therefore use the result that[184]

$$J_\nu'(\nu+z\nu^{\frac{1}{3}}) \sim -\frac{2^{2/3}}{\nu^{2/3}} Ai'(-2^{1/3}z) +$$

$$+\frac{2^{2/3}}{\nu^{4/3}}\left[\frac{4}{5} zAi'(-2^{1/3}z) + \frac{1}{2^{1/3}}\left(\frac{3}{5}z^3 - \frac{1}{5}\right)Ai'(-2^{1/3}z)\right] + 0(\nu^{-2}) \tag{6.18}$$

when z is fixed and $|\nu|$ is large. In this expression Ai(-x) is an Airy function.[185] Thus, if we set

$$\frac{\omega}{c_t} R = \nu + z\nu^{\frac{1}{3}} , \tag{6.19}$$

then

$$\frac{\omega}{c_t} = k\left[1 + \frac{z}{(kR)^{2/3}}\right] . \tag{6.20}$$

To obtain ω/c_t to lowest nonzero order in $(kR)^{-1}$, we see from Eq. (6.20) that we need only the approximation to z that is given by the solutions of

$$Ai'(-2^{1/3}z) = 0 . \tag{6.21}$$

If we denote the zeros of $Ai'(x)$ by $-x_i$, the first few of them are given by[186]

$$x_i = 1.019,\ 3.248,\ 4.820,\ldots \tag{6.22}$$

The dispersion relation (6.20) thus finally takes the form

$$\left(\frac{\omega}{c_t}\right)_i = k\left[1 + \frac{x_i}{2^{1/3}(kR)^{2/3}}\right], \qquad i = 1, 2, 3, \ldots \tag{6.23}$$

From these results we see that a homogeneous, isotropic elastic cylinder with a stress-free surface can support an infinite number of shear horizontal surface acoustic waves that propagate circumferentially along its surface. These waves are dispersive, i.e. their phase velocities are functions of their wavelength. This is due to the presence of a characteristic length in the problem, viz. the radius R of the cylinder. The phase velocities of these waves are also seen from Eqs. (6.22) and (6.23) to be greater than that (ω/c_t) of bulk shear waves in the medium. We have seen earlier that such waves are not possible if the surface is planar. Indeed, we see from Eq. (6.23) that in the limit $R \to \infty$ the phase velocities of all the waves approach ω/c_t, i.e. they become skimming bulk transverse waves.

The surface acoustic waves just obtained are the whispering gallery waves first studied by Lord Rayleigh[180] and which have been the objects of many subsequent investigations.[187-190]

Some physical insight into this solution can be gained in the following way. Let us introduce the new variable

$$z = R\,\ell n\,\frac{R}{r}, \tag{6.24}$$

so that $z > 0$ corresponds to the interior of the solid ($0 < r < R$), while $z < 0$ corresponds to the vacuum outside the solid. In terms of this variable Eq. (6.11b) takes the form

$$\left[\frac{d^2}{dz^2} - k^2 + \frac{\omega^2}{c_t^2(z)}\right]\hat{R}(z) = 0, \tag{6.25}$$

where $k = \nu/R$ and

$$c_t^2(z) = c_t^2\, e^{2\frac{z}{R}}. \tag{6.26}$$

In writing Eq. (6.25) we have defined $\hat{R}(z) \equiv R(r)$. The boundary condition satisfied by $\hat{R}(z)$ is

$$\frac{d}{dz}\hat{R}(z)\Big|_{z=0} = 0. \tag{6.27}$$

From Eqs. (6.25) we see that the equation for $\hat{R}(z)$ resembles that for shear horizontal waves on a planar surface of a medium whose transverse speed of sound increases exponentially with increasing distance into the medium. The analogy is not exact because, as we have seen in subsection 2.1.2.2, in the presence of spatially varying elastic moduli there are additional terms in the equations of motion associated with the spatial derivatives of the elastic moduli. Nevertheless, the analogy is a good one for slowly varying elastic moduli. Since $c_t(z)$ is smaller at the surface of the medium ($z = 0$) than in the interior, this is just the kind of situation that gives rise to Love waves. Thus, we can say that shear horizontal surface acoustic waves propagating circumferentially around the stress-free surface of an elastic cylinder that is convex toward vacuum are analogous to Love waves on a planar stress-free surface in which c_t increases exponentially with increasing distance into the medium from the surface.

In the immediate vicinity of the surface, i.e. for z/R small, we can write Eq. (6.25) in the approximate form

$$\left[\frac{d^2}{dz^2} - k^2 + \frac{\omega^2}{c_t^2} - 2\frac{\omega^2}{c_t^2}\frac{z}{R}\right]\hat{R}(z) = 0, \tag{6.28}$$

which has the solution

$$\hat{R}(z) = \mathrm{Ai}\left(\left(\frac{Rc_t^2}{2\omega^2}\right)^{2/3}\left(2\frac{\omega^2}{c_t^2}\frac{z}{R} + k^2 - \frac{\omega^2}{c_t^2}\right)\right). \tag{6.29}$$

The dispersion relation obtained from the boundary condition (6.27) is

$$\mathrm{Ai}'\left(\left(\frac{Rc_t^2}{2\omega^2}\right)^{2/3}\left(k^2 - \frac{\omega^2}{c_t^2}\right)\right) = 0. \tag{6.30}$$

Thus, we have that

$$\left(\frac{Rc_t^2}{2\omega^2}\right)^{2/3}\left(k^2-\frac{\omega^2}{c_t^2}\right) = -x_i. \tag{6.31}$$

With some rearrangement the solution of this equation is found to be

$$\left(\frac{\omega}{c_t}\right)_i = k\left[1+\frac{x_i}{2^{1/3}(kR)^{2/3}}\right], \tag{6.32}$$

which coincides with the result obtained earlier (Eq. (6.23)).

ii) solid concave to vacuum

When the cylinder is concave to the vacuum we choose for the solution of Eq. (6.11b)

$$R(r) = H_\nu^{(1)}\left(\frac{\omega}{c_t} r\right), \tag{6.33}$$

where $H_\nu^{(1)}(x) = J_\nu(x) + iY_\nu(x)$ is the Hankel function of the first kind of order ν. Its asymptotic behavior for large $|x|$ and fixed ν is

$$H_\nu^{(1)}(x) \sim \left(\frac{2}{\pi x}\right)^{1/2} e^{i(x-\frac{1}{2}\nu\pi-\frac{1}{4}\pi)}. \tag{6.34}$$

Thus this choice for R(r) describes a wave that radiates energy away from the surface into the interior of the solid. It must therefore be attenuated as it propagates along the surface. This means that the wave vector $k = \nu/R$ that is the solution of the dispersion relation

$$H_\nu^{(1)}\left(\frac{\omega}{c_t} R\right)' = 0, \tag{6.35}$$

that follows from Eq. (6.9) must be complex for real ω, $k = k_R + ik_I$, with both k_R and k_I positive.

Solutions of the equation $H_\nu^{(1)}(ka)' = 0$ in the limit as $ka \to \infty$ have been obtained by Franz[191]. The first three of them are

$$\nu = ka + e^{i\frac{\pi}{3}}(ka)^{\frac{1}{3}}\, 0.808617 - e^{-i\frac{\pi}{3}}(ka)^{-\frac{1}{3}}\, 0.145463 + \ldots \tag{6.36a}$$

$$\nu = ka + e^{i\frac{\pi}{3}}(ka)^{\frac{1}{3}}\, 2.578096 - e^{-i\frac{\pi}{3}}(ka)^{-\frac{1}{3}}\, 0.260341 + \ldots \tag{6.36b}$$

$$\nu = ka + e^{i\frac{\pi}{3}}(ka)^{\frac{1}{3}}\, 3.825715 - e^{-i\frac{\pi}{3}}(ka)^{-\frac{1}{3}}\, 0.514009 + \ldots \tag{6.36c}$$

With the replacements $\nu \to kR$ and $ka \to \omega R/c_t$ Eq. (6.36a) yields

$$kR = \frac{\omega}{c_t} R + e^{i\frac{\pi}{3}}\left(\frac{\omega}{c_t} R\right)^{\frac{1}{3}} 0.808617 -$$

$$-e^{-i\frac{\pi}{3}}\left(\frac{\omega}{c_t} R\right)^{-\frac{1}{3}} 0.145463 + \ldots \tag{6.37}$$

so that

$$k_R = \frac{\omega}{c_t}\left[1 + \frac{0.4043}{\left(\frac{\omega}{c_t} R\right)^{2/3}} + O\left(\left(\frac{\omega}{c_t} R\right)^{-\frac{4}{3}}\right)\right] \tag{6.38a}$$

$$k_I = \frac{\omega}{c_t}\left[\frac{0.7003}{\left(\frac{\omega}{c_t} R\right)^{2/3}} + O\left(\left(\frac{\omega}{c_t} R\right)^{-\frac{4}{3}}\right)\right] . \tag{6.38b}$$

Similar results are obtained for the remaining modes.

Thus an infinite number of leaky, shear horizontal surface acoustic waves can propagate circumferentially around the cylindrical boundary of a medium that is concave to the vacuum. They are attenuated as they propagate due to the radiation of energy into the interior of the solid. The wave number of each mode is greater than that of the surface skimming bulk transverse acoustic wave, and approaches the latter as the radius R of the cylinder tends to infinity. In this limit the attenuation of the wave vanishes. Such waves do not exist on a planar surface.

6.2. Grating Surfaces

The emphasis in subsections 6.2 and 6.3 is on the interaction of surface acoustic waves with surface roughness. We will preface a discussion of the more difficult case of statistical surface roughness with a description of what is known for the case of periodically corrugated surfaces, and in the latter context will emphasize nonperturbative formulations of the problems considered.

In considering periodically corrugated surfaces we will limit our discussion to one-dimensional grating profiles. This is a reflection of the state of the art in this field, and all the calculations that will be described in the context of one-dimensional gratings can be repeated for doubly periodic gratings as well. It is hoped that this will be done in the near future.

We begin by considering the propagation of surface acoustic waves across a one-dimensional grating. We consider Rayleigh acoustic waves first, and obtain their dispersion curves and attenuation by methods that are not limited to small-amplitude gratings. We then show that a periodically corrugated surface can "trap" surface acoustic waves of a kind that have no counterpart on a flat surface. These problems are then re-examined for the case in which the rough surface is randomly rough.

6.2.1 Rayleigh Waves

The propagation of surface acoustic waves, in particular Rayleigh waves, across periodically corrugated surfaces has been the object of considerable interest in recent years. In part, this interest arises from the practical, technological applications of such wave propagation in the acoustoelectrical field. It is characteristic of periodic structures, such as a grating ruled on a surface, that they cause wave slowing and create band-gaps or stop bands. The extent of this slowing and the positions and widths of the band-gaps are useful characteristics in the design of Rayleigh wave delay lines, filters, and resonators. In addition the periodic roughness can convert, in a controllable fashion, surface wave energy into bulk waves, and vice versa, thus serving as a surface-bulk, or bulk-surface transducer.

A second reason for the study of Rayleigh waves on periodically roughened surfaces arises from the more general and basic interest in the effects of random roughness on surface wave propagation. Surface roughness, for example, appears to be the most important mechanism for Rayleigh wave attenuation[124]. However, large amplitude random roughness is difficult to treat theoretically and still not easy to characterize experimentally. It is therefore reasonable to investigate a deterministic, and in particular periodic, surface profile as a first, and easier step.

The earliest theoretical work on the propagation of Rayleigh waves across a grating was carried out by Brekhovskikh[192], who used his results to obtain the attenuation of these surface waves due to their scattering into bulk waves by the grating perturbation. The theory of Brekhovskikh was subsequently corrected by Sabine[193], and the results compared to the experimental results of Rischbieter[194] for the attenuation of Rayleigh waves by a grating with a symmetric sawtooth profile on aluminum (see subsection 6.2.1.2).

The theory of Brekhovskikh and Sabine (see also Refs. (195,196)) is a first order scattering theory, and is therefore limited to cases where: (1) the grating amplitude is much smaller than the wavelength of the Rayleigh wave; (2) the slope of the surface profile is everywhere small; and (3) the amplitude of the scattered field is small compared to that of the incident field[192,193]. Moreover, it is valid only when coherent reflection of the Rayleigh wave from the grating can be neglected. Coherent reflection occurs when the period of the grating a is equal to an integral number of half-wavelengths of the Rayleigh wave incident on the grating: $a = n\lambda_R/2$. The cases $n = 1$ and 2 are of the greatest practical interest. In the former case the grating is a "Bragg mirror," and acts as a reflector for the surface acoustic wave. A theory of this kind of Bragg reflection of surface acoustic waves has been developed by several authors[197-202]. The latter case, viz. $a \approx \lambda_R$ is important in connection with the possibility of exciting Rayleigh waves by the conversion of a bulk acoustic wave incident on a grating surface into a surface acoustic wave[203-204].

In subsequent theoretical work it has been shown that in the study of the propagation of a Rayleigh wave across a grating when the wavelength of the Rayleigh wave is close to the period of the grating, multiple wave reflection must be taken into account[205-207] for a correct description of the scattering process. In this and related work[208,209] the amplitudes of the

Rayleigh waves scattered by the corrugations of the grating were calculated to second order in the surface profile function.

In contrast with these essentially perturbation-theoretic calculations, a nonperturbative theory of the propagation of Rayleigh waves across a grating surface of arbitrary, periodic profile has recently been formulated, and applied to the determination of the dispersion curves of such waves[210], and to their attenuation[65]. This theory is capable of yielding accurate results even for gratings of such large corrugation strengths that the low order perturbation calculations break down completely.

The theory was formulated originally on the basis of both Rayleigh's method and the extinction theorem form of Green's theorem (the extended boundary condition)[210]. However, we outline here the simpler theory based on Rayleigh's method, and refer the interested reader to the original work for a description of the theory based on the extended boundary condition.

In the remainder of this subsection and the next subsection we outline the method of calculation employed and describe some of the results obtained. The application of this theory to the attenuation of Rayleigh waves by a grating surface will be described in subsection 6.2.1.2.

The common starting point for both discussions are the assumptions that an isotropic elastic medium occupies the region $x_3 > \zeta(x_1)$, that we are interested in a straightcrested wave, polarized in the x_1x_3-plane, and propagating in the x_1-direction, and that the surface $x_3 = \zeta(x_1)$ is stress-free, and that $\zeta(x_1)$ is a periodic function of x_1 with period a. With these assumptions the elastic displacement field of frequency ω in the medium has the form

$$\vec{u}(\vec{x},t) = (u_1(x_1x_3|\omega),\ 0,\ u_3(x_1x_3|\omega))e^{-i\omega t}, \qquad (6.39)$$

where the amplitude functions $u_1(x_1x_3|\omega)$ and $u_3(x_1x_3|\omega)$ satisfy the pair of coupled partial differential equations

$$\left[\omega^2 + c_\ell^2 \frac{\partial^2}{\partial x_1^2} + c_t^2 \frac{\partial^2}{\partial x_3^2}\right]u_1 + (c_\ell^2 - c_t^2)\frac{\partial^2}{\partial x_1 \partial x_3} u_3 = 0 \qquad (6.40a)$$

$$(c_\ell^2 - c_t^2) \frac{\partial^2}{\partial x_1 \partial x_3} u_1 + [\omega^2 + c_\ell^2 \frac{\partial^2}{\partial x_3^2} + c_t^2 \frac{\partial^2}{\partial x_1^2}] u_3 = 0, \quad (6.40b)$$

that are obtained when Eq. (6.39) is combined with Eqs. (2.7) and (2.20).

These equations of motion have to be supplemented by the boundary conditions on the surface $x_3 = \zeta(x_1)$. These conditions express the requirement that this surface be stress-free, and take the form

$$\sum_\beta T_{\alpha\beta} \hat{n}_\beta \Big|_{x_3 = \zeta(x_1)} = 0, \quad (6.41a)$$

where $\hat{n}$ is the unit vector normal to the surface at each point. In the present case n is given by

$$\hat{n} = [1 + (\zeta'(x_1))^2]^{-1/2} [-\zeta'(x_1), 0, 1], \quad (6.41b)$$

where the prime denotes differentiation with respect to x_1. When we combine Eq. (6.39) with Eqs. (2.2) and (2.20), we find that the only nonzero stresses are

$$T_{11} = \rho c_\ell^2 \frac{\partial u_1}{\partial x_1} + \rho(c_\ell^2 - 2c_t^2) \frac{\partial u_3}{\partial x_3} \quad (6.42a)$$

$$T_{13} = \rho c_t^2 (\frac{\partial u_1}{\partial x_3} + \frac{\partial u_3}{\partial x_1}) = T_{31} \quad (6.42b)$$

$$T_{33} = \rho c_\ell^2 \frac{\partial u_3}{\partial x_3} + \rho(c_\ell^2 - 2c_t^2) \frac{\partial u_1}{\partial x_1} . \quad (6.42c)$$

It follows from Eqs. (6.41)-(6.42) that the boundary conditions are

$$[-\zeta'(x_1)T_{11} + T_{13}]_{x_3 = \zeta(x_1)} = 0 \tag{6.43a}$$

$$[-\zeta'(x_1)T_{31} + T_{33}]_{x_3 = \zeta(x_1)} = 0, \tag{6.43b}$$

Since we seek waves localized to the surface $x_3 = \zeta(x_1)$, we will also require that u_1 and u_3 vanish as $x_3 \to +\infty$.

The solutions of Eqs. (6.40) in the region $x_3 > \zeta(x_1)_{max}$ that satisfy the boundary conditions at infinity are

$$u_1(x_1x_3|\omega) = \sum_{m=-\infty}^{\infty} e^{ik_mx_1}\,[A_m^{(\ell)}e^{-\alpha_\ell(k_m\omega)x_3} + A_m^{(t)}e^{-\alpha_t(k_m\omega)x_3}] \tag{6.44a}$$

$$u_3(x_1x_3|\omega) = \sum_{m=-\infty}^{\infty} e^{ik_mx_1}\,i[\frac{\alpha_\ell(k_m\omega)}{k_m}A_m^{(\ell)}e^{-\alpha_\ell(k_m\omega)x_3} + $$

$$+ \frac{k_m}{\alpha_t(k_m\omega)}A_m^{(t)}e^{-\alpha_t(k_m\omega)x_3}] \tag{6.44b}$$

with

$$k_m \equiv k + \frac{2\pi m}{a}, \tag{6.45}$$

where k is the wave vector of the wave, and where

$$\alpha_{\ell,t}(k_m\omega) = \left(k_m^2 - \frac{\omega^2}{c_{\ell,t}^2}\right)^{1/2} \qquad k_m^2 > \frac{\omega^2}{c_{\ell,t}^2} \tag{6.46a}$$

$$= -i\left(\frac{\omega^2}{c_{\ell,t}^2} - k_m^2\right)^{1/2} \qquad k_m^2 < \frac{\omega^2}{c_{\ell,t}^2}. \tag{6.46b}$$

The solutions given by Eqs. (6.44) possess the Bloch property

$$u_\alpha(x_1 + a, x_3|\omega) = e^{ika}\, u_\alpha(x_1 x_3|\omega), \quad \alpha = 1,3 \tag{6.47}$$

required by the periodicity of our system in the x_1-direction.

When Eqs. (6.44) are substituted into the boundary conditions given by Eqs. (6.42)-(6.43), which constitutes the Rayleigh hypothesis[211] for the present problem, we obtain a pair of coupled equations for the expansion coefficients $\{A_m^{(\ell,t)}\}$,

$$\sum_{m=-\infty}^{\infty} e^{ik_m x_1}\{[ik_m \frac{2c_t^2\alpha_{\ell m}^2 + \omega^2}{c_t^2 k_m^2}\zeta'(x_1) + 2\alpha_{\ell m}]e^{-\alpha_{\ell m}\zeta(x_1)} A_m^{(\ell)} +$$

$$+ [2ik_m\zeta'(x_1) + \frac{\alpha_{tm}^2 + k_m^2}{\alpha_{tm}}]e^{-\alpha_{tm}\zeta(x_1)} A_m^{(t)}\} = 0 \tag{6.48a}$$

$$\sum_{m=-\infty}^{\infty} e^{ik_m x_1}\{[2\alpha_{\ell m}\zeta'(x_1) - ik_m \frac{k_m^2 + \alpha_{tm}^2}{k_m^2}]e^{-\alpha_{\ell m}\zeta(x_1)} A_m^{(\ell)} +$$

$$+ [\frac{\alpha_{tm}^2 + k_m^2}{\alpha_{tm}}\zeta'(x_1) - 2ik_m]e^{-\alpha_{tm}\zeta(x_1)} A_m^{(t)}\} = 0, \tag{6.48b}$$

where we have abbreviated $\alpha_{\ell,t}(k_m\omega)$ by $a_{\ell,tm}$.

It is now convenient to introduce the expansion

$$e^{-\alpha\zeta(x_1)} = \sum_{n=-\infty}^{\infty} \mathcal{J}_n(\alpha) e^{i\frac{2\pi n}{a}x_1} \tag{6.49}$$

from which, by differentiating both sides with respect to x_1, we obtain another useful expansion, viz.

$$\zeta'(x_1)e^{-\alpha\zeta(x_1)} = -\sum_{n=-\infty}^{\infty} \frac{(i2\pi n/a)}{\alpha}\mathcal{J}_n(\alpha) e^{i\frac{2\pi n}{a}x_1}. \tag{6.50}$$

The coefficient $\mathcal{J}_n(\alpha)$ is given by

$$\mathcal{J}_n(\alpha) = \frac{1}{a} \int_{-\frac{a}{2}}^{\frac{a}{2}} dx_1 e^{-i \frac{2\pi n}{a} x_1} e^{-\alpha\zeta(x_1)} . \tag{6.51}$$

When Eqs. (6.49)-(6.50) are substituted into Eqs. (6.48), and the pth Fourier coefficient in each of the resulting Fourier series is equated to zero, we obtain a pair of infinite, homogeneous linear equations for the coefficients $\{A_m^{(\ell,t)}\}$,

$$\sum_{m=-\infty}^{\infty} [\mathcal{J}_{p-m}(\alpha_{\ell m}) \frac{\omega^2(k_p-k_m)+2c_t^2\alpha_{\ell,m}^2 k_p}{c_t^2 k_m \alpha_{\ell m}} A_m^{(\ell)} +$$

$$+ \mathcal{J}_{p-m}(\alpha_{tm}) \frac{2c_t^2 k_p k_m - \omega^2}{c_t^2 \alpha_{tm}} A_m^{(t)}] = 0 \tag{6.52a}$$

$$p = 0,\pm1,\pm2,\pm3,\ldots$$

$$\sum_{m=-\infty}^{\infty} [\mathcal{J}_{p-m}(\alpha_{\ell m}) \frac{2c_t^2 k_p k_m - \omega^2}{c_t^2 k_m} A_m^{(\ell)} +$$

$$+ \mathcal{J}_{p-m}(\alpha_{tm}) \frac{c_t^2(k_m^2+\alpha_{tm}^2)k_p - k_m\omega^2}{c_t^2\alpha_{tm}^2} A_m^{(t)}] = 0 . \tag{6.52b}$$

$$p = 0,\pm1,\pm2,\pm3,\ldots$$

Two different surface profile functions have been used in the calculations of the dispersion curve for Rayleigh waves propagating across a grating and of their attenuation, on the basis of Eqs. (6.52). The first is the sinusoidal profile given by

$$\zeta(x_1) = \zeta_o \cos \frac{2\pi x_1}{a} , \tag{6.53}$$

for which

$$\mathcal{J}_n(\alpha) = (-1)^n I_n(\zeta_o\alpha) , \tag{6.54}$$

where $I_n(x)$ is a modified Bessel function. The second is the symmetric sawtooth profile,

$$\zeta(x_1) = \begin{cases} h + \frac{4h}{a} x_1 & -\frac{a}{2} \leqslant x_1 \leqslant 0 \\ h - \frac{4h}{a} x_1 & 0 \leqslant x_1 \leqslant \frac{a}{2} , \end{cases} \tag{6.55a,b}$$

for which

$$\mathcal{J}_n(\alpha) = \begin{cases} \frac{4h\alpha}{\pi^2 n^2 + 4h^2\alpha^2} \sinh(h\alpha) & n \text{ even} \quad (6.56a) \\ \frac{-4h\alpha}{\pi^2 n^2 + 4h^2\alpha^2} \cosh(h\alpha) & n \text{ odd} . \quad (6.56b) \end{cases}$$

These profiles were chosen not only because they allow the coefficients $\{\mathcal{J}_n(\alpha)\}$ to be obtained analytically rather than numerically, but also because as we will see, experimental data exist for the attenuation of Rayleigh waves on gratings with a symmetric sawtooth profile.

The determination of the dispersion curves and the attenuation of Rayleigh waves on a grating on the basis of Eqs. (6.52) proceed along somewhat different lines, and we consider these two problems in turn.

6.2.1.1. Dispersion

The dispersion curve for Rayleigh waves propagating across a grating, the relation between ω and k for which such waves exist, is obtained by equating to zero the determinant of the coefficients in Eqs. (6.52). Even without a numerical solution several useful general properties of the frequency $\omega(k)$ of these waves can be obtained directly from the matrix of the coefficients in these equations. The first is that $\omega(k)$ is a periodic function of the wave vector k with a period of $2\pi/a$. The second is that $\omega(k)$ is an even function of k. These two properties, which are independent of the grating profile, have the consequence that we need to obtain the dispersion curve only for values of k in the interval $0 < k < \pi/a$. A third useful property of $\omega(k)$ for the profiles (6.53) and (6.55) is that, for given values of c_ℓ and c_t, $\omega(k)a/2\pi$ is a function only of the reduced wave vector $ka/2\pi$ and of the corrugation strength ζ_0/a or h/a.

In actual calculations of the dispersion curve the infinite determinant formed from the coefficients in Eqs. (6.52) was transformed into the determinant of an N x N matrix by restricting p and m to run from $-(N/4)+1$ to $(N/4)$, where N is a multiple of four. The zeros $\{\omega_s(k)\}$ of this truncated determinant were found numerically, and the convergence of the solutions was tested by increasing N (in increments of four), and seeing if they approached stable limiting values.

For the sinusoidal profile, for ratios of ζ_0/a up to 0.25 convergence was found as N increased to 52 (the largest determinant used), although the rate of convergence decreased with increasing ζ_0/a. Thus, for $\zeta_0/a = 0.016$ the frequencies $\{\omega_s(k)\}$ were found with six-figure accuracy already with $N = 12$, while for $\zeta_0/a = 0.25$ a value of $N = 36$ was required to obtain the frequencies with six-figure accuracy. For $N = 32$ and $N = 36$ the frequencies obtained with four- to five-figure accuracy for $\zeta_0/a =$ 0.35 and 0.40, respectively, but when N was increased from 36 to 40 and from 32 to 36, respectively, the results began to diverge. For the symmetric sawtooth profile the calculations were found to converge for $h/a = 0.016$ and 0.04, but when h/a was increased to 0.064 an almost immediate trend towards divergence was seen.

Before proceeding to the results for the dispersion curves it is useful to consider the dispersion curves for Rayleigh waves on a flat surface in both the extended and reduced Brillouin zone schemes. In the extended zone scheme of Fig. 6.2(a) the straight line $\omega = c_R k$ (drawn with a solid and then a dotted segment) is the dispersion curve for a Rayleigh wave on a flat surface, and the dashed lines are the curves $\omega = c_t|k+(2\pi/a)m|$, where either $m = 0$ (the so-called transverse sound line)or $m = \pm 1, \pm 2, \ldots$ Since any point in the (ω,k)-plane in the region above the dashed lines, $\omega > c_t|k+(2\pi/a)m|$, corresponds to imaginary $\alpha_t(k_m\omega)$, the displacement field at such points will have radiative components. Thus a Rayleigh wave, initially on a flat surface with its ω and k lying on the dotted part of the dispersion curve in Fig. 6.2(a), will begin to radiate into the bulk the moment the grating, however weak, is turned on. If a Rayleigh wave with ω,k situated on this radiative part of the dispersion curve is launched across a corrugated surface, it will decay with some characteristic lifetime. The Bloch-type surface waves that are true eigenmodes of the corrugated structure (i.e. possess an infinite lifetime) will thus be found only for (ω,k) beneath the dashed lines in Fig. 6.2(a), originating out of the solid part of the flat surface Rayleigh wave dispersion curve as the grating is turned on.

The flat surface dispersion curve with its solid and dotted segments is folded back into the domain $-\pi/a < k < \pi/a$ (the first Brillouin zone for the grating) in the reduced zone scheme of Fig. 6.2(b), by displacing the portions of this dispersion curve lying outside the first Brillouin zone to the right and to the left by suitable multiples of $2\pi/a$. The nonradiative region is now beneath the single dashed line $\omega = c_t k$.

Points such as those labeled by A, B, C,... in Fig. 6.2 are of particular interest because they correspond to degenerate Rayleigh waves on a flat surface, e.g. $\omega(k_A) = \omega(-k_A)$, whose wave vectors differ by a translation vector of the reciprocal lattice of the grating, i.e. by a multiple of $2\pi/a$. If the Fourier series expansion of the grating profile function $\zeta(x_1)$ has nonzero Fourier components for the wave vectors $2k_A$, $2k_B$, $2k_C$,..., then an application of degenerate perturbation theory shows that as the grating is turned on the degeneracy of the flat surface dispersion curve at points A, B, C,... will be split, and gaps will open up in the dispersion curve at these points.

In fact, however, only a single gap is expected to be seen in the dispersion curve for a Rayleigh wave propagating across the grooves of a grating, viz. the one associated with the point A. This is due to the following reasons. Points of degeneracy such as C, that correspond to $k = 0$ in the reduced zone scheme, are always in the radiative region and therefore cannot lie on the dispersion curves of the Bloch-type surface waves. Thus a second gap in the dispersion curve would have to develop from a point of degeneracy on the flat surface dispersion curve at $k = \pi/a$. The lowest frequency point of this type after the point A is the point C in Fig. 6.2(b). However, this point will always lie quite far into the radiative region also, i.e. its frequency $\omega_c = c_R(3\pi/a)$ must always be greater than the frequency at the boundary point

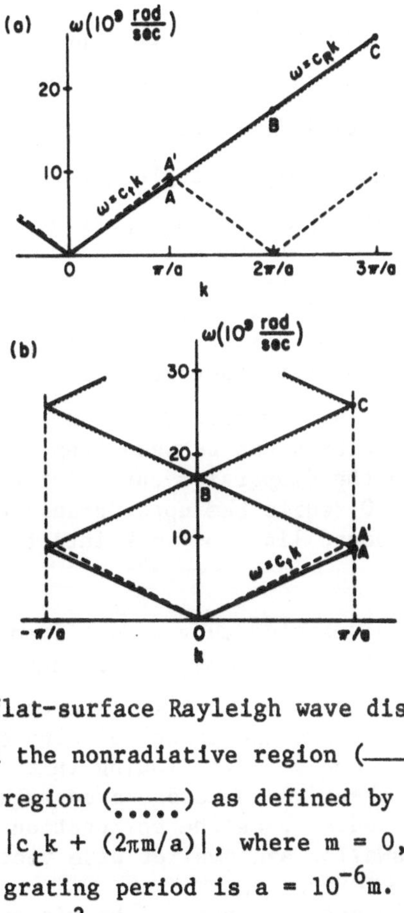

Fig. 6.2. The flat-surface Rayleigh wave dispersion curve $\omega = c_R k$ in the nonradiative region (——) and in the radiative region (·····) as defined by the boundary lines (---) $\omega = |c_t k + (2\pi m/a)|$, where $m = 0, \pm 1, \pm 2, \ldots$, and where the grating period is $a = 10^{-6}$m. $c_\ell = 5 \times 10^3$ m/sec and $c_t = 3 \times 10^3$m/sec. (a) The extended zone and (b) the reduced zone scheme [Ref. 210].

A′, $\omega_{A'} = c_t(\pi/a)$, that is the upper limit on the nonradiative region at $k = \pi/a$. This is because for an isotropic elastic medium the requirements of elastic stability restrict the ratio c_t/c_ℓ to lie in the range $0 < c_t^2/c_\ell^2 = \frac{3}{4}$ (18), that in turn restricts the speed of Rayleigh waves to the range $0.69\ c_t < c_R < 0.96\ c_t$, which implies $\omega_C > 2.1\ \omega_{A'}$ for an isotropic solid. Even for an elastic medium of cubic symmetry the speed of a Rayleigh wave propagating in any direction on a (001) surface must be at least 0.6 times the speed of bulk transverse waves, which implies $\omega_c > 1.8\ \omega_{A'}$. Thus, only a single gap is expected in the dispersion curve for a Rayleigh wave on a grating at $k = \pi/a$.

This is just what is observed in the numerical solutions of Eqs. (6.52). In Figs. 6.3 and 6.4 we show the dispersion curves for Rayleigh waves propagating across a sinusoidal grating (6.53) with $\zeta_o/a = 0.30$, and a symmetric sawtooth grating (6.55) with $h/a = 0.4$, respectively. In each case two branches to the dispersion curve are observed, with a gap between them at $k = \pi/a$. The lower branch merges with the dispersion curve for Rayleigh waves on a flat surface as $k \to 0$, while the upper branch cuts off as it crosses the transverse "sound line" $\omega = c_t k$ into the radiative region.

The bending of the dispersion curves away from the straight line $\omega = c_R k$, the dispersion curve for a flat surface, observed in the results depicted in Figs. 6.3 and 6.4, is accompanied by a reduction in both the group and the phase velocity of the Rayleigh wave. This is the phenomenon of wave slowing that is associated with the propagation of these waves across a corrugated surface. The wave is slowed more and more as the corrugation strength increases, starting at smaller and smaller wave vectors.

The shape of the dispersion curve in the vicinity of the gap at $k = \pi/a$ has been obtained analytically by Glass <u>et al.</u>[210], by perturbation theory, starting from Eqs. (6.52). If we set

$$\omega = c_R(\pi/a) + \Delta\omega, \qquad k = (\pi/a) - \Delta k \tag{6.57}$$

then to lowest nonzero order in the surface profile function, it is found that

$$\Delta\omega = \pm \{[\tfrac{1}{2}\omega_G]^2 + c_R^2(\Delta k)^2\}^{1/2}. \tag{6.58}$$

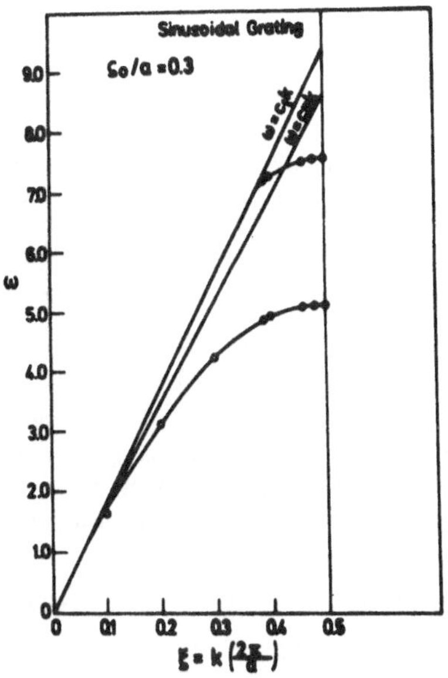

Fig. 6.3. Dispersion curve for Rayleigh waves propagating across a sinusoidal grating with corrugation strength $\zeta_0/a = 0.3$ on an isotropic medium characterized by speeds of sound $c_\ell = 5 \times 10^3$m/sec and $c_t = 3 \times 10^3$m/sec.

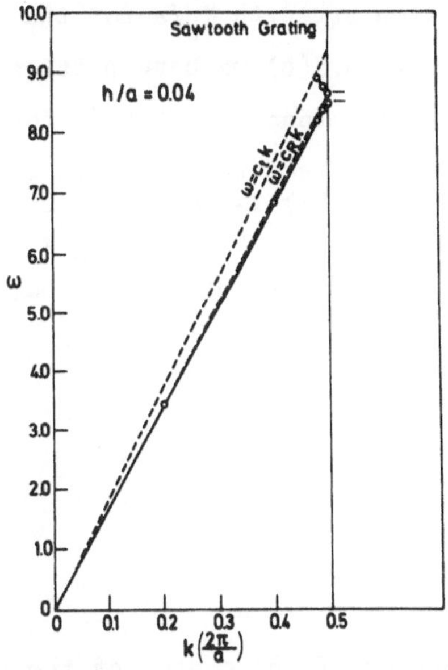

Fig. 6.4. Dispersion curve for Rayleigh waves propagating across a symmetric sawtooth grating with corrugation strength h/a = 0.04 on an isotropic medium characterized by speeds of sound $c_\ell = 5 \times 10^3$m/sec and $c_t = 3 \times 10^3$m/sec. [Ref. 210].

The plus sign corresponds to the upper branch and the minus sign to the lower branch. The magnitude of the gap that opens up in the dispersion curve at $k = \pi/a$ is given by

$$\omega_G = {}^1\!/_2 \, \frac{c_R^3 c_\ell^2}{c_t^3} \, |\hat{\zeta}(+1)| \left(\frac{\pi}{a}\right)^2 \frac{(c_t^2 - c_R^2)^{1/2} (2c_t^2 - c_R^2)^2}{c_R^4 c_\ell^2 - 6c_R^2 c_\ell^2 c_t^2 + 6c_\ell^2 c_t^4 + 4c_R^2 c_t^4 - 6c_t^6} \, . \tag{6.59}$$

In this expression $\hat{\zeta}(+1)$ is the +1 Fourier coefficient of the surface profile function, and is assumed to be nonzero. If we have a profile for which $\hat{\zeta}(+1)$ vanishes, a gap can still open up in the dispersion curve at $k = \pi/a$, but a higher order perturbation calculation is required to obtain its magnitude.

From the results given by Eqs. (6.57)-(6.59) we see that to lowest order in the surface profile function the center of the gap is always at $\omega_{Gc} = c_R(\pi/a)$, and the width of the gap is a linear function of $|\hat{\zeta}(+1)|$. It is seen from the results presented in Figs. 6.3 and 6.4, however, that already for moderate corrugation strengths the center of the gap lies well below the frequency $c_R(\pi/a)$. In Fig. 6.5(a) we have plotted the width of the gap as a function of the corrugation strength ζ_o/a for the sinusoidal profile (6.53), and in Fig. 6.5(b) we have plotted the frequency of the center of the gap as a function of ζ_o/a for the same profile. In both figures the exact result obtained from the numerical solution of Eqs. (6.52) is compared with the prediction of Eqs. (6.57)-(6.59). It is seen that for $\zeta_o/a < 0.03$ the small roughness approximation for the gap width is good, but it is in error by

41% at $\zeta_o/a = 0.2$

46% at $\zeta_o/a = 0.25$

48% at $\zeta_o/a = 0.3$.

Similarly the frequency of the center of the gap is accurately predicted by the small roughness approximation for $\zeta_o/a < 0.03$, but is in error by

3.9% at $\zeta_0/a = 0.1$

16% at $\zeta_0/a = 0.2$

25% at $\zeta_0/a = 0.25$

36% at $\zeta_0/a = 0.3$.

Thus the use of Rayleigh's method can yield the dispersion curves for Rayleigh waves propagating across the grooves of a grating not only for values of the corrugation strength far beyond the limit of validity of the Rayleigh hypothesis(212-215), for an analytic surface profile such as the sinusoidal one, but also for nonanalytic surface profiles, such as the symmetric sawtooth profile, for which the Rayleigh method in principle should not work at all.(216) This convergence is found numerically to have an asymptotic nature, at least for the larger corrugation strengths, but the results obtained nonetheless possess a high degree of accuracy. It is not yet known what the limits on the corrugation strength, and on the profile function, are at which the Rayleigh method breaks down for the calculation of the dispersion curves of surface acoustic waves. What is clear, however, even if the reasons for this are not understood at the present time, is that the Rayleigh method can yield accurate results in "bound state" calculations, i.e. in the calculation of the dispersion curves of surface waves, for considerably larger corrugation strengths than those at which it breaks down in "scattering" calculations, i.e. in the calculation of the diffraction of waves from rough surfaces.

6.2.1.2. Attenuation

In the preceding subsection the dispersion relation for Rayleigh waves propagating across a grating was solved numerically in the non-radiative region. Here we discuss solutions in the radiative region, for k,ω above the transverse sound-line, $\omega = c_t k$, where $\alpha_t(k_m\omega)$ given by Eqs. (6.46) is imaginary, and for k,ω inside the zone boundary gap that results from the grating periodicity. To account for the loss of surface-wave energy due to radiation into the bulk, we consider: (1) real k with

$\omega = \omega_R - i\omega_I$, where $\omega_R > 0$ and $\omega_I > 0$, and (2) real ω with $k = k_R + ik_I$, where $k_I > 0$ or < 0 for a group velocity > 0 or < 0. In the case of complex ω, $\alpha^2_{\ell,t}(k_m\omega) = (k+2\pi m/a)^2 - (\omega_R^2 - \omega_I^2)/c^2_{\ell,t} + i2\omega_R\omega_I/c^2_{\ell,t}$, so we can have waves radiating outward from the surface ($\alpha_I < 0$) only if the branch-cut defining the square-root of α^2 is along the positive imaginary axis.* This choice gives an exponential increase of the wave into the bulk, typical of leaky-waves (see subsection 2.1.1.3). This is a consequence of retardation; at a distance x_3 into the bulk at time t, the outward radiating plane-wave has an amplitude proportional to the amplitude of the decaying surface wave at the time when it, the radiated wave, first left the surface, namely, at a retarded time $t - x_3/\tilde{c}$, where $\tilde{c} = \omega_I/|\alpha_R|$ acts as the propagation speed, i.e., the amplitude is proportional to $\exp(-\omega_I(t-x_3/[\omega_I/\alpha_R])$. for a surface excitation begun at $t = 0$ there can be no radiated wave for $x_3 > \tilde{c}t$; hence there is no real divergence at $x_3 = \infty$. For complex k, we use the same branch-cut to define the α's, obtaining an outwardly propagating, exponentially increasing wave; and the physical interpretation is the same.

The dispersion relation has been solved with complex ω and with complex k, both for (1) the sinusoidal profile (6.53) with ζ_o/a up to 0.25, and (2) the symmetric sawtooth profile (6.55) with h/a up to 0.1166. Values of c_t/c_ℓ from 0.3 to 0.6 were used. The dispersion curves are similar for complex $\omega(\omega_R$ vs. k) and for complex k (ω vs. k_R); they are also very similar for the sawtooth and sinusoidal gratings (through numerical convergence of the results is faster for the latter, which permits larger corrugation strengths ζ_o/a to be studied. Figure 6.6, showing the dispersion curves in the reduced zone scheme, for a sawtooth grating with h/a = 0.1166 and $c_t/c_\ell = 0.3$, is typical. The Rayleigh wave and also

---

* Equivalently, the branch cut can be chosen to coincide with the negative real axis, provided one then seeks solutions on the lower sheet of the corresponding Riemann surface.

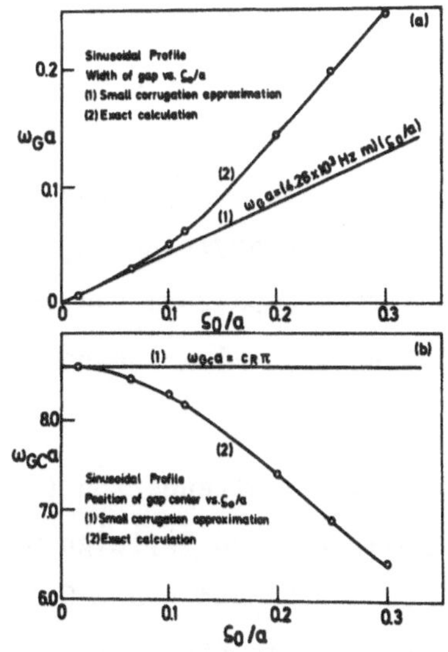

Fig. 6.5.(a) The gap width (times the grating period) $\omega_G a$ as a function of the corrugation strength ζ_o/a, and (b) the position of the center of the gap (times the grating period) $\omega_{Gc}a$ as a function of ζ_o/a - both for a sinusoidal profile - calculated from (1) the small roughness approximation and (2) the numerical solution of Eqs. (6.52). $c_\ell = 5 \times 10^3$ m/sec and $c_t = 3 \times 10^3$m/sec [Ref. 210].

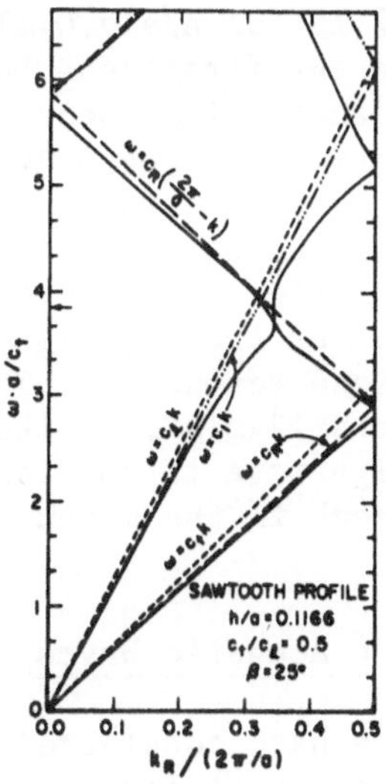

Fig. 6. 6. Dispersion curves: bulk waves (---); flat-surface Rayleigh waves (— —); flat-surface resonances (– •• –); and Rayleigh waves and surface resonances on a sawtooth grating for h/a = 0.1166 (——). For $c_t/c_\ell = 0.5$ [Ref. 65].

the high frequency, intrinsically leaky mode discussed in subsection 2.1.1.3, for the flat surface and for the grating surface are indicated. Gaps at the zone center and zone boundary appear. Otherwise, the curves for the grating depart from those for the flat surface most strongly where the Rayleigh wave branch crosses the new resonance branch, and this departure was found to increase as a function of h/a, indicating a grating-induced interference between the two.

Figure 6.7 shows the acoustical attenuation ($20 \log(e)\lambda_s k_I$) versus λ_s/a, where $\lambda_s = 2\pi c_t/\omega a$, for the Rayleigh wave on sawtooth gratings of $h/a = 0.044$ ($c_t/c_\ell = 0.3$) and $h/a = 0.1166$ (c_t/c_ℓ = 0.3 and 0.5). The surface profiles in Fig. 6.7 were chosen to be identical to those in Fig. 6.8 for the experimental results of Rischbieter[(194)] and the first-order scattering theory of Sabine[(193)] (which follows the work of Brekhovskikh[(192)]). Comparison of Figs. 6.7 and 6.8 shows that the present theory gives the peak heights more accurately than the first order scattering theory, and finds all the experimental peaks, including (1) the sharp peaks at $\lambda_s/a = 2.15$ for $h/a = 0.1166$, which corresponds to solutions inside the first zone boundary gap, which are missed in any perturbative scattering theory and (2) the higher frequency peaks, associated with the edges of the successively higher order gaps, also missed in previous theories. The frequency corresponding to the principal peak is marked by an arrow on the dispersion curve of Fig. 6.7, where it is thus seen that this is the frequency at which the Rayleigh wave branch crosses the branch for the surface resonance. The grating, by mixing reciprocal lattice vectors $\pm 2\pi m/a$ into the Rayleigh wavevector k, can couple the Rayleigh wave to the leaky wave, and thus can account for the principal peak in the measured acoustical attenuation.

6.2.2. Shear Horizontal Surface Acoustic Waves

Thus far in this section we have considered only surface acoustic waves polarized in the sagittal plane. Such waves exist on a planar surface, where they are known as Rayleigh waves, and the presence of a grating largely only modifies properties they already possess on a flat surface.

We have seen that surface acoustic waves polarized perpen-

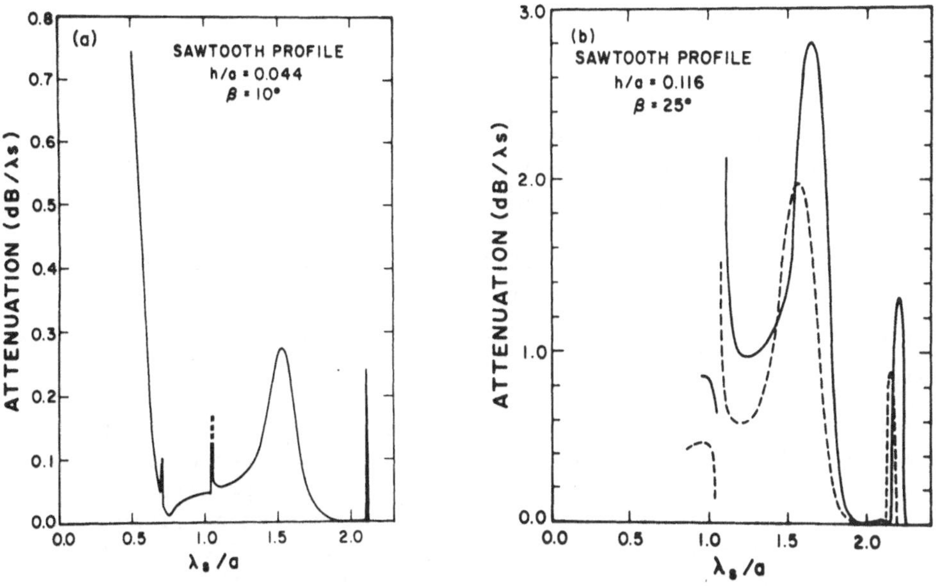

Fig. 6.7. Attenuation as a function of shear wavelength λ_s for Rayleigh waves on sawtooth gratings of corrugation strengths (a) h/a = 0.044 [c_t/c_ℓ = 0.3] and (b) h/a = 0.1166 [c_t/c_ℓ = 0.3 (- -) and c_t/c_ℓ = 0.5 (——) [Ref. 65].

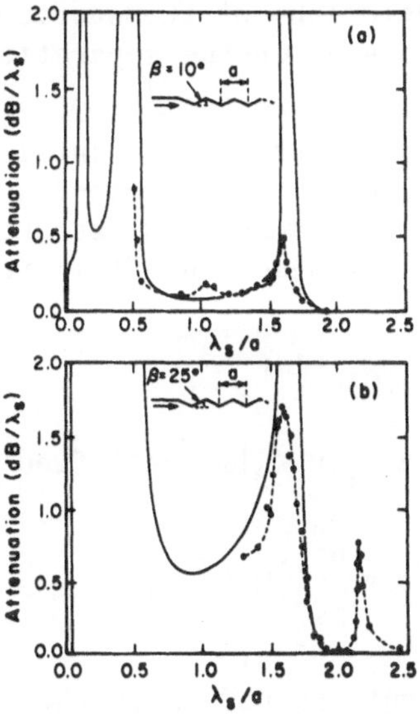

Fig. 6.8. Attenuation as a function of shear wavelength λ_s for Rayleigh waves on an Al sawtooth grating, as measured by Rischbieter (—•—) and calculated by Sabine (——), for a Poisson ratio σ = 0.33, and (a) h/a = 0.44 or $\beta = 10^o$ and (b) h/a = 0.1166 or $\beta = 25^o$ [Ref. 193].

dicular to the sagittal plane (called shear horizontal surface acoustic waves) cannot exist on a stress-free planar surface. The situation is entirely different when a grating is ruled on the surface of the solid. Several years ago Auld et al.(217) calculated, and also measured, the dispersion curves for such shear horizontal surface acoustic waves propagating across the grooves of a rectangular grating. Such waves owe their existence to the corrugations of the surface, and represent a third example of effects induced by surface roughness that have no counterpart for a flat surface.

The work of Auld et al. was followed by similar calculations by Gulyaev and Plesskii(218).

The method of calculation used by Auld, et al., however, is limited to special grating profiles, such as the rectangular one, for which the normal elastic modes in the teeth of the structure can be found, in terms of which the elastic displacement field in the selvedge region of the grating can be expanded. The latter expansion is then matched to a Rayleigh expansion of the displacement field outside the selvedge region. Moreover, they made the approximation of keeping only the dominant mode in the teeth of the grating (i.e. the displacement field was constant across a tooth). Similar approximations were made in the work of Gulyaev and Plesskii, who in addition used a single wave approximation for the displacement field outside the selvedge region. Their results are therefore valid for wavelengths of the surface acoustic wave much larger than the thickness of or the separation between the teeth of the grating.

The dispersion relation for shear horizontal surface acoustic waves propagating across the grooves of a one-dimensional grating has been calculated nonperturbatively by Glass and Maradudin(219). In this case the form of the displacement field is

$$\vec{u}(\vec{x},t) = (0,\ u_2(x_1x_3|\omega),0)e^{-i\omega t}, \tag{6.60}$$

where the amplitude function $u_2(x_1x_3|\omega)$ satisfies

$$-\omega^2 u_2(x_1x_3|\omega) = c_t^2\left(\frac{\partial^2}{\partial x_1^2} + \frac{\partial^2}{\partial x_3^2}\right)u_2(x_1x_3|\omega) . \tag{6.61}$$

The stress-free boundary condition now takes the form

$$\left[-\zeta'(x_1)\frac{\partial}{\partial x_1} + \frac{\partial}{\partial x_3}\right]u_2(x_1x_3|\omega)\Big|_{x_3=\zeta(x_1)} = 0 . \tag{6.62}$$

The Rayleigh expansion

$$u_2(x_1x_3|\omega) = \sum_{m=-\infty}^{\infty} A_m(k\omega)e^{ik_mx_1-\alpha_t(k_m\omega)x_3} \qquad (6.63)$$

satisfies the equation of motion (6.61) for $x_3 > \zeta_{max}$. Continuing this expansion in to the surface, and using it to satisfy the boundary condition (6.62), yields the following equation for the $\{A_p(k\omega)\}$:

$$\sum_{m=-\infty}^{\infty} \frac{k_pk_m-(\omega^2/c_t^2)}{\alpha_t(k_m\omega)} \mathcal{J}_{p-m}(\alpha_t(k_m\omega))A_m(k\omega) = 0 \ . \qquad (6.64)$$

The dispersion relation is obtained by equating to zero the determinant of the coefficients in this equation.

The dispersion relation was solved numerically for two surface profile functions: (1) the sinusoidal profile (6.53); and (2) the symmetric sawtooth profile (6.55), for which the functions $\mathcal{J}_n(\alpha)$ have been given in Eqs. (6.54) and (6.56).

It follows directly from Eq. (6.64) that the frequency $\omega(k)$ that satisfies the dispersion relation is a periodic function of k with period $2\pi/a$. It is also an even function of k. Consequently, we can restrict the wave vector k to the range $0 \leqslant k \leqslant \pi/a$.

The zeros of an N-dimensional determinant, obtained by allowing p and m in Eq. (6.64) to range from $-(n/2)+1$ to $N/2$, with N even, were found, and convergence was studied by increasing N. With the sinusoidal profile convergence was very rapid for values of ζ_0/a up to 0.6: five figure accuracy for the frequencies was achived with N as small as 20, and the results were still found to converge as N was increased beyond this value. For $\zeta_0/a = 0.7$ four-figure accuracy was achieved with N = 24, but oscillations in the values of the frequencies set in as N was increased beyond 24. For the sawtooth profile with h/a equal to 0.1 four-figure accuracy for the frequencies was found for $N \leqslant 20$, but for h/a = 0.2 three-figure accuracy was achieved with N = 20, at which point the results began to diverge with increasing N.

Some of the dispersion curves thus obtained are shown in Fig. 6.9. They are qualitatively similar to those obtained by Auld et al.[217]. There is only one branch to the dispersion curve. It approaches the dispersion curve for bulk transverse waves

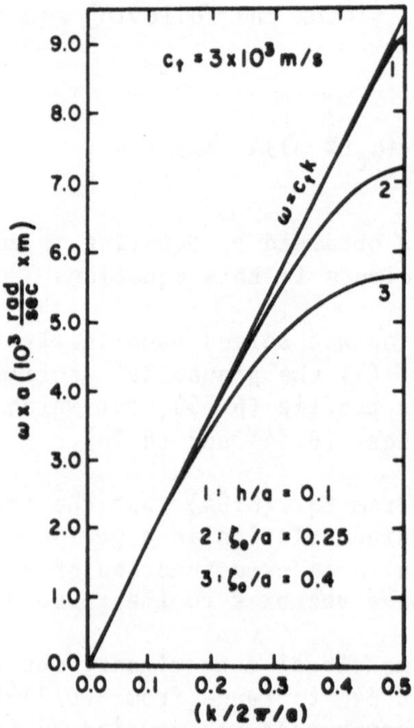

Fig. 6.9. Dispersion curves for shear horizontal surface acoustic waves propagating across (1) a symmetric sawtooth grating for which $h/a = 0.1$; (2) a sinusoidal grating for which $\zeta_0/a = 0.25$; and (3) a sinusoidal grating for which $\zeta_0/a = 0.4$ [Ref. 219].

$\omega = c_t k$, in the limit as $k \to 0$, and its distance from this limiting curve and the domain in which it is dispersive (wave slowing), increase with increasing corrugation strength. The maximum frequency of these waves, i.e. the width of the pass-band, thus also decreases with increasing corrugation strength.

These shear horizontal surface acoustic waves are weakly bound, however, at least in the limit of small k. A perturbation calculation starting from Eq. (6.64) shows that in this limit

$$\alpha_t(k\omega) = \pi(\zeta_0/a)^2(ka)^2 k, \tag{6.65}$$

for a sinusoidal grating. Since the decay length of the wave into the elastic medium is $\alpha_t^{-1}(k\omega)$, we see that for $\zeta_0/a \cong 0.05$ and $ka = 0.1\pi$, the decay length is $(4 \times 10^4/\pi^4)$ a $\sim$ 4000a. Thus it penetrates far into the medium. As the wave vector increases, however, the wave becomes more tightly bound to the surface.

6.3. Randomly Rough Surfaces

An important class of curved surface is the class of randomly rough surfaces. The interest in such surfaces arises because surface roughness is found in varying degrees on all solid surfaces, even carefully prepared ones, and it is important to know in what ways properties of surface acoustic waves obtained on the assumption of a perfectly flat surface are modified in the presence of surface roughness. In particular, surface roughness has the ability to attenuate a surface acoustic wave as it propagates along a randomly rough surface, by scattering energy out of the incident beam into other surface acoustic waves and into bulk waves. Indeed, at the present time surface roughness appears to be the most important extrinsic mechanism responsible for the attenuation of Rayleigh waves.

The first study of the propagation of Rayleigh waves across a randomly rough surface was carried out by Urazakov and Fal'kovskii(220) on the basis of Rayleigh's method(211). In this method the surface localized elastic displacement field obtained from the equations of motion in the region beyond the maximum amplitude of the surface profile is continued in to the rough surface itself, and used in satisfying the stress-free boundary conditions on that surface. Subsequently Maradudin and Mills(221) used a Green's function method to solve the same problem. In two recent papers Eguiluz and Maradudin(222,223) have used two different methods to study the propagation of Rayleigh waves across a randomly rough surface, and corrected some errors in the earlier work of Maradudin and Mills[221] in the process. All of these studies predict that the attenuation rate for a Rayleigh wave is proportional to the fifth power of its frequency, in the limit that the wavelength of the Rayleigh wave is longer than the transverse correlation length of the surface roughness. The latter is a measure of the mean distance between consecutive peaks and valleys on the surface. The frequency dependence of the attenuation rate then becomes much slower than the ω_R^5 law as the wavelengths of the Rayleigh wave becomes comparable to, or shorter than, the transverse correlation length. In the work of Eguiluz and Maradudin[222,223] the change in the frequency of the Rayleigh wave caused by the surface roughness was also determined.

The propagation of a shear horizontal surface acoustic wave across a randomly rough surface was studied theoretically by Hardouin Duparc and Maradudin[224]. Since a periodically corrugated surface can "trap" a surface acoustic wave of this polarization it is not surprising that a randomly rough surface can as well. The frequency and attenuation of such a roughness-trapped surface acoustic wave were obtained by Hardouin Duparc and Maradudin. The propagation of a shear horizontal surface acoustic wave across a random grating had been investigated earlier by Bulgakov and Khankina[225].

In this subsection we sketch out a theory of the propagation of a surface acoustic wave across a randomly rough surface that is based on, but simplifies, the treatment given by Eguiluz and Maradudin[223]. It permits the separation of the sagittally polarized wave (Rayleigh wave) from the shear horizontal wave in a very explicit way.

We assume that an isotropic elastic medium occupies the region $x_3 > \zeta(\vec{x}_\parallel)$. Since the surface is randomly rough we do not know the surface profile function, and are forced to characterize it by its statistical properties. Underlying this description is the notion that there is not a single rough surface but rather an ensemble of such surfaces. Physical observables are to be averaged with respect to this ensemble, and it is assumed that the ensemble average coincides with the average over the surface in the limit of a very large surface area.

Thus, we assume that $\zeta(\vec{x}_\parallel)$ is a stationary stochastic process characterized by the following two properties

$$\langle\zeta(\vec{x}_\parallel)\rangle = 0 \tag{6.66a}$$

$$\langle\zeta(\vec{x}_\parallel)\zeta(\vec{x}'_\parallel)\rangle = \delta^2 W(|\vec{x}_\parallel - \vec{x}'_\parallel|), \tag{6.66b}$$

where the angular brackets denote the average over the ensemble of realizations of the surface profile. The result expressed by Eq. (6.66a) simply states that the mean surface is the plane $x_3 = 0$. In the second property expressed by Eq. (6.66b) $\delta^2 = \langle\zeta^2(\vec{x}_\parallel)\rangle$ is the mean square deviation of the surface from flatness. The fact that the correlation function $W(|\vec{x}_\parallel - \vec{x}'_\parallel|)$ depends on $\vec{x}_\parallel$ and $\vec{x}'_\parallel$ only through their difference (the stationarity assumption) has the consequence that averaging restores infinitesimal translational invariance in the plane $x_3 = 0$ to our system; the fact that it depends on the magnitude of $\vec{x}_\parallel - \vec{x}'_\parallel$ has the consequence that if our system in the absence of roughness is isotropic in the plane $x_3 = 0$ (as it is), then averaging will restore isotropy in this plane to our system.

When an explicit form for $W(|\vec{x}_\parallel - \vec{x}'_\parallel)$ will be needed below we will use the Gaussian form

$$W(|\vec{x}_\parallel - \vec{x}'_\parallel|) = e^{-|\vec{x}_\parallel - \vec{x}'_\parallel|^2/a^2}, \tag{6.67}$$

which combines computational simplicity with physical reasonableness. The characteristic length a appearing in this expression is called the <u>transverse correlation length</u>.

Since surface roughness is a <u>static</u> perturbation of our system, the solutions of the equations of motion of the elastic medium in the region $x_3 > \zeta(\vec{x}_\parallel)_{max}$ that satisfy the boundary conditions at infinity can be written as $\vec{u}(\vec{x},t) = \vec{u}(\vec{x}|\omega)\exp(-i\omega t)$, where

$$\vec{u}(x|\omega) = \int \frac{dq_\parallel}{(2\pi)^2} e^{i\vec{q}_\parallel \cdot \vec{x}_\parallel} \{(\hat{q}_\parallel + i\hat{x}_3 \frac{\alpha_\ell(q_\parallel\omega)}{q_\parallel})A_1(\vec{q}_\parallel\omega) \times$$

$$\times e^{-\alpha_\ell(q_\parallel\omega)x_3} + (\hat{q}_\parallel \times \hat{x}_\parallel)_3 A_2(\vec{q}_\parallel\omega)e^{-\alpha_t(q_\parallel\omega)x_3} +$$

$$+ (\hat{q}_{\parallel} + i\hat{x}_3 \frac{q_{\parallel}}{\alpha_t(q_{\parallel}\omega)})A_3(\vec{q}_{\parallel}\omega)e^{-\alpha_t(q_{\parallel}\omega)\hat{x}_3}\}. \tag{6.68}$$

The coefficients $A_{\alpha}(\vec{q}_{\parallel}\omega)$ (α = 1, 2, 3) are to be determined from the stress-free boundary conditions on the surface $x_3 = \zeta(\vec{x}_{\parallel})$. These are given by

$$\sum_{\beta} T_{\alpha\beta}(\vec{x}|\omega)\hat{n}_{\beta}\Big|_{x_3 = \zeta(\vec{x}_{\parallel})} = 0 \quad \alpha = 1, 2, 3, \tag{6.69}$$

where the stress tensor $T_{\alpha\beta}(\vec{x}|\omega)$ is given by

$$T_{\alpha\beta}(\vec{x}|\omega) = \rho(c_{\ell}^2 - 2c_t^2)\delta_{\alpha\beta}\nabla\cdot\vec{u}(\vec{x}|\omega) +$$

$$+ \rho c_t^2(\frac{\partial u_{\alpha}(\vec{x}|\omega)}{\partial x_{\beta}} + \frac{\partial u_{\beta}(\vec{x}|\omega)}{\partial x_{\alpha}}), \tag{6.70}$$

while the unit normal to the surface at each point is given by

$$\hat{n} = [1 + (\nabla\zeta(\vec{x}_{\parallel}))^2]^{-1/2}(-\frac{\partial\zeta(\vec{x}_{\parallel})}{\partial x_1}, -\frac{\partial\zeta(\vec{x}_{\parallel})}{\partial x_2}, 1). \tag{6.71}$$

Equations (6.68) and (6.70) are now used for the determination of the components of the stress tensor, that are then used together with Eq. (6.71) in the boundary conditions (6.69). When the latter are expanded to first order in $\zeta(\vec{x}_{\parallel})$ (the small roughness limit), the equations for $A_{\alpha}(\vec{q}_{\parallel}\omega)$ take the form

$$\int\frac{d^2q_{\parallel}}{(2\pi)^2} e^{i\vec{q}_{\parallel}\cdot\vec{x}_{\parallel}} \sum_{\beta} \{M_{\alpha\beta}^{(o)}(\vec{q}_{\parallel}|\omega) + \int\frac{d^2Q_{\parallel}}{(2\pi)^2} e^{i\vec{Q}_{\parallel}\cdot\vec{x}_{\parallel}} \times$$

$$\times \hat{\zeta}(\vec{Q}_{\parallel})X_{\alpha\beta}(\vec{Q}_{\parallel};\vec{q}_{\parallel}|\omega)\}A_{\beta}(\vec{q}_{\parallel}\omega) = 0, \tag{6.72}$$

where we have introduced the Fourier transform of $\zeta(\vec{x}_\parallel)$

$$\zeta(\vec{x}_\parallel) = \int \frac{d^2Q_\parallel}{(2\pi)^2} \hat{\zeta}(\vec{Q}_\parallel) e^{i\vec{Q}_\parallel \cdot \vec{x}_\parallel} \quad . \tag{6.73}$$

The elements of the matrices $\overleftrightarrow{M}^{(o)}(\vec{q}_\parallel|\omega)$ and $\overleftrightarrow{X}(\vec{Q}_\parallel;\vec{q}_\parallel|\omega)$ are given in Ref. 223. At this point we do two things. We equate to zero the $\vec{q}_\parallel$th Fourier coefficient of the expression on the left hand side of Eq. (6.80), and multiply the resulting equation from the left by the matrix $\overleftrightarrow{S}(\hat{q}_\parallel)$ defined by

$$\overleftrightarrow{S}(\hat{q}_\parallel) = \begin{pmatrix} \hat{q}_1 & \hat{q}_2 & 0 \\ -\hat{q}_2 & \hat{q}_1 & 0 \\ 0 & 0 & 1 \end{pmatrix} \quad . \tag{6.74}$$

The result is an equation for the $\{A_\alpha(\vec{q}_\parallel\omega)\}$ that can be written as

$$\sum_\beta \overline{M}^{(o)}_{\alpha\beta}(q_\parallel|\omega)A_\beta(\vec{q}_\parallel\omega) - \sum_\beta \int \frac{d^2k_\parallel}{(2\pi)^2} \hat{\zeta}(\vec{q}_\parallel - \vec{k}_\parallel)\overline{M}^{(1)}_{\alpha\beta}(\vec{q}_\parallel;\vec{k}_\parallel|\omega) \times$$

$$\times A_\beta(\vec{k}_\parallel\omega) = 0. \tag{6.75}$$

The elements of the matrices $\overline{M}^{(o)}_{\alpha\beta}$ and $\overline{M}^{(1)}_{\alpha\beta}$ are

$$\overline{M}^{(o)}_{11}(\vec{q}_\parallel|\omega) = 2\alpha_\ell(q_\parallel\omega) \tag{6.76a}$$

$$\overline{M}^{(o)}_{13}(q_\parallel|\omega) = \frac{q_\parallel^2 + \alpha_t^2(q_\parallel\omega)}{\alpha_t(q_\parallel\omega)} \tag{6.76b}$$

$$\overline{M}^{(o)}_{22}(q_\parallel|\omega) = \alpha_t(q_\parallel\omega) \tag{6.76c}$$

$$\overline{M}_{31}^{(o)}(q_{\parallel}|\omega) = \frac{i}{q_{\parallel}}\left(q_{\parallel}^2 + \alpha_t^2(q_{\parallel}\omega)\right) \tag{6.76d}$$

$$\overline{M}_{33}^{(o)}(q_{\parallel}|\omega) = 2iq_{\parallel} \tag{6.76e}$$

$$\overline{M}_{12}^{(o)}(q_{\parallel}|\omega) = \overline{M}_{21}^{(o)}(q_{\parallel}|\omega) = \overline{M}_{23}^{(o)}(q_{\parallel}|\omega) = \overline{M}_{32}^{(o)}(q_{\parallel}|\omega) = 0 \tag{6.76f}$$

$$\overline{M}_{11}^{(1)}(\vec{q}_{\parallel};\vec{k}_{\parallel}|\omega) = \frac{q_{\parallel}}{k_{\parallel}}\frac{\omega^2}{c_t^2}(1-2\lambda^2) - \frac{\omega^2}{c_t^2}(\hat{q}_{\parallel}\cdot\hat{k}_{\parallel}) + 2q_{\parallel}k_{\parallel}(\hat{q}_{\parallel}\cdot\hat{k}_{\parallel})^2 \tag{6.77a}$$

$$\overline{M}_{12}^{(1)}(\vec{q}_{\parallel};\vec{k}_{\parallel}|\omega) = (\hat{q}_{\parallel}\times\hat{k}_{\parallel})_3\left[\frac{\omega^2}{c_t^2} - 2q_{\parallel}k_{\parallel}(\hat{q}_{\parallel}\cdot\hat{k}_{\parallel})\right] \tag{6.77b}$$

$$\overline{M}_{13}^{(1)}(\vec{q}_{\parallel};\vec{k}_{\parallel}|\omega) = -\frac{\omega^2}{c_t^2}(\hat{q}_{\parallel}\cdot\hat{k}_{\parallel}) + 2q_{\parallel}k_{\parallel}(\hat{q}_{\parallel}\cdot\hat{k}_{\parallel})^2 \tag{6.77c}$$

$$\overline{M}_{21}^{(1)}(\vec{q}_{\parallel};\vec{k}_{\parallel}|\omega) = (\hat{q}_{\parallel}\times\hat{k}_{\parallel})_3\left[-\frac{\omega^2}{c_t^2} + 2q_{\parallel}k_{\parallel}(\hat{q}_{\parallel}\cdot\hat{k}_{\parallel})\right] \tag{6.77d}$$

$$\overline{M}_{22}^{(1)}(\vec{q}_{\parallel};\vec{k}_{\parallel}|\omega) = -\frac{\omega^2}{c_t^2}(\hat{q}_{\parallel}\cdot\hat{k}_{\parallel}) + q_{\parallel}k_{\parallel}(\hat{q}_{\parallel}\cdot\hat{k}_{\parallel})^2 - q_{\parallel}k_{\parallel}(\hat{q}_{\parallel}\times\hat{k}_{\parallel})_3^2 \tag{6.77e}$$

$$\overline{M}_{23}^{(1)}(\vec{q}_{\parallel};\vec{k}_{\parallel}|\omega) = (\hat{q}_{\parallel}\times\hat{k}_{\parallel})_3\left[-\frac{\omega^2}{c_t^2} + 2q_{\parallel}k_{\parallel}\;(\hat{q}_{\parallel}\cdot\hat{k}_{\parallel})\right] \tag{6.77f}$$

$$\overline{M}_{31}^{(1)}(\vec{q}_{\parallel};\vec{k}_{\parallel}|\omega) = i\left[-\frac{\omega^2}{c_t^2} + 2q_{\parallel}k_{\parallel}(\hat{q}_{\parallel}\cdot\hat{k}_{\parallel})\right]\frac{\alpha_{\ell}(k_{\parallel}\omega)}{k_{\parallel}} \tag{6.77g}$$

$$\overline{M}_{32}^{(1)}(\vec{q}_{\parallel};\vec{k}_{\parallel}|\omega) = -iq_{\parallel}\alpha_t(k_{\parallel}\omega)(\hat{q}_{\parallel}\times\hat{k}_{\parallel})_3 \tag{6.77h}$$

$$\overline{M}_{33}^{(1)}(\vec{q}_{\parallel};\vec{k}_{\parallel}|\omega) = = i\left[-\frac{\omega^2}{c_t^2}\frac{k_{\parallel}}{\alpha_t(k_{\parallel}\omega)} + q_{\parallel}\frac{k_{\parallel}^2+\alpha_t^2(k_{\parallel}\omega)}{\alpha_t(k_{\parallel}\omega)}(\hat{q}_{\parallel}\cdot\hat{k}_{\parallel})\right], \tag{6.77i}$$

where $\lambda = c_t/c_\ell$. We see that $\overline{M}^{(0)}_{\alpha\beta}(q_\parallel|\omega)$ depends on $\vec{q}_\parallel$ only through its magnitude, while $\overline{M}^{(1)}_{\alpha\beta}(\vec{q}_\parallel;\vec{k}_\parallel|\omega)$ depends on $\vec{q}_\parallel$ and $\vec{k}_\parallel$ only through their magnitudes and through the cosine $(\hat{q}_\parallel\cdot\hat{k}_\parallel)$ and sine $(\hat{q}_\parallel \times \hat{k}_\parallel)_3$ of the angle between them.

The coefficients $\{A_\alpha(\vec{q}_\parallel\omega)\}$ are random quantities because they satisfy a matrix integral equation (6.75) with a random kernel. Just as we have defined $\zeta(\vec{x}_\parallel)$ through its moments (6.66), we define $A_\alpha(\vec{q}_\parallel\omega)$ by its moments. A particularly important moment of $A_\alpha(\vec{q}_\parallel\omega)$ is the first, $\langle A_\alpha(\vec{q}_\parallel\omega)\rangle$, since it describes the propagation of the mean surface acoustic wave across the rough surface.

To obtain from Eq. (6.75) the equation satisfied by $\langle A_\alpha(\vec{q}_\parallel\omega)\rangle$ we introduce the smoothing operator P that averages everything it acts on over the ensemble of realizations of the surface profile: $PA_\alpha(\vec{q}_\parallel\omega) \equiv \langle A_\alpha(\vec{q}_\parallel\omega)\rangle$. We also introduce the complementary operator $Q = 1 - P$ that projects out the fluctuating part of anything it acts on. We apply P and Q to Eq. (6.75) in turn and obtain the pair of equations

$$\sum_\beta \overline{M}^{(0)}_{\alpha\beta}(q_\parallel|\omega)PA_\beta(\vec{q}_\parallel\omega) - \sum_\beta \int\frac{d^2k_\parallel}{(2\pi)^2}\, P\hat{\zeta}(\vec{q}_\parallel-\vec{k}_\parallel)\overline{M}^{(1)}_{\alpha\beta}(\vec{q}_\parallel;\vec{k}_\parallel|\omega) \times$$

$$\times\, QA_\beta(\vec{k}_\parallel\omega) \tag{6.78a}$$

$$\sum_\beta \overline{M}^{(0)}_{\alpha\beta}(q_\parallel|\omega)QA_\beta(\vec{q}_\parallel\omega) - \sum_\beta \int\frac{d^2k_\parallel}{(2\pi)^2}\, Q\hat{\zeta}(\vec{q}_\parallel-\vec{k}_\parallel)\overline{M}^{(1)}_{\alpha\beta}(\vec{q}_\parallel;\vec{k}_\parallel|\omega) \times$$

$$\times\, PA_\beta^{\rightarrow}(k_\parallel\omega). \tag{6.78b}$$

In writing these equations we have used the identity $A_\beta(\vec{k}_\parallel\omega) = PA_\beta(\vec{k}_\parallel\omega) + QA_\beta(\vec{k}_\parallel\omega)$; we have also used the fact that $P\hat{\zeta}(\vec{q}_\parallel - \vec{k}_\parallel) = 0$, which follows from Eq. (6.66a); and we have used the result that $QA_\alpha(\vec{q}_\parallel\omega)$ is of $O(\zeta)$ to drop a term containing $QA_\beta(\vec{k}_\parallel\omega)$ from the right hand side of Eq. (6.78b) as of second order in ζ. We now solve Eq. (6.78b) for $QA_\beta(\vec{k}_\parallel\omega)$ and substitute the result into Eq. (6.78a). In this way we obtain the equations satisfied by $\langle A_\alpha(\vec{q}_\parallel\omega)\rangle$ in the form

$$\sum_\beta \overline{M}^{(0)}_{\alpha\beta}(q_\parallel|\omega)\langle A_\beta(\vec{q}_\parallel\omega)\rangle - \sum_{\beta\mu\nu}\int\frac{d^2k_\parallel}{(2\pi)^2}\int\frac{d^2p_\parallel}{(2\pi)^2}\langle\hat{\zeta}(\vec{q}_\parallel-\vec{k}_\parallel)\zeta(\vec{k}_\parallel-\vec{p}_\parallel)\rangle \times$$

$$\times\, \overline{M}^{(1)}_{\alpha\mu}(\vec{q}_\parallel;\vec{k}_\parallel|\omega)\overline{M}^{(0)-1}_{\mu\nu}(k_\parallel|\omega)\overline{M}^{(1)}_{\nu\beta}(\vec{k}_\parallel;\vec{p}_\parallel|\omega)\langle A_\beta(\vec{p}_\parallel|\omega)\rangle = 0. \tag{6.79}$$

It follows from Eqs. (6.66b) and (6.73) that

$$\langle\hat{\zeta}(\vec{q}_\parallel-\vec{k}_\parallel)\hat{\zeta}(\vec{k}_\parallel-\vec{p}_\parallel)\rangle = (2\pi)^2\,\delta(\vec{q}_\parallel-\vec{p}_\parallel)\,\delta^2 g(|\vec{q}_\parallel-\vec{k}_\parallel|), \tag{6.80}$$

where

$$g(k_\parallel) = \int d^2x_\parallel\, e^{-ik_\parallel\cdot x_\parallel}W(|\vec{x}_\parallel|) \tag{6.81a}$$

$$= \pi a^2 \exp\left(-\tfrac{1}{4}k_\parallel^2 a^2\right), \tag{6.81b}$$

where the second form follows from the particular choice for $W(|\vec{x}_\parallel|)$ given by Eq. (6.65). Thus we obtain finally the homogeneous equations for $\langle A_\alpha(\vec{q}_\parallel\omega)\rangle$,

$$\sum_\beta \{\overline{M}^{(0)}_{\alpha\beta}(q_\parallel|\omega) - \Delta M_{\alpha\beta}(q_\parallel|\omega)\}\langle A_\beta(\vec{q}_\parallel\omega)\rangle = 0\ , \tag{6.82}$$

where

$$\Delta M_{\alpha\beta}(q_\parallel|\omega) = \delta^2\int\frac{d^2k_\parallel}{(2\pi)^2}\,g(|\vec{q}_\parallel-\vec{k}_\parallel|)\sum_{\mu\nu}\overline{M}^{(1)}_{\alpha\mu}(\vec{q}_\parallel;\vec{k}_\parallel|\omega)\times$$

$$\times\,\overline{M}^{(0)-1}_{\mu\nu}(k_\parallel|\omega)\overline{M}^{(1)}_{\nu\beta}(\vec{k}_\parallel;\vec{q}_\parallel|\omega)\ . \tag{6.83}$$

The fact that we obtain a set of algebraic equations for $\langle A_\alpha(\vec{q}_\parallel\omega)\rangle$ instead of a set of integral equations is due to the restoration of infinitesimal translational invariance to our system as a result of the averaging process. The dispersion relation for surface acoustic waves on a randomly rough surface is obtained by equating to zero the determinant of the coefficients of $\langle A_\beta(\vec{q}_\parallel\omega)\rangle$ in this set of equations.

The inverse matrix $\overline{M}^{(o)-1}_{\alpha\beta}(q_\parallel|\omega)$ appearing in Eqs. (6.79) and (6.83) has the simple form

$$\vec{M}^{(o)-1}(q_\parallel|\omega)$$

$$= \frac{1}{D(q_\parallel\omega)}\begin{pmatrix} 2q_\parallel^2\alpha_t(q_\parallel\omega) & 0 & iq_\parallel(q_\parallel^2+\alpha_t^2(q_\parallel\omega) \\ 0 & \dfrac{D(q_\parallel\omega)}{\alpha_t(q_\parallel\omega)} & 0 \\ -\alpha_t(q_\parallel\omega)(q_\parallel^2+\alpha_t^2(q_\parallel\omega)) & 0 & -2iq_\parallel\alpha_t(q_\parallel\omega)\alpha_\ell(q_\parallel\omega) \end{pmatrix} \tag{6.84}$$

where

$$D(q_\parallel\omega) = 4q_\parallel^2\alpha_t(q_\parallel\omega)\alpha_\ell(q_\parallel\omega) - (q_\parallel^2+\alpha_t^2(q_\parallel\omega))^2. \tag{6.85}$$

The vanishing of $D(q_\parallel\omega)$ yields the frequency of a Rayleigh wave on a flat surface.

The set of equations (6.82) is actually simpler than appears to be the case at first glance. When the angular integration on the right hand side of Eq. (6.83) is carried out the 12, 21, 23, and 32 off-diagonal elements of the matrix $\Delta M_{\alpha\beta}(q_\parallel|\omega)$ vanish. This is due to the restoration of isotropy in the plane $x_3 = 0$ to our system by the averaging process. This is also the reason that the nonzero elements of the matrix $\Delta M_{\alpha\beta}(q_\parallel|\omega)$ depend on the wave vector $\vec{q}_\parallel$ only through its magnitude. Equations (6.82) thus break up into two sets of equations that have the forms

$$\begin{pmatrix} \overline{M}^{(o)}_{11}(q_\parallel|\omega) - \Delta M_{11}(q_\parallel|\omega) & \overline{M}^{(o)}_{13}(q_\parallel|\omega) - \Delta M_{13}(q_\parallel|\omega) \\ \overline{M}^{(o)}_{31}(q_\parallel|\omega) - \Delta M_{31}(q_\parallel|\omega) & \overline{M}^{(o)}_{33}(q_\parallel|\omega) - \Delta M_{33}(q_\parallel|\omega) \end{pmatrix} \begin{pmatrix} \langle A_1(q_\parallel\omega)\rangle \\ \langle A_3(q_\parallel\omega)\rangle \end{pmatrix} = 0 \tag{6.86a}$$

$$[\overline{M}^{(o)}_{22}(q_\parallel|\omega) - \Delta M_{22}(q_\parallel|\omega))]\langle A_2(q_\parallel\omega)\rangle = 0. \tag{6.86b}$$

The first of these equations is associated with a wave polarized in the sagittal plane (a Rayleigh wave); the second is associated with a wave polarized perpendicular to the sagittal plane (shear horizontal surface acoustic wave).

We note from the definitions of the matrices entering the matrix $\Delta M_{\alpha\beta}(q_\parallel|\omega)$ that the latter will be complex in general. This is because $\alpha_\ell(q_\parallel\omega)$ changes from being purely real for $k_\parallel > \omega/c_\ell$ to being purely imaginary for $0 < k_\parallel < \omega/c_\ell$, while $\alpha_t(q_\parallel\omega)$ changes from being purely real for $k_\parallel > \omega/c_t$ to being purely imaginary for $0 < k_\parallel < \omega/c_t$. In addition, for $k_\parallel > \omega/c_t$, where $\alpha_\ell(k_\parallel\omega)$ and $\alpha_t(k_\parallel\omega)$ are both real, the matrix $\overline{M}^{(o)-1}_{\mu\nu}(k_\parallel|\omega)$ contributes a complex pole to the integrand on the right hand side of Eq. (6.83) in view of Eq. (6.84). The correct way of dealing with it is to give ω a positive infinitesimal imaginary part, since this ensures the correct analytic continuations of $\alpha_\ell(k_\parallel\omega)$ and $\alpha_t(k_\parallel\omega)$. Thus this pole also makes a complex contribution to $\Delta M_{\alpha\beta}(q_\parallel|\omega)$. The solvability conditions for the equations (6.86) therefore yield complex values of ω even if $q_\parallel$ is real. The imaginary part of $\omega(q_\parallel)$ describes the damping of the Rayleigh wave due to its conversion into bulk waves or into other Rayleigh waves, depending on whether it arises from the integration over the interval $0 < k_\parallel\ \omega/c_t$ or the interval $k_\parallel > \omega/c_t$, respectively.

We now turn to a consideration of the solutions of Eqs. (6.86a) and (6.86b) in turn.

6.3.1. Rayleigh Waves

We recall that the elements of the matrix $\Delta M_{\alpha\beta}(q_\parallel|\omega)$ are of $O(\delta^2)$. In expanding the determinant of the coefficients in Eq. (6.86a) we keep terms of only up to this order in δ. In this way we obtain for the dispersion relation for Rayleigh waves on a randomly rough surface:

$$\begin{aligned} D(q_\parallel\omega) = -iq_\parallel\{ & \overline{M}_{11}^{(o)}(q_\parallel|\omega)\Delta M_{33}(q_\parallel|\omega) + \\ & + \overline{M}_{33}^{(o)}(q_\parallel|\omega)\Delta M_{11}(q_\parallel|\omega) - \\ & - \overline{M}_{13}^{(o)}(q_\parallel|\omega)\Delta M_{31}(q_\parallel|\omega) - \\ & - \overline{M}_{31}^{(o)}(q_\parallel|\omega)\Delta M_{13}(q_\parallel|\omega)\} \\ \equiv \Delta D(q_\parallel\omega). \end{aligned} \tag{6.87}$$

We use the fact that $D(q_\parallel\omega)$ vanishes for $\omega = \omega_R(q_\parallel) = C_R q_\parallel$, to write the solution of Eq. (6.87) in the form

$$\omega = \omega_R(q_\parallel) + \Delta\omega_R(q_\parallel), \tag{6.88}$$

where to lowest order in δ^2

$$\Delta\omega_R(q_\parallel) = \frac{\Delta D(q_\parallel,\omega_R(q_\parallel))}{\frac{\partial}{\partial\omega} D(q_\parallel\omega)\Big|_{\omega = \omega_R(q_\parallel)}} . \tag{6.89}$$

The frequency shift $\Delta\omega_R(q_\parallel)$ is conveniently written in the form

$$\Delta\omega_R(q_\parallel) = \omega_R(q_\parallel)\frac{\delta^2}{a^2}[\omega_1(aq_\parallel) - i\omega_2(aq_\parallel)], \tag{6.90}$$

where $\omega_{1,2}(aq_\parallel)$ are universial dimensionless functions of

$aq_\parallel$. These functions have been calculated by Eguiluz and Maradudin,(223) and are plotted in Figs. 6.10 and 6.11, respectively. From Fig. 6.10 we see that the roughness-induced shift in the frequency of the Rayleigh wave is always negative. From Fig. 6.11 we see that the scattering of the Rayleigh wave into bulk waves is a much more effective attenuation mechanism than is scattering into other Rayleigh waves. The attenuation length of the Rayleigh wave, $\ell(q_\parallel)$, i.e. the distance over which the energy of the Rayleigh wave decays to 1/e of its initial value, is given in terms of $\omega_2(aq_\parallel)$ by

$$\ell^{-1}(q_\parallel) = 2\frac{\delta^2}{a^2} q_\parallel \, \omega_2(aq_\parallel). \tag{6.91}$$

If we assume the value $\delta/a = 0.3$, and the typical values $\omega = 10^8$ rad/sec and $c_R = 3 \times 10$ cm/sec, for which $q_\parallel = 333.33\ \mathrm{cm}^{-1}$ and the wavelength of the Rayleigh wave along the surface $\lambda_R = 2\pi/q_\parallel \cong 0.02$ cm, we find that $\ell = 2.6 \times 10^3$ cm for $aq_\parallel = 0.1$, $\ell = 0.42$ cm for $aq_\parallel = 1$, and $\ell = 0.02$ for $aq_\parallel = 10$. (Note that while in the first case $\ell \gg \lambda_R$, in the last one $\ell \cong \lambda_R$.) The relative downward shift for the frequency of the Rayleigh wave in these three cases is 0.4%, 2%, and 5.8%, respectively.

6.3.2. Shear Horizontal Surface Acoustic Waves

From Eqs. (6.74c) and (6.86b) the dispersion relation for a shear horizontal wave on a randomly rough surface can be written as

$$\alpha_t(q_\parallel \omega) = \Delta M_{22}(q_\parallel | \omega). \tag{6.92}$$

In the absence of surface roughness the solution of this equation is $\omega = c_t q_\parallel$. In the presence of roughness the solution to the lowest nonzero order in δ is

$$\omega_{SH}(q_\parallel) = c_t q_\parallel - \frac{1}{2} c_t \frac{\Delta M_{22}^2(q_\parallel | c_t q_\parallel)}{q_\parallel} . \tag{6.93a}$$

$$= \omega_{SH}^{(1)}(q_\parallel) - i\omega_{SH}^{(2)}(q_\parallel). \tag{6.93b}$$

The second term on the right hand side of Eq. (6.93a) is of $O(\delta^4)$, so that the departure of $\omega_{SH}(q_\parallel)$ from $c_t q_\parallel$ is very small. In Fig. 6. 12 is plotted the result of Hardouin Duparc and

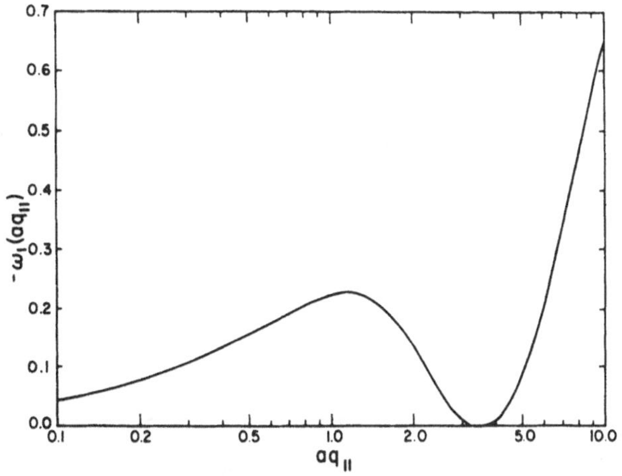

Fig. 6.10. Shift in the frequency of a Rayleigh wave due to surface roughness, as a function of the product $aq_\parallel$. Note that the actual shift is given by $c_R q_\parallel (\delta^2/a^2)\omega_1(aq_\parallel)$ [Ref. 223].

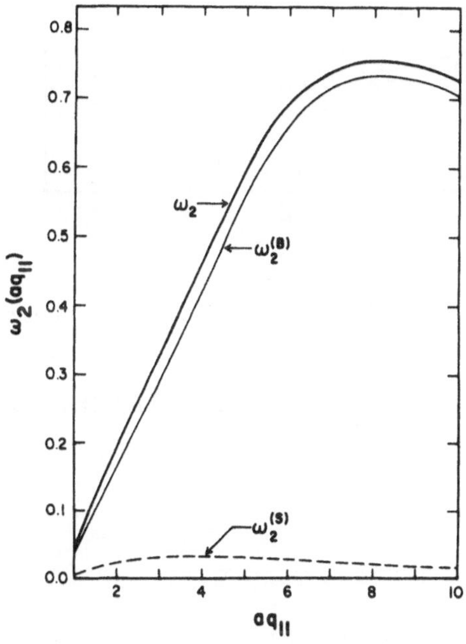

Fig. 6.11. Imaginary part of the change in the frequency of a Rayleigh wave caused by surface roughness. The actual imaginary part is given by $c_R q_\parallel (\delta^2/a^2)\omega_2(aq_\parallel)$ [Ref. 223]. The figure shows that the contribution to $\omega_2(aq_\parallel)$ from the bulk channels ($\omega_2^{(B)}$) dominates that due to the surface channel ($\omega_2^{(S)}$).

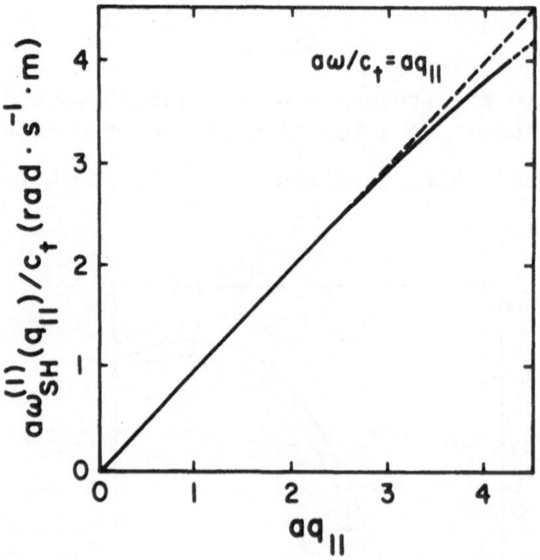

Fig. 6.12. The real part of the frequency of a shear horizontal surface acoustic wave propagating across a randomly rough aluminum surface ($c_t = 2.9 \times 10^3 m\ s^{-1}$) with $\delta/a = 0.05$. [Ref. 224].

Maradudin[224] for the real part of $a\omega_{SH}(q_{\parallel})/c_t$ as a function of $aq_{\parallel}$ for the choices $\delta/a = 0.05$, $c_t = 2.9 \times 10^5$ cm/sec. The value of $\omega_{SH}^{(2)}(q_{\parallel})$ obtained is a factor of 5 to 10 smaller than the real part of the second term on the right hand side of Eq. (6.93a).

The shear horizontal surface acoustic wave discussed in this subsection is only weakly bound to the randomly rough surface, i.e. it penetrates deeply into the medium. This can be seen from Eq. (6.92) and the fact that $\Delta M_{22}(q_{\parallel}/c_t q_{\parallel})$ is of $O(\delta^2)$, so that $\alpha_t(q_{\parallel},\omega_{SH}(q_{\parallel}))$ is also of this order, and is therefore small.

Thus, random surface roughness gives rise to a surface acoustic wave of a kind that cannot exist on a planar surface.

7. Transmission and Reflection of Surface Acoustic Waves by Transverse Discontinuities

The transmission and reflection of surface acoustic waves by transverse geometric and material discontinuities has been an important problem for many years in the fields of geophysics[226-232] and surface acoustic wave technology.[126,233-264]) Most of the recent interest in this problem has centered on the transverse discontinuities typically found in surface acoustic wave devices.

The transmission and reflection of a surface acoustic wave at any boundary is a much more complex problem than the analogous plane wave case. The usual acoustic boundary conditions cannot be satisfied by just a single transmitted and reflected surface acoustic wave. In addition, bulk acoustic waves must be generated by the transmission and reflection processes.

This class of surface acoustic wave problems has not yet yielded to analytical solution. However, numerical techniques utilizing both finite element[232,235,238,241,246,249,250-252 254,256,257] and normal mode analysis[126] (of a finite thickness plate) have proven successful. The normal mode approach is attractive[126] because it provides information about the mode-converted bulk waves in an easier way than does a finite element analysis.

We present two typical examples here: a surface acoustic wave launched along the top surface of an isotropic, elastic plate incident on (a) the edge of a plate(126) (Fig.7.1a) (see also Refs. 232, 234, 235, 238, 240-243, 245, 247, 248, 252, 253, 255, 258, 259-263); or (b) a transverse interface with a second isotropic plate(265) (Fig. 7.1b) (see also Refs. 266-272). The boundary conditions for the physical system shown in Fig. 1a require the stress components T_{11} and T_{13} to vanish at the end of the plate. The boundary conditions for Fig. 7.1b require the continuity of the displacement components u_1 and u_3, as well as of the stress components T_{11} and T_{13} across the interface.

We use two numerical methods to satisfy the boundary conditions. They are (a) the point-matching (collocation) method(273) and (b) the least-squares boundary residual method(274,275). In both methods the displacement components, and hence the stress components, are expanded in terms of the normal modes of the corresponding infinite plate. N terms are kept in each expansion. The convergence criterion imposed for both methods is, that the elastic energy be conserved with an error of less than 1%.

The point-matching method is favored for its simplicity. However, the least-squares boundary residual method can satisfy the same convergence criterion for much smaller values of N compared to the former method, and its convergence is guaranteed as N increases.

The normal modes for an isotropic elastic plate of thickness 2d are well known(276). For wave propagation at a frequency ω in the sagittal x_1x_3-plane, the symmetric (s) and antisymmetric (a) displacement fields are

$$u_1^{(s)} = i\beta A_m^{(s)} \left(\frac{\cosh(\alpha_\ell x_3)}{\sinh(\alpha_\ell d)} - 2 \frac{\alpha_\ell \alpha_t}{\beta^2 + \alpha_t^2} \frac{\cosh(\alpha_t x_3)}{\sinh(\alpha_t d)} \right) e^{i(\beta x_1 - \omega t)} , \tag{7.1}$$

$$u_3^{(s)} = -\alpha_\ell A_m^{(s)} \left(\frac{\sinh(\alpha_\ell x_3)}{\sinh(\alpha_\ell d)} - 2 \frac{\beta^2}{\beta^2 + \alpha_t^2} \frac{\sinh(\alpha_t x_3)}{\sinh(\alpha_t d)} \right) e^{i(\beta x_1 - \omega t)} , \tag{7.2}$$

$$u_1^{(a)} = -i\beta A_m^{(a)}\left(\frac{\sinh(\alpha_\ell x_3)}{\sinh(\alpha_\ell d)} - 2\frac{\alpha_\ell\alpha_t}{\beta^2+\alpha_t^2}\frac{\sinh(\alpha_t x_3)}{\sinh(\alpha_t d)}\right)e^{i(\beta x_1-\omega t)}, \tag{7.3}$$

$$u_3^{(a)} = -\alpha_\ell A_m^{(a)}\left(\frac{\cosh(\alpha_\ell x_3)}{\cosh(\alpha_\ell d)} - 2\frac{\beta^2}{\beta^2+\alpha_t^2}\frac{\cosh(\alpha_t x_3)}{\cosh(\alpha_t d)}\right)e^{i(\beta x_1-\omega t)}, \tag{7.4}$$

$$\omega^2/c_t^2 = \beta^2 - \alpha_t^2, \qquad \omega^2/c_\ell^2 = \beta^2 - \alpha_\ell^2, \tag{7.5}$$

where c_t and c_ℓ are the shear (transverse) and longitudinal wave velocities, respectively. All quantities in Eqs. (7.1)-(7.5) can be real, imaginary, or complex. The modes occur for discrete values of β that obey the dispersion relations (+1 for s and -1 for a)

$$\frac{\tanh(\alpha_t d)}{\tanh(\alpha_\ell d)} = \left(\frac{4\beta^2\alpha_\ell\alpha_t}{(\beta^2+\alpha_t^2)^2}\right)^{\pm 1}. \tag{7.6}$$

The allowed values of β are complex in general, i.e. $\beta = \beta_1 + i\beta_2$. The normal modes used here are ordered with the propagating modes ($\beta_2 = 0$) first in the order of decreasing $|\beta_1|$ and then evanescent modes in the order of increasing $|\beta_2|$. For m = 1, the lowest order symmetric and antisymmetric modes become degenerate in the propagation vector β for thick plates, i.e. for $2d > 5\lambda_R$. The degenerate β are equal to $\beta_R(= 2\pi/\lambda_R)$, i.e. the Rayleigh wave vector.

To test energy conservation, it is necessary to evaluate the energy of the incident, reflected and transmitted surface waves as well as the reflected and transmitted bulk waves carried away from the interface along the x_1-axis. The power P per unit length along the wave front, i.e. the x_2-axis, is given by

$$P = \int_{-d}^{d} dx_3 \, \langle P_1(x_3)\rangle. \tag{7.7}$$

The time averaged elastic Poynting vector $\langle P_1(x_3)\rangle$ is

$$\langle P_1(x_3)\rangle = -\frac{1}{2}\,\mathrm{Re}\{i\omega(T_{11}(x_3)u_1^*(x_3) + T_{13}(x_3)u_3^*(x_3))e^{i(\beta^* - \beta)x_1}\}, \tag{7.8}$$

where a harmonic time dependence is assumed, and T_{11} and T_{13} are stress components.

The selected numerical results to be presented are the elastic energy distribution carried by the bulk waves. It is advantageous to decompose the bulk mode energies into their transverse and longitudinal components. Strictly speaking this is impossible, since the plate mode energies contain interference terms between the transverse and longitudinal components. In practice, these interference terms are small for thick plates, i.e. for $d > 5\lambda_R$.

The bulk wave energy pattern for the reflection of a surface wave at the edge of a silica plate ($\sigma = 0.17$) is shown in Fig. 7.2. The total energy in the diffracted bulk waves in Fig. 7.2 equals 77% of the incident surface energy.

The bulk wave energy patterns for the reflection and transmission of a surface acoustic wave across the interface of two isotropic media with same transverse sound velocity c_t, and different mass densities ρ, ρ' as well as different Poisson's ratios σ, σ' are shown in Figs. 7.3a and 7.3b. In both Figs. 7.3a and 7.3b about 6% and 3% of the incident energy is converted into reflected and transmitted bulk waves, respectively.

The acoustic mismatches between the transmission and reflection sides of the structure in Fig. 7.1b are of equal magnitude and opposite sign. Their bulk radiation patterns are obviously different. The dependence on the sign of the acoustic mismatches indicates the nonreciprocity of the results with respect to the launching side of an incident surface wave for a given system. The electromagnetic analogy of the nonreciprocity has been discussed(277) previously.

8. Focusing of Rayleigh Waves

In an elastically isotropic solid the phonon phase and group velocities do not point in the same direction in general. As a result the energy propagation away from a point source may be focused in certain directions. For bulk acoustic waves there has been considerable experimental(278-280) as well as theoretical work(280-283) on such phonon focusing, and the agreement between theory and experiment is good.

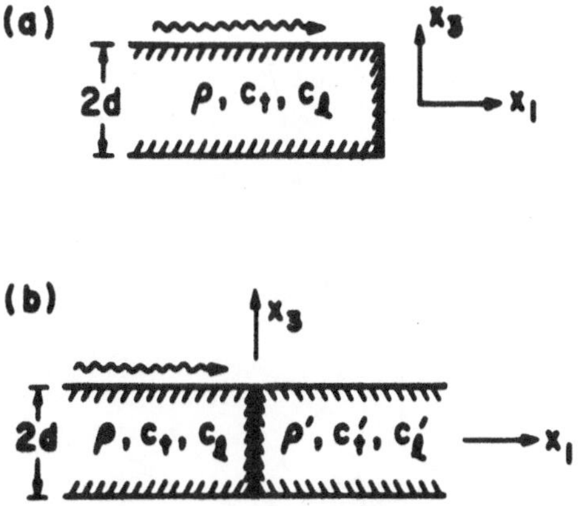

Fig. 7.1. Refraction of a Rayleigh wave from (a) the free end of a semi-infinite plate, and (b) the transverse interface between two different elastic plates.

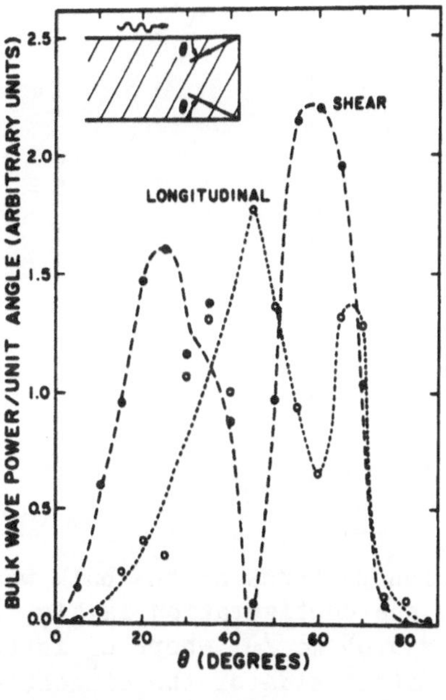

Fig. 7.2. Bulk wave power emitted in the x_1-direction per unit angle (• shear waves, o longitudinal waves) vs. angle θ for $2d = 20\lambda_R$. The points are averages over 10°, and the smooth curves are extrapolations based on $2d = 10\lambda_R$, $20\lambda_R$ [Ref. 126].

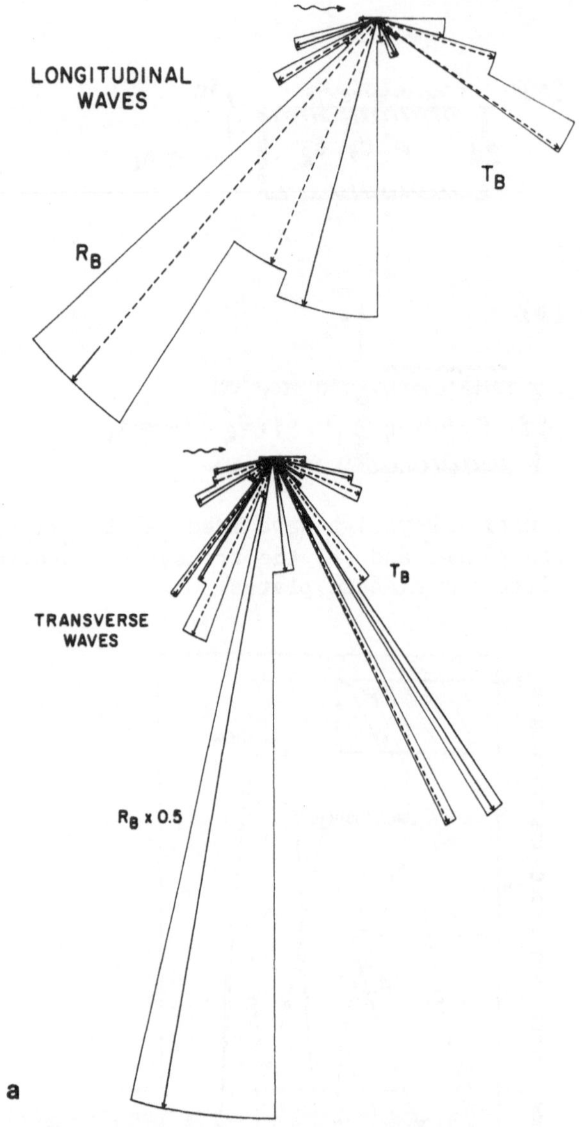

Fig. 7.3(a). The radiation patterns of the bulk waves excited by a Rayleigh wave in the configuration in Fig. 7.1 (b), for a plate thickness $d = 5.05\ \pi c_t/\omega$, where c_t is the transverse sound velocity on either side of the discontinuity. The Poisson ratios (σ, σ') and mass ratio ρ/ρ' are (0.3, 0.1) and 0.5, respectively. The wavy arrow indicates the incident Rayleigh wave. The solid and dotted arrows denote the propagation directions of the symmetric and antisymmetric modes, respectively. The area of each sector enclosing a given arrow is proportional to the energy carried by the given mode. All the figures are in the same arbitrary units [Ref. 265].

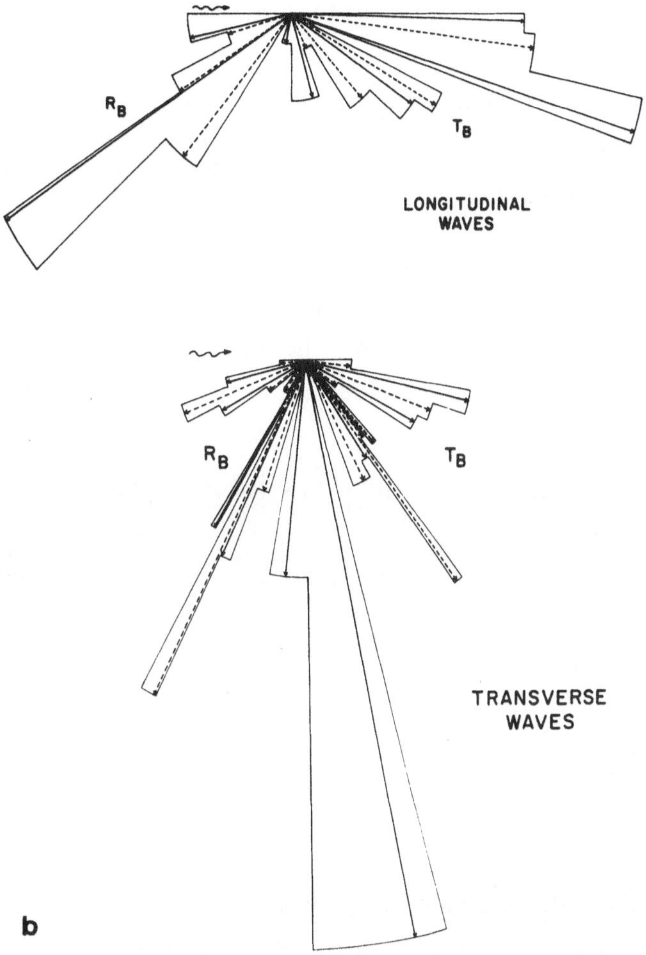

Fig. 7.3 (b). The same as Fig. 7.3(a) except that the Poisson ratios (σ, σ') and mass ratio ρ/ρ' are (0.1, 0.3) and 2.0, respectively [Ref. 265].

The focusing of surface acoustic waves was first studied theoretically by Buchwald.[284,285] However, there are errors in his work in that he found that surface acoustic waves existed only in a restricted range of directions and only on some surfaces. This was shown to be untrue later by Lim and Farnell.[19,42] Additional errors were pointed out by other authors.[286] Subsequently, designers of surface acoustic wave devices began to include focusing effects in calculations of diffraction on very anisotropic surfaces.[287-289]

We outline in this section a theory of the focusing of surface acoustic waves on the stress-free surfaces of nonpiezoelectric media.[290] For a theory of the focusing of surface acoustic waves on piezoelectric media see Ref. 286.

We study the focusing of surface acoustic waves by examining the dynamical Green's tensor $D_{\alpha\beta}(\vec{x},\vec{x}'|\omega)$ for a semi-infinite, anisotropic elastic medium occupying the region $x_3 > 0$.[291] It satisfies the time-independent equations of motion,

$$\sum_{\mu} [\omega^2\delta_{\alpha\mu} - \frac{1}{\rho}\sum_{\beta\nu} C_{\alpha\beta\mu\nu} \frac{\partial^2}{\partial x_\beta \partial x_\nu}] D_{\mu\gamma}(\vec{x},\vec{x}'|\omega) = \delta_{\alpha\gamma}\delta(\vec{x}-\vec{x}')$$

$$x_3,\ x_3' > 0, \quad \alpha,\gamma = 1,\ 2,\ 3, \tag{8.1a}$$

and the boundary conditions

$$\sum_{\mu} C_{\alpha 3\mu\nu} \frac{\partial}{\partial x_\nu} D_{\mu\gamma}(\vec{x},\vec{x}'|\omega)\Big|_{x_3 = 0} = 0$$

$$x_3' > 0, \quad \alpha,\gamma = 1,\ 2,\ 3 \tag{8.1b}$$

at the surface $x_3 = 0$, and vanishing or outgoing wave boundary conditions as $x_3 \to \infty$. The function $D_{\alpha\beta}(\vec{x},\vec{x}'|\omega)$ gives the amplitude of the α-component of the displacement at the point $\vec{x}$ in the medium when a distribution of body force per unit mass given by $\delta(\vec{x}-\vec{x}')\exp(-i\omega t)$ is applied in the β-direction at the point $\vec{x}'$. With no significant loss of generality we will assume that $\vec{x}' = 0$ in what follows.

The resulting Green's tensor can be written quite generally in the form[292]

$$D_{\alpha\beta}(\vec{x},\vec{0}|\omega) = \int \frac{d^2k_{\parallel}}{(2\pi)^2} \{d_{\alpha\beta}^{(\infty)}(\vec{k}_{\parallel}\omega|x_3 0) + \sum_{j=1}^{3} \frac{A_{\alpha\beta}^{(j)}(\vec{k}_{\parallel}\omega|0)}{\omega^2 - \omega_R^2(\vec{k}_{\parallel})} \times$$

$$\times e^{-\alpha_j(\vec{k}_{\parallel}\omega)x_3}\} e^{i\vec{k}_{\parallel}\cdot\vec{x}_{\parallel}}. \qquad (8.2)$$

The first term on the right hand side of this equation is the Green's tensor for an infinitely extended elastic medium. It is the particular integral of Eq. (8.1a) that satisfies exponentially decaying or outgoing wave boundary conditions at infinity. The second term, that is a solution of the corresponding homogeneous equation, represents the addition to the first term that ensures that the sum satisfies the boundary conditions (8.1b), as well as exponentially decaying or outgoing wave boundary conditions at $x_3 = +\infty$. The integrand of this term has a simple pole at $\omega = \omega_R(\vec{k}_{\parallel})$, where $\omega_R(\vec{k}_{\parallel})$ is the frequency of the Rayleigh wave on the surface $x_3 = 0$. It lies below the continuum of bulk wave frequencies corresponding to the wave vector $\vec{k}_{\parallel}$. Neither the function $d_{\alpha\beta}^{(\infty)}(\vec{k}_{\parallel}\omega|x_3x_3')$ nor the function $A_{\alpha\beta}^{(j)}(\vec{k}_{\parallel}\omega|x_3')$ has a pole in the region of $(\omega,\vec{k}_{\parallel})$-space in which the Rayleigh wave can exist.

The $\vec{k}_{\parallel}$-plane over which the integral in Eq. (8.2) is evaluated can be divided into two regions by a closed curve C for each value of ω. Each ray drawn from the origin of $\vec{k}_{\parallel}$-space intersects this curve at the value of $k_{\parallel} = |\vec{k}_{\parallel}|$ for which the minimum bulk wave frequency equals ω. For the given value of ω no Rayleigh waves can exist for values of $\vec{k}_{\parallel}$ lying inside C; they can exist only for values of $\vec{k}_{\parallel}$ lying outside this curve. In the latter region Re $\alpha_j(\vec{k}_{\parallel}\omega) > 0$ for j = 1, 2, 3.

The physical significance of the contribution to $D_{\alpha\beta}(\vec{x},\vec{0}|\omega)$ from each of these two regions of the $\vec{k}_{\parallel}$-plane is that the integration over the interior of the curve C describes a displacement field propagating away from the oscillating point source at $\vec{x}' = 0$ into the interior of the medium, while the integration throughout the region outside the curve describes a displacement field propagating away from the source along the surface of the medium. It is with the latter contribution that we are interested.

Since the asymptotic behavior for large $|\vec{x}_{\parallel}|$ of a function defined by a Fourier integral, such as $D_{\alpha\beta}(\vec{x},\vec{0}|\omega)$, Eq. (8.2), is determined by the singularities of the integrand,[293] and since only the second term on the right hand side of Eq. (8.2) has a pole for values of $\vec{k}_{\parallel}$ lying outside C, the asymptotic form of the surface wave contribution to $D_{\alpha\beta}(\vec{x},\vec{0}|\omega)$ is given by

$$D^{(+)}_{\alpha\beta}(\vec{x},\vec{0}|\omega)_s \sim \sum_{j=1}^{3} \int_{C^>} \frac{d^2k_{\parallel}}{(2\pi)^2} \frac{A^{(j)}_{\alpha\beta}(\vec{k}_{\parallel}\omega|0)}{\omega^2+i\eta-\omega_R^2(\vec{k}_{\parallel})} e^{i\vec{k}_{\parallel}\cdot\vec{x}_{\parallel}-\alpha_j(\vec{k}_{\parallel}\omega)x_3}. \tag{8.3}$$

In this expression $C^>$ denotes the region of the $\vec{k}_{\parallel}$-plane outside the curve C, and the superscript "+" refers to the addition of the term $+i\eta$, where η is a positive infinitesimal, to the denominator of the integrand to define the manner in which the pole is to be dealt with. Our criterion is that the integral in Eq. (8.3) should represent outgoing waves.

With the aid of the integral representation

$$\frac{1}{x+i\eta} = \frac{1}{i}\int_0^{\infty} dt\; e^{ixt-\eta t}, \tag{8.4}$$

we can rewrite Eq. (8.3) as

$$D^{(+)}_{\alpha\beta}(\vec{x},\vec{0}|\omega)_s \sim \sum_{j=1}^{3} \int_{C^>} \frac{d^2k_{\parallel}}{(2\pi)^2 i} \int_0^{\infty} ds\; A^{(j)}_{\alpha\beta}(\vec{k}_{\parallel}\omega|0)\; e^{-\alpha_j(\vec{k}_{\parallel}\omega)x_3} \times$$

$$\times \exp iF(\vec{k}_{\parallel}s;\vec{x}_{\parallel}) , \tag{8.5}$$

where

$$F(\vec{k}_{\parallel}s;\vec{x}_{\parallel}) = \vec{k}_{\parallel}\cdot\vec{x}_{\parallel} + s(\omega^2 - \omega_R^2(\vec{k}_{\parallel})). \tag{8.6}$$

The method of stationary phase[294] tells us that the dominant contribution to the integral for large $|\vec{x}_{\parallel}|$ comes from the neighborhood of those points in s- and $\vec{k}_{\parallel}$-space where the variation of $F(\vec{k}_{\parallel}s;\vec{x}_{\parallel})$ is smallest. These points, called stationary points, are determined from the equations

$$\frac{\partial F}{\partial s} = 0 = \omega^2 - \omega_R^2(\vec{k}_{\parallel}); \ \nabla_{\vec{k}_{\parallel}} F = 0 = \vec{x}_{\parallel} - s\nabla_{\vec{k}_{\parallel}} \omega_R^2(\vec{k}_{\parallel}). \qquad (8.7)$$

We denote the solutions of Eqs. (8.7) by s_ν and $\vec{k}_\nu$. If there is more than one solution for given $\vec{x}_{\parallel}$ and ω^2, we have to sum our result over the contributions associated with each $(s_\nu,\vec{k}_\nu)$.

We see from Eqs. (8.7) that the principal contribution to the asymptotic behavior of $D_{\alpha\beta}^{(+)}(\vec{x},\vec{0}|\omega)_s$ comes from the neighborhood of the points on the constant frequency curve in $\vec{k}_{\parallel}$-space defined by $\omega_R^2(\vec{k}_{\parallel}) = \omega^2$ at which the gradient $\nabla_{\vec{k}_{\parallel}} \omega_R^2(\vec{k}_{\parallel})$ is parallel to $\vec{x}_{\parallel}$. If we rewrite the second of Eqs. (8.7) as $\vec{x}_{\parallel} = 2s_\nu\omega_R(\vec{k}_\nu)\vec{v}_R(\vec{k}_\nu)$, where $\vec{v}_R(\vec{k}_\nu) = \nabla_{\vec{k}_{\parallel}} \omega_R(\vec{k}_{\parallel})|_{\vec{k}_{\parallel}=\vec{k}_\nu}$ is the group velocity of the Rayleigh wave, we see that the principal contribution to the large $|\vec{x}_{\parallel}|$ behavior of $D_{\alpha\beta}^{(+)}(\vec{x},\vec{0}|\omega)_s$ comes from the neighborhood of the point $\vec{k}_\nu$ on the constant frequency curve $\omega_R^2(\vec{k}_{\parallel}) = \omega^2$ at which the group velocity of the Rayleigh wave is parallel to $\vec{x}_{\parallel}$. Since s_ν is positive, $\vec{v}_R(\vec{k}_\nu)$ and $\vec{x}_{\parallel}$ must not only be parallel, they must point in the same direction as well. Now, it can be shown that, due to time reversal symmetry, $\omega_R(-\vec{k}_{\parallel}) = \omega_R(\vec{k}_{\parallel})$. Consequently, we must have that $\vec{v}_R(-\vec{k}_\nu) = -\vec{v}_R(\vec{k}_\nu)$. Thus, for every solution $\vec{k}_\nu$ we also have a second, $-\vec{k}_\nu$, but only one of them corresponds to a group velocity that is pointed in the direction of $\vec{x}_{\parallel}$ (Fig. 8.1). Of the pair of solutions it is only the latter we can retain.

We now expand $F(\vec{k}_{\parallel}s;\vec{x}_{\parallel})$ about the point $(s_\nu,\vec{k}_\nu)$,

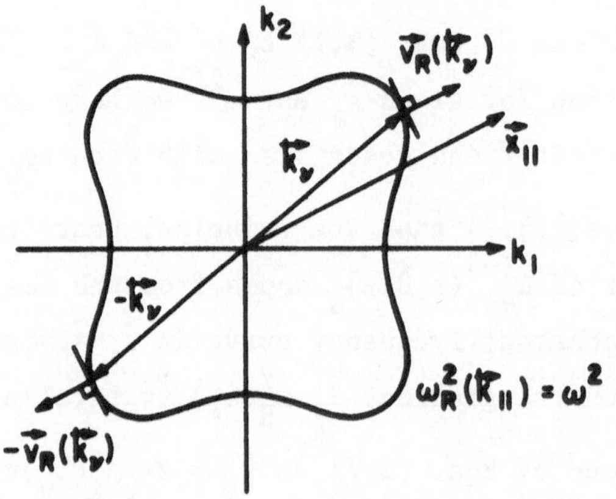

Fig. 8.1. Illustration of the determination of the wave vector $\vec{k}_\nu$ whose neighborhood yields the dominant contribution to the asymptotic behavior of the Green's tensor $D^{(+)}_{\alpha\beta}(\vec{x},\vec{0}|\omega)_s$.

$$F(\vec{k}_\parallel s;\vec{x}_\parallel) = \vec{k}_\nu\cdot\vec{x}_\parallel - \tau\sum_\alpha \kappa_\alpha \frac{\partial}{\partial k_\alpha}\omega_R^2(\vec{k}_\nu) - $$

$$- \frac{1}{2} s_\nu \sum_{\alpha\beta} \kappa_\alpha\kappa_\beta \frac{\partial^2}{\partial k_\alpha \partial k_\beta}\omega_R^2(\vec{k}_\nu) - \ldots \tag{8.8}$$

where $\tau = s - s_\nu$, $\vec{\kappa} = \vec{k}_\parallel - \vec{k}_\nu$. Having made a small τ and $\vec{\kappa}$ expansion of $F(\vec{k}_\parallel s;\vec{x}_\parallel)$ we make little error by extending the integrations in Eq. (8.5) over all $\vec{k}_\parallel$-space. In this way we obtain

$$D_{\alpha\beta}^{(+)}(\vec{x},\vec{0}|\omega)_s \sim \sum_\nu \frac{e^{i\vec{k}_\nu\cdot\vec{x}_\parallel}}{(2\pi)^2 i} \sum_{j=1}^{3} A_{\alpha\beta}^{(j)}(\vec{k}_\nu\omega|0)e^{-\alpha_j(\vec{k}_\nu\omega)x_3} \times$$

$$\times \int d^2\kappa \int_{-\infty}^{\infty} d\tau \exp\{-i\tau\vec{\kappa}\cdot\nabla_{\vec{k}_\parallel}\omega_R^2(\vec{k}_\nu) - \frac{1}{2} is_\nu\vec{\kappa}:\nabla_{\vec{k}_\parallel}\nabla_{\vec{k}_\parallel}\omega_R^2(\vec{k}_\nu):\vec{\kappa} + \ldots\} \tag{8.9}$$

In writing this expression we have replaced the comparatively slowly varying function $A_{\alpha\beta}^{(j)}(\vec{k}_\parallel\omega|0)\exp(-\alpha_j(\vec{k}_\parallel\omega)x_3)$ by its value at $\vec{k}_\parallel = \vec{k}_\nu$, and have taken it outside the integral over $\vec{\kappa}$. The error incurred by making this approximation is of higher order in $|\vec{x}_\parallel|^{-1}$ than the terms retained.

It is now convenient to rotate the coordinate axes in $\vec{\kappa}$-space according to $\vec{\kappa} = \hat{n}\xi_1 + \hat{t}\xi_2$, where $\hat{n}$ is the unit vector along the outward normal to the constant frequency curve $\omega_R^2(\vec{k}_\parallel) = \omega^2$ at the point $\vec{k}_\parallel = \vec{k}_\nu$, while $\hat{t}$ is the unit vector tangent to this curve at the same point, and oriented in such a way that $\hat{n}$ is rotated into $\hat{t}$ by a counterclockwise rotation through $\pi/2$:

$$\hat{n} = \frac{1}{|\nabla_{\vec{k}_\parallel}\omega_R^2(\vec{k}_\nu)|}\left[\frac{\partial\omega_R^2(\vec{k}_\nu)}{\partial k_1}, \frac{\partial\omega_R^2(\vec{k}_\nu)}{\partial k_2}\right] \tag{8.10a}$$

$$\hat{t} = \frac{1}{|\nabla_{\vec{k}_\parallel} \omega_R^2(\vec{k}_\nu)|} \left[- \frac{\partial \omega_R^2(\vec{k}_\nu)}{\partial k_2} , \frac{\partial \omega_R^2(\vec{k}_\nu)}{\partial k_1} \right] . \tag{8.10b}$$

The Jacobian for this transformation is unity. The integral over τ in Eq. (8.9) gives just $2\pi\delta(\xi_1)|\nabla_{\vec{k}_\parallel} \omega_R^2(\vec{k}_\nu)|^{-1}$, and we are left with

$$D_{\alpha\beta}^{(+)}(\vec{x},\vec{0}|\omega)_s \sim \sum_\nu \frac{e^{i\vec{k}_\nu \cdot \vec{x}_\parallel}}{2\pi i} \sum_{j=1}^{3} \frac{A_{\alpha\beta}^{(j)}(\vec{k}_\nu\omega|0)}{|\nabla_{\vec{k}_\parallel} \omega_R^2(\vec{k}_\nu)|} e^{-\alpha_j(\vec{k}_\nu\omega)x_3} \times$$

$$\times \int_{-\infty}^{\infty} d\xi_2 \exp\{-is_\nu[\tfrac{1}{2} a\xi_2^2 + \tfrac{1}{6} b\xi_2^3 + \ldots], \tag{8.11}$$

where

$$a = \sum_{\alpha\beta} \hat{t}_\alpha \hat{t}_\beta \left(\frac{\partial^2 \omega_R^2(\vec{k}_\nu)}{\partial k_\alpha \partial k_\beta}\right) \tag{8.12a}$$

$$b = \sum_{\alpha\beta\gamma} \hat{t}_\alpha \hat{t}_\beta \hat{t}_\gamma \left(\frac{\partial^3 \omega_R^2(\vec{k}_\nu)}{\partial k_\alpha \partial k_\beta \partial k_\gamma}\right). \tag{8.12b}$$

If $a \neq 0$, we can neglect the higher order terms in the exponent over ξ_2 in the integral over ξ_2 in Eq. (8.11). The result of the subsequent integration is

$$D_{\alpha\beta}^{(+)}(\vec{x},\vec{0}|\omega)_s \sim \sum_\nu \frac{e^{i\vec{k}_\nu\cdot\vec{x}_\parallel}}{(2\pi|\vec{x}_\parallel|)^{1/2}} \sum_{j=1}^{3} \frac{A_{\alpha\beta}^{(j)}(\vec{k}_\nu\omega|0)}{|\nabla_{\vec{k}_\parallel}\omega_R^2(\vec{k}_\nu)|} e^{-\alpha_j(\vec{k}_\nu\omega)x_3} \times$$

$$\times \frac{1}{|a|^{1/2}} e^{-i\frac{\pi}{4}\mathrm{sgn}\,a - i\frac{\pi}{2}}, \tag{8.13}$$

where we have used the fact that $s_\nu = |\vec{x}_\parallel||\nabla_{\vec{k}_\parallel}\omega_R^2(\vec{k}_\nu)|^{-1}$. For most values of $\vec{x}_\parallel$, $a \neq 0$ and Eq. (8.13) will hold. From Eq. (8.13) we see that the displacement at $\vec{x}_\parallel$ is proportional to $|\vec{x}_\parallel|^{-1/2}|a|^{-1/2}$. This is in contrast to the result for bulk waves where the displacement at point $\vec{x}$ is proportional to $|\vec{x}|^{-1}$.

At points where $a = 0$ we must keep the term in b in the exponent in Eq. (8.11), and find in this case that

$$D_{\alpha\beta}^{(+)}(\vec{x},\vec{0}|\omega)_s \sim \sum_\nu \frac{e^{i\vec{k}_\nu\cdot\vec{x}_\parallel}}{|\vec{x}_\parallel|^{1/3}} \sum_{j=1}^{3} \frac{A_{\alpha\beta}^{(j)}(\vec{k}_\nu\omega|0)}{|\nabla_{\vec{k}_\parallel}\omega_R^2(\vec{k}_\nu)|^{2/3}} e^{-\alpha_j(\vec{k}_\nu\omega)x_3} \times$$

$$\times \frac{1}{\Gamma(\frac{2}{3})}\left(\frac{2}{9|b|}\right)^{1/3} e^{-i\frac{\pi}{2}}. \tag{8.14}$$

In this case, we see that the displacement at $\vec{x}_\parallel$ is proportional to $|\vec{x}_\parallel|^{-1/3}|b|^{-1/3}$.

If neither a nor b can be neglected, we have the result that

$$\int_{-\infty}^{\infty} d\xi\; e^{-is_\nu(\frac{1}{2}a\xi^2 + \frac{1}{6}b\xi^3)}$$

$$= \int_{-\infty}^{\infty} d\xi\, e^{-is_\nu \frac{b}{6}(\xi^3 + 3\frac{a}{b}\xi^2)}$$

$$= e^{-i\frac{s_\nu}{3}\frac{a^3}{b^2}} \int_{-\infty}^{\infty} d\xi\, e^{-is_\nu \frac{b}{6}[(\xi + \frac{a}{b})^3 - 3\frac{a^2}{b^2}(\xi + \frac{a}{b})]}$$

$$= 2e^{-i\frac{s_\nu}{3}\frac{a^3}{b^2}} \int_0^{\infty} d\xi \cos s_\nu \frac{b}{6}(\xi^3 - \frac{3a^2}{b^2}\xi)$$

$$= 2\pi \left(\frac{2}{|b|}\right)^{\frac{1}{3}} \frac{e^{-i\frac{s_\nu}{3}\frac{a^3}{b^2}}}{s_\nu^{1/3}} A_i\left(-\left(\frac{s_\nu}{2b^2}\right)^{2/3} a^2\right), \tag{8.15}$$

where Ai(-z) is an Airy function, so that

$$D_{\alpha\beta}^{(+)}(\vec{x},\vec{0}|\omega)_s \sim \sum_\nu e^{i\vec{k}_\nu\cdot\vec{x}_\parallel} \sum_{j=1}^{3} A_{\alpha\beta}^{(j)}(\vec{k}_\nu\omega|0) e^{-\alpha_j(\vec{k}_\nu\omega)x_3} \times$$

$$\times \left(\frac{2}{|b|}\right)^{1/3} \frac{e^{-i\frac{s_\nu}{3}\frac{a^3}{b^2} - i\frac{\pi}{2}}}{|\nabla_{\vec{k}_\parallel}\omega_R^2(\vec{k}_\nu)|^{2/3}} \frac{1}{|\vec{x}_\parallel|^{1/3}} Ai\left(-\frac{a^2}{(2b^2)^{2/3}} \frac{|\vec{x}_\parallel|^{2/3}}{|\nabla_{\vec{k}_\parallel}\omega_R^2(\vec{k}_\nu)|^{2/3}}\right). \tag{8.16}$$

We present results illustrating Rayleigh waves on the (110) and (111) surfaces of Cu, Ge and NaCℓ. For these three materials the equations for the Rayleigh wave dispersion relation

$$\omega(\vec{k}_\parallel) = c_R(\hat{k}_\parallel)|\vec{k}_\parallel|, \tag{8.17}$$

where $c_R(\hat{k}_\parallel)$ is the speed of Rayleigh waves in the direction $\hat{k}_\parallel$, were solved numerically.[295] From this dispersion relation the slowness surfaces can be obtained by plotting a constant frequency curve in $\vec{k}_\parallel$-space

$$k_{\parallel}(\hat{k}_{\parallel}) = \frac{\omega}{c_R(\vec{k}_{\parallel})} \,. \tag{8.18}$$

The slowness surfaces obtained in this way for the different materials are presented in the upper parts of Figs. 8.2 and 8.3.[290]

In our theoretical analysis we found that the amplitude of the displacement field at a large distance from a point source should be proportional to $|a|^{-1/2}$. Thus by traveling along the slowness surface and at each point finding the direction of the outward normal and $|a|^{-1/2}$, one obtains a quantity proportional to the amplitude of the displacement field far from a point source. This has been done numerically. The results, a polar plot of the amplitude versus direction of observation, are presented in the lower portions of Figs. 8.2 and 8.3. We see several directions along which the amplitude is strongly enhanced. These directions correspond to places on the slowness surface where $a \to 0$, i.e. where the Gaussian curvature vanishes. These directions in general are not the directions of enhancement for bulk phonons.

The slowness surfaces and focusing patterns do reflect the symmetry of the crystal surface. Thus the slowness surfaces and focusing patterns for the (110) and (111) surfaces can be rotated by 180° and 120°, respectively, without change. This corresponds to the fact that the normal to the surface in the former case is a two-fold rotation axis, while it is a three-fold rotation axis in the latter case.

We see that on the (111) surface the slowness surfaces and focusing patterns for Ge and NaCℓ are very similar despite the fact that the anisotropy factor $A = 2c_{44}/(c_{11} - c_{12})$ for Ge is greater than unity, while that for NaCℓ is smaller than unity. The results presented show that for Rayleigh waves on cubic crystals a large variety of focusing patterns can occur, depending on the crystal surface considered and on the values of the elastic moduli.

The method of analysis employed here should be applicable to the theoretical study of phonon focusing on piezoelectric and magnetoelastic surfaces.

The effects described in this section have yet to be studied experimentally.

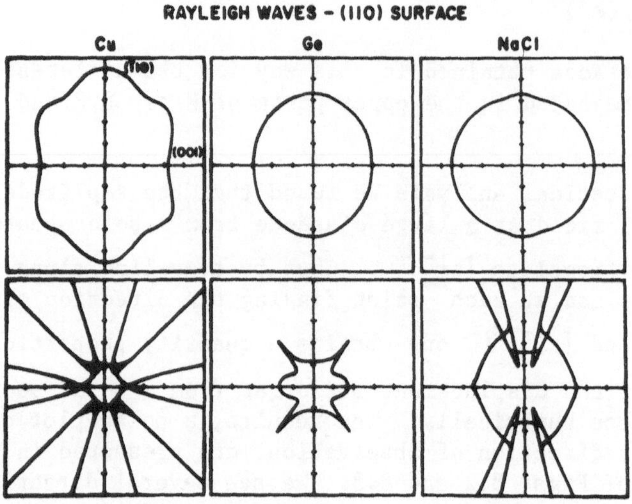

Fig. 8.2. The upper curves are the slowness surfaces for Rayleigh waves on the (110) surfaces of Cu, Ge, and NaCℓ. The lower curves are the focussing patterns derived from the slowness surfaces. The orientation of the crystal axes for the upper and lower curves is given in the upper Cu box [Ref. 290].

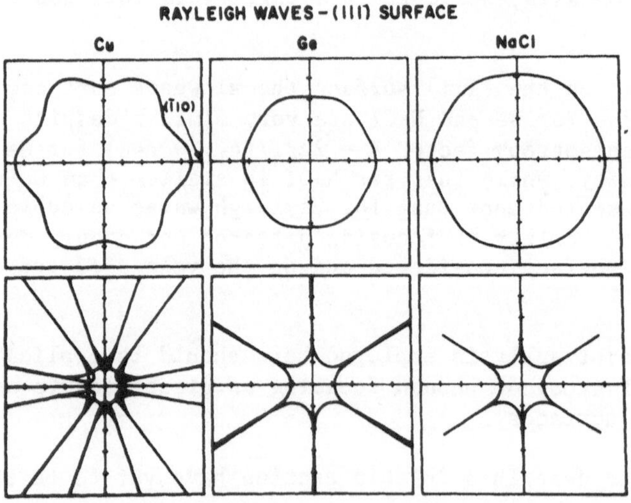

Fig. 8.3. The same as Fig. 8.3, but for the (111) surfaces of Cu, Ge, and NaCℓ [Ref. 290].

9. Effects of Spatial Dispersion on Surface Acoustic Waves

The term spatial dispersion refers to the spatially nonlocal relation between two physical quantities. Perhaps the best known example of spatial dispersion is provided by the most general linear relation between the electric displacement vector and the macroscopic electric field in a dielectric medium:

$$D_\alpha(\vec{x};t) = \sum_\beta \int d^3x' \int dt' \; \varepsilon_{\alpha\beta}(\vec{x},\vec{x}';t,t')E_\beta(\vec{x}';t'), \tag{9.1}$$

where $\varepsilon_{\alpha\beta}(\vec{x},\vec{x}';t,t')$ is the dielectric tensor of the medium, and is nonlocal both in space and in time. The integration over $\vec{x}'$ extends throughout the volume occupied by the medium, and causality requires that $\varepsilon_{\alpha\beta}(\vec{x},\vec{x}';t,t')$ vanish for $t' > t$.

For a spatially homogeneous medium characterized by a time-independent Hamiltonian, $\varepsilon_{\alpha\beta}(x,x';t,t')$ depends on $\vec{x}$ and $\vec{x}'$ only through their difference, and also depends on t and t′ only through their difference. In this case, if we write $D_\alpha(\vec{x};t)$ and $E_\alpha(\vec{x};t)$ in the forms

$$D_\alpha(\vec{x};t) = D_\alpha(\vec{k}\omega)\exp i(\vec{k}\cdot\vec{x} - \omega t) \tag{9.2a}$$

$$E_\alpha(\vec{x};t) = E_\alpha(\vec{k}\omega)\exp i(\vec{k}\cdot\vec{x} - \omega t), \tag{9.2b}$$

the relation between the amplitudes $D_\alpha(\vec{k}\omega)$ and $E_\alpha(\vec{k}\omega)$ becomes

$$D_\alpha(\vec{k}\omega) = \sum_\beta \varepsilon_{\alpha\beta}(\vec{k}\omega)E_\beta(\vec{k}\omega), \tag{9.3}$$

where

$$\varepsilon_{\alpha\beta}(\vec{k}\omega) = \int d^3x \int dt \; \varepsilon_{\alpha\beta}(\vec{x};t)e^{-i\vec{k}\cdot\vec{x}+i\omega t}. \tag{9.4}$$

The dependence of the dielectric tensor $\varepsilon_{\alpha\beta}(\vec{k},\omega)$ in Eq. (9.3) on the wave vector $\vec{k}$ of the electromagnetic field in the medium is referred to as spatial dispersion. (The dependence of $\varepsilon_{\alpha\beta}(\vec{k}\omega)$ on the frequency ω of the field in the medium is referred to as frequency dispersion, although a more consistent term is temporal dispersion.)

If we expand $\varepsilon_{\alpha\beta}(\vec{k}\omega)$ in powers of the components of the wave vector $\vec{k}$, the result is[296]

$$\varepsilon_{\alpha\beta}(\vec{k}\omega) = \varepsilon_{\alpha\beta}(\omega) + \sum_{\gamma} ig_{\alpha\beta\gamma}(\omega)k_{\gamma} + O(k^2), \tag{9.5}$$

where $g_{\alpha\beta\gamma}(\omega)$ is the optical gyrotropic tensor. If we then combine Eqs. (9.2) and (9.3), we obtain the result that

$$D_{\alpha}(\vec{x};\omega) = \sum_{\beta} \varepsilon_{\alpha\beta}(\omega)E_{\beta}(\vec{x};\omega) +$$

$$+ \sum_{\beta\gamma} g_{\alpha\beta\gamma}(\omega) \frac{\partial}{\partial x_{\gamma}} E_{\beta}(\vec{x};\omega) + \ldots, \tag{9.6}$$

where $D_{\alpha}(\vec{x};\omega)$ and $E_{\alpha}(\vec{x};\omega)$ are defined by

$$D_{\alpha}(\vec{x};t) = D_{\alpha}(\vec{x};\omega)e^{-i\omega t} \tag{9.7a}$$

$$E_{\alpha}(\vec{x};t) = E_{\alpha}(\vec{x};\omega)e^{-i\omega t}. \tag{9.7b}$$

Thus, another manifestation of spatial dispersion is the presence of higher spatial derivatives in the constitutive equation relating $\vec{D}$ and $\vec{E}$ than are present when spatial dispersion is neglected.

After this introduction to spatial dispersion we now proceed to demonstrate a central point of this Section, viz. that due to the atomic nature of a crystalline solid, elastic continua also display spatial dispersion when their equations of motion and associated boundary conditions are obtained by going to the long wavelength limit in the corresponding equations of lattice dynamics. We will then show some consequences of this fact for surface acoustic waves.

Rather than maintaining complete generality in what follows, we will proceed from a specific lattice model. While the results obtained have, therefore, a degree of model dependence, this is compensated to some extent by the fact that they automatically possess all the properties required by infinitesimal translational and rotational invariance, that are so important for a correct description of surface acoustic waves[297], as well as by crystal symmetry.

The model we study is a simple cubic crystal, with nearest and next nearest neighbor central and noncentral force interactions among atoms, that is bounded by a stress-free (001) surface. The equilibrium positions of the atoms are given by the vectors

$$\vec{x}(\ell) = a_0(\ell_1,\ell_2,\ell_3), \quad \ell_1,\ell_2 = 0, \pm1, \pm2,\ldots \qquad (9.8)$$
$$\ell_3 = 0,1,2,\ldots$$

where a_0 is the lattice parameter. The equations of motion of the atom (ℓ_1,ℓ_2,ℓ_3) are given by[29]

$$M\ddot{u}_1(\ell_1,\ell_2,\ell_3) = \alpha[u_1(\ell_1+1, \ell_2, \ell_3) -2u_1(\ell_1,\ell_2,\ell_3) +$$
$$+ u_1(\ell_1-1,\ell_2,\ell_3)] +$$
$$+ \beta[u_1(\ell_1+1,\ell_2+1,\ell_3) + u_1(\ell_1-1,\ell_2+1,\ell_3) + u_1(\ell_1+1,\ell_2-1,\ell_3) +$$
$$+ u_1(\ell_1-1,\ell_2-1,\ell_3) + u_1(\ell_1+1,\ell_2,\ell_3+1) + u_1(\ell_1-1,\ell_2,\ell_3+1) +$$
$$+ u_1(\ell_1+1,\ell_2,\ell_3-1) + u_1(\ell_1-1,\ell_2,\ell_3-1) - 8u_1(\ell_1,\ell_2,\ell_3)] +$$
$$+ (\beta+\gamma)[u_2(\ell_1+1,\ell_2+1,\ell_3) + u_2(\ell_1-1,\ell_2-1,\ell_3) -$$
$$- u_2(\ell_1-1,\ell_2+1,\ell_3) - u_2(\ell_1+1,\ell_2-1,\ell_3) + u_3(\ell_1+1,\ell_2,\ell_3+1) +$$

$$+ u_3(\ell_1-1,\ell_2,\ell_3-1) - u_3(\ell_1+1,\ell_2,\ell_3-1) - u_3(\ell_1-1,\ell_2,\ell_3+1)]$$

$$+ 4\gamma[u_1(\ell_1,\ell_2+1,\ell_3) + u_1(\ell_1,\ell_2-1,\ell_3) + u_1(\ell_1,\ell_2,\ell_3+1) +$$

$$+ u_1(\ell_1,\ell_2,\ell_3-1) - 4u_1(\ell_1,\ell_2,\ell_3)], \tag{9.9}$$

with corresponding similar equations for the 2- and 3-components of the displacement field. Here, M is the mass of an atom, $u_\alpha(\ell_1,\ell_2,\ell_3)$ is the α-component of the displacement of the atom at $\vec{x}(\ell)$, α and β are the nearest and next nearest neighbor central force constants, respectively, and γ is an angle bending force constant. The latter characterizes the forces due to the angular stiffness of a system of three consecutive nearest neighbors that form a right angle in the equilibrium configuration.

The equations of motion (9.9) apply to atoms in all planes of the crystal, including the plane $\ell_3 = 0$. This means that they include forces acting on atoms in the plane $\ell_3 = 0$ from atoms in the plane $\ell_3 = -1$, that in fact are missing from the semi-infinite system. Equations (9.9) therefore must be supplemented by boundary conditions that correct for the presence of these forces, and at the same time ensure that the requirements of rotational invariance are satisfied, in the presence of the three-body angle-bending forces. These boundary conditions are[29]

$$\beta[u_1(\ell_1+1,\ell_2,-1) + u_1(\ell_1-1,\ell_2,-1) - 2u_1(\ell_1,\ell_2,0)] +$$

$$+ \beta[i_3(\ell_1-1,\ell_2,-1)-u_3(\ell_1+1,\ell_2,-1)]+4\gamma[u_1(\ell_1,\ell_2,-1)-u_1(\ell_1,\ell_2,0)] +$$

$$+\gamma[u_3(\ell_1-1,\ell_2,0)-u_3(\ell_1+1,\ell_2,0)+u_3(\ell_1-1,\ell_2,-1)-u_3(\ell_1+1,\ell_2,-1)] = 0 \tag{9.10a}$$

$$\beta[u_2(\ell_1,\ell_2+1,-1) + u_2(\ell_1,\ell_2-1,-1) - 2u_2(\ell_1,\ell_2,0)] +$$

$$\beta[u_3(\ell_1,\ell_2-1,-1)-u_3(\ell_1,\ell_2+1,-1)+4\gamma[u_2(\ell_1,\ell_2,-1)-u_2(\ell_1,\ell_2,0)] +$$

$$+\gamma[u_3(\ell_1,\ell_2-1,0)-u_3(\ell_1,\ell_2+1,0)+u_3(\ell_1,\ell_2-1,-1)-u_3(\ell_1,\ell_2+1,-1)] = 0 \tag{9.10b}$$

$$\alpha[u_3(\ell_1,\ell_2,-1)-u_3(\ell_1,\ell_2,0)]+\beta[u_3(\ell_1+1,\ell_2,-1)+u_3(\ell_1-1,\ell_2,-1) +$$

$$+ u_3(\ell_1,\ell_2,-1)-u_3(\ell_1,\ell_2-1,-1)-4u_3(\ell_1,\ell_2 0)] +$$

$$+\beta[u_1(\ell_1-1,\ell_2,-1)-u_1(\ell_1+1,\ell_2,-1)+u_2(\ell_1,\ell_2-1,-1)-u_2(\ell_1,\ell_2+1,-1)] +$$

$$+ 2\gamma[u_3(\ell_1+1,\ell_2,0) + u_3(\ell_1-1,\ell_2,0) + u_3(\ell_1,\ell_2+1,0) +$$

$$+ u_3(\ell_1,\ell_2-1,0) - 4u_3(\ell_1,\ell_2,0)] +$$

$$+\gamma[u_1(\ell_1+1,\ell_2,0)-u_1(\ell_1+1,\ell_2,-1)+u_1(\ell_1-1,\ell_2,-1)-u_1(\ell_1-1,\ell_2,0)] +$$

$$+\gamma[u_2(\ell_1,\ell_2+1,0)-u_2(\ell_1,\ell_2+1,-1)+u_2(\ell_1,\ell_2-1,-1)-u_2(\ell_1,\ell_2-1,0)] = 0. \tag{9.10c}$$

We now pass to the long wavelength limit with the aid of expansions like

$$u_1(\ell_1+1,\ell_2,\ell_3) = u_1(\vec{x}) + a_o \frac{\partial}{\partial x_1} u_1(\vec{x}) + \frac{1}{2} a_o^2 \frac{\partial^2}{\partial x_1^2} u_1(\vec{x}) +$$

$$+ \frac{1}{6} a_o^3 \frac{\partial^3}{\partial x_1^3} u_1(\vec{x}) + \frac{1}{24} a_o^4 \frac{\partial^4}{\partial x_1^4} u_1(\vec{x}) + \dots \tag{9.11}$$

We also divide Eq. (9.9) by the volume of a primitive unit cell, a_o^3, and introduce the mass density $\rho = M/a_o^3$. The equations of motion that are obtained are

$$\begin{aligned}
\rho\ddot{u}_1 = & \left[\frac{\alpha+4\beta}{a_o}\frac{\partial^2}{\partial x_1^2} + \frac{2(\beta+2\gamma)}{a_o}\left(\frac{\partial^2}{\partial x_2^2} + \frac{\partial^2}{\partial x_3^2}\right)\right]u_1 + \\
& + \frac{4(\beta+\gamma)}{a_o}\frac{\partial^2}{\partial x_1 \partial x_2}u_2 + \frac{4(\beta+\gamma)}{a_o}\frac{\partial^2}{\partial x_1 \partial x_3}u_3 + \\
& + \left[\frac{a_o}{12}(\alpha+4\beta)\frac{\partial^4}{\partial x_1^4} + \frac{a_o}{6}(\beta + 2\gamma)\left(\frac{\partial^4}{\partial x_2^4} + \frac{\partial^4}{\partial x_3^4}\right) + \right. \\
& \left. + a_o\beta\left(\frac{\partial^4}{\partial x_1^2 \partial x_2^2} + \frac{\partial^4}{\partial x_1^2 \partial x_3^2}\right)\right]u_1 + \frac{2}{3}a_o(\beta+\gamma)\left(\frac{\partial^4}{\partial x_1^3 \partial x_2} + \frac{\partial^4}{\partial x_1 \partial x_2^3}\right)u_2 \\
& + \frac{2}{3}a_o(\beta+\gamma)\left(\frac{\partial^4}{\partial x_1^3 \partial x_3} + \frac{\partial^4}{\partial x_1 \partial x_3^3}\right)u_3
\end{aligned} \tag{9.12a}$$

$$\begin{aligned}
\rho\ddot{u}_2 = & \frac{4(\beta+\gamma)}{a_o}\frac{\partial^2}{\partial x_2 \partial x_1}u_1 + \left[\frac{\alpha+4\beta}{a_o}\frac{\partial^2}{\partial x_2^2} + \frac{2(\beta+2\gamma)}{a_o}\left(\frac{\partial^2}{\partial x_3^2} + \frac{\partial^2}{\partial x_1^2}\right)\right]u_2 + \\
& + \frac{4(\beta+\gamma)}{a_o}\frac{\partial^2}{\partial x_2 \partial x_3}u_3 + \frac{2}{3}a_o(\beta+\gamma)\left(\frac{\partial^4}{\partial x_2^3 \partial x_1} + \frac{\partial^4}{\partial x_2 \partial x_1^3}\right)u_1 + \\
& + \left[\frac{a_o}{12}(\alpha+4\beta)\frac{\partial^4}{\partial x_2^4} + \frac{a_o}{6}(\beta+2\gamma)\left(\frac{\partial^4}{\partial x_3^4} + \frac{\partial^4}{\partial x_1^4}\right) + \right. \\
& \left. + a_o\beta\left(\frac{\partial^4}{\partial x_2^2 \partial x_3^2} + \frac{\partial^4}{\partial x_2^2 \partial x_1^2}\right)\right]u_2 + \frac{2}{3}a_o(\beta+\gamma)\left(\frac{\partial^4}{\partial x_3^3 \partial x_2} + \frac{\partial^4}{\partial x_2 \partial x_3^3}\right)u_3
\end{aligned}$$

(9.12b)

$$\rho\ddot{u}_3 = \frac{4(\beta+\gamma)}{a_o}\frac{\partial^2}{\partial x_3 \partial x_1}u_1 + \frac{4(\beta+\gamma)}{a_o}\frac{\partial^2}{\partial x_3 \partial x_2}u_2 +$$

$$+ \left[\frac{\alpha+4\beta}{a_o}\frac{\partial^2}{\partial x_3^2} + \frac{2(\beta+2\gamma)}{a_o}\left(\frac{\partial^2}{\partial x_1^2} + \frac{\partial^2}{\partial x_2^2}\right)\right]u_3 +$$

$$+ \frac{2}{3}a_o(\beta+\gamma)\left(\frac{\partial^4}{\partial x_3^3 \partial x_1} + \frac{\partial^4}{\partial x_3 \partial x_1^3}\right)u_1 +$$

$$+ \frac{2}{3}a_o(\beta+\gamma)\left(\frac{\partial^4}{\partial x_3^3 \partial x_2} + \frac{\partial^4}{\partial x_3 \partial x_2^3}\right)u_2 +$$

$$+ \left[\frac{a_o}{12}(\alpha+4\beta)\frac{\partial^4}{\partial x_3^4} + \frac{a_o}{6}(\beta+2\gamma)\left(\frac{\partial^4}{\partial x_1^4} + \frac{\partial^4}{\partial x_2^4}\right) +\right.$$

$$\left. + a_o\beta\left(\frac{\partial^4}{\partial x_3^2 \partial x_1^2} + \frac{\partial^4}{\partial x_3^2 \partial x_2^2}\right)\right]u_3 , \qquad (9.12c)$$

where we have retained terms containing fourth order spatial derivatives. The latter describe the effects of spatial dispersion. If we compare the non-spatially dispersive terms with the equations of motion of elasticity theory for a cubic medium, we can make the following identifications

$$c_{11} = \frac{\alpha+4\beta}{a_o}, \quad c_{12} = \frac{2\beta}{a_o}, \quad c_{44} = \frac{2\beta+4\gamma}{a_o} . \qquad (9.13)$$

The boundary conditions (9.10) at the surface $x_3 = 0$ take the following forms

$$\frac{2(\beta+2\gamma)}{a_o}\left(\frac{\partial u_1}{\partial x_3}+\frac{\partial u_3}{\partial x_1}\right) - \left[\beta\frac{\partial^2}{\partial x_1^2} + (\beta+2\gamma)\frac{\partial^2}{\partial x_3^2}\right]u_1 -$$

$$- 2(\beta+\gamma)\frac{\partial^2 u_3}{\partial x_1 \partial x_3} = 0 \qquad (9.14a)$$

$$\frac{2(\beta+2\gamma)}{a_o}\left(\frac{\partial u_2}{\partial x_3}+\frac{\partial u_3}{\partial x_2}\right) - \left[\beta\frac{\partial^2}{\partial x_2^2} + (\beta+2\gamma)\frac{\partial^2}{\partial x_3^2}\right]u_2 -$$

$$- 2(\beta+\gamma)\frac{\partial^2 u_3}{\partial x_2 \partial x_3} = 0 \qquad (9.14b)$$

$$\frac{2\beta}{a_o}\left(\frac{\partial u_1}{\partial x_1}+\frac{\partial u_2}{\partial x_2}\right) + \frac{\alpha+4\beta}{a_o}\frac{\partial u_3}{\partial x_3}$$

$$- 2(\beta+\gamma)\left(\frac{\partial^2 u_1}{\partial x_1 \partial x_3}+\frac{\partial^2 u_2}{\partial x_2 \partial x_3}\right) - \left[(\beta+2\gamma)\left(\frac{\partial^2}{\partial x_1^2}+\frac{\partial^2}{\partial x_2^2}\right) +\right.$$

$$\left. + \frac{1}{2}(\alpha+4\beta)\frac{\partial^2}{\partial x_3^2}\right]u_3 = 0. \qquad (9.14c)$$

where, again, we have kept the leading spatially dispersive terms.

We now turn to an application of Eqs. (9.12) and (9.14) in the study of surface acoustic waves.

If one applies the preceding results to the study of surface acoustic waves polarized in the sagittal plane, it is found that the effect of spatial dispersion is to make Rayleigh waves, which already exist in the absence of spatial dispersion, dispersive. In what follows we display a much more dramatic effect of spatial dispersion, viz. its role in binding shear horizontal surface acoustic waves that cannot exist in the absence of spatial dispersion.

For this purpose we consider a shear horizontal surface acoustic wave propagating in the [110] direction on the (001) surface of our cubic medium. (For symmetry reasons shear horizontal surface acoustic waves can propagate only in the [100] and [110] directions on the (001) surface of a cubic medium. We will see below that for the model of a cubic crystal underlying the present discussion, no shear horizontal surface acoustic wave can propagate in the [100] direction.)

A shear horizontal surface acoustic wave propagating in the [100] direction on a (001) surface of a cubic medium has a displacement field of the form

$$\vec{u}(\vec{x},t) = (A, -A, 0)e^{i\frac{k}{\sqrt{2}}(x_1 + x_2) - \alpha x_3 - i\omega t} . \quad (9.15)$$

When this expression is substituted into Eqs. (9.12), we obtain a single independent equation, viz.

$$\omega^2 = \frac{1}{a_0\rho}\left[\frac{1}{2}(\alpha + 2\beta)k^2 - 2(\beta+2\gamma)\alpha^2\right] - \frac{a_0}{\rho}\left[\frac{1}{48}(\alpha + 2\beta-12\gamma)k^4 - \frac{1}{2}\beta k^2\alpha^2 + \frac{1}{6}(\beta+2\gamma)\alpha^4\right] . \quad (9.16)$$

Similarly, the boundary conditions (9.14) yield only a single equation, viz.

$$\frac{2(\beta+2\gamma)}{a_0}\alpha = \frac{\beta}{2}k^2 - (\beta + 2\gamma)\alpha^2 . \quad (9.17)$$

It follows from Eq. (9.17) that if α vanishes as $k \to 0$, the solution possessing this property is

$$\alpha = \frac{a_0\beta}{4(\beta+2\gamma)}k^2 - \frac{a_0^3}{32}\frac{\beta^2}{(\beta+2\gamma)^2}k^4 + 0(k^6) , \quad (9.18)$$

or, using Eqs. (9.13)

$$a_o\alpha = \frac{1}{4}\frac{c_{12}}{c_{44}} a_o^2 k^2 - \frac{1}{32}\left(\frac{c_{12}}{c_{44}}\right)^2 a_o^4 k^4 + O((a_o k)^6). \tag{9.19}$$

Consequently, α is positive, as it must be for a surface acoustic wave. When this result is substituted in Eq. (9.16), and only terms of up to $O(k^4)$ are retained, we find that the frequency of the shear horizontal surface acoustic wave is obtained from

$$\omega_{SH}^2(k) = \frac{1}{2\rho}(c_{11} - c_{12})k^2 - \frac{1}{16\rho}\frac{c_{12}^2}{c_{44}} a_o^2 k^4 - $$

$$- \frac{1}{48\rho}(c_{11} + 2c_{12} - 3c_{44})a_o^2 k^4 + O(k^6). \tag{9.20}$$

We can simplify this result somewhat if we note that the squared frequency obtained on setting the decay constant $\alpha = 0$ in Eq. (9.16) is just $\omega_{110}^2(k)$, the square of the bulk transverse acoustic wave propagating in the [110] direction (in the presence of spatial dispersion):

$$\omega_{110}^2(k) = \frac{1}{2\rho}(c_{11} - c_{12})k^2 - \frac{1}{48\rho}(c_{11} + 2c_{12} - 3c_{44})a_o^2 k^4 + $$

$$+ O(k^6) \tag{9.21}$$

Thus, Eq. (9.20) can be rewritten in the form

$$\omega_{SH}^2(k) = \omega_{110}^2(k) - \frac{1}{16\rho}\frac{c_{12}^2}{c_{44}} a_o^2 k^4 + O(k^6) . \tag{9.22}$$

We can rewrite this last result as

$$\frac{\omega_{SH}(k) - \omega_{110}(k)}{\omega_{110}(k)} = -\frac{\eta}{32}\left(\frac{c_{12}}{c_{44}}\right)^2 a_o^2k^2 + O((a_ok)^4) , \qquad (9.23)$$

where

$$\eta = \frac{2c_{44}}{c_{11} - c_{12}} \qquad (9.24)$$

is the anisotropy factor for a cubic crystal. The results given by Eqs. (9.19) and (9.23) were first obtained by Alldredge[298] by a somewhat different, but equivalent, approach.

Thus we see that for the same value of k the shear horizontal surface acoustic wave lies lower in frequency than the bulk transverse wave with frequency $\omega_{110}(k)$, i.e. in a stop band for bulk waves of the same polarization, and differs in frequency from the latter by terms of $O(k^2)$. It is a deeply penetrating wave, because α is of $O(k^2)$ rather than of $O(k)$ as in the case of Rayleigh waves.

If we consider, finally, a shear horizontal wave propagating in the [100] direction on the (001) surface of a cubic medium, the displacement field is given by

$$u(\vec{x};t) = (0, A, 0)e^{ikx_1-\alpha x_3-i\omega t} . \qquad (9.25)$$

From Eqs. (9.12) and (9.14) we obtain the following pair of equations, respectively.

$$\omega^2 = \frac{2\beta+4\gamma}{a_o\rho}(k^2 - \alpha^2) - \frac{a_o}{6}(\beta+2\gamma)(k^4 + \alpha^4) \qquad (9.26a)$$

$$(\beta+2\gamma)a_o\alpha(2+a_o\alpha) = 0 . \qquad (9.26b)$$

From the second of these equations we see that either $\alpha = 0$ or $2a_o^{-1}$, neither of which choices corresponds to a surface acoustic wave.

This result, however, is a consequence of the simplicity of the lattice model on which we based the continuum theory in this section. For example, it can be shown(299) that shear horizontal surface acoustic waves can propagate in the [100] direction on the (001) surface of a body centered cubic crystal with nearest and next nearest neighbor central and noncentral interatomic forces.

Future work on the effects of spatial dispersion on surface acoustic waves will undoubtedly be based on phenomenological equations of motion and boundary conditions that are not based on specific crystal models. Nevertheless, the study of theories based on such models can provide very useful insights into the form such phenomenological theories must take.

10. Surface Acoustic Waves at a Solid-Liquid Interface

In subsection 2.1.2.1.2 we considered the Stoneley waves, polarized in the sagittal plane, that can propagate along the planar interface between two different, semi-infinite, isotropic elastic media. In the present section we study the related, but simpler, problem of the surface acoustic waves, polarized in the sagittal plane, that can propagate along the planar interface between a semi-infinite, isotropic solid and a semi-infinite, inviscid liquid. We will see that this system supports waves not possible in a semi-infinite, isotropic elastic medium alone.

We assume the elastic medium occupies the region $x_3 > 0$, while the liquid occupies the region $x_3 < 0$. In view of the isotropy of this system in the plane $x_3 = 0$, we can assume, with no loss of generality, a wave that propagates in the x_1-direction. The solutions of the equations of motion in the solid then have the form

$$\vec{u}(\vec{x},t) = (u_1(k\omega|x_3),\ 0,\ u_3(k\omega|x_3))e^{ikx_1-i\omega t}, \tag{10.1}$$

where

$$u_1(k\omega|x_3) = A_\ell e^{-\alpha_\ell(k\omega)x_3} + A_t e^{-\alpha_t(k\omega)x_3} \tag{10.2a}$$

$$u_3(k\omega|x_3) = i[\frac{\alpha_\ell(k\omega)}{k} A_\ell e^{-\alpha_\ell(k\omega)x_3} + \frac{k}{\alpha_t(k\omega)} A_t e^{-\alpha_t(k\omega)x_3}], \tag{10.2b}$$

with

$$\alpha_\ell^2(k\omega) = k^2 - \frac{\omega^2}{c_\ell^2}, \quad \alpha_t^2(k\omega) = k^2 - \frac{\omega^2}{c_t^2}. \tag{10.3}$$

The inviscid liquid will be characterized by a velocity potential $\phi(\vec{x},t)$ such that the velocity field in the liquid, $\vec{v}(\vec{x},t)$ is given by

$$\vec{v}(\vec{x},t) = \nabla\phi(\vec{x},t) . \tag{10.4}$$

The velocity potential satisfies the wave equation

$$\frac{\partial^2\phi(\vec{x},t)}{\partial t^2} = c_o^2 \nabla^2\phi(\vec{x},t), \tag{10.5}$$

where c_o is the speed of (longitudinal) sound in the liquid. The pressure $p(\vec{x},t)$ in the liquid is given in terms of the velocity potential by

$$p(\vec{x},t) = - \rho_o \frac{\partial\phi(\vec{x},t)}{\partial t}, \tag{10.6}$$

where ρ_o is the mass density of the fluid.

The solution of Eq. (10.5) that vanishes as $x_3 \to -\infty$ is

$$\phi(\vec{x},t) = \frac{\omega}{k} A_o e^{ikx_1 + \alpha_o(k\omega)x_3 - i\omega t} \tag{10.7}$$

where

$$\alpha_o^2(k\omega) = k^2 - \frac{\omega^2}{c_o^2}. \tag{10.8}$$

The factor (ω/k) on the right hand side of Eq. (10.7) has been introduced so that A_o has the same dimensions as A_1 and A_3.

The boundary conditions at the plane $x_3 = 0$ are: the continuity of the normal components of the velocity,

$$\dot{u}_3(\vec{x},t)\Big|_{x_3=0} = \frac{\partial}{\partial x_3}\phi(\vec{x},t)\Big|_{x_3=0}, \qquad (10.9)$$

and (ii) the continuity of the stresses acting on the plane $x_3 = 0$,

$$\rho c_t^2\left(\frac{\partial u_1}{\partial x_3} + \frac{\partial u_3}{\partial x_1}\right)\Big|_{x_3=0} = 0 \qquad (10.10a)$$

$$\left[\rho c_\ell^2 \frac{\partial u_3}{\partial x_3} + \rho(c_\ell^2 - 2c_t^2)\frac{\partial u_1}{\partial x_1}\right]_{x_3=0} = -p\Big|_{x_3=0}. \qquad (10.10b)$$

In writing Eqs. (10.10) we have used the result that the stress tensor for an inviscid liquid is

$$T_{\alpha\beta} = -p\delta_{\alpha\beta}. \qquad (10.11)$$

The use of the solutions (10.2) and (10.7) in the boundary conditions (10.9)-(10.10) yields the following set of homogeneous equations for the coefficients:

$$\begin{pmatrix} \dfrac{\alpha_\ell}{k} & \dfrac{k}{\alpha_t} & -\dfrac{\alpha_o}{k} \\ 1 & \dfrac{\alpha_t^2 + k^2}{2\alpha_t\alpha_\ell} & 0 \\ 1 & \dfrac{2k^2}{\alpha_t^2 + k^2} & -\dfrac{\rho_o}{\rho}\dfrac{\omega^2}{c_t^2(\alpha_t^2+k^2)} \end{pmatrix} \begin{pmatrix} A_1 \\ A_3 \\ A_o \end{pmatrix} = 0 . \qquad (10.12)$$

When we equate to zero the determinant of the coefficients on the left hand side of this equation, we obtain the dispersion relation for waves localized at a solid liquid interface:

$$4k^2\alpha_t(k\omega)\alpha_\ell(k\omega) - (\alpha_t^2(k\omega) + k^2)^2 = \frac{\rho_o}{\rho}\frac{\omega^4}{c_t^4}\frac{\alpha_\ell(k\omega)}{\alpha_o(k\omega)} \qquad (10.13)$$

The vanishing of the left hand side of this equation is the dispersion relation for Rayleigh waves at the planar, stress-free boundary of a semi-infinite, isotropic elastic medium. The term on the right hand side accounts for the presence of the liquid.

Several types of solutions of Eq. (10.13) have been discovered so far. Equation (10.13) always has a solution for which $\omega = vk$ with $v < c_o < c_t < c_\ell$. In this case $\alpha_o(k\omega)$, $\alpha_t(k\omega)$, $\alpha_\ell(k\omega)$ are all real, and the resulting wave is localized to the interface on both the liquid and solid sides. If we define

$$\frac{v}{c_t} = s, \quad \lambda = \frac{c_t}{c_\ell}, \quad \mu = \frac{c_t}{c_o}, \qquad (10.14)$$

Eq. (10.13) takes the form

$$4(1-s^2)^{1/2}(1-\lambda^2 s^2)^{1/2} - (2-s^2)^2 = \frac{\rho_o}{\rho} s^4 \frac{(1-\lambda^2 s^2)^{1/2}}{(1-\mu^2 s^2)^{1/2}} . \qquad (10.15)$$

If we assume that $\rho_o/\rho \ll 1$, and $1/\mu \ll 1$, then $s \ll 1$ also, since we seek a solution with $v < c_o$. We expand the left hand side of this equation in powers of s^2 and transpose terms to obtain

$$(1-\mu^2 s^2)^{1/2} - \frac{1}{4}\frac{3-2\lambda+\lambda^2}{1-\lambda^2}(1-\mu^2 s^2)^{1/2} s^2 + \ldots = \frac{s^2}{2}\frac{\rho_o}{\rho}\frac{(1-\lambda^2 s^2)^{1/2}}{1-\lambda^2} \qquad (10.16)$$

This equation is solved by successive approximations starting with $s = 1/\mu$, with the result that

$$v \cong c_o\left[1 - \frac{1}{8\mu^8}\left(\frac{\rho_o}{\rho}\right)^2 \frac{\mu^2 - \lambda^2}{(1-\lambda^2)^2}\right]. \qquad (10.17)$$

Thus the velocity of the interface mode is somewhat slower than the speed of sound in the liquid. The result of a numerical solution of Eq. (10.13) for v is plotted in Fig. 10.1.

The expression for $\alpha_o(k\omega)$ with the present assumptions becomes

$$\alpha_o(k\omega) \cong \frac{1}{2} k \frac{\rho_o}{\rho} \frac{(\mu^2 - \lambda^2)}{\mu^3 (1-\lambda^2)} . \tag{10.18}$$

This is small compared to k, and the wave consequently penetrates a great distance into the liquid. A calculation of the 1-component of the power flow vector W_1 (Eq. (3.15)) shows that as a result most of the power in this interface wave is carried by the liquid, not the solid.

More interesting than this wave are the leaky surface acoustic waves that can exist at the solid-liquid interface. We set $k^2 = u$ and choose the branch cuts in the complex u-plane, necessitated by the square roots in the definitions of $\alpha_\ell(k\omega)$, $\alpha_t(k\omega)$, and $\alpha_o(k\omega)$, as shown in Fig. 10.2. We then search for solutions of the equation

$$(u-c)^{1/2}\left[4u(u-b)^{1/2}(u-a)^{1/2} - (2u-b)^2\right] - \frac{\rho_o}{\rho} b^2 (u-a)^{1/2} = 0 \tag{10.19}$$

on the eight sheets of the Riemann surface on which the left hand side is single-valued. In writing Eq. (10.19) we have defined

$$a = \frac{\omega^2}{c_\ell^2}, \quad b = \frac{\omega^2}{c_t^2}, \quad c = \frac{\omega^2}{c_o^2} . \tag{10.20}$$

If we denote these sheets by

$$(\operatorname{sgn} \operatorname{Re}(u-a)^{1/2}, \ \operatorname{sgn} \operatorname{Re}(u-b)^{1/2}, \ \operatorname{sgn} \operatorname{Re}(u-c)^{1/2}), \tag{10.21}$$

then leaky surface acoustic waves are found on the sheets (++-), (+-+), (+--). The interface wave discussed above lies on the sheet (+++). From the form of Eq. (10.19) we see that no distinct additional roots are found on the remaining four sheets. If we set

$$k = k_R + ik_I \tag{10.22a}$$

with k_R, $k_I > 0$, and in turn express k_R and k_I in the forms

$$k_R = \frac{\omega}{c_1}, \; k_I = \frac{\omega}{c_2}, \tag{10.22b}$$

plots of c_1/c_t and c_2/c_t against c_t/c_o and ρ_0/ρ for each of these leaky waves are shown in Figs. 10.3, 10.4, and 10.5, respectively.

The wave on the sheet (++−) is essentially the Rayleigh wave at the solid-vacuum interface. However, because its phase velocity is greater than c_0, it can radiate energy into the liquid. Both its longitudinal and transverse components in the solid are localized to the interface, however.

The mode on the sheet (+−+) is essentially the mode studied by Glass and Maradudin(65) modified, however, by the presence of the liquid.

Solutions of Eq. (10.19) have been carried out by Ansell(64) who, however, did not present his results in the form given by Eqs. (10.22). Some of his results were corrected by Frisk et al.(300), who studied surface acoustic waves propagating circumferentially around a cylindrical solid-liquid interface and then let the radius of the cylinder become infinite.

Thus, liquid-solid interfaces offer a rich variety of waves associated with them. The leaky waves in particular await study by experimental techniques, such as Brillouin scattering, that have proved to be so useful in the study of surface acoustic waves on solid surfaces.

Acknowledgements

I am grateful to Professor A. McGurn for the numerical solution of Eqs. (2.64)and (10.19); to Dr. L. Dobrzynski for pointing out the simplification that resulted in Eq. (6.83); to Professor R. E. Camley for a prepublication copy of his work with Dr. F. Nizzoli on high frequency leaky surface acoustic waves on anisotropic surfaces; to Dr. J. E. Gubernatis for the calculation of the dispersion curves of shear horizontal surface acoustic waves on an inhomogeneous medium; and to Tsai Pyng Shen for preparing Section 7.

This work was supported in part by the National Science Foundation Grant INT-81-15141.

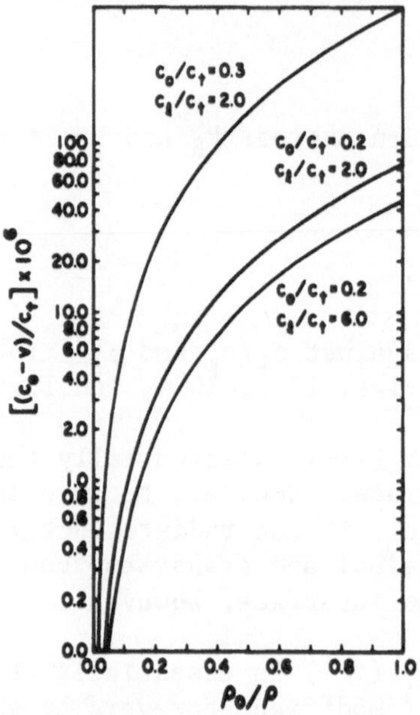

Fig. 10.1. The phase velocity v of a wave localized to the planar interface between a semi-infinite liquid and a semi-infinite isotropic elastic medium.

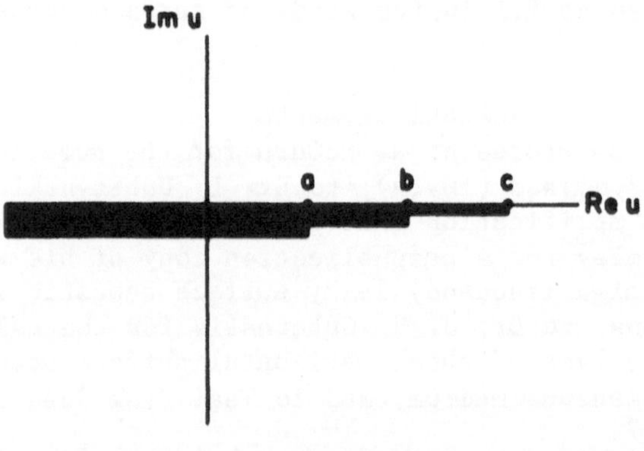

Fig. 10.2. The branch cuts in the complex $u = k^2$ plane that make the function on the left hand side of Eq. (10.19) single-valued. In this figure $a = \omega^2/c_\ell^2$, $b = \omega^2/c_t^2$, and $c = \omega^2/c_o^2$.

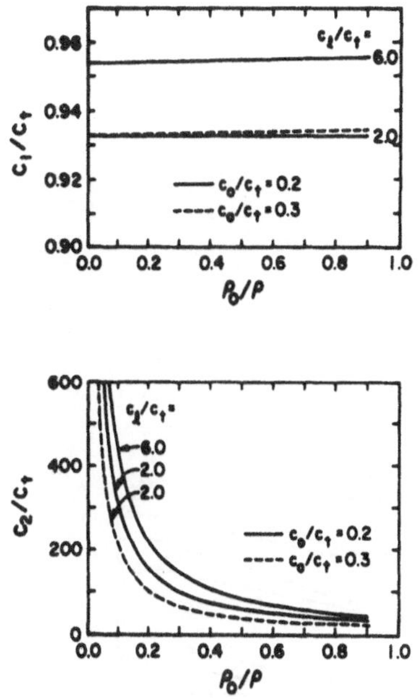

Fig. 10.3. The phase velocities c_1 and c_2 for leaky surface acoustic waves at a liquid-solid interface on the sheet (++−) of the Riemann surface on which the dispersion relation is single-valued.

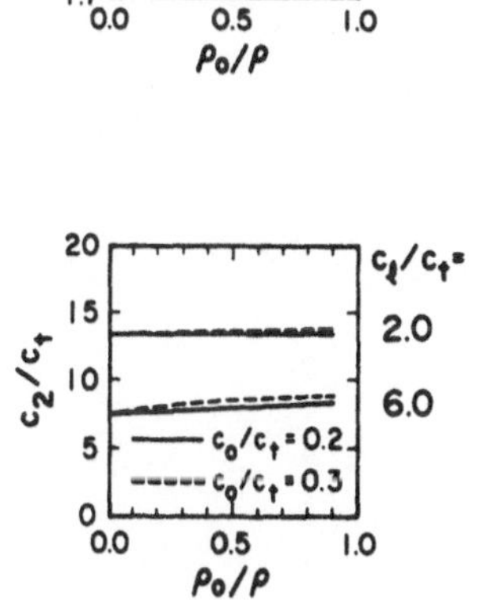

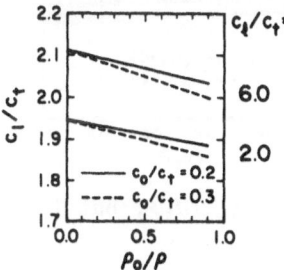

Fig. 10.4. The same as Fig. 10.3, but for the sheet (+−+).

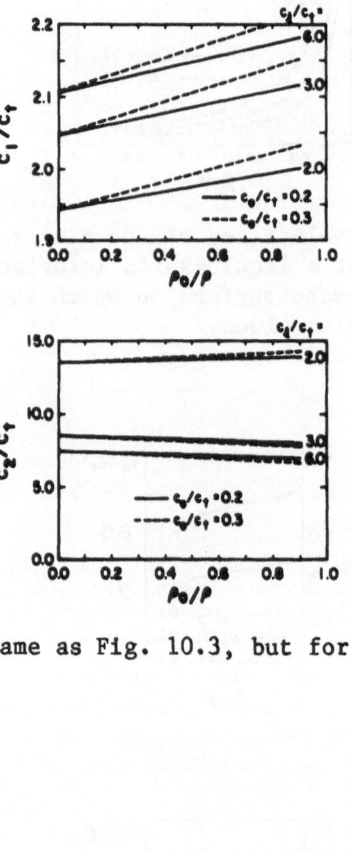

Fig. 10.5. The same as Fig. 10.3, but for the sheet (+--).

REFERENCES

1. Lord Rayleigh, Proc. Lond. Math. Soc. 17, 4 (1887).
2. F. A. Firestone and J. R. Frederick, J. Acoust. Soc. Am. 18, 200 (1946).
3. I. A. Viktorov, Rayleigh and Lamb Waves (Plenum, New York, 1967), Chap. III.
4. J. E. May, Jr., IEEE Spectrum 2, 73 (1965).
5. B. A. Auld, Acoustic Fields and Waves in Solids, vol. II (John Wiley and Sons, New York, 1973).
6. J. D. Maines and E. G. S. Paige, Proc. IEEE 64, 639 (1976).
7. A. A. Oliner, in Acoustic Surface Waves, ed. A. A. Oliner (Springer-Verlag, New York, 1978), p.1.
8. H. F. Tiersten, J. Appl. Phys. 40, 770 (1969).
9. D. L. White, IEEE Trans. Sonics and Ultrasonics, SU-15, 76 (1968).
10. R. M. White, IEEE Trans. Elect. Dev. ED-14, 181 (1967).
11. K. Yoshida and M. Yamanishi, Japan. J. Appl. Phys. 7, 1143 (1968).
12. J. H. Collins and P. J. Hagon, Electronics 42, 97 (1969); 42, 102 (1969); 43, 110 (1970).
13. A. P. van den Heuvel, Sci. Technol. 6, 52 (1969).
14. Many articles on applications of surface acoustic waves are contained in the November, 1969, special issue of the IEEE Trans. Microwave Theory Tech. on microwave acoustics.
15. The March, 1970, issue of the Microwave Journal is devoted to microwave acoustics.
16. R. M. White, Proc. IEEE 58, 1238 (1970).
17. F. Nizzoli and G. I. Stegeman, in Surface Excitations, eds. V. M. Agranovich and R. Loudon (North-Holland, Amsterdam, 1985).
18. See for example, L. D. Landau and E. M. Lifshitz, Mechanics (Pergamon Press, New York, 1969), p. 76.
19. T. C. Lim and G. W. Farnell, J. Appl. Phys. 39, 4319 (1968).
20. G. W. Farnell, In Physical Acoustics, eds. W. P. Mason and R. N. Thurston, Vol. VI, Chapter 3 (Academic Press, New York, (1970).
21. See, for example, W. Prager, Introduction to Mechanics of Continua (Ginn and Co., Boston, 1961), p. 91.
22. A. E. H. Love, A Treatise on the Mathematical Theory of Elasticity (Dover, New York, 1944), p. 294.
23. See, for example, K. H. Yen, K. F. Lau, and R. S. Kagiwada, Electronics Lett. 15, 206 (1979), and references contained therein.
24. I. A. Viktorov, Akust. Zh. 25, 1 (1979) [Soviet Physics - Acoustics 25, 1 (1979)].
25. See, for example, L. D. Landau and E. M. Lifshitz, Theory of Elasticity (Pergamon Press, New York, 1959), p. 99.
26. Ref. 5, p. 92.

27. L. Knopoff, Bull. Seis. Soc. Am. 42, 307 (1952).
28. D. Royer, J. M. Bonnet, and E. Dieulesaint, in The Mechanical Behavior of Electromagnetic Solid Continua, ed. G. A. Maugin (North-Holland, Amsterdam, 1984), p. 23.
29. D. C. Gazis, R. Herman, and R. F. Wallis, Phys. Rev. 119, 533 (1960).
30. D. A. Tursunov, Akust. Zhur. 13, 100 (1967) [Soviet Physics-Acoustics 13, 78 (1967)].
31. R. Stoneley, Proc. Roy. Soc. (London) A232, 447 (1955).
32. J. L. Synge, J. Math. Phys. 35, 323 (1957).
33. A. N. Stroh, J. Math. Phys. 41, 77 (1962).
34. P. K. Currie, Quart. J. Mech. Appl. Math. XXVII, 489 (1974).
35. R. Burridge, Quart. J. Mech. Appl. Math. XXIII, 217 (1970).
36. L. D. Landau, J. Phys. USSR 10, 25 (1946).
37. H. Engan, K. A. Ingebrigtsen, and A. Tonning, Appl. Phys. Lett. 10, 311 (1967).
38. K. A. Ingebrigtsen and A. Tonning, Phys. Rev. 184, 942 (1968).
39. F. R. Rollins, Jr., J. Acoust. Soc. Am. 44, 431 (1968).
40. F. R. Rollins, Jr., T. C. Lim, and G. W. Farnell, Appl. Phys. Lett. 12, 236 (1968).
41. F. W. Voltmer, E. I. Ippen, R. M. White, T. C. Lim, and G. W. Farnell, Proc. IEEE 56, 1634 (1968).
42. T. C. Lim and G. W. Farnell, J. Acoust. Soc. Am. 45, 845 (1969).
43. R. G. Pratt and T. C. Lim, Appl. Phys. Lett. 15, 403 (1969).
44. Ref. 20, pp. 160-163.
45. A. Takayanagi, K. Yamanouchi, and K. Shibayama, Appl. Phys. Lett. 17, 225 (1970).
46. E. Strick, Phil. Trans. Roy. Soc. (London) 251, 488 (1959).
47. R. A. Phinney, Bull. Seis. Soc. Am. 51, 527 (1961).
48. L. M. Brekhovskikh, Waves in Layered Media, Second Edition (Academic Press, New York, 1980), Chapter I, Secs. 7,8; Chapter V, Sec. 36.
49. Ref. 3, p. 48.
50. Ref. 3, p. 118.
51. Ref. 48, Secs. 9-11.
52. Ref. 48, pp. 70-74.
53. G. S. Podlyapol′skii and Y. I. Vasil′ev, Bull. Acad. Sci. USSR, Geophys. Ser. English Transl. 859 (1961).
54. D. A. Lee and D. M. Corbly, IEEE Trans. on Sonics and Ultrasonics SU-24, 206 (1977).
55. I. A. Viktorov, Akust. Zh. 4, 131 (1958) [Soviet Physics - Acoustics 4, 131 (1958)].
56. B. Rulf, J. Acoust. Soc. Am. 45, 493 (1969).
57. I. A. Viktorov, Akust. Zh. 7, 21 (1961) [Soviet Physics - Acoustics 7, 13 (1961)].
58. R. A. Phinney, J. Geophys. Res. 66, 1445 (1961).
59. F. Gilbert, Revs. Geophysics 2, 123 (1964).

60. I. A. Viktorov, Doklady Akad. Nauk SSSR 228, 579 (1976) [Soviet Physics - Doklady 21, 272 (1976)].
61. Ref. 20, p. 164.
62. L. Knopoff (private communication).
63. M. Hayes and R. Rivlin, Zeit. Angew. Math. Phys. 13, 80 (1962).
64. J. H. Ansell, Pure Appl. Geophys. 94, 172 (1972).
65. N. E. Glass and A. A. Maradudin, J. Appl. Phys. 54, 796 (1983).
66. R. E. Camley and F. Nizzoli (J. Phys. C (to appear).
67. J. R. Sandercock, Solid State Comm. 26 547 (1978).
68. R. Loudon, Phys. Rev. Lett. 40, 581 (1978).
69. A. E. H. Love, Some Problems of Geodynamics (Cambridge Univ. Press, London, 1911).
70. Ref. 5, p. 101.
71. I. M. Gel'fgat and E. S. Syrkin, Akust. Zh. 29, 19 (1983) [Soviet Physics - Acoustics 29, 10 (1983)].
72. R. Stoneley Proc. Roy. Soc. (London) A106, 416 (1924).
73. K. Sezawa and K. Kanai, Bull. Earthquake Res. Inst. Tokyo Univ. 16, 504 (1938); 17, 1 (1939).
74. L. Cagniard, Reflexion et Refraction des Ondes Seismiques Progressives (Gauthier-Villars, Paris, 1939).
75. J. G. Scholte, Roy. Astron. Soc. London, Monthly Notices Geophys. Suppl. 5, 120 (1947).
76. H. Koppe, Z. Angew. Math. Mech. 28, 355 (1948).
77. S. Yamaguchi and Y. Saito, Bull. Earthquake Res. Inst. Tokyo Univ. 33, 549 (1955).
78. A. S. Ginzbarg and E. Strick, Bull. Seis. Soc. Am. 48, 51 (1958).
79. T. W. Owen, Progr. Appl. Math. Res. 6, 69 (1964).
80. G. W. Farnell and E. L. Adler, in Physical Acoustics, eds. W. P. Mason and R. N. Thurston, Vol. IX, Chapter 2 (Academic Press, New York, 1972), pp. 72-76.
81. W. L. Pilant, Bull. Seism. Soc. Am. 62, 285 (1972).
82. Ref. 5, p. 97.
83. Ref. 5, p. 102.
84. H. Lamb, Phil. Trans. Roy. Soc. (London) A203, 1 (1904).
85. W. M. Ewing, W. S. Jardetzky, and F. Press, Elastic Waves in Layered Media (McGraw-Hill, New York, 1957), pp. 281-288.
86. Ref. 3, Chap. II.
87. Ref. 5, p. 77.
88. K. Sezawa, Bull. Earthquake Res. Inst. 3, 1 (1927).
89. I. A. Viktorov, Akust. Zh. 24, 780 (1978) [Soviet Physics - Acoustics 24, 441 (1978)].
90. Ref. 25, pp. 109-112.
91. H. F. Tiersten, J. Appl. Phys. 40, 770 (1969).
92. I. Murdoch, J. Mech. Phys. Solids 24, 137 (1976).
93. V. R. Velasco and F. Garcia-Moliner, Physica Scripta 20, 111 (1979).

94. V. R. Velasco, B. Djafari-Rouhani, L. Dobrzynski, and F. Garcia-Moliner, Proc. IVth International Conference on Solid Surfaces, Cannes, 1980, Le Vide, Les Couches Minces 201, (suppl. vol. II), 774, (1980).
95. B. Djafari-Rouhani, L. Dobrzynski, V. R. Velasco, and F. Garcia-Moliner (preprint, 1981).
96. J. E. Gubernatis and A. A. Maradudin (unpublished work).
97. J. M. Richardson, J. Appl. Phys. 48, 498 (1977).
98. C. C. Tseng and R. M. White, J. Appl. Phys. 38, 4274 (1967).
99. C. C. Tseng, J. Appl. Phys. 38, 4281 (1967).
100. J. J. Campbell and W. R. Jones, IEEE Trans. Sonics and Ultrasonics SU-15, 209, (1968).
101. M. I. Kaganov and I. L. Sklovskaya, Fiz. Tver. Tela 8, 3840 (1966) [Soviet Physics - Solid State 8, 2789 (1967)].
102. M. I. Kaganov (private communication).
103. J. L. Bleustein, Appl. Phys. Lett. 13, 412 (1968).
104. Yu. V. Gulyaev, Zh. Eksper i Teor. Fiz. Pis. v Red. 9, 63 (1969) [Soviet Physics - JETP Letters 9, 37 (1969)].
105. R. G. Curtis and M. Redwood, J. Appl. Phys. 46, 2406 (1975).
106. C. Maerfeld, F. Gires, and P. Tournois, Appl. Phys. Lett. 18, 269.
107. G. Koerber and R. F. Vogel, IEEE Trans. Sonics Ultrasonics SU-19, 3 (1972).
108. C. C. Tseng, Appl. Phys. Lett. 16, 253 (1970).
109. G. G. Kessenikh, V. N. Lyubimov, and D. G. Sannikov, Kristallografia 17, 591 (1972) [Soviet Physics - Crystallography 17, 512 (1972)].
110. G. G. Kessenikh and L. A. Shuvalov, Kristallografia 21, 5 (1976) [Soviet Physics - Crystallography 21, 1 (1976)].
111. (a) R. Q. Scott and D. L. Mills, Solid State Comm. 18, 849 (1976), (b) Phys. Rev. B15, 3545 (1977).
112. R. E. Camley and R. Q. Scott, Phys. Rev. B17, 4327 (1978).
113. A. A. Maradudin, Surface Waves, in Modern Problems of Surface Physics, ed. I. J. Lalov (Bulgarian Academy of Sciences, Sofia, 1981), p. 11.
114. A. A. Maradudin, Surface Waves, in Festkorperprobleme, Volume XXI, ed. J. Treusch (Vieweg, Braunschweig, 1981), p. 25.
115. C. Kittel, Phys. Rev. 110, 836 (1958).
116. L. D. Landau and E.M. Lifshitz, Phys. Z. Sowjetunion 8, 153 (1935).
117. C. Kittel, Rev. Mod. Phys. 21, 541 (1949).
118. C. Kittel, Quantum Theory of Solids (J. Wiley, New York, 1963), Chapter 4.
119. M. Matthews and H. van de Vaart, Appl. Phys. Lett. 15, 373 (1969).
120. J. P. Parekh, Electron. Lett. 5, 323 (1969).
121. R. E. Camley and A. A. Maradudin, Phys. Rev. B24, 1255 (1981).

122. R. E. Camley and R. Q. Scott, Phys. Rev. B17, 4327 (1978).
123. R. E. Camley, J. Appl. Phys. 50, 5272 (1979).
124. R. F. Wallis, D. L. Mills, and A. A. Maradudin, Phys. Rev. B19, 3981 (1979).
125. See, for example, R. Normandin, M. Fukui, and G. I. Stegeman, J. Appl. Phys. 50, 81 (1979).
126. K. Portz, G. I. Stegeman, and A. A. Maradudin, Appl. Phys. Lett. 38, 856 (1981).
127. Ref. 3, section 1.1.
128. E. G. Cook and H. E. Valkenburg, Am. Soc. Testing Mater. Bull. 198, 81 (1954).
129. C. Minton, Nondestructive Testing 12, 13 (1954).
130. D. S. Shraiber, Ultrasonic Flaw Detection, in Flaw Detection in Metals, (Gos. Izd. Oboron. Prom., Moscow, 1959), p. 241.
131. L. C. Lynnworth, IEEE Trans. Sonics and Ultrasonics SU-12, 23 (1965).
132. R. M. White and F. W. Voltmer, Appl. Phys. Lett. 7, 314 (1965).
133. R. M. White and F. W. Voltmer, Appl. Phys. Lett. 8, 40 (1966).
134. D. B. Armstrong, IEEE Trans. Sonics and Ultrasonics SU-16, 20 (1969).
135. G. A. Coquin and H. F. Tiersten, J. Acoust. Soc. Am. 41, 921 (1967).
136. C. C. Tseng, IEEE Trans. Electron Devices ED-15, 586 (1968).
137. S. G. Joshi and R. M. White, J. Acoust. Soc. Am. 47, 17 (1969).
138. P. R. Emtage, J. Acoust. Soc. Am. 51, 1142 (1972).
139. W. S. Goruk and G. I. Stegeman, Phys. Lett. 51A, 419 (1975).
140. S. Mishra, G. D. Holah, and R. Bray, Proc. 3rd Int. Conf. Light Scattering in Solids (Flammarion, Paris, 1975), p. 198.
141. K. Dransfeld and E. Salzmann, in Physical Acoustics, eds. W. P. Mason and R. N. Thurston, vol. VII, Chapter 4 (Academic Press, New York, 1970).
142. A. F. Bessonov, L. N. Deryugin, and V. A. Komotskii, Opt. Spectrosk. 49, 382 (1980) [Optics and Spectroscopy 49 207 (1980).
143. V. A. Komotskii and T. D. Black, J. Appl. Phys. 52, 129 (1981).
144. M. Schmidt and K. Dransfeld (preprint, 1981).
145. D. L. Mills, A. A. Maradudin, and E. Burstein, Phys. Rev. Lett. 21, 1178 (1968).
146. D. L. Mills, A. A. Maradudin, and E. Burstein, in Light Scattering Spectra of Solids, ed. G. B. Wright (Springer-Verlag, New York, 1969), p. 399.
147. D. L. Mills, A. A. Maradudin, and E. Burstein, Ann. Phys. (N.Y.) 56, 504 (1970).

148. B. I. Bennett, A. A. Maradudin, and L. R. Swanson, in Proc. 2nd Int. Conf. on Light Scattering in Solids, ed. M. Balkanski (Flammarion, Paris, 1971), p. 443.
149. B. I. Bennett, A. A. Maradudin, and L. R. Swanson, Ann. Phys. (N.Y.) 71, 357 (1972).
150. J. R. Sandercock, Phys. Rev. Lett. 28, 237 (1972).
151. J. R. Sandercock, Phys. Rev. Lett. 29, 1735 (1972).
152. G. Dresselhaus and A. S. Pine, Solid State Commun. 16, 1001 (1975).
153. A. Dervisch and R. Loudon, J. Phys. C9, L669 (1976).
154. J. Dil and E. Brody, Phys. Rev. B14, 5218 (1976).
155. R. Loudon, J. Phys. C11, 403 (1978).
156. K. R. Subbaswamy and A. A. Maradudin, Indian J. Pure Appl. Phys. 16, 282 (1978).
157. A. Dervisch and R. Loudon, J. Phys. C11, L291 (1978).
158. R. Loudon, J. Phys. C11, 2623 (1978).
159. N. L. Rowell and G. I. Stegeman, Phys. Rev. Lett. 41, 970 (1978).
160. N. L. Rowell and G. I. Stegeman, Phys. Rev. B18, 2598 (1978).
161. K. R. Subbaswamy and A. A. Maradudin, Phys. Rev. B18, 4181 (1978).
162. V. Bortolani, F. Nizzoli, and G. Santoro, Phys. Rev. Lett. 41, 39 (1978).
163. V. Bortolani, F. Nizzoli, and G. Santoro, J. Phys. F8, L215 (1978).
164. V. Bortolani, F. Nizzoli, and G. Santoro, Phys. Lett. 68A, 342 (1978).
165. V. Bortolani, F. Nizzoli, G. Santoro, and E. Tosatti, Solid State Comm. 26, 507 (1978).
166. A. M. Marvin, V. Bortolani, F. Nizzoli, and G. Santoro, J. Phys. C13, 299 (1980).
167. A. M. Marvin, V. Bortolani, F. Nizzoli, and G. Santoro, J. Phys. C13, 1607 (1980).
168. E. L. Albuquerque, R. Loudon, and D. R. Tilley, J. Phys. C13, 1775 (1980).
169. J. Sandercock (private communication).
170. J. G. Dil, N. C. J. A. van Hijningen, F. van Dorst, and R. M. Aarts, Appl. Optics 20, 1374 (1981).
171. V. Bortolani, F. Nizzoli, G. Santoro, A. Marvin, and J. R. Sandercock, Phys. Rev. Lett. 43, 224 (1979).
172. J. R. Sandercock, F. Nizzoli, V. Bortolani, G. Santoro, and A. M. Marvin, Proc. IVth International Conference on Solid Surfaces, Cannes, 1980, Le Vide, Les Couches Minces 201 (suppl. vol. II), 754 (1980).
173. R. T. Harley and P. A. Fleury, J. Phys. C12, L863 (1979).
174. J. R. Sandercock, Proc. 7th Int. Conf. on Raman Spectroscopy, ed. W. F. Murphy (Amsterdam, North-Holland, 1980), p. 364.

175. L. Bjerkan and K. Fossheim, Solid State Comm. 21, 1147 (1977).
176. G. Gorodetsky and S. Shaft, Phys. Rev. B23, 6755 (1981).
177. E. Kratzig, Solid State Comm. 9, 1205 (1971).
178. J. Poliquen, M. Depoorter, and A. Defebvre, Proc. IVth International Conference on Solid Surfaces, Cannes, 1980, Le Vide, Le Couches Minces, 201, 201 (suppl. vol. I), 116 (1980).
179. M. I. Borisov and L. L. Konstantinov, Surface Acoustic Waves in Solid Surface Investigations, in Modern Problems of Surface Physics, ed. I. J. Lalov (Bulgarian Academy of Sciences, Sofia, 1981), p. 400.
180. Lord Rayleigh, The Theory of Sound vol. II, 2nd Ed. (Dover, New York, 1945), Section 287.
181. Reference 85 contains a good bibliography of work on this subject.
182. I. S. Sokolnikoff, Mathematical Theory of Elasticity (McGraw-Hill, New York, 1946), p. 202.
183. M. Abramowitz and I. Stegun, Handbook of Mathematical Functions, Fifth Printing (Dover, New York, 1968), p. 359.
184. Ref. 183, p. 367.
185. Ref. 183, p. 446.
186. Ref. 183, p. 478.
187. A. I. Lanin, Abstracts of Papers for the Third All-Union Symposium on Wave Diffraction (in Russian) (Nauka, Moscow, 1964), pp. 108-109.
188. V. S. Buldyrev and A. I. Lanin, in Numerical Methods of Solving Problems in Mathematical Physics (in Russian) (Nauka, Moscow, 1966), pp. 131-143.
189. V. S. Buldyrev, Vestn. Leningrad Univ. Ser. Fiz. Khim. 22, 38 (1965).
190. L. M. Brekhovskikh, Akust. Zh. 13, 541 (1967) [Soviet Physics - Acoustics 13, 462 (1968)].
191. W. Franz, Z. fur Naturforschung 9a, 705 (1954).
192. L. M. Brekhovskikh, Akust. Zh. 5, 288 (1959) [Soviet Physics - Acoustics 5, 288 (1960)].
193. P. V. H. Sabine, Electron. Lett. 6, 149 (1970).
194. F. Rischbieter, Acustica 16, 75 (1965).
195. A. Ronnekleiv and J. Souquet, IEEE Trans. Sonics and Ultrasonics SU-23, 188 (1976).
196. S. Datta and B. J. Hunsinger, IEEE Trans. Sonics and Ultrasonics SU-27, 333 (1980).
197. H. S. Tuan and J. P. Parekh, IEEE Trans. Sonics and Ultrasonics SU-24, 384 (1977).
198. H. A. Haus, IEEE Trans. Sonics and Ultrasonics SU-24, 259 (1977).
199. S. R. Seshadri, J. Acoust. Soc. Am. 65, 687 (1979).
200. A. D. Lapin, Akust. Zh. 25, 766 (1979) [Soviet Physics - Acoustics 25, 432 (1979)].

201. A. D. Lapin, Akust. Zh. 26, 104 (1980) [Soviet Physics - Acoustics 26, 55 (1980)].
202. V. I. Grigor´evskii and V. P. Plesskii, Pis´ma Zh. Tekh. Fiz. 5, 1398 (1979) [Sov. Tech. Phys. Lett. 5, 588 (1979)].
203. R. F. Humphryes and E. A. Ash, Electron. Lett. 5, 175 (1969).
204. I. D. Akhromeeva and V. V. Krylov, Akust. Zh. 23, 510 (1977) [Soviet Physics - Acoustics 23, 292 (1977).
205. Yu. V. Gulyaev, T. N. Kurach, and V. P. Plesskii, Pisma Zh. Tekh. Fiz. 5, 272 (1979) [Sov. Tech. Phys. Lett. 5, 111 (1979)].
206. Yu. V. Gulyaev, T. N. Kurach, and V. P. Plesskii, Zh. Ekspr. i Teor. Fiz. Pis. v Red. 29, 563 (1979) [Soviet Physics - JETP Letters 29, 513 (1979).
207. Yu. V. Gulyaev, T. N. Kurach, and V. P. Plesskii, Akust. Zh. 26, 540 (1980) [Soviet Physics - Acoustics 26, 296 (1980)].
208. Yu. V. Gulyaev and V. P. Plesskii, Fiz. Tver. Tela 21, 3479 (1979) [Soviet Physics - Solid State 21, 2010 (1979)].
209 A. N. Avdeev and V. P. Plesskii, Akust. Zh. 28, 289 (1982) [Soviet Physics - Acoustics 28, 173 (1982).
210. N. E. Glass, R. Loudon, and A. A. Maradudin, Phys. Rev. B24, 6843 (1981).
211. Lord Rayleigh, Philos. Mag. 14, 70 (1907); Ref. 180, p. 89.
212. The Rayleigh hypothesis is known to be valid for the scattering of a scalar plane wave from a sinusoidally corrugated hard wall defined by $x_3 = \zeta_0 \cos(2\pi x_1/a)$, provided $\zeta_0/a < 0.071$. See Refs. 213-215.
213. R. Petit and C. R. Cadilhac, Acad. Sc. B262, 468 (1966).
214. R. F. Millar, Proc. Camb. Philos. Soc. 69, 175, 217 (1971).
215. N. R. Hill and V. Celli, Phys. Rev. B17, 2478 (1978).
216. R. F. Millar, Proc. Camb. Philos. Soc. 65, 773 (1969).
217. B. A. Auld, J. J. Gagnepain, and M. Tan, Electronics Lett. 12, 650 (1976).
218. Yu. V. Gulyaev and V. P. Plesskii, Zh. Tekh. Fiz. 48, 447 (1978) [Soviet Physics - Technical Physics 23, 266 (1978)].
219. N. E. Glass and A. A. Maradudin, Electronics Lett. 17, 773 (1981).
220. E. I. Urazakov and L. A. Fal´kovskii, Zh. Ekspr. i Teor. Fiz. 63, 2297 (1972) [Soviet Physics - JETP 36, 1214 (1973)].
221. A. A. Maradudin and D. L. Mills, Ann. Phys. (N.Y.) 100, 262 (1976).
222. A. G. Eguiluz and A. A. Maradudin, Phys. Rev. B28, 711 (1983).
223. A. G. Eguiluz and A. A. Maradudin, Phys. Rev. B28, 728 (1983).
224. O. Hardouin Duparc and A. A. Maradudin, J. Electron Spectr. and Rel. Phenom. 30, 145 (1983).

225. A. A. Bulgakov and S. I. Khankina, Solid State Comm. 44, 55 (1982).
226. For example, J. C. de Bremaecker, Geophysics 23, 253 (1958).
227. L. Knopoff and A. F. Gangi, Geophysics 25, 1203 (1960).
228. E. R. Lapwood, Geophys. J. Roy. Astron. Soc. 4, 174 (1961).
229. J. Kane and J. Spence, Geophysics 28, 715 (1963).
230. J. A. Hudson and L. Knopoff, J. Geophys. Res. 69 281 (1964).
231. K. Viswanathan, J. T. Kuo, and E. R. Lapwood, J. Geophys. R. Astron. Soc. 24, 401 (1971).
232. M. Munasinghe and G. W. Farnell, J. Geophys. Res. 78, 2454 (1973).
233. I. A. Viktorov, Dokl.Akad. Nauk U.S.S.R. 119, 463 (1958) [Soviet Physics - Doklady 3, 304 (1958)].
234. Ref. 3, pp. 42-46.
235. R. C. M. Li, IEEE Trans. Sonics and Ultrasonics SU-20 63 (1973).
236. M. T. Wauk and R. L. Zimmerman, Electronics Lett. 8, 439 (1972).
237. L. E. Alsop and A. S. Goodman, IBM J. Res. Dev. 16, 365 (1972).
238. M. Munasinghe and G. W. Farnell, IEEE Trans. Sonics and Ultrasonics SU-20, 63 (1973).
239. M. T. Wauk, IEEE Trans. Sonics and Ultrasonics SU-21, 78 (1974).
240. M. Munasinghe and G. W. Farnell, J. Appl. Phys. 44, 2025 (1973).
241. M. Munasinghe and G. W. Farnell, IEEE Trans. Sonics and Ultrasonics SU-21, 76 (1974).
242. W. H. Haydl, Appl. Phys. Lett. 22, 284 (1973); IEEE Trans. Sonics and Ultrasonics SU-21, 78 (1974).
243. W. S. Goruk and G. I. Stegeman, Phys. Lett. 51A, 419 (1973).
244. T. Yoneyama and S. Nishida, J. Acoust. Soc. Am. 55, 738 (1974).
245. H. S. Tuan and R. C. M. Li, J. Acoust. Soc. Am. 55, 1212 (1974).
246. E. Cambiaggio and F. Cuozzo, IEEE Trans. Sonics and Ultrasonics SU-23, 189 (1976).
247. O. W. Otto, IEEE Trans. Sonics and Ultrasonics SU-22, 251 (1975).
248. O. W. Otto, Appl. Phys. Lett. 20, 215 (1975).
249. M. Munasinghe, Ultrasonics 14, 9 (1976).
250. E. Cambiaggio, F. Cuozzo, and E. Rivier, Appl. Phys. Lett. 28, 71 (1976).
251. E. Cambiaggio, F. Cuozzo, and E. Rivier, IEEE Trans. Sonics and Ultrasonics SU-24, 133 (1977).
252. F. Cuozzo, E. Cambiaggio, J. P. Damiano, and E. Rivier, IEEE Trans. Sonics and Ultrasonics, SU-24, 280 (1977).
253. O. W. Otto, J. Appl. Phys. 48, 5105 (1977).

254. E. Cambiaggio, F. Azan, and A. Lantz, IEEE Trans. Sonics and Ultrasonics SU-26, 176 (1979).
255. W. S. Goruk and G. I. Stegeman, Appl. Phys. Lett. 32, 265 (1978).
256. E. Cambiaggio and F. Cuozo, IEEE Trans. Sonics and Ultrasonics SU-26, 340 (1979).
257. F. Cuozzo and E. Cambiaggio, IEEE Trans. Sonics and Ultrasonics SU-27, 141 (1980).
258. W. S. Goruk and G. I. Stegeman, J. Appl. Phys. 50, 6719 (1979).
259. S. Datta and B. J. Hunsinger, IEEE Trans. Sonics and Ultrasonics SU-27, 141 (1980).
260. S. Datta and B. J. Hunsinger, J. Appl. Phys. 50, 3370 (1979).
261. A. D. Lapin, Akust. Zh. 25, 766 (1979) [Soviet Physics - Acoustics 25, 432 (1979).
262. S. Datta and B. J. Hunsinger, IEEE Trans. Sonics and Ultrasonics SU-27, 333 (1980).
263. J. Temmyo, Y. Sakakibara, K. Komatsu, M. Oda, and S. Yoshikawa, IEEE Trans. Sonics and Ultrasonics SU-27, 383 (1980).
264. Yu. V. Gulyaev, V. I. Grigor´evskii, and V. P. Plesski, Zh. Tekh. Fiz. 51, 1338 (1981) [Soviet Physics - Technical Physics 26, 768 (1981)].
265. T. P. Shen, A. A. Maradudin, R. F. Wallis, and G. I. Stegeman (unpublished work).
266. L. E. Alsop, J. Geophys. Res. 71, 3969 (1966).
267. A. McGarr and L. E. Alsop, J. Geophys. Res. 72, 2169 (1967).
268. S. Gregersen and L. E. Alsop, Bull. Seism. Soc. Am. 64, 535 (1974).
269. P. Malischewsky, Pure and Appl. Geophys. 117, 1045 (1979).
270. T. C. Chen and L. E. Alsop, Bull. Seism. Soc. Am. 69, 1409 (1979).
271. A. I. Lutikov, Izv. Earth Phys. 14, 250 (1978).
272. P. Malischewsky, Izv. Akad. Nauk SSSR, Fiz. Zemlyi 11, 87 (1980).
273. D. H. Norrie and G. de Vries, The Finite Element Method (Academic Press, New York, 1973).
274. J. B. Davies, IEEE Trans. Microwave Theory and Techniques MTT-21, 99 (1973).
275. W. M. Visscher, Wave Motion 3, 49 (1981).
276. Ref. 3, p. 69.
277. G. I. Stegeman, A. A. Maradudin, and T. S. Rahman, Phys. Rev. B23, 2576 (1981).
278. G. A. Northrup and J. P. Wolfe, Phys. Rev. B22, 6196 (1980).
279. J. C. Hensel and R. C. Dynes, Phys. Rev. Lett. 43, 1424 (1979).
280. P. Taborek and D. Goodstein, Solid State Comm. 33, 1191 (1980).

281. B. Taylor, H. J. Maris, and C. Elbaum, Phys. Rev. Lett. 23, 416 (1969).
282. B. Taylor, H. J. Maris, and C. Elbaum, Phys. Rev. B3, 1462 (1971).
283. V. T. Buchwald, Proc. Roy. Soc. (London) A253, 563 (1959).
284. V. T. Buchwald, Quart. J. Mech. Appl. Math. 14, 293 (1960).
285. V. T. Buchwald and A. Davis, Quart. J. Mech. Appl. Math. 16, 283 (1963).
286. H. Shirasaki and T. Makimoto, J. Appl. Phys. 49, 658, 651 (1978); 50, 2795 (1979).
287. I. M. Mason and E. A. Ash, J. Appl. Phys. 42, 5343 (1971).
288. M. S. Kharusi and G. W. Farnell, IEEE Trans. Sonics and Ultrasonics SU-18, 34 (1971).
289. T. L. Szabo and A. J. Slobodnick, Jr., IEEE Trans. Sonics and Ultrasonics SU-20, 240 (1973).
290. R. E. Camley and A. A. Maradudin, Phys. Rev. B27, 1959 (1983).
291. This method has been used for an infinite medium in Refs. 280 and 283. For its use in the case of a semi-infinite medium, see Refs. 284 and 286.
292. Explicit determinations of the elements of the tensor $D_{\alpha\beta}(\vec{x},\vec{x}'|\omega)$ for the cases of semi-infinite isotropic and hexagonal media have been given by A. A. Maradudin and D. L. Mills, Ann. Phys. (N.Y.) 100, 262 (1976) and by L. Dobrzynski and A. A. Maradudin, Phys. Rev. B14, 2200 (1976), respectively.
293. M. J. Lighthill, *Introduction to Fourier Analysis and Generalized Functions* (Cambridge University Press, Cambridge, 1958), Chap. IV.
294. G. F. Koster, Phys. Rev. 95, 1436 (1954); A. A. Maradudin, in *Phonons and Phonon Interactions*, ed. T. A. Bak (Benjamin, New York, 1964), p. 424. See, in particular, Appendix D.
295. The methods used here are similar to those of Refs. 29, 19, and 42.
296. See, for example, M. Born and K. Huang, *Dynamical Theory of Crystal Lattices* (Clarendon Press, Oxford, 1954), pp. 336-337.
297. W. Ludwig and B. Lengeler, Solid State Comm. 2, 83 (1964).
298. G. P. Alldredge, Phys. Lett. 41A, 281 (1972).
299. V. L. Zoth, M. A. Thesis, University of Texas at Austin, 1973 (unpublished).
300. G. V. Frisk, J. W. Dickey, and H. Uberall, J. Acoust. Soc. Am. 58, 996 (1975).

VIBRATIONAL ENERGY EXCHANGE BETWEEN GASES AND SOLIDS

Giorgio Benedek

Gruppo Nazionale di Struttura della Materia del C.N.R.
Dipartimento di Fisica dell'Università
Via Celoria 16, I-20133 Milano, Italy

INTRODUCTION

The interest in the subject of gas-surface collisions and energy transfer between gases and solids has much increased during the last two decades in connection with aerospace research. Under rarified-gas conditions, where the mean free path of gas molecules is much larger than any characteristic surface length scale, the gas-surface interaction is no longer governed by boundary layer or bulk flow effects, but depends strictly on individual collisions between the gas molecules and the solid surface.[1] More recently, the inelastic scattering of atoms from solid surfaces has become a powerful tool for surface phonon spectroscopy. A by-product of such spectroscopic studies is the direct measurement of the energy exchanges with individual modes of the vibrational spectrum under selected kinematic and temperature conditions.

As long as the overall energy transfer is concerned, a useful concept is that of energy accommodation coefficient, which represents the thermalization efficiency of gas-surface collisions. The first section of this lecture is devoted to the definition and experimental determination of the energy accommodation coefficient. A short description of the Knudsen cell method[2] is given, with an illustration of the existing phenomenology, particularly the experimental data obtained for different rare gases in contact with a metal surface. These data have received a satisfactory interpretation in the framework of the classical theory in the intermediate and high temperature regimes. The classical theory works well also at low temperatures for heavy rare-gas atoms, whereas significant deviations are noted for He below 50 K. A quantum treatment is needed for the scattering of light atoms at low energy. The second section of the lecture deals with the quantum theory of energy accommodation. The differential reflection coefficient for inelastic atom scattering is derived in a general form in terms of atom-phonon coupling forces. The coupling forces are then expressed by means of the distorted wave Born approximation (DWBA), and, for sake of comparison, in the framework of a hard-corrugated surface model. Advantages and limitations of the DWBA are illustrated in connection with the experimental behaviour of the accommodation coefficient in the low temperature limit. Many early and recent works prove that the attractive molecule-surface potential plays an important role in giving a nonvanishing accommodation coefficient at low temperature ($T\to 0$). It appears, however, that the low-temperature behaviour also depends upon the experimental procedure, particularly on whether equilibrium or non-equilibrium methods are used.

Many classical and quantum theories of the accommodation coefficient have reached a comparatively high level of sophistication as regards the scattering theory and the interaction potential, but not always the specific features of surface dynamics have been properly considered. The nature and classification of surface phonons will be introduced by Prof. Maradudin in his lecture. The theoretical predictions of inelastic scattering intensities based on realistic surface dynamical models can now be tested by means of high-resolution nozzle beam scattering experiments.

ENERGY ACCOMMODATION

The concept of accommodation coefficient was introduced by Knudsen[2] as a quantitative measure of the thermalization efficiency of collisions between the gas molecules and a solid surface. In a nonequilibrium condition, we define the accommodation coefficient as the ratio of the average energy transfer in a single collision to the average energy transfer required to bring a gas molecule to thermal equilibrium with the surface[3,4]:

$$\alpha(T_g,T_s) = \frac{\bar{E}_f-\bar{E}_g}{\bar{E}_s-\bar{E}_g} \tag{1}$$

Here $\bar{E}_g$ is the average energy of a gas molecule at temperature T_g; $\bar{E}_f$ its average energy after a single collision and $\bar{E}_s$ the average energy of the gas molecule at a temperature T_s equal to the (constant) temperature of the solid surface. For a monoatomic gas we use the relations[3]

$$\bar{E}_g = 2kT_g \tag{2}$$

$$\bar{E}_s = 2kT_s \; . \tag{2'}$$

Since for $T_s \to T_g$ the average energy exchange must vanish, it makes sense to define the equilibrium accommodation coefficient as

$$\alpha(T) = \lim_{T_g \to T_s = T} \alpha(T_g, T_s) \tag{3}$$

The equilibrium accommodation coefficient, which depends on a single temperature, contains most of the interesting physics involved in the gas-solid energy exchange. The experimental studies on the energy accommodation of rare gases on tungsten[5-10] have shown that the accommodation coefficient depends essentially on the gas temperature, whereas the surface temperature has only a weak influence on $\alpha(T_g, T_s)$, particularly for light gases (He, Ne).[8-10] Thus, many theoretical investigations have considered indifferently either $\alpha(T) = \alpha(T_g, T_g)$ or $\alpha(T_g, 0)$[4].

The experimental determination of $\alpha(T_g, T_s)$ is based on the direct measure of the net energy transferred in an average collision between a gas molecule and the surface. This can be achieved by means of the Knudsen cell, consisting in a cylindrical glass tube of radius R, where the gas is held at a sufficiently low pressure p. A thin electrically conducting filament of radius r made from the material (e.g. tungsten) whose surface has to be investigated , lies along the cylinder axis. The walls of the tube are in contact with thermostat at temperature T_g. The filament temperature T_s is raised and held above the gas temperature by means of an electric current. The power loss per unit filament area to the gas molecules is directly measurable in terms of the filament current and change in resistance. In order to keep the gas at a uniform temperature T_g equal to the wall temperature the filament surface has to be very small, i.e. $r \ll R$, and the gas pressure low enough to give a mean free path much larger than the linear dimensions of the cell. In this way, the excess energy which the molecules carry from the filament is given to the thermostat, without affecting the average energy of the incident stream. Typical parameters of the Knudsen cells used in low-pressure experiments are $R = 3$ cm, $r = 2 \times 10^{-3}$ cm and $p < 0.05$ mm_{Hg}.[3]

The experimental data[5-9] on the equilibrium accomodation coefficient as function of T for He, Ne, Ar Kr and Xe on tungsten are shown in fig.1. For light atoms, $\alpha(T)$ decreases rapidly at low temperature to a minimum, then shows a slow increase. The minimum occurs at about 40 K for He, $\sim$250 K for Ne and $\sim$600 for Ar. No

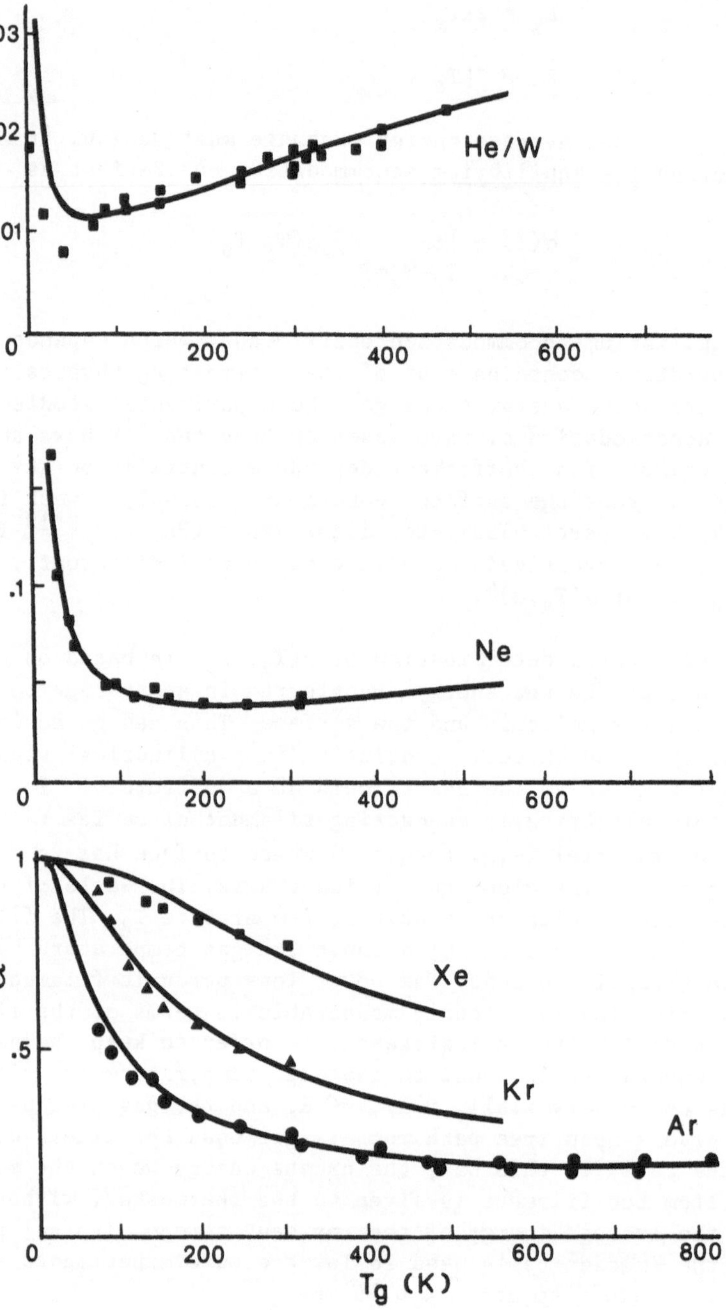

Fig. 1. Experimental energy accommodation coefficient as function of the gas temperature for rare gases on tungsten (from Thomas and Roach, refs.5,6; Wachman and Goodman, refs.7 and 9; Kouptsidis and Menzel, ref.8) and calculated curves from a classical lattice theory (Goodman, ref.17).

minima are found for Kr and Xe in the investigated range, but a linear extrapolation of T_{min} as function of the atomic weight suggests possible minima at $\sim$1500 K and $\sim$2500 K, respectively. On the other hand, $\alpha(T)$ for Kr and Xe reaches saturation (α=1), i.e., a perfect accommodation efficiency below a certain temperature; saturation is not found for light atoms, but according to Goodman theoretical arguments[3], is likely to occur also for Ar and Ne at sufficiently low temperature.

Recent investigations by Sinvani, Cole and Goodstein[10] on the sticking of He on solid surfaces below 4 K give a sticking probability between 2/3 and 1. Similar figures are expected for the accommodation coefficient.

A Phenomenological Model

The apparently universal behavior of the accommodation coefficient as function of T descends from the existence of two regimes, according to whether the incident energy E_g is smaller or larger than the depth D of the surface attractive well. We consider first the case $E_g \ll D$: the kinetic energy of the gas atom during the interaction is $\sim$D. Thus the energy loss ε does not depend on E_g and is approximately constant. Setting T_s=0, $\alpha(T_g,0)$ is of the order ε/E_g and decreases as T_g^{-1}. However, for an incident energy smaller than the energy loss, $E_g \leq \varepsilon$, the atom is trapped in the attractive well (sticking), and $\alpha(T_g,0)$ saturates to unity. In the high energy case, $E_g \gg D$, the energy-loss is proportional to the energy itself and $\alpha(T_g,0)$ tends to a finite limit $\alpha(\infty,0)$. Furthermore, the classical theories[3,4] as well as the quantum theory including many-phonon processes[11] yield an increasing $\alpha(T_g,0)$ for $T_g \to \infty$, so that $\alpha(T_g,0)$ must have a minimum at the crossover $T_g \sim D/2k$. Such qualitative temperature dependence holds also for the equilibrium $\alpha(T)$, by virtue of the approximation $\alpha(T) \sim \alpha(T,0)$.

The high-temperature values of $\alpha(T)$ are roughly proportional to the atomic weight of the gas (Fig.2) showing in that a pure isotopic effect. In general the observation of isotope effects draws immediately our attention to phonon processes. However, at high T, this is easily understood on a classical basis, e.g., using the old hard-spheres model proposed by Baule[12].

Consider an elastic collision between a hard sphere of mass m, radius r and incident energy E_g with another sphere of mass M and radius R, at rest. For an impact parameter (distance of the trajectory of m from the center of M) equal to (R+r) sin ψ, where ψ is the angle between the trajectory and the radii ending at the impact point, the energy transfer is

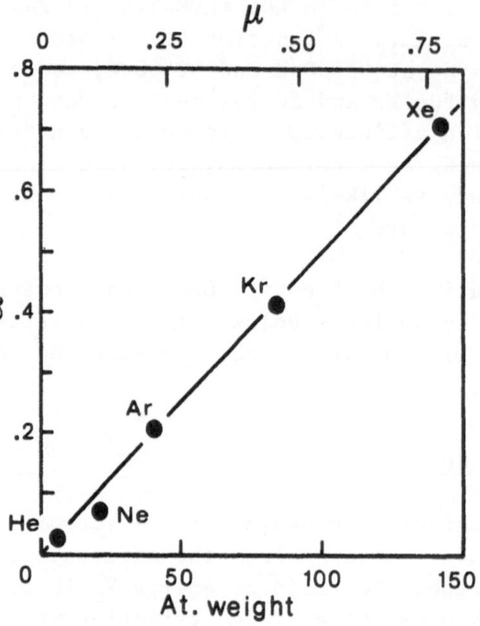

Fig. 2. "High"-temperature $\alpha(T)$ of rare gases on W as function of the atomic weight and of the mass ratio.

$$\varepsilon = E_g \frac{4mM}{(m+M)^2} \cos^2 \psi \tag{4}$$

Assuming that the surface atoms form an exagonal lattice of tangent hard spheres, we perform the average of ε over the cross section πR^2, i.e., for ψ ranging from 0 to ψ_o=arsin $R/(r+R)$. We find

$$\overline{\varepsilon} = E_g \frac{2\mu}{(1+\mu)^2} \left[1+\nu \frac{\nu+2}{(1+\nu)^2} \right], \tag{5}$$

where $\mu \equiv m/M$ and $\nu \equiv r/R$ and $\alpha(\infty,0)=\overline{\varepsilon}/E_g$. (6)

The result of eqs.(5) and (6) gives a description of the effect of masses and sizes of colliding atoms on the accommodation coefficient. As simple as it is, this model correctly predicts, for μ and $\nu \ll 1$, an accommodation coefficient proportional to μ, as found in experiment (Fig.2). However, the values of $\alpha(\infty,0)$ calculated from (5) and (6) for rare gases on tungsten are considerably larger than the experimental values. A possible explanation is that in Baule model the surface atoms work as independent oscillators, whereas in a solid they are connected each other by interatomic forces. Thus surface atoms are to be considered as having an effective mass, which may be quite larger than that of an isolated atom. The lower the oscillator energy (frequency) involved, the larger the effective mass, which explains alone the decrease of $\alpha(T,0)$ below the asymptotic value as T decreases. In the absence of the attractive well effect discussed above, yielding an increase of $\alpha(T,0)$ up to saturation as $T \to 0$, the accommodation coefficient would go to zero with T, as a simple effect of the effective mass increase (and of the quantization of the atomic vibrational motion). Indeed, this result in evident disagreement with experiment, is found in quantum theories of accommodation[4,13-15] when the attractive potential is neglected. (The effect of the attractive well on the quantum accommodation is rather complex and will be discussed below).

Classical Mechanics of Energy Accommodation

A quantitative description of the effective mass effect requires a rigorous treatment of surface dynamics. In which manner surface dynamics enters the accommodation coefficient is perhaps best understood by the classical theory developed by Frank Goodman[3,4]. Two drastic assumptions are made, which do not affect the basic physical content of the theory: firstly, only the component normal to the surface of the atomic motion is considered, so that the dynamical problem is reduced to one dimension. Secondly, only head-on collisions involving a single surface atom are considered, which allows one to project all the dynamical structure onto a single lattice site L and a single cartesian component z. The atom position r(t) and the surface atom displacement u(t) obey the set of coupled equations of motion

$$r(t) = r_0 + v_0 t - m^{-1} \int_0^t dt' \int_0^{t'} d\tau\, f(\tau) \,, \tag{7}$$

$$u(t) = M^{-1} \int_0^t d\tau\, \tilde{g}(t-\tau)\, f(\tau), \tag{8}$$

where

$$f(t) = \Big[dV(z)/dz \Big]_{z=r(t)-u(t)} \tag{9}$$

is the force acting on the surface atom when the gas atom is at distance z, and V(z) is the gas-surface potential; r_0 and v_0 are initial values. The displacement u(t) is that induced by the external force f(t), to which is linearly related via the response function g(t): g(t), also known as surface vibrational Green function depends exclusively on (and bears all information about) the dynamics of the solid surface. More precisely, it is the diagonal element L=L', $\alpha=\beta=z$ of the Green function matrix $\tilde{g}_{\alpha\beta}(L-L';t)$.

For a single oscillator of frequency ω_0 and infinitesimal damping η, $\tilde{g}(t)$ at T=0 is given by[16]

$$\tilde{g}(t) = \frac{-1}{\omega_0} e^{-\eta t} \sin(\omega_0 t)\theta(t) \qquad (10)$$

where $\theta(t)$ is the Heaviside step function. Its Fourier tranform

$$\tilde{g}(\omega^2) \equiv \frac{1}{2\omega}\tilde{g}(\omega) = \int_{-\infty}^{+\infty} dt\ e^{-i\omega t}\tilde{g}(t)$$

$$= \frac{1}{2\omega}\left\{\frac{1}{\omega-\omega_0-i\eta} + \frac{1}{\omega+\omega_0-i\eta}\right\} \qquad (11)$$

shows the spectral features of the system in an explicit way.

For a semiinfinite lattice we need (and have derived elsewhere) the Green function $\tilde{g}_{\alpha\beta}(\mathbf{K},\omega)$, Fourier transform of $\tilde{g}_{\alpha\beta}(\mathbf{L}-\mathbf{L}',\omega)$, where K and ω are the parallel wavevector and angular frequency of the surface phonons and of the surface-projected bulk phonons.

The energy loss in a single collision is given by

$$\varepsilon = \int_0^{\infty} d\tau\ \dot{u}(\tau)f(\tau); \qquad (12)$$

hence $\alpha(T,0)$ is readily obtained. For the single oscillator

$$\varepsilon = \frac{1}{M}\int_0^{\infty} dt \int_0^{\infty} dt'\ \theta(t-t')\cos\left[\omega_0(t-t')\right]f(t)f(t') \qquad (13)$$

is the sum of products of infinitesimal impulses f(t)dt and f(t')dt' at successive times (t>t'), multiplied by the cosine of a phase shift produced by the vibrational motion of the potential. Of course eq.(13) is still rather implicit because f(t) depends on the potential via the distance r(t)-u(t), and has to be calculated from the solution of the coupled equations. Numerical solutions of (7-9) for a Morse potential and a single dynamical model for the solid have been obtained by Goodman.[17] The results for rare gases on W reproduce quite well the experimental data (broken lines in Fig. 1) with the only two adjustable parameters: the potential well depth D and the ratio θ/a, where θ is the temperature equivalent of the maximum phonon frequency and a is the Morse parameter.

The fitted values of D/k are 105 K for He, 250 K for Ne, 1200 K for Ar, 2000 K for Kr and 3200 K, while θ/a is of the order of of 250 AK for all gases.[3] In all cases saturation is reached at low T (there is some disagreement, however, for He in this region), and a minimum is predicted also for heavy gases (not shown in Fig. 1).

The parameter a is related to the repulsive parameter $\rho(=\frac{1}{2}a)$ of the Born-Mayer form, which takes, for closed shells, nearly the same value in many different systems: e.g. in Fumi and Tosi fit of alkali halide interionic potentials[18], $\rho=0.3394$ Å. This gives $a=1.47$ Å^{-1}, which coincides with the average of the values used by Goodman (see Table 13.I in ref.3). This leads to an important remark: the corresponding value of $\theta(=380$ K) is close to the temperature equivalent of the maximum frequency for the bulk tungsten (348 K)[19] and not to the reduced value one would expect for the surface. Actually the peak frequency of W is reduced from 4.4 THz for the bulk to 3.0 THz for the surface.[19] This suggest that bulk phonons, rather than surface phonons, are most efficient in producing accommodation of energy. This result seems to be in conflict with the experimental data on the inelastic scattering of He from various surfaces reported by Toennies and coworkers[20-23], where Rayleigh waves give the dominant contribution. On the other hand, the inelastic scattering of Ne from Ni(111), measured by Feuerbacher and Willis[14], involves a large amount of bulk phonons, according to the interpretation given by Bortolani et al.[25]. Also the time of flight spectra of He from NaF at high incident energy[26] show , beneath the sharp features due to single surface phonons, a large uniform multiphonon background where bulk phonons are predominant. From the point of view of transport, surface phonons should be less efficient in accommodating the energy because of their localized character.

QUANTUM THEORY OF ENERGY ACCOMMODATION

The energy accommodation at low temperature is subject to quantization. A quantum theory of the accommodation coefficient is needed although quantum effects are probably important only for light atoms, specifically for He. Here, they may account for the observed discrepancy between experimental and theoretical $\alpha(T)$ based on the classical theory below 50 K. The quantum theory of atom inelastic scattering and energy accommodation has been approached as early as fifty years ago by Zener, Jackson and Mott, Lennard-Jones and Devonshire and others (for an introduction see Logan's review article[1]). Still the quantum calculations in the low temperature limit were not as successful as the classical theory in the intermediate and high temperature regimes.

A quantum theory of inelastic scattering with a correct many-body treatment of the phonon field has formulated in the late sixties within the distorted wave Born approximation[27-29] and subsequently extended to better approximations for corrugated surfaces[30-33]. However, only few quantum calculations combining an accurate description of surface dynamics with adequate potentials are available to date.[34-38,22,25]

As explained by Goodman[3-4], the attractive potential plays a crucial role in quantum α (T). We shall see below, that also the approximations made in scattering theory play a specific role in quantum accomodation for $T \to 0$.

<u>Inelastic Differential Reflection Coefficient</u>

We consider a transition from an initial state of the system where the atom has an incident wavevector k_i and the solid has the set of phonon numbers $\{n_i\}$, to a final state of atom wavevector k_f and phonon numbers $\{ n_f \}$. The transition rate w_{fi} has the general form[39]

$$w_{fi} = \frac{2\pi}{\hbar} |T_{fi}|^2 \delta(\mathcal{E}_f - \mathcal{E}_i) \tag{14}$$

where $\mathcal{E}_f$ and $\mathcal{E}_i$ are the respective <u>total</u> energies, i.e.,

$$\mathcal{E}_c = E_c + \Sigma_\lambda \hbar\omega_\lambda (n_{\lambda,c} + \tfrac{1}{2}), \qquad (c=f,i) \tag{15}$$

where ω_λ is the frequency of the λ-th phonon and $n_{\lambda,c}$ its occupation number in the state c. The elements T_{fi} of the transition matrix operator[39] are formal solutions of the matrix equation

$$T_{fi} = V_{fi} + \Sigma_c V_{fc}(E_i - E_c + i\eta)^{-1} T_{ci} \tag{16}$$

where $V_{cc'}$ are <u>known</u> matrix elements of the scattering potential between unperturbed (free) atom wavefunctions; index c spans all intermediate atom states and η is $\hbar$ times the inverse lifetime of the state c (possibly $\eta \to 0^+$). The reflection coefficient $R(k_f,k_i)$ is defined as the transition rate per unit incident flux integrated over all final phonon states and averaged over all initial phonon states. The incident atom flux is

$$J_i = \frac{\hbar |k_{iz}| L^2}{m L^3} = \frac{\hbar |k_{iz}|}{mL} \tag{17}$$

where L is a quantization length, alias the edge of a cubic volume above the solid surface confining the gas. Thus

$$R(k_f,k_i) = \frac{2\pi mL}{\hbar^2 |k_{iz}|} \sum_{\{n_f\}\{n_i\}} P_{\{n_i\}} |T_{fi}|^2 \, \delta(E_f-E_i) \tag{18}$$

where $P_{\{n_i\}}$ is the thermal distribution of phonons in the initial state.In inelastic scattering experiments one considers the differential reflection, given by the reflection coefficient times the increment of phase space volume occupied by the final states within a solid angle of $d\Omega_f$ with momentum between k_f and k_f+dk_f:

$$\begin{aligned} d^2R &= R \cdot (L/2\pi)^3 \, k_f^2 \, dk_f \, d\Omega_f \\ &= R \cdot (L/2\pi)^3 (mk_f/\hbar^2) \, dE_f d\Omega_f \, . \end{aligned} \tag{19}$$

The ratio

$$\frac{d^2R}{dE_f d\Omega_f} = \frac{k_f}{|k_{iz}|} \left(\frac{mL^2}{2\pi\hbar^2}\right)^2 \sum_{\{n_f\}\{n_i\}} P_{\{n_i\}} |T_{fi}|^2 \delta(E_f-E_i) \tag{20}$$

is currently called differential reflection coefficient.

The interaction potential V and the operator T, which have energy dimension, are conveniently expressed in units of the quantum $(2\pi\hbar^2/mL^2)$, in order to deal with reduced dimensionless quantities

$$\begin{aligned} \mathsf{V} &\equiv V(mL^2/2\pi\hbar^2) \\ \mathsf{T} &\equiv T(mL^2/2\pi\hbar^2) \end{aligned} \tag{21}$$

As a consequence, the linear atom-phonon coupling constant, which is physically a force, takes the dimension of an inverse length in its reduced form.

By using in eq.(20) the Fourier representation of the δ-function and applying the van Hove transformation[40], we have

$$\frac{d^2R}{dE_f d\Omega_f} = \frac{1}{2\pi\hbar} \frac{k_f}{|k_{iz}|} \int_{-\infty}^{+\infty} dt \, e^{-i\omega t} \langle \mathsf{T}_{fi}(0) \, \mathsf{T}_{fi}^{+}(t) \rangle \tag{22}$$

with

$$\omega \equiv (E_f - E_i)/\hbar .$$

The time evolution of $\mathsf{T}_{fi}(t)$ is driven by the free solid Hamiltonian only, and the brackets <...> denote the thermal average over the lattice vibrations at temperature T_s.

Equation (22) holds for coupling to the phonon system to any order of approximation. However the phonon dynamics governs the evolution of T_{fi} through its field operators in an exponential form. A formal Taylor expansion conceptually yields a jerarchy of processes involving zero, one, two, ... many phonons, but this is much complicated by the time-ordering and commutation rules of phonon operators. It is at this stage that a few simplifying assumptions are necessary.

Firstly, we assume a linear coupling, which means that the phonon-induced modulation of the atom-surface potential V is expanded with respect to the phonon displacements and only the first order terms are considered. Secondly, the linear coupling is treated to first order in the perturbation theory,which corresponds to the one-phonon approximation. Only the linear term in the exponent multiphonon expansion is held. With these approximations, the one-phonon differential reflection coefficient $d^2R^{(1)}/dE_f d\Omega_f$ always factorizes into a scalar product of the coupling forces with the vibrational spectral response of the solid, represented by the imaginary part of the surface-projected Green function matrix. For a monoatomic surface

$$\frac{d^2R^{(1)}}{dE_f d\Omega_f} = \frac{1}{4\pi M} \frac{k_f}{|k_{iz}|} \frac{n(\omega)}{\omega} \sum_{\alpha\beta} F_\alpha^* F_\beta \, \mathrm{Im}\, \tilde{g}_{\alpha\beta}(\mathbf{K},\omega) \tag{23}$$

where F_α are the reduced coupling forces, and

$$n(\omega) = (e^{\hbar\omega/kT_s} - 1)^{-1} \tag{24}$$

is the equilibrium phonon occupation number at temperature T_s. In eq.(23), ω and the parallel wavevector transfer $\mathbf{K}$ are mutually connected through the kinematic relationship

$$\hbar\omega/E_i = -1+\left[(k_i \sin\theta_i - K_p)^2 + K_{np}^2\right]/k_i^2 \sin^2\theta_f \tag{25}$$

where K_p and K_{np} are the components of $\mathbf{K}$ parallel and normal to the incidence plane, respectively.

The factorization yields two completely independent problems: the calculation of $g(\mathbf{K},\omega)$ by solving the surface dynamical problem, and the calculation of the coupling forces from the gas-surface potential model and scattering theory.

We reproduce here two examples relevant to our discussion on quantum accommmodation. We consider the two body reduced potential $V(\mathbf{r})$ between He and a surface atom.

In Manson and Celli DWBA theory[29] the specular part of the atom-surface potential is treated exactly in the scattering theory, whereas the corrugated components of the static potential (producing diffraction) and the phonon-induced modulations of the potential (producing inelastic scattering) are treated as perturbations in the first Born approximation. Thus, He atom waves, distorted by the specular potential, serve as a basis for the calculation of the coupling forces. At low surface temperature, it is found[34]

$$F_\alpha = \int dz\ \chi^*(k_{fz},z)\chi(-k_{iz},z)Q_\alpha V(\mathbf{K},z). \tag{26}$$

where $V(\mathbf{K}, z)$ is the two-dimensional Fourier transform of the reduced interatomic potential $V(\mathbf{r})$ for a surface phonon wavevector $\mathbf{K}$, Q_α is the vector gradient operator

$$Q_\alpha = (i\mathbf{K}, -\frac{\partial}{\partial z}); \tag{27}$$

the distorted wavefunctions $\chi(k_z,z)$ describe the motion of the atom along z in the specular part of the surface potential:

$$\left[-\frac{\hbar^2}{2m}(\frac{\partial^2}{\partial z^2} + k_z^2) + V(\mathbf{0},z)\right]\chi(k_z,z) = 0. \tag{28}$$

In general the z-component of the force, F_z, gives the strongest coupling to lattice vibration. For vanishing incident energy, the matrix element (26) for $\alpha=z$ can be worked out by part integration:

$$F_z = \int dz\ V(\mathbf{K},z)\left[\frac{\partial\chi^*(k_{fz})}{\partial z}\chi(-k_{iz}) + \chi^*(k_{fz})\frac{\partial\chi(-k_{iz})}{\partial z}\right]$$

$$= \int dz\left[V(\mathbf{K},z) - V(0,z)\right]\partial\chi^*(k_{fz})\chi(0)/\partial z$$

$$+ (L^2k_{fz}^2/4)\int dz\ \chi^*(k_{fz})\ \partial\chi(0)/\partial z, \tag{29}$$

where use is made of (28) and of the condition $\lim_{z\to\infty}\chi(0,z)=0$. We note that F_z does not vanish in general for $E_i=0$ alone, but it does if also $E_f\to 0$, because $\mathbf{K}, k_{fz}\to 0$.

The second interesting case is that of scattering from a hard-corrugated surface, whose potential is defined as

$$V(\mathbf{r}) = \begin{cases} \infty & z < D_0(\mathbf{R}) \\ 0 & z > D_0(\mathbf{R}) \end{cases} \tag{30}$$

and $D_0(\mathbf{R})$ is the static surface corrugation function. Here no attractive well is considered. In this case the static corrugation of the potential can be treated exactly[35,22] and we find, for the monoatomic case,

$$F_\alpha = \frac{k_{fz}-k_{iz}}{2\pi L} \int d^2R \; e^{i\{\mathbf{K}\cdot\mathbf{R}+(k_{fz}-k_{iz})D_0(\mathbf{R})\}} f_0(\mathbf{R}) \frac{\partial D_0(\mathbf{R})}{\partial u_\alpha}, \tag{31}$$

where $\mathbf{u}$ is the displacement of any arbitrary surface atom and $f_0(\mathbf{R})$ is the source function for the static surface, related to the "static" reduced T-matrix by the generalized Fourier transform:[22]

$$T_{0,fi} = \frac{1}{2L} \int d^2R \; e^{i\{\mathbf{K}\cdot\mathbf{R}+(k_{fz}-k_{iz})D_0(\mathbf{R})\}} f_0(\mathbf{R}) \tag{32}$$

The static reduced T-matrix $T_{0,fi}$ is that given by (16) for the static corrugated potential. The source function $f_0(\mathbf{R})$ is an inverse length. Levi has shown that a good approximation for not too corrugated surfaces is $f_0(\mathbf{R}) \sim ik_{iz}$ (eikonal approximation).[30] In this case $F_z \to 0$ for $k_{iz} \to 0$, but not necessarily for $\omega \to 0$. In eq.(31) the Debye-Waller factor has been dropped for the reasons explained in the next paragraph.

Unitarity

We note that the intensity of the elastic scattering is obviously reduced by the occurrence of inelastic processes since the total integrated reflection coefficient is equal to unity. Such reduction is expressed by the Debye-Waller factor e^{-2W} (which automatically arises from the evaluation of the T-matrix correlation function) and multiplies all terms of the integrated multiphonon expansion $e^{2W}=1+2W+2W^2+\ldots$. Since in our approximation the one-phonon term represents the entire inelastic scattering, the Debye-Waller factor is conveniently replaced by the unitarity factor $1/(1+R_I/4)^2$, where

$$R_I = \iint (d^2R^{(1)}/dE_f d\Omega_f) dE_f d\Omega_f \tag{33}$$

in the integrated one-phonon reflection coefficient.[29] Thus we write the unitary inelastic reflection coefficient as

$$\frac{d^2R}{dE_f d\Omega_f} = \frac{1}{(1+R_I/4)^2} \frac{d^2R^{(1)}}{dE_f d\Omega_f} . \tag{34}$$

Unitarity is obviously important in the calculation of integral properties such as the accommodation coefficient.

Energy Accommodation at Low Temperatures

Once we know the differential reflection coefficient, we can express the accommodation coefficient as

$$\alpha(T_g,T_s) = \int_0^\infty dE_i (kT_g)^{-1} e^{-E_i/kT_g} \int_0^\infty dE_f [k(T_s-T_g)]^{-1}(E_f-E_i)$$
$$\times \iint_{2\pi} d\Omega_i d\Omega_f (d^2R/dE_f d\Omega_f). \qquad (35)$$

The expression of $\alpha(T_g,T_s)$ has the important property of being invariant against the interchange of initial and final states, as a consequence of the reciprocity theorem, alias detailed balancing.[1] Reciprocity holds if the solid is at thermal equilibrium, which is implicit in the definition of $n(\omega)$, eq.(24). We can take advantage of reciprocity by setting

$$\iint_{2\pi} d\Omega_i d\Omega_f \frac{d^2R}{dE_f d\Omega_f} \equiv \frac{n(\omega)}{\hbar\omega} I(\hbar\omega,E_i) \quad , \qquad (37)$$

where $I(\hbar\omega,E_i) = I(-\hbar\omega,E_F)$ is invariant for $E_i \leftrightarrows E_f$ and writing

$$\alpha(T_g,T_s) = \{4k^2T_g(T_s-T_g)\}^{-1} \iint_0^\infty dE_i dE_f$$
$$\times \{e^{-E_i/kT_g} n(\omega) I(\hbar\omega,E_i) + e^{-E_f/kT_g} n(-\omega) I(-\hbar\omega,E_f)\}$$
$$= \{4k^2T_g(T_s-T_g)\}^{-1} \iint_0^\infty dE_i dE_f \; e^{-E_i/kT_g} n(\omega)$$
$$\times \; I(\hbar\omega,E_i)(1-e^{\hbar\omega(T_g-T_s)/kT_gT_s}) \quad . \qquad (38)$$

It may be convenient substituting the second integration variable E_f with $\varepsilon = E_f - E_i$ and re-write (38) as

$$\alpha(T_g,T_s) = \{4k^2T_g(T_s-T_g)\}^{-1} \int_0^\infty dE_i \; e^{-E_i/kT_g}$$
$$\times \int_{-E_i}^\infty d\varepsilon \; n(\tfrac{\varepsilon}{\hbar}) I(\varepsilon,E_i)(1-e^{(\varepsilon/k)(T_s^{-1}-T_g^{-1})}) . \qquad (39)$$

The low-temperature limits ($T_g,T_s \to 0$) have to be taken with some care because of the complicate analytical behaviour of the double integral. Different situations are found according to whether we consider the equilibrium case ($T_g=T_s \to 0$) or the non-equilibrium cases with $T_g \to 0$ and $T_s > 0$, or viceversa $T_s \to 0$ and $T_g > 0$.

(a) Nonequilibrium with $T_g \ll T_s$: this seems to be unavoidably the situation of Knudsen cell experiments, where the filament current has to be held above a certain threshold in order to observe its oscillations during the energy release from the gas molecules. Set $E_i/kT_g = x$ and $\varepsilon/kT_s = y$; equation (38) becomes

$$\alpha(T_g,T_s) = \frac{1}{4}\int_0^\infty dx\ e^{-x}\int_{-xT_g/T_s}^\infty dy\ (e^y-1)^{-1} \times I(kT_s y, kT_g x)\ (1 - e^{y(1-T_s/T_g)})\ . \tag{40}$$

The limit $T_g \to 0$, if exists, gives

$$\alpha(0,T_s) = \frac{1}{4}\int_0^\infty dy\ (e^y-1)^{-1}\ I(kT_s y,0)\ . \tag{41}$$

For F_z given by eq.(29) (potential well treated with DWBA), $I(kT_s y,\ kT_g x)$ would diverge slowly as its incident-flux prefactor $|k_{iz}|^{-1}$ (see eq.(23)) until the unitarity correction becomes effective and kills the divergence. This argument is qualitative because unitarity correction is valid as long as it is small. However, in this case we expect that $I(kT_s y,0)$ and therefore $\alpha(0,T_s)$ are not manifestly zero, whereas they vanish for the case of a hard corrugated surface, eq.(32). Of course $\alpha(0,T_s) \to 0$ for $T_s \to 0$ in all cases.

(b) Nonequilibrium with $T_s \ll T_g$. The limit $T_s = 0$ has been considered by Goodman in classical lattice theories. Setting $E_i/kT_g = x$ and $\varepsilon/kT_g = y$, one finds

$$\alpha(T_g,0) = \frac{1}{4}\int_0^\infty dx\ e^{-x}\int_{-x}^\infty dy\ I(kT_g x,\ kT_g y)\ . \tag{42}$$

Clearly $\alpha(T_g,0) \to 0$ for $T_f \to 0$ in all cases, as both arguments of I vanish simultaneously.

(c) Equilibrium. Taking the limit $T_s \to T_g \equiv T$ in (38) we have

$$\alpha(T) = \frac{1}{4(kT)^3}\int_0^\infty dE_i\ e^{-E_i/kT}\int_{-E_i}^\infty d\varepsilon\ n(\tfrac{\varepsilon}{\hbar})\ \varepsilon\, I(\varepsilon,E_i), \tag{43}$$

and, setting $E_i/kT = x$, $\varepsilon/kT = y$, one gets

$$\alpha(T) = \frac{1}{4}\int_0^\infty dx\ e^{-x}\int_{-x}^\infty dy\ \frac{y}{e^y - 1}\ I(kTy\ ,\ kTx)\ . \tag{44}$$

Again it is predicted that $\alpha(T)\to 0$, for $T\to 0$. Note the similarity of the equilibrium accommodation coefficient with the expression of $\alpha(T_g,0)$, eq.(41). The modulus of the argument kTy is actually bound by the maximum lattice frequency, i.e., $|kTy|\leq k\theta$. For $T>\Theta$ the factor $y/(e^y-1)$ in (43) is approximately one and $\alpha(T)\simeq\alpha(T_g,0)$, as argued by Goodman and supported by experiment, but at low temperature, say for $T\ll\theta$, this approximation is not evident.

Quantum theory of energy accommodation always predicts a vanishing coefficient when both T_g and T_s tend simultaneously to absolute zero, regardless whether equilibrium ($T_g=T_s$) or nonequilibrium ($T_g\neq T_s$) conditions are maintained in the experiment. However, $\alpha(T_g,T_s)$ remains finite if only one of the two temperatures vanishes.

The experimental conditions in low-temperature measurements have probably more resemblance with the latter nonequilibrium situation, than with the theoretical equilibrium case. On the other hand, recent accurate investigations of the low-energy limit in the quantum theory of sticking, performed by Brenig and coworkers[41-44] and by Brivio and Grimley,[45-46] show that the coupling matrix elements evaluated in the semiclassical WKB approximation for wavefunctions yield a finite sticking (and therefore a finite accommodation) at zero incident energy.

These results qualitatively agree with experiment in that the sticking coefficient increases with decreasing energy. Actually the WKB approximation fails at very low energies, much below those considered in the existing experiments. The exact result tends ultimately to zero at T=0, as expected from the general, but qualitative, discussion reported above for $\alpha(T)$. Clearly, new experiments at very low energies, like those accomplished by Sinvami, Cole and Goodstein,[10] are required in order to prove such theoretical predictions.

The attractive potential plays obviously an indispensable role in the stiking problem. However, Goodman's statement that the range of the attractive well determines the limiting value of $\alpha(T)$ is still matter of discussion.

THE ROLE OF BOUND STATES AND PHONON DENSITY

The attractive well plays a fundamental role in the energy accommodation and sticking problems in that it hosts a certain number of bound states. Manson pointed out long ago[11] that bound states could by themselves explain the finite value of $\alpha(0)$. Sticking obviously requires bound states at the surface, but also energy accomodation is increased if low energy atoms can reside at the surface for a sufficiently long time. What seems to be physically obvious not always is straightforward in theoretical calculations, nor in experimental demonstration. The existence of inelastic resonances with bound states was proved only in 1976 by Cantini, Felcher and Tatarek in angular distributions[47], and more recently by Evans et al.[36] in time-of-flight spectra. The quantum theory of inelastic scattering incorporating inelastic resonant processes was developed in recent years as well[48] and only recently close-coupling calculations became possible[36,49]. However no close-coupling calculation of $\alpha(T)$ has been performed so far.

It is known that bound states are accesible in elastic scattering only on corrugated surfaces, because the lattice has to provide a parallel reciprocal lattice vector for parallel momentum conservation. Thus, on flat metal surfaces, bound states do not take part in elastic scattering. On the contrary, in inelastic scattering phonons can provide sufficient parallel momentum for a bound state to be coupled also on flat metal surfaces. In this case one has to go beyond Manson and Celli DWBA and take into consideration in eq.(16) also the sum over the intermediate bound states, V_{fc} being the matrix element inelastic coupling potential. This is proportional to the force given by eq. (29), where $\chi(0)$ is replaced by $\chi(ik_{nz})$ and $E_n=-(\hbar k_{nz})^2/2m$ is the energy of the n-th bound state. In this case, the force does not vanish for $f_{fz}\to 0$, which ensures a finite contribution to $\alpha(0)$.

The effect of bound states has been carefully reviewed by Goodman in a recent paper devoted to the low-energy accomodation[50]. It is shown that, unlike the continuum-state accommodation coefficient α_c that vanishes at T=0, the bound-state contribution α_b tends to a finite limit at absolute zero. Furthermore, the low temperature analytic behavior of the two contributions scales with the dimensionality d of the phonon density as $\alpha_b \propto (E_i-E_n)^{d-2}$ and $\alpha_c \propto E_i^{d-1}$. Thus for either d=3 (bulk phonons) or d=2 (surface phonons) both α_b and α_c are nonincreasing functions, in contrast with experiment, whereas for d=1 α_b is increasing and α_c constant and for d<1 both terms increase for $T\to 0$. This seems to imply that at very low temperature, when the phonon energies are comparable or smaller than $|E_n|$, the He atom experiences a local dynamics, rather than the dynamics extended to the whole surface or bulk.

We have so far discussed the low-temperature accommodation coefficient in the light of the coupling forces and their analytical behavior. It is clear, however, that in those particular limits and approximations where the interaction part is independent of energy, the analytical structure of $\alpha(T)$ reflects that of the phonon density.

At low temperature the specific features of the vibrational spectrum loose their importance, since only the extreme acoustic spectrum is populated. Manson has shown that the contributions of Rayleigh waves and surface-projected bulk modes, including resonances ("mixed" modes)[11] are different in magnitude but similar in their analytical behavior as function of temperature.

The simplest form of Im $\tilde{g}_{zz}$ we can conceive includes Rayleigh wave with isotropic velocity v_R and a surface-projected bulk phonon band of transverse polarization in the sagittal plane with isotropic velocity v,[51] namely

$$\mathrm{Im}\ \tilde{g}(K,\omega) = (R\,\omega/v_R)\ \{\ \delta(\omega-v_R K) + \delta(\omega+v_R K)\ \} + (R\,\omega/v)\ \mathrm{Re}\ (\omega^2 - v^2K^2)^{-\frac{1}{2}} \qquad (45)$$

for small K. Use of the kinematic condition (25) implies the substitution (assuming planar scattering):

$$K = k_i\ (\ (\hbar\omega/E_i + 1)^{\frac{1}{2}} \sin\theta_f - \sin\theta_i\ ) \qquad (46)$$

in eq.(45). It is an interesting exercise to insert these expressions into the integrals (39-43) and to analyze the limiting behavior of $\alpha(T)$.

Of course a quantitative analysis of $\alpha(T)$ requires a rigorous calculation of the Green function. This problem is the subject of Prof. Maradudin's lectures at the present course, and has been treated by myself in a recent NATO-ASI on collective excitations in solids.[51]

ACKNOWLEDGEMENTS

I am grateful to Dr. G.P. Brivio for useful discussions and to Dr. F.O. Goodman, for keeping me informed about his recent contributions in the area of gas-surface interaction.

REFERENCES

1. R.M. Logan, Theory of Gas-Surface Scattering and Accommodation, in "Solid State Surface Science", M. Green ed., vol.3, Dekker, New York (1973). p.1.
2. M. Knudsen, Ann. der Phys. 34:593 (1911).
3. F.O. Goodman and H.Y. Wachman <u>in</u> "Dynamic Aspects of Surface

Physics", F.O. Goodman ed., Compositori, Bologna (1974), p. 347 -529.
4. F.O. Goodman, J. Phys. Chem. 84:1431 (1980); and F.O. Goodman in "Rarified Gas Dynamics", S.S. Fisher ed., Progr. Astron. Aeron. 74:3 (1981).
5. L.B. Thomas in "Rarified Gas Dynamics", C.L. Brundin, ed., Vol.1, Academic Press, New York (1955) p.155.
6. D.V. Roach and LB. Thomas, ibidem, p.163.
7. H.Y. Wachman, J. Chem. Phys., 45:1532 (1966).
8. J. Kouptsidis and D. Menzel, Ber. Bunsen Gesell. f. Physik Chemie 74:512 (1970).
9. F.O. Goodman and H.Y. Wachman, "Dynamics of Gas-Surface Scattering", Academic Press, New York, (1976).
10. M. Sinvani, M.W. Cole and D.L. Goodstein, Phys.. Rev. Lett. 51:188 (1983).
11. J.R. Manson, J. Chem. Phys. 56:3451 (1972); and J.R. Manson and J. Tompkins in "Rarified Gas Dynamics", J.L. Potter ed., Progr. Astron. Aeron. 51:603 (1977).
12. B. Baule, Ann. der Phys. 44:145 (1914).
13. F.O. Goodman and N. Garcia, Phys. Rev. B. 20:813 (1979).
14. F.O. Goodman, Surf. Sci. 92:185 (1980).
15. N. Garcia, V. Celli and J.R. Manson, J. Chem. Phys. 72:3436 (1980).
16. E.N. Economou, "Green's Functions for Solid State Physicists", Springer, Heidelberg (1979).
17. F.O. Goodman in "Rarified Gas Dynamics", J.H. De Leeuw ed., Vol.II, Academic Press, New York (1966) p.366.
18. F. Fumi and M.P. Tosi, J. Phys. Chem. Solids 25:31 (1964).
19. V. Bortolani, F. Nizzoli and G. Santoro in "Lattice Dynamics", M. Balkanski ed., Flammarion, Paris (1977) p.302.
20. G. Brusdeylins, R.B. Doak and J.P. Toennies, Phys. Rev. Lett. 44:1417 (1980); 16:437 (1981); G. Brusdeylins, R.B. Doak and J.P. Toennies, Phys. Rev. B27:3662 (1983).
21. R.B. Doak, U. Harten and J.P. Toennies, Phys. Rev. Lett. 51: 578 (1983).
22. G. Benedek, J.P. Toennies and R.B. Doak, Phys. Rev. B 28:7276 (1983).
23. J.P. Toennies, J. Vac. Technol. (1984) in press.
24. B. Feuerbacher and R.F. Willis, Phys. Rev. Lett. 47:526 (1981).
25. V. Bortolani, A. Franchini, F. Nizzoli, G. Santoro, G. Benedek and V. Celli, Surf. Sci. 128:249 (1983).
26. G. Benedek, G. Brusdeylins, L. Miglio, J. Skofronick and J.P. Toennies, Phys. Rev. Lett. (1984) in press.
27. N. Cabrera, V. Celli and R. Manson, Phys. Rev. Lett. 22:346 (1969).
28. N. Cabrera, V. Celli, F.O. Goodman and R. Manson, Surf. Sci. 19:67 (1970).

29. R. Manson and V. Celli, Surf. Sci. 24:495 (1971).
30. A.C. Levi, Nuovo Cim. B 54:357 (1979), and references therein.
31. A.C. Levi and H. Suhl, Surf. Sci. 88:221 (1979).
32. G. Armand and J.R. Manson, Surf. Sci. 88:532 (1979).
33. H.D. Meyer, Surf. Sci.104:117 (1981), and references therein.
34. G. Benedek and G. Seriani, Jpn. j. Appl. Phys., Suppl. 2, Pf.2: 545 (1974): He/LiF, in DWBA with Lennar-Jones potentials.
35. G. Benedek and N. Garcia, Surf. Sci. 103:L143 (1981): He/LiF for a hard-corrugated surface with exact numerical treatment of static corrugation.
36. D. Evans, V. Celli, G. Benedek,R.B. Doak and J.P. Toennies, Phys. Rev. Lett. 50:1854 (1983): He/LiF as ref.35 with effect of resonances due to bound states in the attractive well.
37. V. Bortolani, A. Franchini, F. Nizzoli and G. Santoro, Phys. Rev. Lett. 52:429 (1984): He/Ag with DWBA and Morse potential.
38. A.C. Levi, G. Benedek, L. Miglio, G. Platero, V.R. Velasco and F.G. Moliner, Surf. Sci. in press: He/NaF in eikonal approximation with Morse potential.
39. H. Schiff, "Quantum Mechanics" 3rd ed., McGraw-Hill, New York (1966).
40. L. van Hove, Phys. Rev. 95:249 (1954).
41. W. Brenig, Z. Physik B 36:81 (1979) and 36:227 (1980).
42. H. Böheim, W. Brenig and J. Stutzki, Z. Physik B 48:43 (1982).
43. R. Sedlmeir and W. Brenig, Z. Physik B 36:245 (1980).
44. J. Böheim and W. Brenig, Z. Phys. B 41:243 (1981).
45. G.P. Brivio, J. Phys. C: Solid State Physics, 16:L131 (1983).
46. G.P. Brivio and T.B. Grimley, Surf. Sci. 131: 475 (1983).
47. P. Cantini, G.P. Felcher and R. Tatarek, Phys. Rev. Letters 37: 606 (1976); Surf. Sci. 63: 104 (1977).
48. P. Cantini and R. Tatarek, Phys. Rev. B23: 3030 (1981).
49. D. Eichenauer and J.P. Toennies, in "Dynamics at Surfaces", B. Pullmann and J. Jortner, eds. Reidel, Dordrecht (1984).
50. F.O. Goodman, Surf. Sci. 111: 279 (1981).
51. G. Benedek in "Collective Excitations in Solids", B. di Bartolo ed., Plenum Publishing Co., New York (1983) p. 523-558.

THE GAS/PHONON INTERFACE: DESORPTION AND OTHER PHENOMENA

David L. Goodstein
California Institute of Technology
Pasadena, California USA 91125

INTRODUCTION TO ADSORPTION AND DESORPTION

The phenomenon of desorption has fired the imaginations of scientists for decades, leading them, at times, to the very frontiers of knowledge. For example, the first splitting of the atom - J. J. Thomson's discovery of the electron in 1896 - was accomplished when he realized he could improve the vacuum in his cathode ray tubes by baking them before they were sealed. Thomson's insight may have been the first important application of thermal desorption kinetics, a subject we shall explore in some detail. On a more prosaic level, any scientist who has spent lonely and frustrating hours waiting for an experimental system to reach high vacuum was most likely contending with the kinetics of desorption.

When an atom or a molecule approaches a surface, it will, generally speaking, experience an attractive well of some depth followed by a repulsive barrier (see for example Fig. 1 below). Depending on the nature and temperature of the surface and the incident molecule, there is always some chance that the projectile will find a way to dispose of excess energy and become bound at least temporarily in the attractive potential well. This is the process of adsorption. In the reverse case, a molecule already stuck to a surface acquires energy in one of a variety of possible ways, and escapes. This process is called desorption.

Traditionally, adsorption and desorption have been studied more by chemists, who seem to thrive on the complexities of the problem, than by physicists. Nevertheless, we shall concentrate below on the physicist's point of view, pretending to the best of our ability that the world is made up of helium atoms and chemically inert substrates. For the sake of orientation, however, we begin with a few words about the role of desorption in modern surface chemistry.

Contemporary surface science begins with an ultrahigh vacuum (UHV) system, inside of which clean, fresh surfaces may be prepared by ion bombardment, annealing, *in situ* cleaving, or simply by epitaxial growth forming a fresh surface. The state of the surface that results may then be analyzed by a variety of surface-specific analytic techniques. For example, the periodicity of the surface may be determined by diffraction, i.e., Bragg scattering of a beam of electrons (LEED, or low energy electron diffraction) which have the appropriate deBroglie wavelengths, and which do not penetrate more than two or three layers beneath the surface. Beams of atoms, often helium, are sometimes used because they sample the first layer only. The chemical nature of the surface may be detected by Auger spectroscopy, a process in which an incident high energy electron ejects a core electron from a surface atom. This hole in the core is filled by decay from a higher level ΔE above the hole, accompanied by emission of a photon (real or virtual) with $\hbar\omega = \Delta E$, which in turn finally ejects a (typically) valence electron from the material. The energy of the ejected electron, a characteristic combination of ΔE and the valence band energy, reveals the chemical species present on the surface.

A UHV system with a combined LEED and Auger spectrometer is now a more or less off-the-shelf item, *de rigeur* for modern surface science. A variety of other techniques are also used (Prutton, 1983).

Once a clean surface is prepared and characterized, overlayers may be adsorbed to it, either physically or chemically (chemisorption differs from physisorption mainly in that the strength of the bond to the surface (of order eV) is much stronger than the physisorption bond ($<$ 0.1 eV)). The adsorbed species are then studied in their turn using many of the same techniques applied to bare surfaces, and others as well, such as infrared spectroscopy, electron energy loss spectroscopy (EELS), and even inelastic atomic beam scattering. In these techniques, the loss (or gain) of energy of a photon, electron or atom scattered from a surface reveals the vibrational states of adsorbed molecules.

The study of desorption is also an important tool of surface chemistry. Often, desorption is accomplished by changing the chemical state of the molecules on the surface from one that prefers to stick to one that does not. These changes may be induced by bombardment with electrons, ions, photons, application of dc electric fields, or just by heating. Each of these processes adds its own contribution to the acronymic zoo of contemporary surface science (see, e.g., Menzel, 1975).

Of particular interest to us is thermal desorption (TD). Here, either the chemical change that leads to desorption of some part of the adsorbate, or simply desorption of the molecule itself from the surface is caused by heating. A standard technique is to warm the surface such that its temperature is a predetermined function of time (often, but not always, linear). As desorption proceeds, peaks in the desorption rate are observed. They are a combined result of increasingly favorable conditions for desorption of a species, and depletion of that species from the surface. From the results of this technique (sometimes called temperature programmed desorption, or TPD), one tries to deduce the rate constants and activation energies that govern the processes involved.

This brief description can do no more than give a bit of the flavor of a vigorous and growing field. It is of interest to us here mainly by contrast to the radically different approach to desorption that forms the central focus of this chapter.

There is an almost entirely separate (and much older) tradition in the physics and physical chemistry of adsorption which is mainly concerned with films in thermodynamic equilibrium with their own vapors, principally adsorbed on high surface area adsorbents such as powders. (The word adsorbent means the same as substrate; the adsorbed species is called the adsorbate.) The principal experimental method of this field was the adsorption isotherm, a constant temperature measurement of the amount adsorbed versus gas pressure, and the usual objective was to determine the area of the adsorbent by interpreting adsorption isotherms according to theories by Langmuir, Brunauer, Emmett, and Teller (BET), and others (de Boer, 1968; Young and Crowell, 1962).

In more recent times, the discovery of adsorbates (mainly forms of graphite) combining large area and great uniformity has opened a new field: the study of two-dimensional matter. Submonolayer films of various adsorbates on graphite (and a few other) substrates are found to have two-dimensional gas and liquid phases as well as solid phases that are either commensurate or incommensurate with the underlying substrate lattice. The study of these phases and the phase transitions between them by means of LEED, neutron scattering, X-ray scattering, nuclear magnetic resonance (NMR), thermodynamic measurements, atomic beam scattering, and other techniques is a lively field of research today (Dash, 1975; Sinha, 1980; Cole et al., 1982). Most recently, attention has centered on the growth of thicker films with special reference to the question of whether a given substance spreads out smoothly ("wets") or beads up on a given substrate (see, e.g., Pandit et al., 1982; Sequin et al., 1983; Goodstein et al., 1985a).

Regardless of what particular phenomenon may be of interest, it seems fair to say that the central idea of physical adsorption is the adsorption potential (Bruch, 1983). It is the theoretical starting point for describing any aspect of adsorption or, as we shall see, desorption. Sketched in Fig. 1, its general form is, assuming the substrate occupies the half space $z < 0$,

$$V(z)=ae^{-bz} - \frac{C_3}{z^3} \tag{1}$$

Here the repulsive term is due to Pauli repulsion between the electron clouds of the adsorbate molecule and the substrate. The latter falls off exponentially for $z > 0$, giving the first term on the right its form. The z^{-3} dependence of the attractive term is perhaps most easily understood by recalling that the long-range attraction between atoms or molecules due to mutually induced dipole moments falls off with distance as r^{-6}. Summing (or

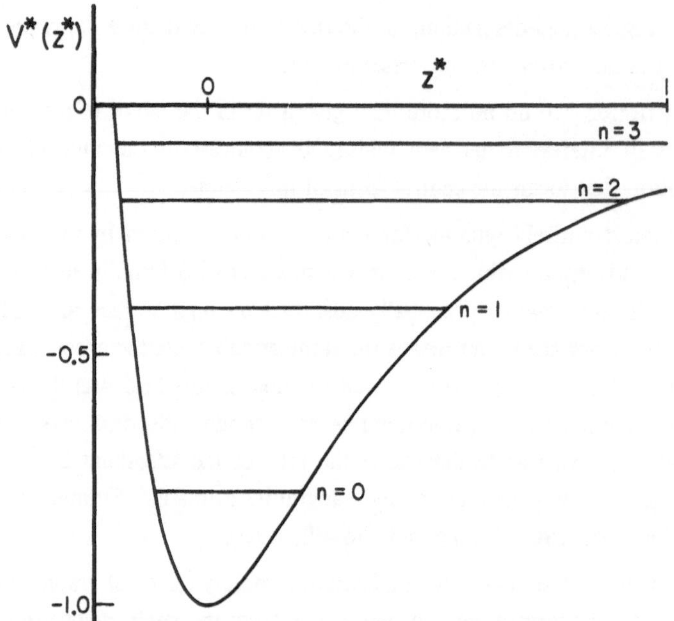

Figure 1. The adsorption potential, Eq. (1), in dimensionless form, as explained in the text. The first four bound states are shown.

integrating) over the three-dimensional half space $z < 0$ yields the dependence z^{-3}.

In some applications it is important to realize that the substrate potential is corrugated. In other words, V(z) is not independent of x and y. For our purposes, however, we will be able to ignore that complication considering our potential to be suitably averaged over the two-dimensional unit cell of the substrate.

What is important for our application is the fact that V(z) has a number of bound states, ε_n, n = 0, 1, 2 ...., found by solving the Schrödinger equation for Eq. (1) and also shown in Fig 1. Here the ground state energy, ε_0, is what is normally meant by the term adsorption binding energy, but the excited states also play a role, for example in determining the thermodynamic heats of adsorption that are generally measured.

There are only two instances in which the actual ground state binding energy, ε_0, has been measured by purely thermodynamic means: ^3He and ^4He adsorbed on graphite (Elgin and Goodstein, 1974; Elgin et al., 1978). The nature of the measurement is worth recounting because related ideas and manipulations will be used later in discussing desorption. It proceeds as follows:

At some convenient temperature (3-4K) a measurement is made of the helium gas pressure in equilibrium with a small amount of the same isotope adsorbed on the graphite surface (data in the range 0.1 to 0.6 monolayer are needed for this purpose). Knowing the temperature T and the pressure P, one finds the chemical potential

$$\mu = -k_B T \log \frac{k_B T}{P\Lambda^3} \tag{2}$$

where Λ is the thermal deBroglie wavelength,

$$\Lambda = \frac{2\pi\hbar}{\sqrt{2\pi m k_B T}} \tag{3}$$

m being the helium atomic mass. Virial corrections to (2) may be applied if necessary. Then, since the chemical potential of the gas is equal to that of the adsorbed film in equilibrium, one can use these data to construct $\mu(N, T)$ of the film, where N is the amount adsorbed per unit area, in the range of T where the measurements are made.

Now these data must be combined with measurements of the isosteric (constant N) heat capacity of the film between this temperature and 0K (in practice, ~50mk),

$$C_N = T\left(\frac{\partial S}{\partial T}\right)_N \tag{4}$$

The heat capacity data are inverted to construct the functional dependences of the entropy, S(N, T). Then, using a Maxwell relation,

$$\left(\frac{\partial S}{\partial N}\right)_T = -\left(\frac{\partial \mu}{\partial T}\right)_N \tag{5}$$

one can, starting with $\mu(N, T)$, find $\mu(N, O)$. In the range 0.1 to 0.6 monolayers, $\mu(N, O)$ turns out to be constant, independent of N. This quantity, the low coverage, 0K chemical potential, technically the change in the energy of the system when one atom is added at 0K, is equal to ε_o (a small correction is made to account for the fact that the last atom added adsorbs onto a two-dimensional island of preadsorbed helium (Novaco and Campbell, 1975)).

There is, however, a more fruitful way of finding ε_o and the other excited bound state energies as well. The method involves observing the elastic scattering of a beam of atoms or molecules from a surface. Under precisely determined conditions of incident energy and direction, it is possible for the projectile to fall resonantly into one of the bound states while conserving energy (it picks up kinetic energy of motion along the surface in compensation for its loss of potential energy in being captured) and also conserving the component of its momentum parallel to the surface to within a reciprocal lattice vector of the substrate lattice. In other words, the final energy of the captured projectile,

$$E_{K,n} = \frac{\hbar^2 K^2}{2m} - \varepsilon_n \tag{6}$$

where $\hbar \mathbf{K}$ is its momentum along the surface, is equal to its incident energy

$$E_i = \frac{p_{11}^{\;2}}{2m} + \frac{p_z^{\;2}}{2m}$$

where $\mathbf{p}_{11}$ is the component of incident momentum parallel to the surface, subject to

$$\mathbf{p}_{11} = \hbar\,(\mathbf{K} \pm \mathbf{G})$$

where G is a reciprocal lattice vector in the plane of the substrate. Such an event, called a bound state resonance, usually shows up as a minimum or dip in the elastic reflection as a function of incident energy and direction. Measurements of incident energy and angles at which these dips occur may be combined with the above kinematical equations to give the bound state energies, ε_n.

In the cases of ^{3}He and ^{4}He on graphite, the values of ε_o determined from these bound state resonances in atomic beam scattering experiments (Boato et al., 1978; Derry et al., 1979) have turned out to agree with the thermodynamically measured values to within 1%. This is one of the real success stories of recent surface physics (Cole et al., 1981).

Atomic beam experiments have also given bound state energies for helium, H_2, HD, and D_2 on a variety of other substrates, both metals and insulators. Those values can be inverted to find the parameters in V(z) for any given system. Better yet, a single form has been found which fits all systems in a kind of law of corresponding states (Vidali et al., 1983). That is the form plotted in Fig. 1.

The starting point of the law of corresponding states is the long-range attractive part of Eq. (1),

$$C_3 = \frac{\hbar}{4\pi}\int_0^\infty \alpha(i\omega)\,\frac{\varepsilon(i\omega)-1}{\varepsilon(i\omega)+1}\,d\omega$$

where $\alpha(i\omega)$ and $\varepsilon(i\omega)$, the polarizability of the adsorbate molecule and dielectric function of the substrate, respectively, are obtained from optical absorption measurements (Bruch and Watanabe, 1977) or, in practice, suitably approximated from inadequate data. In any case, C_3 is regarded as a known parameter.

The potential V(z) shown in Fig. 1 is made universal by giving the energy scale in units of the well depth D, and the length scale in units of a characteristic length l, which is formed from D and C_3, $l^3 = (C_3/D)$. In addition, the origin of the z axis is shifted to z_{min}, the point where V(z) has its minimum by defining

$$z^* = \frac{z - z_{\min}}{l}$$

The dimensionless universal potential is then

$$V^*(z^*) = \frac{V(lz^* + z_{\min})}{D}$$

where by definition, $V^*(0) = -1$.

The idea then, is that for each adsorbate-substrate system, one finds the two scaling parameters D and l (or equivalently D and C_3) and the adsorption potential (and its bound states) are known. This procedure has been shown to work remarkably well for the long list of observed bound state resonances, and also some thermodynamic data for the heavier noble gases adsorbed on graphite (Vidali et al., 1983). Like any law of corresponding states, it is extremely useful (regardless of its theoretical merit) because it correlates a great many otherwise confusing experimental data, and permits sensible estimates to be made in new situations. Unfortunately, as we shall see, the data needed for calculating l and D are not yet always available.

As we saw earlier, the binding energy of a single adsorbate molecule on a bare surface can be deduced from the low coverage chemical potential at 0K. As more adsorbate is added to the surface, however, interactions between adsorbate atoms start to become increasingly important. After a number of layers are formed, if the film wets the substrate (i.e., grows continuously to infinite thickness) then the chemical potential approaches that of the bulk material of the adsorbate, μ_o. A useful approximation for many purposes is that the chemical potential of a multilayer film is equal to that of the bulk material, modified only by the long-range part of V(z),

$$\mu = \mu_o - \frac{\Delta C_3}{d^3}$$

where d is the thickness of the film, and ΔC_3 is the difference between the coefficient C_3 of the substrate, and the value C_3 would have if the substrate were replaced by bulk adsorbate. Applying the same idea at finite temperature, with $\mu_o(T)$ the chemical potential of bulk adsorbate at coexistence with its own vapor,

$$\mu(N,T) = \mu_o(T) - \frac{\Delta C_3}{d^3} \tag{7}$$

where d may be related to N via the number density, n, $N \approx nd$. The implicit assumption here is that the partial molar entropy of the film $(\partial S/\partial N)_T$ is equal to the specific entropy of the bulk material, i.e., that the film is a thin slab of bulk. This may be seen from Eq. (5). Since the film and bulk are each in equilibrium with their own vapor, Eq. (2) may be combined with Eq. (7) to predict that the vapor pressure of the film P(N,T) will approach the bulk vapor pressure $P_o(T)$ according to

$$P = P_o e^{-\frac{\Delta C_3}{k_B T d^3}} \tag{8}$$

Equation (8) is known as the Frenkel-Halsey-Hill (FHH) isotherm. Applied with care, it can be a useful starting point for reasoning about thick films.

The equilibrium between a film and its vapor is a dynamical process with adsorption and desorption simultaneously occuring in exact balance. Molecules of the gas (assumed ideal) strike a unit area of the surface at the rate

$$\frac{dN_{inc}}{dt} = \frac{P}{\sqrt{2\pi m k_B T}} \tag{9}$$

bringing with them a flux of energy,

$$\frac{dQ_{inc}}{dt} = P\sqrt{2k_B T/\pi m} \tag{10}$$

Obviously, atoms must desorb or reflect from the film so as to produce exactly the same outgoing fluxes. However, the condition for thermodynamic equilibrium is much more demanding than that.

True thermodynamic equilibrium must satisfy the principle of detailed balance. The principle can be expressed as follows: Imagine a reference plane far enough above the surface to escape even the long-range part of Eq. (1), but close enough so that it is well within one mean free path in the gas (these conditions are easily met in many experiments). Thus, every molecule passing through the reference plane from above will strike the surface, and every molecule passing through it from below came directly from the surface. The principle of detailed balance requires that the flux of molecules through the reference plane from below be identical in every detail to that from above. Any departure from this behavior in equilibrium would be equivalent to a Maxwell demon, making it possible in principle to devise a means of violating the Second Law of Thermodynamics (Goodstein, 1975).

To illustrate the point with a specific example, we shall see below that if a single substrate phonon causes a single molecule to desorb while conserving energy and momentum, the desorbed molecule will tend to emerge in the direction perpendicular to the surface. If this is the dominant mechanism of desorption, then the desorbed flux will be nearly perpendicular to the plane of the surface. But the incident flux is isotropic, molecules crossing the reference frame from above at all angles. In these circumstances, the only way the principle of detailed balance can be obeyed is if only those atoms incident normal to the surface stick to it, while those incident at other angles are reflected. We thus see that the probability of sticking, or the sticking coefficient α_s, is a key parameter in the kinetics of adsorption and desorption (Comsa, 1982).

PHONON PROPAGATION, REFLECTION AND TRANSMISSION

If a temperature gradient is present in a crystal, heat will flow to reestablish equilibrium. Usually, in metals, electrons carry most of the heat while in nonconducting matter the job is done by phonons. Phonons generally suffer frequent collisions with other phonons (due to anharmonicity and Umklapp processes), impurities, defects, and so on. As a result, heat spreads diffusively. However, at low temperatures ($T << \Theta_D$) Umklapp processes cannot occur, and there are few other phonons to scatter from. Consequently, phonon mean free paths in pure single crystals at low temperature may be larger than the size of the sample. In the decade of the 1970's, a number of groups set out to make use of the ballistic propagation of thermal phonons through crystals, largely in the hope of discovering why phonons pass so easily through interfaces, a phenomenon called the Anomalous Kapitza Conductance.

In the heroic early days of low temperature physics Peter Kapitza (1941) discovered that if heat were made to flow through a solid into a bath of superfluid helium there would be a jump or discontinuity in temperature at the interface between the two media. The (mysterious) cause of that discontinuity came to be known as the Kapitza Thermal Boundary Resistance. However, it was soon shown that such an effect was to be expected on the basis of simple elastic theory (Little, 1959; Khalatnikov, 1965). Sound waves incident at a boundary should be refracted or reflected in accordance with the degree of acoustic impedance mismatch between the two media (the analogy to power transmission and electric impedance mismatch between circuits is evident). The acoustic impedance is $Z = \rho c$ where ρ is the density and c the speed of sound. Thus, a boundary resistance producing a temperature discontinuity in heat flow is not only unsurprising, it is absolutely required at all real interfaces, a simple consequence of applying continuous stress and continuous displacement boundary conditions. Moreover, at the interface between liquid helium and anything else, it must be particularly large, since liquid helium has both the smallest ρ and the smallest c of any condensed medium. The great surprise, then, turned out not to be that there is a boundary resistance, but instead that it is much too small. Measured resistances between various solids and liquid helium were consistently a hundred or even a thousand times smaller than could be accounted for by Khalatnikov theory. The experiments typically consisted of measuring the temperature gradient in a heat flux in each of the two media, and extrapolating to the interface to find the temperature jump, $\Delta T = \dot{Q}/R_k$, where $\dot{Q}$ is the heat flux and R_k the Kapitza resistance of the interface (Wyatt, 1980).

The anomaly, then, is not in the resistance but in the conductance. This strange phenomenon, still unsolved after nearly half a century, became the focus of most of the early experiments with ballistic thermal phonons.

The results of a typical ballistic phonon experiment are shown in Fig. 2. A heater and a bolometer (detector of heat), about which more later, are attached to one face of a crystal. With the far surface in vacuum, the bolometer signal as a function of time after the pulse shows three peaks due to longitudinal, transverse, and mode-converted phonons (again, more about this below). When the vacuum at the interface is replaced by liquid helium, the signal that returns to the bolometer drops well below its original size, the remainder of the energy escaping into the helium instead of being reflected. This substantial transmission is the Kapitza anomaly; acoustic mismatch theory would predict a change of a fraction of 1%, much too small to detect. This vivid method of detecting the effect represents a substantial technical advance over the traditional method of measuring and extrapolating temperature gradients (Kinder et al., 1979).

These experiments, known as phonon reflection experiments, have led to a number of important insights into the nature of the Kapitza problem. Perhaps the most significant is that there is no Kapitza anomaly at all (i.e., the vacuum and helium signals are indistinguishable) when the reflecting surface is freshly cleaved or laser annealed *in situ* (Weber et al., 1978; Basso et al., 1983). Another, particularly relevant to our concerns in this chapter, is the result of reflections from a surface in neither vacuum nor liquid helium, but rather the intermediate case where the surface has adsorbed on it a thin film of helium. It is found that, as the film thickness increases from zero, the reflected phonon signal decreases from the vacuum value, reaching the bulk signal at about 3 layers adsorbed (as judged by a Frenkel-Halsey-Hill isotherm fit) (Guo and Maris, 1974; Kinder and Dietsche, 1974; Long et al., 1974). It was this observation, leading to the idea that the Kapitza effect was somehow related to desorption, that inspired much of the investigations to follow

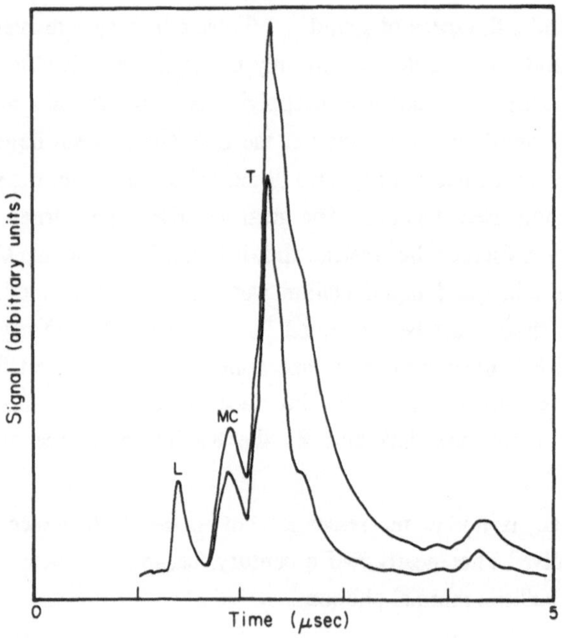

Figure 2. A phonon reflection signal. Three peaks are due to reflected longitudinal (L), transverse (T), and mode converted (L → T or T → L) phonons. Upper curve for reflecting face in vacuum; lower curve in liquid helium.

below. However, the explanation of the three-layer effect turned out to be related to film kinetics (Goodstein and Weimer, 1983; Weimer and Goodstein, 1983). Essentially, the point is that very thin films take much longer than the duration of a typical experimental pulse to react and absorb energy. As the film gets thicker it reacts more quickly until by three layers it is capable of absorbing all the available energy in the pulse. In any case, once three atomic layers (or so) are adsorbed, the reflected phonon signal becomes independent of the medium at the interface, not only as the film grows to infinite thickness, but even as the bulk liquid is pressurized and even solidified (Goodstein et al., 1980).

In a sense, these phonon reflection experiments are analogous to the surface spectroscopies using electrons, photons, atoms and so on, discussed in the last section. An obvious advance in the technique would be to improve its phonon frequency resolution, and a number of valiant attempts to do so have been made (Kinder, 1972, 1985; Bron, 1980). It has also turned out to be possible to improve greatly the resolution of the signal in both time and space. Those improvements opened an avenue of research that has proved very fruitful indeed, and is the central subject of this chapter.

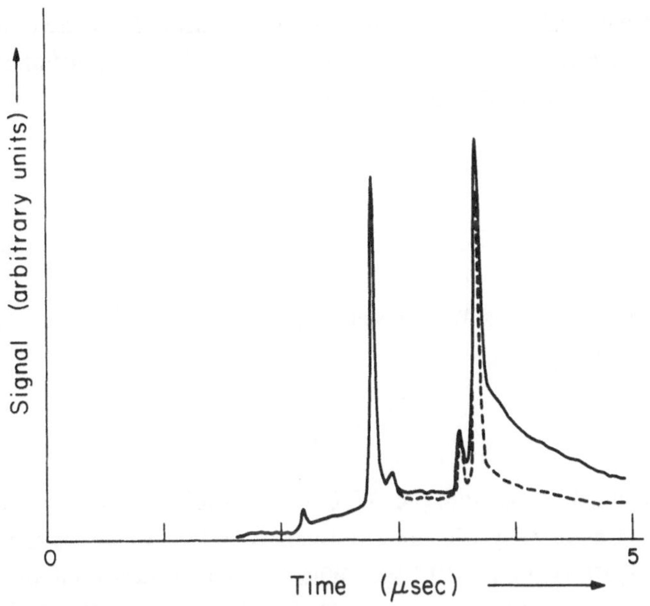

Figure 3. High resolution phonon-reflection spectrum. Solid curve is due to reflection from the sapphire-vacuum interface, dashed curve from the sapphire-liquid helium interface.

Figure 3 shows a high resolution phonon reflection spectrum. Suddenly, instead of three broad bumps, we see five narrow peaks. As many as nine peaks are possible depending on the geometry of the experiment. This result was entirely unexpected before it was observed. The same very small, very fast heaters and detectors responsible for these results form the basis of the desorption experiments to be discussed in the next section. Let us pause briefly to describe them (Taborek and Goodstein, 1979).

The high resolution measurements are made using thin film evaporated ohmic heaters and superconducting transition bolometers. These are standard devices; the high resolution is achieved by making them very small. The typical area of a high resolution heater or bolometer is of order 0.1mm^2. The time resolution in detecting the arrival of a pulse of phonons depends on the width of the electrical input pulse to the heater, the thermal relaxation times of heater and bolometer, and the resolution of the readout electronics. All of these are easily made to be less than about 50nsec. However, it is also limited by $(l/c)\sin\theta$ where l is the size of the device, c the speed of sound, and θ the angle between the plane of the device and the direction of propagation. This is the difference in arrival times on opposite sides of the device. Since c $\sim 10^6$ cm/sec (in sapphire), we must have $l \lesssim 0.5$mm to make l/c comparable to the other times in the problem.

The heaters, made of constantan, nichrome or (above its superconducting transition temperature) aluminum, are 250 to 1000Å thick, designed to have 50Ω resistance matching the impedance of the coaxial transmission lines attached to them. If a pulse of current dissipates power W in a heater whose initial temperature was T_o, its temperature after the onset of the pulse is governed by the conservation of energy:

$$W = C\frac{\partial T}{\partial t} + \xi\,(T^4 - T_o^4). \tag{11}$$

The power either warms the heater via its heat capacity C, or radiates phonons with a known coefficient ξ (related to the Kapitza effect). In a typical case, a heater made of 5 x 10^{-8}gm of constantan, dissipating W = 1 watt starting with T_o = 2K, will reach a final steady state temperature, T_h,

$$T_h = [\frac{W}{\xi} + T_o{}^4]^{1/4} \tag{12}$$

of about 16K. The initial temperature rise, $\partial T/\partial t \approx W/C \approx 10^{11}$K/sec is very rapid. Steady state is approached exponentially with time constant $\tau_h = R_k C$ where R_k is the Kapitza resistance $R_k = (4\xi T_h^3)^{-1}$, giving $\tau_h \approx 10^{-9}$ sec. When the pulse is turned off, the same happens in reverse, except that the final decay to T_o has time constant $C(4\xi T_o{}^3)^{-1} \approx 40 \times 10^{-9}$ sec. Similar considerations apply at the bolometer which has a thermal time constant of the same order of magnitude. The bolometer is made of 2000Å thick films of tin which become superconducting at 3.7K. Held at the edge of its superconducting transition, the electrical resistance of the bolometer changes sharply when it is heated; this is the detection mechanism. Experiments may be performed at temperatures below 3.7K by applying a magnetic field to the bolometer, thereby reducing its transition temperature. Commercially available pulse generators provide square pulses of width $\gtrsim$30nsec to the heater, and standard readout devices (e.g., a boxcar integrator) make it possible to resolve pulses of that width in the detector. Thus, if the heater and bolometer are sufficiently small laterally, all the characteristic times are of the same order ($\lesssim$50nsec) and high resolution phonon reflection experiments become possible.

The complex time-of-flight spectra revealed by high resolution phonon reflections came as a surprise because the low resolution result was in accord with expectations from standard isotropic elastic theory. In an isotropic crystal, there would be three phonon modes, one longitudinal (L the fastest) and two transverse (T) modes with the same speed (about half that of the longitudinal). Either an L or T wave incident at a surface can excite both L and T reflections, provided the surface parallel component of the wavevector **k** is conserved and the incident polarization vector has some projection on the outgoing one. The result is three reflection peaks due to L → L, L → T or T → L, and T → T processes.

The difference between this expectation and the high resolution result is due to crystal anisotropy (Taborek and Goodstein, 1979). One would not expect this to be a very important effect in sapphire, the most isotropic of the commonly used crystals (the transverse sound speeds differ by only a few percent), but, rather than 3 peaks, as many as 8 are resolved in some experimental spectra. Clearly, even small anisotropies can have important effects on reflection.

Because of anisotropy, the energy flow associated with a sound wave is not generally in the same direction as **k**. The reason is that, no matter in what direction the planewave fronts propagate, energy tends to get dragged in the direction of the stronger elastic constants. Conservation of the parallel component of **k** is still the condition for reflection at the interface, but now a **k** vector from the heater that reflects into a **k** vector toward the bolometer does not necessarily send any energy to be detected at the bolometer. In general, the energy will follow a completely different path, arriving at a different point on the lower surface after reflection. Moreover, the speed of each mode now varies with direction through the crystal so that each possible path up to the surface and back that does carry energy to the bolometer will generally produce a different time-of-flight than all others. Thus, there are nine possible processes, each with its own arrival time, capable of carrying energy from heater to bolometer by means of specular reflection.

To test these ideas, calculations and experiments were done for ten different heater-bolometer combinations on a sapphire crystal. A total of 54 peaks were predicted to occur, each separated from all others by at least the 25nsec resolution of the instruments. Every one was observed to arrive at the predicted time, fully validating the analysis.

The foregoing discussion is concerned entirely with the arrival times of specular peaks. An analysis of the shapes, widths and heights of the peaks tells a quite different and equally important story. There are three principal conclusions:

1. Only part of the phonon energy incident at the vacuum interface is reflected specularly. The rest is scattered randomly, or diffusely, back into the crystal.
2. The diffuse contribution to the spectrum is, in some directions, strongly affected or even dominated by phonon focusing, another effect of crystal anisotropy (Maris, 1971).
3. When the vacuum at the interface is replaced by liquid helium the portion of the spectrum due to specular reflections is unaffected, but the diffusely reflected part vanishes almost entirely from the signal. In other words, that part of the signal in vacuum which is due to diffuse reflection is responsible for the Anomalous Kapitza Conductance (Taborek and Goodstein, 1980a, 1980b; Marx and Eisenmenger, 1982; Northrop and Wolfe, 1984).

The first and last of these points is most easily seen by examining a spectrum which (for reasons that will become clearer later) is little affected by phonon focusing. Figure 3 is a case-in-point, showing a vacuum spectrum and also one with liquid helium. As others have also observed (Folinsbee and Harrison, 1978), the effect of the helium is not so much to reduce the peaks as to narrow them, in this case to a width of around 50nsec, consistent with the original pulse width and (as we have seen) the resolution of the experiment. The long, straggling tail that the helium has eliminated is due to phonons whose time of arrival cannot be accounted for by any specular process.

Moreover, the difference between the vacuum and helium spectra in Fig. 3 is very much what one should expect from diffuse reflections. For a given process (say, slow transverse → slow transverse, responsible for the last peak), diffuse reflection will lead to a sudden onset in the signal at precisely the specular reflection time since specular reflection is the time-of-flight minimizing process. After the sudden steplike onset, due to diffuse reflections from the same place on the surface responsible for the specular reflections, the straggling tail, due to diffuse reflections from points increasingly farther away on the upper surface, is just what one expects.

A simple quantitative analysis of diffuse scattering of a delta function heat pulse, $\dot{Q}(t) = Q_o\delta(t)$, gives precisely that result. The diffuse contribution to the signal S(t) is a sudden jump from zero at the specular arrival time followed by a decay that goes $\sim t^{-5}$ (the straggling tail). By the same analysis, if we assume the fraction of nonspecular scattering at the interface to be η, then the integrated area under the specular and nonspecular portions of S(t) gives the ratio of energies returning to the bolometer,

$$\frac{E_{specular}}{E_{diffuse}} = \frac{1}{2\pi}\,\frac{1-\eta}{\eta} \qquad (13)$$

This formula, together with a measurement of the areas in Fig. 3 (and many others like it) yields the estimate $\eta \approx 0.5$ for the crystals used in these experiments. The reason for this value is almost surely connected to the quality of the surface. In fact, in an analogous recent experiment (Northrop and Wolfe, 1984) the fraction of specular reflection was reduced from 80% to zero by roughening the sapphire surface.

Figures 4 and 5 show two more experimental spectra, both taken with heater and bolometer about 2mm apart but along different directions in the plane of the lower surface. Figure 5 is rather similar to the data we have been discussing, but Fig. 4 is dramatically different. Superimposed on a large background (the reason for which will become clear later), it has five narrow specular peaks at the predicted arrival times followed by a much larger bump at about 3.5 μsec, a time that does not correspond to the arrival of energy by any specular process. Moreover, when liquid helium is added at the reflecting interface the large bump nearly vanishes. If the conclusions we have reached above are correct, the large bump must somehow be due to diffuse reflections.

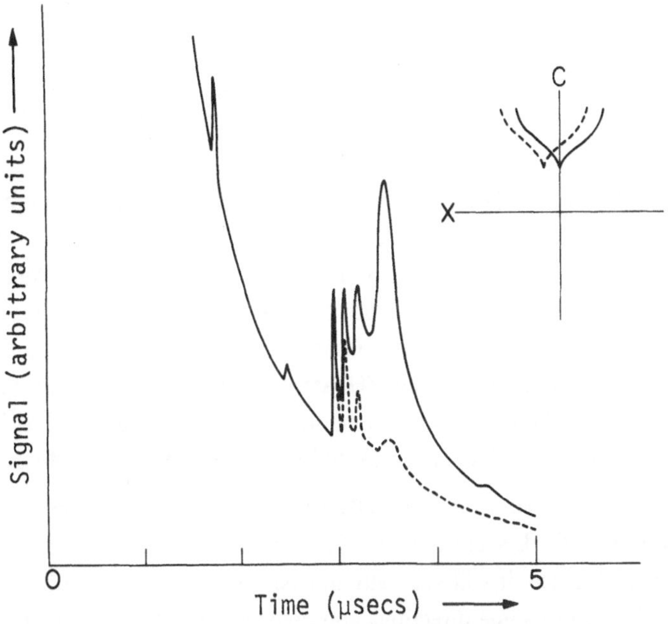

Figure 4. Spectra due to phonons reflected from the sapphire-vacuum interface (solid curve) and the sapphire-liquid helium interface (dashed curve). As explained in the text, the inset shows the intersection of a caustic surface from the heater (solid curve) and bolometer (dashed curve) with the reflecting interface. The fact that these curves intersect each other near their cusps is responsible for the large, broad bump at 3.5μsec in the vacuum spectrum.

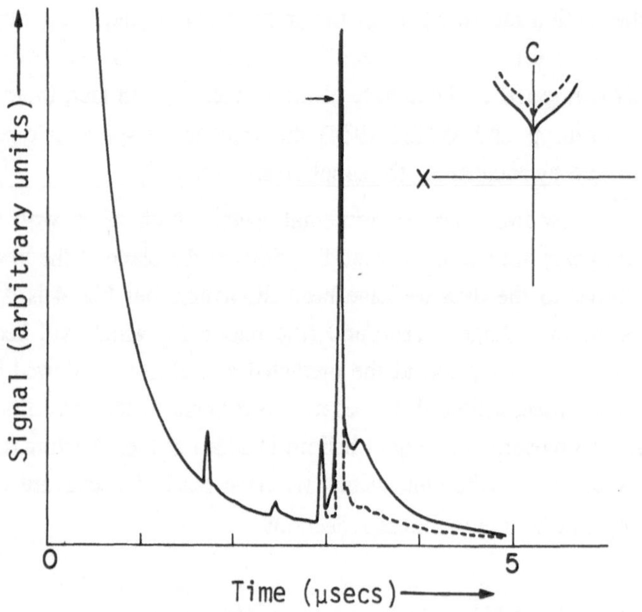

Figure 5. Same as Fig. 4., except that in this case the heater and bolometer caustics do not intersect.

The large bump is in fact a consequence of diffuse scattering in combination with phonon focusing. Phonon focusing is a consequence of crystal anisotropy that was well known before these experiments were performed, but the theory then available was inadequate to account for the effect. By a kind of historical accident, a number of groups became interested in phonon focusing singularities at almost the same time, developing various ways to describe these phenomena (Hensel and Dynes, 1979; Northrop and Wolfe, 1979). We will make use here of a catastrophe theory approach which is helpful in understanding the phonon reflection data (Taborek and Goodstein, 1980a, 1980b).

As we have already seen, in a real, anisotropic crystal, phonon energy does not flow in the same direction as the **k** vector. Thus, if a point source of phonons sends out a burst which is isotropic in k-space, it will generally not be isotropic in its distribution of energy, the energy being focused in some directions and defocused in others. A particularly useful way of thinking about the problem is in terms of the so-called slowness surface, the surface in k-space of constant ω for each sound mode.

In an isotropic crystal, the slowness surface would be spherical, with two concentric sheets, one for the longitudinal mode, and the other (larger) one for the two degenerate transverse modes. For each **k** vector from the origin to a point on the slowness surface, the associated energy flow is in the direction of the group velocity, $\nabla_k\omega$, in other words perpendicular to the slowness surface (just as an electric field, $\mathbf{E} = -\nabla\varphi$ is perpendicular to an equipotential surface). As shown in Fig. 6a, for an isotropic crystal with a spherical slowness surface, **k** and $\nabla_k\omega$ are always in the same direction.

To understand what happens when the slowness surface is not spherical, we can consider a bizarre but illustrative example, a slowness surface in the form of a cube (Fig. 6b). Then all the **k** vectors from the origin to each face of the cube are associated with group velocities $\nabla_k\omega$ in the same direction, perpendicular to the face. In real space, that would correspond to all the energy from the point source being concentrated into six beams perpendicular to the six faces of the cube.

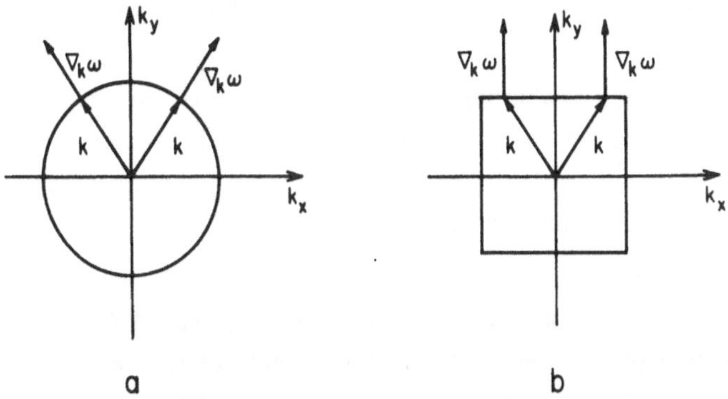

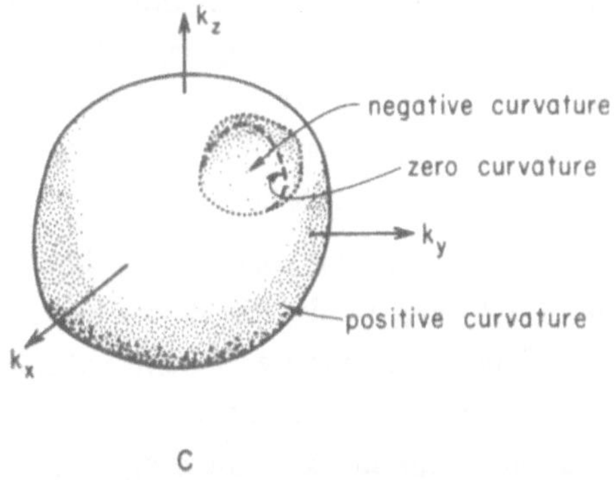

Figure 6. Some examples of slowness surfaces: a) Spherical slowness surface, corresponding to an isotropic crystal; k-vectors and associated group velocities, $\nabla_k\omega$, are in the same direction. b) Cubic slowness surface (not corresponding to any real crystal). All k-vectors to any surface have group velocities in a single direction. c) Distorted spherical slowness surface, having regions of positive and negative curvature joined at a closed curve of zero curvature.

In conventional phonon focusing theory, if a wavepacket of **k** vectors subtends solid angle $d\Omega_k$, and the associated packet of group velocities subtends $d\Omega p$, then the phonon focusing factor is defined as $d\Omega_k/d\Omega p$ since that measures the degree to which energy has been concentrated. For a spherical slowness surface, $d\Omega_k/d\Omega p = 1$ everywhere, whereas on the cube it is infinite along the six directions perpendicular to the faces. It is easy to see what has caused the infinity: the flat surfaces have refracted the energy from a number of **k** vectors into the same direction. In fact, the focusing factor turns out always to be inversely proportional to the curvature of the slowness surface, having infinities wherever the curvature is zero. As we can see from the example of the cube, however, an infinite focusing factor does not mean infinite intensity. In the cube, one sixth of the available energy goes off into each of the six singular directions. The energy flux density is genuinely singular in those directions, however, in the sense that it does not fall off as r^{-2} (it is independent of r instead).

Real crystals do not have cubic slowness surfaces, of course, but they do have many of the qualitative features of that extreme example. The phonon focusing factor is infinite wherever the curvature of the surface is zero, those directions being mathematically singular in the sense that the energy flux falls off more slowly than r^{-2}.

The shape of the slowness surface of a real crystal is determined by the numerical values of its elastic constants (sapphire, for example, has six independent elastic constants). It is usually a perturbed sphere, a typical perturbation being a depression, as one obtains by pushing a thumb into the surface of a balloon. The result (as shown in Fig. 6c) is a region of negative curvature, connected to the remaining positively curved surface at a closed curve of points of zero curvature. The wavevectors from the origin to those points of zero curvature on the slowness surface sweep out a topologically conical surface in k-space, with a corresponding topologically conical surface in real space on which there is infinite (i.e., singular) phonon focusing. In classical wave theory this is known as a caustic surface; in catastrophy theory it is called a fold catastrophe (Gilmore, 1981).

The role of catastrophe theory in this analysis is to show that this kind of behavior is typical, or perhaps one should say, stable. Small changes in the elastic constants will change the precise shape of the caustic surface, but will not make it go away; its existence is not a lucky accident. Moreover, the mathematics of catastrophe theory yields two more bonuses: it tells us the distance dependence of the energy flux along the caustic surface (it goes as $r^{-5/3}$) and it tells us that somewhere on the causitc surface there is typically a cusp which is an even higher order singularity ($r^{-3/2}$). It is this cusp which, as we shall see, is responsible for the big bump in the diffuse reflection spectrum.

The existence and importance of caustics in phonon transmission and reflection, beautifully illustrated in phonon imaging experiments, is discussed by Wolfe and Northrop elsewhere in this volume (see also Northrop and Wolfe, 1979, 1984). Since the exact geometry of the caustic depends on the numerical values of the elastic constants, its direction in the crystal is not generally along symmetry directions or planes, and must be found numerically. The result of a numerical calculation for the sapphire crystals used in producing Figs. 4 and 5 is that a caustic surface complete with cusp, originating from the heater, intersects the upper reflecting surface.

For orientation, the sapphire crystal is trigonal, the unit cell formed by elongating one diagonal of a cube. The elongated axis called C is a threefold axis of symmetry. In the plane perpendicular to C is a twofold axis called X, and a Y axis chosen to form an orthogonal right handed (XYC) system. The samples used in all the experiments reported in this chapter are cylinders whose axis is parallel to the Y axis of the lattice (also known as the (I$\bar{1}$OO) direction). The top and bottom surfaces thus lie in the C-X plane.

Given this orientation, numerical calculation showed that the caustic surface due to the slower transverse mode from a point source (the heater) at the origin would intersect the upper surface with a cusp in the C direction as shown by the solid curves in the insets of Figs. 4 and 5. For diffuse scattering back into the crystal, each point on the upper surface is a source of heat. Points along the intersection of the caustic and the upper surface are the strongest sources, and the cusp is strongest of all.

A caustic surface emanating from the bolometer will intersect the upper surface in a similar curve, displaced from the heater caustic just as the bolometer is displaced from the heater. The intersection of the bolometer caustic with the upper surface is a curve of points from which diffusely reflected phonons will be strongly focused back to the bolometer. These are shown by dashed curves in Figs. 4 and 5. In the case of Fig. 5, the bolometer is displaced from the heater by 2mm along the C axis. Consequently (as shown in the inset), the heater and bolometer caustics do not intersect, and the resulting reflection spectrum is not strongly affected by phonon focusing. However, in the case of Fig. 4, the bolometer is displaced 2mm from the heater along the X axis, and the two caustics do intersect as shown in the inset. The point of intersection on the upper surface is a hot spot from which diffusely radiated phonons are strongly focused back to the bolometer. It is this intersection that produces the big bump in Fig. 4.

This interpretation of the origin of the big bump is supported by a number of corroborating observations. For example, the arrival time of the bump corresponds precisely to the time-of-flight of the path from heater to intersection to bolometer. A series of experiments also showed that the size of the bump could be increased or diminished by moving the intersection nearer to or further from the cusp which is the direction of maximum focusing (conventional phonon focusing theory could not explain this observation since it merely gives infinite focusing everywhere on the caustic). Northrop and Wolfe (1984 and this volume) have also reported that sharp, stable structures like the big bump can be producd by diffuse reflection in combination with phonon focusing. Incidentally, the large background in Fig. 4 is explained by the intersection of the caustics within the crystal; the phonon scattering that occurs in the crystal is strongly focused back to the bolometer.

Having explained the origin of the big bump, it becomes possible to make effective use of it. The big bump is entirely due to diffusely scattered phonons, and those phonons are responsible for the anomalous Kapitza conductance. In effect, the big bump concentrates many of those phonons into a small portion of the time-of-flight spectrum affording a very sensitive probe of the Kapitza anomaly. For example, as we see in Fig. 4, that very prominent feature of the spectrum vanishes when there is liquid helium at the interface. This circumstance has been used in an experiment that helped to upset an important misconception about the nature of the Kapitza problem (Taborek and Goodstein, 1981).

For "classical" interfaces (i.e., any interface not involving liquid helium, and to a lesser extent, hydrogen and neon), the observed temperature discontinuities in a heat flux were in satisfactory agreement with the expectations of acoustic mismatch theory (Anderson, 1980). It was thus widely believed that the anomalous conduction was a quantum mechanical effect related to the low mass of helium, hydrogen and neon. However, spectra, comparing vacuum reflections to those at a surface onto which a thick film of argon has been flash evaporated, are indistinguishable from Fig. 4. If, as we have argued, the difference between the solid and dashed spectra in Fig. 4 is a supremely sensitive illustration of the Kapitza anomaly in liquid helium, precisely the same anomaly occurs in solid argon. The anomaly consists of complete transmission of what would otherwise have been diffusely reflected phonons. Companion experiments showed exactly the same result when the helium or argon was replaced by neon, krypton or xenon. Whatever the nature of anomaly, it occurs equally at all these interfaces. Marx and Eisenmenger (1982) have reached similar conclusions about the anomaly at the interface between silicon and a variety of other materials.

It is easy to see in retrospect why previous investigators were misled into associating the anomaly chiefly with helium. The older heat flux experiments were difficult and subject to substantial errors in finding the temperature discontinuity by extrapolating temperature gradients. The results, regardless of the interface in question, were consistent with roughly half the energy being transmitted. For most interfaces, the acoustic mismatch is modest and one expects roughly half the energy to be transmitted. Only for the lighter elements, and especially helium, was there a clear discrepancy between observation (half the flux transmitted) and expectation (less than 1%). However, the phonon reflection experiments make it manifestly obvious that essentially the same phenomenon is occuring at all the interfaces studied because now the experiments at different interfaces are compared to each other rather than to theoretical expectations.

DESORPTION OF HELIUM

Now that we have mastered the necessary background in surface science, adsorbed films, and the behavior of thermal phonons in sapphire crystals, we are prepared to put these elements together to study the subject at hand: the desorption of helium. This is a story of rewarding progress in the face of formidable obstacles with, in many cases, the obstacles almost as interesting as the progress.

Three kinds of experiments stand out as particularly significant. In one, the time required for a film to desorb, τ, has been measured directly in real time (rather than being inferred, e.g., from TPD experiments as mentioned above). The second, called the phonoatomic effect by analogy to the photoelectric effect, demonstrates directly the existence of events in which a single phonon causes the desorption of a single atom. The third, a scattering experiment, measures the sticking coefficient α_s. However, before getting to the experiments, let us try to set the problem in its theoretical context.

As mentioned in section II, adsorption-desorption equilibrium must obey the principle of detailed balance. In effect, that means that the flux of atoms leaving the surface must be identical in every respect to the flux that would result if the crystal were replaced by a half space of gas. A similar statement can be made about the phonons inside the crystal; the equilibrium flux that leaves the surface must be identical to that which would result from a half space of sapphire crystal. Thus, the system may be viewed as two gases, one of atoms and the other of phonons, with mutual annihilation and creation producing equilibrium at the interface. Yet, at the pressures and temperatures of these experiments, the gas of atoms is in its extreme classical limit (i.e., the number density $n << \Lambda^{-3}$ where Λ is the thermal deBroglie wavelength), whereas the phonon gas is in its extreme quantum limit ($T << \Theta_D$). Unless both gases simply reflect from the interface, incident phonons must create just the right classical distribution of desorbing atoms, and adsorbing atoms must create just the right Bose distribution of phonons. But the gases do not simply reflect. We have already seen that on the order of half the phonons disappear at a typical interface (if the adsorbed film is three layers thick or more), and we will see below that most of the incident atoms stick to a typical interface. We thus have an intriguing question to investigate: what are the mechanisms that achieve equilibrium at the interface between the two systems?

In most recent theoretical work, attention is centered on the microscopic mechanisms of desorption and on predicting desorption kinetics, for which at least some experimental data exist. The usual approach is to assume an atom-surface potential (Eq. (1) for example, although various others are used). An adsorbed atom is in a bound state of this potential. Vibrations of the substrate due to the phonons induce transitions in the state of the atom, some of which are transitions into the unbound continuum, i.e., desorption. Matrix elements are calculated, densities of states assumed, and transition rates are computed. Within this scheme there are so many assumptions, approximations and choices of parameters to be made that by the end of the calculation it is hard to tell whether either agreement or disagreement with experiment is significant. (For a sampling of this literature, see Ying and Bendow, 1973; Bendow and Ying, 1973; Goodman and Romero, 1978; Gortel et al., 1980; Jedrzejk et al., 1981; Leuthäuser, 1981, 1983; Goldys et al., 1982; Sommer and

Kreuzer, 1982; Goodman, 1982; Cole and Toigo, 1982a, 1982b.)

Nevertheless, it is possible to extract common features from these competing theories which do help us to understand what ought to be expected in the desorption of helium. One point, of course, is that desorption is not complicated by the chemical reactions studied in surface chemistry. Another advantage of helium is that, since its adsorption binding energy (typically between 20 and 90 k_B in the experiments below) is much less than the Debye temperature of the substrate ($\sim 10^3$K for sapphire, for example), single phonons exist that are capable of desorbing atoms. It is generally either argued or assumed that, since multiphonon processes are of higher order, single phonon processes will be the dominant mechanism of desorption. But single phonon desorption has kinematical features that are independent of the details of any particular theory. The phonon-atom interaction must conserve both energy and the component of momentum parallel to the surface. Suppose (to take a typical example) a phonon of energy $40k_B$ desorbs a helium atom bound with $30k_B$, so that it emerges with $10k_B$ of kinetic energy, $p^2/2m$. Then the atom has momentum $|\mathbf{p}| = \sqrt{20k_B m} \simeq 1 \times 10^{-19}$ $gm\ cm\ sec^{-1}$, but the phonon only had momentum $|\hbar \mathbf{k}| = 40k_B/c \approx 6 \times 10^{-21}$ in the same units (c $\approx 10^6$cm/sec is the speed of sound). Thus, no matter what the direction of the incident phonon, no more than a few percent of the atoms's momentum can be in the conserved component parallel to the surface. In other words, the atom must desorb very nearly perpendicular to the surface. The prediction that desorption will occur in a narrow cone perpendicular to the surface is a general feature of single-phonon theories. This feature, of course, does not help us to understand how detailed balance is maintained - quite the opposite in fact, since detailed balance requires an isotropic flux from the surface.

Theories differ in the extent to which they take into account the possibility of thermalization within the film. The most extreme example is a phenomenalogical model which assumes that all desorption is due only to thermalization (Weimer and Goodstein, 1983; Goodstein and Weimer, 1983). It was this model that explained the three-layer effect in phonon reflection mentioned above. However, this model does not help us understand how detailed balance works either; it merely assumes it is obeyed.

The simplest kind of helium desorption experiment is shown in Fig. 7. The elements are still the heaters, bolometers and crystals of the previous section, but now they are arranged so that a film of helium adsorbed onto the heater may be rapidly desorbed by a heat pulse and detected by the bolometer. The time-of-flight signal that results has a characteristic Maxwellian shape; the desorbed atoms arrive at the bolometer with a thermal distribution of kinetic energies, $\overline{(1/2)mv^2} \approx (3/2)k_B T_a$ where T_a is the temperature of the atoms, and since the time-of-flight $t \sim v^{-1}$, the arrival time of the signal maximum, $t_m \sim (T_a)^{-1/2}$. Thus, the spectrum in Fig. 7 is a thermometer for the arriving helium beam. Careful analysis of much experimental data shows that generally $T_a \approx T_h$ where T_h (as in Eq. (12)) is the steady state temperature to which the heater has been pulsed (Cohen and King, 1973; Andres et al., 1977; Sinvani et al., 1984).

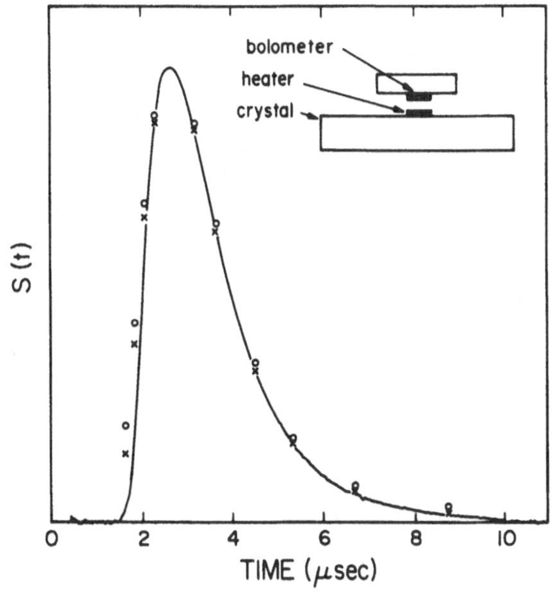

Figure 7. Simple desorption experiment. Geometry, as shown in the inset, with heater and bolometer separated by 1.2mm. Solid curve is the experimental spectrum.Open circles and crosses are the predictions respectively of a single-phonon desorption model and a thermal desorption model (Sinvani et al., 1984).

Much effort has been expended trying by means of theoretical analysis to learn more than just the temperature T_a from simple desorption spectra like Fig. 7. These efforts, although helpful in some respects such as sorting out the effects of heater-bolometer geometry, have not led to a deeper understanding of the desoption process. One reason is that a number of radically different models of flash desorption tend to yield indistinguishable predictions for S(t). Another is that all such analyses are based on the assumption of ballistic propagation from heater to bolometer (Sinvani et al., 1984). The experiments are done under conditions in which the desorbing atoms do not suffer collisions with preexisting gas, but it is harder to eliminate collisions between desorbing atoms.

A simple analysis (Cowin et al., 1978) shows that if ΔN atoms per unit area are desorbed in time τ, then the number of collisions per atom after time t will be $\nu \approx \sigma\Delta N \log(t/\tau)$, where σ is the cross section. The factor $\sigma\Delta N$ is roughly one per layer desorbed, and experience shows that the logrithmic factor is never very far from one. Thus, unless ΔN is much less than a monolayer, the experiment is being performed in that embarrassing regime of very few collisions where the behavior is neither ballistic nor hydrodynamic. In a typical case like Fig. 7, $\sigma\Delta N$ and ν are both of order one.

Since the effect of a small number of collisions defies theoretical analysis, data like Fig. 7 are best handled empirically. It is found that the apparent temperature of the beam T_a, measured normal to the desorbing surface (as in Fig. 7) is not much affected by the collisions, but at an angle θ to the normal, both the mean energy of the desorbed beam and its intensity decrease (Taborek, 1982). The decrease of intensity can be fitted to the form $\cos^{\gamma}\theta$, where γ must be measured in each experimental situation (for isotropic desorption and ballistic flight, $\gamma = 1$). Thus, although the results of simple desorption experiments can be well characterized and described, they reveal little about the underlying mechanisms of desorption.

Nevertheless, experiments in this configuration have been used to measure the desorption time τ as well as the time needed for readsorption. This is done not by studying just S(t), but rather how it changes as a function of the duration of the heater pulse, and of the time between heater pulses. With the first of these (pulse duration), one studies the kinetics of desorption. With the second (time between pulses), one studies the kinetics of adsorption.

Let us first consider adsorption (Sinvani and Goodstein, 1983). It is found that S(t) as in Fig. 7 is independent of the pulsing rate provided the time between pulses is longer than a minimum value called t_{rc}, the critical repetition rate. Typically, t_{rc} is between 10^{-3} and 10^{-1} sec, compared to pulse widths of 10^{-8} to 10^{-5} sec. As the time between pulses decreases below t_{rc}, the height of S(t) decreases monotonically. The reason is, obviously, that for intervals less than t_{rc}, the film is not able to recover to its initial equilibrium state.

Over a very wide range of conditions this recovery is limited by the rate at which atoms arrive from the gas phase to be readsorbed into the film. Using Eq. (9), we can write the rate of change of the amount adsorbed as

$$\frac{dN}{dt} = \frac{\alpha_s P}{\sqrt{2\pi m k_B T}} - \dot{N}_{des} \tag{14}$$

where $\dot{N}_{des}$ is the rate of desorption. Just after a heat pulse, the film is far from equilibrium (in many experiments, in fact, the surface is stripped bare), and $\dot{N}_{des}$ is negligible compared to the sticking term. Then, since the gas can be taken to be an infinite reservoir with constant P and T, dN/dt is simply constant and N grows linearly in time. Eventually, N approaches its equilibrium value, the two terms on the right become comparable, and the final value of N is approached exponentially. This last phase occupies such a small fraction of t_{rc} that it can be ignored for practical purposes, however, so that Eq. (14) can be integrated to give

$$t_{rc} = \frac{\Delta N}{\alpha_s} \frac{\sqrt{2\pi m k_B T}}{P} \tag{15}$$

In a regime where the gas pressure P was high enough to be measured externally, it was confirmed that $t_{rc} \propto P^{-1}$, with $\Delta N/\alpha_s \approx 1\times10^{15}$ atoms/cm^2, independent of pulse power. Since, as we shall see below, $\alpha_s \approx 1$, close to a full monolayer ($\Delta N \approx 1\times10^{15}$) was being desorbed in these experiments, roughly the entire initial coverage. When the pressure is too low to be measured externally, this technique can be turned around and used as a manometer; provided sufficient power is used to desorb the entire film, gas pressure is measured (to within ~30%) by measuring t_{rc}. The errors and uncertainties in this procedure are tolerable because modest changes in N and T will cause P to vary over 5 or 6 orders of magnitude. The measurement of P is used via Eq. (2) to find μ, the chemical potential which is the real quantity of interest. A 30% error in P translates into a 1% error in μ.

Thus, adsorption kinetics are used in these experiments to find the preexisting equilibrium state (i.e., μ) of the gas and hence the film. This is very convenient experimentally but tells us little about the adsorption and desorption processes themselves.

We do, however, learn a great deal about the kinetics of desorption by studying the signal S(t) as a function of the heater pulse duration t_p (Sinvani et al., 1982). Figure 8 shows a series of spectra for different t_p (all other conditions including pulse power are kept

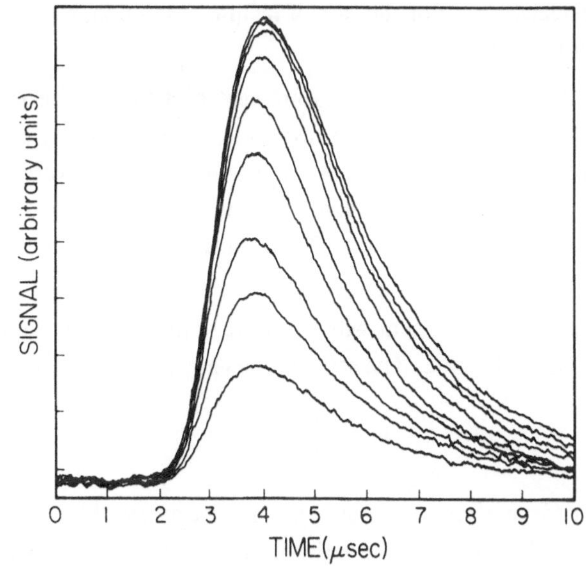

Figure 8. Desorption spectra for various pulse widths at constant heater power. In ascending order, t_p = 0.03, 0.06, 0.08, 0.15, 0.22, 0.5, 1, 1.5, and 2.5 μsec. Experimental geometry as in Fig. 7.

constant). As t_p increases from 30nsec to 60nsec, the signal (roughly) doubles in size. But as t_p increases from 1 μsec to 1.5 to 2.5 μsec it hardly changes at all. The reason is that by the time 1 μsec is past, desorption is complete; leaving the heater on for a longer time has no effect on the signal. A more quantitative result is obtained by assuming that the signal maximum $S_m(t_p)$ approaches its saturation value exponentially

$$S_m(t_p) = S_m(\infty)\,(1 - e^{-t_p/\tau})$$

The data of Fig. 8 fit this formula with $\tau = 110$nsec.

The desorption time constant τ, a quantity of considerable practical importance, has long proved elusive despite much experimental and theoretical attention. It is generally deduced indirectly using temperature programmed desorption (see above) and other techniques, sometimes with very misleading results. The experiments shown here are, to our knowledge, the first direct determination of this key quantity.

Estimates of τ have been made using kinetic theory (Frenkel, 1955), chemical reaction rate theory (Menzel, 1975), the quantum theories of desorption outlined above, and others. The result is generally described in activated form,

$$\tau = \tau_o e^{E/k_B T} \tag{16}$$

where E is some heat or binding energy of adsorption. Theories differ from each other in their estimates of the prefactor τ_o. For the case of helium desorption, estimates of τ_o have ranged from 10^{-5} down to 10^{-14}sec.

The data in Fig. 8 represent a single measurement of τ for a given initial chemical potential and temperature, and a given final temperature, T_h, governed by the pulse power (Eq. (12)). A series of such measurements holding μ and T fixed but varying T_h permits the form of Eq. (16) to be tested. The dependence of the parameters E and τ_o on μ and T can also be tested by additional experiments. The results of extensive measurements can be summarized briefly:

The data are in excellent agreement with Eq. (16), with T_h in the exponential. The initial temperature has little influence. The activation energy is found to fit

$$E = \frac{2}{3}\mu \tag{17}$$

over a wide range of μ (from about 30 to 90k_B). The reason for this strange but simple dependence is unknown. The prefactor τ_o varies from 10^{-9} to 10^{-10} sec, decreasing as E increases, a phenomenon known as the compensation effect.

The most surprising thing about these results is that they are so simple. The experiments themselves are rather violent; the temperature is raised from $\lesssim$3K to 10K or more in a matter of nanoseconds, desorbing nearly the entire film. By contrast, theories usually assume small perturbations such as $\Delta N << N$ and $\Delta T << T$. It is thus unfair from the outset to compare these experiments to theory. To make matters worse, as we have seen, the theories themselves are subject to large uncertainties.

We neverthless can get a feeling for the situation as follows. To stand in for the various single-phonon theories, we adopt a simple model in which all phonons incident at the surface with energy $\hbar\omega > \mu$ succeed in desorbing atoms, and all others have no effect. Alternatively, we can assume that the incident flux of phonons thermalizes in the film, warming it to temperature T_h, which causes desorption because it raises the vapor pressure of the film. Putting in the parameters of the experiment, both of these models make essentially the same prediction (Goodstein, 1984)

$$\tau \approx 10^{-12} e^{|\mu|/k_B T_h} sec$$

This form differs from the one that fits the data, but the smaller prefactor and larger exponent combine to give values of τ in reasonable (i.e., order of magnitude) agreement with the data.

This state of affairs, in which theory and experiment are not quite comparable and not quite in either agreement or disagreement, is hardly satisfactory, but perhaps little more should be expected of the relatively crude experiments we have been discussing in which the film is adsorbed directly onto a heater. Further progress can be made, but only by means of more subtle experiments.

An experiment of the requisite subtlety is shown in Fig. 9 (Sinvani et al., 1983a; Sinvani et al., 1984a). Here, the heater is on the bottom of the crystal, but desorption takes place from the upper surface, to be detected by the bolometer. A mask limits the region from which desorption will be detected. A burst of phonons from the heater spreads out in the crystal. Some of these (about 5 x 10^{-4} of them) reach the desorption region. These are hot phonons (thermally distributed at $T_h \approx 10$K) but there are very few of them. If thermal desorption occurs, the total energy available is not enough to heat the film very much, so the time-of-flight signal will give a temperature close to ambient. On the other hand, if single phonon desorption occurs, the time-of-flight will correspond to the temperature of the heater, which determines the energies of individual phonons. This experiment is thus the phonon analog of the photoelectric effect except that, lacking a tunable monochromatic source of phonons, one cannot expect a work function type threshold effect.

The result, to be seen in Fig. 9, is that both thermal and single phonon processes occur. There is both a thermal or ambient temperature bump and an earlier bump due to single phonons. The earlier bump, whose arrival time can be changed as expected by changing the heater temperature (i.e., the pulse power), is the phonoatomic effect.

In separate, carefully designed experiments, it has even been possible to resolve two phonoatomic effect bumps due to the separate arrival at the upper surface of longitudinal and transverse phonons. L and T phonons are about equally efficient in desorbing atoms. If the crystal were isotropic, only L phonons would be expected to produce desorption at

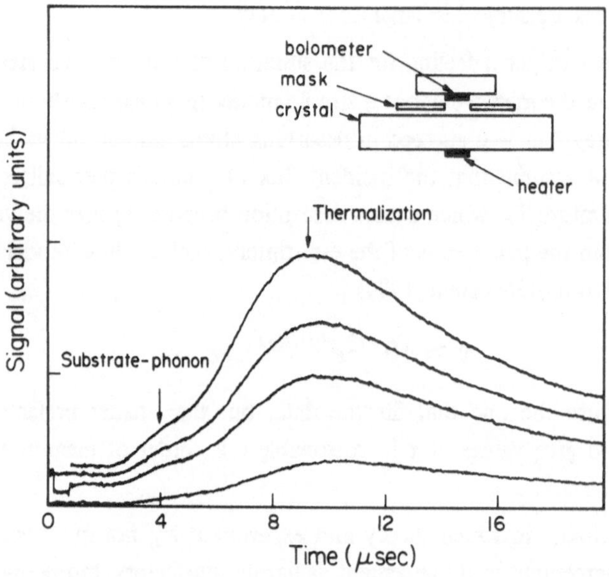

Figure 9. The phonoatomic effect. Experimental geometry shown in the inset. The bumps marked substrate-phonon are due to desorption of atoms by single phonons (the phonoatomic effect). The larger, later bumps, marked thermalization, are due to a slight warming of the film. The different curves are due to different heater powers.

normal incidence, but in an anisotropic crystal, this result should have been expected.

The phonoatomic effect has most recently been studied as a function of initial coverage (Goodstein et al., 1985). For relatively thick films (about 2 layers), it is found that desorption is mostly thermal, the area under the phonoatomic bump amounting to about 5% of the total. As the initial film thickness is reduced however, single-phonon desorption becomes increasingly important.

The final category of experiments is that which examines the fate of an atom incident on the surface. These are atom-surface scattering experiments in which the temperatures of both the atom-beam and the target surface are far lower than have ever been reported before (Sinvani et al., 1983b; Sinvani et al., 1984b). In fact, they differ from previous atom surface scattering experiments precisely in that the atomic beam deBroglie wavelengths are so long (~4Å) that there can be no diffraction effects of the kind that are the purpose of previous work. In these experiments we are interested instead in the sticking coefficient α_s, in how big it is, and in what it depends on. Quantum theories tend to predict it will approach zero as temperature approaches zero (Knowles and Suhl, 1977; Gumhalter and Crljen, 1984).

The basic experimental arrangement is shown in Fig. 10. The beam to be scattered is created by flash desorption from the heater h on the lower surface. As we have seen, the beam produced that way is well characterized experimentally. A bolometer b_2 on the lower surface detects the scattered atoms. Either a heater or a bolometer may be placed on the upper surface at the point from which specularly scattered atoms would reach b_2, or the upper surface may be left bare. The heaters and bolometers are just as described above.

In the experiment shown in Fig. 10, time-of-flight analysis of the reflected signal at b_2 shows that what reflection occurs is close to specular (any other process would lead to later arrival times). Hurst et al. (1979) have observed similar results for Xe scattered from Pb (111). Comparison of the size of that signal to the one at a bolometer on the target surface yields a sensitive measure of the sticking coefficient. The dependence of α_s on angle of incidence is determined by varying the distance between the upper and lower surfaces, thereby changing the angle of incidence. Replacing the bolometer by a heater on the upper

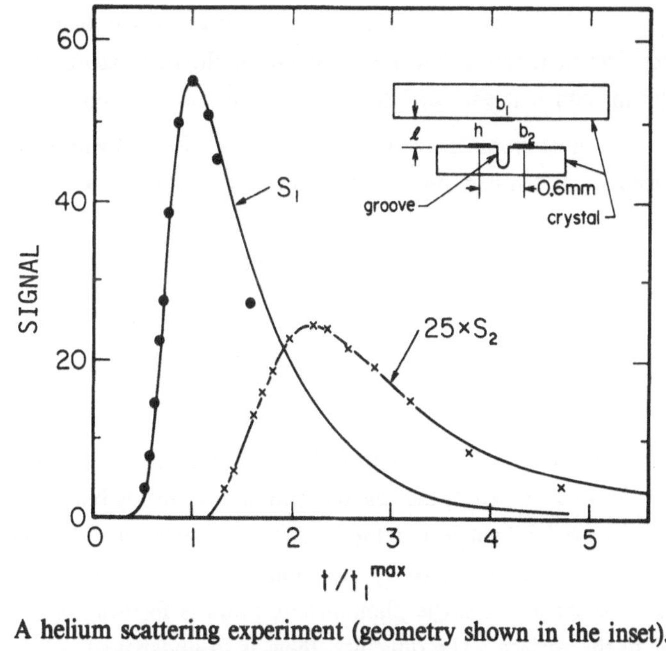

Figure 10. A helium scattering experiment (geometry shown in the inset). S_1, the signal at b_1, is characterized empirically (black dots). Given this characterization, the arrival times of S_2 (signal at b_2) are predicted assuming specular reflection. The fit to S_2 (the crosses) then depends only on the height of the signal maximum, which determines α_s. In other experiments, b_1 may be replaced by a heater, or the surface left bare as described in the text.

surface makes it possible to study α_s as a function of substrate temperature, and also of helium coverage (by pulsing the heater and allowing it to cool in time much less than t_{rc} before the incident beam arrives). When the upper heater is pulsed after a scattering event, a redesorption signal decisively confirms that most of the incident atoms had stuck.

The result of these experiments is that α_s is quite large: 0.94 at 14° from normal incidence, decreasing only to 0.77 at 45° incidence. Substrate coverage, substrate temperature, and beam temperature (in the range ~8 to 19K) do not appear to influence the results. It also does not seem to matter whether the target surface is a constantan heater, a tin bolometer, or bare sapphire. The experiments also show why a previous measurement (Thomas, 1967) gave $\alpha_s \approx 0.02$ at 19K: as the substrate temperature is increased, redesorption becomes increasingly rapid (consistent with Eq. (16)) until at 19K it is effectively instantaneous.

Let us summarize briefly some of the results of these experiments. If a thermal (~10K) pulse of phonons is incident at the surface, desorption is dominated by single phonon events at low coverage (< 1 layer), but becomes increasingly thermal at higher coverages (95% thermal at ~2 layers). Of the incident phonons, about 50% are specularly reflected regardless of helium coverage. The remainder are diffusely scattered if there is no helium, but increasingly transmitted across the interface with growing coverage until all are absorbed at ~3 layers. This 3-layer effect is correctly described by the thermal model of desorption which predicts that, as the film thickness grows, thermal desorption will occur with a short enough time constant τ to carry the pulse of heat away promptly.

On the other hand, most atoms incident on the film are adsorbed with high probability α_s nearly independent of temperature (3-19K), preexisting film thickness (0-1 layer), and substrate material. Those atoms that do not stick are reflected specularly. Most importantly, α_s is large, being already close to 0.8 at 45° incidence and rising from there as the normal is approached.

DISCUSSION

The results outlined in the last section appear to present us with a dilemma. To summarize briefly: theory leads us to expect that the desorption of a weakly bound species such as helium will be dominated by single phonon events. The existence of single phonon events has been convincingly demonstrated by the phonoatomic effect experiments. However, a kinematical consequence of single phonon desorption is focusing of the desorbed beam perpendicular to the surface. The difficulty, then, is to understand how desorption, proceeding principally by this mechanism, can give rise to the isotropic flux from the surface required when desorption and adsorption are in equilibrium. The problem could be reconciled by a sticking coefficient that falls rapidly to zero for non-normal incidence, but direct measurements rule out that possibility.

This dilemma is not restricted to adsorption/desorption equilibrium, but arises also in the case of condensation and evaporation at the interface of bulk liquid helium. Elegant measurements of the scattering of helium atoms from the surface of bulk liquid helium have shown that α_s is even larger than that at a solid interface, being essentially equal to unity except in the limit of grazing incidence (Nayak et al., 1983). To complete the picture, recent experiments have yielded results analogous to the phonoatomic effect, evaporated atoms having energies and angular distributions characteristic of evaporation by single phonons and rotons (Hope et al., 1984).

To understand how, in principle, the dilemma might be resolved for the case of adsorption and desorption, we consider in elementary form a model that has been analyzed extensively by H. Kreuzer and his colleagues (see, e.g., Goldys et al., 1982). The basic idea is that the adsorbed atoms are bound in the potential well shown in Fig. 1. Thus, confined in their motion normal to the surface in states with energy ε_n, they are nevertheless free to move parallel to the surface with momentum $\hbar$ K. The possible energies of an adsorbed atom, $E_{K,n}$, are thus the same as those in the analysis of bound state resonances in Eq. (6). A sketch of these states is shown in Fig. 11. We are assuming an ideal plane surface in the low coverage limit where adsorbed atoms do not interact with each other, but the basic ideas also apply at higher coverage (Sommer et al., 1982).

In thermodynamic equilibrium, all of these states are populated with Boltzmann probabilities, $e^{-E_{K,n}/k_BT}$. Motion in the plane of the surface is purely classical and thus, according to the equipartition theorem, the distributions of those components of momentum is the same as they would be if the atoms were free in the gas phase. Now, when an atom is desorbed by a phonon in equilibrium, it is true that the phonon has negligible momentum to contribute parallel to the surface, but the atoms already have all the momentum they need in that plane, and so they emerge, not focused normal to the surface, but nearly isotropically.

In the phonoatomic effect experiments, however, superimposed on the phonon background at ambient temperature ($T \sim 2K$), there is a very weak beam of phonons with a Bose distribution at $\sim$10K. A small number of these have enough energy to desorb atoms directly from the states $E_{K,n}$, and some of them do so, but they do not have enough momentum to promote atoms into states of high K. The remaining phonons thermalize in the film, but they only succeed in raising the film temperature slightly. The atoms desorbed by the high energy phonons thus emerge with $\sim$10K characteristic energy as observed, but they should have only $\sim$2K characteristic momentum parallel to the surface (Taborek, 1982).

Although this simple picture shows clearly that there is no conflict in principle between the experimental results and the requirements of detailed balance in equilibrium, there is still much to be resolved. One particular embarrassment is that, although μ and T are measured in this work, film thicknesses are only guessed at. Not even the parameters

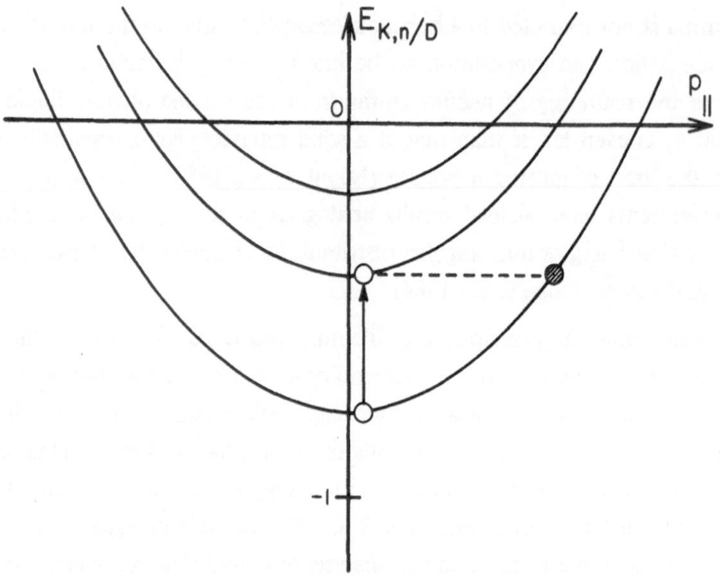

Figure 11. The states of an adsorbed atom, Eq. (6), with $p_{\parallel} = \hbar K$. The arrow shows a possible excitation of an atom by a phonon; the excited state shown by the shaded circle should have the same Boltzmann probability of occupation as the unshaded excited state, but it cannot be reached by a single-phonon event. The flux of high energy phonons in the phonoatomic effect experiment thus produces a nonequilibrium distribution of excited states in the adsorbed film.

necessary for applying the universal adsorption potential are known for the sapphire-helium system. However, the most important difficulties are related to the nature of the substrates used in the experiments.

As we saw above, modern surface science has developed sophisticated techniques for preparing and characterizing surfaces. Those techniques generally require large volume UHV systems with ample room for diagnostic devices, and seldom involve very low temperature surfaces. The substrates used in the desorption and scattering experiments reported here could not be subject to that kind of treatment for compelling technical reasons. Aside from the problems posed by low temperatures, normal cleaning procedures (heating, sputtering, etc.) would destroy the delicate heaters and bolometers needed for these experiments, and the space needed for diagnostic apparatus would be difficult to provide. For example, in the scattering experiments, the source heater, the target surface and the detecting bolometer are all contained in a volume of less than one cubic millimeter.

The sapphire crystals used in these experiments were optically polished, usually on known crystallographic planes, and were treated with reasonable laboratory care, but by the standards of modern surface science they are basically uncharacterized.

It is nevertheless possible to make some comments about the influence of the substrate on the outcome of the experiments. To begin with, for example, the principle of detailed balance is just as applicable at a dirty or rough surface as it is at a flat, clean surface. Thus, all the arguments guided by that principle remain valid. Moreover, in some instances, a little is known about the effect on these experiments of preparing a more ideal surface. As mentioned above, phonon reflection experiments have been performed at freshly cleaved and laser annealed surfaces. The result was the vanishing of the anomalous Kapitza conductance. At the very least, that means that experiments such as the phonoatomic effect would be much more difficult with ideal surfaces since the coupling of phonons across the interface and the resulting desorption signals would be orders of magnitude smaller. However, the question of surface imperfections is obviously much more important than a mere matter of experimental convenience.

The surface theorist wishes to apply the machinery of theoretical physics to ideal, flat, chemically pure surfaces where there is some hope of making clear and testable predictions. The young but growing field of experimental surface science, sketched above, is quite properly doing its best to produce and examine such surfaces. Yet, with regard to the phenomena we have discussed in this chapter, it may be that the most interesting physics occurs precisely because real surfaces are seldom perfect. Moreover, those phenomena are of very considerable practical importance.

Consider what one would expect to be the results of the experiments we have described here if they were performed at perfect surfaces. First of all, we have seen that there would be no anomalous Kapitza effect; incident phonons would nearly all reflect. The converse of that is the expectation that nearly all incident atoms would reflect; α_s should be close to zero. The problem of detailed balance at an interface between an atom gas and a phonon gas would be solved largely by the two systems nearly ignoring one another entirely.

A world made up of surfaces acting that way would lack more than just the Kapitza anomaly. There would be no cryopumping, which works because atoms stick to cold surfaces. Probably, hydrodynamic flows, at least low temperature ones, would be subject to perfect slip rather than non-slip boundary conditions (Haff, 1984) since the mechanisms of adsorption of atoms must be similar to those which transmit momentum from a hydrodynamic flow to a wall. Moreover, there is evidence that surface damage promotes adhesion between materials (Werner et al., 1982), and so on. In other words, a world of perfect surfaces might be radically different from the one we are familiar with. It would therefore seem wise not only to aim at studying more perfect surfaces, but also to begin to think about systematic study of the effects of making surfaces more like those that shape the common phenomena of the real world.

The experiments we have discussed offer some clues as to how a program of describing the roughness of surfaces might be organized. As mentioned above, the surfaces used in them are optically polished, that is to say, flat on all scales larger than $\sim 10^3$Å. The phonon reflection experiments used as probes phonons whose typical wavelength was $\sim \Theta_D/T_h \sim 100$ lattice spaces. Fifty percent of them were specularly reflected, meaning (in some sense) that on that scale, about half the surface is flat and in the proper plane. Moreover, the atoms in the scattering experiments had thermal wavelengths ~4Å, and about 20% of them were specularly reflected. Thus, at least 20% of the surface was flat on that scale. In other words, it is possible by these and other techniques to have knowledge of the roughness of the surface on a variety of length scales. This fact suggests that it might be useful to develop a language for discussing surface roughness analogous to that used for percolation theory and fractal surfaces.

No matter what directions are taken by future research, the experiments discussed above have brought us a long way in a short time. Just a few years ago, the crudest relevant fact about helium desorption, the desorption time τ, was unknown even in order of magnitude. Today, not only is that parameter much better understood, but we are able to consider delicate questions concerning disequilibrium between the internal degrees of freedom of an adsorbed film. At the very least, the experiments we have discussed point the way to a new technology for the study of surfaces and films.

ACKNOWLEDGEMENT

This work was supported by the U.S. Office of Naval Research, Contract No. N0014-80-C-0447. I would like to thank past, present and honorary members of the Caltech group who have participated in this work: B. Axan, M. W. Cole, R. Maboudian, F. Scaramuzzi, M. Sinvani, P. Taborek, G. Vidali, and M. Weimer.

REFERENCES

Anderson, A. C., 1980, The Kapitza thermal boundary resistance between two solids, in "Nonequilibrium Superconductivity, Phonons, and Kapitza Boundaries," K. E. Gray, ed., Plenum, New York.

Andres, K., Dynes, R. C., and Narayanamurti, V., 1977, Velocity spectrum of atoms evaporating from a liquid He surface at low temperatures, Phys. Rev. A, 8:2501.

Basso, H. C., Dietsche, W., and Kinder, H., 1983, Kapitza resistance of laser-annealed surfaces, in "Phonon Scattering in Condensed Matter," Eisenmenger and Lassmann.

Bendow, B., and Ying, S.-C., 1973, Phonon-induced desorption of adatoms from crystal surfaces I, Phys. Rev. B, 7:622.

Boato, G. P., Cantini, P., and Tatarek, R., 1978, Phys. Rev. Lett., 40:887.

Bron, W. E., 1980, Spectroscopy of high-frequency phonons, Rep. Prog. Phys., 43:302.

Bruch, L. W., and Watanabe, H., 1977, Polarization energy in the atom-surface interaction from optical data, Surf. Sci., 65:619.

Bruch, L. W., 1983, Theory of physisorption interactions, Surf. Sci., 125:194.

Cohen, S. A., and King, J. G., 1973, Measurement of lifetimes and binding energies of atoms adsorbed on surfaces at low temperatures by a rapid-flash-desorption technique, Phys. Rev. Lett., 31:703.

Cole, M. W., Frankl, D. R., and Goodstein, D. L., 1981, Probing the helium-graphite interaction, Rev. Mod. Phys., 53:199.

Cole, M. W., Toigo, F., and Tosatti, E., 1982, "Statistical Mechanics of Adsorption," North Holland, Amsterdam [Surf. Sci., 125:1 (1983)].

Cole, M. W., and Toigo, F., 1982a, Kinetics of elementary processes at surfaces in "Interfacial Aspects of Phase Transformations," B. Mutaftschiev, p 223, Riedel, Dordecht, Holland.

Cole, M. W., and Toigo, F., 1982b, Atom-phonon interaction at a surface, Surf. Sci., 119:L346.

Comsa, G., 1982, The dynamical parameters of desorbing molecules, in "Dynamics of Gas-Surface Interactions," G. Benedek and V. Valbusa, ed., p. 117, Springer Series in Chemical Physics, Vol. 21, Springer-Verlag, New York.

Cowin, J. P., Auerbach, D. J., Becker, C., and Wharton, L., 1978, Measurement of fast desorption kinetics of D_2 from tungsten by laser induced thermal desorption, Surf. Sci., 78:545.

Dash, J. G., 1975, "Films on Solid Surfaces," Academic, New York.

de Boer, J. H., 1968, "The Dynamical Character of Adsorption," Oxford, Clarendon Press.

Derry, G. D., Wesner, D., Carlos, W.E., and Frankl, D. R., 1979, Selective adsorption of ^3He and ^4He on the basal plane surface of graphite, Surf. Sci., 87:629.

Elgin, R. L., and Goodstein, D. L., 1974, Thermodynamic study of the ^4He monolayer adsorbed on grafoil, Phys. Rev. A, 9:2657.

Elgin, R. L., Greif, J. M., and Goodstein, D. L., 1978, The ground state of the helium atom-graphite surface potential, Phys. Rev. Lett., 41:1723.

Folinsbee, J. T., and Harrison, J. P., 1978, Phonon reflection at a silicon/^3He interface, J. Low Temp. Phys., 32:469.

Frenkel, J., "Kinetic Theory of Liquids," Dover, New York.

Gilmore, R., 1981, "Catastrophe Theory for Scientists and Engineers," Wiley, New York.

Goldys, E., Gortel, Z. W., and Kreuzer, H. J., 1982, Desorption kinetics mediated by surface phonon modes, Surf. Sci., 116:33.

Goodman, F. O., and Garcia, N., 1982, Nonequilibrium desorption of atoms and molecules from surfaces, Surf. Sci., 120:251.

Goodman, F. O., and Romero, I., 1978, One-phonon scattering of atoms in three dimensions by a simplified continuum model of a surface: thermal desorption, J. Chem. Phys., 69:1086.

Goodstein, D. L., 1975, "States of Matter," pp 87-89, Prentice-Hall, Englewood Cliffs, NJ.

Goodstein, D. L., Paterno, G., Scaramuzzi, F., and Taborek, P., 1980, Phonon reflections from solid-solid interfaces, in "Nonequilibrium Superconductivity, Phonons, and Kapitza Boundaries, K. E. Gray, ed., p 665, Plenum, New York.

Goodstein, D. L., and Weimer, M., 1983, A simple model of helium desorption kinetics, Surf. Sci., 125:227.

Goodstein, D. L., 1984, The adsorption and desorption of helium films, in "Many Body Phenomena at Surfaces," D. Langreth and H. Suhl, ed., p 277, Academic, San Fransisco.

Goodstein, D. L., Hamilton, J. J., Lysek, M. J., and Vidali, G., 1985a, Phase diagrams of multilayer adsorbed methane, Surf. Sci., in press.

Goodstein, D. L., Maboudian, R., Scaramuzzi, F., Sinvani, M., and Vidali, G., 1985b, in preparation.

Gortel, Z. W., Kreuzer, H. J., and Spaner, D., 1980, Quantum statistical theory of flash desorption, J. Chem. Phys., 72:234.

Gumhalter, B., and Crljen, Z., 1984, The effect of electronic surface response on sticking of atoms on metallic substrates, Surf. Sci., 139:231.

Guo, C. J., and Maris, H. J., 1974, Experimental study of the reflection of phonons at an interface between dielectric crystals and liquid helium, gaseous helium and solid neon, Phys. Rev. A, 10:960.

Haff, P., 1984, private communication.

Hamilton, J. J., and Goodstein, D. L., 1983, Thermodynamic study of methane multilayers adsorbed on graphite, Phys. Rev. B, 28:3838.

Hensel, J. C., Dynes, R. C., 1979, Observation of singular behavior in the focusing of ballistic phonons in Ge, Phys. Rev. Lett., 43:1033.

Hope, F. R., Baird, M. J., and Wyatt, A. F. G., 1984, Quantum evaporation from liquid ^{4}He by rotons, Phys. Rev. Lett., 52:1528.

Hurst, J. E., Becker, C. A., Cowin, J. P., Janda, K. C., and Wharton, L., 1979, Observation of direct inelastic scattering in the presence of trapping desorption scattering: Xe on Pt(III), Phys. Rev. Lett., 43:1175.

Jedrzejk, C., Freed, K. F., Efrima, S., and Metiu, H., 1981, A one-dimensional microscopic model for the rate of thermal desorption of an atom. The role of multiphonon processes, Chem. Phys. Lett., 79:227.

Kapitza, P. L., 1941, The study of heat transfer in helium II, J. Phys. USSR, 4:181.

Khalatnikov, I. M., 1965, Heat exchange between a solid and helium II, in "An Introduction to the Theory of Superfluidity," W. A. Benjamin Inc., New York.

Kinder, H., 1972, Spectroscopy with phonons on $Al_2O_3 : V^{3+}$ using the phonon bremsstrahlung of a superconducting tunnel junction, Phys. Rev. Lett., 28:1564.

Kinder, H., and Dietsche, W., 1974, Strong phonon conversion at the helium-solid interface, Phys. Rev. Lett., 33:578.

Kinder, H., Weber, J., and Dietsche, W., 1979, Kapitza resistance studies using phonon pulse studies, in "Phonon Scattering in Condensed Matter, H. J. Maris, ed., p 173, Plenum, New York.

Knowles, T. R., and Suhl, H., 1977, Sticking coefficient of atoms on solid surfaces at low temperatures, Phys. Rev. Lett., 39:1417.

Leuthäusser, V., 1981, Kinetic theory of adsorption and desorption, Z. Phys. B, 44:101.

Leuthäusser, V., 1983, Kinetic theory of desorption: energy and angular distributions, Z. Phys. B, 50:65.

Little, W. A., 1959, The transport of heat between dissimilar solids at low temperatures, Can. J. Phys., 37:334.

Long, A. R., Sherlock, R. A., and Wyatt, A. F. G., 1974, Phonon reflection at a cleaved sodium fluoride-helium film interface, J. Low Temp. Phys., 15:523.

Maris, H. J., 1971, Enhancement of heat pulses in crystals due to elastic anisotropy, J. Acoust. Soc. Am., 50:812.

Marx, D., and Eisenmenger, W., 1982, Phonon scattering at silicon crystal surfaces, Z. Phys. B, 48:277.

Menzel, D., 1975, Desorption phenomena, in "Topics in Applied Physics," Vol. 4, R. Gomer, ed., p 101, Springer-Verlag, New York.

Nayak, V. U., Edwards, D. O., and Masuhara, N., 1983, Scattering of ^{4}He atoms grazing the liquid-^{4}He surface, Phys. Rev. Lett., 50:990

Northrop, G. A., and Wolfe, J. P., 1979, Ballistic phonon imaging in solids - a new look at phonon focusing, Phys. Rev. Lett., 43:1424.

Northrop, G. A., and Wolfe, J. P., 1984, Phonon reflection imaging: a determination of specular versus diffuse boundary scattering, Phys. Rev. Lett., 52:2156.

Novaco, A. D., and Campbell, C. E., 1975, Effects of crystalline substrate potentials on quasi-two-dimensional liquid helium, Phys. Rev. B, 11:2525.

Pandit, R., Schick, M., and Wortis, M., 1982, Systematics of multilayer adsorption phenomena on attractive substrates, Phys. Rev. B, 26:5112.

Prutton, M., 1983, "Surface Physics," Clarendon Press, Oxford.

Sequin, J. L., Suzanne, J., Bienfait, M., Dash, J. G., and Venables, J. A., 1983, Complete and incomplete wetting in multilayer adsorption: high energy electron diffraction studies of Xe, Ar, N_2 and Ne films on graphite, Phys. Rev. Lett., 51:122.

Sinha, S., 1980, "Ordering in Two Dimensions," North Holland, New York.

Sinvani, M., Taborek, P., and Goodstein, D., 1982, Direct measurement of desorption kinetics of ^{4}He at low temperatures, Phys. Rev. Lett., 48:1259.

Sinvani, M., and Goodstein, D., 1983, Relaxation time of an adsorbing ^{4}He film, Surf. Sci., 125:291.

Sinvani, M., Taborek, P., and Goodstein, D., 1983a, Direct and thermal desorption of ^{4}He films, Phys. Lett., 95A:59.

Sinvani, M., Cole, M. W., and Goodstein, D. L., 1983b, Sticking probability of ^{4}He on solid surfaces at low temperature, Phys. Rev. Lett., 51:188.

Sinvani, M., Goodstein, D. L., Cole, M. W., and Taborek, P., 1984a, Desorption of atoms from helium films, Phys. Rev. B, to be published 1 Sept.

Sinvani, M., Goodstein, D. L., and Cole, M. W., 1984b, Scattering of low-energy helium atoms from a low-temperature solid surface, Phys. Rev. B, 29:3905.

Sommer, E. and Kreuzer, H. J., 1982, Physisorption kinetics from mean field theory: compensation effect near monolayer coverage, Phys. Rev. Lett., 49:61.

Taborek, P., and Goodstein, D., 1979, Phonon reflection at a sapphire-vacuum interface, J. Phys. C.: Solid State Phys., 12:4737.

Taborek, P., and Goodstein, D., 1980a, Diffuse reflection of phonons and the anomalous Kapitza resistance, Phys. Rev. B, 22:1550.

Taborek, P., and Goodstein, D., 1980b, Phonon focusing catastrophes, Solid State Communications, 33:1191.

Taborek, P., and Goodstein, D., 1981, Phonon reflection at noble gas interfaces, Solid State Communications, 38:215.

Taborek, P., 1982, Critical cone in phonon-induced desorption of helium, Phys. Rev. Lett., 48:1737.

Thomas, L. B., 1967, A collection of some controlled surface thermal accomodation coefficient measurements, in "Rarified Gas Dynamics," C. L. Brundin, ed., Vol. 1, p 155, Academic, New York.

Vidali, G., Cole, M. W., and Klein, J. R., 1983, Shape of physical adsorption potentials, Phys. Rev., 28:3064.

Weber, J., Sandermann, W., Dietsche, W., and Kinder, H., 1978, Absence of anomalous Kapitza conductance on freshly cleaned surfaces, Phys. Rev. Lett., 40:1469.

Weimer, M., and Goodstein, D., 1983, Continuum model of helium desorption kinetics, Phys. Rev. Lett., 50:193.

Werner, B. T., Vreeland, T., Mendenhall, M. H., Qui, Y., and Tombrello, T. A., 1982, Enhanced adhesion from high energy ion irradiation, Thin Solid Films, 104:163.

Wyatt, A. F. G., 1980, Kapitza conductance of solid-liquid He interfaces, in "Nonequilibrium Superconductivity, Phonons, and Kapitza Boundaries, K. E. Gray, ed., p 31, Plenum, New York.

Ying, S.-C., and Bendow, B., 1973, Phonon-induced desorption of adatoms from crystal surfaces II, Phys. Rev. B, 7:637.

Young, D. M., and Crowell, A. D., 1962, "Physical Adsorption of Gases," Butterworths, London.

PARTICIPANTS OF THE NATO ADVANCED STUDY INSTITUTE ON
NONEQUILIBRIUM PHONON DYNAMICS
August 27-September 7, 1984
Les Arcs, France

(by R.J. Goosens, The Netherlands)

INDEX